FORMULAS FROM ALGEBRA

Exponents

$$a^m a^n = a^{m+n}$$
$$(a^m)^n = a^{mn}$$
$$\frac{a^m}{a^n} = a^{m-n}$$
$$(ab)^n = a^n b^n$$
$$\left(\frac{a}{b}\right)^n = \frac{a^n}{b^n}$$

Radicals

$$(\sqrt[n]{a})^n = a$$
$$\sqrt[n]{a^n} = a, \text{ if } a \geqslant 0$$
$$\sqrt[n]{ab} = \sqrt[n]{a}\sqrt[n]{b}$$
$$\sqrt[n]{\frac{a}{b}} = \frac{\sqrt[n]{a}}{\sqrt[n]{b}}$$

Logarithms

$$\log_a MN = \log_a M + \log_a N$$
$$\log_a (M/N) = \log_a M - \log_a N$$
$$\log_a (N^p) = p \log_a N$$

Factoring Formulas

$$x^2 - y^2 = (x-y)(x+y)$$
$$x^3 - y^3 = (x-y)(x^2 + xy + y^2)$$
$$x^3 + y^3 = (x+y)(x^2 - xy + y^2)$$
$$x^2 + 2xy + y^2 = (x+y)^2$$
$$x^2 - 2xy + y^2 = (x-y)^2$$
$$x^3 + 3x^2y + 3xy^2 + y^3 = (x+y)^3$$

Binomial Formula

$$(x+y)^n = {}_nC_0 x^n y^0 + {}_nC_1 x^{n-1} y^1 + \cdots + {}_nC_{n-1} x^1 y^{n-1} + {}_nC_n x^0 y^n$$

Quadratic Formula

The solutions to $ax^2 + bx + c = 0$ are $x = \dfrac{-b \pm \sqrt{b^2 - 4ac}}{2a}$

Complex Numbers

Multiplication: $(a + bi)(c + di) = (ac - bd) + (ad + bc)i$

Polar form: $a + bi = r(\cos\theta + i\sin\theta)$ where $r = \sqrt{a^2 + b^2}$

Powers: $[r(\cos\theta + i\sin\theta)]^n = r^n(\cos n\theta + i\sin n\theta)$

Roots: $\sqrt[n]{r}\left[\cos\left(\frac{\theta + k\cdot 360^\circ}{n}\right) + i\sin\left(\frac{\theta + k\cdot 360^\circ}{n}\right)\right]$ $k = 0, 1, 2, \cdots, n-1$

Algebra and Trigonometry

Algebra and

WALTER FLEMING

Hamline University

DALE VARBERG

Hamline University

Trigonometry

Prentice-Hall Inc., Englewood Cliffs, New Jersey 07632

Library of Congress Cataloging in Publication Data

FLEMING, WALTER.
Algebra and trigonometry.

Includes index.
1. Algebra. 2. Trigonometry. I. Varberg, Dale E., joint author. II. Title.
QA154.2.F52 512'.13 79-18439
ISBN 0-13-021824-3

Other titles by the same authors:
College Algebra
Plane Trigonometry

Algebra and Trigonometry
Walter Fleming/Dale Varberg

Printed in the United States of America

10 9 8 7 6 5 4 3 2 1

Editorial/production supervision by Ellen W. Caughey and Reynold R. Rieger
Interior Design by Mark A. Binn
Cover Design by Mark A. Binn
Manufacturing buyer: Edmund W. Leone

Prentice-Hall International, Inc., *London*
Prentice-Hall of Australia Pty. Limited, *Sydney*
Prentice-Hall of Canada, Ltd., *Toronto*
Prentice-Hall of India Private Limited, *New Delhi*
Prentice-Hall of Japan, Inc., *Tokyo*
Prentice-Hall of Southeast Asia Pte. Ltd., *Singapore*
Whitehall Books Limited, *Wellington, New Zealand*

To our wives

Dorothy and Idella

Contents

Preface xiii

CHAPTER ONE

Numbers and Their Properties 1

1-1 What is Algebra? 2

1-2 The Integers and the Rational Numbers 7

1-3 The Real Numbers 13

1-4 Fundamental Properties of the Real Numbers 20

1-5 Order and Absolute Value 27

1-6 The Complex Numbers 33

Chapter Summary 38

CHAPTER TWO

Exponents and Polynomials 41

2-1 Integral Exponents 42

2-2 Scientific Notation 50

2-3 Pocket Calculators 56

2-4 Polynomials 61

2-5 Factoring Polynomials 67

2-6 Rational Expressions 75

Chapter Summary 81

CHAPTER THREE

Equations and Inequalities 85

3-1 Equations 86

3-2 Applications Using One Unknown 93

3-3 Two Equations in Two Unknowns 99

3-4 Quadratic Equations 106

3-5 Inequalities 114

3-6 More Applications (Optional) 122

Chapter Summary 128

CHAPTER FOUR

Coordinates and Curves 131

4-1 The Cartesian Coordinate System 132

4-2 Algebra and Geometry United 138

4-3 The Straight Line 145

4-4 The Parabola 153
4-5 Ellipses and Hyperbolas (Optional) 162
Chapter Summary 169

CHAPTER FIVE

Functions and Their Graphs 173

5-1 Functions 174
5-2 Graphs of Functions 181
5-3 Graphing Rational Functions 189
5-4 Putting Functions Together 196
5-5 Inverse Functions 203
Chapter Summary 210

CHAPTER SIX

Exponential and Logarithmic Functions 213

6-1 Radicals 214
6-2 Exponents and Exponential Functions 220
6-3 Exponential Growth 226
6-4 Logarithms and Logarithmic Functions 233
6-5 Applications of Logarithms 240
6-6 Common Logarithms (Optional) 247
6-7 Calculations with Logarithms (Optional) 252
Chapter Summary 256

CHAPTER SEVEN

The Trigonometric Functions 259

7-1 Right-triangle Trigonometry 260

7-2 Angles and Arcs 266

7-3 The Sine and Cosine Functions 272

7-4 Four More Trigonometric Functions 278

7-5 Finding Values of the Trigonometric Functions 283

7-6 Graphs of the Trigonometric Functions 288

Chapter Summary 294

CHAPTER EIGHT

Trigonometric Identities and Equations 297

8-1 Identities 298

8-2 More Identities 304

8-3 Trigonometric Equations 312

8-4 Inverse Trigonometric Functions 318

Chapter Summary 325

CHAPTER NINE

Applications of Trigonometry 329

9-1 Oblique Triangles: Law of Sines 330

9-2 Oblique Triangles: Law of Cosines 335

9-3 Simple Harmonic Motion 340

9-4 The Polar Coordinate System 347

9-5 Polar Representation of Complex Numbers 353
9-6 Powers and Roots of Complex Numbers 360
Chapter Summary 366

CHAPTER TEN

Theory of Polynomial Equations 369

10-1 Division of Polynomials 370
10-2 Factorization Theory for Polynomials 376
10-3 Polynomials Equations with Real Coefficients 384
10-4 The Method of Successive Approximations 392
Chapter Summary 399

CHAPTER ELEVEN

Systems of Equations and Inequalities 401

11-1 Equivalent Systems of Equations 402
11-2 Matrix Methods 409
11-3 The Algebra of Matrices 416
11-4 Multiplicative Inverses 424
11-5 Second- and Third-Order Determinants 431
11-6 Higher-Order Determinants 439
11-7 Systems of Inequalities 445
Chapter Summary 453

CHAPTER TWELVE

Sequences and Counting Problems 457

12-1 Number Sequences 458

12-2 Arithmetic Sequences 465

12-3 Geometric Sequences 471

12-4 Mathematical Induction 478

12-5 Counting Ordered Arrangements 486

12-6 Counting Unordered Collections 494

12-7 The Binomial Formula 501

Chapter Summary 505

Appendix 509

Use of Tables 510

Table A. Natural Logarithms 512

Table B. Common Logarithms 514

Table C. Trigonometric Functions (degrees) 516

Table D. Trigonometric Functions (radians) 519

Answers A1

Index I1

Preface

Of making many books there is no end, and much study is a weariness of the flesh.

Ecclesiastes 12:12

One imagines that King Solomon was looking at algebra and trigonometry texts as he wrote the above words. There are at least 30 such books in print, all filled with good mathematics. Most of the books we've tried in our combined 60 years of teaching, however, make mathematics taste like unsalted cardboard. Generally, theory has been overemphasized and problem solving has been shortchanged.

A fresh wind is blowing in the mathematical community. Intuition and insight are equally respected with logical reasoning. Skill at calculation and algebraic manipulation is demanded. Problems that arise from the world of experience are welcomed. Word problems are stressed. With this fresh outlook in mind, a number of special features have been incorporated into our text.

Lively opening displays Each section begins with a challenging problem, a historical anecdote, a famous quotation, or an appropriate cartoon. These displays are designed to spark the readers' curiosity, to draw them into the section.

Informal writing style The text avoids the ponderous theorem-proof style of most mathematics books, rather giving simple illustrations illustrated by carefully chosen examples. Technical jargon is avoided except where clarity and precision of thought require it; slices of humor are injected where possible.

Problem solving Greater deference is shown to George Polya than to Nickolas Bourbaki (see Chapter 3). The text poses good problems and gives clear guidance in how to solve them. There are more than 3600 problems for the student to work, including numerous "word" problems. If readers develop a taste for and skill at problem solving, we will have succeeded in one of our chief aims.

Examples There are many examples in the text, but they are not placed exclusively within the discussion. A few examples are used to highlight the major points of each section. Then, in the problem sets, several more examples designed to elucidate and to expand upon the text are added. Each of these examples is accompanied by a number of related exercises for the student. To avoid the old danger that students will adopt the cookbook approach to problem solving, following the example without thinking, each problem set concludes with miscellaneous problems designed to test all the skills needed in the section and perhaps to challenge the very best students.

Calculators The advent of inexpensive, electronic calculators is good for mathematics and cannot be ignored. The use of the scientific calculator is woven into the fabric of the book wherever appropriate. However, so that the text can be used successfully by a student who does not own one of these marvels, we identify with a [c] those problems which are difficult to do with paper and pencil. Such problems can be omitted without loss of continuity.

Flexibility This book has been designed for use in either one- or two-term courses. In two terms, most of the book can be covered at a pace of one section per class period. Well-prepared students can omit or quickly review Chapters 1 and 2. At Hamline University, we teach the following one-semester course (see Dependence Chart, page xv).

CHAPTER	TITLE	SECTIONS COVERED	DAYS REQUIRED
1	Numbers and Their Properties	1-6	2
2	Exponents and Polynomials	1-6	4
3	Equations and Inequalities	1-5	4
4	Coordinates and Curves	1-4	3
5	Functions and Their Graphs	1-5	3
6	Exponential and Logarithmic Functions	1-5	4
7	The Trigonometric Functions	1-6	5
8	Trigonometric Identities and Equations	1-4	4
9	Applications of Trigonometry	1-3, 5, 6	5
10	Theory of Polynomial Equations	1, 2, 4	3
			37

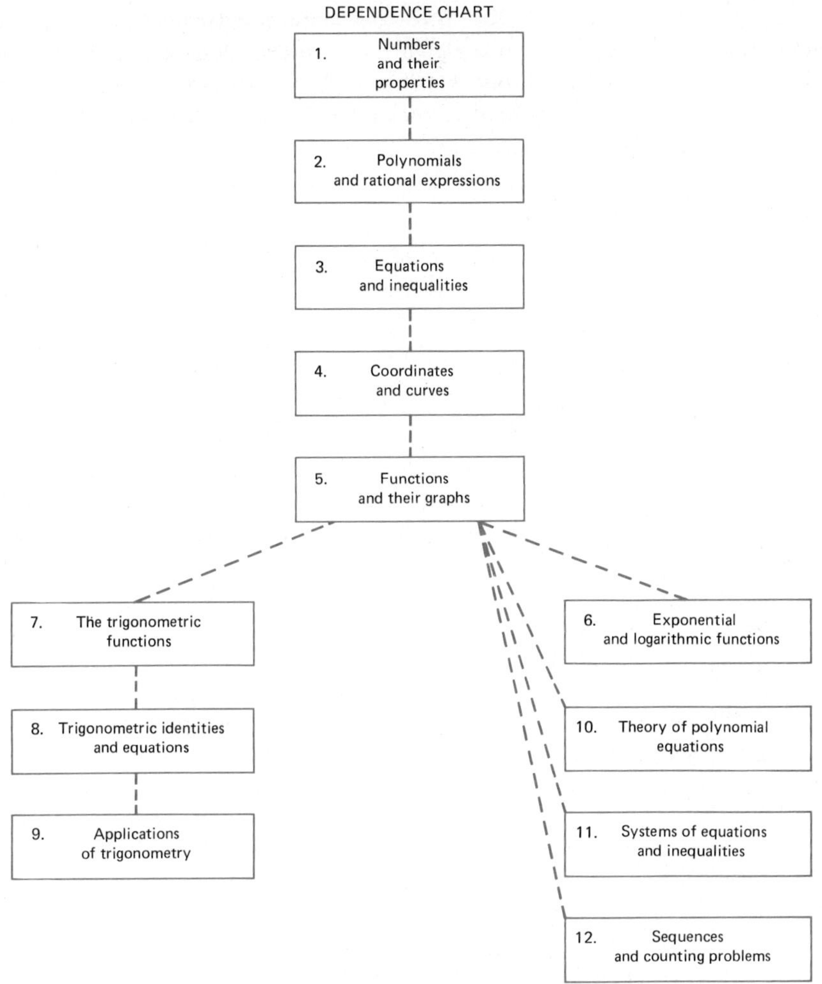

Supplementary materials An Instructor's Manual is available upon adoption of the text. Advice on how to use the text, answers to the even-numbered problems, four versions of chapter tests for each chapter, and an answer key for these tests are included in the manual.

Marilyn Mays Gilchrist, North Lake College, Irving, Texas, has written a Student Study Guide to accompany the text. It contains behavioral objectives for each chapter, notes expanding on difficult material, additional examples and problems, and a concepts test for each chapter.

Acknowledgments

A book, like a painting, goes through several stages. There is the initial vision of something great and wonderful. There is the long and often tedious task of putting down word upon word, sentence after sentence, like brush-strokes on canvas. Finally, there is that climactic day when the completed work is unveiled for public scrutiny.

But unlike painters, textbook authors seldom wait for some final grand occasion to expose their work to criticism. They know that the best books are subjected to the critical examination of strong reviewers at every stage of development. Thus, it is with considerable gratitude and pride that we acknowledge the continuing advice, helpful criticism, and occasional praise of a large corps of reviewers. Each of the following people served as a reviewer for at least part of the manuscript. Four of them (Hale, J. Hall, Smith, Wells) not only critiqued the entire book but also met with the authors for a face-to-face reviewer conference.

Dean Alders, Mankato State University

Wayne Andrepont, The University of Southwestern Louisiana

Alfred Borm, Southwest Texas State University

James Calhoun, Western Illinois University

Donald Coram, Oklahoma State University

Leonard Deaton, California State University, Los Angeles

Gary Fowler, Louisiana State University

Margaret Gessaman, The University of Nebraska at Omaha

Mark Hale, Jr., University of Florida

D. W. Hall, Michigan State University

James E. Hall, University of Wisconsin, Madison

Eldon Miller, University of Mississippi

John Minnick, DeAnza College

Wallace E. Parr, University of Maryland, Baltimore County

H. D. Perry, Texas A & M University

Jean Smith, Middlesex Community College

Shirley Sorensen, University of Maryland, College Park

Monty Strauss, Texas Tech University

Carroll Wells, Western Kentucky University

We congratulate the staff at Prentice-Hall for an expert production job. Our final accolade goes to acquisitions editor Harry Gaines for his help in putting the book together.

WALTER FLEMING

DALE VARBERG

Numbers are an indispensable tool of civilization, serving to whip its activities into some sort of order . . . The complexity of a civilization is mirrored in the complexity of its numbers.

Philip J. Davis

CHAPTER ONE

Numbers and Their Properties

1-1 What is Algebra?

1-2 The Integers and the Rational Numbers

1-3 The Real Numbers

1-4 Fundamental Properties of the Real Numbers

1-5 Order and Absolute Value

1-6 The Complex Numbers

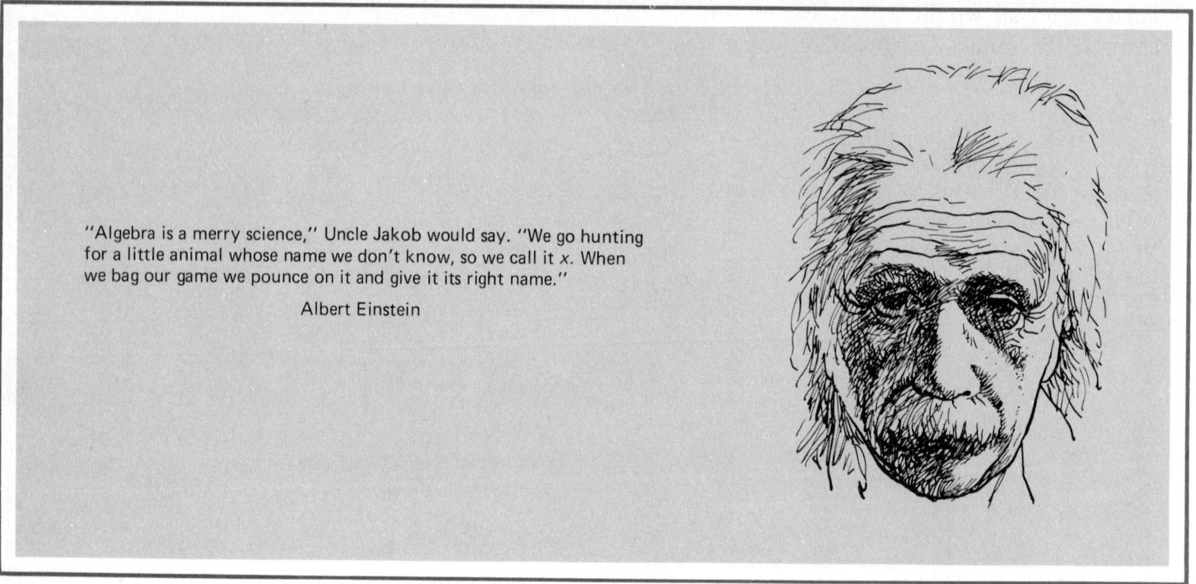

1-1 What is Algebra?

Sometimes the simplest questions seem the hardest to answer. One frustrated ninth grader responded, "Algebra is all about x and y, but nobody knows what they are." Albert Einstein was fond of his Uncle Jakob's definition, which is quoted above. A contemporary mathematician, Morris Kline, refers to algebra as generalized arithmetic. There is some truth in all of these statements, but perhaps Kline's statement is closest to the heart of the matter. What does he mean?

In arithmetic we are concerned with numbers and the four operations of addition, subtraction, multiplication, and division. We learn to understand and manipulate expressions like

$$16 - 11 \qquad \frac{3}{24} \qquad (13)(29)$$

In algebra we do the same thing, but we are more likely to write

$$a - b \qquad \frac{x}{y} \qquad mn$$

without specifying precisely what numbers these letters represent. This determination to stay uncommitted (not to know what x and y are) offers some tremendous advantages. Here are two of them.

Generality and Conciseness All of us know that $3 + 4$ is the same as $4 + 3$ and that $7 + 9$ equals $9 + 7$. We could fill pages and books, even libraries, with the corresponding facts about other numbers. All of them would be correct and all would be important. But we can achieve the same effect much more economically by writing

$$a + b = b + a$$

This simple formula says all there is to be said about adding two numbers in opposite order. It states a general law and does it on one-fourth of a line.

Or take the well-known facts that if I drive 30 miles per hour for 2 hours, I will travel 60 miles, and that if I fly 120 miles per hour for 3 hours, I will cover 360 miles. These and all other similar facts are summarized in the general formula

$$D = RT$$

which is an abbreviation for

$$\text{distance} = \text{rate} \times \text{time}$$

Problem Solving Uncle Jakob's definition of algebra hinted at something else that is very important. In algebra, as in life, there are many problems. Often they involve finding a number which is initially unknown but which must satisfy certain conditions. If these conditions can be translated into the symbols of algebra, it may take only a simple manipulation to find the answer or, as Uncle Jakob put it, to bag our game. Here is an illustration.

> Roger Longbottom has rented a motorboat for 5 hours from a river resort. He was told that the boat will travel 6 miles per hour upstream and 12 miles per hour downstream. How far upstream can he go and still return the boat to the resort within the allotted 5-hour time period?

We recognize immediately that this is a distance–rate–time problem; the formula $D = RT$ is certain to be important. Now what is it that we want to know? We want to find a distance, namely, how far upstream Roger dares to go. Let us call that distance x miles. Next we summarize the information that is given, keeping in mind that, since $D = RT$, it is also true that $T = D/R$.

	GOING	RETURNING
Distance (miles)	x	x
Rate (miles per hour)	6	12
Time (hours)	$x/6$	$x/12$

There is one piece of information we have not used; it is the key to the whole problem. The total time allowed is 5 hours, which is the sum of the time going and the time returning. Thus

$$\frac{x}{6} + \frac{x}{12} = 5$$

After multiplying both sides by 12, we have

$$2x + x = 60$$
$$3x = 60$$
$$x = 20$$

Roger can travel 20 miles upstream and still return within 5 hours.

We intend to emphasize problem solving in this book. To be able to read a mathematics book with understanding is important. To learn to calculate accurately and to manipulate symbols with ease is a worthy goal. But to be able to solve problems, easy problems and hard ones, practical problems and abstract ones, is a supreme achievement.

It is time for you to try your hand at some problems.

Problem Set 1-1

Example A (Writing phrases in algebraic notation) Use the symbols x and y to express the following phrases in algebraic notation.
(a) A number divided by the sum of twice that number and another number.
(b) The sum of 32 and $\frac{9}{5}$ of a Celsius temperature reading.

SOLUTION. (a) If x is the first number and y is the second, then the given phrase can be expressed by

$$\frac{x}{2x + y}$$

(b) If x represents the Celsius reading, then we can write the given phrase as

$$32 + \tfrac{9}{5}x$$

Perhaps you recognize this as the Fahrenheit reading corresponding to a Celsius reading of x.

Express each of the following phrases in algebraic notation using the symbols x and y. Be sure to indicate what x and y represent unless this is already stated in the problem.

1. One number plus one-third of another number.
2. The average of two numbers.
3. Twice one number divided by three times another.
4. The sum of a number and its square.
5. Ten percent of a number added to that number.
6. Twenty percent of the amount by which a number exceeds 50.
7. The sum of the squares of two sides of a triangle.
8. One-half the product of the base and the height of a triangle.
9. The distance in miles that a car travels in x hours at y miles per hour.
10. The time in hours it takes to go x miles at y miles per hour.
11. The rate in miles per hour of a boat that traveled y miles in x hours.

12. The total distance a car traveled in 8 hours if its rate was x miles per hour for 3 hours and y miles per hour for 5 hours.
13. The time in hours it took a boat to travel 30 miles upstream and back if its rate upstream was x miles per hour and its rate downstream was y miles per hour.
14. The time in hours it took a boat to travel 30 miles upstream and back if its rate in still water was x miles per hour and the rate of the stream was y miles per hour. Assume that x is greater than y.

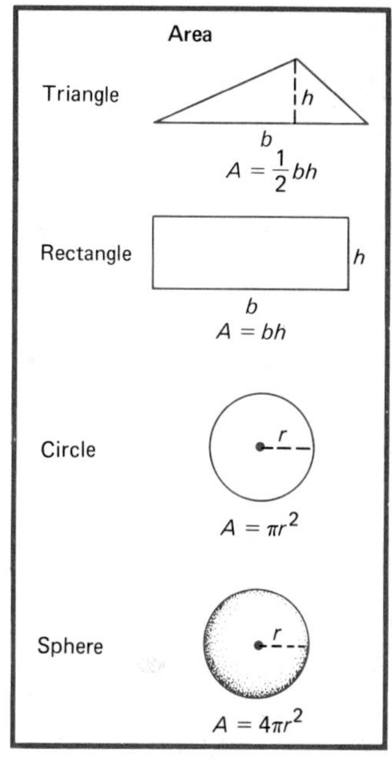

In the margin are some formulas from geometry. Use them to express each of the following in algebraic symbols. In each problem, assume that all dimensions are given in terms of the same unit of length (such as a centimeter).

15. The area of a square of side x.
16. The area of a triangle whose height is $\frac{1}{3}$ the length of its base.
17. The surface area of a cube of side x.
18. The area of the surface of a rectangular box whose dimensions are x, $2x$, and $3x$.
19. The surface area of a sphere whose diameter is x.
20. The area of a Norman window whose shape is that of a square of side x topped with a semicircle.
21. The volume of a box with square base of length x and height 10.
22. The volume of a cylinder whose radius and height are both x.
23. The volume of a sphere of diameter x.
24. The volume of what is left of a cylinder of radius 10 and altitude 12 when a hole of radius x is drilled along the center axis of the cylinder. (Assume x is less than 10.)
25. The volume of what is left when a round hole of radius 2 is drilled through a cube of side x. (Assume that the hole is drilled perpendicular to a side and that x is greater than 4.)
26. The volume of what is left of a cylinder when a square hole of width x is drilled around the center axis of the cylinder of radius $3x$ and height y.
27. The area and perimeter of a running track in the shape of a square with semicircular ends if the square has side x. (Recall that the circumference of a circle satisfies $C = 2\pi r$, where r is the radius.)
28. The area of what is left of a circle of radius r after removing an isosceles triangle which has a diameter as a base and the opposite vertex on the circumference of the circle.

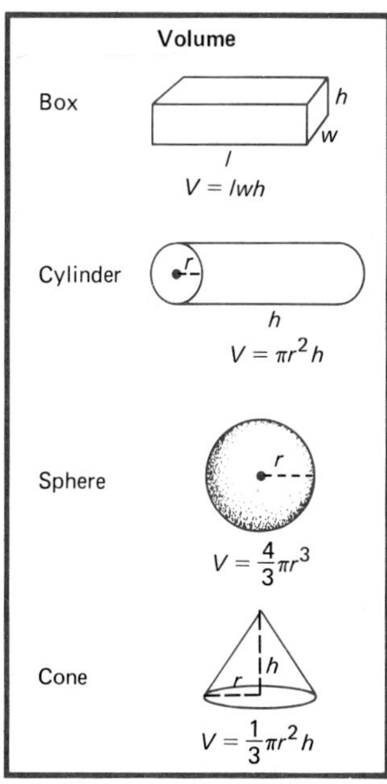

Example B (Stating sentences as algebraic equations) Express each of the following sentences as an equation using the symbol x. Then solve for x.

(a) The sum of a number and three-fourths of that number is 21.

(b) A rectangular field which is 125 meters longer than it is wide has a perimeter of 650 meters.

SOLUTION. (a) Let x be the number. The sentence can be written as

$$x + \tfrac{3}{4}x = 21$$

If we multiply both sides by 4, we have

$$\begin{aligned} 4x + 3x &= 84 \\ 7x &= 84 \\ x &= 12 \end{aligned}$$

It is always wise to make at least a mental check on the solution. The sum of 12 and three-fourths of 12, that is, the sum of 12 and 9 does equal 21.

(b) Let the width of the field be x meters. Then its length is $x + 125$ meters. Since the perimeter is twice the width plus twice the length, we write

$$2x + 2(x + 125) = 650$$

To solve, we remove parentheses and simplify.

$$\begin{aligned} 2x + 2x + 250 &= 650 \\ 4x + 250 &= 650 \\ 4x &= 400 \\ x &= 100 \end{aligned}$$

The field is 225 meters long and 100 meters wide.

Express each of the following sentences as an algebraic equation in x and then solve for x. Start by writing down what x represents. (*Note*: Chapter 3 covers this kind of problem in more detail.)

29. The sum of a number and one-half of that number is 45.
30. Fifteen percent of a number added to that number is 10.35.
31. The sum of two consecutive odd numbers is 168.
32. The sum of three consecutive even numbers is 180.
33. A car going at x miles per hour travels 252 miles in $4\frac{1}{2}$ hours.
34. A circle of radius x centimeters has an area of 64π square centimeters.

Miscellaneous Problems

35. A rectangle 3 meters longer than it is wide has a perimeter of 48 meters. How wide is the rectangle?
36. The area of a rectangle is 54 square feet and its length is $1\frac{1}{2}$ times its width. Find the length.
37. A cube has side length 4 inches. Find its volume and its surface area.
38. The surface area of a certain sphere is the same as the area of a circle of radius 8 centimeters. Find the radius of the sphere.
39. If the radius of a sphere is doubled, what happens to its volume? Its surface area?
40. A boat can go 6 miles per hour in still water. If the rate of the current is 2 miles per hour, how far upstream can this boat go and return in 4 hours?
41. Thomas Goodroad knows that his small plane can cruise 150 miles per hour in still air and that his tank holds enough gas for 4 hours of flying. If he takes off on patrol against a north wind of 50 miles per hour, how long can he fly in that direction and then return safely?

42. On a 200-mile trip, Janet Creighton drove 40 miles per hour the first 100 miles and 60 miles per hour the rest of the way. What is her average speed for the trip?

43. On a 400-kilometer trip, Jonathan drove at 50 kilometers per hour the first 200 kilometers and at 70 kilometers per hour the rest of the way. What was his average speed?

44. Ken Woodward drove from Oregon City to Cozy Harbor, a distance of 200 kilometers, in $2\frac{1}{2}$ hours. What speed must he average on the return trip in order to average 90 kilometers per hour for the entire trip?

"And how many hours a day did you do lessons?" said Alice, in a hurry to change the subject.

"Ten hours the first day," said the Mock Turtle; "nine the next, and so on."

"What a curious plan!" exclaimed Alice.

"That's the reason they're called lessons," the Gryphon remarked, "because they lessen from day to day."

This was quite a new idea to Alice, and she thought it over a little before she made her next remark. "Then the eleventh day must have been a holiday?"

"Of course it was," said the Mock Turtle.

"And how did you manage on the twelfth?" Alice went on eagerly.

"That's enough about lessons," the Gryphon interrupted in a very decided tone; "tell her something about the games now."

From *Alice's Adventures in Wonderland* by Lewis Carroll, mathematician, clergyman, and storyteller

1-2 The Integers and the Rational Numbers

If algebra is generalized arithmetic, then arithmetic is going to be important in this course. That is why we first review the familiar number systems of mathematics.

The human race learned to count before it learned to write; so did most of us. The **whole numbers** 0, 1, 2, 3, . . . entered our vocabularies before we started school. We used them to count our toys, our friends, and our money. Counting forward to larger and larger numbers was no problem; counting backwards was a different matter: 5, 4, 3, 2, 1, 0. But what comes after 0? It bothered Alice in Wonderland, it bothered mathematicians as late as 1300, and it still bothers some people. However, if we are to talk about debts, cold temperatures, and even moon launch countdowns, we must have an answer. Faced with this problem, mathematicians invented a host of new numbers, $-1, -2, -3, \ldots$, called **negative integers.** Together with the whole numbers, they form the system of integers.

The Integers The **integers** are the numbers ..., −5, −4, −3, −2, −1, 0, 1, 2, 3, 4, 5, ... This array is familiar to anyone who has looked at a thermometer lying on its side. In fact, that is an excellent way to think of the integers. There they label points, equally spaced along a line, as in the following diagram.

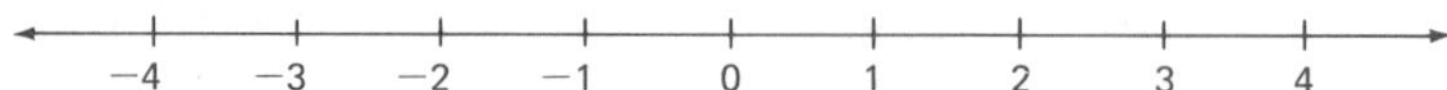

If we are talking about temperatures, −4 means 4 degrees below zero; if we are referring to money, −4 might represent a debt of 4 dollars.

Numbers, used as labels for points, debts, or football players, are significant. But what really makes them useful is our ability to combine them, that is, to add, subtract, and multiply them. All of this is familiar to you so we shall only remind you of two properties or rules that sometimes cause trouble.

(i) $$(-a)(b) = (a)(-b) = -(ab)$$

(ii) $$(-a)(-b) = ab$$

	Day	Amount gained each day	My worth
	−4 ·	−4	= $16
Three days ago	−3 ·	−4	= 12
	−2 ·	−4	= 8
	−1 ·	−4	= 4
Today	0 ·	−4	= 0
	1 ·	−4	= −4
	2 ·	−4	= −8
Three days from now	3 ·	−4	= −12
	4 ·	−4	= −16

These rules actually make perfectly good sense. For example, if I am flat broke today and am losing 4 dollars each day (that is, gaining −4 dollars each day), then 3 days from now I will be worth $3(-4) = -(3 \cdot 4) = -12$ dollars. On the other hand, 3 days ago (that is, on day −3), I must have been worth 12 dollars; that is, $(-3)(-4) = 3 \cdot 4 = 12$. The chart in the margin shows my worth each day starting 4 days ago and continuing to 4 days from now. Can you tell how much I was worth 6 days ago and how much I will be worth 10 days from now?

The problem set will give you practice in working with the integers. In the meantime, we note that with all their beauty and usefulness, the integers are plagued by a serious flaw: You cannot always divide them—that is, if you want an integer as the answer. For, while $\frac{4}{2}$, $\frac{9}{3}$, and $\frac{12}{2}$ are perfectly respectable integers, $\frac{1}{2}$, $\frac{3}{4}$, and $\frac{5}{3}$ are not. To make sense of these symbols requires a further enlargement of the number system.

> The informal use of fractions (ratios) is very old. We know from the Rhind papyrus that the Egyptians were quite proficient with fractions by 1650 B.C. However, they used a different notation and considered only ratios of positive integers.

The Rational Numbers In a certain sense, mathematicians are master inventors. When new kinds of numbers are needed, they invent them. Faced with the need to divide, mathematicians simply decided that all quotients (or ratios) of two integers a/b, with $b \neq 0$, were to be regarded as numbers. That meant that

$$\frac{3}{4} \quad \frac{7}{8} \quad \frac{-2}{3} \quad \frac{14}{-16} \quad \frac{8}{2}$$

and all similar ratios were numbers, numbers with all the rights and privileges of the integers and even a bit more: division, except by zero, was always possible. Naturally, these numbers were called **rational numbers** (ratio numbers).

The rational numbers are admirably suited for certain very practical measurement problems. Take a piece of string of length 1 unit and divide it into two parts of equal length. We say that each part has length $\frac{1}{2}$. Take the same piece of string and divide it into 4 equal parts. Then each part has length $\frac{1}{4}$ and two of them together have length $\frac{2}{4}$. Thus $\frac{1}{2}$ and $\frac{2}{4}$ must stand for the same number, that is,

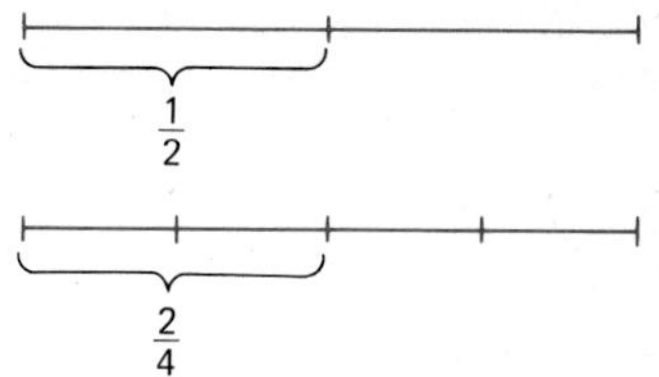

$$\frac{1}{2} = \frac{2 \cdot 1}{2 \cdot 2} = \frac{2}{4}$$

Considerations like this suggest an important agreement. We agree that

$$\boxed{\frac{a}{b} = \frac{k \cdot a}{k \cdot b}}$$

for any nonzero number k. Thus $\frac{1}{2}, \frac{2}{4}, \frac{3}{6}, -4/-8, \ldots$ are all treated as symbols for the same rational number.

We should learn to read equations backwards as well as forwards. Read backwards, the boxed equation tells us that we can divide numerator and denominator (top and bottom) of a quotient by the same nonzero number k. Or, in language that may convey the idea even better, it says that we can cancel a common factor from numerator and denominator.

$$\frac{24}{32} = \frac{\not{8} \cdot 3}{\not{8} \cdot 4} = \frac{3}{4}$$

Among the many symbols for the same rational number, one is given special honor, the reduced form. If numerator a and denominator b of the rational number a/b have no common integer divisors (factors) greater than 1 and if b is positive, we say a/b is in **reduced form**. Thus, $\frac{3}{4}$ is the reduced form of $\frac{24}{32}$ and $-2/3$ is the reduced form of $50/-75$.

We call attention to another rather obvious fact. Notice that $\frac{2}{1}$ is technically the reduced form of $\frac{4}{2}, \frac{6}{3}$, and so on. However, we almost never write $\frac{2}{1}$, since the ordinary meaning of division implies that $\frac{2}{1}$ is equal to the integer 2. In fact, for any integer a,

$$\boxed{\frac{a}{1} = a}$$

Thus the class of rational numbers contains the integers as a subclass.

Let us go back to that horizontal thermometer, the calibrated line we looked at earlier. Now we can label many more points. In fact, it seems that we can fill the line with labels.

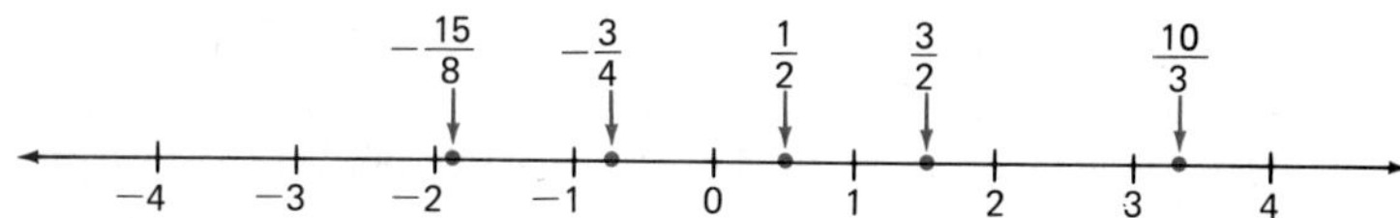

No discussion of the rational numbers is complete without mention of how to add, subtract, multiply, and divide them. You are familiar with these operations but perhaps you need to review them. Such a review is one of the major purposes of the following problem set.

Problem Set 1-2

Example A (Removing grouping symbols) Simplify $-3[2 - (6 + x)]$.

SOLUTION. Keep these two things in mind. First, always begin work with the innermost parentheses. Second, remember that a minus sign preceding a set of parentheses (or brackets) means all that is between them must be multiplied by -1, and so the sign of each term must be changed when the parentheses are removed. Thus

$$-3[2 - (6 + x)] = -3[2 - 6 - x] = -3[-4 - x] = 12 + 3x$$

Simplify each of the following.

1. $4 - 2(8 - 12)$
2. $-5 + 2(3 - 18)$
3. $-3 + 2[-5 - (12 - 3)]$
4. $-2[1 + 3(6 - 8) + 4] + 3$
5. $-4[3(-6 + 13) - 2(7 - 5)] + 1$
6. $5[-(7 + 12 - 16) + 4] + 2$
7. $-3[4(5 - x) + 2x] + 3x$
8. $-4[x - (3 - x) + 5] + x$
9. $2[-t(3 + 5 - 11) + 4t]$
10. $y - 2[-3(y + 1) + y]$

Example B (Reducing fractions) Reduce (a) $\frac{24}{36}$; (b) $\frac{3 + 9}{3 + 6}$; (c) $\frac{12 + 3x}{3x}$.

SOLUTION. You can cancel common factors (which are multiplied) from numerator and denominator. Do not make the common mistake of trying to cancel common terms (which are added).

(a) $\frac{24}{36} = \frac{\cancel{12}\cdot 2}{\cancel{12}\cdot 3} = \frac{2}{3}$

(b) $\frac{3 + 9}{3 + 6} = \frac{\cancel{3} + 9}{\cancel{3} + 6} = \frac{9}{6} = \frac{3}{2}$ Wrong

$\frac{3 + 9}{3 + 6} = \frac{12}{9} = \frac{\cancel{3}\cdot 4}{\cancel{3}\cdot 3} = \frac{4}{3}$ Right

(c) $\frac{12 + 3x}{3x} = \frac{12 + \cancel{3x}}{\cancel{3x}} = 12$ Wrong

$\frac{12 + 3x}{3x} = \frac{\cancel{3}(4 + x)}{\cancel{3}\cdot x} = \frac{4 + x}{x}$ Right

Reduce each of the following, leaving your answer with a positive denominator.

11. $\frac{24}{27}$
12. $\frac{16}{36}$
13. $\frac{45}{-60}$

14. $\dfrac{63}{-81}$ 15. $\dfrac{3 - 9x}{6}$ 16. $\dfrac{4 + 6x}{4 - 8}$

17. $\dfrac{4x - 6}{4 - 8}$ 18. $\dfrac{4x - 12}{8}$

Example C (Adding and subtracting fractions) Simplify (a) $\frac{3}{4} + \frac{5}{4}$; (b) $\frac{3}{5} + \frac{5}{4}$; (c) $\frac{8}{9} - \frac{5}{12}$.

SOLUTION. We add fractions with the same denominator by adding numerators. If the fractions have different denominators, we first rewrite them as equivalent fractions with the same denominator and then add. Similar rules apply for subtraction.

(a) $\dfrac{3}{4} + \dfrac{5}{4} = \dfrac{3 + 5}{4} = \dfrac{8}{4} = 2$

(b) $\dfrac{3}{5} + \dfrac{5}{4} = \dfrac{12}{20} + \dfrac{25}{20} = \dfrac{37}{20}$

(c) $\dfrac{8}{9} - \dfrac{5}{12} = \dfrac{32}{36} - \dfrac{15}{36} = \dfrac{17}{36}$

Now simplify each of the following.

19. $\dfrac{5}{6} + \dfrac{11}{12}$ 20. $\dfrac{8}{10} - \dfrac{3}{20}$

21. $\dfrac{4}{5} - \dfrac{3}{20} + \dfrac{3}{10}$ 22. $\dfrac{11}{24} + \dfrac{2}{3} - \dfrac{5}{12}$

23. $\dfrac{5}{12} + \dfrac{7}{18} - \dfrac{1}{6}$ 24. $\dfrac{11}{15} + \dfrac{3}{4} - \dfrac{5}{6}$

25. $\dfrac{-5}{27} + \dfrac{5}{12} + \dfrac{3}{4}$ 26. $\dfrac{23}{30} + \dfrac{2}{25} - \dfrac{3}{5}$

Example D (Multiplying and dividing fractions) Simplify (a) $\frac{3}{4}\cdot\frac{5}{7}$; (b) $\frac{3}{4}\cdot\frac{16}{27}$; (c) $\dfrac{\frac{3}{4}}{\frac{9}{16}}$.

SOLUTION. We multiply fractions by multiplying numerators and multiplying denominators. To divide fractions, invert (that is, take the reciprocal of) the divisor and multiply.

(a) $\dfrac{3}{4}\cdot\dfrac{5}{7} = \dfrac{3\cdot 5}{4\cdot 7} = \dfrac{15}{28}$

(b) $\dfrac{3}{4}\cdot\dfrac{16}{27} = \dfrac{3\cdot 16}{4\cdot 27} = \dfrac{\cancel{3}\cdot\cancel{4}\cdot 4}{\cancel{4}\cdot\cancel{3}\cdot 9} = \dfrac{4}{9}$

(c) $\dfrac{\frac{3}{4}}{\frac{9}{16}} = \dfrac{3}{4}\cdot\dfrac{16}{9} = \dfrac{\cancel{3}\cdot\cancel{4}\cdot 4}{\cancel{4}\cdot\cancel{3}\cdot 3} = \dfrac{4}{3}$

Simplify.

27. $\frac{5}{6} \cdot \frac{9}{15}$

28. $\frac{4}{13} \cdot \frac{5}{12}$

29. $\frac{3}{4} \cdot \frac{6}{15} \cdot \frac{5}{2}$

30. $\frac{9}{11} \cdot \frac{33}{5} \cdot \frac{15}{18}$

31. $\frac{\frac{5}{6}}{\frac{8}{12}}$

32. $\frac{\frac{9}{24}}{\frac{15}{12}}$

33. $\frac{\frac{3}{4}}{2}$

34. $\frac{\frac{6}{7}}{9}$

35. $\frac{6}{\frac{7}{9}}$

36. $\frac{3}{\frac{4}{5}} \cdot \frac{7}{5}$

Example E (Complicated fractions) Simplify (a) $\frac{\frac{2}{3} + \frac{1}{5}}{\frac{5}{7} - \frac{1}{2}}$; (b) $\frac{\frac{5}{6} - \frac{2}{15}}{\frac{11}{30} + \frac{3}{5}}$.

SOLUTION. We show two methods of attacking four-story expressions.
(a) Start by working with the top portion and bottom portion separately.

$$\frac{\frac{2}{3} + \frac{1}{5}}{\frac{5}{7} - \frac{1}{2}} = \frac{\frac{10}{15} + \frac{3}{15}}{\frac{10}{14} - \frac{7}{14}} = \frac{\frac{13}{15}}{\frac{3}{14}} = \frac{13}{15} \cdot \frac{14}{3} = \frac{182}{45}$$

(b) Multiply the top and bottom portions by a common denominator of all the simple fractions.

$$\frac{\frac{5}{6} - \frac{2}{15}}{\frac{11}{30} + \frac{3}{5}} = \frac{30(\frac{5}{6} - \frac{2}{15})}{30(\frac{11}{30} + \frac{3}{5})} = \frac{25 - 4}{11 + 18} = \frac{21}{29}$$

Simplify.

37. $\frac{\frac{2}{3} + \frac{3}{4}}{\frac{7}{12}}$

38. $\frac{\frac{3}{5} - \frac{3}{4}}{\frac{9}{20}}$

39. $\frac{\frac{2}{3} + \frac{3}{4}}{\frac{2}{3} - \frac{3}{4}}$

40. $\frac{\frac{8}{9} - \frac{2}{27}}{\frac{8}{9} + \frac{2}{27}}$

41. $\frac{\frac{5}{6} - \frac{1}{12}}{\frac{3}{4} + \frac{2}{3}}$

42. $\frac{\frac{3}{50} - \frac{1}{2} + \frac{4}{5}}{\frac{4}{25} + \frac{7}{10}}$

Miscellaneous Problems

Perform the indicated operations and simplify.

43. $\frac{5}{6} - \frac{1}{9} + \frac{2}{3}$

44. $\frac{3}{4} - \frac{7}{12} + \frac{2}{9}$

45. $-\frac{2}{3}\left(\frac{5}{4} - \frac{1}{12}\right)$

46. $\frac{3}{2}\left(\frac{8}{9} - \frac{5}{6}\right)$

47. $\frac{1}{3}\left[\frac{1}{2}\left(\frac{1}{4} - \frac{1}{3}\right) + \frac{1}{6}\right]$

48. $-\frac{1}{3}\left[\frac{2}{5} - \frac{1}{2}\left(\frac{1}{3} - \frac{1}{5}\right)\right]$

49. $\frac{14}{33}\left(\frac{2}{3} - \frac{1}{7}\right)^2$

50. $\left(\frac{5}{7} + \frac{7}{9}\right) \div \frac{3}{2}$

51. $1 \div (2 + \frac{3}{4})$

52. $15(2 - \frac{3}{5})$

53. $\frac{\frac{11}{49} - \frac{3}{7}}{\frac{11}{49} + \frac{3}{7}}$

54. $\frac{\frac{1}{2} - \frac{3}{4} + \frac{7}{8}}{\frac{1}{2} + \frac{3}{4} - \frac{7}{8}}$

55. $2 + \dfrac{\frac{3}{4} + \frac{5}{12}}{\frac{2}{3}}$

56. $\dfrac{11}{12} - \dfrac{\frac{2}{3} - \frac{3}{4}}{\frac{5}{6} + \frac{1}{12}}$

57. $1 - \dfrac{2}{2 + \frac{3}{4}}$

58. $2 + \dfrac{3}{1 + \frac{5}{2}}$

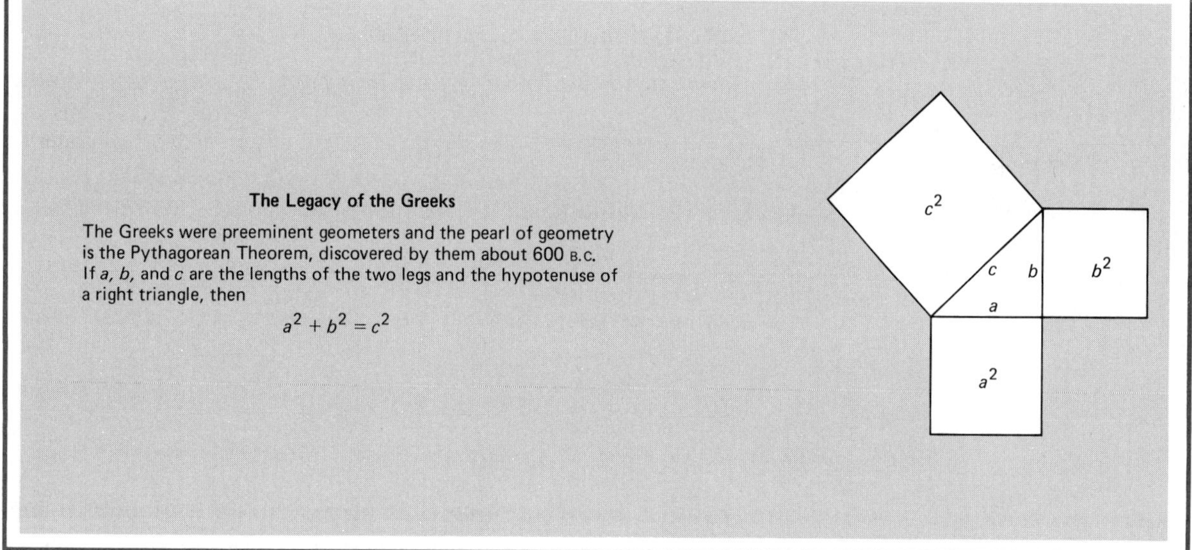

1-3 The Real Numbers

The Pythagorean Theorem, that beautiful gem of geometry, came to be the great nemesis of Greek mathematics. For hidden in that elegant formula is a consequence that seemed to destroy the Greek conception of numbers. Here is a paraphrased version of an old story reported by Euclid.

"Consider," said Euclid, "a right triangle with legs of unit length and hypotenuse of length C. By the Theorem of Pythagoras,

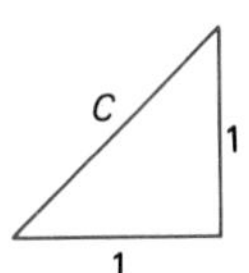

$$C^2 = 1^2 + 1^2 = 2$$

That is, $C = \sqrt{2}$. But then, as I will show by a logical argument, the quantity $\sqrt{2}$, whatever it is, is not a rational number. To put it bluntly, if only rational numbers exist, then the simplest right triangle we know has a hypotenuse whose length cannot be measured."

Euclid's argument (soon to be presented) is flawless; it is also devilishly clever. But to understand it, we need to know some basic facts about prime numbers.

A Digression on Prime Numbers A **prime number** is a whole number with exactly two whole-number divisors, itself and 1. The first few primes are

$$2, 3, 5, 7, 11, 13, 17, 19, 23, 29$$

Prime numbers are the building blocks of other whole numbers. For example,

$$18 = 2 \cdot 3 \cdot 3 \qquad 40 = 2 \cdot 2 \cdot 2 \cdot 5$$

This type of factorization is possible for all nonprime whole numbers greater than 1.

FUNDAMENTAL THEOREM OF ARITHMETIC

Any nonprime whole number (greater than 1) can be written as the product of a unique set of prime numbers.

This theorem is important in many parts of mathematics (see the problem set), but here we want to point out one simple consequence. When the square of any whole number is written as a product of primes, each prime occurs as a factor an even number of times. For example,

$$(18)^2 = 18 \cdot 18 = 2 \cdot 3 \cdot 3 \cdot 2 \cdot 3 \cdot 3 = \underbrace{2 \cdot 2}_{\text{two 2's}} \cdot \underbrace{3 \cdot 3 \cdot 3 \cdot 3}_{\text{four 3's}}$$

$$(40)^2 = 40 \cdot 40 = 2 \cdot 2 \cdot 2 \cdot 5 \cdot 2 \cdot 2 \cdot 2 \cdot 5 = \underbrace{2 \cdot 2 \cdot 2 \cdot 2 \cdot 2 \cdot 2}_{\text{six 2's}} \cdot \underbrace{5 \cdot 5}_{\text{two 5's}}$$

The Proof that $\sqrt{2}$ is Irrational Suppose that $\sqrt{2}$ is a rational number, that is, suppose that $\sqrt{2} = m/n$, where m and n are whole numbers (necessarily greater than 1). Then

$$2 = \frac{m^2}{n^2}$$

and so

$$2n^2 = m^2$$

Now imagine that both n and m are written as products of primes. As we saw above, both n^2 and m^2 must then have an even number of 2's as factors. Thus in the above equation, the prime 2 appears on the left an odd number of times but on the right an even number of times. This is clearly impossible. What can be wrong? The only thing that can be wrong is our supposition that $\sqrt{2}$ was a rational number.

> Reductio ad absurdum, which Euclid loved so much, is one of a mathematician's finest weapons. It is a far finer gambit than any chess gambit: a chess player may offer the sacrifice of a pawn or even a piece, but a mathematician offers the game.
>
> G. H. Hardy

To let one number $\sqrt{2}$ through the dike was bad enough. But, as Euclid realized, a host of others came pouring through with it. Exactly the same proof shows that $\sqrt{3}$, $\sqrt{5}$, $\sqrt{7}$, and, in fact, the square roots of all primes are irrational. The Greeks, who steadfastly insisted that all measurements must be based on whole numbers and their ratios, could not find a satisfactory way out of this dilemma. Today we recognize that the only adequate solution is to enlarge the number system.

The Real Numbers Let us take a bold step and simply declare that every line segment shall have a number that measures its length. The set of all numbers that can measure lengths, together with their negatives and zero, constitute the **real numbers.** Thus the rational numbers are automatically real numbers; the positive rational numbers certainly measure lengths.

Consider again the thermometer on its side, the calibrated line. We may have thought we had labeled every point. Not so; there were many holes corresponding to what we now call $\sqrt{2}$, $\sqrt{5}$, π, and so on. But with the introduction of the real numbers, all the holes are filled in; every point has a number label. Because of this, we often call this calibrated line the **real line**.

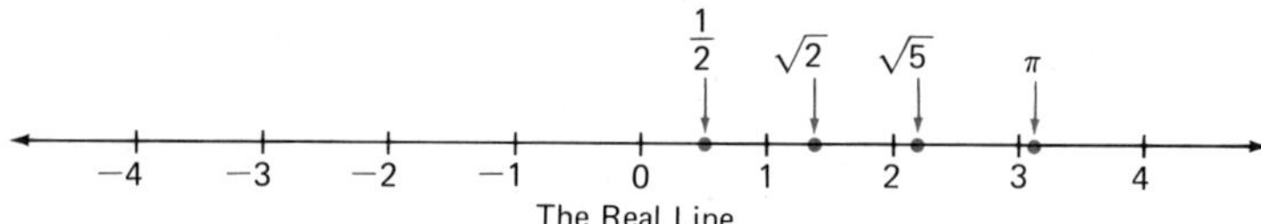

The Real Line

Decimals There is another important way to describe the real numbers. It calls for review of a basic idea. Recall that

$$.4 = \frac{4}{10} \qquad .7 = \frac{7}{10}$$

Similarly,

$$.41 = \frac{4}{10} + \frac{1}{100} = \frac{40}{100} + \frac{1}{100} = \frac{41}{100}$$

$$.731 = \frac{7}{10} + \frac{3}{100} + \frac{1}{1000} = \frac{700}{1000} + \frac{30}{1000} + \frac{1}{1000} = \frac{731}{1000}$$

It is a simple matter to locate a decimal on the number line. For example, to locate 1.4, we divide the interval from 1 to 2 into ten equal parts and pick the fourth point of division.

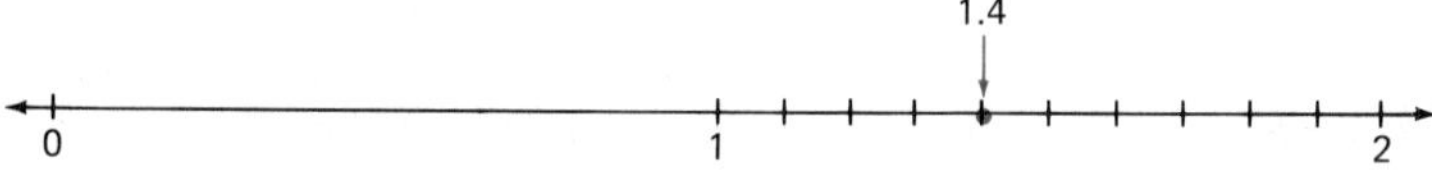

If the interval from 1.4 to 1.5 is divided into ten equal parts, the second point of division corresponds to 1.42.

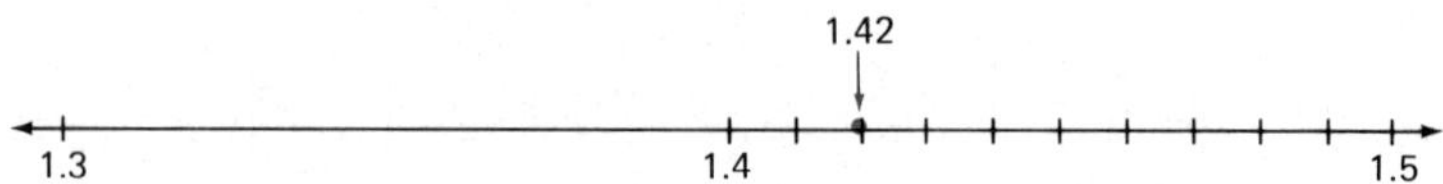


```
    .875
8)7.000
   64
    60
    56
     40
     40
```

We can find the decimal corresponding to a rational number by long division. For example, the division in the margin shows that $\frac{7}{8} = .875$. If we try the same procedure on $\frac{1}{3}$, something different happens. The decimal just keeps on going; it is an **unending decimal**. Actually, the terminating decimal .875 can be thought of as unending if we annex zeros. Thus

```
   .33333
3)1.00000
    9
    10
     9
     10
      9
      10
       9
       10
        9
        1
```

$$\frac{7}{8} = .875 = .8750000\ldots$$
$$\frac{1}{3} = .3333333\ldots$$
$$\frac{1}{6} = .16666\ldots$$

Let us take an example that is a bit more complicated, $\frac{2}{7}$.

```
   .285714
7)(2.0)000000
   14
    60
    56
     40
     35
      50
      49
       10
        7
        30
        28
        (20)
```

If we continue the division, the pattern must repeat (note the circled 20's). Thus

$$\frac{2}{7} = .285714285714285714\ldots$$

which can also be written

$$\frac{2}{7} = \overline{.285714}$$

The bar indicates that the pattern of digits 285714 repeats indefinitely.

In fact, the decimal expansion of any rational number must inevitably start repeating, because there are only finitely many different possible remainders in the division process (at most as many as the divisor). It is a remarkable coincidence that the converse statement is also true. A repeating decimal must inevitably represent a rational number (see Problems 25–31). Thus the rational numbers are precisely those numbers that can be represented by repeating decimals.

What about the nonrepeating, unending decimals like

$$.12112111211112\ldots$$

They represent the **irrational numbers**. And they, together with the rational numbers, constitute the real numbers.

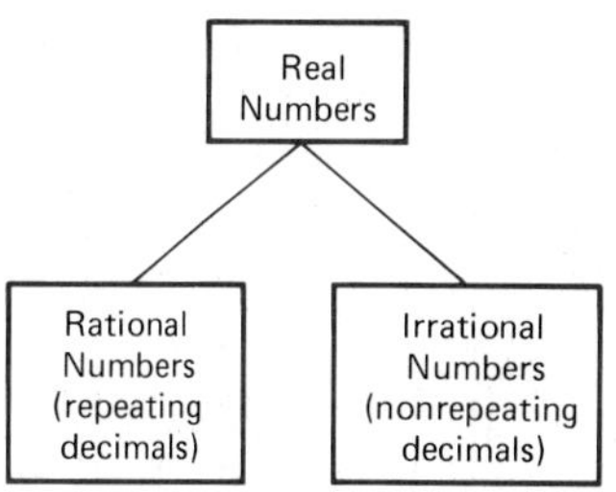

We showed that $\sqrt{2}$ is not rational (that is, it is irrational). It too has a decimal expansion.

$$\sqrt{2} = 1.414214\ldots$$

Actually, the decimal expansion of $\sqrt{2}$ is known to several thousand places. It does not repeat. It cannot. It is a fact of mathematics.

Problem Set 1-3

Example A (Prime factorization) Find the prime factorizations of 168 and 420.

SOLUTION. Factor out as many 2's as possible, then 3's, then 5's, and so on.

$$168 = 2\cdot84 = 2\cdot2\cdot42 = 2\cdot2\cdot2\cdot21 = 2\cdot2\cdot2\cdot3\cdot7$$

$$420 = 2\cdot210 = 2\cdot2\cdot105 = 2\cdot2\cdot3\cdot35 = 2\cdot2\cdot3\cdot5\cdot7$$

Write the prime factorization of each number.

1. 250
2. 504
3. 200
4. 2079
5. 2100
6. 1650

Example B (Least common multiple) The least common multiple (lcm) of several positive integers is the smallest positive integer that is a multiple of all of them. Find the least common multiple of 168 and 420.

SOLUTION. We found the prime factorizations of these two numbers in Example A. To find the least common multiple of 168 and 420, write down the product of all factors that occur in either number, repeating a factor according to the greatest number of times it occurs in either number.

$$\text{lcm}(168, 420) = 2\cdot2\cdot2\cdot3\cdot5\cdot7 = 840$$

Find each of the following using the information you obtained in Problems 1–6.

7. lcm(250, 200)
8. lcm(504, 2079)
9. lcm(250, 2100)
10. lcm(504, 1650)
11. lcm(250, 200, 2100)
12. lcm(504, 2079, 1650)

Example C (Least common denominator) Calculate $\frac{5}{168} + \frac{13}{420}$.

SOLUTION. Write both fractions with a common denominator. The best choice of common denominator is the least common multiple of 168 and 420, namely 840, obtained in Example B. We call it the least common denominator.

$$\frac{5}{168} + \frac{13}{420} = \frac{5 \cdot 5}{840} + \frac{13 \cdot 2}{840} = \frac{25 + 26}{840} = \frac{51}{840} = \frac{17}{280}$$

Calculate each of the following, using the results of Problems 7–12.

13. $\frac{3}{250} + \frac{17}{200}$

14. $\frac{5}{504} - \frac{1}{2079}$

15. $\frac{7}{250} - \frac{1}{2100}$

16. $\frac{13}{504} + \frac{13}{1650}$

17. $\frac{3}{250} - \frac{17}{200} + \frac{11}{2100}$

18. $\frac{13}{504} + \frac{13}{1650} - \frac{17}{2079}$

Example D (Rational numbers as repeating decimals) Write $\frac{68}{165}$ as a repeating decimal.

SOLUTION.

```
       .412
165/68.0000
    660
    ---
     200
     165
     ---
      350
      330
      ---
       200
```

Answer. $\frac{68}{165} = .41212\ldots = .4\overline{12}$.

Find the repeating decimal expansion for each number. Use the bar notation for your answer.

19. $\frac{2}{3}$

20. $\frac{3}{5}$

21. $\frac{5}{8}$

22. $\frac{13}{11}$

23. $\frac{6}{13}$

24. $\frac{4}{13}$

Example E (Repeating decimals as rational numbers) Write $.\overline{24}$ as the ratio of two integers.

SOLUTION. Let $x = .\overline{24} = .242424\ldots$. Then $100x = 24.242424\ldots$. Subtract x from $100x$ and simplify.

$$\begin{aligned} 100x &= 24.2424\ldots \\ x &= .2424\ldots \\ 99x &= 24 \\ x &= \tfrac{24}{99} = \tfrac{8}{33} \end{aligned}$$

Note that we multiplied x by 100 because x is a decimal that repeats in a two-digit group. If the decimal had repeated in a three-digit group, we would have multiplied by 1000.

Write each of the following as a ratio of two integers.

25. $.\overline{7}$ 26. $.\overline{123}$ 27. $.\overline{235}$

28. $.875$ 29. $.325$ 30. $.5\overline{21}$

31. $.3\overline{21}$

Miscellaneous Problems

32. Find the prime factorization of 420 and 750.
33. Find the least common multiple of 420 and 750. Use the result of Problem 32.
34. Calculate each of the following.

 (a) $\frac{17}{420} + \frac{11}{750}$ (b) $\frac{17}{420} \div \frac{11}{750}$

35. Find the repeating decimal expansion of each of the following.
 (a) $\frac{5}{16}$ (b) $\frac{3}{11}$
36. Write as a ratio of two integers.
 (a) $.\overline{27}$ (b) $.\overline{207}$
37. Write the sum $.\overline{23} + .\overline{405}$ as a repeating decimal.
38. Is the sum of two rational numbers necessarily rational?
39. Is the sum of two irrational numbers necessarily irrational?
40. Answer the questions 38 and 39 with the word *sum* replaced by *product*.
41. Show that $\sqrt{2} + \frac{3}{4}$ is irrational. (*Hint*: Let $\sqrt{2} + \frac{3}{4} = r$. By adding $-\frac{3}{4}$ to both sides, show that it is impossible for r to be rational.)
42. Show that if x is irrational and y is rational, then $x + y$ is irrational.
43. Show that $\frac{3}{4}\sqrt{2}$ is irrational.
44. Show that if x is a nonzero rational number and y is an irrational number, then xy is irrational.
45. Which of the following are rational numbers?

 (a) $\sqrt{\frac{8}{18}}$ (b) $(3\sqrt{2})(\sqrt{2})$

 (c) $\sqrt{2}(\sqrt{2} + 1)$ (d) $(3 + 2\sqrt{2}) + (4 - 2\sqrt{2})$

 (e) $\dfrac{4\sqrt{2} + 2}{2\sqrt{2} + 1}$ (f) $(.12)\sqrt{2}$

 (g) $.\overline{12}$ (h) $.12112111211112\ldots$
46. What is the smallest positive integer?
47. Write a positive rational number that is smaller than .00000001. Is there a smallest positive rational number?
48. The number $(1/10{,}000{,}000)\sqrt{2}$ is irrational (see Problem 44 of this set). Write a smaller positive irrational number. Is there a smallest positive irrational number?
49. Mimic the proof that $\sqrt{2}$ is irrational (see the first part of this section) to show that $\sqrt{3}$ is irrational.
50. Show that the square root of any prime number is irrational.

51. It is known that π is irrational. What does this mean about its decimal expansion?

52. Show that $\sqrt{\pi}$ is irrational.

53. The beginning of the decimal expansion of π is

$$\pi = 3.14159\ldots$$

Is $\pi - \frac{22}{7}$ positive, negative, or zero?

Playing by the Rules

Chess has rules; so does mathematics. You can't move a pawn backwards; you must not divide by zero. Chess without rules is no game at all. Numbers without definite properties are worthless curiosities. A good chess player and a good mathematician play by the rules.

1-4 Fundamental Properties of the Real Numbers

Now we face an important question. Are the real numbers adequate to handle all the applications we are likely to encounter? Or will some problems arise that will force us to enlarge our number system again? The answer is that the real numbers are sufficient for most purposes. Except for one type of problem, to be described in Section 1-6, we will do our work within the context of the real numbers. From now on, when we say number with no qualifying adjective, we mean real number. You can count on it.

All of this suggests that we ought to pay some attention to the fundamental properties of the real numbers. By a fundamental property, we mean something so basic that we must understand it, and to understand means more than to memorize. Understanding a property means to see the purpose the property serves, to recognize its implications, and to be able to derive other things from it.

Associative Property Addition and multiplication are the fundamental operations; subtraction and division are offshoots of them. These operations are binary operations, that is, they work on two numbers at a time. Thus $3 + 4 + 5$ is technically meaningless. We ought to write either $3 + (4 + 5)$ or $(3 + 4) + 5$. But, luckily, it really does not matter; we get the same answer either way. Addition is **associative**.

$$a + (b + c) = (a + b) + c$$

Thus we can write $3 + 4 + 5$ or even $3 + 4 + 5 + 6 + 7$ without ambiguity. The answer will be the same regardless of the order in which the addition is done.

Addition and multiplication are like Siamese twins: What is true for one is quite likely to be true for the other. Thus multiplication, too, is associative.

$$a \cdot (b \cdot c) = (a \cdot b) \cdot c$$

If we wish, we can write $a \cdot b \cdot c$ with no parentheses at all.

Commutative Property It makes some difference whether you first put on your slippers and then take a bath or vice versa—the difference between wet and dry slippers. But for addition or multiplication, order does not matter. Both operations are **commutative**.

$$a + b = b + a$$

$$a \cdot b = b \cdot a$$

Thus

$$3 + 4 = 4 + 3$$

$$3 + 4 + 6 = 4 + 3 + 6 = 6 + 4 + 3$$

and

$$7 \cdot 8 \cdot 9 = 8 \cdot 7 \cdot 9 = 9 \cdot 8 \cdot 7$$

Neutral Elements To be neutral is to sit on the sidelines and refuse to do battle. A neutral party can be ignored; the outcome is not affected by its presence. Thus we call 0 the **neutral element** for addition; its presence can be ignored in an addition.

$$\boxed{a + 0 = 0 + a = a}$$

Similarly, 1 is the neutral element for multiplication since

$$\boxed{a \cdot 1 = 1 \cdot a = a}$$

The numbers 0 and 1 are also called the **identity elements** for addition and multiplication, respectively.

While 0 is inactive in addition, its effect in multiplication is overwhelming; any number multiplied by 0 is completely wiped out.

$$a \cdot 0 = 0 \cdot a = 0$$

We have not put this property in a box; it is not quite as basic as the others since it can be derived from them (see Problem 37). However, you should still know it.

Inverses The numbers 3 and -3 are like an acid and a base: When you add them together, they neutralize each other. We refer to them as inverses of each other. In fact, every number a has an **additive inverse** $-a$ (also called the negative of a). It satisfies

$$\boxed{a + (-a) = (-a) + a = 0}$$

Similarly, every number a different from 0 has a **multiplicative inverse** a^{-1} satisfying

$$\boxed{a \cdot a^{-1} = a^{-1} \cdot a = 1}$$

Thus, $3^{-1} = \frac{1}{3}$ since $3 \cdot \frac{1}{3} = 1$. In fact, for any $a \neq 0$, $a^{-1} = 1/a$, and for this reason we often say the "reciprocal of a" rather than the "multiplicative inverse of a."

Distributive Property When we indicate an addition and a multiplication in the same expression, we face a problem. For example, what do we mean by $3 + 4 \cdot 2$? If we mean $(3 + 4) \cdot 2$, the answer is 14; if we mean $3 + (4 \cdot 2)$, the answer is 11. Most of us would not give the first answer. We are so familiar with a convention that we use it without thinking, but now is a good time to emphasize it. In any expression involving additions and multiplications which has no parentheses, we agree to do all the multiplications first. Thus

$$4 \cdot 5 + 3 = 20 + 3 = 23$$

and

$$4 \cdot 5 + 6 \cdot 2 = 20 + 12 = 32$$

We can always overrule this agreement by inserting parentheses. For example,

$$4 \cdot (5 + 3) = 4 \cdot 8 = 32$$

The agreement just described is a matter of convenience; no law forces it upon us. However, the **distributive property**,

$$a \cdot (b + c) = a \cdot b + a \cdot c$$
$$(b + c) \cdot a = b \cdot a + c \cdot a$$

is not a matter of choice or convenience. Rather, it is another of the fundamental properties of numbers. That it must hold for the positive integers is almost obvious, as may be seen by examining the diagram in the margin. But we assert that it is just as true that

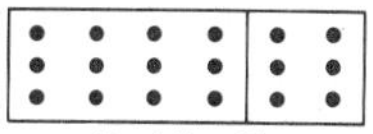

$$\tfrac{1}{2} \cdot (\sqrt{2} + \pi) = \tfrac{1}{2} \cdot \sqrt{2} + \tfrac{1}{2} \cdot \pi$$

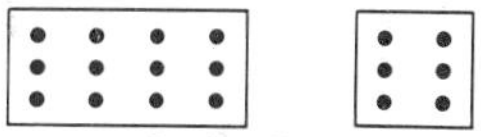

$3 \cdot (4 + 2)$

$3 \cdot 4 + 3 \cdot 2$

Actually, we use the distributive property all the time, often without realizing it. The familiar calculation in the margin is really a shorthand version of

$$\begin{aligned} 65 \cdot 34 = (5 + 60)34 &= 5 \cdot 34 + 60 \cdot 34 \\ &= 170 + 2040 = 2210 \end{aligned}$$

$$\begin{array}{r} 34 \\ \times\ \underline{65} \\ 170 \\ +\ \underline{204} \\ 2210 \end{array}$$

Subtraction and Division Addition and multiplication are the basic operations; subtraction and division are dependent on them. Subtraction is the addition of an additive inverse and division is multiplication by a reciprocal. Thus

$$a - b = a + (-b)$$

and

$$a \div b = a \cdot b^{-1} = a \cdot \frac{1}{b}$$

Clearly $3 - 5 \neq 5 - 3$ and $3 \div 5 \neq 5 \div 3$, which tell us that subtraction and division are not commutative. Nor are they associative. In spite of these drawbacks, these operations are important to us.

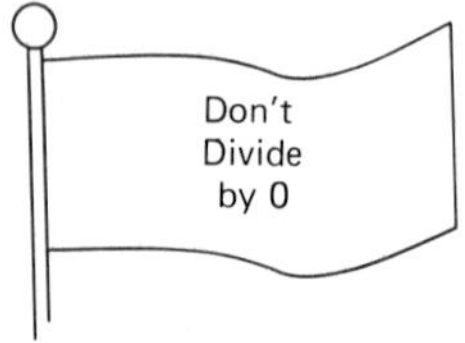

There is one restriction: we never divide by zero. Why do we exclude $\frac{4}{0}$, $\frac{6}{0}$, $\frac{10}{0}$, and similar expressions from our consideration? We exclude them to maintain consistency. If $\frac{4}{0}$ were a number q, that is, if $\frac{4}{0} = q$, then $4 = 0 \cdot q = 0$, which is nonsense. The symbol $\frac{0}{0}$ is meaningless for a different reason: If $\frac{0}{0} = p$, then $0 = 0 \cdot p$, which is true for any number p. We choose to avoid such ambiguities by excluding division by zero.

Problem Set 1-4

Example A (Simple calculations) Calculate $31.9 + 45 + 68.1 + 155 + 43.2$.

SOLUTION. Intelligent use of the associative and commutative properties makes this a breeze.

$$(31.9 + 68.1) + (45 + 155) + 43.2 = 100 + 200 + 43.2 = 343.2$$

Find the following sums or products using the basic properties we have introduced. If you can do it all in your head, that will be fine.

1. $420 + 431 + 580$
2. $99{,}985 + 67 + 15$
3. $983 + 400 + 300 + 17$
4. $8.75 + 14 + 36 + 1.25$
5. $\frac{11}{13} + 43 + \frac{2}{13} + 17$
6. $\frac{15}{8} + \frac{5}{6} + \frac{1}{6} + \frac{9}{8}$
7. $6 \cdot \frac{3}{4} \cdot \frac{1}{6} \cdot 4$
8. $99 + 98 + 97 + 3 + 2 + 1$
9. $5 \cdot \frac{1}{3} \cdot \frac{2}{5} \cdot 6 \cdot \frac{1}{2}$
10. $(.25)(363)(400)(\frac{1}{3})$

Example B (Number properties) What properties justify each of the following?
(a) $6 + [5 + (-6)] = [6 + (-6)] + 5$ (b) $-\frac{1}{3}(\frac{6}{7} - \frac{9}{11}) = -\frac{2}{7} + \frac{3}{11}$

SOLUTION. (a) Commutative and associative properties for addition:

$$6 + [5 + (-6)] = 6 + [(-6) + 5] = [6 + (-6)] + 5$$

(b) The definition of subtraction and the distributive property:

$$-\tfrac{1}{3}(\tfrac{6}{7} - \tfrac{9}{11}) = -\tfrac{1}{3}(\tfrac{6}{7} + (-\tfrac{9}{11})) = -\tfrac{1}{3} \cdot \tfrac{6}{7} + (-\tfrac{1}{3})(-\tfrac{9}{11}) = -\tfrac{2}{7} + \tfrac{3}{11}$$

Name the properties that justify each of the following.

11. $\frac{5}{6}(\frac{3}{4} \cdot 12) = (12 \cdot \frac{5}{6})\frac{3}{4}$
12. $4 + (.52 - 2) = (4 - 2) + .52$
13. $(2 + 3) + 4 = 4 + (2 + 3)$
14. $9(48) = 9(40) + 9(8)$
15. $6 + [-6 + 5]$
    $= [6 + (-6)] + 5 = 5$
16. $6[\frac{1}{6} \cdot 4] = [6 \cdot \frac{1}{6}]4 = 4$
17. $-(\sqrt{5} + \sqrt{3} - 5)$
    $= -\sqrt{5} - \sqrt{3} + 5$
18. $(b/c)(b/c)^{-1} = 1$
19. $(x + 4)(x + 2)$
    $= (x + 4)x + (x + 4)2$
20. $(a + b)a^{-1} = 1 + b/a$

Example C (More on properties) Which of the following are true for all real numbers a, b, and c?
(a) $(a - b) - c = a - (b - c)$ (b) $(a + b) \div c = (a \div c) + (b \div c)$

SOLUTION. (a) Not true; subtraction is not associative. For example, if $a = 12$, $b = 9$, and $c = 5$, then

$$(a - b) - c = (12 - 9) - 5 = 3 - 5 = -2$$
$$a - (b - c) = 12 - (9 - 5) = 12 - 4 = 8$$

(b) True, provided $c \neq 0$:

$$(a + b) \div c = (a + b)(1/c) = a(1/c) + b(1/c) = (a \div c) + (b \div c)$$

Which of the following are true for all choices of a, b, and c? If false, provide an example. If true, provide a demonstration similar to that in part (b) above.

21. $a - (b - c) = a - b + c$
22. $a + bc = ac + bc$
23. $a \div (b + c)$ $= (a \div b) + (a \div c)$
24. $(a + b)^{-1} = a^{-1} + b^{-1}$
25. $ab(a^{-1} + b^{-1}) = b + a$
26. $a(a + b + c) = a^2 + ab + ac$
27. $(a + b)(a^{-1} + b^{-1}) = 1$
28. $(a + b)(a + b)^{-1} = 1$
29. $(a + b)(a + b) = a^2 + b^2$
30. $(a + b)(a - b) = a^2 - b^2$
31. $a \div (b \div c) = (a \div b) \div c$
32. $a \div (b \div c) = (a \cdot c) \div b$

Example D (Some new binary operations) Define binary operations $\oplus$ and $\odot$ for the real numbers by

$$a \oplus b = \frac{a + b}{2} \quad \text{and} \quad a \odot b = ab + 1$$

(a) Are these operations commutative? (b) Are they associative? (c) Does the distributive property $a \odot (b \oplus c) = (a \odot b) \oplus (a \odot c)$ hold?

SOLUTION. (a) $\oplus$ is commutative since $(a + b)/2 = (b + a)/2$; $\odot$ is commutative since $ab + 1 = ba + 1$. (b) $\oplus$ is not associative since

$$(a \oplus b) \oplus c = \frac{a + b}{2} \oplus c = \frac{(a + b)/2 + c}{2} = \frac{a + b + 2c}{4}$$

$$a \oplus (b \oplus c) = a \oplus \frac{b + c}{2} = \frac{a + (b + c)/2}{2} = \frac{2a + b + c}{4}$$

$\odot$ is not associative since

$$(a \odot b) \odot c = (ab + 1) \odot c = (ab + 1)c + 1 = abc + c + 1$$
$$a \odot (b \odot c) = a \odot (bc + 1) = a(bc + 1) + 1 = abc + a + 1$$

(c) The distributive property holds since

$$a \odot (b \oplus c) = a \odot \frac{b + c}{2} = a\left(\frac{b + c}{2}\right) + 1 = \frac{ab + ac + 2}{2}$$

$$(a \odot b) \oplus (a \odot c) = (ab + 1) \oplus (ac + 1) = \frac{ab + ac + 2}{2}$$

33. Define $\oplus$ and $\odot$ by $a \oplus b = a + 2b$ and $a \odot b = 3ab$.
    (a) Is $\oplus$ commutative? Associative?
    (b) Is $\odot$ commutative? Associative?
    (c) Is $a \odot (b \oplus c) = (a \odot b) \oplus (a \odot c)$?

34. Let $a \oplus b = a$ and $a \odot b = \text{lcm}(a, b)$ (see page 17).
    (a) Is $\oplus$ commutative? Associative?
    (b) Is $\odot$ commutative? Associative?
    (c) Is $a \odot (b \oplus c) = (a \odot b) \oplus (a \odot c)$? Is $a \oplus (b \odot c) = (a \oplus b) \odot (a \oplus c)$?

35. Let $a \oplus b$ be the minimum of a and b and $a \odot b$ be the maximum of a and b, these operations being considered for the set $\{1, 2, 3, 4, 5\}$. Thus $3 \oplus 5 = 3$, $3 \odot 5 = 5$, and $3 \oplus 3 = 3$.
    (a) Is $\oplus$ commutative? Associative?
    (b) Is $\odot$ commutative? Associative?
    (c) Is there a neutral element for $\oplus$? What is it?
    (d) What is the neutral element for $\odot$?
    (e) Does each element have an additive inverse?
    (f) Does each element have a multiplicative inverse?
    (g) Does $a \odot (b \oplus c) = (a \odot b) \oplus (a \odot c)$?

36. Consider the collection of all subsets of the integers and let $\oplus$ denote union and $\odot$ denote intersection (normally denoted by $\cup$ and $\cap$). Answer the questions of Problem 35 for these operations.

37. We claimed that we could show that $a \cdot 0 = 0$ for all a using only the properties displayed in boxes. Justify each of the following equalities by one of those properties.

$$\begin{aligned} 0 &= -(a \cdot 0) + a \cdot 0 \\ &= -(a \cdot 0) + a \cdot (0 + 0) \\ &= -(a \cdot 0) + (a \cdot 0 + a \cdot 0) \\ &= [-(a \cdot 0) + a \cdot 0] + a \cdot 0 \\ &= 0 + a \cdot 0 \\ &= a \cdot 0 \end{aligned}$$

38. Demonstrate that $(-a) \cdot b = -(a \cdot b)$ and $(-a) \cdot (-b) = a \cdot b$ using only the properties displayed in boxes and the result of Problem 37.

39. Let $\#$ represent the exponentiation operation, that is, $a \# b = a^b$. Is $\#$ commutative? Is $\#$ associative?

"I wonder, sir, if you would be willing to move to the end?"

1-5 Order and Absolute Value

We may question the bartender's tact, but not his mathematical taste. He has a deep feeling for an important notion that we call order; it is intimately tied up with the real number system. To describe this notion, we introduce a special symbol $<$; it stands for the phrase "is less than."

We begin by recalling that every real number (except 0) falls into one of two classes. Either it is positive or it is negative. Then, given two real numbers a and b, we say that

$$a < b \text{ if } b - a \text{ is positive}$$

Thus $-3 < -2$ since $-2 - (-3)$ is positive. Similarly, $3 < \pi$ since

$$\pi - 3 = 3.14159\ldots - 3 = .14159\ldots$$

which is a positive number.

Another and more intuitive way to think about the symbol $<$ is to relate it to the real number line. To say that $a < b$ means that a is to the left of b on the real line.

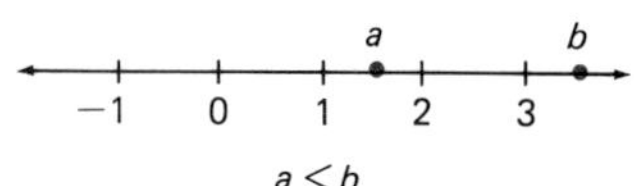

$a < b$

The symbol $<$ has a twin sister, denoted by $>$, which is read "is greater than." If you know how $<$ behaves, you automatically know how $>$ behaves. Thus there is no need to say much about $>$. It is enough to note that $b > a$ means exactly the same thing as $a < b$. In particular, $b > 0$ and $0 < b$ mean the same thing; both say that b is a positive number. Relations like $a < b$ and $b > a$ are called **inequalities**.

Properties of < If Homer is shorter than Ichabod and Ichabod is shorter than Jehu, then of course Homer is shorter than Jehu. This and other properties of the "less than" relation seem almost obvious. The following is a formal statement of the three most important properties.

PROPERTIES OF INEQUALITIES

1. (Transitivity). If $a < b$ and $b < c$, then $a < c$.
2. (Addition). If $a < b$, then $a + c < b + c$.
3. (Multiplication). If $a < b$ and $c > 0$, then $a \cdot c < b \cdot c$
   If $a < b$ and $c < 0$, then $a \cdot c > b \cdot c$

Property 2 says that you can add the same number to both sides of an inequality. It also says that you can subtract the same number from both sides, since c can be negative. Property 3 has a catch. Notice that if we multiply both sides by a positive number, we preserve the direction of the inequality; however, if we multiply by a negative number, we reverse the direction of the inequality. Thus

$$2 < 3$$

and

$$2 \cdot 4 < 3 \cdot 4$$

but

$$2(-4) > 3(-4)$$

These facts are shown on the following number line.

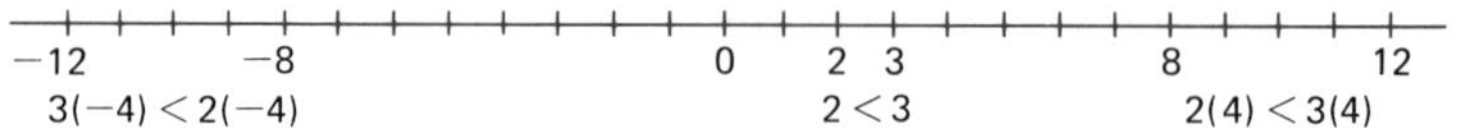

Division, which is equivalent to multiplication by the reciprocal, also satisfies Property 3.

The ≤ Relation In addition to its twin sister >, the symbol < has a half sister. It is denoted by ≤ and is read "is less than or equal to." We say

$a \leq b$ if $b - a$ is either positive or zero

For example, it is correct to say $2 \le 3$; it is also correct to say $2 \le 2$. This new relation behaves very much like $<$. In fact, if we put a bar under every $<$ and $>$ in the *Properties of Inequalities* displayed above, the resulting statements will be correct. Naturally, $b \ge a$ means the same thing as $a \le b$.

Intervals We can use the order symbols $<$ and $\le$ to describe intervals on the real line. When we write $-1.5 < x \le 3.2$, we mean that x is simultaneously greater than -1.5 and less than or equal to 3.2. The set of all such numbers is the interval shown below.

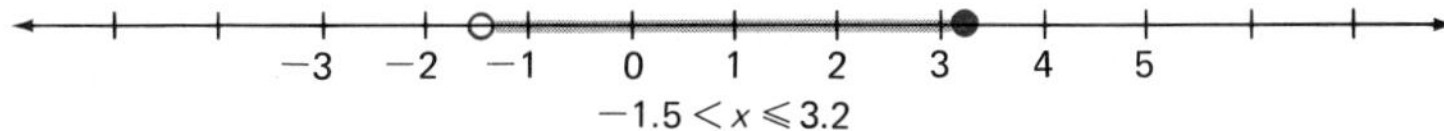

The small circle at the left indicates that -1.5 is left out; the heavy dot at the right indicates that 3.2 is included. The following diagram illustrates other possibilities.

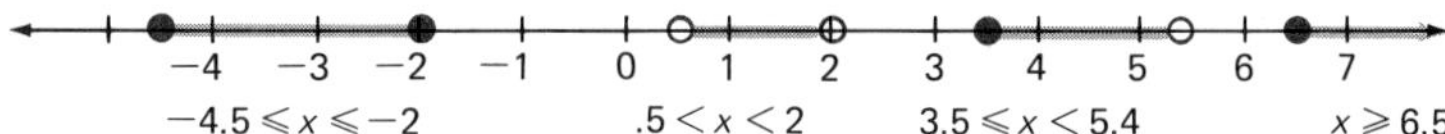

Sometimes we use set notation to describe an interval. For example, to denote the interval at the far left above, we could write $\{x: -4.5 \le x \le -2\}$, which is read "the set of all x such that x is greater than or equal to -4.5 and less than or equal to -2."

We should not write nonsense such as

$$3 < x < 2 \quad \text{or} \quad 2 > x < 3$$

The first is simply a contradiction. There is no number both greater than 3 and less than 2. The second says that x is both less than 2 and less than 3. This should be written simply as $x < 2$.

Absolute Value Often we want to describe the size of a number, not caring whether it is positive or negative. To do this, we introduce the concept of absolute value, symbolized by two vertical bars $|\ |$. It is defined by

$$|a| = \begin{cases} a & \text{if } a \ge 0 \\ -a & \text{if } a < 0 \end{cases}$$

This two-pronged definition can cause confusion. It says that if a number is positive or zero, its absolute value is itself. But if a number is negative, then its absolute value is its additive inverse (which is a positive number). For example,

$$|7| = 7$$

since $a = 7$ is positive. On the other hand,

$$|-7| = -(-7) = 7$$

since $a = -7$ is negative. Note that $|0| = 0$.

We may also think of $|a|$ geometrically. It represents the distance between a and 0 on the number line. More generally, $|a - b|$ is the distance between a and b.

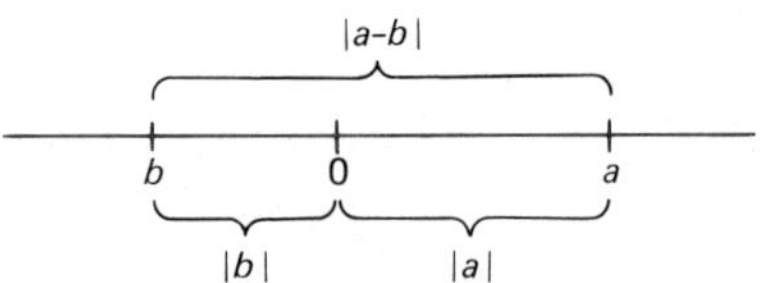

You should satisfy yourself that this is true for arbitrary choices of a and b, for example, that $|7 - (-3)|$ really is the distance between 7 and -3.

The properties of absolute values are straightforward and easy to remember.

PROPERTIES OF ABSOLUTE VALUES

1. $|a \cdot b| = |a| \cdot |b|$

2. $\left|\dfrac{a}{b}\right| = \dfrac{|a|}{|b|}$

3. $|a + b| \leq |a| + |b|$

Problem Set 1-5

In Problems 1–12, replace the symbol # by the appropriate symbol $<$, $>$, or $=$.

1. $1.5 \# -1.6$
2. $-2 \# -3$
3. $\sqrt{2} \# 1.4$
4. $\pi \# 3.15$
5. $\frac{1}{5} \# \frac{1}{6}$
6. $-\frac{1}{5} \# -\frac{1}{6}$
7. $5 - \sqrt{2} \# 5 - \sqrt{3}$
8. $\sqrt{2} - 5 \# \sqrt{3} - 5$
9. $-\frac{3}{16}\pi \# -\frac{3}{17}\pi$
10. $(\frac{16}{17})^2 \# (\frac{17}{18})^2$
11. $|-\pi + (-2)| \# |-\pi| + |-2|$
12. $|\pi - 2| \# \pi - 2$
13. Order the following numbers from least to greatest.

$$\tfrac{3}{4},\ -2,\ \sqrt{2},\ -\pi/2,\ -\tfrac{3}{2}\sqrt{2},\ \tfrac{43}{24}$$

14. Order the following numbers from least to greatest.

$$5 - 5,\ .37,\ -\sqrt{3},\ \tfrac{14}{33},\ -\tfrac{7}{4},\ -\tfrac{49}{35},\ \tfrac{3}{8}$$

Use a real number line to show the set of numbers that satisfy each given inequality.

15. $x < -4$
16. $x < 3$
17. $x \geq -2$
18. $x \leq 3$
19. $-1 < x < 3$
20. $2 < x < 5$

21. $0 < x \leq 3$ 22. $-3 \leq x < 2$ 23. $-\frac{1}{2} \leq x \leq \frac{3}{2}$

24. $-\frac{7}{4} \leq x \leq -\frac{3}{4}$

Write an inequality for each interval.

25. −2 −1 0 1 2 3

26. −1 0 1 2 3

27. −2 −1 0 1 2 3

28. −1 0 1 2 3

29. −2 −1 0 1 2 3

30. −1 0 1 2 3

31. −2 −1 0 1 2 3

32. −1 0 1 2 3

Example (Removing absolute values) Write each of the following without the absolute value symbol. Then show the corresponding interval(s) on the real number line.

(a) $|x| < 3$ (b) $|x - 2| < 3$ (c) $|x - 4| \geq 2$

SOLUTION. (a) Since $|x|$, the distance between x and 0 on the number line, is less than 3, x must be between -3 and 3. We write $-3 < x < 3$.

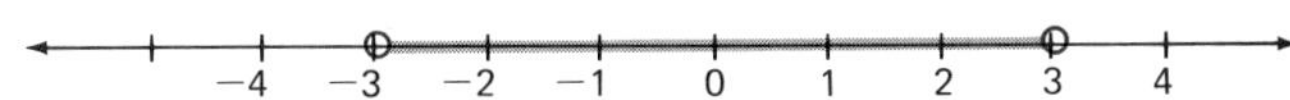

(b) Here $x - 2$, instead of x, must be between -3 and 3, that is,

$$-3 < x - 2 < 3$$

Adding 2 to each quantity gives

$$-1 < x < 5$$

(c) The distance between x and 4 is greater than or equal to 2. This means that $x - 4 \leq -2$ or $x - 4 \geq 2$. That is,

$$x \leq 2 \quad \text{or} \quad x \geq 6$$

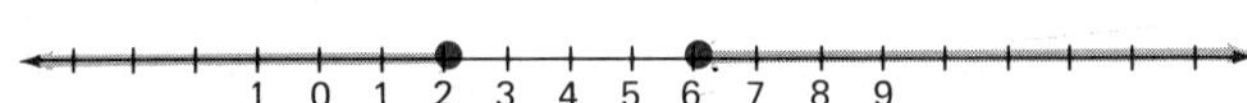

In set notation, we may write the solution set for the inequality in part (c) as $\{x : x \leq 2 \quad \text{or} \quad x \geq 6\}$ or equivalently as $\{x : x \geq 2\} \cup \{x : x \geq 6\}$.

Write each of the following without the absolute value symbol. Show the corresponding interval(s) on the real number line.

33. $|x| \leq 4$ 34. $|x| \geq 2$ 35. $|x - 3| < 2$

36. $|x - 5| < 1$ 37. $|x + 1| \leq 3$ 38. $|x + \frac{3}{2}| \leq \frac{1}{2}$

39. $|x - 5| > 5$ 40. $|x + 3| < 3$

Miscellaneous Problems

In Problems 41–48, show on the number line the set of numbers that satisfy the given inequality (or inequalities).

41. $x \leq -2$
42. $x > 9$
43. $-4 < x \leq 2$
44. $2 \leq x < 7$
45. $|x + 3| < -4$
46. $|x - 2| \leq 3$
47. $|x - 5| > 2$
48. $|x - 2| \geq 3$
49. If $a < b$, does it follow that $a^2 < b^2$?
50. If $0 < a < b$, does it follow that $a^2 < ab$? That $ab < b^2$? That $a^2 < b^2$?
51. If $0 < a < b$, how are $1/a$ and $1/b$ related to each other?
52. If $0 < a < b$, how are $\sqrt{a}$ and $\sqrt{b}$ related to each other?
53. Note that $-1 < x < 7$ can be rewritten as $|x - 3| < 4$. Rewrite each of the following double inequalities as a single inequality involving an absolute value.
    (a) $2 < x < 10$ (b) $-6 \leq x \leq 4$
    (c) $3 < x < 6$ (d) $-3 < x < 6$

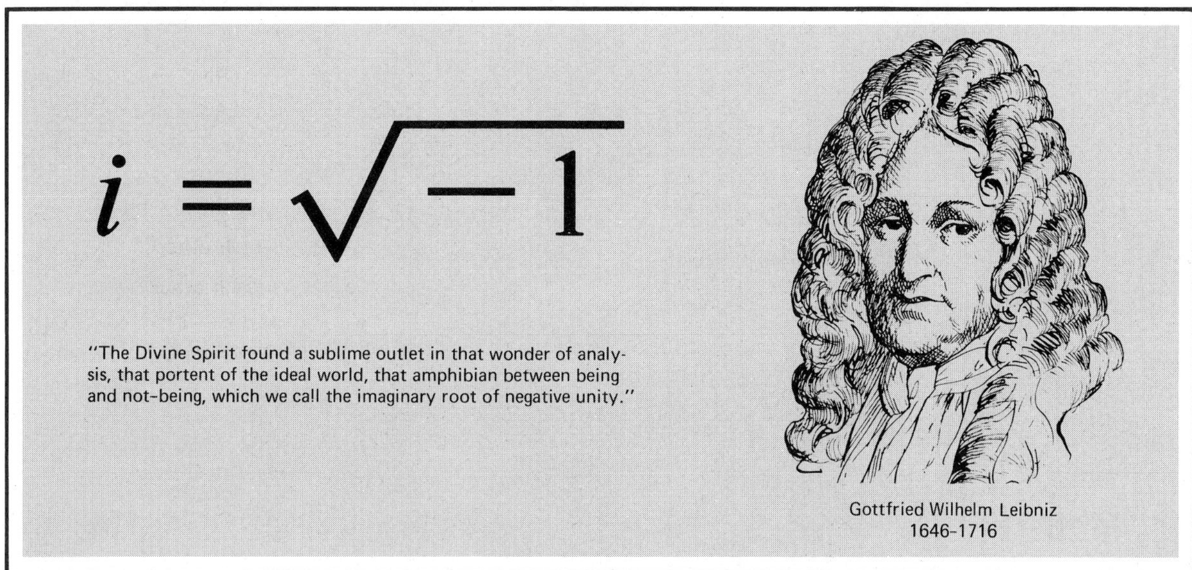

Gottfried Wilhelm Leibniz
1646–1716

1-6 The Complex Numbers

As early as 1550, the Italian mathematician Raffael Bombelli had introduced numbers like $\sqrt{-1}$, $\sqrt{-2}$, and so on, to solve certain equations. But mathematicians had a hard time deciding whether they should be considered legitimate numbers. Even Leibniz, who ranks with Newton as the greatest mathematician of the late seventeenth century, called them amphibians between being and not being. He wrote them down, he used them in calculations, but he carefully covered his tracks by calling them imaginary numbers. Unfortunately that name (which was actually first used by Descartes) has stuck, though we now prefer the term complex numbers. Let us see what these new numbers are and why they are needed.

Go back to the whole numbers 0, 1, 2, 3, ... We can easily solve the equation $x + 3 = 7$ within this system ($x = 4$). On the other hand, the equation $x + 7 = 3$ has no whole number solution. To solve it, we need the negative integer -4. Similarly, we cannot solve $3x = 2$ in the integers. We can say that the solution is $\frac{2}{3}$ only after the rational numbers have been introduced. To their dismay, the Greeks discovered that $x^2 - 2 = 0$ had no rational solution. We conquered that problem by enlarging our family of numbers to the real numbers. But there are still simple equations without solutions. Consider $x^2 + 1 = 0$. Try as you will, you will never solve it within the real number system.

By now, our procedure is well established. When we need new numbers, we invent them. This time we invent a number denoted by i (or by $\sqrt{-1}$) which satisfies $i^2 = -1$. However, we cannot get by with just one new number. For after we have adjoined it to the real numbers, we still must be able to multiply and add. Thus with i, we also need numbers such as

$$2i \qquad -4i \qquad (\tfrac{3}{2})i, \ldots$$

which are called pure imaginary numbers. We also need

$$3 + 2i \qquad 11 + (-4i) \qquad \tfrac{3}{4} + \tfrac{3}{2}i, \ldots$$

and it appears that we need even more complicated things such as

$$(3 + 8i + 2i^2 + 6i^3)(5 + 2i)$$

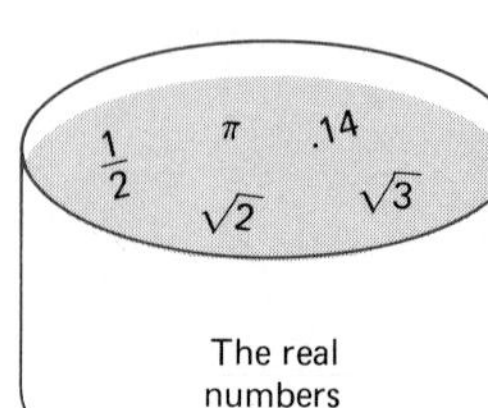

Actually, this last number can be simplified to $1 + 12i$ as we shall see later. In fact, no matter how many additions and multiplications we do, after the expressions are simplified we shall never have anything more complicated than a number of the form $a + bi$. Such numbers, that is, numbers of the form $a + bi$ where a and b are real, are called **complex numbers**. We refer to a as the **real part** and b as the **imaginary part** of $a + bi$. We shall agree that $0 \cdot i = 0$. Then every real number is a complex number, for $a = a + 0 = a + 0i$.

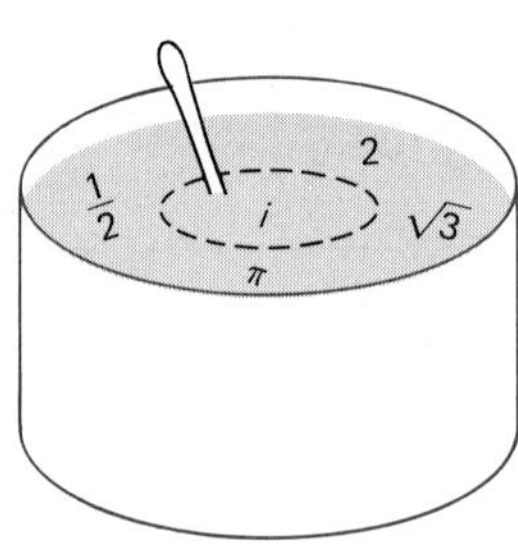

Addition and Multiplication We cannot say anything sensible about operations for complex numbers until we agree on the meaning of equality. The definition that seems most natural is this: $a + bi = c + di$ if and only if $a = c$ and $b = d$. That is, two complex numbers are equal if and only if their real parts and their imaginary parts are equal.

Now we can consider addition. Actually we have already used the plus sign in $a + bi$. That was like trying to add apples and bananas. The addition can be indicated but no further simplification is possible. We do not get apples and we do not get bananas; we get fruit salad.

When we have two numbers of the form $a + bi$, we can actually perform an addition. We just add the real parts and the imaginary parts separately, that is, we add the apples and we add the bananas. Thus

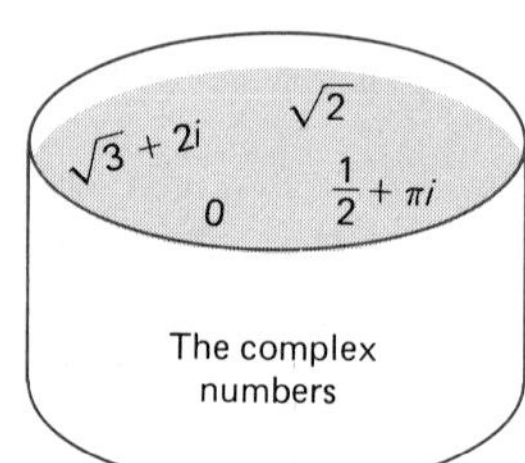

$$(3 + 2i) + (6 + 5i) = 9 + 7i$$

and more generally,

$$(a + bi) + (c + di) = (a + c) + (b + d)i$$

When we consider multiplication, our desire to maintain the properties of Section 1.4 leads to a definition that looks complicated. Thus let us first look at some examples.

(i) $2(3 + 4i) = 6 + 8i$ (distributive property)

(ii) $(3i)(-4i) = -12i^2 = 12$ (commutative and associative properties and $i^2 = -1$)

(iii) $(3 + 2i)(6 + 5i) = (3 + 2i)6 + (3 + 2i)5i$ (distributive property)
$= 18 + 12i + 15i + 10i^2$ (distributive, commutative, and associative properties)
$= 18 + (12 + 15)i - 10$ (distributive property)
$= 8 + 27i$ (commutative property)

The same kind of reasoning applied to the general case leads to

$$(a + bi)(c + di) = (ac - bd) + (ad + bc)i$$

which we take as the definition of multiplication for complex numbers.

Actually there is no need to memorize the formula for multiplication. Just do what comes naturally (that is, use familiar properties) and then replace i^2 by -1 wherever it arises, as in the following example.

$$\begin{aligned}(2 - 3i)(5 + 4i) &= (10 - 12i^2) + (8i - 15i) \\ &= (10 + 12) + (-7i) \\ &= 22 - 7i\end{aligned}$$

Consider the more complicated expression mentioned earlier. After noting that $i^3 = i^2 i = -i$, we have

$$\begin{aligned}(3 + 8i + 2i^2 + 6i^3)(5 + 2i) &= (3 + 8i - 2 - 6i)(5 + 2i) \\ &= (1 + 2i)(5 + 2i) \\ &= (5 + 4i^2) + (2i + 10i) \\ &= 1 + 12i\end{aligned}$$

Subtraction and Division Subtraction is easy; we simply subtract corresponding real and imaginary parts. For example,

$$\begin{aligned}(3 + 6i) - (5 + 2i) &= (3 - 5) + (6i - 2i) \\ &= -2 + 4i\end{aligned}$$

and

$$\begin{aligned}(5 + 2i) - (3 + 7i) &= (5 - 3) + (2i - 7i) \\ &= 2 + (-5i) \\ &= 2 - 5i\end{aligned}$$

Division is somewhat more difficult. We first note that $a - bi$ is called the **conjugate** of $a + bi$. Thus $2 - 3i$ is the conjugate of $2 + 3i$ and $-2 + 5i$ is the conjugate of $-2 - 5i$. Next, we observe that a complex number times its conjugate is a real number. For example,

$$(3 + 4i)(3 - 4i) = 9 + 16 = 25$$

and in general

$$(a + bi)(a - bi) = a^2 + b^2$$

Now to simplify the quotient of two complex numbers, multiply the numerator and denominator by the conjugate of the denominator, as illustrated below.

$$\frac{2 + 3i}{3 + 4i} = \frac{(2 + 3i)(3 - 4i)}{(3 + 4i)(3 - 4i)} = \frac{18 + i}{9 + 16} = \frac{18}{25} + \frac{1}{25}i$$

The effect of this multiplication is to replace a complex denominator by a real one. Notice that the result of the division is a number in the form $a + bi$.

A Genuine Extension We assert that the complex numbers constitute a genuine enlargement of the real numbers. This means first of all that they include the real numbers, since any real number a can be written as $a + 0i$. Second, the complex numbers satisfy all the properties we discussed in Section 1-4 for the real numbers. The order properties of Section 1-5, however, do not apply to the complex numbers.

The diagram below summarizes our development of the number systems.

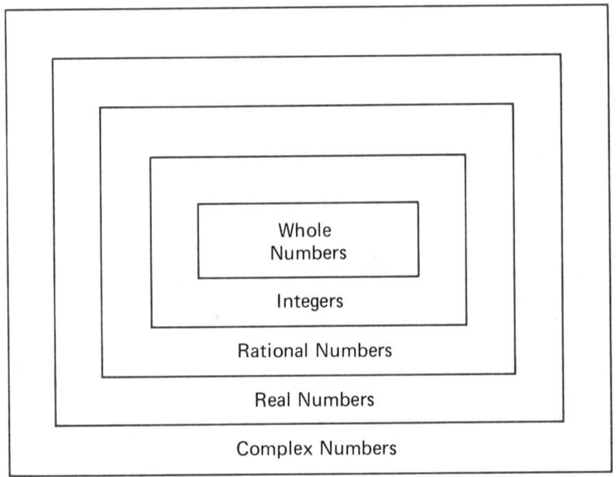

Is there any need to enlarge the number system again? The answer is no, and for a good reason. With the complex numbers, we can solve any equation that arises in algebra. Right now, we expect you to take this statement on faith.

Problem Set 1-6

Carry out the indicated operations and write the answer in the form $a + bi$.

1. $(2 + 3i) + (-4 + 5i)$
2. $(3 - 4i) + (-5 - 6i)$
3. $5i - (4 + 6i)$
4. $(-3i + 4) + 8i$
5. $(3i - 6) + (3i + 6)$
6. $(6 + 3i) + (6 - 3i)$
7. $4i^2 + 7i$
8. $i^3 + 2i$
9. $i(4 - 11i)$
10. $(3 + 5i)i$

11. $(3i + 5)(2i + 4)$

12. $(2i + 3)(3i - 7)$

13. $(3i + 5)^2$

14. $(-3i - 5)^2$

15. $(5 + 6i)(5 - 6i)$

16. $(\sqrt{3} + \sqrt{2}i)(\sqrt{3} - \sqrt{2}i)$

17. $\dfrac{5 + 2i}{1 - i}$

18. $\dfrac{4 + 9i}{2 + 3i}$

19. $\dfrac{5 + 2i}{i}$

20. $\dfrac{-4 + 11i}{-i}$

21. $\dfrac{(2 + i)(3 + 2i)}{1 + i}$

22. $\dfrac{(4 - i)(5 - 2i)}{4 + i}$

Example A (Multiplicative inverses) Find $(5 + 4i)^{-1}$, the multiplicative inverse of $5 + 4i$.

SOLUTION. The multiplicative inverse is just the reciprocal. Thus,

$$(5 + 4i)^{-1} = \frac{1}{5 + 4i} = \frac{1(5 - 4i)}{(5 + 4i)(5 - 4i)} = \frac{5 - 4i}{25 - 16i^2}$$

$$= \frac{5 - 4i}{25 + 16} = \frac{5 - 4i}{41} = \frac{5}{41} - \frac{4}{41}i$$

Find:

23. $(2 - i)^{-1}$

24. $(3 + 2i)^{-1}$

25. $(\sqrt{3} + i)^{-1}$

26. $(\sqrt{2} - \sqrt{2}i)^{-1}$

Use the results above to perform the following divisions. For example, division by $2 - i$ is the same as multiplication by $(2 - i)^{-1}$.

27. $\dfrac{2 + 3i}{2 - i}$

28. $\dfrac{1 + 2i}{3 + 2i}$

29. $\dfrac{4 - i}{\sqrt{3} + i}$

30. $\dfrac{\sqrt{2} + \sqrt{2}i}{\sqrt{2} - \sqrt{2}i}$

Example B (Complex roots) We say that a is a 4th root of b if $a \cdot a \cdot a \cdot a = b$ (more briefly, $a^4 = b$). Thus 3 is a 4th root of 81 because $3 \cdot 3 \cdot 3 \cdot 3 = 81$. Show that $1 + i$ is a 4th root of -4.

SOLUTION.

$$\begin{aligned}(1 + i)(1 + i)(1 + i)(1 + i) &= (1 + i)^2(1 + i)^2 \\ &= (1 + 2i - 1)(1 + 2i - 1) \\ &= (2i)(2i) \\ &= 4i^2 \\ &= -4\end{aligned}$$

31. Show that i is a 4th root of 1. Can you find three other 4th roots of 1?

32. Show that $-1 - i$ is a 4th root of -4.

33. Show that $1 - i$ is a 4th root of -4.

34. Show that $-1 + i$ is a 4th root of -4.

Miscellaneous Problems

35. Perform the indicated operations and write your answers in the form $a + bi$.
(a) $(5 - 3i) - (7 + 2i) + (6 - 7i)$
(b) $(5 - 3i)(5 + 3i) - (4 + 2i)(3 - 4i)$
(c) $\dfrac{5 - 2i}{3 + 4i} + i^3$

36. Simplify, giving your answer in the form $a + bi$.
(a) $\dfrac{(3 + 2i)(2 - 3i)}{5 - i}$ (b) $\dfrac{2}{3 - i} - \dfrac{5}{2 + 2i} + \dfrac{1}{i}$

37. Show that $1 + \sqrt{3}i$ is a cube root of -8. (You need to demonstrate that $(1 + \sqrt{3}i)(1 + \sqrt{3}i)(1 + \sqrt{3}i) = -8$.)

38. Show that $1 + 2i$ and $2 - i$ are both 4th roots of $-7 - 24i$. Can you guess two more 4th roots of $-7 - 24i$? Check your answers.

39. Solve the equation $3ix - 7i = 2i + 1$.

40. Show that $2 + i$ and $-2 - i$ are solutions of $x^2 - 3 - 4i = 0$.

CHAPTER SUMMARY

Algebra is generalized arithmetic. In it, we use letters to stand for numbers and then manipulate these letters according to definite rules.

The number systems used in algebra are the **whole numbers**, the **integers**, the **rational numbers**, the **real numbers**, and the **complex numbers**. Except for the complex numbers, we can visualize all these numbers as labels for points along a line, called the **real line**. Rational numbers can be expressed as ratios of integers. Real numbers (and therefore rational numbers) can be expressed as **decimals**. In fact, the rational numbers can be expressed as **repeating decimals** and the **irrational** (not rational) numbers as **nonrepeating decimals**.

The fundamental properties of numbers include the **commutative**, **associative**, and **distributive** laws. Because of their special properties, 0 and 1 are called the **neutral elements** for addition and multiplication. Within the rational number system, the real number system, and the complex number system, all numbers have **additive inverses** and all but 0 have **multiplicative inverses**.

The real numbers are **ordered** by the relation $<$ (is less than); its properties should be well understood as should those of its relatives $\leq$, $>$, and $\geq$. The symbol $|\ |$, read **absolute value,** denotes the magnitude of a number regardless of whether it is positive or negative.

CHAPTER REVIEW PROBLEM SET

1. An open box with a square base of length x centimeters has height y centimeters. Express its volume V and surface area S in terms of x and y.

2. Suppose that an airplane can fly x miles per hour in still air and that the wind is blowing from east at y miles per hour. Express in terms of x and y the time it will take for the plane to fly 100 miles due east and then return.

3. Perform the indicated operations and simplify.

   (a) $\dfrac{13}{24} - \dfrac{7}{12} + \dfrac{2}{3}$ (b) $\dfrac{9}{24} \cdot \dfrac{15}{18} \cdot \dfrac{8}{5}$

   (c) $\dfrac{1}{3}\left(\dfrac{1}{2} - \left(\dfrac{5}{6} - \dfrac{2}{3}\right)\right)$ (d) $\left(\dfrac{11}{30}\right) \Big/ \left(\dfrac{33}{18}\right)$

   (e) $\dfrac{\frac{3}{4} - \frac{1}{12} + \frac{3}{8}}{\frac{3}{4} + \frac{5}{12} - \frac{7}{8}}$ (f) $3 + \dfrac{\frac{3}{4} - \frac{7}{8}}{\frac{5}{12}}$

4. Find the prime factorizations of 1000 and 180.

5. Find the least common multiple of 1000 and 180.

6. Express $\frac{5}{13}$ and $\frac{11}{7}$ as repeating decimals.

7. Express $.\overline{257}$ and $1.\overline{23}$ as ratios of two integers.

8. Is $(a \div b) \times c = (a \times c) \div b$? Is $(a + b)^{-1} = a^{-1} + b^{-1}$?

9. State the associative law of multiplication and the commutative law of addition.

10. Order the following numbers from least to greatest: $\frac{29}{20}$, 1.4, $1.\overline{4}$, $\sqrt{2}$, $\frac{13}{8}$.

11. Write the inequalities $|x + 2| < 6$ and $|2x - 1| \leq 3$ without using the absolute value symbol. Then show the corresponding intervals on the real line.

12. Is $|-x| = x$? Explain.

13. Write each of the following in the form $a + bi$.
    (a) $(3 - 2i) + (-7 + 6i) - (2 - 3i)$
    (b) $(3 - 2i)(3 + 2i) - 4i^2 + (2i)^3$
    (c) $2 + 3i + (3 - i)/(2 + 3i)$
    (d) $(2 + i)^3$
    (e) $(5 - 3i)^{-1}$

14. Assuming that $\sqrt{5}$ is irrational, show that $5 + \sqrt{5}$ and $5\sqrt{5}$ are irrational.

15. A rectangle with a perimeter of 72 centimeters is 4 centimeters longer than it is wide. Find its area.

Algebra is the intellectual instrument for rendering clear the quantitative aspects of the world.

Alfred North Whitehead

CHAPTER TWO

Exponents and Polynomials

2-1 Integral Exponents

2-2 Scientific Notation

2-3 Pocket Calculators

2-4 Polynomials

2-5 Factoring Polynomials

2-6 Rational Expressions

A Long Sermon Leads To A Good Problem

The sermon was long; the sanctuary was hot. It was even hard to stay awake. For awhile, young Franklin Figit amused himself by doodling on the Sunday bulletin. Growing tired of that, he tried folding the bulletin in more and more complicated ways. That's when a bright idea hit him. "I wonder what would happen," he said to himself, "if I took a mammoth bulletin, folded it in half, then in half again, and so on until I had folded it 40 times. How high a stack of paper would that make?"

2-1 Integral Exponents

Young Franklin has posed a very interesting question. What do you think the answer is? 10 inches? 3 feet? 500 feet? Make a guess and write it in the margin. When we finally work out the solution late in this section, you are likely to be surprised at the answer.

Let us make a start on the problem right away. If the bulletin were c units thick ($c = .01$ inches would be a reasonable value), then after folding once, it would be $2c$ units thick. After two folds, it would measure $2 \cdot 2c$ units thick, and after 40 folds it would have a thickness of

$$2\cdot c$$

Nobody with a sense of economy and elegance would write a product of forty 2's in this manner. To indicate such a product, most ordinary people and all mathematicians prefer to write 2^{40}. The number 40 is called an **exponent**; it tells you how many 2's to multiply together. The number 2^{40} is called a **power** of 2 and we read it "2 to the 40th power."

In the general case, if b is any number ($\frac{3}{4}$, π, $\sqrt{5}$, i, . . .) and n is a positive integer, then

$$b^n = \underbrace{b \cdot b \cdot b \cdots b}_{n \text{ factors}}$$

Thus

$$b^3 = b \cdot b \cdot b \qquad b^5 = b \cdot b \cdot b \cdot b \cdot b$$

How do we write the product of 1000 b's? (Honey is not the answer we have in mind.) The product of 1000 b's is written as b^{1000}.

Rules for Exponents The behavior of exponents is excellent, being governed by a few simple rules that are easy to remember. Consider multiplication first. If we multiply 2^5 by 2^8, we have

$$
\begin{aligned}
2^5 \cdot 2^8 &= \underbrace{(2\cdot 2\cdot 2\cdot 2\cdot 2)}_{5}\underbrace{(2\cdot 2\cdot 2\cdot 2\cdot 2\cdot 2\cdot 2\cdot 2)}_{8} \\
&= \underbrace{2\cdot 2\cdot 2\cdot 2\cdot 2\cdot 2\cdot 2\cdot 2\cdot 2\cdot 2\cdot 2\cdot 2\cdot 2}_{13} \\
&= 2^{13} \\
&= 2^{5+8}
\end{aligned}
$$

It suggests that to find the product of powers of 2, you should add the exponents. There is nothing special about 2; it could just as well be 5, $\frac{2}{3}$, or π. We can put the general rule in a nutshell by using the symbols of algebra.

$$\boxed{b^m \cdot b^n = b^{m+n}}$$

Here b can stand for any number, but (for now) think of m and n as positive integers. Be careful with this rule. If you write

$$3^4 \cdot 3^5 = 3^9$$

or

$$\pi^9 \cdot \pi^{12} \cdot \pi^2 = \pi^{23}$$

that is fine. But do not try to use the rule on $2^4 \cdot 3^5$ or $a^2 \cdot b^3$; it just does not apply.

Next consider the problem of raising a power to a power. By definition, $(2^{10})^3$ is $2^{10} \cdot 2^{10} \cdot 2^{10}$, which allows us to apply the rule above. Thus

$$(2^{10})^3 = 2^{10} \cdot 2^{10} \cdot 2^{10} = 2^{10+10+10} = 2^{10\cdot 3}$$

It appears that to raise a power to a power we should multiply the exponents; in symbols

$$\boxed{(b^m)^n = b^{m\cdot n}}$$

Try to convince yourself that this rule is true for any number b and for any positive integer exponents m and n.

Sometimes we need to simplify quotients like

$$\frac{8^6}{8^6} \qquad \frac{2^9}{2^5} \qquad \frac{10^4}{10^6}$$

The first one is easy enough; it equals 1. Furthermore,

$$\frac{2^9}{2^5} = \frac{2^5 \cdot 2^4}{2^5} = 2^4 = 2^{9-5}$$

and

$$\frac{10^4}{10^6} = \frac{10^4}{10^4 \cdot 10^2} = \frac{1}{10^2} = \frac{1}{10^{6-4}}$$

These illustrate the general rules.

$$\frac{b^m}{b^n} = 1 \qquad \text{if } m = n$$

$$\frac{b^m}{b^n} = b^{m-n} \qquad \text{if } m > n$$

$$\frac{b^m}{b^n} = \frac{1}{b^{n-m}} \qquad \text{if } n > m$$

In each case, we assume $b \neq 0$.

We did not put a box around these rules simply because we are not happy with them. It took three lines to describe what happens when you divide powers of the same number. Surely we can do better than that, but first we shall have to extend the notion of exponents to numbers other than positive integers.

Zero and Negative Exponents So far, symbols like 4^0 and 10^{-3} have not been used. We want to give them meaning and do it in a way that is consistent with what we have already learned. For example, 4^0 must behave so that

$$4^0 \cdot 4^7 = 4^{0+7} = 4^7$$

This can happen only if $4^0 = 1$. More generally, we require that

$$\boxed{b^0 = 1}$$

Here b can be any number except 0 (0^0 will be left undefined).

What about 10^{-3}? If it is to be admitted to the family of powers, it too must abide by the rules. Thus we insist that

$$10^{-3} \cdot 10^3 = 10^{-3+3} = 10^0 = 1$$

This means that 10^{-3} has to be the reciprocal of 10^3. Consequently, we are led to make the definition

$$\boxed{b^{-n} = \frac{1}{b^n} \qquad b \neq 0}$$

Now the complicated law for quotients of powers can be put in a simple form, as the following examples illustrate.

$$\frac{b^4}{b^9} = \frac{b^4}{b^4 \cdot b^5} = \frac{1}{b^5} = b^{-5} = b^{4-9}$$

$$\frac{b^5}{b^5} = 1 = b^0 = b^{5-5}$$

$$\frac{b^{-3}}{b^{-9}} = \frac{1/b^3}{1/b^9} = \frac{b^9}{b^3} = b^6 = b^{-3-(-9)}$$

In fact, for any choice of integers m and n, we find that

$$\frac{b^m}{b^n} = b^{m-n} \qquad b \neq 0$$

What about the two rules we learned earlier? Are they still valid when m and n are arbitrary (possibly negative) integers? The answer is yes. A few illustrations may help convince you.

$$b^{-3} \cdot b^7 = \frac{1}{b^3} \cdot b^7 = \frac{b^7}{b^3} = b^4 = b^{-3+7}$$

$$(b^{-5})^2 = \left(\frac{1}{b^5}\right)^2 = \frac{1}{b^5} \cdot \frac{1}{b^5} = \frac{1}{b^{10}} = b^{-10} = b^{(-5)(2)}$$

A summary of the main rules of exponents follows. We include the three rules already discussed, plus two new rules that will be described in detail in the problem set (see Example B).

RULES FOR EXPONENTS

If a and b are (real or complex) numbers and if m and n are any integers,

1. $b^m b^n = b^{m+n}$
2. $(b^m)^n = b^{mn}$
3. $\dfrac{b^m}{b^n} = b^{m-n} \qquad b \neq 0$
4. $(ab)^n = a^n b^n$
5. $\left(\dfrac{a}{b}\right)^n = \dfrac{a^n}{b^n} \qquad b \neq 0$

The Paper Folding Problem Again It is now a simple matter to solve Franklin Figit's paper folding problem, especially if we are satisfied with a reasonable approximation to the answer. To be specific, let us approximate 1 foot by 10 inches, 1 mile by 5000 feet, and 2^{10} (which is really 1024) by 1000. Then a bulletin of thickness .01 inch will make a stack of the following height when folded 40 times ($\approx$ means "is approximately equal to").

$$\begin{aligned}(.01)2^{40} &= (.01)\cdot 2^{10}\cdot 2^{10}\cdot 2^{10}\cdot 2^{10} \text{ inches}\\ &\approx \frac{1}{10^2}\cdot 10^3\cdot 10^3\cdot 10^3\cdot 10^3 \text{ inches}\\ &= 10^{10} \text{ inches}\\ &\approx 10^9 \text{ feet}\\ &\approx \frac{10\cdot 10^8}{5\cdot 10^3} \text{ miles}\\ &= 2\cdot 10^5 \text{ miles}\\ &= 200{,}000 \text{ miles}\end{aligned}$$

That is a stack of paper that would reach almost to the moon.

Problem Set 2-1

Example A (Removing negative exponents) Rewrite without negative exponents and simplify.

(a) -4^{-2} (b) $(-4)^{-2}$ (c) $((\frac{3}{4})^{-1})^2$ (d) $2^5 3^{-2} 2^{-3}$

SOLUTION. (a) The exponent -2 applies just to 4.

$$-4^{-2} = -\frac{1}{4^2} = -\frac{1}{4\cdot 4} = -\frac{1}{16}$$

(b) The exponent -2 now applies to -4.

$$(-4)^{-2} = \frac{1}{(-4)^2} = \frac{1}{16}$$

(c) First apply the rule for a power of a power.

$$\left(\left(\frac{3}{4}\right)^{-1}\right)^2 = \left(\frac{3}{4}\right)^{-2} = \frac{1}{(\frac{3}{4})^2} = \frac{1}{\frac{9}{16}} = \frac{16}{9}$$

(d) Note that the two powers of 2 can be combined.

$$2^5 3^{-2} 2^{-3} = 2^5 2^{-3} 3^{-2} = 2^2\cdot\frac{1}{3^2} = \frac{4}{9}$$

Write without negative exponents and simplify.

1. 5^{-2}
2. $(-5)^{-2}$
3. -5^{-2}
4. 2^{-5}
5. $(-2)^{-5}$
6. $\left(\frac{1}{5}\right)^{-2}$
7. $\left(\frac{-2}{3}\right)^{-3}$
8. $-\frac{2^{-3}}{3}$
9. $\frac{2^{-2}}{3^{-3}}$
10. $\left(\left(\frac{2}{3}\right)^{-2}\right)^2$
11. $\left(\left(\frac{3}{2}\right)^{-2}\right)^{-2}$
12. $\frac{4^0 + 0^4}{4^{-1}}$

13. $\dfrac{2^{-2} - 4^{-3}}{(-2)^2 + (-4)^0}$ 14. $\dfrac{3^{-1} + 2^{-3}}{(-1)^3 + (-3)^2}$ 15. $3^3 \cdot 2^{-3} \cdot 3^{-5}$

16. $4^2 \cdot 4^{-4} \cdot 3^0$

Example B (Rules for products and quotients) Expressions like $(ab)^n$ and $(a/b)^n$ often arise; we need rules for handling them. Notice that

$$(ab)^n = \underbrace{(ab)(ab)\ldots(ab)}_{n \text{ times}} = \underbrace{a \cdot a \cdots a}_{n \text{ times}} \cdot \underbrace{b \cdot b \cdots b}_{n \text{ times}} = a^n b^n$$

$$\left(\frac{a}{b}\right)^n = \underbrace{\left(\frac{a}{b}\right)\left(\frac{a}{b}\right)\cdots\left(\frac{a}{b}\right)}_{n \text{ times}} = \frac{a \cdot a \cdots a}{b \cdot b \cdots b} = \frac{a^n}{b^n}$$

Our demonstrations are valid for any positive integer n, but the results are correct even if n is negative or zero. Thus for any integer n,

$$(ab)^n = a^n b^n$$

$$\left(\frac{a}{b}\right)^n = \frac{a^n}{b^n}$$

Use these rules to simplify (a) $(2x)^6$; (b) $(2x/3)^4$; (c) $(x^{-1}y^2)^{-3}$.

SOLUTION.

(a) $(2x)^6 = 2^6x^6 = 64x^6$

(b) $\left(\dfrac{2x}{3}\right)^4 = \dfrac{(2x)^4}{3^4} = \dfrac{2^4x^4}{3^4} = \dfrac{16x^4}{81}$

(c) $(x^{-1}y^2)^{-3} = (x^{-1})^{-3}(y^2)^{-3} = x^3y^{-6} = x^3 \cdot \dfrac{1}{y^6} = \dfrac{x^3}{y^6}$

Simplify, writing your answer without negative exponents.

17. $(3x)^4$ 18. $\left(\dfrac{2}{y}\right)^5$ 19. $(xy^2)^6$

20. $\left(\dfrac{y^2}{3z}\right)^4$ 21. $\left(\dfrac{2x^2y}{w^3}\right)^4$ 22. $\left(\dfrac{\sqrt{2}x}{3}\right)^4$

23. $\left(\dfrac{3x^{-1}y^2}{z^2}\right)^3$ 24. $\left(\dfrac{2x^{-2}y}{z^{-1}}\right)^2$ 25. $\left(\dfrac{\sqrt{5}i}{x^{-2}}\right)^4$

26. $(i\sqrt{3}x^{-2})^6$

Example C (Simplifying complicated expressions) Simplify

(a) $\dfrac{4ab^{-2}c^3}{a^{-3}b^3c^{-1}}$; (b) $\left[\dfrac{(2xz^{-2})^3(x^{-2}z)}{2xz^2}\right]^4$; (c) $(a^{-1} + b^{-2})^{-1}$.

SOLUTION.

(a) $\dfrac{4ab^{-2}c^3}{a^{-3}b^3c^{-1}} = \dfrac{4a(1/b^2)\cdot c^3}{(1/a^3)b^3(1/c)} = \dfrac{4ac^3/b^2}{b^3/(a^3c)} = \dfrac{4ac^3}{b^2}\cdot\dfrac{a^3c}{b^3} = \dfrac{4a^4c^4}{b^5}$

We point out that in simplifying expressions like the one above, *a factor can be moved from numerator to denominator or vice versa by changing the sign of its exponent.* That is important enough to remember. Let us do part (a) again using this fact.

$$\frac{4ab^{-2}c^3}{a^{-3}b^3c^{-1}} = \frac{4aa^3c^3c}{b^2b^3} = \frac{4a^4c^4}{b^5}$$

(b) $$\left[\frac{(2xz^{-2})^3(x^{-2}z)}{2xz^2}\right]^4 = \left[\frac{8x^3z^{-6}x^{-2}z}{2xz^2}\right]^4$$

$$= \left[\frac{8xz^{-5}}{2xz^2}\right]^4$$

$$= \left[\frac{4}{z^2z^5}\right]^4 = \frac{256}{z^{28}}$$

(c) $$(a^{-1} + b^{-2})^{-1} = \left(\frac{1}{a} + \frac{1}{b^2}\right)^{-1} = \left(\frac{b^2 + a}{ab^2}\right)^{-1} = \frac{ab^2}{b^2 + a}$$

Note the difference between a product and a sum.

$$(a^{-1}\cdot b^{-2})^{-1} = a\cdot b^2$$

but

$$(a^{-1} + b^{-2})^{-1} \neq a + b^2$$

Simplify, leaving your answer free of negative exponents.

27. $\dfrac{2x^{-3}y^2z}{x^3y^4z^{-2}}$

28. $\dfrac{3x^{-5}y^{-3}z^4}{9x^2yz^{-1}}$

29. $\left(\dfrac{-2xy}{z^2}\right)^{-1}(x^2y^{-3})^2$

30. $(4ab^2)^3\left(\dfrac{-a^3}{2b}\right)^2$

31. $\dfrac{ab^{-1}}{(ab)^{-1}}\cdot\dfrac{a^2b}{b^{-2}}$

32. $\dfrac{3(b^{-2}d)^4(2bd^3)^2}{(2b^2d^3)(b^{-1}d^2)^5}$

33. $\left[\dfrac{(3b^{-2}d)(2)(bd^3)^2}{12b^3d^{-1}}\right]^5$

34. $\left[\dfrac{(ab^2)^{-1}}{(ba^2)^{-2}}\right]^{-1}$

35. $(a^{-2} + a^{-3})^{-1}$

36. $a^{-2} + a^{-3}$

37. $\dfrac{x^{-1}}{y^{-1}} - \left(\dfrac{x}{y}\right)^{-1}$

38. $(x^{-1} - y^{-1})^{-1}$

Miscellaneous Problems

Simplify the expressions in Problems 39–49, leaving your answer free of negative exponents.

39. $(\sqrt{3}x^2y^{-2})^2$

40. $(\frac{1}{3}x^{-1}z^3)^3$

41. $\dfrac{(3x^{-1}y)^2}{3xy^{-2}}$

42. $\left(\dfrac{3x^{-1}y}{6xy^{-2}}\right)^2$

43. $[(2x^{-2})^3(3z^{-1}x)^2]^2$

44. $\left(\dfrac{5x^2z^{-3}}{(2x^{-1}z^2)^3}\right)^{-1}$

45. $x^2 + x^{-2}$

46. $(x + x^{-1})^2$

47. $(x + x^{-1})^{-1}$

48. $\left(1 - \frac{x - 1}{x + 1}\right)^{-1}$

49. $[1 - (1 + x^{-1})^{-1}]^{-1}$

50. Express each of the following as a power of 5.

(a) $(25)^{-2}$ (b) $\left(\frac{1}{5^3}\right)^{-2}$ (c) $(.2)(5^{-4})^{-2}$

51. Express each of the following as a power of 2.

(a) $\frac{1}{32}$ (b) 1 (c) $\frac{1}{2} \cdot \frac{1}{16} \cdot \frac{1}{8} \cdot \frac{1}{4}$

52. Which is larger, 2^{1000} or $(1000)^{100}$?

53. G. P. Jetty has agreed to pay his private secretary according to the following plan: 1¢ the first day, 2¢ the second day, 4¢ the third day, and so on.
   (a) About how much will the secretary make on the 20th day? (Recall that 2^{10} is approximately 1000).
   (b) About how much will the secretary make on the 50th day?

54. Refer to Problem 53.
   (a) How much will the secretary make during the first 3 days? 4 days? 5 days?
   (b) How do your answers in part (a) compare with the amounts the secretary made on the 4th day? 5th day? 6th day?
   (c) Guess at a formula for the total amount the secretary will make in n days.
   (d) About how much will the secretary make altogether during the first 30 days? 40 days?

55. Refer to Problem 54. If G. P. Jetty is worth 2 billion dollars and his secretary starts work on January 1, about when will G. P. Jetty go broke?

56. Consider Franklin Figit's paper folding problem with which we began this section. If the pile of paper covers 1 square inch after 40 folds, approximately how much area did it cover at the beginning?

Numbers Large and Small

The numbers of modern science range from the mammoth to the minuscule. For example the speed of light is

29,979,000,000 centimeters per second

while the mass of the proton is

.00000000000000000000000167 gram

How can we simplify calculations with numbers like these?

2-2 Scientific Notation

The difficulty in working with very large or very small numbers is that there are so many zeros, zeros which serve only to place the decimal point. It is easy enough to calculate

$$\frac{(32)(284)}{128} = 71$$

But to calculate

$$P = \frac{(3{,}200{,}000{,}000)(.0000000284)}{.00000000128}$$

seems much harder, even though the answers in both cases have the same nonzero digits. The only difference is in the placement of the decimal point. To simplify calculations like this, we introduce a method of writing numbers called scientific notation. It is extensively used in all of modern science.

Scientific Notation Defined A positive number is in **scientific notation** when it is written in the form

$$c \times 10^n$$

where n is an integer and c is a real number satisfying the inequality $1 \leq c < 10$. For example, scientists would normally write the following.

Speed of light: 2.9979×10^{10} centimeters per second
Mass of proton: 1.67×10^{-24} grams

Schroeder could have said that a googol is 1×10^{100}, but perhaps that would not have helped Lucy feel better.

To calculate the number P introduced above, we write

$$3{,}200{,}000{,}000 = 3.2 \times 10^{9}$$
$$.0000000284 = 2.84 \times 10^{-8}$$
$$.00000000128 = 1.28 \times 10^{-9}$$

Then

$$P = \frac{(3.2 \times 10^{9})(2.84 \times 10^{-8})}{1.28 \times 10^{-9}}$$
$$= \frac{(3.2)(2.84)}{1.28} \times 10^{9-8-(-9)}$$
$$= 7.1 \times 10^{10}$$

Significant Digits When working with measurements (which are always approximate), it is important to know what rules to follow in making calculations and reporting answers. Suppose, for example, that we wish to calculate the area of a rectangular metal strip whose length and width are recorded as 21.52 centimeters and 1.6 centimeters, respectively. Multiplying these numbers, we get 34.432. Can we honestly claim 34.432 square centimeters as the area of the metal strip? Hardly. To say that 21.52 is the length really means that the length is between 21.515 and 21.525. Similarly, the width is between 1.55 and 1.65. Thus the area is somewhere between $(21.515)(1.55) = 33.34825$ and $(21.525)(1.65) = 35.51625$. If we are to report one number as the area, what should it be? To answer, we need the concept of significant digits.

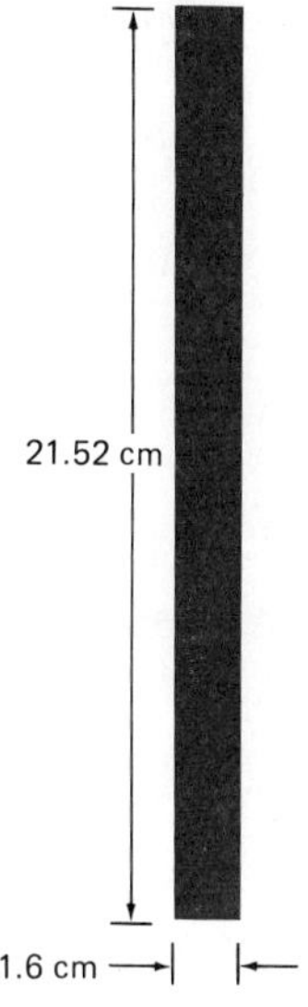

We illustrate what we mean by **significant digits** with four examples.

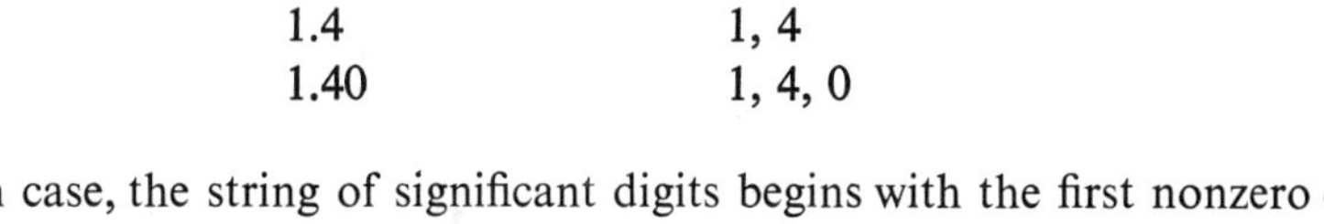

NUMBER	SIGNIFICANT DIGITS
.0024	2, 4
1.205	1, 2, 0, 5
1.4	1, 4
1.40	1, 4, 0

In each case, the string of significant digits begins with the first nonzero digit and ends with the last digit definitely specified.

Now we can state a rule to use in calculations involving approximate numbers.

RULE 1

In any calculation involving multiplication, division, finding powers, or extracting roots of numbers obtained as measurements, the answer should be given with the same number of significant digits as the measurement with fewest significant digits.

In using this rule, we suggest carrying one more significant digit than is present in the least accurate measurement through all intermediate calculations and then rounding off at the final step. Incidentally, we round up if the first neglected digit is 5 or more and down otherwise.

Here is how we use the rule in the example involving the area of a metal strip.

	LENGTH	WIDTH
Measurement	21.52	1.6
No. of significant digits	4	2
Measurement rounded off	21.5	1.6

$$\text{Area:}\quad (21.5)(1.6) = 34.40$$
$$\text{Area rounded off:}\quad 34$$

Suppose that the width had been measured as 1.61. Then we would have calculated

$$(21.52)(1.61) = 34.6472$$

and reported 34.6 as the area.

Though it will be needed less often in this book, there is also a rule for addition and subtraction.

RULE 2

In a calculation involving addition or subtraction of numbers obtained as measurements, the reported answer should not show more decimal places than does the measurement with fewest decimal places.

For example, suppose we are to find the perimeter of a triangle whose sides were measured as 4.123, 5.2, and 3.49. While these numbers sum to 12.813, we should round this off to 12.8 in reporting the answer since the least precise of the data, 5.2, shows only one decimal place.

Whether the final zeros of a number are significant digits or whether they simply serve to place the decimal point must sometimes be determined from the context. If the population of a city is given as 490,000 in a list which clearly gives populations to the nearest thousand, then the first zero is significant but the others are not. We can avoid any ambiguity by always writing approximate numbers in scientific notation. The population of that city would then be written as 4.90×10^5. If it were correct to the nearest hundred, we would write 4.900×10^5.

The Metric System Several of the examples of this section have involved metric units; it is time we said a word about the metric system of measurement. It has long been recognized by most countries that the metric system, with its emphasis on 10 and powers of 10, offers an attractive way to measure length, volume, and weight. Only in the United States do we hang on to our hodge-podge of inches, feet, miles, pounds, and quarts. Even here, however, it appears that the metric system will gradually win acceptance.

The following chart summarizes the metric system. It highlights the relationship to powers of 10 within the metric system and gives some of the conversion factors that relate the metric system to our English system. Using the table, you should be able to answer questions such as the following.

1. How many kilometers per hour are equivalent to 65 miles per hour?
2. How many grams are equivalent to 200 pounds?

THE METRIC SYSTEM

LENGTH	VOLUME	WEIGHT
kilometer = 10^3 meter	kiloliter = 10^3 liter	kilogram = 10^3 gram
hectometer = 10^2 meter	hectoliter = 10^2 liter	hectogram = 10^2 gram
dekameter = 10 meter	dekaliter = 10 liter	dekagram = 10 gram
meter = 1 meter	liter = 1 liter	gram = 1 gram
decimeter = 10^{-1} meter	deciliter = 10^{-1} liter	decigram = 10^{-1} gram
centimeter = 10^{-2} meter	centiliter = 10^{-2} liter	centigram = 10^{-2} gram
millimeter = 10^{-3} meter	milliliter = 10^{-3} liter	milligram = 10^{-3} gram
1 kilometer $\approx$ .62 miles	1 liter $\approx$ 1.057 quarts	1 kilogram $\approx$ 2.20 pounds
1 inch $\approx$ 2.54 centimeters	1 liter = 10^3 cubic centimeters	1 pound $\approx$ 453.6 grams

The answer to the two questions we posed are:

1. 65 miles per hour = 65/.62, or 105 kilometers per hour
2. 200 pounds = 200(453.6) grams = 9.07×10^4 grams

Problem Set 2-2

Write each of the following numbers in scientific notation.

1. 341,000,000
2. 25 billion
3. .0000000513
4. .00000000012
5. .0000000001245
6. .0000000000012578

Calculate, assuming that all given numbers are exact.

7. $\dfrac{(1.2 \times 10^9)(2.47 \times 10^5)}{3.2 \times 10^6}$
8. $\dfrac{(3.3 \times 10^{11})(1.4 \times 10^{-4})}{2.1 \times 10^3}$
9. $\dfrac{(1.32 \times 10^{-5})(42,000)}{6.6 \times 10^{15}}$
10. $\dfrac{(1.43 \times 10^6)(2.3 \times 10^{14})}{4,100,000}$

Calculate, assuming that each of the given numbers is the result of a measurement.

11. $(1.2 \times 10^5)(3.4325 \times 10^{-6})$
12. $(2.3 \times 10^{-6})(2.135 \times 10^{-2})$
13. $(2.41 \times 10^4)(3.2 \times 10^5)(1.555 \times 10^{-2})$
14. $(3.2456 \times 10^3)(4.32 \times 10^4)$

Example (Conversions between units)

(a) Convert 2.56×10^4 kilometers to centimeters.
(b) Convert 3.42×10^2 kilograms to ounces.
(c) Convert 43.8 cubic meters to liters.

SOLUTION.

(a)
$$\begin{aligned} 2.56 \times 10^4 \text{ kilometers} &= (2.56 \times 10^4)(10^3) \text{ meters} \\ &= (2.56 \times 10^4)(10^3)(10^2) \text{ centimeters} \\ &= 2.56 \times 10^9 \text{ centimeters} \end{aligned}$$

(b)
$$\begin{aligned} 3.42 \times 10^2 \text{ kilograms} &= (3.42 \times 10^2)(2.2) \text{ pounds} \\ &= (3.42 \times 10^2)(2.2)(16) \text{ ounces} \\ &= 1.2 \times 10^4 \text{ ounces} \end{aligned}$$

(c)
$$\begin{aligned} 43.8 \text{ cubic meters} &= (43.8)(100)^3 \text{ cubic centimeters} \\ &= \frac{(43.8)(100)^3}{10^3} \text{ liters} \\ &= 4.38 \times 10^4 \text{ liters} \end{aligned}$$

Express each of the following in scientific notation.

15. The number of centimeters in 413.2 meters.
16. The number of millimeters in 1.32×10^4 kilometers.
17. The number of kilometers in 4×10^{15} millimeters.
18. The number of meters in 1.92×10^8 centimeters.
19. The number of millimeters in one yard (36 inches).
20. The number of inches in 150 decimeters.
21. The number of deciliters in a kiloliter.
22. The number of milliliters in a hectoliter.
23. The number of liters in 4 cubic meters.
24. The number of milliliters in 3.6×10^2 cubic dekameters.
25. The number of kilograms in 4.63×10^{15} centigrams.
26. The number of grams in 4.1×10^3 pounds.

Miscellaneous Problems

Perform the following calculations, leaving your answers in scientific notation. Assume all numbers to be exact.

27. $\dfrac{(.000021)(240{,}000)}{7000}$

28. $\dfrac{(36{,}000{,}000)(.000011)}{.0000033}$

29. $(54)(.00005)(12{,}000{,}000)^2$

30. $\dfrac{(3400)^2(400{,}000)^3}{(.017)^2}$

31. $\dfrac{(.0003)^{-4}(3000)^2}{(.09)^2}$

32. $\dfrac{(.05)^{-3}(5000)^3}{(200)^{-4}}$

One mole of any substance is an amount equal to its molecular weight in grams. Avogadro's number (6.02×10^{23}) *is the number of molecules in a mole. The atomic weights of hydrogen, carbon, and oxygen are* 1, 12, *and* 16, *respectively. Thus the molecular weight of water* (H_2O) *is* $2 \cdot 1 + 1 \cdot 16 = 18$.

33. How many molecules are there in 36 grams of water? (*Hint*: Thirty-six grams of water equal 2 moles, since the molecular weight of water is 18.)
34. How many molecules are there in 56.3 grams of water?
35. How many molecules are there in 51.2 grams of carbon dioxide (CO_2)? (Note that the molecular weight of carbon dioxide is $1 \cdot 12 + 2 \cdot 16 = 44$.)
36. How many molecules are there in 36 grams of cholesterol ($C_{27}H_{46}O$)?
37. One mole of a gas occupies 22.414 liters under standard conditions. Find the volume occupied by 51.2 grams of the gas carbon dioxide under these conditions.
38. How much does one molecule of water weigh?

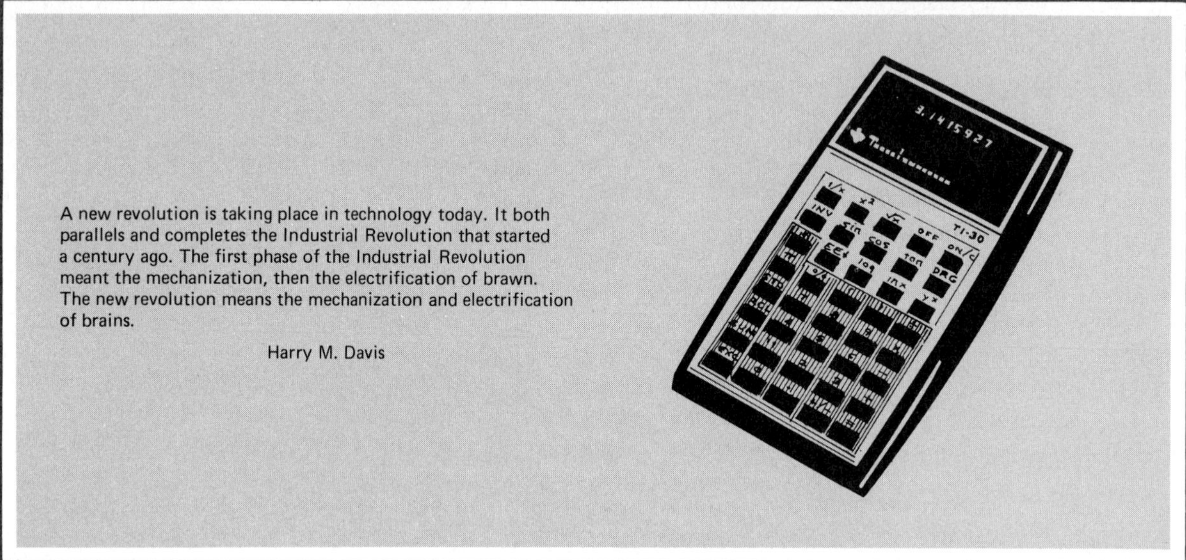

A new revolution is taking place in technology today. It both parallels and completes the Industrial Revolution that started a century ago. The first phase of the Industrial Revolution meant the mechanization, then the electrification of brawn. The new revolution means the mechanization and electrification of brains.

Harry M. Davis

2-3 Pocket Calculators

Most people can do arithmetic if they have to. But few do it with either enthusiasm or accuracy. Frankly, it is a rather dull subject. The spectacular sales of pocket electronic calculators demonstrate both our need to do arithmetic and our distaste for it.

Pocket calculators vary greatly in what they can do. The simplest perform only the four arithmetic operations and sell for under $10. The most sophisticated are programmable and can automatically perform dozens of operations in sequence. For this course, a standard scientific calculator is ideal. In addition to the four arithmetic operations, it will calculate values for the exponential and logarithmic functions and the trigonometric and inverse trigonometric functions.

Two kinds of logic are used in pocket calculators, **reverse Polish** logic and **algebraic** logic. The former avoids the use of parentheses but is slightly tricky to learn. Algebraic logic uses parentheses and mimics the procedures of ordinary algebra. The Texas Instruments Model 30 shown above uses algebraic logic and will serve to illustrate our discussion of calculators. You will need to study the instruction book for any other calculator to understand the minor modifications required to use that calculator.

Entering Data To enter 238.75, simply press the keys 238 · 75 in order. If the negative of this number is desired, press the same keys and then press the change sign key [+/−]. Numbers larger than 10^8 or smaller than 10^{-8} must be entered in scientific notation. For example, 2.3875×10^{19} would be entered as follows.

2 [·] 3875 [EE] 19

The key [EE] (which stands for *enter exponent*) controls the two places at the extreme right of the display. They are reserved for the exponent on 10. After pressing the indicated keys, the display will read

2.3875 19

If you press

2 [·] 3875 [+/−] [EE] 19 [+/−]

the display will read

−2.3875 −19

In making a complicated calculation, you may enter some numbers in standard notation and others in scientific notation. The calculator understands either form and makes the proper translations. If any of the entered data is in scientific notation, it will display the answer in this format. Also, if the answer is too large or too small for standard format, the calculator will automatically convert the result of a calculation to scientific notation.

Doing Arithmetic The five keys [+], [−], [×], [÷], and [=] are the work horses for arithmetic in any calculator using algebraic logic. To perform the calculation

$$175 + 34 - 18$$

simply press the keys indicated below.

175 [+] 34 [−] 18 [=]

The answer 191 will appear in the display.

Or consider (175)(14)/18. Press

175 [×] 14 [÷] 18 [=]

and the calculator will display 136.11111.

An expression involving additions (or subtractions) and multiplications (or divisions) may be ambiguous. For example, $2 \times 3 + 4 \times 5$ could have several meanings depending on which operations are performed first.

(i) $2 \times (3 + 4) \times 5 = 70$

(ii) $(2 \times 3) + (4 \times 5) = 26$

(iii) $[(2 \times 3) + 4] \times 5 = 50$

(iv) $2 \times [3 + (4 \times 5)] = 46$

Parentheses are used in mathematics and in calculators to indicate the order in which operations are to be performed. To do calculation (i), press

2 [×] [(] 3 [+] 4 [)] [×] 5 [=]

Similarly to do calculation (iii), press

[(] [(] 2 [×] 3 [)] [+] 4 [)] [×] 5 [=]

Recall that, in arithmetic, we have an agreement that when no parentheses are used, multiplications and divisions are done before additions and subtractions. Thus

$$2 \times 3 + 4 \times 5$$

is interpreted as $(2 \times 3) + (4 \times 5)$. The same convention is used in most calculators. Pressing

2 [×] 3 [+] 4 [×] 5 [=]

will yield the answer 26. Similarly for calculation (iii), pressing

[(] 2 [×] 3 [+] 4 [)] [×] 5 [=]

will yield 50, since within the parentheses, the calculator will do the multiplication first. However, when in doubt use parentheses, since without them it is easy to make errors.

Special Functions Most scientific calculators have keys for finding powers and roots of a number. On our sample calculator, the [y^x] key is used to raise a number y to the xth power. For example, to calculate $2.75^{-.34}$, press

2 [.] 75 [y^x] [.] 34 [+/−] [=]

and the correct result, .70896841, will appear in the display.

Finding a root is the inverse of raising to a power. For example, taking a cube root is the inverse operation of cubing. Thus, to calculate $\sqrt[3]{17}$, press 17 [INV] [y^x] 3 [=] and you will get 2.5712816. In using the [y^x] key, the calculator insists that y be positive. However, x may be either positive or negative.

Square roots occur so often that there is a special key for them on some calculators. To calculate $\sqrt{17}$, simply press 17 [$\sqrt{\ }$] and you will immediately get 4.1231056.

Our introduction to pocket calculators has been very brief. The problem set below will give you practice in using your particular model. To clear up any difficulties and to learn other features of your calculator, consult your instruction book.

Problem Set 2-3

Use your pocket calculator to perform the calculations in this problem set. We suggest that you begin by making a mental estimate of the answer. For example, the answer to Problem 3 might be estimated as $(3 - 6)(14 \times 50) = -2100$. *Similarly, the answer to Problem 17 might be estimated as* $(1.5)(10)/20 = .75$. *This will help you catch errors caused by pressing the wrong keys or failing to use parentheses properly.*

1. $34.1 - 49.95 + 64.2$
2. $7.465 + 3.12 - .0156$
3. $(3.42 - 6.71)(14.3 \times 51.9)$
4. $(21.34 + 2.37)(74.13 - 26.3)$
5. $\dfrac{514 + 31.9}{52.6 - 50.8}$
6. $\dfrac{547.3 - 832.7}{.0567 + .0416}$
7. $\dfrac{(6.34 \times 10^{7})(537.8)}{1.23 \times 10^{-5}}$
8. $\dfrac{(5.23 \times 10^{16})(.0012)}{1.34 \times 10^{11}}$
9. $\dfrac{6.34 \times 10^{7}}{.00152 + .00341}$
10. $\dfrac{3.134 \times 10^{-8}}{5.123 + 6.1457}$
11. $\dfrac{532 + 1.346}{34.91}(1.75 - 2.61)$
12. $\dfrac{39.95 - 42.34}{15.76 - 16.71}(5.31 \times 10^{4})$
13. $(1.214)^{3}$
14. $(3.617)^{-2}$
15. $\sqrt[4]{1.215}$
16. $\sqrt[7]{1.5789}$
17. $\dfrac{(1.34)(2.345)^{3}}{\sqrt{364}}$
18. $\dfrac{(14.72)^{12}(59.3)^{11}}{\sqrt{17.1}}$
19. $\dfrac{\sqrt{130} - \sqrt{5}}{15^{6} - 4^{8}}$
20. $\dfrac{\sqrt{143.2} + \sqrt{36.1}}{(234.1)^{4} - (11.2)^{2}}$

Example (A speed of light problem) How long will it take a light ray to reach the earth from the sun? Assume that light travels 2.9979×10^{10} centimeters per second, that 1 mile is equivalent to 1.609 kilometers, and that it is 9.30×10^{7} miles from the sun to the earth.

SOLUTION. The speed of light in kilometers per second is 2.9979×10^{5}, which we shall round to 2.998×10^{5}. The time required is

$$\frac{(9.30 \times 10^{7})(1.609)}{2.998 \times 10^{5}} = 4.9912 \times 10^{2} \approx 499$$

seconds.

21. How long will it take a light ray to travel from the moon to the earth? (The distance to the moon is 2.39×10^{5} miles.)
22. How long would it take a rocket traveling 4500 miles per hour to reach the sun from the earth.
23. A light year is the distance light travels in one year (365.24 days). Our nearest star is 4.300 light years away. How many meters is that?
24. How long would it take a rocket going 4500 miles per hour to get from the earth to our nearest star (see Problem 23)?
25. What is the area in square meters of a rectangular field 2.3 light years by 4.5 light years?
26. What is the volume in cubic meters of a cube 4.3 light years on a side?

Miscellaneous Problems

27. Calculate $\sqrt[3]{29.64} - \dfrac{1}{(.0387)^2} + \sqrt{51.54}$.

28. Calculate $\dfrac{(4.26)^2 + (9.32)^3}{5.8 \times 10^3}$.

All numbers in the remaining problems are to be considered approximate. In reporting your answer, round off your answers using the two rules given in Section 2.2.

29. Find the average of the following weights, all measured in pounds: 184.3, 212, 112.8, 274.3, 147.
30. Calculate $[(18.6 - 16)^2 + (22.4 - 16)^2 + (49.0 - 16)^2 + (13.3 - 16)^2]/4$. Consider 16 and 4 to be exact.
31. Find the area of a rectangle whose dimensions are 39.22 and 11.83 millimeters.
32. Calculate the volume of a box whose dimensions are 21.92, 18.43, and 12.02 centimeters.
33. How many centimeters are there in a mile?
34. Find the volume of a cone whose base radius is 23.2 centimeters and whose height is 42.8 centimeters. (The volume of a cone is given by $V = \pi r^2 h/3$.)

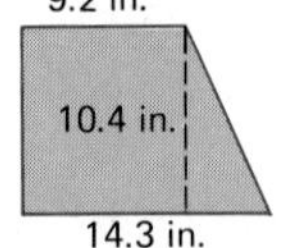

35. How many cubic feet of helium are there in a spherical balloon of radius 40.9 meters? (The volume of a sphere is given by $V = 4\pi r^3/3$.)
36. Find the area of the trapezoid shown in the margin.
37. If I walk 43.6 meters due east and then 97.6 meters due north, how far am I from my starting point?

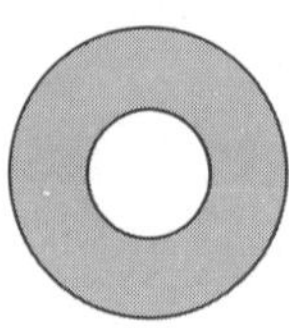

38. Find the area of the ring shown in the margin if the outer circle has radius 13.25 centimeters and the inner circle has radius 6.42 centimeters.

Poly-what?

polygamist A man who has several wives.
polyglot A person who speaks several languages.
polygon A plane figure which has several sides.
polymath A person who knows much about many subjects.
polynomial An expression which is the sum of several terms.
polyphony A musical combination of several harmonizing parts.

Examples of Polynomials

$3x^2 + 11x + 9$

$\frac{3}{2}x - 5$

$\sqrt{3}x^5 + \pi x^3 + 2x^2 + \frac{5}{4}$

$-3x^4 + 2x^3 - 3x$

$5x^{12}$

39

2-4 Polynomials

The dictionary definition of polynomial given above is suggestive, but it is not nearly precise enough for mathematicians. The expression $2x^{-2} + \sqrt{x} + 1/x$ is a sum of several terms, but it is not a polynomial. On the other hand, $3x^4$ is just one term; yet it is a perfectly good polynomial. Fundamentally, a **real polynomial in** x is any expression that can be obtained from the real numbers and x using only the operations of addition, subtraction, and multiplication. For example, we can get $3 \cdot x \cdot x \cdot x \cdot x$ or $3x^4$ by multiplication. We can get $2x^3$ by the same process and then have

$$3x^4 + 2x^3$$

by addition. We could never get $2x^{-2} = 2/x^2$ or $\sqrt{x}$; the first involves a division and the second involves taking a root. Thus $2/x^2$ and $\sqrt{x}$ are not polynomials. Try to convince yourself that all of the expressions in the box above are polynomials.

There is another way to define a polynomial in fewer words. A **real polynomial in** x is any expression of the form

$$a_n x^n + a_{n-1}x^{n-1} + a_{n-2}x^{n-2} + \cdots + a_1 x + a_0$$

where the a's are real numbers and n is a nonnegative integer. We shall not have much to do with complex polynomials in x, but they are defined in exactly the same way, except that the a's are allowed to be complex numbers. In either case, we refer to the a's as **coefficients**. The **degree** of a polynomial is the largest exponent that occurs in the polynomial in a term with a nonzero coefficient. Here are several more examples.

1. $\frac{3}{2}x - 5$ is a first degree (or **linear**) polynomial in x.
2. $3y^2 - 2y + 16$ is a second degree (or **quadratic**) polynomial in y.
3. $\sqrt{3}t^5 - \pi t^3 + 2t^2 - 17$ is a fifth degree polynomial in t.
4. $5x^3$ is a third degree polynomial in x. It is also called a **monomial** since it has only one term.
5. 13 is a polynomial of degree zero. Does that seem strange to you? If it helps, think of it as $13x^0$.

The whole subject of polynomials would be pretty dull and almost useless if it stopped with a definition. Fortunately, polynomials—like numbers—can be manipulated. In fact, they behave something like the integers. Just as the sum, difference, and product of two integers are integers, so the sum, difference, and product of two polynomials are polynomials. Notice that we did not mention division; that subject is discussed in Section 2-6 and in Chapter 10.

Addition Adding two polynomials is a snap. Treat x like a number and use the commutative, associative, and distributive properties freely. When you are all done, you will discover that you have just grouped like terms (that is, terms of the same degree) and added their coefficients. Here, for example, is how we add $x^3 + 2x^2 + 7x + 5$ and $x^2 - 3x - 4$.

$$\begin{aligned}
&(x^3 + 2x^2 + 7x + 5) + (x^2 - 3x - 4) \\
&\quad = x^3 + (2x^2 + x^2) + (7x - 3x) + (5 - 4) \qquad \text{(associative and commutative properties)} \\
&\quad = x^3 + (2 + 1)x^2 + (7 - 3)x + 1 \qquad \text{(distributive property)} \\
&\quad = x^3 + 3x^2 + 4x + 1
\end{aligned}$$

How important are the parentheses in this example? Actually, they are indispensable only in the third line, where we used the distributive property. Why did we use them in the first and second lines? Because they emphasize what is happening. In the first line, they show which polynomials are added; in the second line, they draw attention to the terms being grouped. To shed additional light, notice that

$$x^3 + 3x^2 + 4x + 1$$

is the correct answer not only for

$$(x^3 + 2x^2 + 7x + 5) + (x^2 - 3x - 4)$$

but also for

$$(x^3 + 2x^2) + (7x + 5 + x^2) + (-3x - 4)$$

and even for

$$(x^3) + (2x^2) + (7x) + (5) + (x^2) + (-3x) + (-4)$$

Subtraction How do we subtract two polynomials? We replace the subtracted polynomial by its negative and add. For example, we rewrite

$$(3x^2 - 5x + 2) - (5x^2 + 4x - 4)$$

as

$$(3x^2 - 5x + 2) + (-5x^2 - 4x + 4)$$

Then, after grouping like terms, we obtain

$$\begin{aligned}(3x^2 - 5x^2) + (-5x - 4x) + (2 + 4) &= (3 - 5)x^2 + (-5 - 4)x + (2 + 4) \\ &= -2x^2 - 9x + 6\end{aligned}$$

If you can go directly from the original problem to the answer, do so; we do not want to make simple things complicated. But do not make the common mistake of thinking

$$(3x^2 - 5x + 2) - (5x^2 + 4x - 4) = 3x^2 - 5x + 2 - 5x^2 + 4x - 4$$

The minus sign in front of $(5x^2 + 4x - 4)$ changes the sign of all three terms.

Multiplication The distributive property is the basic tool in multiplication. Here is a simple example.

$$\begin{aligned}(3x^2)(2x^3 + 7) &= (3x^2)(2x^3) + (3x^2)(7) \\ &= 6x^5 + 21x^2\end{aligned}$$

But things can get more complicated.

$$\begin{aligned}(3x - 4)(2x^3 - 7x + 8) &= (3x)(2x^3 - 7x + 8) + (-4)(2x^3 - 7x + 8) \\ &= (3x)(2x^3) + (3x)(-7x) + (3x)(8) + (-4)(2x^3) \\ &\quad + (-4)(-7x) + (-4)(8) \\ &= 6x^4 - 21x^2 + 24x - 8x^3 + 28x - 32 \\ &= 6x^4 - 8x^3 - 21x^2 + 52x - 32\end{aligned}$$

Notice that each term of $3x - 4$ multiplies each term of $2x^3 - 7x + 8$.

The process just illustrated is unwieldy; it is also easy to make mistakes. You may find the format below helpful.

$$\begin{array}{lllll} 2x^3 & - 7x + 8 & & & \\ 3x - 4 & & & & \\ \hline 6x^4 & & - 21x^2 & + 24x & \\ & - 8x^3 & & + 28x & - 32 \\ \hline 6x^4 & - 8x^3 & - 21x^2 & + 52x & - 32 \end{array}$$

When both polynomials are linear (that is, of the form $ax + b$), there is a handy shortcut. For example, just one look at $(x + 4)(x + 5)$ convinces us that the product has the form $x^2 + (\quad) + 20$. It is the middle term that may cause a little trouble. Think of it this way.

$$\overbrace{(x + \underbrace{4)(x}_{4x} + 5)}^{5x} = x^2 + (\overset{5x + 4x}{\downarrow}) + 20$$
$$= x^2 + 9x + 20$$

Similarly,

$$\overbrace{(2x - \underbrace{3)(x}_{-3x} + 5)}^{10x} = 2x^2 + (\overset{10x - 3x}{\downarrow}) - 15$$
$$= 2x^2 + 7x - 15$$

Finally,

$$\overbrace{(3x + \underbrace{2)(5x}_{10x} - 7)}^{-21x} = 15x^2 + (\overset{-21x + 10x}{\downarrow}) - 14$$
$$= 15x^2 - 11x - 14$$

Soon you should be able to find such simple products in your head.

Polynomials in Several Variables So far, we have considered only polynomials in a single variable. That is too restrictive for later work. Expressions such as

$$x^2y + 3xy + y \qquad u^3 + 3u^2v + 3uv^2 + v^3$$

will appear often. They present little in the way of new ideas and so we defer consideration of them to the problem set (see Problems 69–86).

Problem Set 2-4

Decide whether each expression is a polynomial. If it is, give its degree.

1. $3x^2 - x + 2$
2. $4x^5 - x$
3. $\pi s^5 - \sqrt{2}$
4. $3\sqrt{2}t$
5. $16\sqrt{2}$
6. $511/\sqrt{2}$
7. $3t^2 + \sqrt{t} + 1$
8. $t^2 + 3t + 1/t$
9. $3t^{-2} + 2t^{-1} + 5$
10. $5 + 4t + 6t^{10}$

Perform the indicated operations and simplify. Write your answer as a polynomial in descending powers of the variable. It is suggested that a calculator be used to solve problems marked with [c] .

11. $(2x - 7) + (-4x + 8)$
12. $(\frac{3}{2}x - \frac{1}{4}) + \frac{5}{6}x$
13. $(2x^2 - 5x + 6) + (2x^2 + 5x - 6)$
14. $(\sqrt{3}t + 5) + (6 - 4 - 2\sqrt{3}t)$
15. $(5 - 11x^2 + 4x) + (x - 4 + 9x^2)$
16. $(x^2 - 5x + 4) + (3x^2 + 8x - 7)$

17. $(2x - 7) - (-4x + 8)$

18. $(\frac{3}{2}x - \frac{1}{4}) - \frac{5}{6}x$

19. $(2x^2 - 5x + 6)$
$- (2x^2 + 5x - 6)$

20. $y^3 - 4y + 6 - (3y^2 + 6y - 3)$

21. $5x(7x - 11) + 19$

22. $-x^2(7x^3 - 5x + 1)$

23. $(t + 5)(t + 11)$

24. $(t - 5)(t + 13)$

25. $(x + 9)(x - 10)$

26. $(x - 13)(x - 7)$

27. $(2t - 1)(t + 7)$

28. $(3t - 5)(4t - 2)$

29. $(4 + y)(y - 2)$

30. $1 + y(y - 2)$

C 31. $(3.41x - 2.53)(2.34x + 1.77)$

C 32. $(.66x + .87)(.41x - 3.12)$

Example (Two special products) Some products occur so often that they deserve to be highlighted. Here are two product formulas: the first for the square of a binomial and the second for the product of the sum and difference of two terms.

$$(x + a)^2 = x^2 + 2ax + a^2$$

$$(x + a)(x - a) = x^2 - a^2$$

Use these formulas to find each product.

(a) $(x + 7)^2$ (b) $(x + 9)(x - 9)$

(c) $(2x^2 - 3x)^2$ (d) $[(x^2 + 2) + x][(x^2 + 2) - x]$

SOLUTION.

(a) $(x + 7)^2 = x^2 + 2 \cdot 7x + 7^2 = x^2 + 14x + 49$

(b) $(x + 9)(x - 9) = x^2 - 9^2 = x^2 - 81$

(c) We apply the first formula in the box with $2x^2$ and $-3x$ taking the place of x and a, respectively.

$$\begin{aligned}(2x^2 - 3x)^2 &= [2x^2 + (-3x)]^2 = (2x^2)^2 + 2(2x^2)(-3x) + (-3x)^2 \\ &= 4x^4 - 12x^3 + 9x^2\end{aligned}$$

(d) We apply the second formula in the box with $x^2 + 2$ and x playing the roles of x and a, respectively.

$$\begin{aligned}[(x^2 + 2) + x][(x^2 + 2) - x] &= (x^2 + 2)^2 - x^2 \\ &= x^4 + 4x^2 + 4 - x^2 \\ &= x^4 + 3x^2 + 4\end{aligned}$$

Use the formulas in the example to perform the following multiplications. Write your answer as a polynomial in descending powers of the variable.

33. $(x + 10)^2$

34. $(y + 12)^2$

35. $(x + 8)(x - 8)$

36. $(t - 5)(t + 5)$

37. $(2t - 5)^2$

38. $(3s + 11)^2$

39. $(2x^4 + 5x)(2x^4 - 5x)$

40. $(u^3 + 2u^2)(u^3 - 2u^2)$

41. $[(t + 2) + t^3]^2$

42. $[(1 - x) + x^2]^2$

43. $[(t + 2) + t^3][(t + 2) - t^3]$

44. $[(1 - x) + x^2][(1 - x) - x^2]$

C 45. $(2.3x - 1.4)^2$

C 46. $(2.43x - 1.79)(2.43x + 1.79)$

Miscellaneous Problems

Perform the indicated operations and simplify. Write your answer as a polynomial in descending powers of the variable.

47. $2x^2 - 4x + 5$
    $- 3x(x^2 + 7x - 2)$

48. $(2y - 4)(y + 3) + (y^2 - 9)$

49. $(y + 3)(2y - 5)$

50. $(2s - 1)(3s + 2)$

51. $(2y + 5)^2$

52. $(3x - 2)^2$

53. $(2z - 7)(2z + 7)$

54. $(5x + 4)(5x - 4)$

55. $(x^2 + 4)(x^2 - 3)$

56. $(y^3 + 2)(y^3 - 5)$

57. $(2x^2 + x - 5)(3x + 2)$

58. $(3x^2 - 4x + 1)(2x^2 - 5)$

59. $(x + \sqrt{2})(x - \sqrt{2})$

60. $(x + 2i)(x - 2i)$

61. $(x - 2\sqrt{7})^2$

62. $(\sqrt{2}x + 5)^2$

63. $(2x + 1)^2 - (2x - 1)^2$

64. $4(x + 2)^2 - (x + 4)^2$

65. $(2x + 1)^3$

66. $(3y + 2)^3$

[C] 67. $(2.39x + 4.12)^2 - 4.13x$

[C] 68. $(2.39x - 4.12)(2.39x - 4.12)$
    $+ (1.12x)^2$

The remaining problems involve polynomials in several variables. In each case, do the indicated operations and simplify. For example,

$$(3x + 2y)(x - 5y) = 3x^2 - 13xy - 10y^2$$

69. $(5x - y)(5x + y)$

70. $(2x + 3y)(2x - 3y)$

71. $(5x - y)^2 + y^2$

72. $(2x + 3y)^2 + 3x^2$

73. $(2x - 3y)(4x + y)$

74. $(2x + 4y)(x - y)$

75. $(x + y)(x^2 - xy + y^2)$

76. $(x - y)(x^2 + xy + y^2)$

77. $(x + 2y)^3$

78. $(x - 2y)^3$

79. $(x^2 + y^2)(x^2 - y^2)$

80. $(x^2 + 2xy)(x^2 - 2xy)$

81. $(x^2 + 2y^2)(x^2 - 3y^2)$

82. $(2x^2 - 3y^2)(2x^2 + 4y^2)$

83. $(3xy + z)(2xy - 3z)$

84. $(2xy + 3uv)(2xy - 4uv)$

85. $(x^2 + xy + y^2)(x^2 + 2xy - y^2)$

86. $(x^2 - 3xy + y^2)(x^2 + y^2)$

Why Johnny Can't Factor

Polynomial to be Factored	Johnny's Answer	Teacher's Comments
1. $x^6 + 2x^2$	$x^2(x^3 + 2)$	Wrong. Have you forgotten that $x^2x^3 = x^5$? Right answer: $x^2(x^4 + 2)$
2. $x^2 + 5x + 6$	$(x + 6)(x + 1)$	Wrong. You didn't check the middle term. Right answer: $(x + 2)(x + 3)$
3. $x^2 - 4y^2$	$(x + 2y)(x - 2y)$	Right.
4. $x^2 + 4y^2$	$(x + 2y)^2$	Wrong. $(x + 2y)^2 = x^2 + 4xy + 4y^2$ Right answer: $x^2 + 4y^2$ doesn't factor using real coefficients.
5. $x^2y^2 + 6xy + 9$	Impossible	Wrong. x^2y^2 and 9 are squares. You should have suspected a perfect square. Right answer: $(xy + 3)^2$

2-5 Factoring Polynomials

To factor 90 means to write it as a product of smaller numbers; to factor it completely means to write it as a product of primes, that is, numbers that cannot be factored further. Thus we have factored 90 when we write $90 = 9 \cdot 10$, but it is not factored completely until we write

$$90 = 2 \cdot 3 \cdot 3 \cdot 5$$

Similarly to **factor** a polynomial means to write it as a product of simpler polynomials; to **factor** a polynomial **completely** is to write it as a product of polynomials that cannot be factored further. Thus when we write

$$x^3 - 9x = x(x^2 - 9)$$

we have factored $x^3 - 9x$, but not until we write

$$x^3 - 9x = x(x + 3)(x - 3)$$

have we factored $x^3 - 9x$ completely.

Now why can't Johnny factor? He can't factor because he can't multiply. If he doesn't know that

$$(x + 2)(x + 3) = x^2 + 5x + 6$$

he certainly is not going to know how to factor $x^2 + 5x + 6$. That is why we urge you to memorize the special product formulas listed on page 68. Of course, a product formula is also a factoring formula when read from right to left.

Product Formulas ⟶ ⟵ Factoring Formulas
1. $a(x + y + z) = ax + ay + az$
2. $(x + a)(x + b) = x^2 + (a + b)x + ab$
3. $(x + y)^2 = x^2 + 2xy + y^2$
4. $(x + y)(x - y) = x^2 - y^2$
5. $(x + y)(x^2 - xy + y^2) = x^3 + y^3$
6. $(x - y)(x^2 + xy + y^2) = x^3 - y^3$
7. $(x + y)^3 = x^3 + 3x^2y + 3xy^2 + y^3$

To urge memorization may be a bit old-fashioned, but we suggest that a fact, once memorized, becomes a permanent friend. It is best to memorize in words. For example, read formula (3) as follows: the square of a sum of two terms is the first squared plus twice their product plus the second squared.

Taking Out a Common Factor This, the simplest factoring procedure, is based on formula (1) above. You should always try this process first. Take Johnny's first problem as an example. Both terms of $x^6 + 2x^2$ have x^2 as a factor, so we take it out.

$$x^6 + 2x^2 = x^2(x^4 + 2)$$

Always factor out as much as you can. Taking 2 out of $4xy^2 - 6x^3y^4 + 8x^4y^2$ is not nearly enough, though it is a common factor; taking out $2xy$ is not enough either. You should take out $2xy^2$. Then

$$4xy^2 - 6x^3y^4 + 8x^4y^2 = 2xy^2(2 - 3x^2y^2 + 4x^3)$$

Factoring by Trial and Error In factoring, as in life, success often results from trying and trying again. What does not work is systematically eliminated; eventually, effort is rewarded. Let us see how this process works on $x^2 - 5x - 14$. We need to find numbers a and b such that

$$x^2 - 5x - 14 = (x + a)(x + b)$$

Since ab must equal -14, two possibilities immediately occur to us: $a = 7$ and $b = -2$ or $a = -7$ and $b = 2$. Try them both to see if one works.

$$(x + 7)(x - 2) = x^2 + 5x - 14$$

$$(x - 7)(x + 2) = x^2 - 5x - 14 \qquad \text{Success!}$$

The brackets help us calculate the middle term, the crucial step in this kind of factoring.

Here is a tougher factoring problem: Factor $2x^2 + 13x - 15$. It is a safe bet that if $2x^2 + 13x - 15$ factors at all, then

$$2x^2 + 13x - 15 = (2x + a)(x + b).$$

Since $ab = -15$, we are likely to try combinations of 3 and 5 first.

$$(2x + 5)(x - 3) = 2x^2 - x - 15$$
$$(2x - 5)(x + 3) = 2x^2 + x - 15$$
$$(2x + 3)(x - 5) = 2x^2 - 7x - 15$$
$$(2x - 3)(x + 5) = 2x^2 + 7x - 15$$

Discouraging, isn't it? But that is a poor reason to give up. Maybe we have missed some possibilities. We have, since combinations of 15 and 1 might work.

$$(2x - 15)(x + 1) = 2x^2 - 13x - 15$$
$$(2x + 15)(x - 1) = 2x^2 + 13x - 15 \qquad \text{Success!}$$

When you have had a lot of practice, you will be able to speed up the process. You will simply write

$$2x^2 + 13x - 15 = (2x + ?)(x + ?)$$

and mentally try the various possibilities until you find the right one. Of course, it may happen, as in the case of $2x^2 - 4x + 5$, that you cannot find a factorization.

Perfect Squares Certain second degree (quadratic) polynomials are a breeze to factor.

$$x^2 + 10x + 25 = (x + 5)(x + 5) = (x + 5)^2$$
$$x^2 - 12x + 36 = (x - 6)^2$$
$$4x^2 + 12x + 9 = (2x + 3)^2$$

Each is modeled after a special product formula, formula 3.

$$a^2 + 2ab + b^2 = (a + b)^2$$

We look for first and last terms that are squares, say of a and b. Then we ask if the middle term is twice their product.

But we need to be very flexible; a and b might be quite complicated. Consider $x^4 + 2x^2y^3 + y^6$. The first term is the square of x^2 and the last is the square of y^3; the middle term is twice their product.

$$\begin{aligned} x^4 + 2x^2y^3 + y^6 &= (x^2)^2 + 2(x^2)(y^3) + (y^3)^2 \\ &= (x^2 + y^3)^2 \end{aligned}$$

Similarly,

$$y^2z^2 - 6ayz + 9a^2 = (yz - 3a)^2.$$

However,

$$a^4b^2 + 6a^2bc + 4c^2 \neq (a^2b + 2c)^2$$

since the middle term does not check.

Difference of Squares Do you see a common feature in the following polynomials?

$$x^2 - 16 \qquad y^2 - 100 \qquad 4y^2 - 9b^2$$

Each is the difference of two squares. From one of our special product formulas (formula 4), we know that

$$\boxed{a^2 - b^2 = (a + b)(a - b)}$$

Thus

$$\begin{aligned} x^2 - 16 &= (x + 4)(x - 4) \\ y^2 - 100 &= (y + 10)(y - 10) \\ 4y^2 - 9b^2 &= (2y + 3b)(2y - 3b) \end{aligned}$$

Sum and Difference of Cubes Now we are ready for some high-class factoring. Consider $8x^3 + 27$ and $x^3z^3 - 1000$. The first is a sum of cubes and the second is a difference of cubes. The secrets to success are the two special product formulas for cubes.

$$\boxed{\begin{aligned} a^3 + b^3 &= (a + b)(a^2 - ab + b^2) \\ a^3 - b^3 &= (a - b)(a^2 + ab + b^2) \end{aligned}}$$

To factor $8x^3 + 27$, replace a by $2x$ and b by 3 in the first formula.

$$\begin{aligned} 8x^3 + 27 = (2x)^3 + 3^3 &= (2x + 3)[(2x)^2 - (2x)(3) + 3^2] \\ &= (2x + 3)(4x^2 - 6x + 9) \end{aligned}$$

Similarly, to factor $x^3z^3 - 1000$, let $a = xz$ and $b = 10$ in the second formula.

$$x^3z^3 - 1000 = (xz)^3 - 10^3 = (xz - 10)(x^2z^2 + 10xz + 100)$$

Someone is sure to make a terrible mistake and write

$$x^3 + y^3 = (x + y)^3 \qquad \text{Wrong!!}$$

Remember that

$$(x + y)^3 = x^3 + 3x^2y + 3xy^2 + y^3$$

To Factor or Not to Factor Which of the following can be factored?

(i)	$x^2 - 4$
(ii)	$x^2 - 6$
(iii)	$x^2 + 16$

Did you say only the first one? You are correct if we insist on integer coefficients, or as we say, if we **factor over the integers**. But if we factor over the real numbers (that is, insist only that the coefficients be real), then the second polynomial also factors.

$$x^2 - 6 = (x + \sqrt{6})(x - \sqrt{6})$$

If we factor over the complex numbers, even the third polynomial will factor.

$$x^2 + 16 = (x + 4i)(x - 4i)$$

For this reason, we should always spell out what kind of coefficients we permit in the answer. We give specific directions in the following problem set.

Problem Set 2-5

Factor completely over the integers (that is, allow only integer coefficients in your answers).

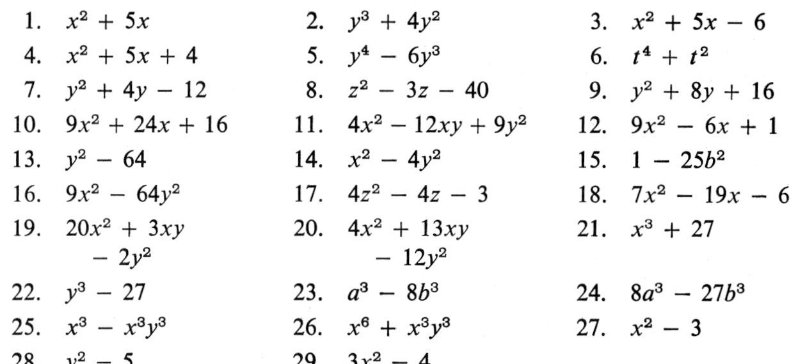

1. $x^2 + 5x$	2. $y^3 + 4y^2$	3. $x^2 + 5x - 6$
4. $x^2 + 5x + 4$	5. $y^4 - 6y^3$	6. $t^4 + t^2$
7. $y^2 + 4y - 12$	8. $z^2 - 3z - 40$	9. $y^2 + 8y + 16$
10. $9x^2 + 24x + 16$	11. $4x^2 - 12xy + 9y^2$	12. $9x^2 - 6x + 1$
13. $y^2 - 64$	14. $x^2 - 4y^2$	15. $1 - 25b^2$
16. $9x^2 - 64y^2$	17. $4z^2 - 4z - 3$	18. $7x^2 - 19x - 6$
19. $20x^2 + 3xy - 2y^2$	20. $4x^2 + 13xy - 12y^2$	21. $x^3 + 27$
22. $y^3 - 27$	23. $a^3 - 8b^3$	24. $8a^3 - 27b^3$
25. $x^3 - x^3y^3$	26. $x^6 + x^3y^3$	27. $x^2 - 3$
28. $y^2 - 5$	29. $3x^2 - 4$	

Factor completely over the real numbers.

30. $x^2 - 3$
31. $y^2 - 5$
32. $3x^2 - 4$
33. $5z^2 - 4$
34. $t^4 - t^2$
35. $t^4 - 2t^2$
36. $x^2 + 2\sqrt{2}x + 2$
37. $y^2 - 2\sqrt{3}y + 3$
38. $x^2 + 4y^2$
39. $x^2 + 9$
40. $4x^2 + 1$

Factor completely over the complex numbers.

41. $x^2 + 9$
42. $4x^2 + 1$

Example A (Factoring by substitution) Factor completely over the integers.

(a) $3x^4 + 10x^2 - 8$ (b) $(x + 2y)^2 - 3(x + 2y) - 10$

SOLUTION. (a) Replace x^2 by u (or some other favorite letter of yours). Then

$$3x^4 + 10x^2 - 8 = 3u^2 + 10u - 8$$

But we know how to factor the latter:

$$3u^2 + 10u - 8 = (3u - 2)(u + 4)$$

Thus, when we go back to x, we get

$$3x^4 + 10x^2 - 8 = (3x^2 - 2)(x^2 + 4)$$

Neither of these quadratic polynomials factors further (using integer coefficients), so we are done.

(b) Here we could let $u = x + 2y$ and then factor the resulting quadratic polynomial. But this time, let us do that step mentally and write

$$\begin{aligned}(x + 2y)^2 - 3(x + 2y) - 10 &= [(x + 2y) + \ ?][(x + 2y) - \ ?] \\ &= (x + 2y + 2)(x + 2y - 5)\end{aligned}$$

Factor completely over the integers. If you can factor without actually making a substitution, that is fine.

43. $x^6 + 9x^3 + 14$
44. $x^4 - x^2 - 6$
45. $4x^4 - 37x^2 + 9$
46. $6y^4 + 13y^2 - 5$
47. $(x + 4y)^2 + 6(x + 4y) + 9$
48. $(m - n)^2 + 5(m - n) + 4$
49. $x^4 - x^2y^2 - 6y^4$
50. $x^4y^4 + 5x^4y^2 + 6x^4$

Example B (Factoring in stages) Factor $x^6 - y^6$ over the integers.

SOLUTION. First think of this expression as a difference of squares. Then factor again.

$$\begin{aligned}x^6 - y^6 &= (x^3 - y^3)(x^3 + y^3) \\ &= (x - y)(x^2 + xy + y^2)(x + y)(x^2 - xy + y^2)\end{aligned}$$

Factor completely over the integers.

51. $x^6 - 64$
52. $x^4 - y^4$
53. $x^8 - x^4y^4$
54. $x^9 - 64x^3$
55. $x^6 + y^6$ (sum of cubes)
56. $a^6 + 64$

Example C (Factoring by grouping) Factor.

(a) $am - an - bm + bn$ (b) $a^2 - 4ab + 4b^2 - c^2$

SOLUTION. To factor expressions involving more than three terms will usually require grouping of some of the terms together.

(a)
$$\begin{aligned} am - an - bm + bn &= (am - an) - (bm - bn) \\ &= a(m - n) - b(m - n) \\ &= (a - b)(m - n) \end{aligned}$$

Note that $m - n$ was a common factor of both terms. Only after removing that factor did we have a factored form.

(b)
$$\begin{aligned} a^2 - 4ab + 4b^2 - c^2 &= (a^2 - 4ab + 4b^2) - c^2 \\ &= (a - 2b)^2 - c^2 \qquad \text{(difference of squares)} \\ &= [(a - 2b) + c][(a - 2b) - c] \\ &= (a - 2b + c)(a - 2b - c) \end{aligned}$$

Factor completely over the integers.

57. $x^3 - 4x^2 + x - 4$
58. $y^3 + 3y^2 - 2y - 6$
59. $4x^2 - 4x + 1 - y^2$
60. $9a^2 - 4b^2 - 12b - 9$
61. $3x + 3y - x^2 - xy$
62. $y^2 - 3y + xy - 3x$
63. $x^2 + 6xy + 9y^2 + 2x + 6y$
64. $y^2 + 4xy + 4x^2 - 3y - 6x$
65. $x^2 + 2xy + y^2 + 3x + 3y + 2$
66. $a^2 - 2ab + b^2 - c^2 + 4cd - 4d^2$

Example D (Factoring by adding and subtracting the same thing) Factor $x^4 + 4$ over the integers.

SOLUTION. Most people would bet money that this one does not factor. We have to admit that it is tricky. But see what happens when we add and subtract $4x^2$.

$$\begin{aligned} x^4 + 4 &= x^4 + 4x^2 + 4 - 4x^2 \\ &= (x^4 + 4x^2 + 4) - 4x^2 \\ &= (x^2 + 2)^2 - (2x)^2 \qquad \text{(difference of squares)} \\ &= (x^2 + 2 + 2x)(x^2 + 2 - 2x) \\ &= (x^2 + 2x + 2)(x^2 - 2x + 2) \end{aligned}$$

Factor completely over the integers.

67. $x^4 + 64$
68. $y^8 + 4$
69. $x^4 + x^2 + 1$
70. $x^8 + 3x^4 + 4$

Miscellaneous Problems

Factor completely over the integers.

71. $5x^2 - 10x$
72. $3x^2 - 18x + 27$
73. $4x^2 + 4x + 1$
74. $9 - 4m^2$
75. $4y^2 - 1$
76. $6x^2 - 5x + 1$
77. $2y^2 - 7y + 5$
78. $6x^2 - 5x - 6$
79. $4x^4 - 16$
80. $4x^4 - 4x$
81. $a^2 + 2ab + b^2 - 25$
82. $b^2 + 4bc + 4c^2 + 3b + 6c$

83. $a^2 + 2ab + b^2 - 2a - 2b$

84. $b^2 + 4bc + 4c^2 - 1$

85. $-2xy^3 + 54x$

86. $x^3y^2 + x^2y^3 + 2xy^4$

87. $x^4 - y^2 + 4y - 4$

88. $a^6 - b^6$

89. $(a + 2b)^2 - (a + 2b) - 6$

90. $x^2 + 3x + 4$

91. $x^2 + x + 1$

92. $x^4 + 3x^2 + 4$

93. $x^4 + x^2 + 1$

94. $4b^4 - 37b^2 + 9$

95. $(x + 3)^2(x - 7)^3 + (x + 3)(x - 7)^2$

96. $(x - 4)(x + 5)^4 - (x - 4)^2(x + 5)^3$

Factor completely over the real numbers.

[C] 97. $x^2 - 3.124$

[C] 98. $x^3 - 5.13$

[C] 99. $x^3 + 4.91$

[C] 100. $2x^2 - 3.22$

2-6 Rational Expressions

Remember that the quotient (ratio) of two integers is a rational number. A quotient (ratio) of two polynomials is called a **rational expression**. Here are some examples involving one variable.

$$\frac{x}{x+1} \qquad \frac{3x^2+6}{x^2+2x} \qquad \frac{x^{15}+3x}{6x^2+11}$$

The example in the riddle is a rational expression in two variables.

We add, subtract, multiply, and divide rational expressions by the same rules we used for rational numbers (see Section 1-2). The result is always a rational expression. That should not surprise us, since we had the same kind of experience with rational numbers.

Reduced Form A rational expression is in **reduced form** if its numerator and denominator have no (nontrivial) common factor. For example, $x/(x + 2)$ is in reduced form, but $x^2/(x^2 + 2x)$ is not, since

$$\frac{x^2}{x^2+2x} = \frac{\cancel{x}\cdot x}{\cancel{x}(x+2)} = \frac{x}{x+2}$$

To reduce a rational expression, we factor numerator and denominator and divide out, or cancel, common factors. Here are two examples.

$$\frac{x^2+7x+10}{x^2-25} = \frac{\cancel{(x+5)}(x+2)}{\cancel{(x+5)}(x-5)} = \frac{x+2}{x-5}$$

$$\frac{2x^2+5xy-3y^2}{2x^2+xy-y^2} = \frac{\cancel{(2x-y)}(x+3y)}{\cancel{(2x-y)}(x+y)} = \frac{x+3y}{x+y}$$

Students have been known to write

$$\frac{x + 6y}{x + 2} = \frac{\cancel{x} + 6y}{\cancel{x} + 2} = \frac{6y}{2} = 3y$$

That is worse than impolite; it is absolutely wrong! You cannot cancel *terms*, only *factors*. As a matter of fact, $x + 6y$ and $x + 2$ have no common factor and therefore $(x + 6y)/(x + 2)$ is already in reduced form.

Addition and Subtraction We add (or subtract) rational expressions by rewriting them so that they have the same denominator and then adding (or subtracting) the new numerators. Suppose we want to add

$$\frac{3}{x} + \frac{2}{x + 1}$$

The appropriate common denominator is $x(x + 1)$. Remember that we can multiply numerator and denominator of a fraction by the same thing. Accordingly,

$$\frac{3}{x} + \frac{2}{x + 1} = \frac{3(x + 1)}{x(x + 1)} + \frac{x \cdot 2}{x(x + 1)} = \frac{3x + 3 + 2x}{x(x + 1)} = \frac{5x + 3}{x(x + 1)}$$

Here is a more complicated example, a subtraction problem. Study each step carefully.

Begin by factoring

$(x - 1)^2\ (x + 1)$ is the lowest common denominator

Forgetting the parentheses around $2x^2 + 3x + 1$ would be a serious blunder

No cancellation is possible since $x^2 - 6x - 1$ doesn't factor over the integers

$$\begin{aligned}
\frac{3x}{x^2 - 1} - \frac{2x + 1}{x^2 - 2x + 1} &= \frac{3x}{(x - 1)(x + 1)} - \frac{2x + 1}{(x - 1)^2} \\
&= \frac{3x(x - 1)}{(x - 1)^2(x + 1)} - \frac{(2x + 1)(x + 1)}{(x - 1)^2(x + 1)} \\
&= \frac{3x^2 - 3x - (2x^2 + 3x + 1)}{(x - 1)^2(x + 1)} \\
&= \frac{3x^2 - 3x - 2x^2 - 3x - 1}{(x - 1)^2(x + 1)} \\
&= \frac{x^2 - 6x - 1}{(x - 1)^2(x + 1)}
\end{aligned}$$

Multiplication We multiply rational expressions in the same manner as we do rational numbers; that is, we multiply numerators and multiply denominators. For example,

$$\frac{3}{x + 5} \cdot \frac{x + 2}{x - 4} = \frac{3(x + 2)}{(x + 5)(x - 4)} = \frac{3x + 6}{x^2 + x - 20}$$

Sometimes we need to reduce the product, if we want the simplest possible answer. Here is an illustration.

$$\frac{2x-3}{x+5}\cdot\frac{x^2-25}{6xy-9y} = \frac{(2x-3)(x^2-25)}{(x+5)(6xy-9y)}$$

$$= \frac{\cancel{(2x-3)}(x-5)\cancel{(x+5)}}{\cancel{(x+5)}(3y)\cancel{(2x-3)}}$$

$$= \frac{x-5}{3y}$$

This example shows that it is a good idea to do as much factoring as possible at the outset. That is what set up the cancellation.

Division There are no real surprises with division, as we simply invert the divisor and multiply. Here is a nontrivial example.

$$\frac{x^2-5x+4}{2x+6} \div \frac{2x^2-x-1}{x^2+5x+6} = \frac{x^2-5x+4}{2x+6}\cdot\frac{x^2+5x+6}{2x^2-x-1}$$

$$= \frac{(x^2-5x+4)(x^2+5x+6)}{(2x+6)(2x^2-x-1)}$$

$$= \frac{(x-4)\cancel{(x-1)}(x+2)\cancel{(x+3)}}{2\cancel{(x+3)}\cancel{(x-1)}(2x+1)}$$

$$= \frac{(x-4)(x+2)}{2(2x+1)}$$

Problem Set 2-6

Reduce each of the following.

1. $\dfrac{x+6}{x^2-36}$

2. $\dfrac{x^2-1}{4x-4}$

3. $\dfrac{y^2+y}{5y+5}$

4. $\dfrac{x^2-7x+6}{x^2-4x-12}$

5. $\dfrac{(x+2)^3}{x^2-4}$

6. $\dfrac{x^3+a^3}{(x+a)^2}$

7. $\dfrac{zx^2+4xyz+4y^2z}{x^2+3xy+2y^2}$

8. $\dfrac{x^3-27}{3x^2+9x+27}$

Perform the indicated operations and simplify.

9. $\dfrac{5}{x-2}+\dfrac{4}{x+2}$

10. $\dfrac{5}{x-2}-\dfrac{4}{x+2}$

11. $\dfrac{5x}{x^2-4}+\dfrac{3}{x+2}$

12. $\dfrac{3}{x}-\dfrac{2}{x+3}+\dfrac{1}{x^2+3x}$

13. $\dfrac{2}{xy} + \dfrac{3}{xy^2} - \dfrac{1}{x^2y^2}$

14. $\dfrac{x + y}{xy^3} - \dfrac{x - y}{y^4}$

15. $\dfrac{x + 1}{x^2 - 4x + 4} + \dfrac{4}{x^2 + 3x - 10}$

16. $\dfrac{x^2}{x^2 - x + 1} - \dfrac{x + 1}{x}$

Example A (The three signs of a fraction) Simplify $\dfrac{x}{3x - 6} - \dfrac{2}{2 - x}$.

SOLUTION.

$$\frac{x}{3x - 6} - \frac{2}{2 - x} = \frac{x}{3(x - 2)} - \frac{2}{2 - x}$$

Now we make a crucial observation. Notice that

$$-(2 - x) = -2 + x = x - 2$$

That is, $2 - x$ and $x - 2$ are negatives of each other. Thus the expression above may be rewritten as

$$\begin{aligned} \frac{x}{3(x - 2)} - \frac{2}{2 - x} &= \frac{x}{3(x - 2)} - \frac{2}{-(x - 2)} \\ &= \frac{x}{3(x - 2)} + \frac{2}{x - 2} \\ &= \frac{x}{3(x - 2)} + \frac{6}{3(x - 2)} \\ &= \frac{x + 6}{3(x - 2)} \end{aligned}$$

When we replaced $-\dfrac{2}{-(x - 2)}$ by $\dfrac{2}{x - 2}$, we used the fact that

$$-\frac{a}{-b} = \frac{a}{b}$$

Keep in mind that a fraction has three sign positions: numerator, denominator, and total fraction. You may change any two of them without changing the value of the fraction. Thus

$$\frac{a}{b} = -\frac{a}{-b} = -\frac{-a}{b} = \frac{-a}{-b}$$

Simplify.

17. $\dfrac{4}{2x - 1} + \dfrac{x}{1 - 2x}$

18. $\dfrac{x}{6x - 2} - \dfrac{3}{1 - 3x}$

19. $\dfrac{2}{6y - 2} + \dfrac{y}{9y^2 - 1} - \dfrac{2y + 1}{1 - 3y}$

20. $\dfrac{x}{4x^2 - 1} + \dfrac{2}{4x - 2} - \dfrac{3x + 1}{1 - 2x}$

21. $\dfrac{m^2}{m^2 - 2m + 1} - \dfrac{1}{3 - 3m}$

22. $\dfrac{2x}{x^2 - y^2} + \dfrac{1}{x + y} + \dfrac{1}{y - x}$

In Problems 23–30, multiply and express in simplest form, as illustrated in the text.

23. $\dfrac{5}{2x-1}\cdot\dfrac{x}{x+1}$

24. $\dfrac{3}{x^2-2x}\cdot\dfrac{x-2}{x}$

25. $\dfrac{x+2}{x^2-9}\cdot\dfrac{x+3}{x^2-4}$

26. $\left(1+\dfrac{1}{x+2}\right)\left(\dfrac{4}{3x+9}\right)$

27. $x^2y^4\left(\dfrac{x}{y^2}-\dfrac{y}{x^2}\right)$

28. $\dfrac{5x^2}{x^3+y^3}\left(\dfrac{1}{xy^2}-\dfrac{1}{x^2y^2}+\dfrac{1}{x^3y}\right)$

29. $\left(\dfrac{x^2+5x}{x^2-16}\right)\left(\dfrac{x^2-2x-24}{x^2-x-30}\right)$

30. $\left(\dfrac{x^3-125}{2x^3-10x^2}\right)$
$$\times\left(\frac{7x}{x^3+5x^2+25x}\right)$$

Express the quotients in Problems 31–37 in simplest form, as illustrated in the text.

31. $\dfrac{\dfrac{5}{2x-1}}{\dfrac{x}{x+1}}$

32. $\dfrac{\dfrac{5}{2x-1}}{\dfrac{x}{4x^2-1}}$

33. $\dfrac{\dfrac{x+2}{x^2-4}}{x}$

34. $\dfrac{\dfrac{x+2}{x^2-3x}}{\dfrac{x^2-4}{x}}$

35. $\dfrac{\dfrac{x^2+a^3}{x^3-a^3}}{\dfrac{x+2a}{(x-a)^2}}$

36. $\dfrac{1+\dfrac{2}{b}}{1-\dfrac{4}{b^2}}$

37. $\dfrac{\dfrac{y^2+y-2}{y^2+4y}}{\dfrac{2y^2-8}{y^2+2y-8}}$

Example B (A quotient arising in calculus) Simplify

$$\frac{\dfrac{2}{x+h}-\dfrac{2}{x}}{h}$$

SOLUTION. This expression may look artificial, but it is one you are apt to find in calculus. It represents the average rate of change in $2/x$ as x changes to $x + h$. We begin by simplifying the complicated numerator.

$$\frac{\dfrac{2}{x+h}-\dfrac{2}{x}}{h}=\frac{\dfrac{2x-2(x+h)}{(x+h)x}}{h}=\frac{\dfrac{2x-2x-2h}{(x+h)x}}{\dfrac{h}{1}}$$

$$=\frac{-2h}{(x+h)x}\cdot\frac{1}{h}=\frac{-2}{(x+h)x}$$

Simplify each of the following.

38. $$\frac{\dfrac{4}{x+h} - \dfrac{4}{x}}{h}$$

39. $$\frac{\dfrac{1}{2x+2h+3} - \dfrac{1}{2x+3}}{h}$$

40. $$\frac{\dfrac{x+h}{x+h+4} - \dfrac{x}{x+4}}{h}$$

41. $$\frac{\dfrac{1}{(x+h)^2} - \dfrac{1}{x^2}}{h}$$

Example C (Four-story fractions) Simplify

$$\frac{\dfrac{x}{x-4} - \dfrac{3}{x+3}}{\dfrac{1}{x} + \dfrac{2}{x-4}}$$

SOLUTION. *Method 1* (Simplify the numerator and denominator separately and then divide.)

$$\frac{\dfrac{x}{x-4} - \dfrac{3}{x+3}}{\dfrac{1}{x} + \dfrac{2}{x-4}} \quad \frac{\dfrac{x(x+3) - 3(x-4)}{(x-4)(x+3)}}{\dfrac{x-4+2x}{x(x-4)}}$$

$$= \frac{\dfrac{x^2+3x-3x+12}{(x-4)(x+3)}}{\dfrac{3x-4}{x(x-4)}}$$

$$= \frac{x^2+12}{\cancel{(x-4)}(x+3)} \cdot \frac{x\cancel{(x-4)}}{3x-4}$$

$$= \frac{x(x^2+12)}{(x+3)(3x-4)}$$

Method 2 (Multiply the fractions in the numerator and denominator by a common denominator, in this case, $(x - 4)(x + 3)x$.)

$$\frac{\dfrac{x}{x-4} - \dfrac{3}{x+3}}{\dfrac{1}{x} + \dfrac{2}{x-4}} = \frac{\left(\dfrac{x}{x-4} - \dfrac{3}{x+3}\right)(x-4)(x+3)x}{\left(\dfrac{1}{x} + \dfrac{2}{x-4}\right)\quad(x-4)(x+3)x}$$

$$= \frac{x^2(x+3) - 3x(x-4)}{(x-4)(x+3) + 2x(x+3)}$$

$$\frac{x[x(x+3) - 3(x-4)]}{(x+3)(x-4+2x)}$$

$$= \frac{x[x^2+12]}{(x+3)(3x-4)}$$

Simplify, using either of the above methods.

42. $\dfrac{\dfrac{1}{x+2} - \dfrac{3}{x^2-4}}{\dfrac{3}{x-2}}$

43. $\dfrac{\dfrac{y}{y+4} - \dfrac{2}{y^2+5y+4}}{\dfrac{4}{y+1} + \dfrac{3}{y+4}}$

44. $\dfrac{\dfrac{1}{x} - \dfrac{1}{x-2} + \dfrac{3}{x^2-2x}}{\dfrac{x}{x-2} + \dfrac{3}{x}}$

45. $\dfrac{\dfrac{a^2}{b^2} - \dfrac{b^2}{a^2}}{\dfrac{a}{b} - \dfrac{b}{a}}$

46. $\dfrac{n - \dfrac{n^2}{n-m}}{1 + \dfrac{m^2}{n^2-m^2}}$

47. $\dfrac{\dfrac{x^2}{x-y} - x}{\dfrac{y^2}{x-y} + y}$

48. $1 - \dfrac{x - (1/x)}{1 - (1/x)}$

49. $\dfrac{y - \dfrac{1}{1 + (1/y)}}{y + \dfrac{1}{y - (1/y)}}$

Miscellaneous Problems

In each of the following, perform the indicated operations and simplify.

50. $\dfrac{1-2x}{x-4} + \dfrac{x+2}{3x-12}$

51. $x + y + \dfrac{x^2}{x-y}$

52. $\dfrac{x+y}{z} - \left(\dfrac{x}{z} + \dfrac{y}{z}\right)$

53. $\dfrac{2x+1}{x^2+4x-60} - \dfrac{3x}{2x-12}$

54. $\dfrac{\dfrac{25-9u^2}{2u+3}}{5u+3u^2}$

55. $\dfrac{\dfrac{x^3-y^3}{2x+3y}}{\dfrac{x-y}{4x^2-9y^2}}$

56. $\dfrac{x^2-4}{x^2-5x} \cdot \dfrac{x^2-10x+25}{x^2+3x-10}$

57. $\dfrac{3x}{4x^2-9} - \dfrac{5}{3-2x}$

58. $\dfrac{1}{x}\left(x + \dfrac{1}{x}\right)^{-1}$

59. $\left(\dfrac{x-2}{x}\right)^2 \div (3x-6)$

CHAPTER SUMMARY

An **exponent** is a numerical superscript placed on a number to indicate that a certain operation is to be performed on that number. Thus

$$b^4 = b \cdot b \cdot b \cdot b \qquad b^0 = 1 \qquad b^{-3} = \frac{1}{b^3} = \frac{1}{b \cdot b \cdot b}$$

Exponents mix together according to five laws called the **rules of exponents**.

A number is in **scientific notation** when it appears in the form $c \times 10^n$, where $1 \leq |c| < 10$ and n is an integer. Very small and very large numbers are commonly written this way. Pocket calculators often use scientific notation in their displays. These electronic marvels are designed to take the drudgery out of arithmetic calculations and are a valuable tool in an algebra-trigonometry course.

An expression of the form

$$a_n x^n + a_{n-1} x^{n-1} + \cdots + a_1 x + a_0$$

is called a **polynomial** in x. The exponent n is its **degree** (provided $a_n \neq 0$); the a's are its **coefficients**. Polynomials can be added, subtracted, and multiplied, the result in each case being another polynomial.

To **factor** a polynomial is to write it as a product of simpler polynomials; to **factor over the integers** is to write a polynomial as a product of polynomials with integer coefficients. Here are five examples.

$x^2 - 2ax = x(x - 2a)$	(Common factor)
$4x^2 - 25 = (2x + 5)(2x - 5)$	(Difference of squares)
$6x^2 + x - 15 = (2x - 3)(3x + 5)$	(Trial and error)
$x^2 + 14x + 49 = (x + 7)^2$	(Perfect square)
$x^3 + 1000 = (x + 10)(x^2 - 10x + 100)$	(Sum of cubes)

A quotient (ratio) of two polynomials is called a **rational expression**. The expression is in **reduced form** if its numerator and denominator have no nontrivial common factors. We add, subtract, multiply, and divide rational expressions in much the same way as we do rational numbers.

CHAPTER REVIEW PROBLEM SET

Simplify, leaving your answer free of negative exponents.

1. $\left(\frac{3}{4}\right)^{-2}$
2. $\left(\frac{5}{6}\right)^{2}\left(\frac{5}{6}\right)^{-4}$
3. $\left(\frac{5}{6} + \frac{1}{3}\right)^{-2}$
4. $\frac{2x^{-2}y^2}{4xy^{-3}}$
5. $(3x^{-2}y^3)^{-3}$
6. $\frac{(a^{-2}b)^2(2ab^{-3})^{-1}}{(ab^{-2})^3}$

Express in scientific notation.

7. 1,382,000
8. $(3.1)10^4(2.2)10^{-7}$
9. $\frac{(6.5)10^4}{(1.3)10^{-3}}$

Decide which of the following are polynomials. Give the degree of each polynomial.

10. $\frac{3}{2}x^2 - \sqrt{2}x + \pi$
11. $10x^3 + 5\sqrt{x} + 6$
12. $4t^{-2} + 5t^{-1} + 6$
13. $\dfrac{4}{x^3 + 11}$

Perform the indicated operations and simplify.

14. $(3x - 5) + (2x^2 - 2x + 11)$
15. $5 - x^3 - (x^2 - 2x^3 + x - 2)$
16. $(2x - 3)(x + 5)$
17. $(2x - 1)^2 - 4x^2$
18. $(z^3 + 4)(z^3 - 4)$
19. $(2x^2 - 3w)(x^2 + 4w)$
20. $(x + 2a)(x^2 - 2ax + 4a^2)$
21. $(2y - 3)(3y^2 - 5y + 6)$
22. $(3t^2 - t + 1)^2$
23. $(a + bcd)(a - bcd)$

In Problems 24–33, factor completely over the integers.

24. $2x^4 - x^3 + 11x^2$
25. $y^2 - 7y + 12$
26. $6z^2 + z - 1$
27. $49a^2 - 25$
28. $9c^2 - 24cd + 16d^2$
29. $a^3 - 27$
30. $8a^3b^3 + 1$
31. $x^8 - x^4y^4$
32. $x^2 + 2xy + y^2 - z^4$
33. $4c^2 - d^2 - 6c - 3d$
34. Factor $9x^2 - 11$ over the real numbers.
35. Factor $x^2 + 16$ over the complex numbers.

Reduce each of the following.

36. $\dfrac{x^3 - 8}{2x - 4}$
37. $\dfrac{2x - 2x^2}{x^3 - 2x^2 + x}$

Perform the indicated operations and simplify.

38. $\dfrac{18}{x^2 + 3x} - \dfrac{4}{x} + \dfrac{6}{x + 3}$
39. $\dfrac{x^2 + x - 6}{x^2 - 1} \cdot \dfrac{x^2 + x - 2}{x^2 + 5x + 6}$
40. $\dfrac{\dfrac{x}{x - 3} - \dfrac{2}{x^2 - 4x + 3}}{\dfrac{5}{x - 1} + \dfrac{5}{x - 3}}$

As the sun eclipses the stars by its brilliancy, so the man of knowledge will eclipse the fame of others in the assemblies of the people if he proposes algebraic problems, and still more if he solves them.

Brahmagupta

CHAPTER THREE

Equations and Inequalities

3-1 Equations

3-2 Applications Using One Unknown

3-3 Two Equations in Two Unknowns

3-4 Quadratic Equations

3-5 Inequalities

3-6 More Applications (Optional)

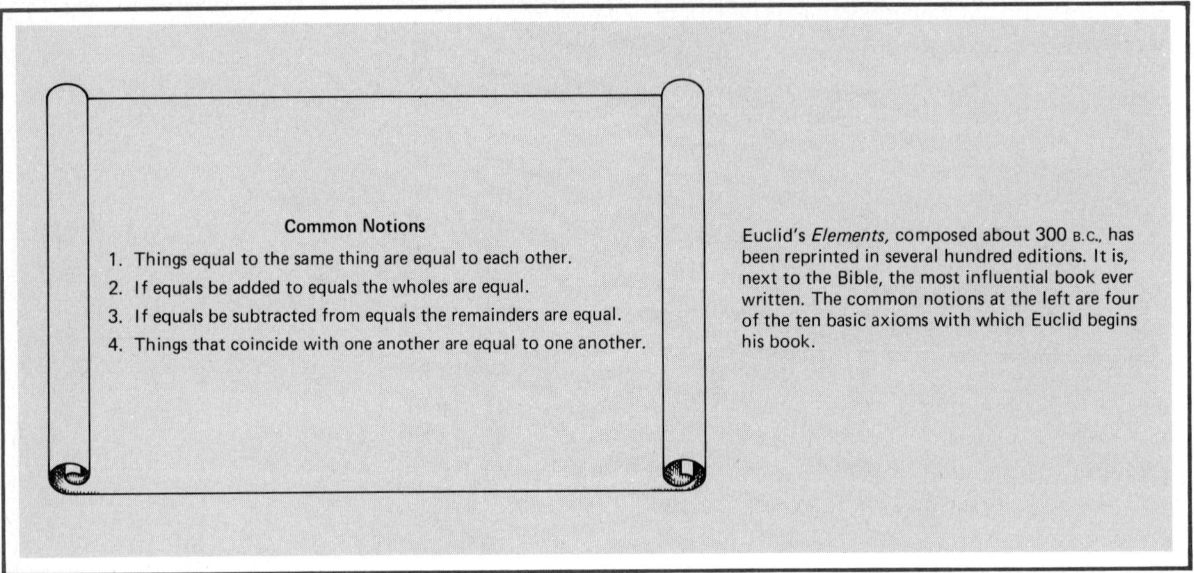

Euclid's *Elements*, composed about 300 B.C., has been reprinted in several hundred editions. It is, next to the Bible, the most influential book ever written. The common notions at the left are four of the ten basic axioms with which Euclid begins his book.

3-1 Equations

It was part of Euclid's genius to recognize that our usage of the word *equals* is fundamental to all that we do in mathematics. But to describe that usage may not be as simple as Euclid thought. When we write

$$.25 + \tfrac{3}{4} + \tfrac{1}{3} - (.3333\ldots) = 1$$

we certainly do not mean that the symbol on the left coincides with the one on the right. Instead, we mean that both symbols, the complicated one and the simple one, stand for (or name) the same number. That is the basic meaning of *equals* as used in this book.

> **Equality?**
>
> "We hold these truths to be self-evident, that all men are created equal ... "

But having said that, we must make another distinction. When we write

$$x^2 - 25 = (x - 5)(x + 5)$$

and

$$x^2 - 25 = 0$$

we have two quite different things in mind. In the first case, we are making an assertion. We claim that no matter what number x represents, the expressions on the left and right of the equality stand for the same number. This certainly cannot be our meaning in the second case. There we are asking a question: What numbers can x symbolize so that both sides of the equality $x^2 - 25 = 0$ stand for the same number?

An equality that is true for all values of the variable is called an **identity**. One that is true only for some values is called a conditional **equation**. And here are the corresponding jobs for us to do. We **prove** identities, but we **solve** (or find the solutions of) equations. Both jobs will be very important in this book; however, it is the second that interests us most right now.

Solving Equations Sometimes we can solve an equation by inspection. It takes no mathematical apparatus and little imagination to see that

$$x + 4 = 6$$

has $x = 2$ as a solution. On the other hand, to solve

$$2x^2 + 8x = 10$$

is quite a different matter. For this kind of equation, we need some machinery. Our general strategy is to modify an equation one step at a time until it is in a form where the solution is obvious. Of course, we must be careful that the modifications we make do not change the solutions. Here, too, Euclid pointed the way.

> RULES FOR MODIFYING EQUATIONS
>
> **1.** Adding the same quantity to (or subtracting the same quantity from) both sides of an equation does not change its solutions.
>
> **2.** Multiplying (or dividing) both sides of an equation by the same nonzero quantity does not change its solutions.

Consider $2x^2 + 8x = 10$ again. One way to solve this equation is to use the following steps.

Given equation:	$2x^2 + 8x = 10$
Divide by 2:	$x^2 + 4x = 5$
Add 4:	$x^2 + 4x + 4 = 9$
Factor left side:	$(x + 2)^2 = 9$

Now we can see that $x + 2$ must equal 3 or -3. In the first case, $x = 1$; in the second, $x = -5$. Thus the solutions of $2x^2 + 8x = 10$ are 1 and -5. We will solve this equation by another method in Section 3-4.

Linear Equations The simplest equation to solve is one in which the *variable* (also called the *unknown*) occurs only to the first power. Consider

$$12x - 9 = 5x + 5$$

Our procedure is to use the rules for modifying equations to bring all the terms in x to one side and the constant terms to the other and then to divide by the coefficient of x. The result is that we have x all alone on one side of the equation and a number (the solution) on the other.

Given equation:	$12x - 9 = 5x + 5$
Add 9:	$12x = 5x + 14$
Subtract $5x$:	$7x = 14$
Divide by 7:	$x = 2$

It is always a good idea to check your answer. In the original equation, replace x by the value that you found to see if a true statement results.

$$12(2) - 9 \stackrel{?}{=} 5(2) + 5$$
$$15 = 15$$

An equation of the form $ax + b = 0$ $(a \neq 0)$ is called a **linear equation**. It has one solution, $x = -b/a$. Many equations not initially in this form can be transformed to it using the rules we have learned.

Equations That Can be Changed to Linear Form Consider

$$\frac{2}{x+1} = \frac{3}{2x-2}$$

If we agree to exclude $x = -1$ and $x = 1$ from consideration, then $(x + 1)(2x - 2)$ is not zero and we may multiply both sides by that expression. We get

$$\frac{2}{x+1}(x+1)(2x-2) = \frac{3}{2x-2}(x+1)(2x-2)$$
$$2(2x - 2) = 3(x + 1)$$
$$4x - 4 = 3x + 3$$
$$x = 7$$

As usual, we check our solution in the original equation.

$$\frac{2}{7+1} \stackrel{?}{=} \frac{3}{14-2}$$
$$\frac{2}{8} = \frac{3}{12}$$

The importance of checking is illustrated by our next example.

$$\frac{3x}{x-3} = 1 + \frac{9}{x-3}$$

To solve, we multiply both sides by $x - 3$ and then simplify.

$$\frac{3x}{x-3}(x-3) = \left(1 + \frac{9}{x-3}\right)(x-3)$$

$$\begin{aligned} 3x &= x - 3 + 9 \\ 2x &= 6 \\ x &= 3 \end{aligned}$$

When we check in the original equation, we get

$$\frac{3 \cdot 3}{3-3} \stackrel{?}{=} 1 + \frac{9}{3-3}$$

This is nonsense, since it involves division by zero. What went wrong? If $x = 3$, then in our very first step we actually multiplied both sides of the equation by zero, a forbidden operation. Thus the given equation has no solution.

The strategy of multiplying both sides by $x - 3$ in this example was appropriate, even though it initially led us to an incorrect answer. We did not worry, because we knew that in the end we were going to run a check. We should always check answers, especially in any situation in which we have multiplied by an expression involving the unknown. Such a multiplication may introduce an *extraneous* solution.

Here is another example of a similar type of equation.

$$\frac{x+4}{(x+1)(x-2)} - \frac{3}{x+1} - \frac{2}{x-2} = \frac{-8}{(x+1)(x-2)}$$

We have no choice but to multiply both sides of the equation by $(x+1)(x-2)$. This gives

$$\begin{aligned} x + 4 - 3(x-2) - 2(x+1) &= -8 \\ x + 4 - 3x + 6 - 2x - 2 &= -8 \\ -4x + 8 &= -8 \\ -4x &= -16 \\ x &= 4 \end{aligned}$$

At this point, $x = 4$ is an apparent solution; however, we are not sure until we check it in the original equation.

$$\frac{4+4}{(4+1)(4-2)} - \frac{3}{4+1} - \frac{2}{4-2} \stackrel{?}{=} \frac{-8}{(4+1)(4-2)}$$

$$\frac{8}{5 \cdot 2} - \frac{3}{5} - \frac{2}{2} \stackrel{?}{=} \frac{-8}{5 \cdot 2}$$

$$\frac{8}{10} - \frac{6}{10} - \frac{10}{10} = \frac{-8}{10}$$

It works, so $x = 4$ is a solution.

Problem Set 3-1

Determine which of the following are identities and which are conditional equations.

1. $2(x + 4) = 8$
2. $2(x + 4) = 2x + 8$
3. $3(2x - \frac{2}{3}) = 6x - 2$
4. $2x - 4 - \frac{2}{3}x = \frac{4}{3}x - 4$
5. $\frac{2}{3}x + 4 = \frac{1}{2}x - 1$
6. $3(x - 2) = 2(x - 3) + x$
7. $(x + 2)^2 = x^2 + 4$
8. $x(x + 2) = x^2 + 2x$
9. $x^2 - 9 = (x + 3)(x - 3)$
10. $x^2 - 5x + 6 = (x - 1)(x - 6)$

Solve each of the following equations.

11. $4x - 3 = 3x - 1$
12. $2x + 5 = 5x + 14$
13. $2t + \frac{1}{2} = 4t - \frac{7}{2} + 8t$
14. $y + \frac{1}{3} = 2y - \frac{2}{3} - 6y$
15. $3(x - 2) = 5(x - 3)$
16. $4(x + 1) = 2(x - 3)$
17. $\sqrt{3}z + 4 = -\sqrt{3}z + 8$
18. $\sqrt{2}x + 1 = x + \sqrt{2}$

C 19. $3.23x - 6.15 = 1.41x + 7.63$

(First obtain $x = \dfrac{7.63 + 6.15}{3.23 - 1.41}$ and then use the calculator.)

C 20. $42.1x + 11.9 = 1.03x - 4.32$

C 21. $(6.13 \times 10^{-8})x + (5.34 \times 10^{-6}) = 0$

C 22. $(5.11 \times 10^{11})x - (6.12 \times 10^{12}) = 0$

Example A (Equations involving fractions) Solve $\frac{2}{3}x - \frac{3}{4} = \frac{7}{6}x + \frac{1}{2}$.

SOLUTION. When an equation is cluttered up with many fractions, the best first step may be to get rid of them. To do this, multiply both sides by the lowest common denominator (in this case, 12). Then proceed as usual.

$$\begin{aligned} 12(\tfrac{2}{3}x - \tfrac{3}{4}) &= 12(\tfrac{7}{6}x + \tfrac{1}{2}) \\ 8x - 9 &= 14x + 6 \quad \text{(distributive property)} \\ 8x &= 14x + 15 \quad \text{(add 9)} \\ -6x &= 15 \quad \text{(subtract } 14x\text{)} \\ x &= \frac{15}{-6} = -\frac{5}{2} \quad \text{(divide by } -6\text{)} \end{aligned}$$

The solution should now be checked in the original equation. We leave the check to the student.

Solve by first clearing the fractions.

23. $\frac{2}{3}x + 4 = \frac{1}{2}x$
24. $\frac{2}{3}x - 4 = \frac{1}{2}x + 4$
25. $\frac{9}{10}x + \frac{5}{8} = \frac{1}{5}x + \frac{9}{20}$
26. $\frac{1}{3}x + \frac{1}{4} = \frac{1}{5}x + \frac{1}{6}$
27. $\frac{3}{4}(x - 2) = \frac{9}{5}$
28. $\dfrac{x}{8} = \dfrac{2}{3}(2 - x)$

The following equations are nonlinear equations that become linear when cleared of fractions. Solve each equation and check your solutions as some might be extraneous (see page 89).

29. $\dfrac{5}{x + 2} = \dfrac{2}{x - 1}$
30. $\dfrac{10}{2x - 1} = \dfrac{14}{x + 4}$

31. $\dfrac{2}{x-3} + \dfrac{3}{x-7} = \dfrac{7}{(x-3)(x-7)}$

32. $\dfrac{2}{x-1} + \dfrac{3}{x+1} = \dfrac{19}{x^2-1}$

33. $\dfrac{x}{x-2} = 2 + \dfrac{2}{x-2}$

34. $\dfrac{x}{2x-4} - \dfrac{2}{3} = \dfrac{7-2x}{3x-6}$

Sometimes an equation that appears to be quadratic is actually equivalent to a linear equation. For example, if we subtract x^2 from both sides of the equation $x^2 + 3x = x^2 + 5$, we see that it is equivalent to $3x = 5$. Use this idea to solve each of the following.

35. $x^2 + 4x = x^2 - 3$

36. $x^2 - 2x = x^2 + 3x + 20$

37. $(x-4)(x+5) = (x+2)(x+3)$

38. $(2x-1)(2x+3) = 4x^2 + 6$

Example B (Solving for one variable in terms of others) Solve $I = nE/(R + nr)$ for n.

SOLUTION. This is a typical problem in science in which an equality relates several variables and we want to solve for one of them in terms of the others. To do this, we proceed as if the other variables were simply numbers, which, after all, is what every variable represents.

$$I = \frac{nE}{R+nr} \qquad \text{(original equality)}$$

$$\begin{aligned}
(R+nr)I &= nE && \text{(multiply by } R+nr\text{)}\\
RI + nrI &= nE && \text{(distributive property)}\\
nrI - nE &= -RI && \text{(subtract } nE \text{ and } RI\text{)}\\
n(rI - E) &= -RI && \text{(factor)}
\end{aligned}$$

$$n = \frac{-RI}{rI - E} \qquad \text{(divide by } rI - E\text{)}$$

Solve for the indicated variable in terms of the remaining variables.

39. $A = P + Prt$ for P

40. $R = \dfrac{E}{L-5}$ for L

41. $I = \dfrac{nE}{R+nr}$ for r

42. $mv = Ft + mv_0$ for m

43. $A = 2\pi r^2 + 2\pi rh$ for h

44. $F = \frac{9}{5}C + 32$ for C

45. $R = \dfrac{R_1R_2}{R_1+R_2}$ for R_1

46. $\dfrac{1}{R} = \dfrac{1}{R_1} + \dfrac{1}{R_2} + \dfrac{1}{R_3}$ for R_2

Miscellaneous Problems

Solve the equations in Problems 47–58.

47. $5x - 11 = 2x + 13$

48. $-2x + 29 = 3x + 74$

49. $\frac{2}{5}y - 4 = \frac{1}{3}y$

50. $\frac{5}{7}a = \frac{2}{3}a - \frac{4}{3}$

[C] 51. $0.081x - 0.123 = 0.152x$

[C] 52. $42.34(x - 1.35) = 432.78$

53. $\dfrac{2}{x-3} = \dfrac{5}{x+2}$

54. $\dfrac{4}{2x+1} = \dfrac{-3}{x-4}$

55. $\dfrac{2x}{x-3} = \dfrac{2x+1}{x+2}$

56. $\dfrac{2}{(x+2)(x-1)} - \dfrac{1}{x+2} = \dfrac{3}{x-1}$

57. $\dfrac{x}{x+3} = -1 - \dfrac{3}{x+3}$

58. $\dfrac{x^2}{x-2} = \dfrac{4}{x-2}$

59. The Celsius and Fahrenheit temperature scales are related by the formula $C = \frac{5}{9}(F - 32)$.
   (a) Solve for F in terms of C.
   (b) What reading on the Fahrenheit scale corresponds to 35 degrees Celsius?
   (c) How warm is it (in Fahrenheit degrees) when the Celsius reading equals the Fahrenheit reading?
   (d) How warm is it (in degrees Fahrenheit) when the Fahrenheit reading is 50 degrees higher than the Celsius reading?
   (e) How warm is it (in Celsius degrees) when the Celsius reading is one-third of the Fahrenheit reading?

60. If a principal P is invested at the simple interest rate r for t years, then the accumulated amount after t years is given by $A = P + Prt$. For example, \$2000 invested at 8 percent simple interest will accumulate to $2000 + 2000(.08)(10)$, or \$3600, in 10 years.
   (a) If \$2000 is invested at 9 percent simple interest, to what sum will it accumulate in 5 years?
   (b) How long will it take \$2000 to grow to \$5000 if invested at 9 percent simple interest?
   (c) A principal of \$2000 grew to \$4000 in 8 years at a certain interest rate r. Find r.

Rules for the Direction of the Mind

1. Reduce any kind of problem to a mathematical problem.
2. Reduce any kind of mathematical problem to a problem of algebra.
3. Reduce any problem of algebra to the solution of a single equation.

René Descartes
1596–1650

3-2 Applications Using One Unknown

Besides being a mathematician, Descartes was a first-rate philosopher. His name appears prominently in every philosophy text. We admit that his rules for the direction of the mind are overstated. Not every problem in life can be solved this way. But they do suggest a style of thinking that has been very fruitful, especially in the sciences. We intend to exploit it.

A Typical Word Problem Sometimes a little story can illustrate some big ideas.

> John plans to take his wife Helen out to dinner. Concerned about their financial situation, Helen asks him point-blank, "How much money do you have?" Never one to give a simple answer when a complicated one will do, John replies, "If I had $12 more than I have and then doubled that amount, I'd be $50 richer than I am." Helen's response is best left unrecorded.

The problem, of course, is to find out exactly how much money John has. Our task is to take a complicated word description, translate it into mathematical symbols, and then let the machinery of algebra grind out the answer. First we introduce a symbol x. It usually will stand for the principal unknown in the problem. But we need to be very precise. It is not enough to let x be John's money, or even to let x be the amount of John's money, though that is better. The symbol x must represent a number. What we should say is

Let x be the number of dollars John has.

Our story puts restrictions on x; x must satisfy a specified condition. That condition must be translated into an equation. We shall do it by bits and pieces.

WORD PHRASE	ALGEBRAIC TRANSLATION
how much John has	x
\$12 more than he has	$x + 12$
double that amount	$2(x + 12)$
\$50 richer than he is	$x + 50$

Read John's answer again. It says that the expressions $2(x + 12)$ and $x + 50$ are equal. Thus

$$\begin{aligned} 2(x + 12) &= x + 50 \\ 2x + 24 &= x + 50 \\ x &= 26 \end{aligned}$$

John has $26, enough for a pretty good dinner for two—even at today's prices.

A Distance-Rate Problem Problems involving rates and distances occur very frequently in physics. Usually their solution involves use of the formula $D = RT$, which stands for "distance equals rate multiplied by time."

> At 2:00 P.M., Slowpoke left Kansas City traveling due east at 45 miles per hour. An hour later, Speedy started after him going 60 miles per hour. When will Speedy catch up with Slowpoke?

Most of us can grasp the essential features in a picture more readily than in a mass of words. That is why all good mathematicians make sketches that summarize what is given.

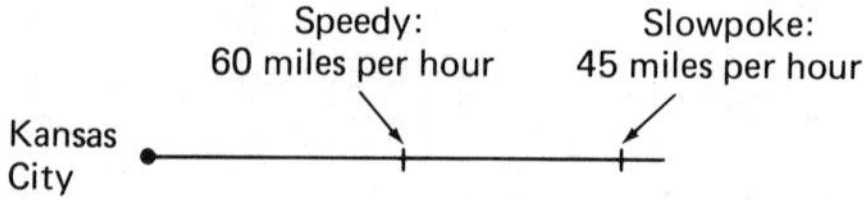

Next assign the unknown. Be precise.
Poor: Let t be time.
Better: Let t be the time when Speedy catches up.
Good: Let t be the number of hours after 2:00 P.M. when Speedy catches up with Slowpoke.

Notice two things.

1. Slowpoke drove t hours; Speedy, starting an hour later, drove only $t - 1$ hours.
2. Both drove the same distance.

Now use the formula $D = RT$ to conclude that Slowpoke drove a distance of $45t$ miles and Speedy drove a distance of $60(t - 1)$ miles. By statement (2), these are equal.

$$\begin{aligned} 45t &= 60(t - 1) \\ 45t &= 60t - 60 \\ 60 &= 15t \\ 4 &= t \end{aligned}$$

Speedy will catch up with Slowpoke 4 hours after 2:00 P.M., or at 6:00 P.M.

A Mixture Problem Here is a problem from chemistry.

How many liters of a 60 percent solution of nitric acid should be added to 10 liters of a 30 percent solution to obtain a 50 percent solution?

We are certain that a picture will help most students with this problem.

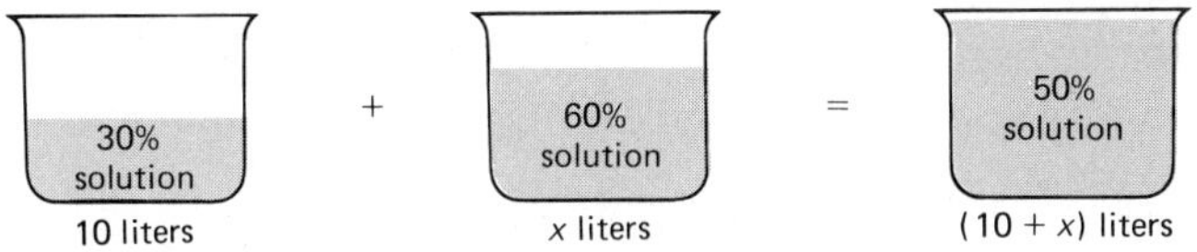

30% solution — 10 liters + 60% solution — x liters = 50% solution — $(10 + x)$ liters

We have indicated on the picture what x represents, but let us be specific.

Let x be the number of liters of 60 percent solution to be added.

Now we make a crucial observation, one that will seem obvious once you think about it.

$$\begin{pmatrix}\text{The amount of}\\ \text{pure acid we}\\ \text{start with}\end{pmatrix} + \begin{pmatrix}\text{The amount of}\\ \text{pure acid we}\\ \text{add}\end{pmatrix} = \begin{pmatrix}\text{The amount of}\\ \text{pure acid we}\\ \text{end with}\end{pmatrix}$$

In symbols, this becomes

$$(.30)(10) + (.60)x = (.50)(10 + x)$$

The big job has been accomplished; we have the equation. After multiplying both sides by 10 to clear the equation of decimal fractions, we can easily solve for x.

$$\begin{aligned}(3)(10) + 6x &= 5(10 + x)\\ 30 + 6x &= 50 + 5x\\ x &= 20\end{aligned}$$

We should add 20 liters of 60 percent solution.

Summary One hears students say that they have a mental block when it comes to "word problems." Yet most of the problems of the real world are initially stated in words. We want to destroy those mental blocks and give you one of the most satisfying experiences in mathematics. Everyone can do word problems. Here are a few simple suggestions.

1. Read the problem very carefully so that you know exactly what it says and what it asks.
2. Draw a picture or make a chart that summarizes the given information.

3. Identify the unknown quantity and assign a letter to it. Be sure it represents a number (for example, of dollars, of miles, or of liters).
4. Note the condition or restriction that the problem puts on the unknown. Translate it into an equation.
5. Solve the equation. Your result is a specific number. The unknown has become known.
6. State your conclusion in words (for example, Speedy will catch up to Slowpoke at 4 hours after 2:00 P.M., or at 6:00 P.M.).
7. Note whether your answer is reasonable. If, for example, you found that Speedy would not catch up with Slowpoke until 4000 hours after 2:00 P.M., you should suspect that you have made a mistake.

Problem Set 3-2

1. The sum of 15 and twice a certain number is 33. Find that number.
2. Tom says to Jerry: I am thinking of a number. When I subtract 5 from that number and then multiply the result by 3, I get 42. What was my number? Jerry figured it out. Can you?
3. Find the number for which twice the number is 12 less than 3 times the number.
4. The result of adding 28 to 4 times a certain number is the same as subtracting 5 from 7 times that number. Find the number.
5. The sum of three consecutive positive integers is 72. Find the smallest one.

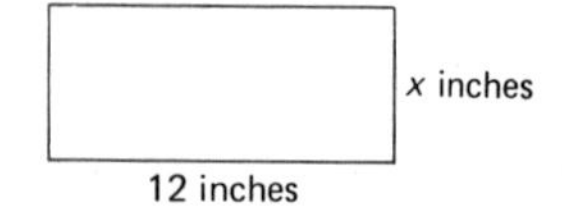

6. The perimeter (distance around) of the rectangle shown in the margin is 31 inches. Find the width x.
7. A wire 130 centimeters long is bent into the shape of a rectangle which is 3 centimeters longer than it is wide. Find the width of the rectangle.
8. A rancher wants to put 2850 pounds of feed into two empty bins. If she wants the larger bin to contain 750 pounds more of the feed than the smaller one, how much must she put into each?
9. Mary scored 61 on her first math test. What must she score on a second test to bring her average up to 75?
10. Henry has scores of 61, 73, and 82 on his first three tests. How well must he do on the fourth and final test to wind up with an average of 75 for the course?
11. A change box contains 21 dimes. How many quarters must be put in to bring the total amount of change to $3.85?
12. A change box contains $3.00 in dimes and nothing else. A certain number of dimes are taken out and replaced by an equal number of quarters, with the result that the box now contains $4.20. How many dimes are taken out?
13. A woman has $4.45 in dimes and quarters in her purse. If there are 25 coins in all, how many dimes are there?
14. Young Amy has saved up $8.05 in nickels and quarters. She has 29 more nickels than quarters. How many quarters does she have?

15. Read the example (in this section) about Slowpoke and Speedy again. When will Speedy be 100 miles ahead of Slowpoke?

16. Two long-distance runners start out from the same spot on an oval track which is $\frac{1}{2}$ mile around. If one runs at 6 miles per hour and the other at 7 miles an hour, when will the faster runner be one lap ahead of the slower runner?

17. The city of Harmony is 455 miles from the city of Dissension. At 12:00 noon Paul Haymaker leaves Harmony traveling at 60 miles per hour toward Dissension. Simultaneously, Nick Ploughman starts from Dissension heading toward Harmony, managing only 45 miles per hour. When will they meet?

18. Suppose in Problem 17, Mr. Ploughman starts at 3:00 P.M. At what time will the two drivers meet?

19. Luella can row 1 mile upstream in the same amount of time that it takes her to row two miles downstream. If the rate of the current is 3 miles per hour, how fast can she row in still water?

20. An airplane flew with the wind for 1 hour and returned the same distance against the wind in $1\frac{1}{2}$ hours. If the speed of the plane in still air is 300 miles per hour, find the speed of the wind.

21. A father is three times as old as his son, but 15 years from now he will be only twice as old as his son. How old is his son now?

22. Jim Warmath was in charge of ticket sales at a football game. The price for general admission was \$3.50, while reserved seat tickets sold for \$5.00. He lost track of the ticket count, but he knew that 110 more general admission tickets had been sold than reserve seat tickets and that total gate receipts were \$980. See if you can find out how many general admission tickets were sold.

23. How many cubic centimeters of a 40 percent solution of hydrochloric acid should be added to 2000 cubic centimeters of a 20 percent solution to obtain a 35 percent solution?

24. In Problem 23, how much of the 40 percent solution would have to be added in order to have a 39 percent solution?

25. A tank contains 1000 liters of 30 percent brine solution. Boiling off water from the solution will increase the percentage of salt. How much water should be boiled off to achieve a 35 percent solution?

26. Sheila Carlson invested \$10,000 in a savings and loan association, some at 7 percent (simple interest) per year and the rest at $8\frac{1}{2}$ percent. How much did she invest at 7 percent if the total amount of interest for one year was \$796?

27. At the Style King shop, a man's suit was marked down 15 percent and sold at \$123.25. What was the original price?

28. Mr. Titus Canby bickered with a furrier over the price of a fur coat he intended to buy for his wife. The furrier offered to reduce the price by 10 percent. Titus was still not satisfied; he said he would buy the coat if the furrier would come down an additional \$200 on the price. The furrier agreed and sold the coat for \$1960. What was the original price?

29. The Conkwrights plan to put in a concrete drive from the street to the garage. The drive is 36 feet long and they plan to make it 4 inches thick.

Since there is a delivery charge for less than 4 cubic yards of ready mixed concrete, the Conkwrights have decided to make the drive just wide enough to use 4 cubic yards. How wide should they make it?

30. Tom can do a certain job in 3 days, Dick can do it in 4 days, and Harry can do it in 5 days. How long will it take them working together? (*Hint*: In one day Tom can do $\frac{1}{3}$ of the job; in x days, he can do $x/3$ of the job.)
31. It takes Jack 5 days to hoe his vegetable garden. Jack and Jill together can do it in 3 days. How long would it take Jill to hoe the garden by herself?

Example (Balancing weight problems) Susan, who weighs 80 pounds, wants to ride on a seesaw with her father, who weighs 200 pounds. The plank is 20 feet long and the fulcrum is at the center. If Susan sits at the very end of her side, how far from the fulcrum should her father sit to achieve balance?

SOLUTION. Let x be the number of feet from the fulcrum to the point where Susan's dad sits. A law of physics demands that *the weight times the distance from the fulcrum* must be the same for both sides in order to have balance. For Susan, weight times distance is $80 \cdot 10$; for her dad, it is $200x$. This gives us the equation

$$200x = 800$$

from which we get $x = 4$. Susan's father should sit 4 feet from the fulcrum.

32. Where should Susan's father sit if Susan moves 2 feet closer to the fulcrum?
33. If Susan sits at one end with Roscoe, a 12-pound puppy, in her arms, where should her father sit?

Find x in each of the following. Assume in each case that the plank is 20 feet long, that the fulcrum is at the center, and that the weights balance.

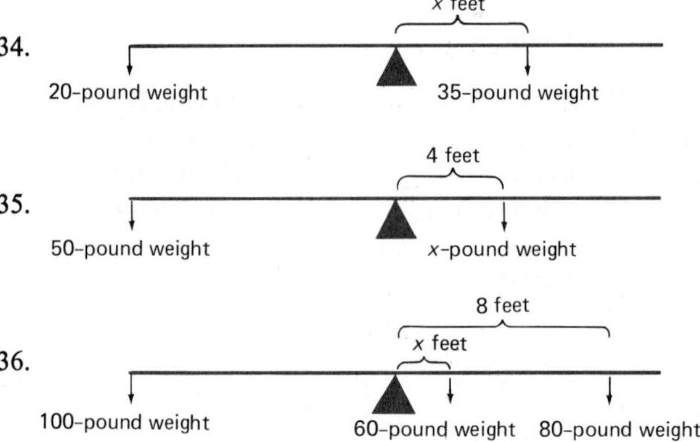

(*Hint*: On the right side, you add the two products.)

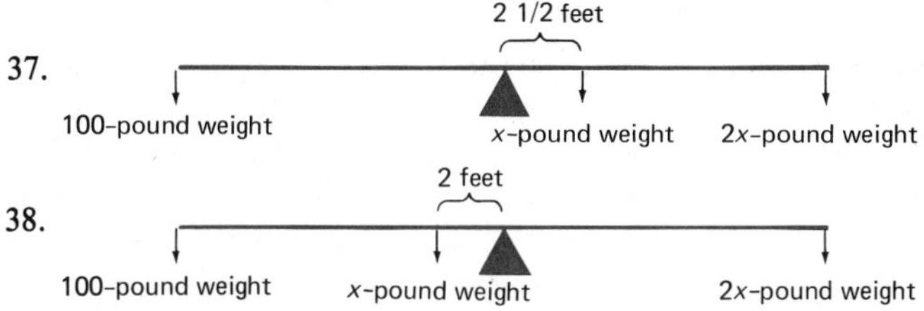

The Farmer's Riddle

I have a collection of hens and rabbits. These animals have 50 heads and 140 feet. How many hens and how many rabbits do I have?

3-3 Two Equations in Two Unknowns

No one has thought more deeply or written more wisely about problem solving than George Polya. In his book, *Mathematical Discovery* (Volume 1, Wiley, 1962), Polya uses the farmer's riddle as the starting point for a brilliant essay on setting up equations to solve problems. He suggests three different approaches that we might take to untangle the riddle.

Trial and Error There are 50 animals altogether. They cannot all be hens as that would give only 100 feet. Nor can all be rabbits; that would give 200 feet. Surely the right answer is somewhere between these extremes. Let us try 25 of each. That gives 50 hen-feet and 100 rabbit-feet, or a total of 150, which is too many. We need more hens and fewer rabbits. Well, try 28 hens and 22 rabbits. It does not work. Try 30 hens and 20 rabbits. There it is! That gives 60 feet plus 80 feet, just what we wanted.

Hens	Rabbits	Feet
50	0	100
0	50	200
25	25	150
28	22	144
30	20	140

Bright Idea Let us use a little imagination. Imagine that we catch the hens and rabbits engaged in a weird new game. The hens are all standing on one foot and the rabbits are hopping around on their two hind feet. In this remarkable situation, only half of the feet, that is 70 feet, are in use. We can think of 70 as counting each hen once and each rabbit twice. If we subtract the total number of animals, namely 50, we will have the number of rabbits. There it is! There have to be $70 - 50 = 20$ rabbits; and that leaves 30 hens.

Algebra Trial and error is time consuming and inefficient, especially in problems with many possibilities. And we cannot expect a brilliant idea to come along for every problem. We need a systematic method that depends neither on guess work nor on sudden visions. Algebra provides such a method. To use it, we must translate the problem into algebraic symbols and set up equations.

ENGLISH	ALGEBRAIC SYMBOLS
The farmer has	
a certain number of hens	x
and	
a certain number of rabbits.	y
These animals have 50 heads	$x + y = 50$
and 140 feet.	$2x + 4y = 140$

"I hope that I shall shock a few people in asserting that the most important single task of mathematical instruction in the secondary schools is to teach the setting up of equations to solve word problems."

G. Polya

Now we have two unknowns, x and y, but we also have two equations relating them. We want to find the values for x and y that satisfy both equations at the same time. There are two standard methods.

Method of Addition or Subtraction We learned two rules for modifying equations in Section 3-1. Here is another rule, especially useful in solving a **system of equations,** that is, a set of several equations in several unknowns.

RULE 3

You may add one equation to another (or subtract one equation from another) without changing the simultaneous solutions of a system of equations.

Here is how this rule is used to solve the farmer's riddle.

Given equations: $\begin{cases} x + y = 50 \\ 2x + 4y = 140 \end{cases}$

Multiply the first equation by (-2): $-2x - 2y = -100$

Write the second equation: $\underline{2x + 4y = 140}$

Add the two equations: $2y = 40$

Multiply by $\frac{1}{2}$: $\boxed{y = 20}$

Substitute $y = 20$ into one of the original equations (in this case, we shall use the first one): $x + 20 = 50$

Add -20: $\boxed{x = 30}$

The key idea is this: Multiplying the first equation by -2 makes the coefficients of x in the two equations negatives of each other. Addition of the two equations eliminates x, leaving one equation in the single unknown y. The resulting equation can be solved by methods learned earlier (see Section 3-1).

Method of Substitution Consider the same pair of equations again.

$$\begin{cases} x + y = 50 \\ 2x + 4y = 140 \end{cases}$$

We may solve the first equation for y in terms of x and substitute the result in the second equation

$$\begin{aligned} y &= 50 - x \\ 2x + 4(50 - x) &= 140 \\ 2x + 200 - 4x &= 140 \\ -2x &= -60 \\ \boxed{x} &\boxed{= 30} \end{aligned}$$

Then substitute the value obtained for x in the expression for y.

$$\boxed{y = 50 - 30 = 20}$$

Naturally, our results agree with those obtained earlier.

Whichever method we use, it is a good idea to check our answer against the original problem. Thirty chickens and 20 rabbits do have a total of 50 heads and they do have $(30)(2) + (20)(4) = 140$ feet in all.

A Distance-Rate Problem An airplane, flying with the help of a strong tail wind, covered 1200 miles in 2 hours. However, the return trip flying into the wind took $2\frac{1}{2}$ hours. How fast would the plane have flown in still air and what was the speed of the wind, assuming both rates to be constant?

Here is the solution. Let

x = speed of the plane in still air in miles per hour
y = speed of the wind in miles per hour

Then

$x + y$ = speed of the plane with the wind
$x - y$ = speed of the plane against the wind

Next, we recall the familiar formula $D = RT$, or distance equals rate multiplied by time. Applying it in the form $TR = D$ to the two trips yields

$$\begin{aligned} 2(x + y) &= 1200 \quad \text{(with wind)} \\ \tfrac{5}{2}(x - y) &= 1200 \quad \text{(against wind)} \end{aligned}$$

or equivalently

$$\begin{aligned} 2x + 2y &= 1200 \\ 5x - 5y &= 2400 \end{aligned}$$

To eliminate y, we multiply the first equation by 5 and the second by 2, and then add the two equations.

$$\begin{aligned} 10x + 10y &= 6000 \\ 10x - 10y &= 4800 \\ \hline 20x &= 10{,}800 \\ x &= 540 \end{aligned}$$

Finally, substituting $x = 540$ in the first of the original equations gives

$$\begin{aligned} 2 \cdot 540 + 2y &= 1200 \\ 1080 + 2y &= 1200 \\ 2y &= 120 \\ y &= 60 \end{aligned}$$

We conclude that the plane's speed in still air was 540 miles per hour and that the wind speed was 60 miles per hour.

A check against the original statement of the problem shows that our answers are correct. With the wind, the plane will travel at 600 miles per hour and will cover 1200 miles in 2 hours. Against the wind, the plane will fly at 480 miles per hour and take $2\frac{1}{2}$ hours to cover 1200 miles.

Problem Set 3-3

In each of the following, find the values for the two unknowns that satisfy both equations. Use whichever method you prefer.

1. $2x + 3y = 13$; $y = 13$
2. $2x - 3y = 7$; $x = -4$
3. $2u - 5v = 23$; $2u = 3$
4. $5s + 6t = 2$; $3t = -4$
5. $7x + 2y = -1$; $y = 4x + 7$
6. $7x + 2y = -1$; $x = -5y + 14$
7. $y = -2x + 11$; $y = 3x - 9$
8. $x = 5y$; $x = -3y - 24$
9. $x - y = 14$; $x + y = -2$
10. $2x - 3y = 8$; $4x + 3y = 16$
11. $2s - 3t = -10$; $5s + 6t = 29$
12. $2w - 3z = -23$; $8w + 2z = -22$
13. $5x - 4y = 19$; $7x + 3y = 18$
14. $4x - 2y = 16$; $6x + 5y = 24$

(*Hint*: In Problem 13, multiply top equation by 3 and the bottom one by 4; then add.)

15. $7x - 4y = 0$
$2x + 7y = 57$

16. $2a + 3b = 0$
$3a - 2b = \frac{13}{2}$

17. $\frac{2}{3}x + y = 4$
$x + 2y = 5$

18. $\frac{3}{4}x - \frac{1}{2}y = 12$
$x + y = -8$

19. $.125x - .2y = 3$
$.75x + .3y = 10.5$

20. $.13x - .24y = 1$
$2.6x + 4y = -2.4$

21. $\frac{4}{x} + \frac{3}{y} = 17$
$\frac{1}{x} - \frac{3}{y} = -7$

22. $\frac{4}{x} - \frac{2}{y} = 12$
$\frac{5}{x} + \frac{1}{y} = 8$

(*Hint*: Let $u = 1/x$ and $v = 1/y$. Solve for u and v and then find x and y.)

23. $\frac{2}{\sqrt{x}} - \frac{1}{\sqrt{y}} = \frac{2}{3}$
$\frac{1}{\sqrt{x}} + \frac{2}{\sqrt{y}} = \frac{7}{6}$

24. $\frac{1}{x-2} - \frac{2}{y+3} = 3$
$\frac{5}{x-2} + \frac{2}{y+3} = 3$

Example (Three unknowns) Find values for x, y, and z that satisfy all three of the following equations.

$$\begin{aligned} 2x + y - 3z &= -9 \\ x - 2y + 4z &= 17 \\ 3x - y - z &= 2 \end{aligned}$$

Solution. The idea is to eliminate one of the unknowns from two different pairs of equations. This gives us just two equations in two unknowns, which we solve as before. Let us eliminate y from the first two equations. To do this, we multiply the first equation by 2 and add to the second.

$$\begin{aligned} 4x + 2y - 6z &= -18 \\ x - 2y + 4z &= 17 \\ \hline 5x \qquad - 2z &= -1 \end{aligned}$$

Next, we eliminate y from the first and third equations by simply adding them.

$$\begin{aligned} 2x + y - 3z &= -9 \\ 3x - y - z &= 2 \\ \hline 5x + \quad - 4z &= -7 \end{aligned}$$

Our problem thus reduces to solving the following system of two equations in two unknowns.

$$\begin{aligned} 5x - 2z &= -1 \\ 5x - 4z &= -7 \end{aligned}$$

We leave this for you to do. You should get $z = 3$ and $x = 1$. If you substitute these values for x and z in any of the original equations, you will find that $y = -2$. Thus

$$x = 1 \qquad y = -2 \qquad z = 3$$

Solve each of these systems for x, y, and z.

25. $\begin{aligned} 4x - y + 2z &= 2 \\ -3x + y - 4z &= -1 \\ x \quad\quad + 5z &= 1 \end{aligned}$

26. $\begin{aligned} x + 4y - 8z &= -10 \\ 3x - y + 5z &= 12 \\ -4x + 2y + z &= -9 \end{aligned}$

27. $\begin{aligned} 2x + 3y + 4z &= -6 \\ -x + 4y - 6z &= 6 \\ 3x - 2y + 2z &= 2 \end{aligned}$

28. $\begin{aligned} 3x \quad\quad + z &= 0 \\ 3x + 2y + z &= 4 \\ 9x + 5y + 10z &= 3 \end{aligned}$

Miscellaneous Problems

29. Find two numbers whose sum is 18 and whose difference is 4.
30. One number is three times as large as another and their sum is 14. Find the two numbers.
31. A man sold two lots at a total price of \$13,000. If he received \$1400 more for one lot than for the other, what was the selling price of each lot?
32. Ian Stockton's estate is valued at \$5000 more than three times as much as his wife's estate. The combined value of their estates is \$185,000. Find the value of each estate.
33. Mrs. Goldthorpe invests $1\frac{1}{2}$ times as much in savings certificates as in government bonds. If her total investment amounts to \$47,500, how much does she invest in each?
34. A ticket manager sold 700 more general admission tickets than reserved seat tickets. How many of each kind did he sell if he sold 2300 tickets in all?
35. The attendance at a professional football game was 45,000 and the total gate receipts were \$385,000. If each person bought either an \$8 ticket or a \$10 ticket, how many tickets of each kind were sold?
36. A change box contains \$30.70 in quarters and dimes. If there are 190 coins in all, how many quarters and how many dimes are there?
37. Johnny boasts that he has 8 more than 4 times as many nickels as he has dimes. When his mother asks him what it all amounts to, his reply is \$1.30. How many nickels and how many dimes does he have?
38. A woman needs to borrow \$80,000 for a business venture. She is able to get an 8 percent loan at a bank and a 9 percent loan at a savings and loan association. In each case, the interest is payable after one year. How much does she borrow from each institution if the total interest due after one year is \$6650?
39. A tourist takes a trip of 700 miles driving at an average speed of 30 miles per hour before lunch and at an average speed of 50 miles per hour after lunch. His time on the road after lunch is one hour more than twice the time on the road before lunch. How many hours does he drive before stopping for lunch? How long is he on the road after lunch?
40. James Carson rows 14 miles downstream on the Mississippi River and then rows back. If it takes him 2 hours going downstream and 7 hours coming back, what is the rate of the current and how fast can he row in still water?

41. A grocer has some coffee worth \$1.69 per pound and some worth only \$1.26 per pound. How much of each kind of coffee should she mix together to get 100 pounds worth \$1.53 per pound?

42. A solution that is 40 percent alcohol is to be mixed with one that is 90 percent alcohol to obtain 100 liters of 60 percent alcohol solution. How many liters of each will be used?

43. Susan Sharp paid \$4800 for some dresses and coats. She paid \$40 for each dress and \$100 for each coat. She sold the dresses at 20 percent profit and the coats at 50 percent profit. Her total profit was \$1800. How many coats and how many dresses did she buy?

44. Workers in a certain factory are classified into two groups, depending upon their skills for their jobs. Group 1 workers are paid \$8.00 per hour and group 2 workers are paid \$5.00 per hour. In negotiations for a new contract, the union demands that workers in the second group have their hourly wages brought up to $\frac{2}{3}$ of those for the workers in the first group. The company has 55 employees in group 1 and 40 in group 2, all of whom work a 40-hour week. If the company is prepared to increase its weekly payroll by \$5760, what hourly wages should be proposed for each class of workers?

Old Stuff

By 2000 B.C. Babylonian arithmetic had evolved into a well-developed rhetorical algebra. Not only were quadratic equations solved, both by the equivalent of substituting in a general formula and by completing the square, but some cubic (third degree) and biquadratic (fourth degree) equations were discussed.

Howard Eves

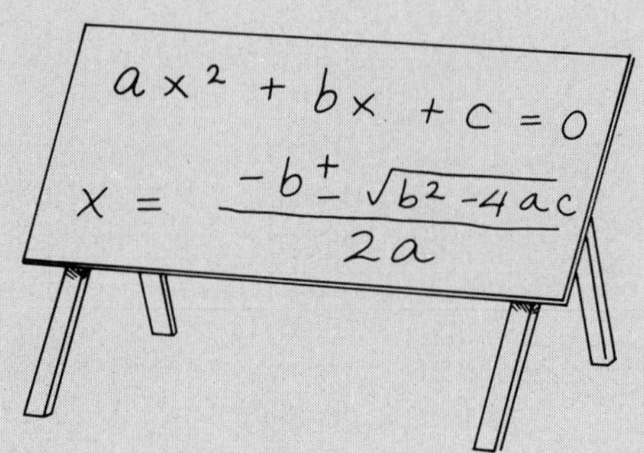

3-4 Quadratic Equations

A linear (first degree) equation may be put in the form $ax + b = 0$. We have seen that such an equation has exactly one solution, $x = -b/a$. That is simple and straightforward; no one is likely to stumble over it. But even the ancient Babylonians knew that equation solving goes far beyond this simple case. In fact, a good part of mathematical history revolves around attempts to solve more and more complicated equations.

The next case to consider is the second degree or **quadratic** equation, that is, an equation of the form

$$ax^2 + bx + c = 0 \qquad (a \neq 0)$$

Here are some examples.

(i) $x^2 - 4 = 0$

(ii) $x^2 - x - 6 = 0$

(iii) $8x^2 - 2x = 1$

(iv) $x^2 = 6x - 2$

While equations (iii) and (iv) do not quite fit the pattern, we accept them because they readily transform to equations in standard form.

(iii) $8x^2 - 2x - 1 = 0$

(iv) $x^2 - 6x + 2 = 0$

Solution by Factoring All of us remember that 0 times any number is 0. Just as important but sometimes forgotten is the fact that if the product of two numbers is 0, then one or both of the factors must be 0.

If $u = 0$ or $v = 0$, then $u \cdot v = 0$.
If $u \cdot v = 0$, then either $u = 0$, or $v = 0$, or both.

This fact allows us to solve any quadratic equation which has 0 on one side provided we can factor its other side. Simply factor, set each factor equal to 0, and solve the resulting linear equations. We illustrate.

(i)
$$x^2 - 4 = 0$$
$$(x - 2)(x + 2) = 0$$
$$x - 2 = 0 \qquad x + 2 = 0$$
$$x = 2 \qquad x = -2$$

(ii)
$$x^2 - x - 6 = 0$$
$$(x - 3)(x + 2) = 0$$
$$x - 3 = 0 \qquad x + 2 = 0$$
$$x = 3 \qquad x = -2$$

(iii)
$$8x^2 - 2x - 1 = 0$$
$$(4x + 1)(2x - 1) = 0$$
$$4x + 1 = 0 \qquad 2x - 1 = 0$$
$$x = -\tfrac{1}{4} \qquad x = \tfrac{1}{2}$$

Equation (iv) remains unsolved; we do not know how to factor its left side. For this equation, we need a more powerful method. First, however, we need a brief discussion of square roots.

Square Roots The number 9 has two square roots, 3 and -3. In fact, every positive number has two square roots, one positive and the other negative. If a is positive, its **positive square root** is denoted by $\sqrt{a}$. Thus $\sqrt{9} = 3$. Do not write $\sqrt{9} = -3$ or $\sqrt{9} = \pm 3$; both are wrong. But you can say that the two square roots of 9 are $\pm\sqrt{9}$ (or ± 3) and that the two square roots of 7 are $\pm\sqrt{7}$.

Here are two important properties of square roots, valid for any positive numbers a and b.

$$\sqrt{ab} = \sqrt{a}\sqrt{b}$$
$$\sqrt{\frac{a}{b}} = \frac{\sqrt{a}}{\sqrt{b}}$$

For example,

$$\sqrt{4 \cdot 16} = \sqrt{4}\sqrt{16} = 2 \cdot 4 = 8$$
$$\sqrt{28} = \sqrt{4 \cdot 7} = \sqrt{4}\sqrt{7} = 2\sqrt{7}$$
$$\sqrt{\frac{4}{9}} = \frac{\sqrt{4}}{\sqrt{9}} = \frac{2}{3}$$

The square roots of a negative number are imaginary. For example, the two square roots of -9 are $3i$ and $-3i$, since

$$(3i)^2 = 3^2 i^2 = 9(-1) = -9$$
$$(-3i)^2 = (-3)^2 i^2 = 9(-1) = -9$$

In fact, if a is positive, the two square roots of $-a$ are $\pm\sqrt{a}i$. And in this case, the symbol $\sqrt{-a}$ will denote $\sqrt{a}i$. Thus $\sqrt{-7} = \sqrt{7}i$.

Completing the Square Consider equation (iv) again.

$$x^2 - 6x + 2 = 0$$

We may rewrite it as

$$x^2 - 6x = -2$$

Now add 9 to both sides, making the left side a perfect square, and factor.

$$x^2 - 6x + 9 = -2 + 9$$
$$(x - 3)^2 = 7$$

This means that $x - 3$ must be one of the two square roots of 7. That is,

$$x - 3 = \pm\sqrt{7}$$

Hence

$$x = 3 + \sqrt{7} \quad \text{or} \quad x = 3 - \sqrt{7}$$

You may ask how we knew that we should add 9. Any expression of the form $x^2 + px$ becomes a perfect square when $(p/2)^2$ is added, since

$$x^2 + px + \left(\frac{p}{2}\right)^2 = \left(x + \frac{p}{2}\right)^2$$

For example, $x^2 + 10x$ becomes a perfect square when we add $(10/2)^2$ or 25.

$$x^2 + 10x + 25 = (x + 5)^2$$

The rule for completing the square (namely, add $(p/2)^2$) works only when the coefficient of x^2 is 1. However, that fact causes no difficulty for quadratic equations. If the leading coefficient is not 1, we simply divide both sides by this coefficient and then complete the square. We illustrate with the equation

$$2x^2 - x - 3 = 0$$

We divide both sides by 2 and proceed as before.

$$\begin{aligned}
x^2 - \tfrac{1}{2}x - \tfrac{3}{2} &= 0 \\
x^2 - \tfrac{1}{2}x \qquad &= \tfrac{3}{2} \\
x^2 - \tfrac{1}{2}x + (\tfrac{1}{4})^2 &= \tfrac{3}{2} + (\tfrac{1}{4})^2 \\
(x - \tfrac{1}{4})^2 &= \tfrac{25}{16} \\
x - \tfrac{1}{4} &= \pm\tfrac{5}{4}
\end{aligned}$$

$$x = \tfrac{1}{4} + \tfrac{5}{4} = \tfrac{3}{2} \quad \text{or} \quad x = \tfrac{1}{4} - \tfrac{5}{4} = -1$$

The Quadratic Formula The method of completing the square works on any quadratic equation. But there is a way of doing this process once and for all. Consider the general quadratic equation

$$ax^2 + bx + c = 0$$

with real coefficients $a \neq 0$, b, and c. First add $-c$ to both sides and then divide by a to obtain

$$x^2 + \frac{b}{a}x = -\frac{c}{a}$$

Next complete the square by adding $(b/2a)^2$ to both sides and then simplify.

$$\begin{aligned}
x^2 + \frac{b}{a}x + \left(\frac{b}{2a}\right)^2 &= -\frac{c}{a} + \left(\frac{b}{2a}\right)^2 \\
\left(x + \frac{b}{2a}\right)^2 &= -\frac{c}{a} + \frac{b^2}{4a^2} \\
\left(x + \frac{b}{2a}\right)^2 &= \frac{b^2 - 4ac}{4a^2}
\end{aligned}$$

Finally take the square root of both sides.

$$x + \frac{b}{2a} = \pm\frac{\sqrt{b^2 - 4ac}}{2a}$$

or

$$x = \frac{-b}{2a} \pm \frac{\sqrt{b^2 - 4ac}}{2a}$$

We call this result the **quadratic formula** and normally write it as follows.

$$x = \frac{-b \pm \sqrt{b^2 - 4ac}}{2a}$$

Let us see how it works on example (iv).

$$x^2 - 6x + 2 = 0$$

Here $a = 1$, $b = -6$, and $c = 2$. Thus

$$\begin{aligned} x &= \frac{-(-6) \pm \sqrt{36 - 4 \cdot 2}}{2} \\ &= \frac{6 \pm \sqrt{28}}{2} \\ &= \frac{6 \pm \sqrt{4 \cdot 7}}{2} \\ &= \frac{6 \pm 2\sqrt{7}}{2} \\ &= \frac{2(3 \pm \sqrt{7})}{2} \\ &= 3 \pm \sqrt{7} \end{aligned}$$

As a second example, consider $2x^2 - 4x + \frac{25}{8} = 0$. Here $a = 2$, $b = -4$, and $c = \frac{25}{8}$. The quadratic formula gives

$$x = \frac{4 \pm \sqrt{16 - 25}}{4} = \frac{4 \pm \sqrt{-9}}{4} = \frac{4 \pm 3i}{4}$$

The expression $b^2 - 4ac$ that appears under the square root sign in the quadratic formula is called the **discriminant**. It determines the character of the solutions.

1. If $b^2 - 4ac > 0$, there are two real solutions.
2. If $b^2 - 4ac = 0$, there is one real solution.
3. If $b^2 - 4ac < 0$, there are two nonreal solutions.

Problem Set 3-4

Example A (Simplifying square roots) Simplify

(a) $\sqrt{54}$; (b) $\dfrac{2 + \sqrt{48}}{4}$; (c) $\dfrac{\sqrt{6}}{\sqrt{150}}$.

SOLUTION. (a) $\sqrt{54} = \sqrt{9 \cdot 6} = \sqrt{9}\sqrt{6} = 3\sqrt{6}$
Here we factored out the largest square in 54, namely 9, and then used the first property of square roots.

(b) $\dfrac{2 + \sqrt{48}}{4} = \dfrac{2 + \sqrt{16 \cdot 3}}{4} = \dfrac{2 + 4\sqrt{3}}{4} = \dfrac{2(1 + 2\sqrt{3})}{4} = \dfrac{1 + 2\sqrt{3}}{2}$

If you are tempted to continue as follows,

$$\frac{1 + \not{2}\sqrt{3}}{\not{2}} = 1 + \sqrt{3}$$

resist the temptation. That cancellation is wrong, because 2 is not a factor of the entire numerator but only of $2\sqrt{3}$.

(c) $\frac{\sqrt{6}}{\sqrt{150}} = \sqrt{\frac{6}{150}} = \sqrt{\frac{1}{25}} = \frac{1}{5}$

Sometimes, as in this case, it is best to write a quotient of two square roots as a single square root and then simplify.

Simplify each of the following.

1. $\sqrt{50}$
2. $\sqrt{300}$
3. $\sqrt{\frac{1}{4}}$
4. $\sqrt{\frac{3}{27}}$
5. $\frac{\sqrt{-45}}{\sqrt{20}}$
6. $\sqrt{.04}$
7. $\sqrt{11^2 \cdot 4}$
8. $\frac{\sqrt{2^3 \cdot 5}}{\sqrt{2 \cdot 5^3}}$
9. $\frac{5 + \sqrt{72}}{5}$
10. $\frac{4 - \sqrt{12}}{2}$
11. $\frac{10 + \sqrt{75}}{20}$
12. $\frac{4 + \sqrt{-4}}{4}$
13. $\frac{18 + \sqrt{-9}}{6}$
14. $\frac{3 + \sqrt{-8}}{3}$

Example B (Quadratics that are already perfect squares) Solve (a) $x^2 = 9$; (b) $(x + 3)^2 = 17$; (c) $(2y - 5)^2 = 16$

SOLUTION. The easiest way to solve these equations is to take square roots of both sides.

(a) $x = 3$ or $x = -3$

(b) $x + 3 = \pm\sqrt{17}$

$x = -3 + \sqrt{17}$ or $x = -3 - \sqrt{17}$

(c) $2y - 5 = \pm 4$

$2y = 5 + 4$ or $2y = 5 - 4$

$y = \frac{9}{2}$ or $y = \frac{1}{2}$

Solve by the method above.

15. $x^2 = 25$
16. $x^2 = 14$
17. $(x - 3)^2 = 16$
18. $(x + 4)^2 = 49$
19. $(2x + 5)^2 = 100$
20. $(3y - \frac{1}{3})^2 = 25$
21. $m^2 = -9$
22. $(m - 6)^2 = -36$

Solve by factoring.

23. $x^2 - 9 = 0$
24. $x^2 - \frac{9}{4} = 0$
25. $m^2 - .0144 = 0$
26. $x^2 - x - 2 = 0$
27. $x^2 - 3x - 10 = 0$
28. $x^2 + 13x + 22 = 0$
29. $3x^2 + 5x - 2 = 0$
30. $3x^2 + x - 2 = 0$
31. $6x^2 - 13x - 28 = 0$
32. $10x^2 + 19x - 15 = 0$

Solve by completing the square.

33. $x^2 + 8x = 9$
34. $x^2 - 12x = 45$
35. $z^2 - z = \frac{3}{4}$
36. $x^2 + 5x = 2\frac{3}{4}$
37. $x^2 + 4x = -9$
38. $x^2 - 14x = -65$

Solve by using the quadratic formula.

39. $x^2 + 8x + 12 = 0$
40. $x^2 - 2x - 15 = 0$
41. $x^2 + 5x + 3 = 0$
42. $z^2 - 3z - 8 = 0$
43. $3x^2 - 6x - 11 = 0$
44. $4t^2 - t - 3 = 0$
45. $x^2 + 4x + 5 = 0$
46. $y^2 - 6y + 10 = 0$
47. $2z^2 - 6z + 11 = 0$
48. $x^2 + x + 1 = 0$

Solve using the quadratic formula. Write your answers rounded to four decimal places.

[C] 49. $2x^2 - \pi x - 1 = 0$

[C] 50. $3x^2 - \sqrt{2}x - 3\pi = 0$

[C] 51. $x^2 + .8235x - 1.3728 = 0$

[C] 52. $5x^2 - \sqrt{3}x - 4.3213 = 0$

Example C (Solving for one variable in terms of the other) Solve for y in terms of x.
(a) $y^2 - 2xy + x^2 - 2 = 0$ (b) $(y - 3x)^2 + 3(y - 3x) - 4 = 0$

SOLUTION. (a) We use the quadratic formula with $a = 1$, $b = -2x$, and $c = x^2 - 2$.

$$y = \frac{2x \pm \sqrt{4x^2 - 4(x^2 - 2)}}{2} = \frac{2x \pm \sqrt{8}}{2} = \frac{2x \pm 2\sqrt{2}}{2} = x \pm \sqrt{2}$$

So $y = x + \sqrt{2}$ or $y = x - \sqrt{2}$.

(b) If we substitute z for $y - 3x$, the equation becomes

$$z^2 + 3z - 4 = 0$$

We solve this equation for z.

$$(z + 4)(z - 1) = 0$$
$$z = -4 \quad \text{or} \quad z = 1$$

Thus

$$y - 3x = -4 \quad \text{or} \quad y - 3x = 1$$
$$y = 3x - 4 \quad \text{or} \quad y = 3x + 1$$

Solve for y in terms of x.

53. $(y - 2)^2 = 4x^2$
54. $(y + 3x)^2 = 9$
55. $(y + 3x)^2 = 9x^2$
56. $4y^2 + 4xy - 5 + x^2 = 0$
57. $(y + 2x)^2 - 8(y + 2x) + 15 = 0$
58. $(x - 2y + 3)^2 - 3(x - 2y + 3) + 2 = 0$

Miscellaneous Problems

Solve equations 59–70 using any method you like.

59. $x^2 = 144$
60. $x^2 = -4$
61. $(x - 2)(x + \frac{3}{2}) = 0$
62. $(2y + 1)^2 = \frac{9}{4}$
63. $2x^2 + 5x - 3 = 0$
64. $x^2 + 12x = -40$
65. $4m^2 + 5m = 0$
66. $4m^2 + 5m + 1 = 0$
67. $x^2 + 2x - 5 = 0$
68. $2x^2 + x = x^2 - 2x + 3$
69. $x^2 - 5ix - 4 = 0$
70. $x^2 + 2ix + 4 = 0$

71. A rectangle has a perimeter of 40 feet and an area of 91 square feet. Find the dimensions of the rectangle.

72. The sum of the squares of three consecutive positive integers is 365. Find the smallest of the three integers.

C 73. A garden in the form of a square is surrounded by a walk 2 feet wide. What are the dimensions of the garden if the combined area of the garden and the walk is 800 square meters?

C 74. Find the radius of a circle whose area is twice the area of a square which is 11 inches on a side.

75. A square piece of cardboard is used to construct a tray by cutting out one-inch squares from the four corners and then turning up the four flaps. Find the size of the original square if the resulting tray has a volume of 128 cubic inches.

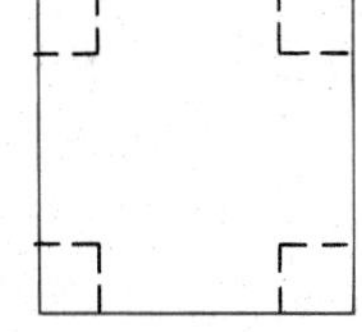

76. One leg of a right triangle is 2 inches longer than the other leg. If the hypotenuse is $\frac{5}{2}\sqrt{10}$ inches long, how long are the legs?

77. A rock is thrown vertically upward at an initial speed of 128 feet per second. Assume that its distance from the ground t seconds later is $(-16t^2 + 128t)$ feet. (This is approximately correct if we neglect air resistance.)
(a) At what time will the rock strike the ground?
(b) When will the rock be 192 feet from the ground?
(c) When will the rock be 400 feet from the ground?

78. The distance from St. Charles to St. Agnes is 360 miles. Jones and Smith, driving in separate cars, start out from St. Charles for St. Agnes at the same time. Jones drives 5 miles per hour faster than Smith and arrives at St. Agnes 1 hour earlier. How fast is each one driving?

In each of the following, find the values for the two unknowns that satisfy both equations.

79. $x^2 + y^2 = 25$
$3x + y = 13$
(*Hint*: Solve the second equation for y in terms of x and substitute this value in the first equation.)

80. $xy = 20$
$y = 2x - 3$

81. $2x^2 - xy + y^2 = 11$
$x - 2y = 4$

82. $x^2 + 4y^2 = 8$
$3x^2 + 2y^2 = 14$

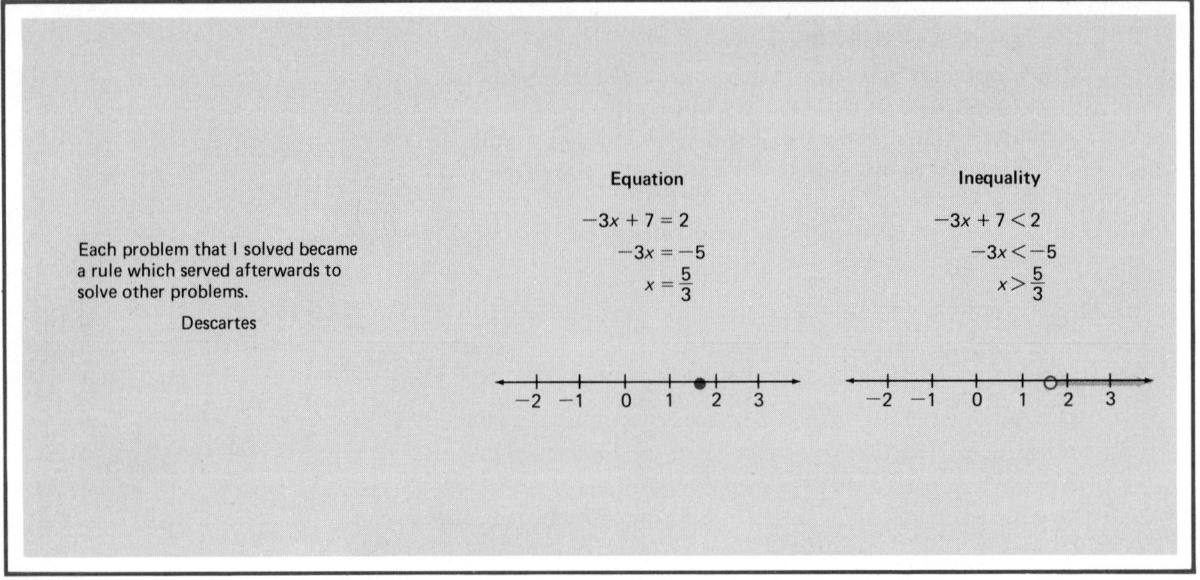

3-5 Inequalities

Solving an inequality is very much like solving an equation, as the example above demonstrates. However, there are dangers in proceeding too mechanically. It will be important to think at every step.

Recall the distinction we made between identities and equations in Section 3-1. A similar distinction applies to inequalities. An inequality which is true for all values of the variables is called an **unconditional inequality**. Examples are

$$(x - 3)^2 + 1 > 0$$

and

$$|x| \leq |x| + |y|$$

Most inequalities (for example, $-3x + 7 < 2$) are true only for some values of the variables; we call them **conditional inequalities**. Our primary task in this section is to solve conditional inequalities, that is, to find all those numbers which make a conditional inequality true.

Linear Inequalities To solve the linear inequality $Ax + B < C$, we try to rewrite it in successive steps until the variable x stands by itself on one side of the inequality (see opening display). This depends primarily on the properties stated in Section 1-5 and repeated here.

PROPERTIES OF INEQUALITIES

1. (Transitivity). If $a < b$ and $b < c$, then $a < c$.
2. (Addition). If $a < b$, then $a + c < b + c$.
3. (Multiplication). If $a < b$ and $c > 0$, then $a \cdot c < b \cdot c$.
   If $a < b$ and $c < 0$, then $a \cdot c > b \cdot c$.

We illustrate the use of these properties, applied to $\leq$ rather than $<$, by solving the following inequality.

	$-2x + 6 \leq 18 + 4x$
Add $-4x$:	$-6x + 6 \leq 18$
Add -6:	$-6x \leq 12$
Multiply by $-\frac{1}{6}$:	$x \geq -2$

By rights, we should check this solution. All we know so far is that any value of x that satisfies the original inequality satisfies $x \geq -2$. Can we go in the opposite direction? Yes, because every step is reversible. For example, starting with $x \geq -2$, we can multiply by -6 to get $-6x \leq 12$. In practice, we do not actually carry out this check as we recognize that Property 2 can be restated:

$a < b$ is equivalent to $a + c < b + c$

There are similar restatements of Property 3.

Quadratic Inequalities To solve

$$x^2 - 2x - 3 > 0$$

we first factor, obtaining

$$(x + 1)(x - 3) > 0$$

Next we ask ourselves when the product of two numbers is positive. There are two cases; either both factors are negative or both factors are positive.

Case 1 (Both negative). We want to know when both factors are negative, that is, we seek to solve $x + 1 < 0$ and $x - 3 < 0$ simultaneously. The first gives $x < -1$ and the second gives $x < 3$. Together they give $x < -1$.

Case 2 (Both positive). Both factors are positive when $x + 1 > 0$ and $x - 3 > 0$, that is, when $x > -1$ and $x > 3$. These give $x > 3$.

The solution set for the original inequality is the union of the solution sets for the two cases. In set notation, it may be written either as

$$\{x: x < -1 \text{ or } x > 3\}$$

or as

$$\{x: x < -1\} \cup \{x: x > 3\}$$

The chart below summarizes what we have learned.

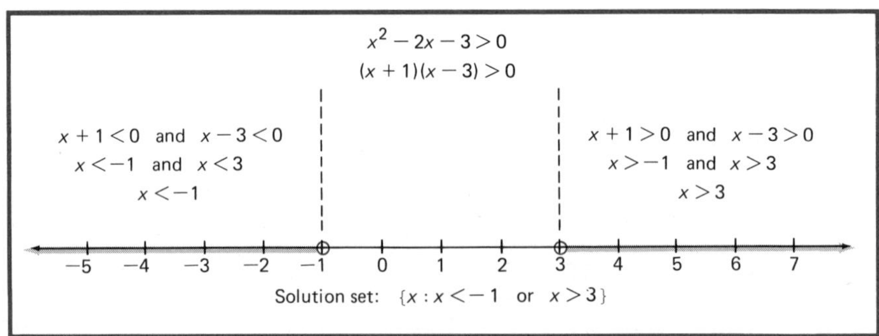

Split-Point Method The preceding example could have been approached in a slightly different way using the notion of split-points. The solutions of the equation $(x + 1)(x - 3) = 0$, which are -1 and 3, serve as split-points that divide the real line into the three intervals: $x < -1$, $-1 < x < 3$, and $3 < x$. Since $(x + 1)(x - 3)$ can change sign only at a split-point, it must be of one sign (that is, be either always positive or always negative) on each of these intervals. To determine which of them make up the solution set of the inequality $(x + 1)(x - 3) > 0$, all we need to do is pick a single (arbitrary) point from each interval and test it for inclusion in the solution set. If it passes the test, the entire interval from which it was drawn belongs to the solution set.

To show how this method works, let us consider the third degree inequality

$$(x + 2)(x - 1)(x - 4) < 0$$

The solutions of the corresponding equation

$$(x + 2)(x - 1)(x - 4) = 0$$

are -2, 1, and 4. They break the real line into the four intervals $x < -2$, $-2 < x < 1$, $1 < x < 4$, and $4 < x$. Suppose we pick -3 as the test point for the interval $x < -2$. We see that -3 makes each of the three factors

$x + 2$, $x - 1$, and $x - 4$ negative, and so it makes their product $(x + 2) \times (x - 1)(x - 4)$ negative. You should pick test points from each of the other three intervals to verify the results shown below.

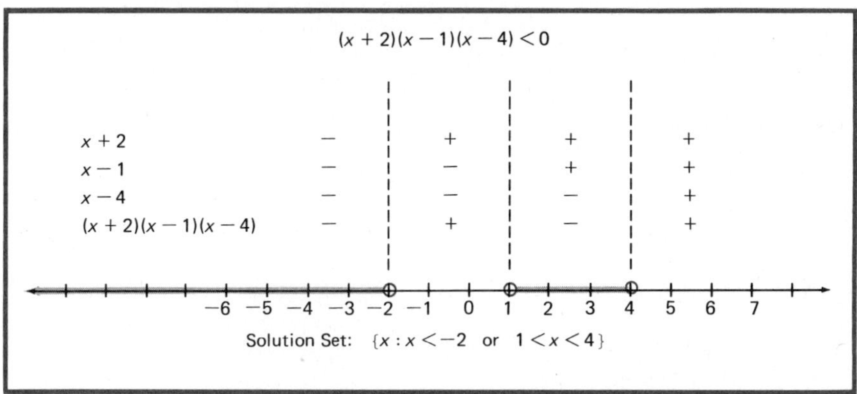

Inequalities With Absolute Values The basic fact to remember is that

$$|x| < a$$

means the same thing as the double inequality

$$-a < x < a$$

To solve the inequality $|3x - 2| < 4$, proceed as follows.

Given inequality:	$|3x - 2| < 4$
Remove absolute value:	$-4 < 3x - 2 < 4$
Add 2:	$-2 < 3x < 6$
Multiply by $\frac{1}{3}$:	$-\frac{2}{3} < x < 2$

Here, as in the preceding examples, we could use the split-point method. The solutions of the equation $|3x - 2| = 4$, which are $-\frac{2}{3}$ and 2, would be the split-points. The solution would again be $-\frac{2}{3} < x < 2$.

An Application A student wishes to get a grade of *B* in Mathematics 16. On the first four tests, he got 82 percent, 63 percent, 78 percent, and 90 percent, respectively. A grade of *B* requires an average between 80 percent and 90 percent, inclusive. What grade on the fifth test would qualify this student for a *B*?

Let x represent the grade (in percent) on the fifth test. The inequality to be satisfied is

$$80 \leq \frac{82 + 63 + 78 + 90 + x}{5} \leq 90$$

This can be rewritten successively as

$$80 \leq \frac{313 + x}{5} \leq 90$$

$$400 \leq 313 + x \leq 450$$
$$87 \leq x \leq 137$$

A score greater than 100 is impossible, so the actual solution to this problem is $87 \leq x \leq 100$.

Problem Set 3-5

Which of the following inequalities are unconditional and which are conditional?

1. $x \geq 0$
2. $x^2 \geq 0$
3. $x^2 + 1 > 0$
4. $x^2 > 1$
5. $x - 2 < -5$
6. $2x + 3 > -1$
7. $x(x + 4) \leq 0$
8. $(x - 1)(x + 2) > 0$
9. $(x + 1)^2 > x^2$
10. $(x - 2)^2 \leq x^2$
11. $(x + 1)^2 > x^2 + 2x$
12. $(x - 2)^2 > x(x - 4)$

Solve each of the following inequalities and show the solution set on the real number line.

13. $3x + 7 < x - 5$
14. $-2x + 11 > x - 4$
15. $\frac{2}{3}x + 1 > \frac{1}{2}x - 3$
(*Hint*: First get rid of the fractions.)
16. $3x - \frac{1}{2} < \frac{1}{2}x + 4$
17. $\frac{3}{4}x - \frac{1}{2} < \frac{1}{6}x + 2$
18. $\frac{2}{7}x + \frac{1}{3} \leq -\frac{2}{3}x + \frac{15}{14}$
19. $(x - 2)(x + 5) \leq 0$
20. $(x + 1)(x + 4) \geq 0$
21. $(2x - 1)(x + 3) > 0$
22. $(3x + 2)(x - 2) < 0$
23. $x^2 - 5x + 4 \geq 0$
(*Hint*: Factor the left side.)
24. $x^2 + 4x + 3 \leq 0$
25. $2x^2 - 7x + 3 < 0$
26. $3x^2 - 5x - 2 > 0$
27. $|2x + 3| < 2$
28. $|2x - 4| \leq 3$
29. $|-2x - 1| \leq 1$
30. $|-2x + 3| > 2$
31. $(x + 4)x(x - 3) \geq 0$
32. $(x + 3)x(x - 3) \geq 0$
33. $(x - 2)^2(x - 5) < 0$
34. $(x + 1)^2(x - 1) > 0$

Example A (Inequalities involving quotients) Solve the following inequality.

$$\frac{3}{x - 2} > \frac{2}{x}$$

SOLUTION. We rewrite the inequality as follows:

$$\frac{3}{x-2} - \frac{2}{x} > 0 \qquad \text{(add } -2/x \text{ to both sides)}$$

$$\frac{3x - 2(x-2)}{(x-2)x} > 0 \qquad \text{(combine fractions)}$$

$$\frac{x+4}{(x-2)x} > 0 \qquad \text{(simplify numerator)}$$

$$(x+4)(x-2)x > 0 \qquad \text{(multiply both sides by } [(x-2)x]^2\text{)}$$

Now we can use -4, 0, and 2 as split-points and proceed as in the text. The following chart shows the result.

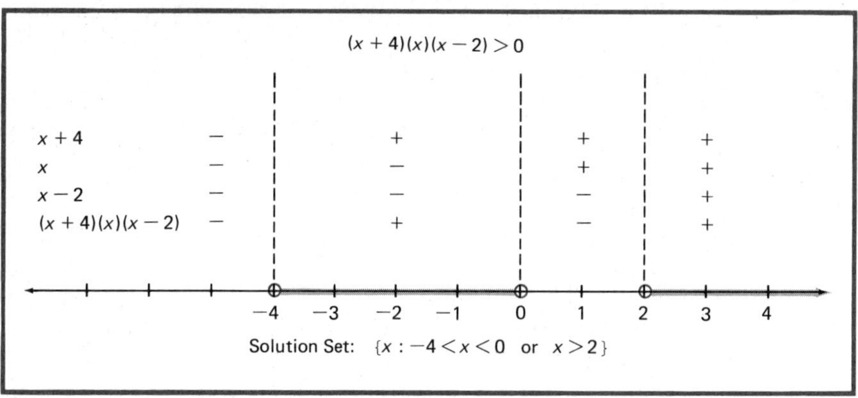

Solve each of the following inequalities.

35. $\dfrac{x-5}{x+2} \leq 0$

36. $\dfrac{x+3}{x-2} > 0$

37. $\dfrac{x(x+2)}{x-5} > 0$

38. $\dfrac{x-1}{(x-3)(x+3)} \geq 0$

39. $\dfrac{5}{x-3} > \dfrac{4}{x-2}$

40. $\dfrac{-3}{x+1} < \dfrac{2}{x-4}$

Example B (Rewriting with absolute values) Write $-2 < x < 8$ as an inequality involving absolute values.

SOLUTION. Look at this interval on the number line. It is 10 units long and its midpoint is at 3. A number x is in this interval provided that it is within a radius of 5 of this midpoint, that is, if

$$|x - 3| < 5$$

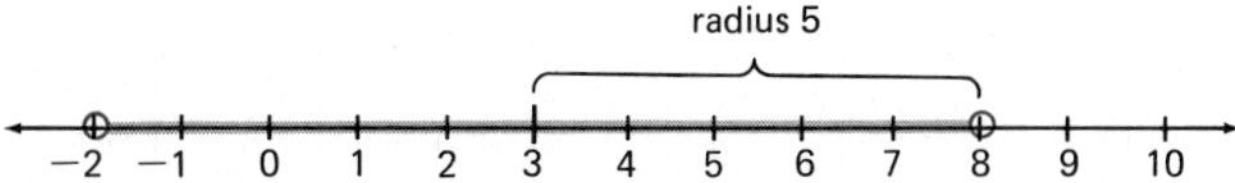

We can check that $|x - 3| < 5$ is equivalent to the original inequality by writing it as

$$-5 < x - 3 < 5$$

and then adding 3 to each member.

Write each of the following as an inequality involving absolute values.

41. $0 < x < 6$
42. $0 < x < 12$
43. $-1 \leq x \leq 7$
44. $-3 \leq x \leq 7$
45. $2 < x < 11$
46. $-10 < x < -3$

Example C (Quadratic inequalities that cannot be factored by inspection) Solve the inequality

$$x^2 - 5x + 3 \geq 0$$

SOLUTION. Even though $x^2 - 5x + 3$ cannot be factored by inspection, we can solve the quadratic equation $x^2 - 5x + 3 = 0$ by use of the quadratic formula. We obtain the two solutions $(5 - \sqrt{13})/2 \approx .7$ and $(5 + \sqrt{13})/2 \approx 4.3$. These two numbers split the real line into three intervals from which the numbers 0, 1, and 5 could be picked as test points. Notice that $x = 0$ makes $x^2 - 5x + 3$ positive, $x = 1$ makes it negative, and $x = 5$ makes it positive. This gives us the solution set

$$\{x : x \leq (5 - \sqrt{13})/2 \quad \text{or} \quad x \geq (5 + \sqrt{13})/2\}$$

which can be pictured as follows.

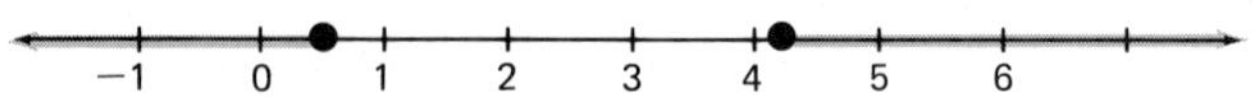

The split-points are included in the solution set because our inequality $x^2 - 5x + 3 \geq 0$ includes equality.

Solve each of the following inequalities and display the solution set on the number line.

47. $x^2 - 7 < 0$
48. $x^2 - 12 > 0$
49. $x^2 - 4x + 2 \geq 0$
50. $x^2 - 4x - 2 \leq 0$
51. [C] $x^2 + 6.32x + 3.49 > 0$
52. [C] $x^2 + 4.23x - 2.79 < 0$

Example D (Finding least values) Find the least value that $x^2 - 4x + 9$ can take on.

SOLUTION. We use the method of completing squares to write $x^2 - 4x + 9$ as the sum of a perfect square and a constant.

$$x^2 - 4x + 9 = (x^2 - 4x + 4) + 5 = (x - 2)^2 + 5$$

Since the smallest value $(x - 2)^2$ can take on is zero, the smallest value $(x - 2)^2 + 5$ can assume is 5.

Find the least value each of the following expressions can take on.

53. $x^2 + 8x + 20$
54. $x^2 + 10x + 40$
55. $x^2 - 2x + 101$
56. $x^2 - 4x + 104$

Miscellaneous Problems

Solve each inequality and show its solution set on the number line.

57. $\frac{1}{2}x - \frac{3}{4} < \frac{2}{3}x$
58. $2(x - \frac{3}{5}) \geq 3(x + \frac{1}{3})$
59. $(x + 4)(x - 2) > 0$
60. $x^2 - 4x - 5 < 0$
61. $|3x - 2| \leq 3$
62. $|2x + 3| > 2$
63. $\frac{5}{x} \leq \frac{2}{x - 3}$
64. $(x - 1)(x + 2)^2(x - 5) \geq 0$

In Problems 65–68, find the set of values of k for which the solutions of the quadratic equation are real. Remember that the discriminant ($b^2 - 4ac$) needs to be greater than or equal to zero.

65. $x^2 + 3x + k = 0$
66. $x^2 - kx + 10 = 0$
67. $x^2 + kx - 2 = 0$
68. $x^2 + kx + k = 0$
69. Tom scored 73, 82, 69, and 94 points on four 100-point tests in Mathematics 30. Suppose that a grade of B requires an average between 75 percent and 85 percent.
    (a) What score on a 100-point final exam would qualify Tom for a B?
    (b) What score on a 200-point final would qualify him for a B?
70. A certain department of a company has a staff of six persons whose average annual salary is $14,500. An additional person will be hired. What salary can be offered if the average salary for the department cannot exceed $18,000?
71. A stone thrown vertically upward from the roof of a building has a height of $(-16t^2 + 64t + 80)$ feet above the ground after t seconds.
    (a) What is the greatest height reached (see Example D of this section)?
    (b) During what time period is the stone higher than 80 feet?
    (c) At what time does the stone strike the ground?
    [C] (d) During what time interval on its way down is the stone less than 40 feet above the ground?
72. Podunk University hires two professors, Smith and Jones. Smith's starting salary is $14,500, while Jones starts at $16,000; however, Smith will get an annual increment of $800 while Jones will receive only $550. How soon will Smith's salary be higher than that of Jones?
73. A company has two garbage trucks to cover the weekly route of 800 miles. Truck A gets 5 miles per gallon of gas, while truck B gets 10 miles per gallon. If gasoline is rationed and the company is allowed only 90 gallons per week, what is the largest number of miles the company will dare drive truck A?
74. Company A will rent a car for $15 per day plus 10¢ per mile, while company B will rent a car for $10 a day plus 12¢ per mile. I need a car for 10 days. At least how many miles must I drive for me to consider renting from company A?

Solve the following inequalities.

75. $|x - 4| \leq |x - 2|$
76. $|x + 3| \geq |x - 3|$
77. $|x| < x + 4$
78. $|x| + |x - 1| > 3$

Johnny's Dilemma

Although he had been warned against it a thousand times, Johnny still walked across the railroad bridge when he was in a hurry—which was most of the time. It was the day of the biggest football game of the season and Johnny was late. He was already one fourth of the way across the bridge when he heard something. Turning to look, he saw the train coming; it was exactly one bridge-length away from him. Which way should he run?

3-6 More Applications (Optional)

We would not suggest that the problem above is a typical application of algebra. Johnny would be well advised to forget about algebra, make an instant decision, and hope for the best. Nevertheless, Johnny's problem is intriguing. Moreover, it can serve to reemphasize the important principles of real life problem solving that we mentioned briefly at the end of Section 3-2.

First, **clarify the question that is asked**. "Which way should he run?" must mean "Which way should Johnny run to have the best chance of surviving?" That is still too vaguely stated for mathematical analysis. We think the question really means, "Which of the two directions allows Johnny to run at the slowest rate and still avoid the train?" Most questions that come to us from the real world are loosely stated. Our first job is always to pin down precisely (at least in our own minds) what the real question is. It is foolish to try to answer a question that we do not understand. That should be obvious, but it is often overlooked.

Johnny's problem appears to be difficult for another reason; there does not seem to be enough information. We do not know how long the bridge is, we do not know how fast the train is traveling, and we do not know how fast Johnny can run. We could easily despair of making any progress on the problem!

This leads us to state our second principle. **Organize the information you have**. The best way to do this is to draw a diagram or picture that somehow captures the essence of the problem. It should be an abstract or idealized picture. We shall represent people and trains by points and train tracks and bridges by line segments. These are only approximations to the real situation, but they are necessary if progress is to be made. After the picture is drawn,

label the key quantities of the problem with letters and write down what they represent. Here is one way to represent Johnny's situation.

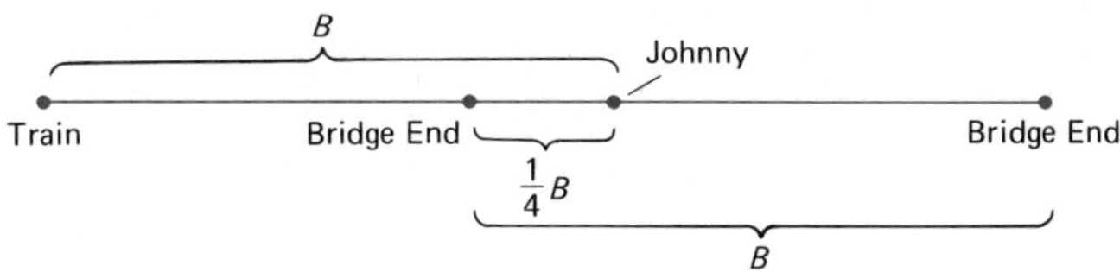

B = length of bridge in feet
S = speed of train in feet per second
Y = speed in feet per second that Johnny must run toward train to escape
Z = speed in feet per second that Johnny must run away from train to escape

Our next principle is this. **Write down the algebraic relationships that exist between the symbols you have introduced**. In Johnny's case, we need to remember the formula $D = RT$ (or $T = D/R$), which relates distance, rate, and time. Now, if Johnny runs toward the train just fast enough to escape, then his time to run the distance $B/4$ must equal the time for the train to go $3B/4$. That gives

(i) $$\frac{B/4}{Y} = \frac{3B/4}{S}$$

On the other hand, if Johnny runs away from the train, he will have to cover $3B/4$ in the same time that the train goes $7B/4$. Thus,

(ii) $$\frac{3B/4}{Z} = \frac{7B/4}{S}$$

Now we have done what is often the hardest part of the problem, translating it into algebraic equations. Our job now is to **manipulate the equations until a solution appears**. In the example we are considering, we must solve the equations for Y and Z in terms of S. When we do this, we get

(i) $$Y = \tfrac{1}{3}S$$
(ii) $$Z = \tfrac{3}{7}S$$

There remains a crucial step: **Interpret the result in the language of the original problem**. In Johnny's case, this step is easy. If he runs toward the train, he will need to run at $\frac{1}{3}$ the speed of the train to escape. If he runs away from the train, he must run at $\frac{3}{7}$ of the speed of the train. Clearly, he will have a better chance of making it if he runs toward the train.

Perhaps by now you need to be reassured. We do not expect you to go through a long-winded analysis like ours for every problem you meet. But we do think that you will have to apply the principles we have stated if you are to be successful at problem solving. The problem set that follows is designed to let you apply these principles to problems from several areas of applied mathematics.

Problem Set 3-6

The problems below are arranged according to type. Only the first set (rate-time problems) relates directly to the example of this section. However, all of them should require some use of the principles we have enunciated. Do not expect to solve all of the problems. But do accept them as challenges worthy of genuine effort.

Rate-Time Problems

1. Sound travels at 1100 feet per second in air and at 5000 feet per second in water. An explosion on a distant ship was recorded by the underwater recording device at a monitoring station. Thirteen seconds later the operator heard the explosion. How far was the ship from the station?
2. The primary wave and secondary wave of an earthquake travel away from the epicenter at rates of 8 and 4.8 kilometers per second, respectively. If the primary wave arrives at a seismic station 15 seconds ahead of the secondary wave, how far is the station from the epicenter?
3. John often flies from Clear Lake to Sun City and returns on the same day. On a windless day, he can average 120 miles per hour for the round trip, while on a windy day, he averaged 140 miles per hour one way and 100 miles per hour the other. It took him 15 minutes longer on the windy day. How far apart are Clear Lake and Sun City?
4. A classic puzzle problem goes like this: If a column of men 3 miles long is marching at 5 miles per hour, how long will it take a courier on a horse traveling at 25 miles per hour to deliver a message from the end of the column to the front and then return?
5. How long after 4:00 P.M. will the minute hand on a clock overtake the hour hand?
6. At what time between 4:00 and 5:00 do the hands of a clock form a straight line?
7. An old machine is able to do a certain job in 8 hours. Recently a new machine was installed. Working together, the two machines did the same job in 3 hours. How long would it take the new machine to do the job by itself?
8. Center City uses fire trucks to fill the village swimming pool. It takes one truck 3 hours to do the job and another 2 hours. How long would it take them working together?
9. An airplane takes off from a carrier at sea and flies west for 2 hours at 600 miles per hour. It then returns at 500 miles per hour. In the meantime, the carrier has traveled west at 30 miles per hour. When will the two meet?
10. A passenger train 480 feet long traveling at 75 miles per hour meets a freight train 1856 feet long traveling on parallel tracks at 45 miles per hour. How long does it take the trains to pass each other? (*Note*: Sixty miles per hour is equivalent to 88 feet per second.)

11. A medical company makes two types of heart valves, standard and deluxe. It takes 5 minutes on the lathe and 10 minutes on the drill press to make a standard valve, but 9 minutes on the lathe and 15 minutes on the drill press for a deluxe valve. On a certain day the lathe will be available for 4 hours and the drill press for 7 hours. How many valves of each kind should the company make that day if both machines are to be fully utilized?

[C] 12. Two boats travel at right angles to each other after leaving the dock at 1:00 P.M. At 3:00 P.M., they are 16 miles apart. If the first boat travels 6 miles per hour faster than the second, what are their rates?

13. A car is traveling at an unknown rate. If it traveled 15 miles per hour faster, it would take 90 minutes less to go 450 miles. How fast is the car going?

14. A boy walked along a level road for awhile, then up a hill. At the top of the hill he turned around and walked back to his starting point. He walked 4 miles per hour on level ground, 3 miles per hour uphill, and 6 miles per hour downhill, with the total trip taking 5 hours. How far did he walk altogether?

15. Jack and Jill live at opposite ends of the same street. Jack wanted to deliver a box at Jill's house and Jill wanted to leave some flowers at Jack's house. They started at the same moment, each walking at a constant speed. They met the first time 300 meters from Jack's house. On their return trip, they met 400 meters from Jill's house. How long is the street? (Assume that neither loitered at the other house nor when they met.)

[C] 16. Two cars are traveling toward each other on a straight road at the same constant speed. A plane flying at 350 miles per hour passes over the second car 2 hours after passing over the first car. The plane continues to fly in the same direction and is 2400 miles from the cars when they pass. Find the speed of the cars.

Science Problems

17. The illumination I in foot-candles on a surface d feet from a light source of c candlepower is given by the formula $I = c/d^2$. How far should an 80-candlepower light be placed from a surface to give the same illumination as a 20-candlepower light at 10 feet?

[C] 18. By experiment, it has been found that a car traveling v miles per hour will require d feet to stop, where $d = 0.044v^2 + 1.1v$. Find the velocity of a car if it took 176 feet to stop.

[C] 19. A bridge 200 feet long was built in the winter with no provision for expansion. In the summer, the supporting beams expanded in length by 8 inches, forcing the center of the bridge to drop. Assuming, for simplicity, that the bridge took the shape of a V, how far did the center drop? First guess at the answer and then work it out.

20. The distance s (in feet) traveled by an object in t seconds when it has an initial velocity v_0 and a constant acceleration a is given by $s = v_0t + \frac{1}{2}at^2$. An object was observed to travel 32 feet in 4 seconds and 72 feet in 6 seconds. Find the initial velocity and the acceleration.

21. A chemist has 5 kiloliters of 20 percent sulfuric acid solution. She wishes to increase its strength to 30 percent by draining off some and replacing it with 80 percent solution. How much should she drain off?

22. How many liters each of a 35 percent alcohol solution and a 95 percent alcohol solution must be mixed to obtain 12 liters of an 80 percent alcohol solution?

23. One atom of carbon combines with 2 atoms of oxygen to form one molecule of carbon dioxide. The atomic weights of carbon and oxygen are 12.0 and 16.0, respectively. How many milligrams of oxygen are required to produce 4.52 milligrams of carbon dioxide?

24. Four atoms of iron (atomic weight 55.85) combine with 6 atoms of oxygen (atomic weight 16.00) to form 2 molecules of rust. How many grams of iron would there be in 79.85 grams of rust?

[C] 25. A sample weighing .5000 grams contained only sodium chloride and sodium bromide. The chlorine and bromine from this sample were precipitated together as silver chloride and silver bromide. This precipitate weighed 1.100 grams. Sodium chloride is 60.6 percent chlorine, sodium bromide is 77.6 percent bromine, silver chloride is 24.7 percent chlorine, and silver bromide is 42.5 percent bromine. Calculate the weights of sodium chloride and sodium bromide in the sample.

Business Problems

26. Sarah Tyler bought stock in the ABC Company on Monday. The stock went up 10 percent on Tuesday and then dropped 10 percent on Wednesday. If she sold the stock on Wednesday for $1000, what did she pay on Monday?

27. If Jane Witherspoon has $4182 in her savings account and wants to buy stock in the ABC Company at $59 a share, how many shares can she buy and still maintain a balance of at least $2000 in her account?

28. Six men plan to take a charter flight to Bear Lake in Canada for a fishing trip, sharing the cost equally. They discover that if they took three more men along, each share of the original six would be reduced by $150. What is the total cost of the charter?

29. Susan has a job at Jenny's Nut Shop. Jenny asks Susan to prepare 25 pounds of mixed nuts worth $1.74 per pound by using walnuts valued at $1.30 per pound and cashews valued at $2.30 per pound. How many pounds of each kind should Susan use?

30. Alec Brown plans to sell toy gizmos at the state fair. He can buy them at 40¢ apiece and will sell them for 65¢ each. It will cost him $200 to rent a booth for the 7-day fair. How many gizmos must he sell to just break even?

31. The cost of manufacturing a product is the sum of fixed plant costs (real estate taxes, utilities, and so on) and variable costs (labor, raw materials, and so on) that depend on the number of units produced. The profit P that the company makes in a year is given by

$$P = TR - (FC + VC)$$

where TR is the total revenue (total sales), FC is the total fixed cost, and VC is the total variable cost. A company that makes one product has \$32,000 in total fixed costs. If the variable cost of producing one unit is \$4 and if units can be sold at \$6 each, find out how many units must be produced to give a profit of \$15,000.

32. The ABC company has total fixed costs of \$100,000 and total variable costs equal to 80 percent of total sales. What must the total sales be to yield a profit of \$40,000 (see Problem 31)?

33. Do Problem 32 assuming that the company pays 30 percent income taxes on its profit and wants a profit of \$40,000 after taxes.

34. The XYZ company has total fixed yearly costs of \$120,000 and last year had total variable costs of \$350,000 while selling 200,000 gizmos at \$2.50 each. Competition will force the manager to reduce the sales price to \$2.00 each next year. If total fixed costs and variable costs per unit are expected to remain the same, how many gizmos will the company have to sell to have the same profit as last year?

35. Podunk University wishes to maintain a student-faculty ratio of 1 faculty member for every 15 undergraduates and 1 faculty member for each 6 graduate students. It costs the university \$600 for each undergraduate student and \$900 for each graduate student over and above what is received in tuition. If the university expects \$2,181,600 in gifts (beyond tuition) next year and will have 300 faculty members, how many undergraduate and how many graduate students should it admit?

36. A department store purchased a number of smoke detectors at a total cost of \$2000. In unpacking them, the stock boys damaged 8 of them so badly that they could not be sold. The remaining detectors were sold at a profit of \$25 each, and a total profit of \$400 was realized when all of them were sold. How many smoke detectors were originally purchased?

Geometry Problems

37. A flag that is 6 feet by 8 feet has a blue cross of uniform width on a white background. The cross extends to all 4 edges of the flag. The area of the cross and the background are equal. Find the width of the cross.

38. If a right triangle has hypotenuse 2 units longer than one leg and 4 units longer than the other, find the dimensions of the triangle.

39. The area and perimeter of a right triangle are both 30. Find its dimensions.

40. Assume that the earth is a sphere of radius 4000 miles. How far is the horizon from an airplane 5 miles high?

41. Three mutually tangent circles have centers A, B, and C and radii a, b, and c, respectively. The lengths of the segments AB, BC, and CA are 13, 15, and 18, respectively. Find the lengths of the radii. Assume that each circle is outside of the other two.

42. A rectangle is inscribed in a circle of radius 5. Find the dimensions of the rectangle if its area is 40.

[C] 43. A ladder is standing against a house with its lower end 10 feet from the house. When the lower end is pulled 2 feet farther from the house, the upper end slides 3 feet down the house. How long is the ladder?

[C] 44. A trapezoid (a quadrilateral with two sides parallel) is inscribed in a square 12 inches on a side. One of the parallel sides is the diagonal of the square. If the trapezoid has area 24 square inches, how far apart are its parallel sides?

45. A 40-inch length of wire is cut in two. One of the pieces is bent to form a square and the other is bent to form a rectangle three times as long as wide. If the combined area of the square and the rectangle is $55\frac{3}{4}$ square inches, where was the wire cut?

CHAPTER SUMMARY

The equalities $(x + 1)^2 = x^2 + 2x + 1$ and $x^2 = 4$ are quite different in character. The first, called an **identity**, is true for all values of x. The second, called a conditional **equation**, is true only for certain values of x, in fact, only for $x = 2$ and $x = -2$. To **solve** an equation is to find those values of the unknown which make the equality true; it is one of the major tasks of mathematics.

The equation $ax + b = 0$ $(a \neq 0)$ is called a **linear equation** and has exactly one solution, $x = -b/a$. Similarly, $ax^2 + bx + c = 0$ $(a \neq 0)$ is a **quadratic equation** and usually has two solutions. Sometimes they can be found by **factoring** the left side and setting both factors equal to zero. Another method that always works is **completing the square**, but the best general method is simply substituting in the **quadratic formula**.

$$x = \frac{-b \pm \sqrt{b^2 - 4ac}}{2a}$$

Here $b^2 - 4ac$, called the **discriminant**, plays a critical role. The equation has two real solutions, one real solution, or two nonreal solutions according as the discriminant is positive, zero, or negative.

Equations arise naturally in the study of word problems. Such problems may lead to one equation in one unknown but often lead to a **system** of several equations in several unknowns. In the latter case, our task is to find the values of the unknowns that satisfy all the equations of the system simultaneously.

Inequalities look like equations with the equal sign replaced by $<$, $\leq$, $>$, or $\geq$. The methods for solving **conditional inequalities** are very similar to those for conditional equations. One difference is that the direction of an inequality sign is reversed upon multiplication or division by a negative number. Another is that the set of solutions consists of one or more **intervals** of numbers, rather than a finite set. For example, the inequality $3x - 2 < 5$ has the solution set $\{x: x < \frac{7}{3}\}$.

CHAPTER REVIEW PROBLEM SET

1. Which are identities and which are conditional equations?
   (a) $3(x - 2) = 3x - 6$ (b) $3x - 2 = x - 6$
   (c) $(x + 2)^2 = x^2 + 4$ (d) $x^2 + 5x + 6 = (x + 3)(x + 2)$
2. Solve the following equations.
   (a) $3(x + \frac{1}{2}) = x - \frac{1}{3}$ (b) $\dfrac{6}{x - 5} = \dfrac{21}{x}$
   (c) $(x - 1)(2x + 1) = (2x - 3)(x + 2)$ (d) $\dfrac{x}{2x + 2} - 1 = \dfrac{8 - 3x}{6x + 6}$
3. In $s = \frac{1}{2}at^2 + v_0t$, solve for v_0 in terms of the other variables.
4. How cold is it in degrees Celsius when the Celsius reading is twice the Fahrenheit reading?
5. Solve the systems below.
   (a) $2x - 3y = 7$, $x + 4y = -2$ (b) $\frac{1}{3}x - \frac{5}{6}y = 2$, $\frac{1}{2}x + y = -\frac{3}{2}$
6. Simplify.
   (a) $\dfrac{6 + \sqrt{18}}{12}$ (b) $\dfrac{10 - \sqrt{300}}{2}$ (c) $\dfrac{-2 + \sqrt{-8}}{6}$

Solve the quadratic equations in Problems 7–16 by any method you choose.

7. $x^2 = 49$
8. $(x - 2)(x + 5) = 0$
9. $(x + 3)^2 = 0$
10. $(x - 2)^2 = 25$
11. $(2x + 1)^2 = 81$
12. $x^2 - 9x + 20 = 0$
13. $x^2 = 4x$
14. $x^2 + 2x - 4 = 0$
15. $3x^2 + x - 1 = 0$
16. $x^2 + mx + 2n = 0$

Solve for y in terms of x.

17. $(y - 2x)^2 = 4$
18. $(2y + 3x)^2 - 4(2y + 3x) + 3 = 0$

Solve the inequalities in Problems 19–22.

19. $-5x + 3 \geq 2x - 9$
20. $x^2 + 5x - 6 \geq 0$
21. $x^2 + x - 3 < 0$
22. $\dfrac{x - 4}{x + 1} > 0$
23. Jill Garcia has $10,000 to invest. How much should she put in the bank at 6 percent interest and how much in the credit union at 8 percent interest if she hopes to have $730 interest at the end of one year?
24. A fast train left Chicago at 6:00 P.M., traveling at 60 miles per hour. At 8:00 A.M., a slower train left St. Paul traveling at a constant rate. If these two cities are 450 miles apart and the two trains crashed head on at 11:00 A.M., what was the rate of the slower train?
25. John Appleseed rowed upstream for a distance of 4 miles in 2 hours. If he had rowed twice as hard and the current had been half as strong, he could have done it in $\frac{4}{7}$ hour. What was the rate of the current?

And so Fermat and Descartes turned to the application of algebra to the study of geometry. The subject they created is called coordinate, or analytic, geometry; its central idea is the association of algebraic equations with curves and surfaces. This creation ranks as one of the richest and most fruitful veins of thought ever struck in mathematics.

Morris Kline

CHAPTER FOUR

Coordinates and Curves

4-1 The Cartesian Coordinate System

4-2 Algebra and Geometry United

4-3 The Straight Line

4-4 The Parabola

4-5 Ellipses and Hyperbolas (Optional)

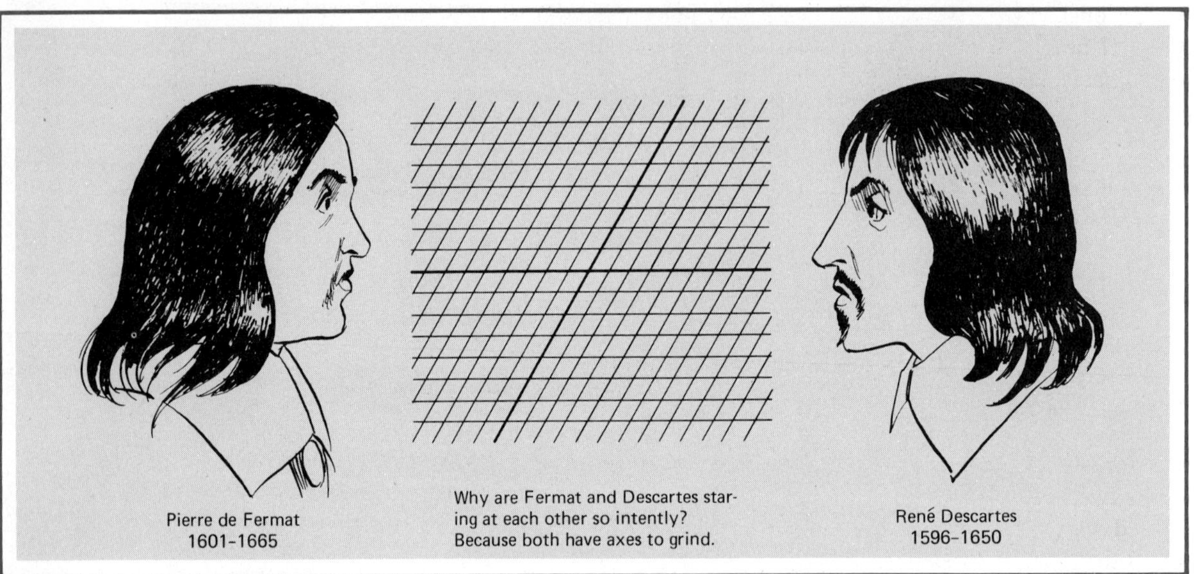
Pierre de Fermat
1601-1665

Why are Fermat and Descartes staring at each other so intently? Because both have axes to grind.

René Descartes
1596-1650

4-1 The Cartesian Coordinate System

Two Frenchmen deserve credit for the idea of a coordinate system. Pierre de Fermat was a lawyer who made mathematics his hobby. In 1629, he wrote a paper which makes explicit use of coordinates to describe points and curves. René Descartes was a philosopher who thought mathematics could unlock the secrets of the universe. He published his *La Géométrie* in 1637. It is a famous book and though it does emphasize the role of algebra in solving geometric problems, one finds only a hint of coordinates there. By virtue of having the idea first and more explicitly, Fermat ought to get the major credit. History can be a fickle friend; coordinates are known as cartesian coordinates, named after René Descartes.

No matter who gets the credit, it was an idea whose time had come. It made possible the invention of calculus, one of the greatest inventions of the human mind. That invention was to come in 1665 at the hands of a 23-year-old genius named Isaac Newton. You will probably study calculus later on. There you will use the ideas of this chapter over and over.

Review of the Real Line Recall the real line, which was introduced in Section 1-3.

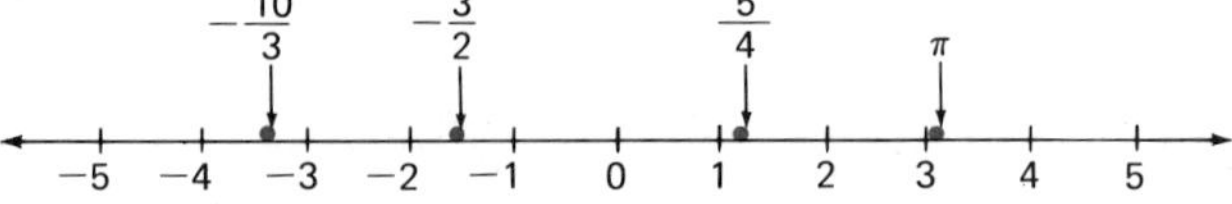

Every point on this line can be given a label, a real number, which specifies exactly where the point is. We call this label the **coordinate** of the point.

Consider now two points A and B with coordinates a and b, respectively. We will need a formula for the distance between A and B in terms of the coordinates a and b. The formula is

$$d(A, B) = |b - a|$$

and it is correct whether A is to the right or to the left of B. Note the two examples in the margin. In the first case,

$$d(A, B) = |3 - (-2)| = |5| = 5$$

In the second,

$$d(A, B) = |-1 - 5| = |-6| = 6$$

Cartesian Coordinates In the plane, produce two copies of the real line, one horizontal and the other vertical, so that they intersect at the zero points of the two lines. The two lines are called **coordinate axes**; their intersection is labeled with O and is called the **origin**. By convention, the horizontal line is called the x-**axis** and the vertical line is called the y-**axis**. The positive half of the x-axis is to the right; the positive half of the y-axis is upward. The coordinate axes divide the plane into four regions called **quadrants**, labeled I, II, III, and IV, as shown in the margin.

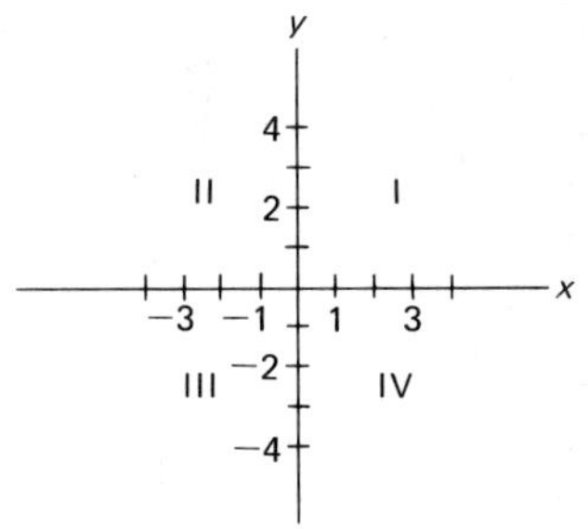

Each point P in the plane can now be assigned a pair of numbers called its **cartesian coordinates**. If vertical and horizontal lines through P intersect the x- and y-axes at a and b, respectively, then P has coordinates (a, b). We

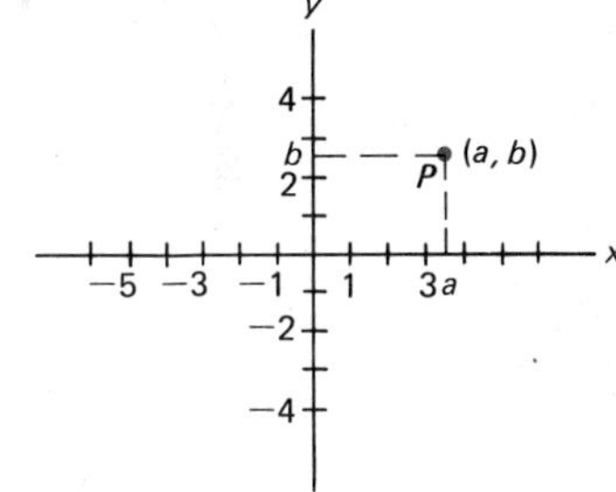

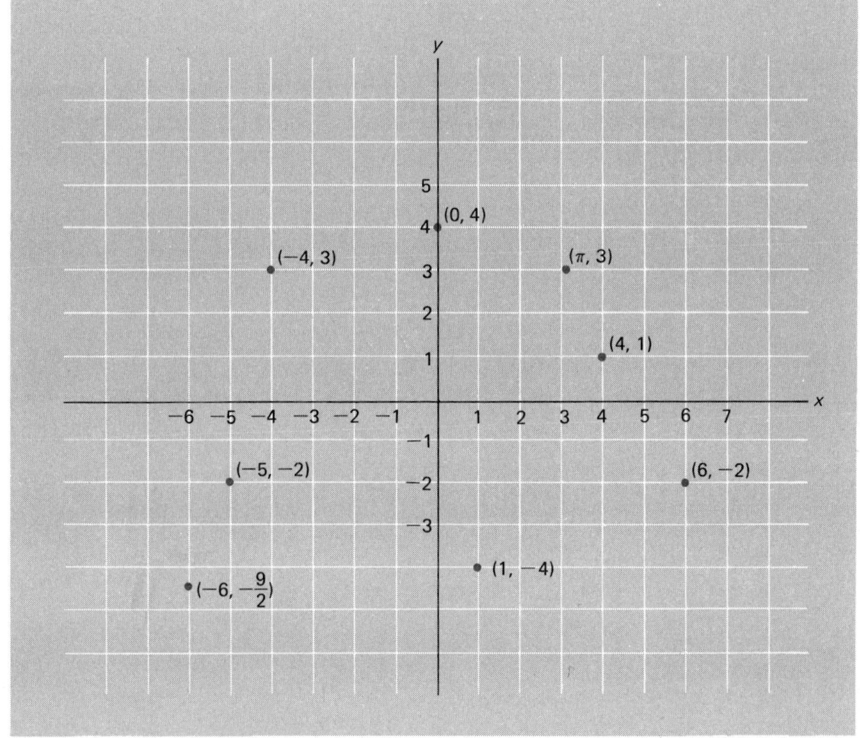

call (a, b) an **ordered pair** of numbers because it makes a difference which number is first. The first number a is the x-**coordinate** (or abscissa); the second number b is the y-**coordinate** (or ordinate).

Conversely, take any ordered pair (a, b) of real numbers. The vertical line through a on the x-axis and the horizontal line through b on the y-axis meet in a point P whose coordinates are (a, b).

Think of it this way: The coordinates of a point are the address of that point. If you have found a house (or a point), you can read its address. Conversely, if you know the address of a house (or a point), you can always locate it. In the diagram at the bottom of page 133, we have shown the coordinates (addresses) of several points.

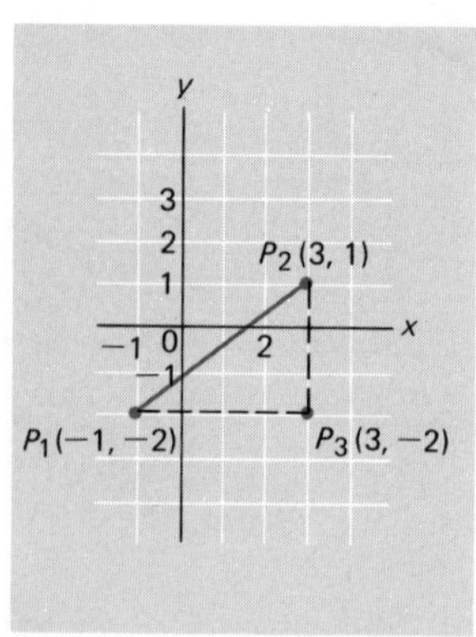

The Distance Formula Consider the points P_1 and P_2 with coordinates $(-1, -2)$ and $(3, 1)$, respectively. The segment joining P_1 and P_2 is the hypotenuse of a right triangle with right angle at $P_3(3, -2)$ (see the diagram in the margin). We easily calculate the lengths of the two legs.

$$d(P_1, P_3) = |3 - (-1)| = 4 \qquad d(P_2, P_3) = |1-(-2)| = 3$$

By the Pythagorean Theorem (see Section 1-3),

$$d(P_1, P_2) = \sqrt{4^2 + 3^2} = \sqrt{25} = 5$$

Next consider two arbitrary points $P_1(x_1, y_1)$ and $P_2(x_2, y_2)$ that are not on the same horizontal or vertical line. They determine a right triangle with legs of length $|x_2 - x_1|$ and $|y_2 - y_1|$. By the Pythagorean Theorem,

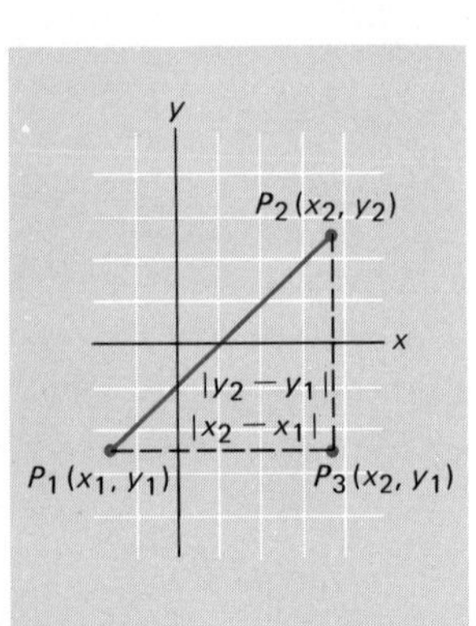

$$d(P_1, P_2) = \sqrt{(x_2 - x_1)^2 + (y_2 - y_1)^2}$$

This formula is known as the **distance formula**. You should check that it is valid even if P_1 and P_2 lie on the same vertical or horizontal line.

The Midpoint Formula Consider two points $A(x_1, y_1)$ and $B(x_2, y_2)$ in the plane and let $P(x, y)$ be the midpoint of the segment joining them. Drop perpendiculars from A, P, and B to the x-axis as shown in the diagram. Then x is midway between x_1 and x_2, so

$$x - x_1 = x_2 - x$$
$$2x = x_1 + x_2$$
$$x = \frac{x_1 + x_2}{2}$$

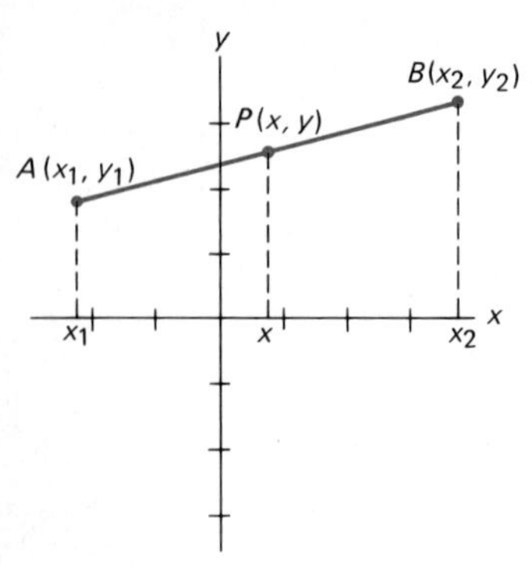

Similar reasoning applies to y. The result, called the **midpoint formula**, says that the coordinates (x, y) of the midpoint P are given by

$$x = \frac{x_1 + x_2}{2} \qquad y = \frac{y_1 + y_2}{2}$$

For example, the midpoint of the segment joining $A(-3, -2)$ and $B(5, 9)$ has coordinates.

$$\left(\frac{-3+5}{2}, \frac{-2+9}{2}\right) = \left(1, \frac{7}{2}\right)$$

Problem Set 4-1

Find $d(A, B)$ where A and B are points on the number line having the given coordinates a and b.

1. $a = -3$, $b = 2$
2. $a = 5$, $b = -6$
3. $a = \frac{11}{4}$, $b = -\frac{5}{4}$
4. $a = \frac{31}{8}$, $b = \frac{13}{8}$
5. $a = 3.26$, $b = 4.96$
6. $a = -1.45$, $b = -5.65$
7. $a = 2 - \pi$, $b = \pi - 3$
8. $a = 4 + 2\sqrt{3}$, $b = -6 + \sqrt{3}$

Again, A and B are points on the number line. Given a and $d(A, B)$, what are the two possible values of b?

9. $a = 5$, $d(A, B) = 2$
10. $a = 9$, $d(A, B) = 5$
11. $a = -2$, $d(A, B) = 4$
12. $a = 3$, $d(A, B) = 7$
13. $a = \frac{5}{2}$, $d(A, B) = \frac{3}{4}$
14. $a = 1.8$, $d(A, B) = 2.4$

Plot A, B, C, and D on a coordinate system. Name the quadrilateral $ABCD$. That is, is it a square, a rectangle, or what?

15. $A(4, 3)$, $B(4, -3)$, $C(-4, 3)$, $D(-4, -3)$
16. $A(-1, 6)$, $B(0, 5)$, $C(3, 2)$, $D(5, 0)$
17. $A(1, 3)$, $B(2, 6)$, $C(4, 7)$, $D(3, 4)$
18. $A(0, 2)$, $B(3, 0)$, $C(2, -1)$, $D(-1, 1)$

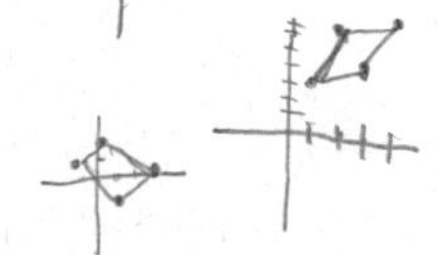

In Problems 19–26, find $d(P_1, P_2)$ where P_1 and P_2 have the given coordinates. Also find the coordinates of the midpoint of P_1P_2.

19. $(2, -1)$, $(5, 3)$
20. $(2, 1)$, $(7, 13)$
21. $(4, 2)$, $(2, 4)$
22. $(-1, 5)$, $(6, 7)$
23. $(\sqrt{3}, 0)$, $(0, \sqrt{6})$
24. $(\sqrt{2}, 0)$, $(0, -\sqrt{7})$
25. [C] $(1.234, -5.132)$, $(6.714, 8.341)$
26. [C] $(-42.1, 16.3)$, $(12.2, -5.3)$
27. The points A, B, C, and D of Problem 17 form a parallelogram.
    (a) Find $d(A, B)$, $d(B, C)$, $d(C, D)$, and $d(D, A)$.
    (b) Find the coordinates of the midpoints of the diagonals AC and BD of the parallelogram $ABCD$.
    (c) Your answers to parts (a) and (b) agree with what facts about a parallelogram?
28. The points $(3, -1)$ and $(3, 3)$ are two vertices of a square. Give two pairs of other possible vertices. Can you give a third pair?
29. Let $ABCD$ be a rectangle whose sides are parallel to the coordinate axes. Find the coordinates of B and D if the coordinates of A and C are:
    (a) $(-2, 0)$ and $(4, 3)$ (Draw a picture.)
    (b) $(2, -1)$ and $(8, 7)$

30. Show that the triangle whose vertices are (5, 3), (−2, 4) and (10, 8) is isosceles.

31. Use the distance formula to show that the triangle whose vertices are (2, −4), (4, 0), and (8, −2) is a right triangle.

32. (a) Find the point on the y-axis that is equidistant from the points (3, 1) and (6, 4). (*Hint*: Let the unknown point be $(0, y)$.)
    (b) Find the point on the x-axis that is equidistant from (3, 1) and (6, 4).

33. Use the distance formula to show that the three points are on a line.
    (a) (0, 0), (3, 4), (−6, −8)
    (b) (−4, 1), (−1, 5), (5, 13)

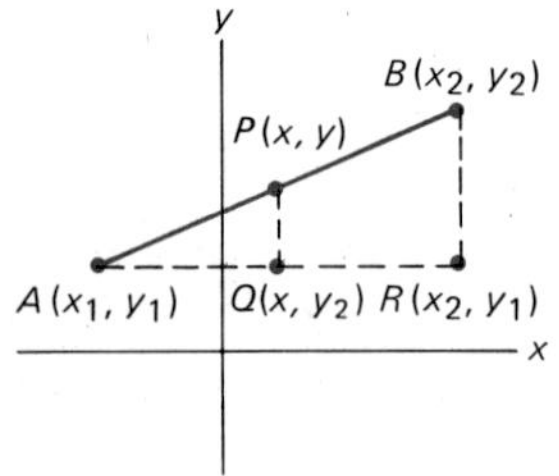

Example (Point of division formula) Let t be a number between 0 and 1 and let $A(x_1, y_1)$ and $B(x_2, y_2)$ be two points in the plane. Find the point $P(x, y)$ on the segment AB for which

$$\frac{d(A, P)}{d(A, B)} = t$$

(For example, if $t = \frac{3}{4}$, we seek the point $P(x, y)$ which is $\frac{3}{4}$ of the way from A to B.)

SOLUTION. Introduce points $Q(x, y_1)$ and $R(x_2, y_1)$ as shown in the diagram. The triangles APQ and ABR are similar, so

$$\frac{d(A, Q)}{d(A, R)} = \frac{d(A, P)}{d(A, B)} = t$$

Now $d(A, Q) = |x - x_1|$ and $d(A, R) = |x_2 - x_1|$. Moreover, $x - x_1$ and $x_2 - x_1$ are either both positive (as in our picture) or both negative (as in the case where A is to the right of B). Thus

$$\frac{d(A, Q)}{d(A, R)} = \frac{|x - x_1|}{|x_2 - x_1|} = \frac{x - x_1}{x_2 - x_1} = t$$

Solving this for x yields

$$\begin{aligned} x - x_1 &= t(x_2 - x_1) \\ x &= x_1 + tx_2 - tx_1 \\ x &= (1 - t)x_1 + tx_2 \end{aligned}$$

In a similar manner, we can find y. The results are summarized below.

$$\begin{aligned} x &= (1 - t)x_1 + tx_2 \\ y &= (1 - t)y_1 + ty_2 \end{aligned}$$

Note that when $t = \frac{1}{2}$, this is the midpoint formula.

Use the formula just developed to find $P(x, y)$ for the given values of t and given points A and B. Plot the points A, P, and B.

34. $A(5, -8)$, $B(11, 4)$, $t = \frac{1}{3}$
35. $A(5, -8)$, $B(11, 4)$, $t = \frac{2}{3}$
36. $A(5, -8)$, $B(11, 4)$, $t = \frac{5}{6}$
37. $A(4, 9)$, $B(104, 209)$, $t = \frac{13}{100}$

Miscellaneous Problems

38. Let A and B be points on the number line with coordinates -3 and 5, respectively.
    (a) Find the coordinate of the midpoint of AB.
    (b) Find the coordinate of C if B is the midpoint of the segment AC.
39. Let A and B be the points on the number line with coordinates -4 and 6, respectively. Find the coordinates of the three points which divide AB into 4 equal parts.
40. Find the lengths of the sides of the triangle with vertices $A(2, -4)$, $B(3, 2)$, and $C(-5, 1)$.

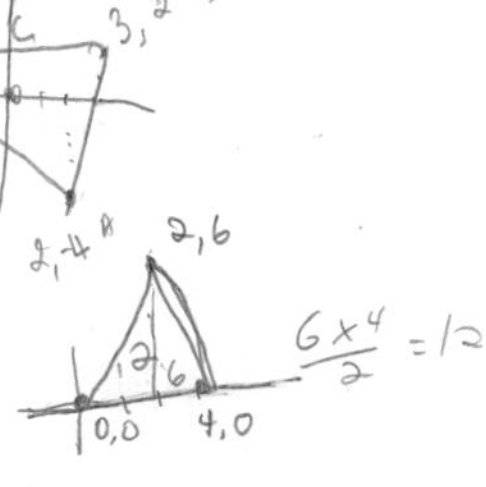

41. Given the points $A(2, 3)$ and $B(-4, 7)$.
    (a) Find the coordinates of the midpoint of the segment AB.
    (b) Find the coordinates of C if B is the midpoint of the segment AC.
42. Find x such that the distance between the points $(3, 0)$ and $(x, 3)$ is 5.
43. Find the area of the triangle with vertices $(0, 0)$, $(4, 0)$, and $(2, 6)$.
44. Consider the triangle with vertices $A(0, 0)$, $B(4, 0)$, and $C(0, 5)$.
    (a) Find the area of this triangle.
    (b) Find the length of BC.
    (c) Find the length of the altitude from A to BC.

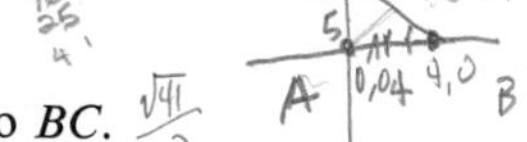

45. Given $A(-4, 3)$ and $B(21, 38)$, find the coordinates of the four points that divide AB into five equal parts. (*Hint*: For the point closest to A, use $t = \frac{1}{5}$.)
46. Consider the figure in the margin.
    (a) Find the coordinates of the midpoints L, M, and N of the sides of the triangle OAB.
    (b) The segment OM is called the median from vertex O. Find the coordinates of the point on OM that is $\frac{2}{3}$ of the way from O to M.
    (c) Answer the question in part (b) for the medians from A to L and from B to N.
    (d) What fact about the medians of a triangle is illustrated by your results in parts (b) and (c)?

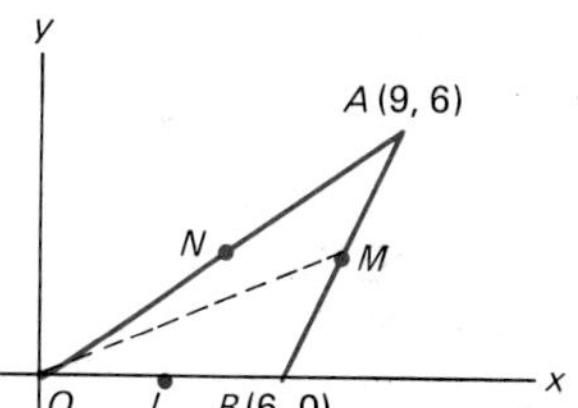

Ⓒ 47. Cities A, B, and C are vertices of a right triangle, as shown in the margin. AB and BC are roads of lengths 214 and 179 miles, respectively. An airline flies route AC, which is not a road. It costs \$3.71 per mile to ship a certain product by truck and \$4.82 per mile by air. Calculate the costs for both methods of transportation in order to see which is cheaper.

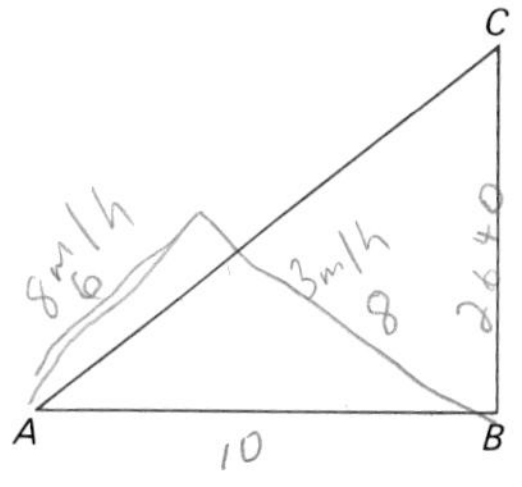

Ⓒ 48. City B is 10 miles downstream from city A and on the opposite side of a river $\frac{1}{2}$ mile wide. Tom O'Shanter will run from city A along the river for 6 miles, then swim diagonally to city B. If he runs at 8 miles per hour and swims at 3 miles per hour, how long will it take him to get from city A to city B?

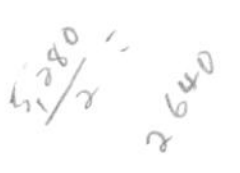

4-2 Algebra and Geometry United

The Greeks were preeminent geometers but poor algebraists. Though they were able to solve a host of geometry problems, their limited algebraic skills kept others beyond their grasp. By 1600, geometry was a mature and eligible bachelor. Algebra was a young woman only recently come of age. Fermat and Descartes were the matchmakers; they brought the two together. The resulting union is called **analytic geometry**, or coordinate geometry.

The Graph of an Equation An equation is an algebraic object. By means of a coordinate system, it can be transformed into a curve, a geometric object. Here is how it is done.

Consider the equation $y = x^2 - 3$. Its set of solutions is the set of ordered pairs (x, y) that satisfy the equation. These ordered pairs are the coordinates of points in the plane. The set of all such points is called the **graph** of the equation. The graph of an equation in two variables x and y will usually be a curve.

To obtain this graph, we follow a definite procedure.

1. Obtain the coordinates of a few points.
2. Plot those points in the plane.
3. Connect the points with a smooth curve in the order of increasing x values.

The best way to do step 1 is to make a **table of values**. Assign values to one of the variables, say x, determine the corresponding values of the other, and then list the pairs of values in tabular form. The whole three-step procedure is illustrated below for the previously mentioned equation, $y = x^2 - 3$.

$y = x^2 - 3$

x	y
-3	6
-2	1
-1	-2
0	-3
1	-2
2	1
3	6

Step 1
Make a table of values

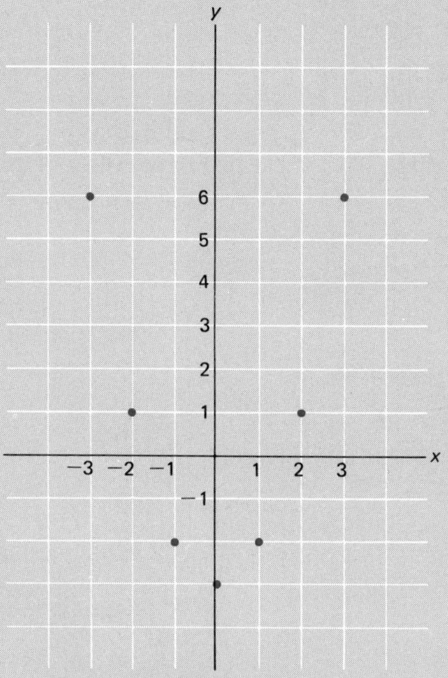

Step 2
Plot those points

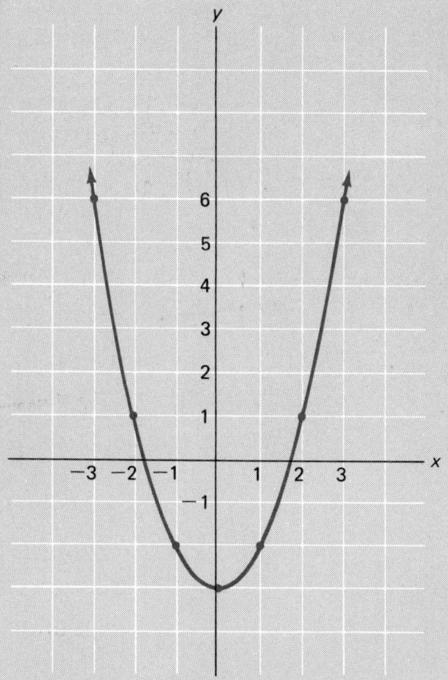

Step 3
Connect those points with a smooth curve

Of course, you need to use common sense and even a little faith. When you connect the points you have plotted with a smooth curve, you are assuming that the curve behaves nicely between consecutive points; that is faith. That is why you should plot enough points so the outline of the curve seems very clear; the more points you plot, the less faith you will need. Also you should recognize that you can seldom display the whole curve. In our example, the curve has infinitely long arms opening wider and wider. But our graph does show the essential features. That is what we always aim to do—show enough of the graph so the essential features are visible.

Symmetry of a Graph The graph of $y = x^2 - 3$, drawn above and again in the margin, has a nice property of symmetry. If the coordinate plane were folded along the y-axis, the two branches would coincide. For example, (3, 6) would coincide with $(-3, 6)$; (2, 1) would coincide with $(-2, 1)$; and, more generally, (x, y) would coincide with $(-x, y)$. Algebraically, this corresponds to the fact that we may replace x by $-x$ in the equation $y = x^2 - 3$ without changing it.

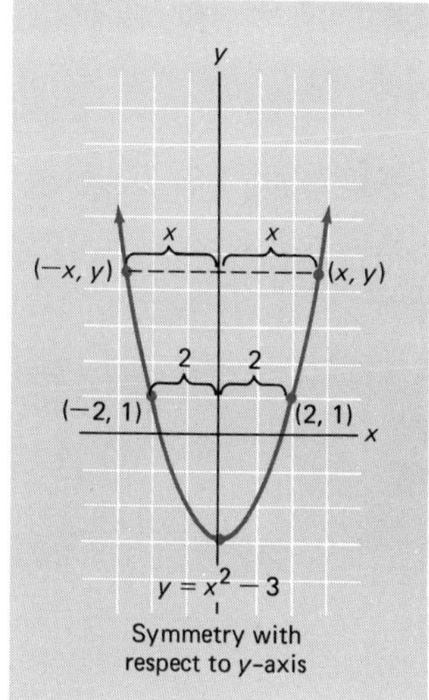

Symmetry with respect to y-axis

Whenever an equation is unchanged by replacing (x, y) with $(-x, y)$, the graph of the equation is **symmetric with respect to the y-axis**. Likewise, if the equation is unchanged when (x, y) is replaced by $(x, -y)$, its graph is **symmetric with respect to the x-axis**. The equation $x = 1 + y^2$ is of the latter type; its graph is shown in the margin.

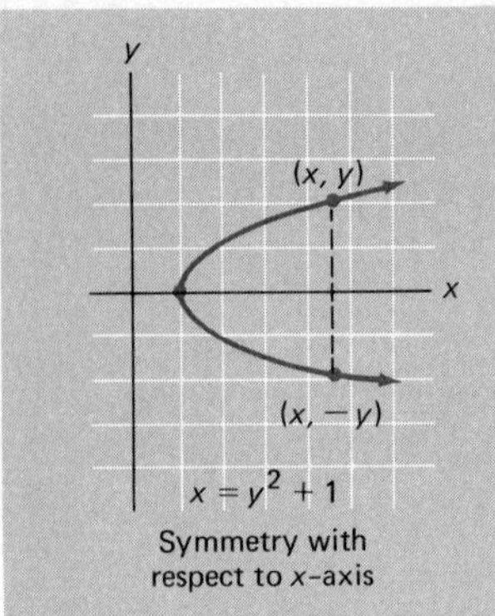

Symmetry with respect to x-axis

A third type of symmetry is **symmetry with respect to the origin**. It occurs whenever replacing (x, y) by $(-x, -y)$ produces no change in the equation. The equation $y = x^3$ is a good example as $-y = (-x)^3$ is equivalent to $y = x^3$. The graph is shown below. Note that the dotted line segment from $(-x, -y)$ to (x, y) is bisected by the origin.

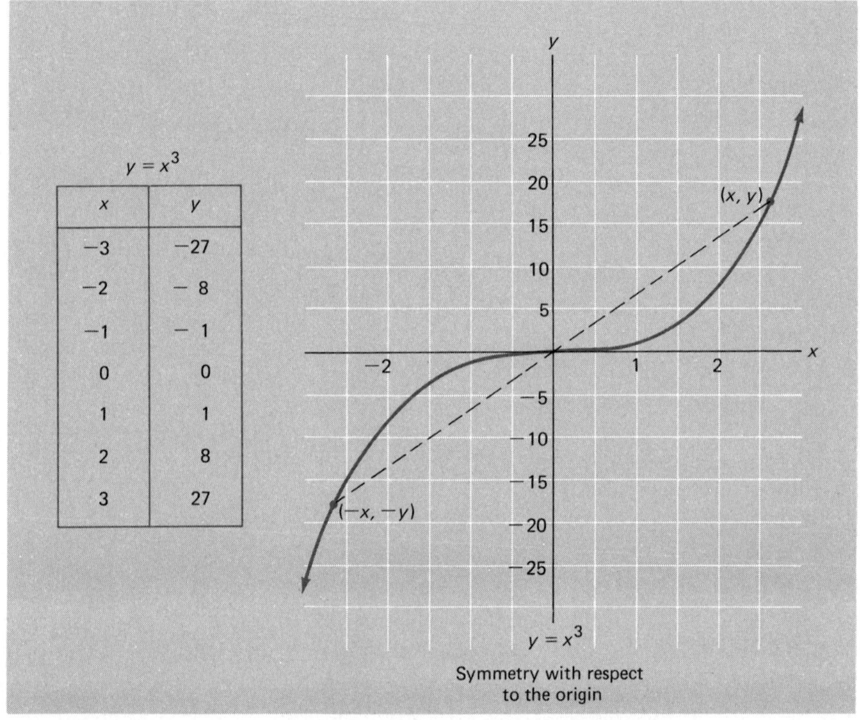

$y = x^3$

x	y
−3	−27
−2	− 8
−1	− 1
0	0
1	1
2	8
3	27

Symmetry with respect to the origin

In graphing $y = x^3$, we used a smaller scale on the y-axis than on the x-axis. This made it possible to show a larger portion of the graph. We suggest that before putting scales on the two axes, you should examine your table of values. Choose scales so that all of your points can be plotted and still keep your graph of reasonable size.

Graphing an equation is an extremely important operation. It gives us a picture to look at. Most of us can absorb qualitative information from a picture much more easily than from symbols. But if we want precise quantitative information, then symbols are better; they are easier to manipulate. That is why we must be able to reverse the process just described, which is our next topic.

The Equation of a Graph A graph is a geometric object, a picture. How can we turn it into an algebraic object, an equation? Sometimes it is easy, but not always. As an example, consider a circle of radius 3. It consists of all points 3 units from a fixed point, called the center. The picture in the margin shows this circle with its center at the origin of a coordinate system.

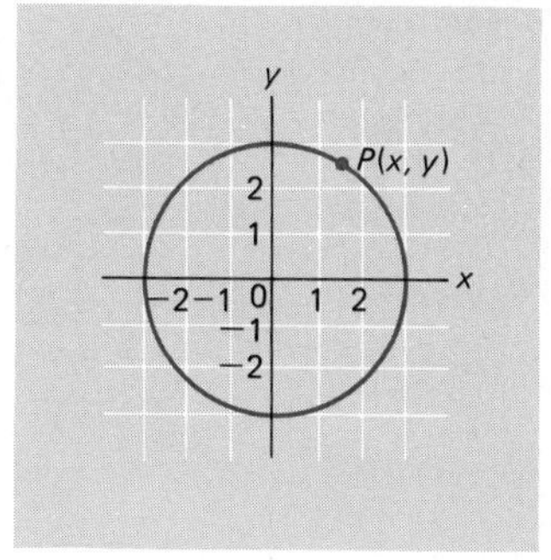

Take *any* point P on the circle and label its coordinates (x, y). It must satisfy the equation

$$d(P, O) = 3$$

From the distance formula of the previous section, we have

$$\sqrt{(x-0)^2 + (y-0)^2} = 3$$

or equivalently (after squaring both sides)

$$x^2 + y^2 = 9$$

This is the equation we sought.

We could move to other types of curves, attempting to find their equations. In fact, we will do exactly that in succeeding sections. Right now, we shall consider more general circles, that is, circles with arbitrary radii and arbitrary centers.

The Standard Equation of a Circle Consider a circle of radius r with center at (a, b). To find its equation, take an arbitrary point on the circle with coordinates (x, y). According to the distance formula, it must satisfy the equation

$$\sqrt{(x-a)^2 + (y-b)^2} = r$$

or, equivalently,

$$(x-a)^2 + (y-b)^2 = r^2$$

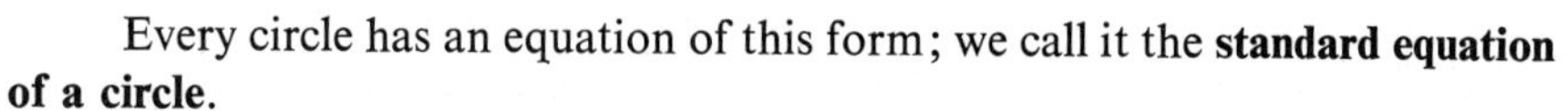

Every circle has an equation of this form; we call it the **standard equation of a circle.**

As an example, let us find the equation of a circle of radius 6 centered at $(2, -1)$. We use the boxed equation with $a = 2$, $b = -1$, and $r = 6$. This gives

$$(x-2)^2 + (y+1)^2 = 36$$

Let us go the other way. What circle has the equation

$$(x+3)^2 + (y-4)^2 = 49$$

This is a circle with center at $(-3, 4)$ and radius 7.

Consider this last equation again. Notice that it could be written as

$$x^2 + 6x + 9 + y^2 - 8y + 16 = 49$$

or equivalently as

$$x^2 + y^2 + 6x - 8y = 24$$

A natural question to ask is whether every equation of the form

$$x^2 + y^2 + Dx + Ey = F$$

is the equation of a circle. Take

$$x^2 + y^2 - 6y + 16x = 8$$

as an example. Recalling a skill we learned in Section 3-4 (completing the square), we may rewrite this as

$$(x^2 + 16x + \quad) + (y^2 - 6y + \quad) = 8$$

or

$$(x^2 + 16x + 64) + (y^2 - 6y + 9) = 8 + 64 + 9$$

or

$$(x + 8)^2 + (y - 3)^2 = 81$$

We recognize this to be the equation of a circle with center at $(-8, 3)$ and radius 9. In fact, we will always have a circle unless the number on the right side in the last step is negative or zero.

Problem Set 4-2 *Graph each of the following equations, showing enough of the graph to bring out its essential features. Note any of the three kinds of symmetry discussed in the text.*

1. $y = 3x - 2$
2. $y = 2x + 1$
3. $y = -x^2 + 4$
4. $y = -x^2 - 2x$
5. $y = x^2 - 4x$
6. $y = x^3 + 2$
7. $y = -x^3$
8. $y = \dfrac{12}{x^2 + 4}$
9. $y = \dfrac{4}{x^2 + 1}$
10. $y = x^3 + x$

Example (More graphing) Graph the equation $x = y^2 - 2y + 4$.

SOLUTION. Assign values to y and calculate the corresponding x.

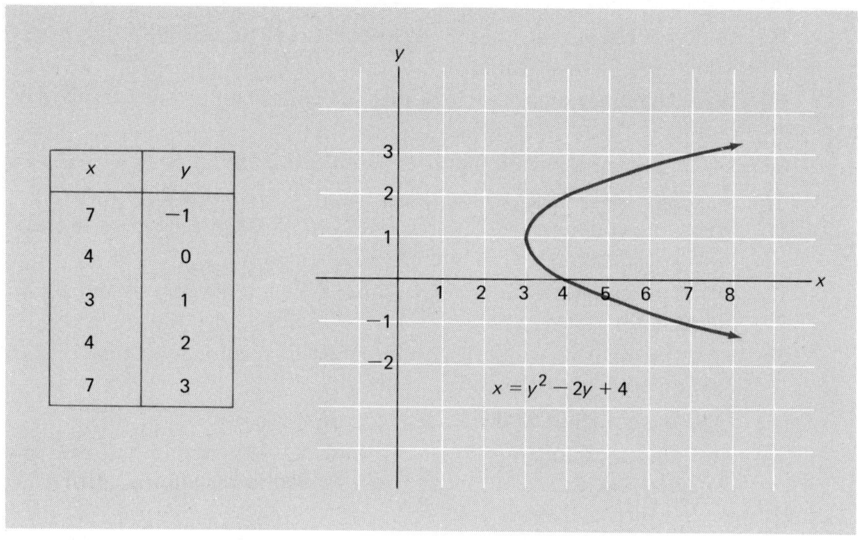

x	y
7	−1
4	0
3	1
4	2
7	3

Note that this graph is symmetric with respect to the line $y = 1$.

Graph each of the following equations.

11. $x = 2y - 1$
12. $x = -3y + 1$
13. $x = -2y^2$
14. $x = 2y - y^2$
15. $x = y^3$
16. $x = 8 - y^3$

Write the equation of the circle with the given center and radius.

17. Center (0, 0), radius 6.
18. Center (2, 3), radius 3.
19. Center (4, 1), radius 5.
20. Center (2, −1) radius $\sqrt{7}$.
21. Center (−2, 1), radius $\sqrt{3}$.
22. Center (π, $\frac{3}{4}$), radius $\frac{1}{2}$.

Graph the following equations.

23. $(x - 2)^2 + y^2 = 16$
24. $(x - 2)^2 + (y + 2)^2 = 25$
25. $(x + 1)^2 + (y - 3)^2 = 64$
26. $(x + 4)^2 + (y + 6)^2 = \frac{49}{4}$

Find the center and radius of each of the following circles. (*Hint*: Complete the squares.)

27. $x^2 + y^2 + 2x - 10y + 25 = 0$
28. $x^2 + y^2 - 6y = 16$
29. $x^2 + y^2 - 12x + 35 = 0$
30. $x^2 + y^2 - 10x + 10y = 0$
31. $4x^2 + 4y^2 + 4x - 12y + 1 = 0$
32. $3x^2 + 3y^2 - 2x + 4y = \frac{20}{3}$

Miscellaneous Problems

Sketch the graphs of the equations in 33–37.

33. $y = 12/x$ 34. $y = 3 + 4/x^2$ 35. $y = 2(x - 1)^2$

36. $x = -y^2 + 8$ 37. $x = 4y - y^2$

38. Which of the graphs in 33–37 are symmetric with respect to the y-axis? The x-axis? The origin?

39. Write the equation of the circle with center at (0, 0) and passing through (4, 5).

40. Write the equation of the circle with center at (2, −3) and passing through (5, 2).

41. Write the equation of the circle with AB as diameter where A and B have coordinates (−3, 7) and (3, −7).

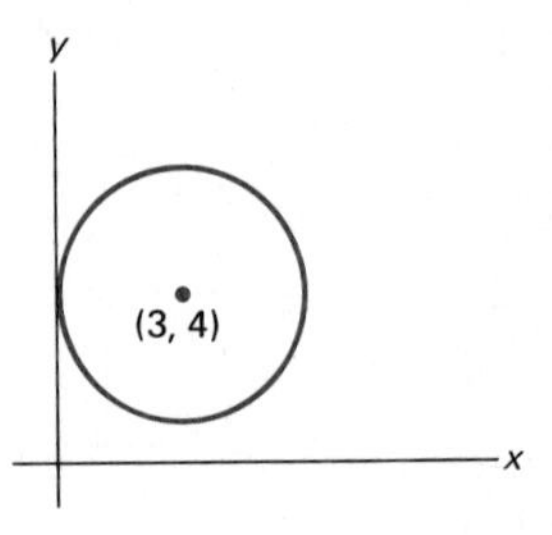

42. Write the equation of the circle with diameter AB where A and B have coordinates (3, −1) and (−5, 7).

43. Find the equation of the circle with center on the line $y = x$ and tangent to the y-axis at (0, 6).

44. Find the equation of the circle pictured in the margin.

45. The points (2, 3), (6, 3), (6, −1), and (2, −1) are the corners of a square. Find the equation of the inscribed circle of the square (the circle which is tangent to all four sides).

46. Find the equation of the circumscribed circle of the square in Problem 45. (This circle passes through all four corner points.)

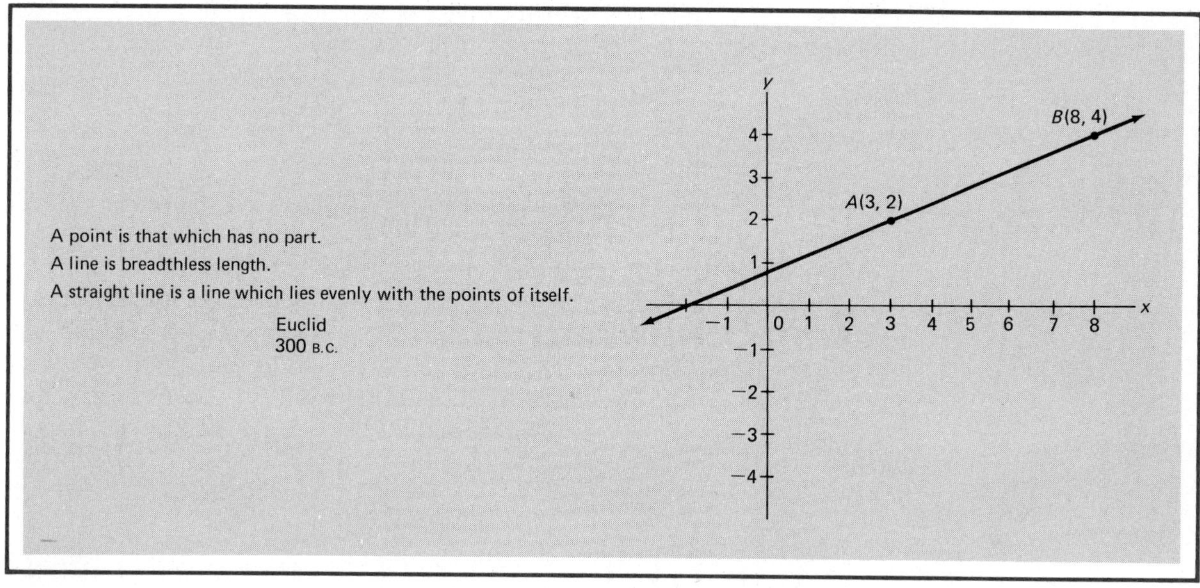

4-3 The Straight Line

Euclid's definition of a straight line is not very helpful, but neither are most of the alternatives we have heard. Fortunately, we all know what we mean by a straight line even if we cannot seem to describe it in terms of more primitive ideas. There is one thing on which we must agree: Given two points (for example, A and B above), there is one and only one straight line that passes through them. And contrary to Euclid, let us agree that the word *line* shall always mean straight line.

A line is a geometric object. When it is placed in a coordinate system, it ought to have an equation just as a circle does. How do we find the equation of a line? To answer this question we will need the notion of slope.

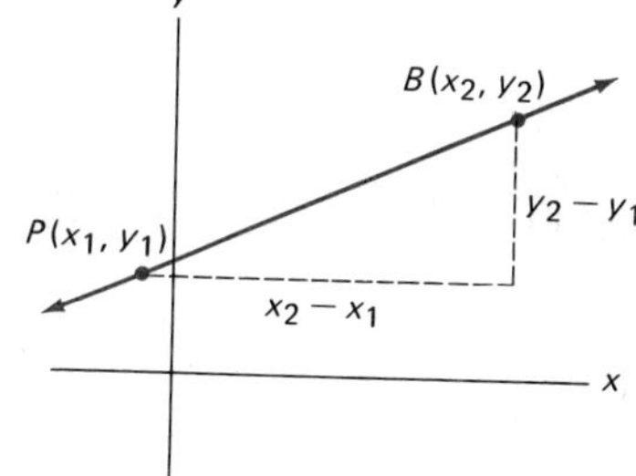

The Slope of a Line Consider the line in our opening diagram. From point A to point B, there is a **rise** (vertical change) of 2 units and a **run** (horizontal change) of 5 units. We say that the line has a slope of $\frac{2}{5}$. In general, for a line through $A(x_1, y_1)$ and $B(x_2, y_2)$, where $x_1 \neq x_2$, we define the **slope** m of that line by

$$m = \frac{\text{rise}}{\text{run}} = \frac{y_2 - y_1}{x_2 - x_1}$$

You should immediately raise a question. A line has many points. Does the value we get for the slope depend on which pair of points we use for A and B? The similar triangles in the marginal diagram show us that

$$\frac{y_2' - y_1'}{x_2' - x_1'} = \frac{y_2 - y_1}{x_2 - x_1}$$

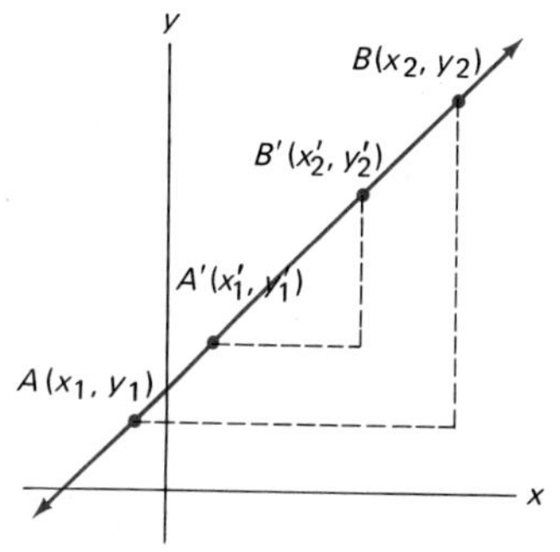

Thus, points A' and B' would do just as well as A and B. It does not even matter whether A is to the left or right of B since

$$\frac{y_1 - y_2}{x_1 - x_2} = \frac{y_2 - y_1}{x_2 - x_1}$$

All that matters is that we subtract the coordinates in the same order in numerator and denominator.

The slope m is a measure of the steepness of a line, as the diagram below illustrates. Notice that a horizontal line has zero slope and a line that rises to the right has positive slope. The larger this positive slope is, the more steeply the line rises. A line that falls to the right has negative slope. The concept of slope for a vertical line makes no sense since it would involve division by zero. Therefore the notion of slope for a vertical line is left undefined.

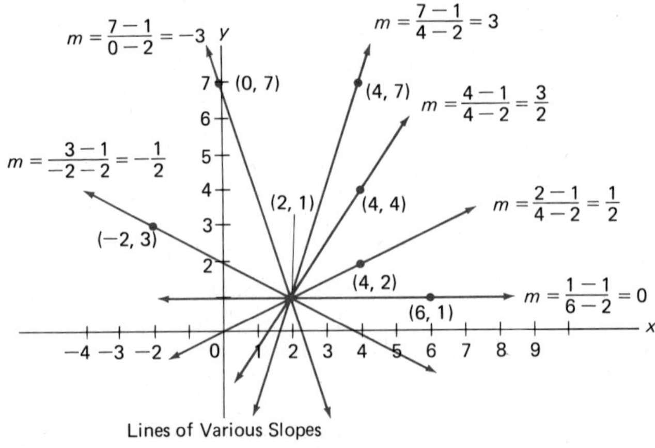

Lines of Various Slopes

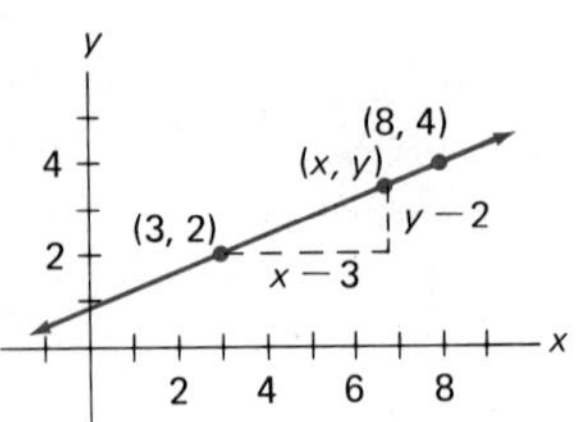

The Point-Slope Form Consider again the line of our opening diagram; it is reproduced in the margin. We know

1. it passes through (3, 2);
2. it has slope $\frac{2}{5}$.

Take any other point on that line, such as one with coordinates (x, y). If we use this point together with (3, 2) to measure slope, we must get $\frac{2}{5}$, that is,

$$\frac{y - 2}{x - 3} = \frac{2}{5}$$

or, after multiplying by $x - 3$,

$$y - 2 = \tfrac{2}{5}(x - 3)$$

Notice that this last equation is satisfied by all points on the line, even by (3, 2). Moreover, no points not on the line can satisfy this equation.

What we have just done in an example can be done in general. The line passing through the (fixed) point (x_1, y_1) with slope m has equation

$$y - y_1 = m(x - x_1)$$

We call it the **point-slope** form of the equation of a line.

Consider once more the line of our example. That line passes through (8, 4) as well as (3, 2). If we use (8, 4) as (x_1, y_1), we get the equation

$$y - 4 = \tfrac{2}{5}(x - 8)$$

which looks quite different from

$$y - 2 = \tfrac{2}{5}(x - 3)$$

However, both can be simplified to $5y - 2x = 4$; they are equivalent.

The Slope-Intercept Form Just as most ideas can be expressed in several different ways, so the equation of a line can be expressed in various forms. Suppose we are given the slope m for a line and the y-intercept b (that is, the line intersects the y-axis at $(0, b)$). Choosing $(0, b)$ as (x_1, y_1) and applying the point-slope form, we get

$$y - b = m(x - 0)$$

which we can rewrite as

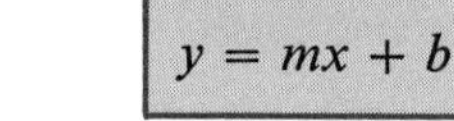

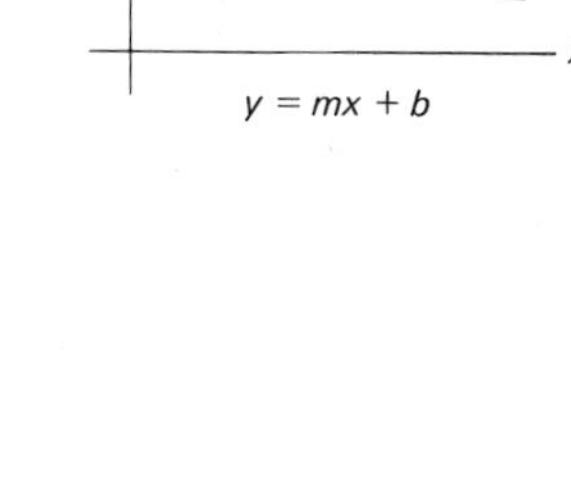

The latter is called the **slope-intercept** form.

Why get excited about that, you ask? Any time we see an equation written this way, we recognize it as a line and can immediately read its slope and y-intercept. For example, consider the equation

$$3x - 2y + 4 = 0$$

If we solve for y, we get

$$y = \tfrac{3}{2}x + 2$$

It is the equation of a line with slope $\frac{3}{2}$ and y-intercept 2.

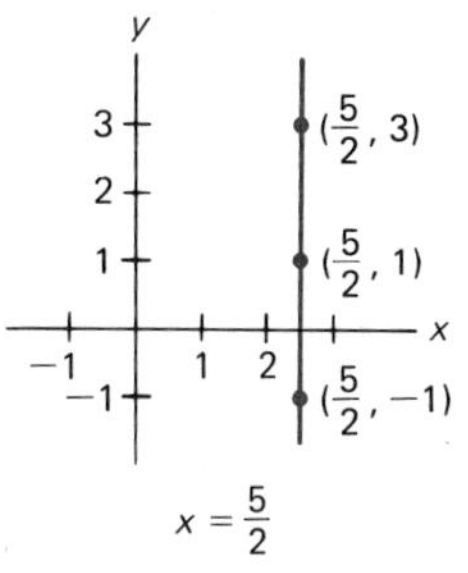

Equation of a Vertical Line Vertical lines do not fit within the discussion above; they do not have slopes. But they do have equations, very simple ones. The line in the marginal picture has equation $x = \frac{5}{2}$, since every point on the line satisfies this equation. The equation of any vertical line can be put in the form

$$x = k$$

where k is a constant. It should be noted that the equation of a horizontal line can be written in the form $y = k$.

The Form $Ax + By + C = 0$ It would be nice to have a form that covered all lines including vertical lines. Consider for example,

$$\begin{array}{ll} \text{(i)} & y - 2 = -4(x + 2) \\ \text{(ii)} & y = 5x - 3 \\ \text{(iii)} & x = 5 \end{array}$$

These can be rewritten (by taking everything to the left side) as follows:

$$\begin{array}{ll} \text{(i)} & 4x + y + 6 = 0 \\ \text{(ii)} & -5x + y + 3 = 0 \\ \text{(iii)} & x + 0y - 5 = 0 \end{array}$$

All are of the form

$$Ax + By + C = 0$$

which we call the **general linear equation**. It takes only a moment's thought to see that the equation of any line can be put in this form. Conversely, the graph of $Ax + By + C = 0$ is always a line (if A and B are not both zero (see Problem 64)).

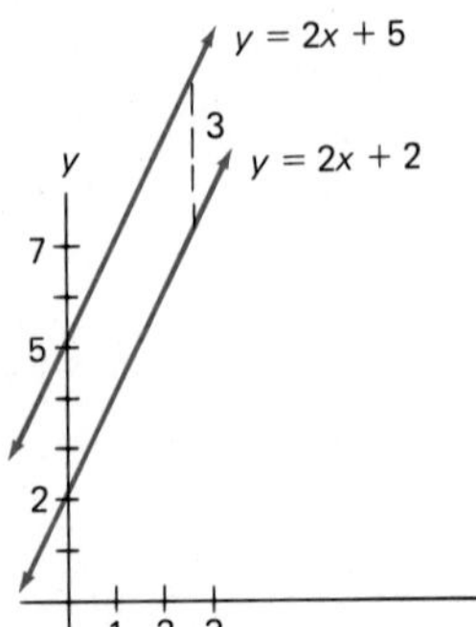

Parallel Lines If two lines have the same slope, they are parallel. Thus $y = 2x + 2$ and $y = 2x + 5$ represent parallel lines; both have a slope of 2. The second line is 3 units above the first for every value of x.

Similarly, the lines with equations $-2x + 3y + 12 = 0$ and $4x - 6y = 5$ are parallel. To see this, solve these equations for y (that is, find the slope-intercept form); you get $y = \frac{2}{3}x - 4$ and $y = \frac{2}{3}x - \frac{5}{6}$, respectively. Both have slope $\frac{2}{3}$; they are parallel.

We may summarize by stating that *two nonvertical lines are parallel if and only if they have the same slope.*

Perpendicular Lines Is there a simple slope condition which characterizes perpendicular lines? Yes; *two nonvertical lines are perpendicular if and only if their slopes are negative reciprocals of each other.* We are not going to prove this, but an example will help explain why it is true. The slopes of the lines $y = \frac{3}{4}x$ and $y = -\frac{4}{3}x$ are negative reciprocals of each other. Both lines pass through the origin. The points (4, 3) and (3, −4) are on the first and second lines, respectively. The two right triangles shown in the diagram in the margin are congruent with $\angle 1 = \angle 4$ and $\angle 2 = \angle 3$. But

$$\angle 1 + \angle 2 = 90°$$

and therefore

$$\angle 1 + \angle 3 = 90°$$

That says the two lines are perpendicular to each other.

The lines $2x - 3y = 5$ and $3x + 2y = -4$ are also perpendicular, since, after solving them for y, we see that the first has slope $\frac{2}{3}$ and the second has slope $-\frac{3}{2}$.

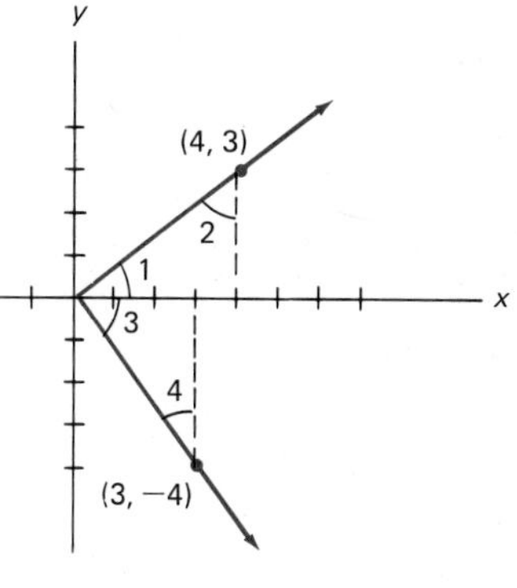

Problem Set 4-3

Find the slope of the line containing the given two points.

1. (2, 3) and (4, 8)
2. (4, 1) and (8, 2)
3. (−4, 2) and (3, 0)
4. (2, −4) and (0, −6)
5. (3, 0) and (0, 5)
6. (−6, 0) and (0, 6)
7. [C] (−1.732, 5.014) and (4.315, 6.175)
8. [C] $(\pi, \sqrt{3})$ and $(1.642, \sqrt{2})$

Find an equation for each of the following lines. Then write your answer in the form $Ax + By + C = 0$.

9. Through (2, 3) with slope 4.
10. Through (4, 2) with slope 3.
11. Through (3, −4) with slope −2.
12. Through (−5, 2) with slope −1.
13. With y-intercept 4 and slope −2.
14. With y-intercept −3 and slope 1.
15. With y-intercept 5 and slope 0.
16. With y-intercept 1 and slope −1.
17. Through (2, 3) and (4, 8).
18. Through (4, 1) and (8, 2).
19. Through (3, 0) and (0, 5).
20. Through (−6, 0) and (0, 6).
21. [C] Through $(\sqrt{3}, \sqrt{7})$ and $(\sqrt{2}, \pi)$.
22. [C] Through $(\pi, \sqrt{3})$ and $(\pi + 1, 2\sqrt{3})$.
23. Through (2, −3) and (2, 5).
24. Through (−5, 0) and (−5, 4).

In Problems 25–32, find the slope and y-intercept of each line.

25. $y = 3x + 5$
26. $y = 6x + 2$
27. $3y = 2x - 4$
28. $2y = 5x + 2$

29. $2x + 3y = 6$ 30. $4x + 5y = -20$

31. $y + 2 = -4(x - 1)$ 32. $y - 3 = 5(x + 2)$

33. Write the equation of the line through $(3, -3)$
    (a) parallel to the line $y = 2x + 5$;
    (b) perpendicular to the line $y = 2x + 5$;
    (c) parallel to the line $2x + 3y = 6$;
    (d) perpendicular to the line $2x + 3y = 6$;
    (e) parallel to the line through $(-1, 2)$ and $(3, -1)$;
    (f) parallel to the line $x = 8$;
    (g) perpendicular to the line $x = 8$.

34. Find the value of k for which the line $4x + ky = 5$
    (a) passes through the point $(2, 1)$;
    (b) is parallel to the y-axis;
    (c) is parallel to the line $6x - 9y = 10$;
    (d) has equal x- and y-intercepts;
    (e) is perpendicular to the line $y - 2 = 2(x + 1)$.

35. Write the equation of the line through $(0, -4)$ that is perpendicular to the line $y + 2 = -\frac{1}{2}(x - 1)$.

36. Find the value of k such that the line $kx - 3y = 10$
    (a) is parallel to the line $y = 2x + 4$;
    (b) is perpendicular to the line $y = 2x + 4$;
    (c) is perpendicular to the line $2x + 3y = 6$.

Example A (Intersection of two lines) Find the coordinates of the point of intersection of the lines $3x - 4y = 5$ and $x + 2y = 5$.

SOLUTION. We simply solve the two equations simultaneously (see Section 3-3). Multiply the second equation by -3 and then add the two equations.

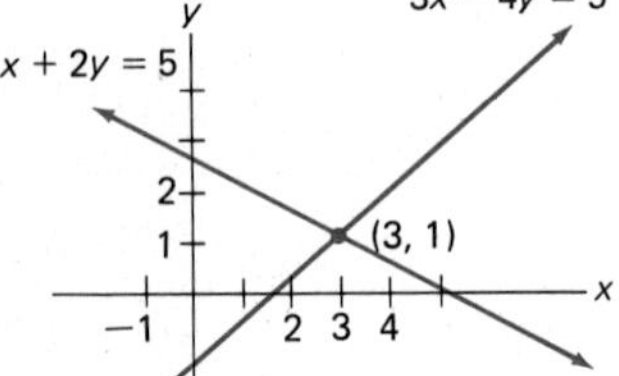

$$\begin{aligned} 3x - 4y &= 5 \\ -3x - 6y &= -15 \\ \hline -10y &= -10 \\ y &= 1 \\ x &= 3 \end{aligned}$$

Find the coordinates of the point of intersection in each problem below. Then write the equation of the line through that point perpendicular to the line given first.

37. $2x + 3y = 4$, $-3x + y = 5$ 38. $4x - 5y = 8$, $2x + y = -10$

39. $3x - 4y = 5$, $2x + 3y = 9$ 40. $5x - 2y = 5$, $2x + 3y = 6$

Example B (Distance from a point to a line) It can be shown that the distance d from the point (x_1, y_1) to the line $Ax + By + C = 0$ is

$$d = \frac{|Ax_1 + By_1 + C|}{\sqrt{A^2 + B^2}}$$

Find the distance from (1, 2) to $3x - 4y = 5$.

SOLUTION. First write the equation as $3x - 4y - 5 = 0$. The formula gives

$$d = \frac{|3\cdot 1 - 4\cdot 2 - 5|}{\sqrt{3^2 + (-4)^2}} = \frac{|-10|}{5} = 2$$

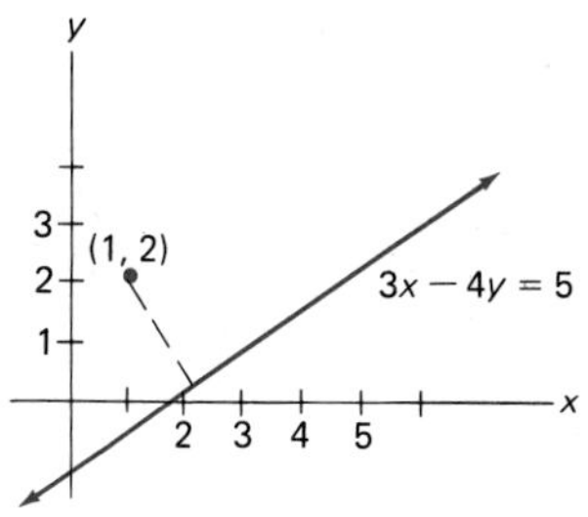

In each case, find the distance from the given point to the given line.

41. $(-3, 2)$, $3x + 4y = 6$
42. $(4, -1)$, $2x - 2y + 4 = 0$
43. $(-2, -1)$, $5y = 12x + 1$
44. $(3, -1)$, $y = 2x - 5$

Find the (perpendicular) distance between the given parallel lines. (*Hint*: First find a point on one of the lines.)

45. $3x + 4y = 6$, $3x + 4y = 12$
46. $5x + 12y = 2$, $5x + 12y = 7$

Miscellaneous Problems

47. Find the slope of the line containing the points (2, 1) and (−3, 5).
48. Find the equation of the line through the point (4, 1) that
    (a) passes through the origin;
    (b) is parallel to the line $y = 3x - 4$;
    (c) is perpendicular to the line $x + 2y = 6$;
    (d) is parallel to the y-axis;
    (e) has y-intercept 2.
49. Find the equation of the line that passes through the intersection of the lines $2x - 3y = 2$ and $5x + 3y = 26$ and is parallel to the line $x + y = 4$.
50. Consider the line segment AB where A and B have coordinates (2, −1) and (6, 5). Find the equation of the line which bisects AB at right angles.
51. Which pairs of lines below are parallel, which are perpendicular, and which are neither?
    (a) $y = 4x - 7$; $y + 3 = 4(x + 1)$
    (b) $x + 2y = 4$; $x - 2y = -4$
    (c) $3x + 2y = 4$; $2x - 3y = 0$
    (d) $x = 4$; $x = -2$
    (e) $2x + 5y = 0$; $5x = 2y$
52. Find the distance between the parallel lines $3x - 4y = 12$ and $3x - 4y = -13$.
53. The Celsius and Fahrenheit temperature scales are related by the equation $F = \frac{9}{5}C + 32$.
    (a) If C increases by 1, by how much does F increase?
    (b) The graph of $F = \frac{9}{5}C + 32$ is a line. Assuming the F-axis to be vertical, what is the slope of the line? What is its F-intercept?
    (c) Solve the system of equations

$$F = \tfrac{9}{5}C + 32$$
$$F = C$$

    Interpret your answer.

54. Suppose that you borrow \$20,000 today at 8 percent simple interest per year. Find a formula for the amount A you will owe after t years (see Problem 60 in Section 3-1).

55. Graph the equation you obtained in Problem 54 using a vertical A-axis. What is the slope of the graph? What is its A-intercept?

56. A heavy piece of road equipment costs \$120,000 and each year it depreciates at 8 percent of its original value. Find a formula for the value V of the item after t years.

57. The graph of the answer to Problem 56 is a straight line. What is its slope, assuming the t-axis to be horizontal? Interpret the slope.

58. Past experience indicates that egg production in Matlin County is growing linearly. In 1960, it was 700,000 cases and in 1970, it was 820,000 cases. Write a formula for the number N of cases produced n years after 1960 and use it to predict egg production in the year 2000.

59. A piece of equipment purchased today for \$80,000 will depreciate linearly to a scrap value of \$2000 after 20 years. Write a formula for its value V after n years.

60. Suppose that the profit P that a company realizes in selling x items of a certain commodity is given by $P = 450x - 2000$.
    (a) Interpret the value of P when $x = 0$
    (b) Find the slope of the graph of the above equation. This slope is called the **marginal profit**. What is its economic interpretation?

61. The cost C of producing x items of a certain commodity is given by $C = .75x + 200$. The slope of its graph is called the **marginal cost**. Find it and give an economic interpretation.

62. Does (3, 9) lie above or below the line $y = 3x - 1$?

63. Show that the equation of the line with x-intercept a and y-intercept b is

$$\frac{x}{a} + \frac{y}{b} = 1$$

64. Show that the graph of $Ax + By + C = 0$ is always a line (provided A and B are not both 0). (*Hint*: Consider two cases: (i) $B = 0$ and (ii) $B \neq 0$.)

65. Find the equation of the line through (2, 3) which has equal x- and y-intercepts.

66. Show that for each value of k, the equation

$$2x - y + 4 + k(x + 3y - 6) = 0$$

represents a line through the intersection of the two lines $2x - y + 4 = 0$ and $x + 3y - 6 = 0$. (*Hint*: It is not necessary to find the point of intersection (x_0, y_0).)

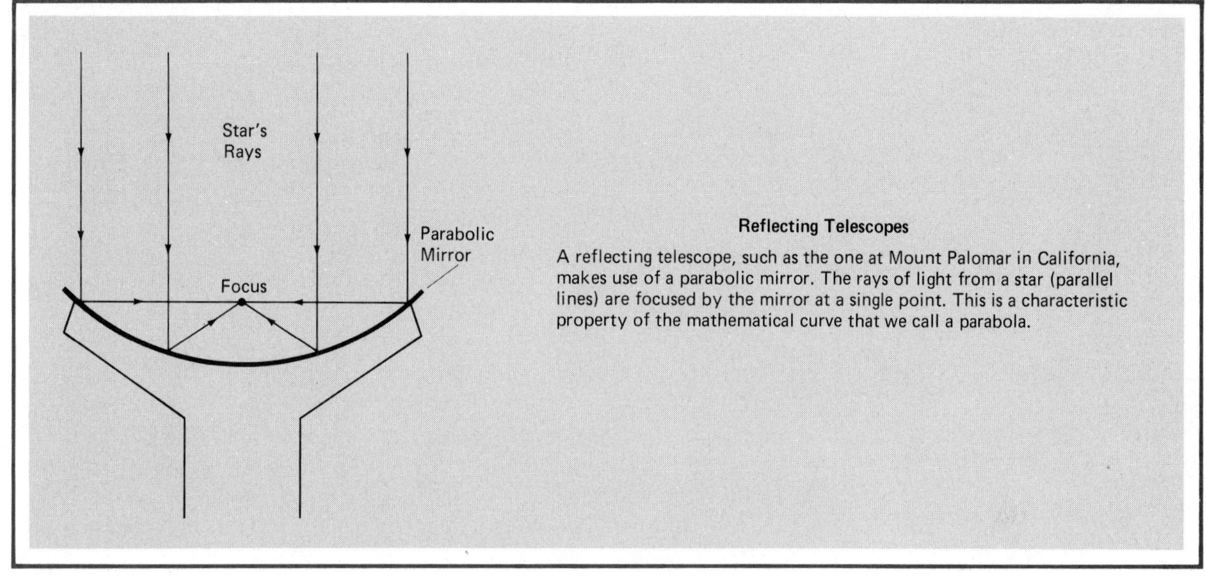

Reflecting Telescopes

A reflecting telescope, such as the one at Mount Palomar in California, makes use of a parabolic mirror. The rays of light from a star (parallel lines) are focused by the mirror at a single point. This is a characteristic property of the mathematical curve that we call a parabola.

4-4 The Parabola

We have seen that the graph of $y = ax + b$ is always a line. Now we want to study the graph of $y = ax^2 + bx + c$ $(a \neq 0)$. As we shall discover, this graph is always a smooth, cup-shaped curve something like the cross section of the mirror shown in the opening diagram. We call it a **parabola**.

Some Simple Cases The simplest case of all is $y = x^2$. The graph of this equation is shown in the margin. Two important features should be noted.

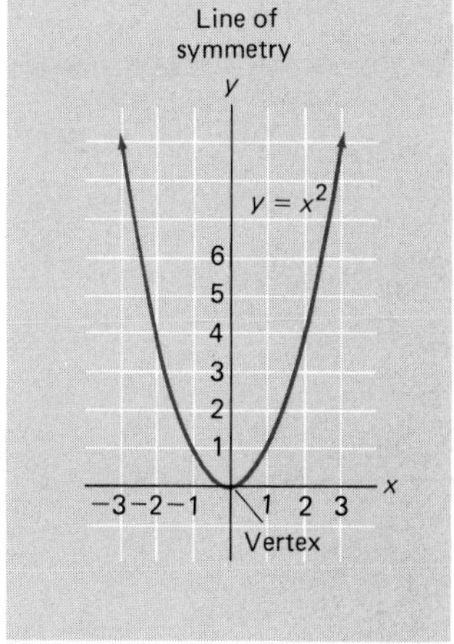

1. The curve is symmetric about the y-axis. This follows from the fact that the equation $y = x^2$ is not changed if we replace (x, y) by $(-x, y)$.
2. The curve reaches its lowest point at $(0, 0)$, the point where the curve intersects the line of symmetry. We call this point the **vertex** of the parabola.

Next we consider how the graph of $y = x^2$ is modified as we look successively at:

A. $y = ax^2$
B. $y = x^2 + k$
C. $y = (x - h)^2 = x^2 - 2hx + h^2$
D. $y = a(x - h)^2 + k$

Several examples of these graphs are shown on page 154. Be sure to study them very carefully.

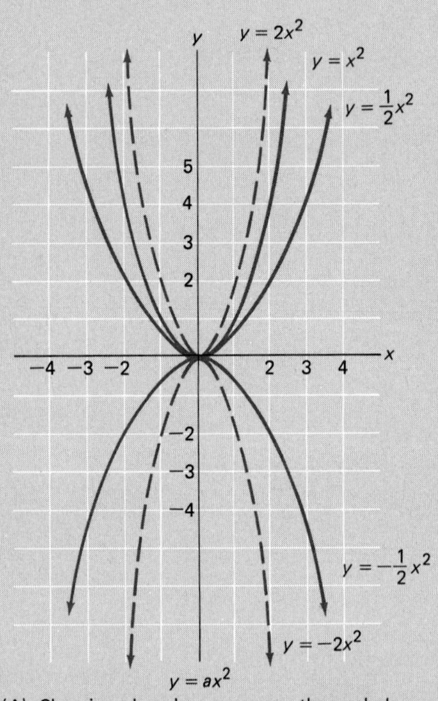

$y = ax^2$

(A) Changing a broadens or narrows the parabola. Making its sign negative turns it downward. Vertex is at (0, 0)

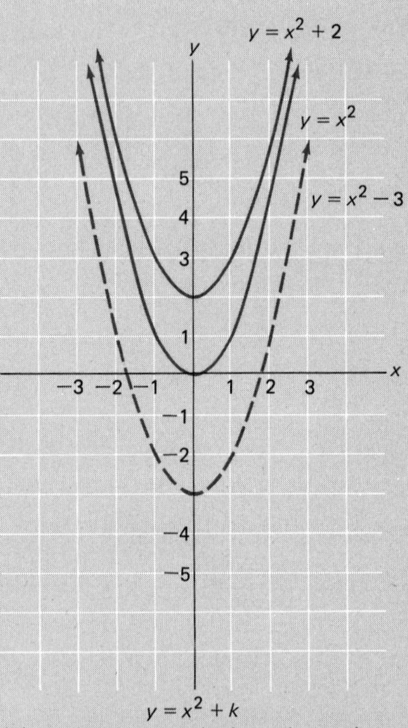

$y = x^2 + k$

(B) Adding a constant k raises or lowers the parabola, depending on whether k is positive or negative. Vertex is at $(0, k)$.

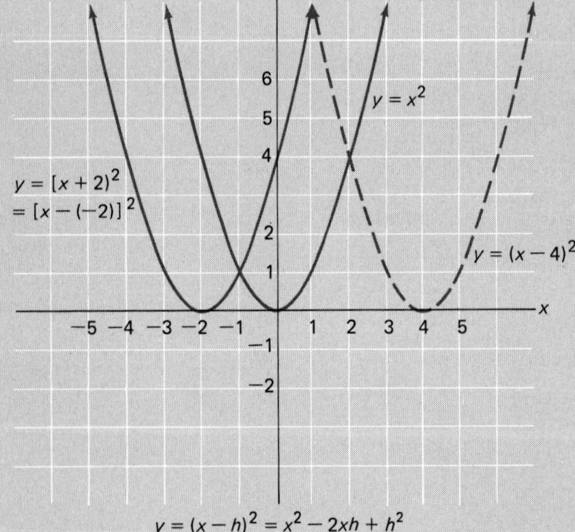

$y = (x - h)^2 = x^2 - 2xh + h^2$

(C) Changing h shifts the parabola right or left depending on whether h is positive or negative. Vertex is at $(h, 0)$.

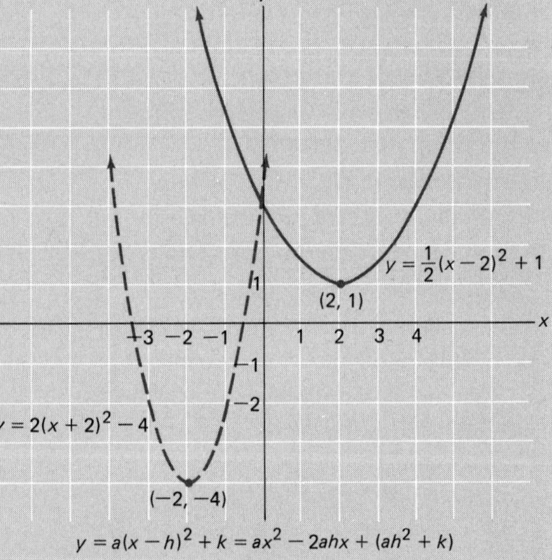

$y = a(x - h)^2 + k = ax^2 - 2ahx + (ah^2 + k)$

(D) Here we have combined all three possibilities. Vertex is at (h, k).

The most general equation considered so far is $y = a(x - h)^2 + k$. The graph of this equation is a parabola with vertex (h, k) and line of symmetry $x = h$. If we expand $a(x - h)^2$ and collect terms on the right side, the equation takes the form $y = ax^2 + bx + c$. Conversely, $y = ax^2 + bx + c$ $(a \neq 0)$ always represents a parabola with a vertical line of symmetry, as we shall now show. We use the method of completing the square to rewrite $y = ax^2 + bx + c$ as follows:

$$\begin{aligned} y &= a\left(x^2 + \frac{b}{a}x + \quad\right) + c \\ &= a\left[x^2 + \frac{b}{a}x + \left(\frac{b}{2a}\right)^2\right] + c - a\left(\frac{b}{2a}\right)^2 \\ &= a\left(x + \frac{b}{2a}\right)^2 + \left(c - \frac{b^2}{4a}\right) \end{aligned}$$

This is the equation of a parabola with vertex at $(-b/2a, c - b^2/4a)$ and line of symmetry $x = -b/2a$.

As an example, consider $y = 2x^2 + 8x + 3$. Its graph is a parabola with vertex at $x = -b/2a = -8/2 \cdot 2 = -2$. The y-coordinate of the vertex can be calculated from the formula $c - (b^2/4a)$ or perhaps better by simply substituting $x = -2$ in the original equation. In either case, we get $y = -5$.

Geometric Definition of a Parabola So far, we have approached the study of parabolas from an algebraic point of view, that is, by means of the equation $y = ax^2 + bx + c$. Now we shall look at the parabolas from a geometric perspective, as the Greeks did. Here is their definition of a parabola: Consider a fixed line L (called the **directrix**) and a fixed point F (called the **focus**). The set of all points equidistant from the line L and the point F is called a **parabola**. The point V midway between F and L is the **vertex** of the parabola (see the diagram in the margin).

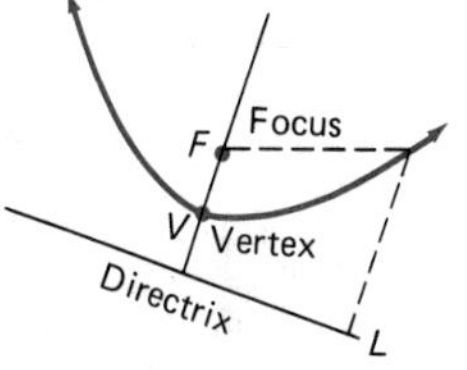

To see that this definition is consistent with our earlier approach, we introduce a coordinate system. Put V at the origin $(0, 0)$, put F at $(0, p)$, and let L be the line $y = -p$. Then the y-axis becomes the line of symmetry of the parabola. Now if $P(x, y)$ is any point on the parabola and $M(x, -p)$ is the corresponding point directly below P on the line $y = -p$, then

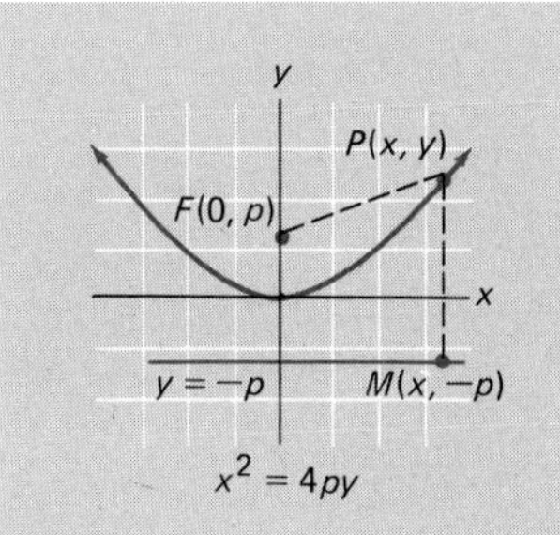

$$d(P, F) = d(P, M)$$

Using the distance formula, we have

$$\sqrt{(x - 0)^2 + (y - p)^2} = \sqrt{(x - x)^2 + (y + p)^2}$$

After squaring both sides, this becomes

$$\begin{aligned} x^2 + y^2 - 2py + p^2 &= y^2 + 2py + p^2 \\ x^2 - 2py &= 2py \\ x^2 &= 4py \end{aligned}$$

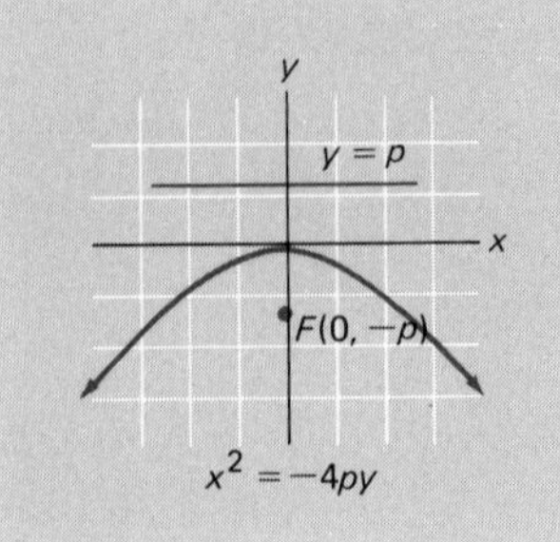

In the same manner, we could show that if the focus is at $(0, -p)$ and the directrix is $y = p$, the equation of the parabola is

$$x^2 = -4py$$

More generally, when the vertex of the parabola is at (h, k), the focus p units above (h, k), and the directrix p units below (h, k), the equation of the parabola is

$$(x - h)^2 = 4p(y - k)$$

Finally, when the relative positions of the focus and directrix are reversed, the equation is

$$(x - h)^2 = -4p(y - k)$$

All of this is summarized in the table below.

VERTEX AT (0, 0)	VERTEX AT (h, k)	PARABOLA OPENS
$x^2 = 4py$	$(x - h)^2 = 4p(y - k)$	upward
$x^2 = -4py$	$(x - h)^2 = -4p(y - k)$	downward

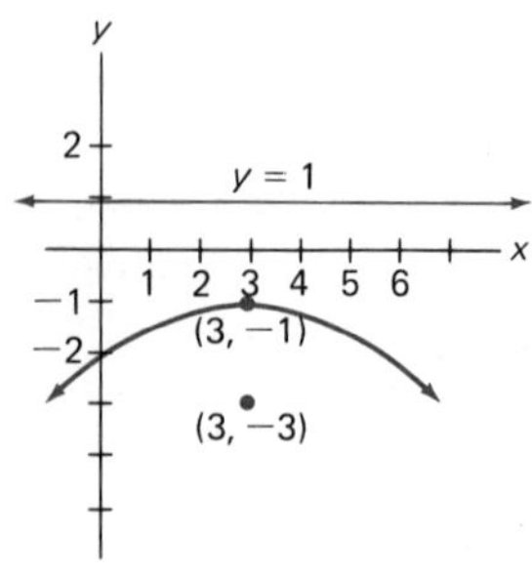

Consider $(x - 3)^2 = -8(y + 1)$ as an example. This is the equation of a parabola that opens downward, with vertex at $(3, -1)$. Since $4p = 8$, $p = 2$. This means the focus is 2 units below the vertex, that is, at $(3, -3)$. The directrix is 2 units above the vertex; its equation is $y = 1$.

It should be clear that an equation of the form

$$(x - h)^2 = \pm 4p(y - k)$$

can also be written in the form

$$y = ax^2 + bx + c$$

For example,

$$(x - 3)^2 = -8(y + 1)$$

can be written successively as

$$x^2 - 6x + 9 = -8y - 8$$
$$x^2 - 6x + 17 = -8y$$
$$-\tfrac{1}{8}(x^2 - 6x + 17) = y$$
$$y = -\tfrac{1}{8}x^2 + \tfrac{3}{4}x - \tfrac{17}{8}$$

Applications The parabola has an important optical property. Consider a cup-shaped mirror with a parabolic cross section. If a light source is placed at the focus, the reflected rays of light are parallel to the axis. This principle is used in flashlight mirrors. Conversely, if parallel light rays (as from a star) hit a parabolic mirror, they will be "focused" at the focus of the parabola, provided they come from a direction parallel to the axis of the mirror. This is the basis for the design of reflecting telescopes (see the opening box of this section).

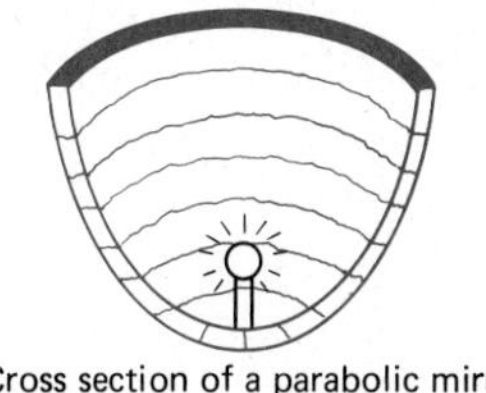

Cross section of a parabolic mirror with light source at focus

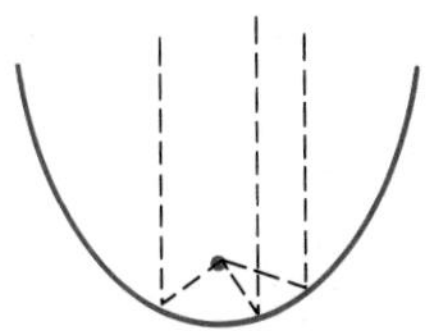

We now turn to a very different kind of application. From physics, we learn that the path of a projectile is a parabola. It is known, for example, that a projectile fired at an angle of 45° from the horizontal with an initial speed of $320\sqrt{2}$ feet per second follows a curve with equation

$$y = -\tfrac{1}{6400}x^2 + x$$

where the coordinate axes are placed as shown in the diagram in the margin. Taking this for granted, we may ask two questions.

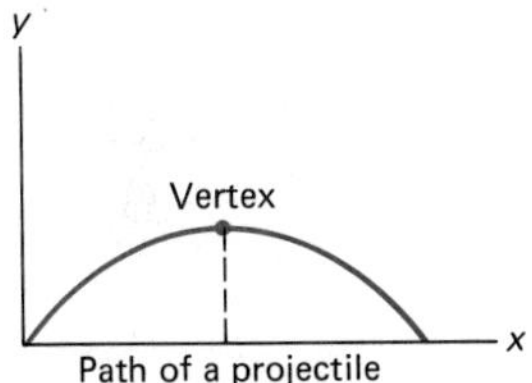

Path of a projectile

1. What is the maximum height attained by the projectile?
2. What is the range (horizontal distance traveled) of the projectile?

To find the maximum height is simply to find the y-coordinate of the vertex. First we find the x-coordinate.

$$x = \frac{-b}{2a} = -\frac{1}{-2/6400} = 3200$$

When we substitute this value in the equation, we get

$$y = -\tfrac{1}{6400}(3200)^2 + 3200 = -1600 + 3200 = 1600$$

The greatest height is thus 1600 feet.

The range of the projectile is the x-coordinate of the point where it lands. By symmetry, this is simply twice the x-coordinate of the vertex; that is,

$$\text{range} = 2(3200) = 6400 \text{ feet}$$

This value could also be obtained by solving the quadratic equation

$$-\tfrac{1}{6400}x^2 + x = 0$$

since the x-coordinate of the landing point is the value of x when $y = 0$.

Problem Set 4-4

The equations in Problems 1–10 represent parabolas. Sketch the graph of each parabola, indicating the coordinates of the vertex.

1. $y = 3x^2$
2. $y = -2x^2$
3. $y = x^2 + 5$
4. $y = 2x^2 - 4$
5. $y = (x - 4)^2$
6. $y = -(x + 3)^2$
7. $y = 2(x - 1)^2 + 5$
8. $y = 3(x + 2)^2 - 4$
9. $y = -4(x - 2)^2 + 1$
10. $y = \frac{1}{2}(x + 3)^2 + 3$

Write each of the following in the form $y = ax^2 + bx + c$.

11. $y = 2(x - 1)^2 + 7$
12. $y = -3(x + 2)^2 + 5$
13. $-2y + 5 = (x - 5)^2$
14. $3y + 6 = (x + 3)^2$

Sketch the graph of each equation. Begin by plotting the vertex and at least one point on each side of the vertex. Recall that the x-coordinate of the vertex for $y = ax^2 + bx + c$ is $x = -b/2a$.

15. $y = x^2 + 2x$
16. $y = 3x^2 - 6x$
17. $y = -2x^2 + 8x + 1$
18. $y = -3x^2 + 6x + 4$

Example A (Horizontal parabolas) Sketch the parabolas (a) $x = 2y^2$; (b) $x = 2(y - 3)^2 - 4$.

SOLUTION. Note that the roles of x and y are interchanged when compared to earlier examples. The line of symmetry will therefore be horizontal. The vertex in (a) is (0, 0); in (b), it is (−4, 3).

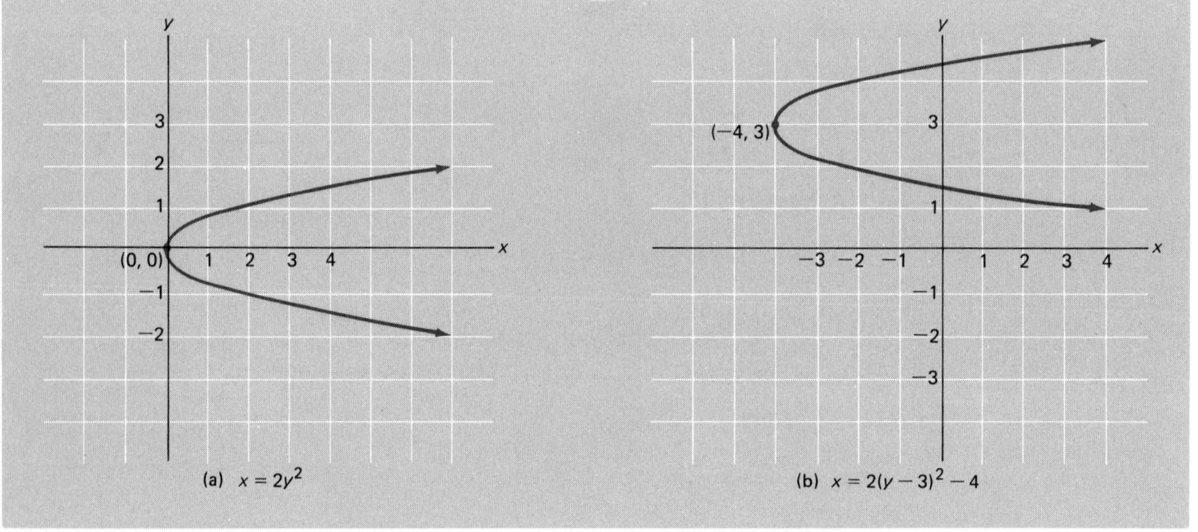

In Problems 19–24, sketch the graph of the equation and indicate the coordinates of the vertex.

19. $x = -2y^2$
20. $x = -2y^2 + 8$
21. $x = -2(y + 2)^2 + 8$
22. $x = 3(y - 1)^2 + 6$
23. $x = y^2 + 4y + 2$. (*Note*: The y-coordinate of the vertex is at $-b/2a$.)
24. $x = 4y^2 - 8y + 10$

Example B (Intersection of a line and a parabola) Find the points of intersection of the line $y = -2x + 2$ and the parabola $y = 2x^2 - 4x - 2$.

SOLUTION. We must solve the two equations simultaneously. This is easy to do by substituting the expression for y from the first equation into the second equation and then solving the resulting equation for x.

$$\begin{aligned} -2x + 2 &= 2x^2 - 4x - 2 \\ 0 &= 2x^2 - 2x - 4 \\ 0 &= 2(x^2 - x - 2) \\ 0 &= 2(x - 2)(x + 1) \\ x &= -1 \qquad x = 2 \end{aligned}$$

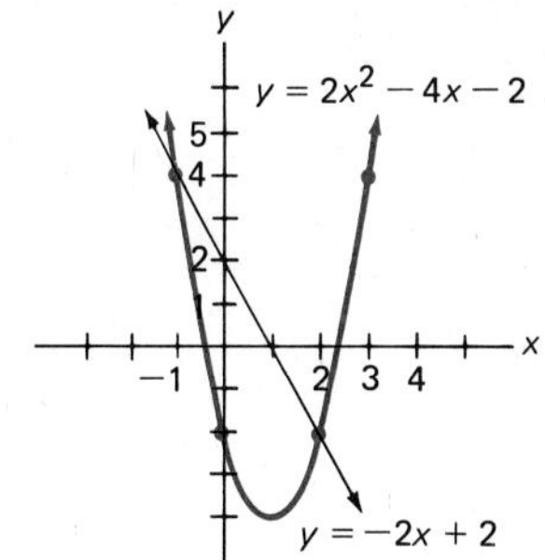

By substitution, we find the corresponding values of y to be 4 and -2; the intersection points are therefore $(-1, 4)$ and $(2, -2)$.

Find the points of intersection for each pair of equations.

25. $y = -x + 1$
    $y = x^2 + 2x + 1$

26. $y = -x + 4$
    $y = -x^2 + 2x + 4$

27. $y = -2x + 1$
    $y = -x^2 - x + 3$

28. $y = -3x + 15$
    $y = 3x^2 - 3x + 12$

[C] 29. $y = 1.5x + 3.2$
    $y = x^2 - 2.9x$

[C] 30. $y = 2.1x - 6.4$
    $y = -1.2x^2 + 4.3$

Example C (Problems involving the focus and directrix)

(a) Find the equation of the parabola with focus at $(-2, 1)$ and directrix $y = 7$.

(b) Find the vertex, focus, directrix, and line of symmetry of the parabola whose equation is $y = x^2 - 4x + 6$.

SOLUTION. (a) Since the focus is below the directrix, the parabola opens downward and has an equation of the form

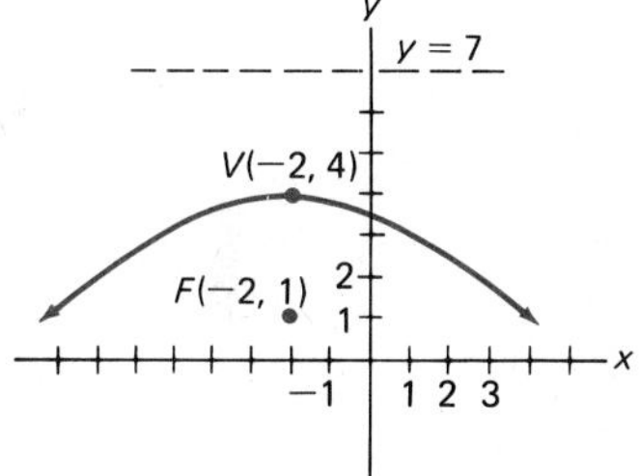

$$(x - h)^2 = -4p(y - k)$$

The distance $2p$ from the focus to directrix is 6 (see the diagram in the margin). Therefore $p = 3$ and the vertex is at $(-2, 4)$, midway between the focus and directrix. The desired equation is

$$(x + 2)^2 = -12(y - 4)$$

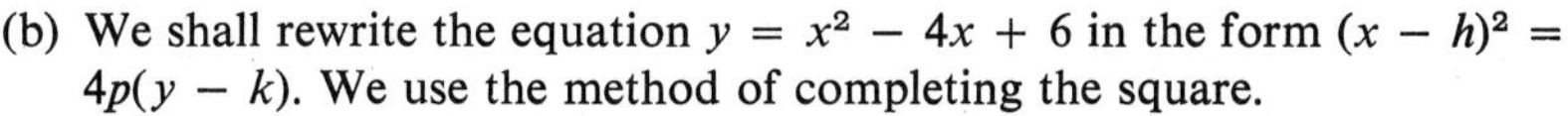
(b) We shall rewrite the equation $y = x^2 - 4x + 6$ in the form $(x - h)^2 = 4p(y - k)$. We use the method of completing the square.

$$\begin{aligned} y &= x^2 - 4x + 6 \\ y &= (x^2 - 4x + 4) + 6 - 4 \\ y - 2 &= (x - 2)^2 \end{aligned}$$

Thus the vertex is at $(2, 2)$, $4p = 1$, and $p = \frac{1}{4}$. The focus is $\frac{1}{4}$ unit above the vertex at $(2, \frac{9}{4})$ and the directrix is $\frac{1}{4}$ unit below the vertex with equation $y = \frac{7}{4}$. The line $x = 2$ is the line of symmetry.

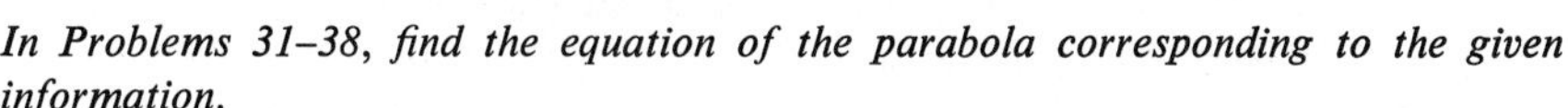
In Problems 31–38, find the equation of the parabola corresponding to the given information.

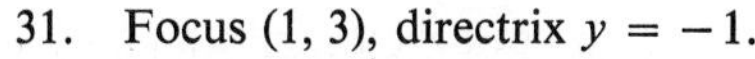
31. Focus $(1, 3)$, directrix $y = -1$.
32. Focus $(2, 0)$, directrix $y = -6$.
33. Focus $(-3, 4)$, directrix $y = 6$.

34. Focus $(4, -6)$, directrix $y = 0$.
35. Focus $(2, 3)$, vertex $(2, -1)$.
36. Focus $(-3, 5)$, vertex $(-3, -11)$.
37. Focus $(0, 0)$, vertex $(0, 6)$.
38. Focus $(4, 2)$, vertex $(4, 7)$.

For each parabola in Problems 39–44, find the vertex, focus, directrix, and line of symmetry.

39. $x^2 = 8y$
40. $x^2 = -12y$
41. $(x - 2)^2 = -16(y - 4)$
42. $(x - 1)^2 = 4(y - 6)$
43. $y = x^2 - 6x + 4$
44. $y = x^2 + 8x - 2$

Miscellaneous Problems

45. Sketch the following parabolas on the same coordinate system.
(a) $y = 2x^2 - 8x + 4$ (b) $y = 2x^2 - 8x + 8$
(c) $y = 2x^2 - 8x + 11$
46. Sketch the following parabolas on the same coordinate system.
(a) $y = -3x^2 + 6x + 9$ (b) $y = -3x^2 + 6x - 3$
(c) $y = -3x^2 + 6x - 9$
47. Calculate the value of the discriminant $b^2 - 4ac$ for each of the parabolas in Problem 45. Make a conjecture about the number of x-intercepts of the parabola $y = ax^2 + bx + c$ if $b^2 - 4ac > 0$, if $b^2 - 4ac = 0$, and if $b^2 - 4ac < 0$.
48. For what values of k does the parabola $y = x^2 - 6x + k$ have two x-intercepts?
49. Find the value of a for which the parabola $y = ax^2$ passes through the point $(2, 5)$.
50. Find the points of intersection of the parabolas $y = 4 - x^2$ and $y = x^2 - 14$.
51. Find the points where the line $y = 2x - 4$ intersects the parabola $y = x^2 - 12$.
52. Use algebra to show that the parabola $y = x^2 - 4x + 6$ has just one point in common with the line $y = 2x - 3$. Then sketch both equations on the same coordinate system.
53. Use algebra to show that the parabola $y = -x^2 + 2x + 4$ and the line $y = -2x + 9$ have no point in common. Then sketch both equations on the same coordinate system.
54. For any $a \neq 0$, the equation $y = a(x - 2)(x - 8)$ represents a parabola.
(a) Find its x-intercepts.
(b) Use part (a) to determine the x-coordinate of the vertex.
(c) Find the value of a if the parabola passes through $(10, 40)$.

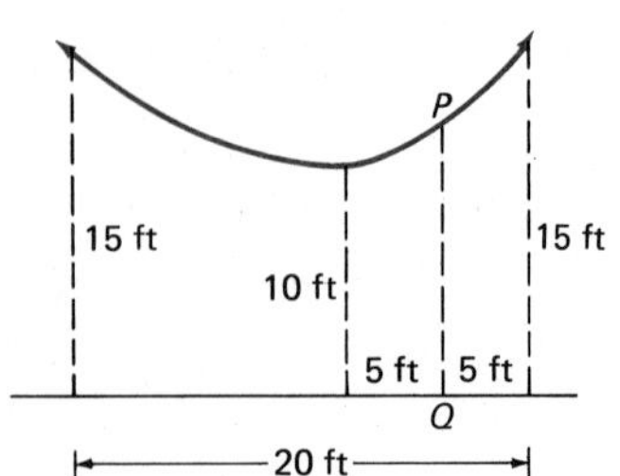

55. If the shape of the hanging rope shown in the margin is part of a parabola, find the distance PQ.
56. Starting at $(0, 0)$, a projectile travels along the path $y = -\frac{1}{256}x^2 + \sqrt{3}x$. Sketch the path, find the maximum height, and find the range.
57. Find the equation of the vertical parabola that passes through $(-1, 2)$, $(1, -1)$, and $(2, 1)$.

58. A retailer has learned from experience that if she charges x dollars apiece for a toy truck, she can sell $300 - 100x$ of them. The trucks cost her \$2 each. Write a formula for her total profit P in terms of x. Then determine what she should charge to maximize her profit.

59. A company that makes fancy golf carts will have overhead of \$12,000 per year and direct costs (labor and materials) of \$80 per cart. It sells all its carts to a certain retailer and has a guaranteed price of \$120 each. The company will give a discount of 1 percent for 100 carts, 2 percent for 200 carts, and in general $x/100$ percent for x carts. Its maximum possible production is 2500 carts.
   (a) Write a formula for C, the cost of producing x carts.
   (b) Show that its total receipts R in dollars is $R = 120x - .012x^2$.
   [C] (c) What is the smallest number of carts it can produce and still break even?
   [C] (d) What number of carts will produce a maximum profit and what is this profit?

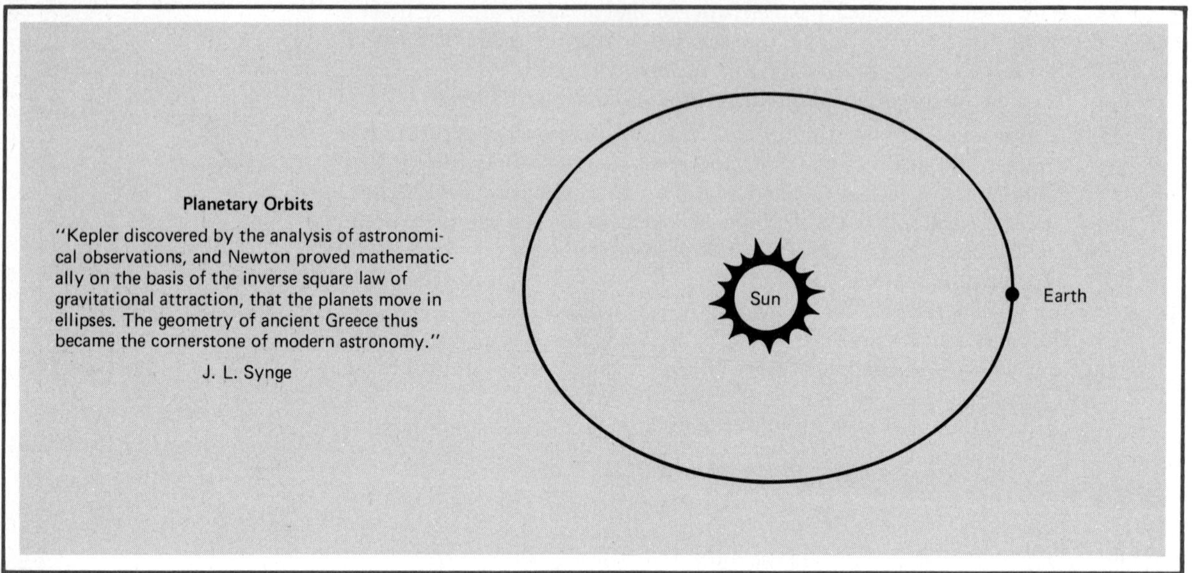

4-5 Ellipses and Hyperbolas (Optional)

Roughly speaking, an ellipse is a flattened circle. In the case of the earth's orbit about the sun, there is very little flattening (less than the opening box indicates). But all ellipses, be they nearly circular or very flat, have important properties in common. As with parabolas, we shall begin algebraically by discussing equations of ellipses.

Equations of Ellipses

Consider the equation

$$\frac{x^2}{25} + \frac{y^2}{16} = 1$$

Because x and y can be replaced by $-x$ and $-y$, respectively, without changing the equation, the graph is symmetric with respect to both axes and the origin. To find the x-intercepts, we let $y = 0$ and solve for x.

$$\frac{x^2}{25} = 1$$

$$x^2 = 25$$
$$x = \pm 5$$

Thus the graph intersects the x-axis at $(\pm 5, 0)$. By a similar procedure (letting $x = 0$), we find that the graph intersects the y-axis at $(0, \pm 4)$. Plotting these points and a few others leads to the graph at the top of page 163.

This curve is an example of an ellipse. The dotted line segment PQ with endpoints on the ellipse and passing through the origin is called a **diameter**. The longest diameter, AB, is the **major diameter** (sometimes called the major

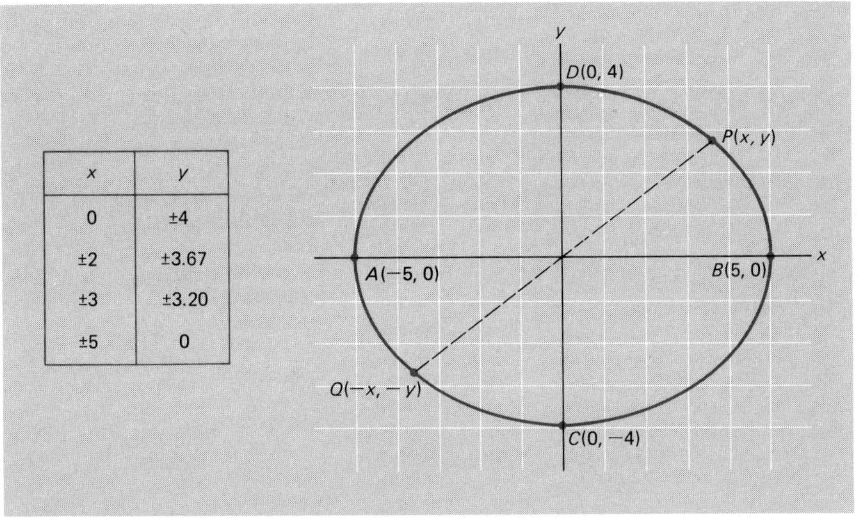

x	y
0	±4
±2	±3.67
±3	±3.20
±5	0

axis) and the shortest one, CD, is the **minor diameter**. The origin, which is a bisector of every diameter, is appropriately called the **center** of the ellipse.

More generally, if a and b are any positive numbers, the equation

$$\frac{x^2}{a^2} + \frac{y^2}{b^2} = 1$$

represents an ellipse with center at the origin and intersecting the x- and y-axes at $(\pm a, 0)$ and $(0, \pm b)$, respectively. If $a > b$, it is called a *horizontal* ellipse; if $a = b$, it is a *circle* of radius a; and if $a < b$, it is called a *vertical* ellipse.

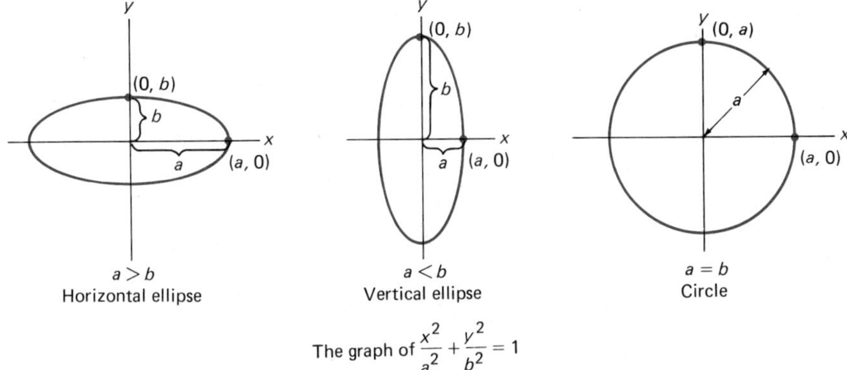

The graph of $\frac{x^2}{a^2} + \frac{y^2}{b^2} = 1$

If we move the ellipse (without turning it) so that its center is at (h, k) rather than the origin, its equation takes the form

$$\boxed{\frac{(x-h)^2}{a^2} + \frac{(y-k)^2}{b^2} = 1}$$

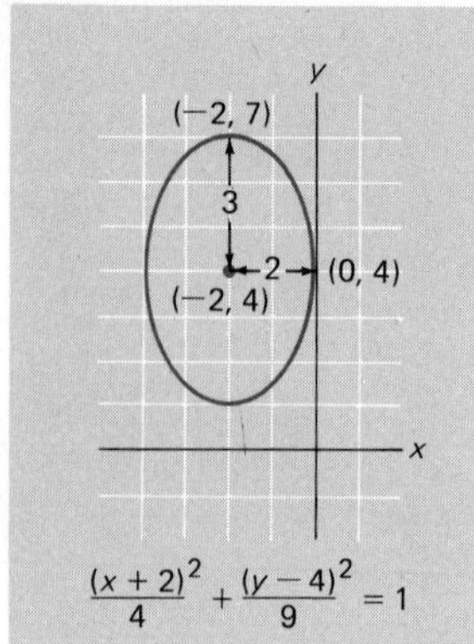

Again, the relative sizes of a and b determine whether it is a horizontal ellipse, a vertical ellipse, or a circle. For example, the graph of

$$\frac{(x+2)^2}{4} + \frac{(y-4)^2}{9} = 1$$

is a vertical ellipse centered at $(-2, 4)$ with major diameter of length $2 \cdot 3 = 6$ and minor diameter of length $2 \cdot 2 = 4$ (see the diagram in the margin).

Equations of Hyperbolas What a difference a change in sign can make! The graphs of

$$\frac{x^2}{25} + \frac{y^2}{16} = 1$$

and

$$\frac{x^2}{25} - \frac{y^2}{16} = 1$$

are as different as night and day. The first is an ellipse; the second is a hyperbola. Let us see what we can find out about this hyperbola.

First, note that it has x-intercepts ± 5, but no y-intercepts. Since x can be replaced by $-x$ and y can be replaced by $-y$ without changing the equation, the graph is symmetric with respect to both axes and the origin. If we solve for y in terms of x, we get

$$y = \pm \tfrac{4}{5}\sqrt{x^2 - 25}$$

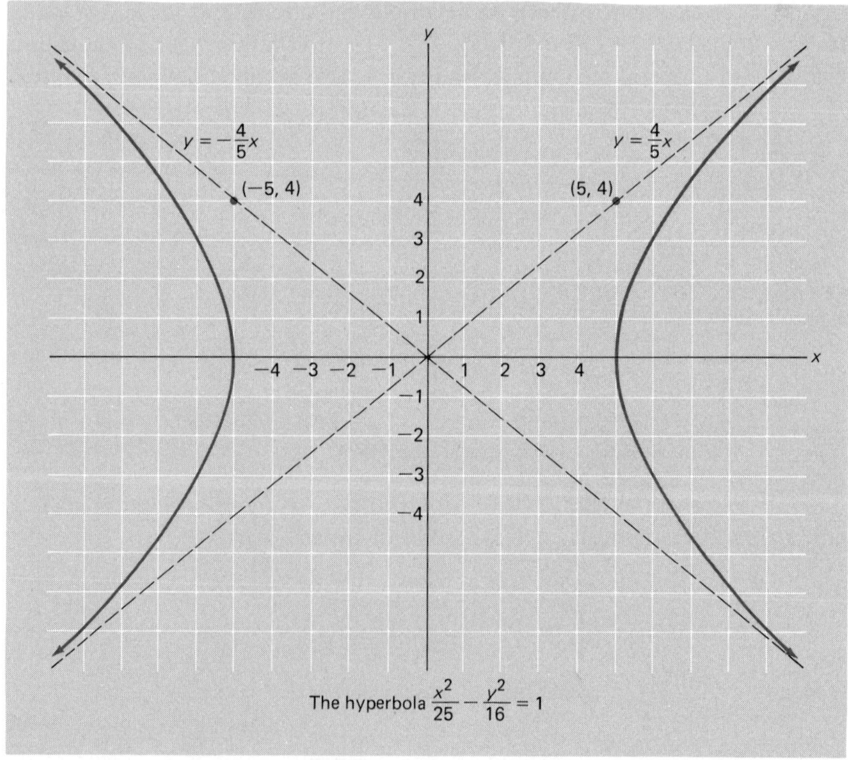

The hyperbola $\frac{x^2}{25} - \frac{y^2}{16} = 1$

This implies that we must have $|x| \geq 5$, so the graph has no points between $x = -5$ and $x = 5$. Moreover, for large x, $\sqrt{x^2 - 25}$ is about the same size as $\sqrt{x^2}$. Thus when x is large, y behaves much like $\frac{4}{5}x$ or $-\frac{4}{5}x$.

When we put all of this information together and plot a few points, we are led to the graph at the bottom of page 164.

The lines $y = \frac{4}{5}x$ and $y = -\frac{4}{5}x$ are called **asymptotes**; the curve snuggles up to them for large $|x|$. In Chapter 5, we will have more to say about asymptotes.

Geometric Definition of an Ellipse Instead of approaching the ellipse through equations as we have done, a geometric condition can be used to define it. Suppose that a point P moves in the plane in such a way that the sum of its distances from two fixed points F_1 and F_2 is a constant. The resulting curve is an **ellipse** with center O midway between F_1 and F_2. The fixed points F_1 and F_2 are called the **foci** (plural of focus) of the ellipse.

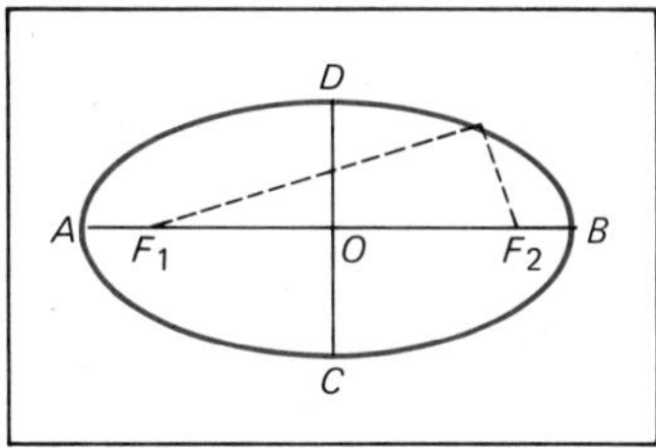

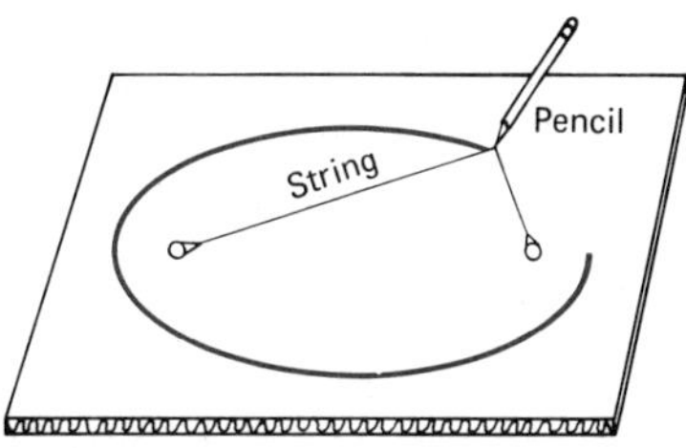

To understand this definition, look at the two pictures above. The one on the right shows how to draw an ellipse. Start with a sheet of cardboard, a piece of string with small loops at each end, two thumbtacks, and a pencil. Stick the two thumbtacks through the small loops and into the cardboard, pull the string taut with the pencil, and trace a curve as shown. The thumbtacks are the foci of the resulting ellipse.

We should be able to derive the equation of an ellipse from the definition just given. To illustrate, consider the ellipse with foci $F_1(-3, 0)$ and $F_2(3, 0)$ having "string length" 10. If $P(x, y)$ is any point on this ellipse,

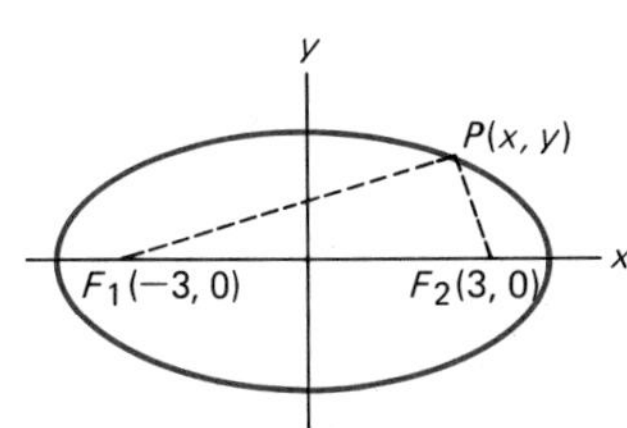

$$d(P, F_1) + d(P, F_2) = 10$$

This is equivalent (by the distance formula) to

$$\sqrt{(x + 3)^2 + y^2} + \sqrt{(x - 3)^2 + y^2} = 10$$

We could stop here and call this the equation of the ellipse, but it does not look much like the simple equations we studied at the beginning of this section.

To simplify, transfer one of the square roots to the right side and then square both sides.

$$\sqrt{(x-3)^2+y^2} = 10 - \sqrt{(x+3)^2+y^2}$$

$$(x-3)^2+y^2 = 100 - 20\sqrt{(x+3)^2+y^2} + (x+3)^2+y^2$$

That trick got rid of one square root, which is progress; perhaps another squaring operation will get rid of the other. But first we should simplify as much as we can, being sure to have the remaining radical alone on one side of the equation.

$$x^2-6x+9+y^2 = 100 - 20\sqrt{(x+3)^2+y^2} + x^2+6x+9+y^2$$

$$-12x-100 = -20\sqrt{(x+3)^2+y^2}$$

$$3x+25 = 5\sqrt{(x+3)^2+y^2}$$

Squaring both sides and simplifying leads to

$$16x^2+25y^2 = 400$$

which is really quite nice. Dividing both sides of the equation by 400 gives the form that we studied earlier.

$$\frac{x^2}{25}+\frac{y^2}{16} = 1$$

Suppose that we had chosen $(-c, 0)$ and $(c, 0)$ as foci and $2a$ as the "string length." A similar derivation would lead to

$$\frac{x^2}{a^2}+\frac{y^2}{a^2-c^2} = 1$$

If we let $b^2 = a^2 - c^2$, we get

$$\frac{x^2}{a^2}+\frac{y^2}{b^2} = 1$$

which can be recognized as an equation studied earlier. Notice that since $a > b$, this is a horizontal ellipse. To get a vertical ellipse from the geometric definition, we would put the foci on the y-axis.

The Conic Sections The Greeks called the four curves—circles, parabolas, ellipses, and hyperbolas—*conic sections.* If you take a cone with two nappes and pass planes through it at various angles, the curve of intersection in each instance is one of these four curves (assuming that the intersecting plane does not pass through the apex of the cone). The diagrams below illustrate this important fact.

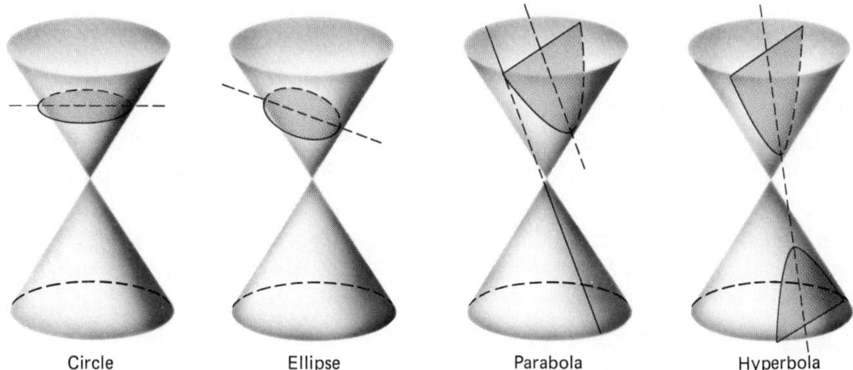

Problem Set 4-5

Each of the following equations represents an ellipse. Find its center and the endpoints of the major and minor diameters and graph the ellipse.

1. $\dfrac{x^2}{25} + \dfrac{y^2}{9} = 1$
2. $\dfrac{x^2}{36} + \dfrac{y^2}{16} = 1$
3. $\dfrac{x^2}{9} + \dfrac{y^2}{25} = 1$
4. $\dfrac{x^2}{4} + \dfrac{y^2}{25} = 1$
5. $\dfrac{(x-2)^2}{25} + \dfrac{(y+1)^2}{9} = 1$
6. $\dfrac{(x-3)^2}{36} + \dfrac{(y-4)^2}{25} = 1$
7. $\dfrac{(x+3)^2}{9} + \dfrac{y^2}{16} = 1$
8. $\dfrac{x^2}{4} + \dfrac{(y-4)^2}{49} = 1$

In Problems 9–14, find the center of the ellipse with AB and CD as major and minor diameters, respectively. Then write the equation of the ellipse.

9. $A(6, 0)$, $B(-6, 0)$, $C(0, 3)$, $D(0, -3)$
10. $A(5, 0)$, $B(-5, 0)$, $C(0, 1)$, $D(0, -1)$
11. $A(0, 6)$, $B(0, -6)$, $C(4, 0)$, $D(-4, 0)$
12. $A(0, 3)$, $B(0, -3)$, $C(2, 0)$, $D(-2, 0)$
13. $A(-4, 3)$, $B(8, 3)$, $C(2, 1)$, $D(2, 5)$
    (*Hint*: Use the midpoint formula to find the center.)
14. $A(3, -4)$, $B(11, -4)$, $C(7, -7)$, $D(7, -1)$

Example (Changing to standard form) Change the equation

$$x^2 + 4y^2 - 8x + 16y = -28$$

to the standard form for the equation of an ellipse. Decide whether the ellipse is horizontal or vertical, identify its center, and give the lengths of the major and minor diameters.

SOLUTION. We use the familiar process of completing the squares.

$$\begin{aligned}(x^2 - 8x + \quad) + 4(y^2 + 4y + \quad) &= -28\\ (x^2 - 8x + 16) + 4(y^2 + 4y + 4) &= -28 + 16 + 4(4)\\ (x - 4)^2 + 4(y + 2)^2 &= 4\\ \frac{(x - 4)^2}{4} + \frac{(y + 2)^2}{1} &= 1\end{aligned}$$

Since the larger denominator is in the x-term, we conclude that the ellipse is horizontal. Its center is at $(4, -2)$, the major diameter has length 4, and the minor diameter has length 2.

For each of the following, change the equation to standard form, decide whether the corresponding ellipse is horizontal or vertical, identify its center, and give the lengths of major and minor diameters.

15. $16x^2 + 9y^2 + 64x - 54y = -1$
16. $4x^2 + y^2 - 8x + 8y = -16$
17. $9x^2 + 4y^2 - 72x = -108$
18. $25x^2 + 4y^2 + 16y = 84$
19. $9x^2 + 16y^2 + 36x - 96y = -36$
20. $x^2 + 4y^2 - 2x + 32y = -61$

Sketch the graphs of the following hyperbolas. Show the asymptotes as dotted lines.

21. $\frac{x^2}{9} - \frac{y^2}{36} = 1$
22. $\frac{x^2}{25} - \frac{y^2}{4} = 1$
23. $\frac{x^2}{25} - \frac{y^2}{9} = 1$
24. $\frac{x^2}{4} - \frac{y^2}{9} = 1$
25. $\frac{(x - 2)^2}{25} - \frac{(y - 1)^2}{4} = 1$
26. $\frac{(x + 1)^2}{4} - \frac{(y - 2)^2}{9} = 1$

Miscellaneous Problems

In Problems 27–34, decide whether the graph of the given equation is a circle, an ellipse, or a hyperbola. Then sketch its graph.

27. $\frac{x^2}{64} + \frac{y^2}{16} = 1$
28. $\frac{x^2}{16} + \frac{y^2}{16} = 1$
29. $\frac{x^2}{64} - \frac{y^2}{16} = 1$
30. $\frac{x^2}{4} + \frac{y^2}{9} = 1$
31. $\frac{(x - 2)^2}{11} + \frac{(y + 3)^2}{11} = 1$
32. $\frac{(x - 2)^2}{4} - \frac{(y + 2)^2}{9} = 1$
33. $x^2 + 2y^2 - 6x + 8y = 1$
34. $x^2 - 2y^2 - 6x + 8y = 1$
35. Find the equation of the ellipse whose major diameter has endpoints $(-4, 2)$ and $(10, 2)$ and whose minor diameter has length 10.
36. Find the equation of the ellipse that passes through the point $(3, 2)$ and has $(\pm 5, 0)$ as endpoints of the major diameter.

37. Find the equation of the ellipse with foci at $(\pm 6, 0)$ and with major diameter of length 20.

38. Find the equation of the ellipse with foci at $(0, \pm 6)$ and with major diameter of length 20.

C 39. Find the y-coordinates of the points for which $x = \pm 2.5$ on the ellipse

$$\frac{x^2}{24} + \frac{y^2}{19} = 1$$

C 40. Find the y-coordinates of the points for which $x = \pm 5$ on the hyperbola

$$\frac{x^2}{17} - \frac{y^2}{11} = 1$$

C 41. The area of the ellipse $x^2/a^2 + y^2/b^2 = 1$ is πab. Find the areas (correct to two decimal places) of the following ellipses.

(a) $\frac{x^2}{7} + \frac{y^2}{11} = 1$ (b) $\frac{x^2}{111} + y^2 = 1$

C 42. Find the radius (correct to 2 decimal places) of the circle which has the same area as each ellipse.

(a) $\frac{x^2}{256} + \frac{y^2}{89} = 1$ (b) $\frac{x^2}{50} + \frac{y^2}{19} = 1$

43. Let $F_1(-5, 0)$ and $F_2(5, 0)$ be two given points and let $P(x, y)$ move so that

$$|d(P, F_1) - d(P, F_2)| = 8$$

Derive and simplify the equation of the path. It should turn out to be a hyperbola.

CHAPTER SUMMARY

Like a city planner, we introduce in the plane two main streets, one vertical (the y-**axis**) and the other horizontal (the x-**axis**). Relative to these axes, we can specify any point by giving its address (x, y). The numbers x and y, called **cartesian coordinates**, measure the directed distances from the vertical and horizontal axes, respectively. And given two points A and B with addresses (x_1, y_1) and (x_2, y_2), we may calculate the distance between them from the **distance formula**:

$$d(A, B) = \sqrt{(x_2 - x_1)^2 + (y_2 - y_1)^2}$$

In **analytic geometry**, we use the notion of coordinates to combine algebra and geometry. Thus we may graph the equation $y = x^2$ (algebra), thereby turning it into a curve (geometry). Conversely, we may take the circle with radius 6 and center $(-1, 4)$ and give it the equation

$$(x + 1)^2 + (y - 4)^2 = 36$$

The simplest of all curves is a **line**. If a line passes through (x_1, y_1) and (x_2, y_2) with $x_1 \neq x_2$, then its **slope** m is given by

$$m = \frac{\text{rise}}{\text{run}} = \frac{y_2 - y_1}{x_2 - x_1}$$

There are two important forms for the equation of a nonvertical line.

point-slope form:	$y - y_1 = m(x - x_1)$
slope-intercept form:	$y = mx + b$

Vertical lines do not have slope; their equations take the form $x = k$. All lines (vertical and nonvertical) can be written in the form of the **general linear equation.**

$$Ax + By + C = 0$$

The distance d from the point (x_1, y_1) to the line $Ax + By + C = 0$ is

$$d = \frac{|Ax_1 + By_1 + C|}{\sqrt{A^2 + B^2}}$$

Nonvertical lines are parallel if their slopes are equal, and they are perpendicular if their slopes are negative reciprocals.

Somewhat more complicated curves are the *conic sections*: **parabolas, circles, ellipses**, and **hyperbolas**. Here are typical equations for them.

parabola:	$(x - h)^2 = 4p(y - k)$
circle:	$(x - h)^2 + (y - k)^2 = r^2$
ellipse:	$\frac{(x - h)^2}{a^2} + \frac{(y - k)^2}{b^2} = 1$
hyperbola:	$\frac{(x - h)^2}{a^2} - \frac{(y - k)^2}{b^2} = 1$

CHAPTER REVIEW PROBLEM SET

1. Name the graph of each of the following.
   (a) $y = -4x + 3$ (b) $x^2 + y^2 = 25$
   (c) $2x - 3y = 0$ (d) $y = x^2$
   (e) $(y - 2)^2 = -3(x + 1)$ (f) $(x - 1)^2 + (y + 2)^2 = 9$
   (g) $\frac{x^2}{9} + \frac{y^2}{49} = 1$ (h) $\frac{x^2}{9} - \frac{y^2}{49} = 1$
   (i) $x^2 - 4y = 16$ (j) $x^2 + 4y^2 = 16$
   (k) $x^2 - 4y^2 = 16$ (l) $x^2 - 4y^2 = -16$
2. Sketch the graph of each of the following equations. For some, you will need to make a table of values.
   (a) $y = -3x + 4$ (b) $x^2 + y^2 = 16$

(c) $y = x^2 + \dfrac{1}{x}$ (d) $y = x^3 - 4x$

(e) $y = \sqrt{x} + 2$ (f) $\dfrac{x^2}{25} + \dfrac{y^2}{9} = 1$

3. Consider the triangle determined by the points $A(-2, 3)$, $B(3, 5)$ and $C(1, 9)$.
   (a) Sketch the triangle.
   (b) Find the lengths of the three sides.
   (c) Find the slopes of the lines which contain the three sides.
   (d) Write equations for each of these lines.
   (e) Find the midpoints of AB, BC, and CA.
   (f) Find the equation of the line through A parallel to BC.
   (g) Find the equation of the line through A perpendicular to BC.
   (h) Find the length of the altitude from A to side BC.
   (i) Calculate the area of the triangle.
   (j) Show that the line segment joining the midpoints of AB and AC is parallel to BC and is one-half the length of BC.
4. Write the equation of the line satisfying the given conditions in the form $Ax + By + C = 0$.
   (a) It is vertical and passes through $(4, -1)$.
   (b) It is horizontal and passes through $(4, -1)$.
   (c) It is parallel to the line $3x + 2y = 4$ and passes through $(4, -1)$.
   (d) It passes through $(2, -5)$ and has x-intercept 3.
   (e) It is perpendicular to the line $2y = -3x$ and passes through $(-2, 3)$.
   (f) It is parallel to the lines $y = 4x + 5$ and $y = 4x - 7$ and is midway between them.
   (g) It passes through $(4, 6)$ and has equal x- and y-intercepts.
   (h) It is tangent to the circle $x^2 + y^2 = 169$ at the point $(12, 5)$.
5. Find the vertex, focus, and directrix of the following parabolas.
   (a) $x^2 = -6y$ (b) $y^2 = 16x$
   (c) $(x + 2)^2 = 10(y - 1)$ (d) $(y + \frac{1}{2})^2 = -3(x - \frac{1}{4})$
   (e) $y = x^2 - 4x$ (f) $2y^2 + 12y = x - 1$
6. Write the equation of the parabola satisfying the following conditions.
   (a) Vertex at $(0, 0)$, focus at $(-5, 0)$.
   (b) Focus at $(0, 8)$, directrix $y = -8$.
   (c) Vertex at $(3, 0)$, axis of symmetry $y = 0$, and passing through $(1, 12)$.
7. Find the points of intersection of the parabola $y = x^2 + x$ and the line $6x - y - 4 = 0$.
8. By completing the squares, identify each of the following conic sections and give pertinent information about it.
   (a) $x^2 + 2x + y^2 - 4y = 2$
   (b) $2x^2 + y^2 + 4x - 4y = 14$
   (c) $2x^2 - y^2 + 4x - 4y = 14$
   (d) $2x^2 + 4x - y = 14$
9. Find the equation of the ellipse satisfying the following conditions.
   (a) Foci $(0, \pm 4)$, $a = 5$.
   (b) Endpoints of major diameter $(0, \pm 13)$ and endpoints of minor diameter $(\pm 5, 0)$.

Mathematicians do not deal in objects, but in relations between objects; thus, they are free to replace some objects by others so long as the relations remain unchanged. Content to them is irrelevant: they are interested in form only.

Henri Poincaré

CHAPTER FIVE

Functions and Their Graphs

5-1 Functions

5-2 Graphs of Functions

5-3 Graphing Rational Functions

5-4 Putting Functions Together

5-5 Inverse Functions

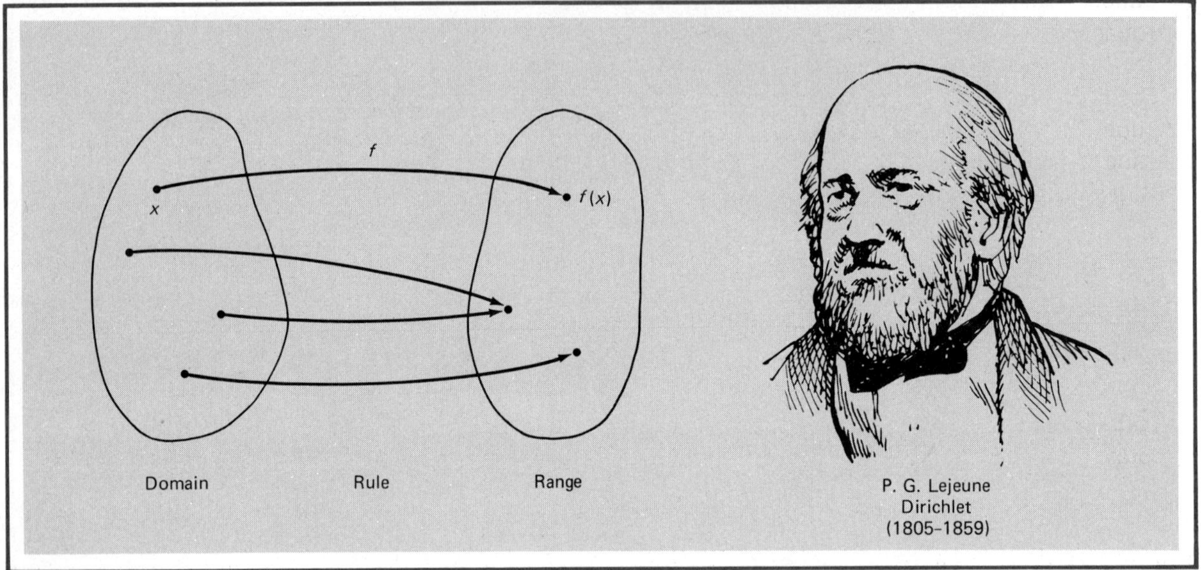

P. G. Lejeune Dirichlet (1805–1859)

5-1 Functions

One of the most important ideas in mathematics is that of a function. For a long time, mathematicians and scientists wanted a precise way to describe the relationships that may exist between two variables. It is somewhat surprising that it took so long for the idea to crystallize into a clear, unambiguous concept. The French mathematician P. G. Lejeune Dirichlet (1805–1859) is credited with the modern definition of function.

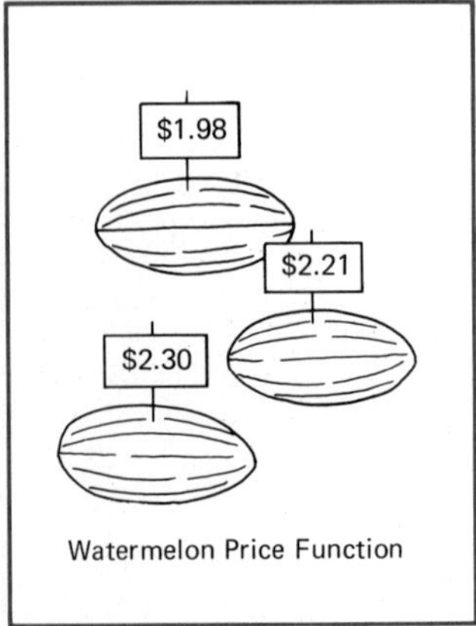

Watermelon Price Function

DEFINITION

A **function** *is a rule which assigns to each element in one set (called the domain of the function) exactly one element of another set (called the range of the function).*

Three examples will help clarify this idea. When a grocer puts a price tag on each of the watermelons for sale, a function is determined. Its domain is the set of watermelons, its range is the set of prices, and the rule is the procedure the grocer uses in assigning prices (perhaps a specified amount per pound).

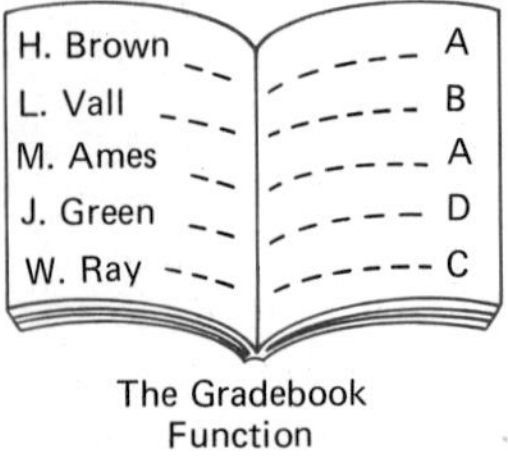

The Gradebook Function

When a professor assigns a grade to each student in a class, he or she is determining a function. The domain is the set of students and the range is the set of grades, but who can say what the rule is? Many professors prefer to keep that a secret.

A much more typical function from our point of view is the *squaring* function displayed in the margin. It takes a number from the domain $\{-2, -1, 0, 1, 2, 3\}$ and squares it, producing a number in the range $\{0, 1, 4, 9\}$. This function is typical for two reasons: Both the domain and range are sets of numbers and the rule can be specified by giving an algebraic formula. Most functions in this book will be of this type.

The Squaring Function

Functional Notation Long ago, mathematicians introduced a special notation for functions. A single letter like f (or g or h) is used to name the function. Then $f(x)$, read *f of x* or *f at x*, denotes the value that f assigns to x. Thus, if f names the squaring function,

$$f(-2) = 4 \qquad f(2) = 4 \qquad f(-1) = 1$$

and, in general,

$$f(x) = x^2$$

We call this last result the *formula* for the function f. It tells us in a concise algebraic way what f does to any number. Notice that the given formula and

$$f(y) = y^2 \qquad f(z) = z^2$$

all say the same thing; the letter used for the domain variable is a matter of no significance, though it does happen that we shall usually use x. Many functions do not have simple formulas (see Problems 17 and 18), but in this book, most of them do.

For a further example, consider the function that cubes a number and then subtracts 1 from the result. If we name this function g, then

$$\begin{aligned} g(2) &= 2^3 - 1 = 7 \\ g(-1) &= (-1)^3 - 1 = -2 \\ g(.5) &= (.5)^3 - 1 = -.875 \\ g(\pi) &= \pi^3 - 1 \approx 30 \end{aligned}$$

and, in general,

$$g(x) = x^3 - 1$$

Few students would have trouble using this formula when x is replaced by a specific number. However, it is important to be able to use it when x is replaced by anything whatever, even an algebraic expression. Be sure you understand the following calculations.

$$\begin{aligned} g(a) &= a^3 - 1 \\ g(y^2) &= (y^2)^3 - 1 = y^6 - 1 \\ g\left(\frac{1}{z}\right) &= \left(\frac{1}{z}\right)^3 - 1 = \frac{1}{z^3} - 1 \\ g(2 + h) &= (2 + h)^3 - 1 = 8 + 12h + 6h^2 + h^3 - 1 \\ &= h^3 + 6h^2 + 12h + 7 \\ g(x + h) &= (x + h)^3 - 1 = x^3 + 3x^2h + 3xh^2 + h^3 - 1 \end{aligned}$$

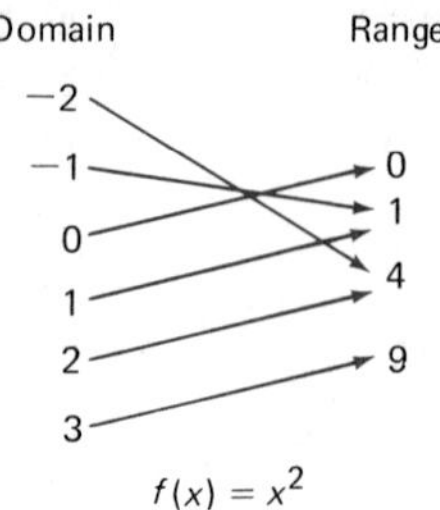

Domain and Range The rule of correspondence is the heart of a function, but a function is not completely determined until its domain is given. Recall that the **domain** is the set of elements to which the function assigns values. In the case of the squaring function f (reproduced in the margin), we gave the domain as the set $\{-2, -1, 0, 1, 2, 3\}$. We could just as well have specified the domain as the set of all real numbers; the formula $f(x) = x^2$ would still make perfectly good sense. In fact, there is a common agreement that if no domain is specified, it is understood to be the largest set of real numbers for which the rule for the function makes sense and gives real number values. We call it the **natural domain** of the function. Thus if no domain is specified for the function with formula $g(x) = x^3 - 1$, it is assumed to be the set of all real numbers. Similarly if

$$h(x) = \frac{1}{x - 1}$$

we would take the natural domain to consist of all real numbers except 1. Here, the number 1 is excluded to avoid division by zero.

Once the domain is understood and the rule of correspondence is given, the **range** of the function is determined. It is the set of values of the function. Here are several examples.

RULE	DOMAIN	RANGE
$F(x) = 4x$	All reals	All reals
$G(x) = \sqrt{x - 3}$	$\{x: x \geq 3\}$	Nonnegative reals
$H(x) = \dfrac{1}{(x - 2)^2}$	$\{x: x \neq 2\}$	Positive reals

Independent and Dependent Variables Scientists like to use the language of variables in talking about functions. Let us illustrate by referring to an object falling under the influence of gravity near the earth's surface. If the object is very dense (so air resistance can be neglected), it will fall according to the formula

$$d = 16t^2$$

where d represents the distance in feet the object falls during the first t seconds. Since time moves along quite independently of everything else, a physicist would call t the **independent variable**. The distance d depends on t; it is called the **dependent variable**. And d is said to be a function of t.

Variation Sometimes we describe the relationship of a dependent variable to one or more independent variables using the language of variation. To say that y **varies directly** as x means that $y = kx$ for some fixed number k. To say that y **varies inversely** as x means that $y = k/x$ for some fixed number k.

Finally, to say that z **varies jointly** as x and y means that $z = kxy$ for a fixed number k.

In variation problems, we are often not only given the form of the relationship, but also some set of corresponding values of the variables involved. Then we can evaluate the constant k. Suppose, for example, that y varies directly as x and that $y = 10$ when $x = -2$. Substituting these values in the equation $y = kx$, we get $10 = k(-2)$. Thus $k = -5$ and we can write the explicit formula $y = -5x$.

Problem Set 5-1

1. If $f(x) = x^2 - 4$, evaluate each expression.
   (a) $f(-2)$ (b) $f(0)$ (c) $f(\frac{1}{2})$ (d) $f(.1)$
   (e) $f(\sqrt{2})$ (f) $f(a)$ (g) $f(1/x)$ (h) $f(x + 1)$
2. If $f(x) = (x - 4)^2$, evaluate each expression in Problem 1.
3. If $f(x) = 1/(x - 4)$, evaluate each expression.
   (a) $f(8)$ (b) $f(2)$ (c) $f(\frac{9}{2})$ (d) $f(\frac{31}{8})$
   (e) $f(4)$ (f) $f(4.01)$ (g) $f(1/x)$ (h) $f(x^2)$
   (i) $f(2 + h)$ (j) $f(2 - h)$
4. If $f(x) = x^2$ and $g(x) = 2/x$, evaluate each expression.
   (a) $f(-7)$ (b) $g(-4)$ (c) $f(\frac{1}{4})$ (d) $1/f(4)$
   (e) $g(\frac{1}{4})$ (f) $1/g(4)$ (g) $g(0)$ (h) $g(1)f(1)$
   (i) $f(g(1))$ (j) $f(1)/g(1)$

Example A (Finding natural domains) Find the natural domain of (a) $f(x) = 4x/[(x + 2)(x - 3)]$; (b) $g(x) = \sqrt{x^2 - 4}$.

SOLUTION. We recall that the natural domain is the largest set of real numbers for which the formula makes sense and gives real number values. Thus in part (a), the domain consists of all real numbers except -2 and 3; in part (b) we must have $x^2 \geq 4$, which is equivalent to $|x| \geq 2$. Notice that if $|x| < 2$, we would be taking the square root of a negative number, so the result would not be a real number.

In Problems 5–16, find the natural domain of the given function.

5. $f(x) = x^2 - 4$
6. $f(x) = (x - 4)^2$
7. $g(x) = \dfrac{1}{x^2 - 4}$
8. $g(x) = \dfrac{1}{9 - x^2}$
9. $h(x) = \dfrac{2}{x^2 - x - 6}$
10. $h(x) = \dfrac{1}{2x^2 + 3x - 2}$
11. $F(x) = \dfrac{1}{x^2 + 4}$
12. $F(x) = \dfrac{1}{9 + x^2}$
13. $G(x) = \sqrt{x - 2}$
14. $G(x) = \sqrt{x + 2}$
15. $H(x) = \dfrac{1}{5 - \sqrt{x}}$
16. $H(x) = \dfrac{1}{\sqrt{x + 1} - 2}$

17. Not all functions arising in mathematics have rules given by simple algebraic formulas. Let $f(n)$ be the nth digit in the decimal expansion of

$$\pi = 3.14159265358979323846\ldots$$

Thus $f(1) = 3$ and $f(3) = 4$. Find (a) $f(6)$; (b) $f(9)$; (c) $f(16)$. What is the natural domain for this function?

18. Let g be the function which assigns to each positive integer the number of factors in its prime factorization. Thus

$$\begin{aligned} g(2) &= 1 \\ g(4) &= g(2\cdot 2) = 2 \\ g(36) &= g(2\cdot 2\cdot 3\cdot 3) = 4 \end{aligned}$$

Find (a) $g(24)$; (b) $g(37)$; (c) $g(64)$; (d) $g(162)$. Can you find a formula for this function?

Example B (Functions of two variables) The formula $f(x, y) = x^2 + 2y^2$ determines a function of two variables. It assigns a number to each ordered pair of numbers (x, y). For example,

$$\begin{aligned} f(2, -3) &= 2^2 + 2(-3)^2 = 22 \\ f(3, 8) &= 3^2 + 2(8)^2 = 137 \end{aligned}$$

Let $g(x, y) = 3xy - 5x$ and $G(x, y) = (5x + 3y)/(2x - y)$. Find each of the following.

19. $g(2, 5)$
20. $g(-1, 3)$
21. $g(5, 2)$
22. $g(3, -1)$
23. $G(1, 1)$
24. $G(3, 3)$
25. $G(\frac{1}{2}, 1)$
26. $G(5, 0)$
27. $g(2x, 3y)$
28. $G(2x, 4y)$
29. $g(x, 1/x)$
30. $G(x - y, y)$

Example C (More on variation) Suppose that w varies jointly as x and the square root of y and that $w = 14$ when $x = 2$ and $y = 4$. Find an explicit formula for w.

SOLUTION. We translate the first statement into mathematical symbols as

$$w = kx\sqrt{y}$$

To evaluate k, we substitute the given values for w, x, and y.

$$14 = k\cdot 2\sqrt{4} = 4k$$

or

$$k = \tfrac{14}{4} = \tfrac{7}{2}$$

Thus the explicit formula for w is

$$w = \tfrac{7}{2}x\sqrt{y}$$

In Problems 31–36, find an explicit formula for the dependent variable.

31. y varies directly as x, and $y = 12$ when $x = 3$.
32. y varies directly as x^2, and $y = 4$ when $x = 0.1$.
33. y varies inversely as x (that is, $y = k/x$), and $y = 5$ when $x = \frac{1}{5}$.

34. V varies jointly as r^2 and h, and $V = 75$ when $r = 5$ and $h = 9$.

35. I varies directly as s and inversely as d^2, and $I = 9$ when $s = 4$ and $d = 12$.

36. W varies directly as x and inversely as the square root of yz, and $W = 5$ when $x = 7.5$, $y = 2$, and $z = 18$.

Ⓒ 37. For Problem 35, find the value of I corresponding to $s = 4.26$ and $d = 13.1$.

Ⓒ 38. For Problem 36, find the value of W corresponding to $x = 3.49$, $y = 13.1$, and $z = 14.6$.

Example D (Applied variation problems) The volume of a certain gas varies directly as the temperature and inversely as the pressure. Suppose that the volume is 462 cubic inches when the temperature is 440°K (absolute or Kelvin units of temperature) and the pressure is 40 pounds per square inch. (a) Write an explicit formula for the volume in terms of temperature and pressure. (b) Use your formula to find the volume when the temperature is 300°K and the pressure is 70 pounds per square inch.

SOLUTION. (a) First we introduce letters for the three variables.

V: the volume in cubic inches
T: the Kelvin temperature reading
P: the pressure in pounds per square inch

Then $V = kT/P$, where k is a constant. Next we substitute the given values for V, T, and P.

$$462 = \frac{k(440)}{40}$$

$$k = \frac{40}{440}(462) = 42$$

Our explicit formula is $V = 42T/P$. (b) If $T = 300$ and $P = 70$, $V = (42)(300)/70 = 180$. The volume of the gas is 180 cubic inches.

39. The maximum range of a projectile varies as the square of the initial velocity. If the range is 16,000 feet when the initial velocity is 600 feet per second,
    (a) write an explicit formula for R in terms of v, where R is the range in feet and v is the initial velocity in feet per second;
    (b) use this formula to find the range when the initial velocity is 800 feet per second.

40. The time of a single vibration of a pendulum varies as the square root of the length of the pendulum. If a pendulum 64 centimeters long beats every 2 seconds, how long must a pendulum be to beat 3 times per second?

41. Suppose that the amount of gasoline used by a car varies jointly as the distance traveled and the square root of the average speed. If a car used 8 gallons on a 100-mile trip going at an average speed of 64 miles per hour, how many gallons would that car use on a 160-mile trip at an average speed of 25 miles per hour?

Miscellaneous Problems

42. If $f(x) = -3x^2 + (12/x)$, evaluate each expression.
(a) $f(2)$ (b) $f(-1)$ (c) $f(\frac{3}{2})$
(d) $f(1/x)$ (e) $f(\sqrt{3})$ (f) $f(a + b)$

43. If $g(x) = [5/(x - 2)^2] + 4x$, evaluate each expression.
(a) $g(0)$ (b) $g(3)$ (c) $g(1)$
(d) $g(\frac{3}{2})$ (e) $g(2.1)$ (f) $g(2 + h)$

44. If $f(x, y) = (2x^2 + xy)/(x - y)$, evaluate each expression.
(a) $f(2, 1)$ (b) $f(3, 0)$ (c) $f(2\sqrt{2}, \sqrt{2})$

In Problems 45–48, find the natural domain of the function.

45. $g(x) = \dfrac{x - 2}{(x^2 + 1)(x - 4)}$

46. $g(x) = \dfrac{3x + 8}{(2 - x)(x^2 + 7)}$

47. $h(x) = \dfrac{2x}{2 - \sqrt{x}}$

48. $h(x) = \dfrac{3x^2 + 2}{1 - \sqrt{2x}}$

In Problems 49 and 50, find an explicit formula for y and evaluate y when x = 4.

49. $y = kx^3$, and $y = 4$ when $x = 0.1$.

50. y varies inversely as x^2, and $y = 48$ when $x = \frac{1}{4}$.

In Problems 51–54, translate the given formula into a verbal statement using the language of variation.

51. $y = \dfrac{k}{x^3}$

52. $y = \dfrac{ks^2}{t}$

53. $z = \dfrac{kxy}{\sqrt{w}}$

54. $u = \dfrac{kr^2}{\sqrt{st}}$

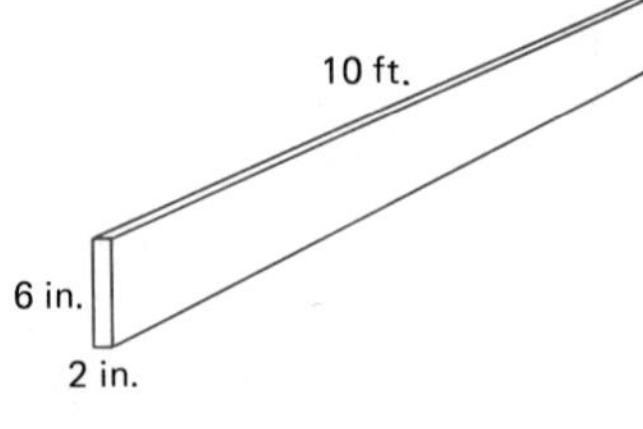

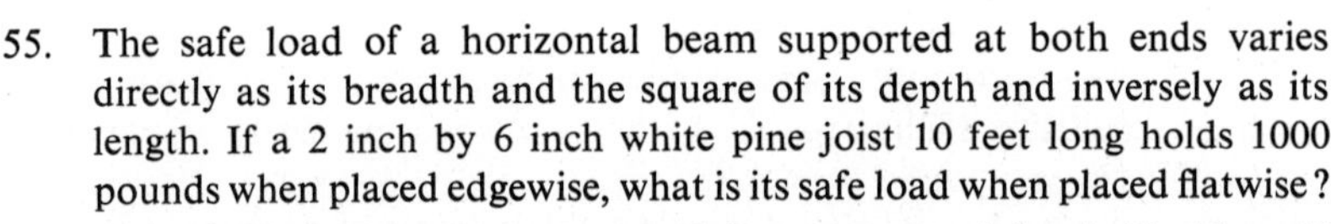

55. The safe load of a horizontal beam supported at both ends varies directly as its breadth and the square of its depth and inversely as its length. If a 2 inch by 6 inch white pine joist 10 feet long holds 1000 pounds when placed edgewise, what is its safe load when placed flatwise?

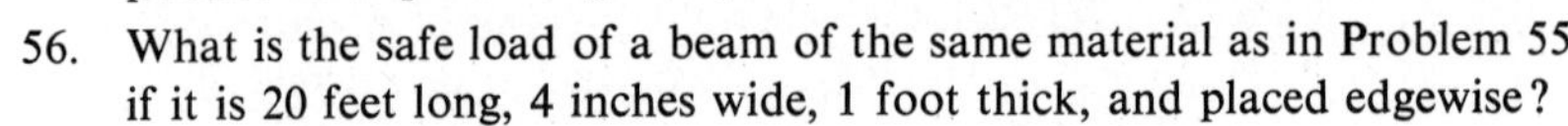

56. What is the safe load of a beam of the same material as in Problem 55 if it is 20 feet long, 4 inches wide, 1 foot thick, and placed edgewise?

[C] 57. The density $D(T)$ of dry air at a pressure of 76 centimeters of mercury and at a temperature of T degrees Celsius is given by

$$D(T) = \frac{.001293}{1 + (.00367)T}$$

(a) What is the density at 13.6 degrees Celsius?
(b) Does the density increase or decrease as the temperature increases?

58. A water tank has the shape of a right circular cone with vertex downward. The radius of the top is 20 feet and the height is 40 feet. Express the volume $V(d)$ of the water as a function of its depth.

59. A plant has the capacity to produce from 1 to 100 refrigerators per day. The daily overhead for the plant is \$2200 and the direct cost (labor and material) of producing one refrigerator is \$151. Write a formula for $T(x)$, the total cost of producing x refrigerators per day, and also the unit cost $U(x)$. What is the domain for these functions?

60. It cost the ABC company $800 + 20\sqrt{x}$ dollars to make x toy ovens. Each oven sells for \$6. Express the total profit $T(x)$ and average profit $A(x)$ as functions of x. What are the domains for these functions?

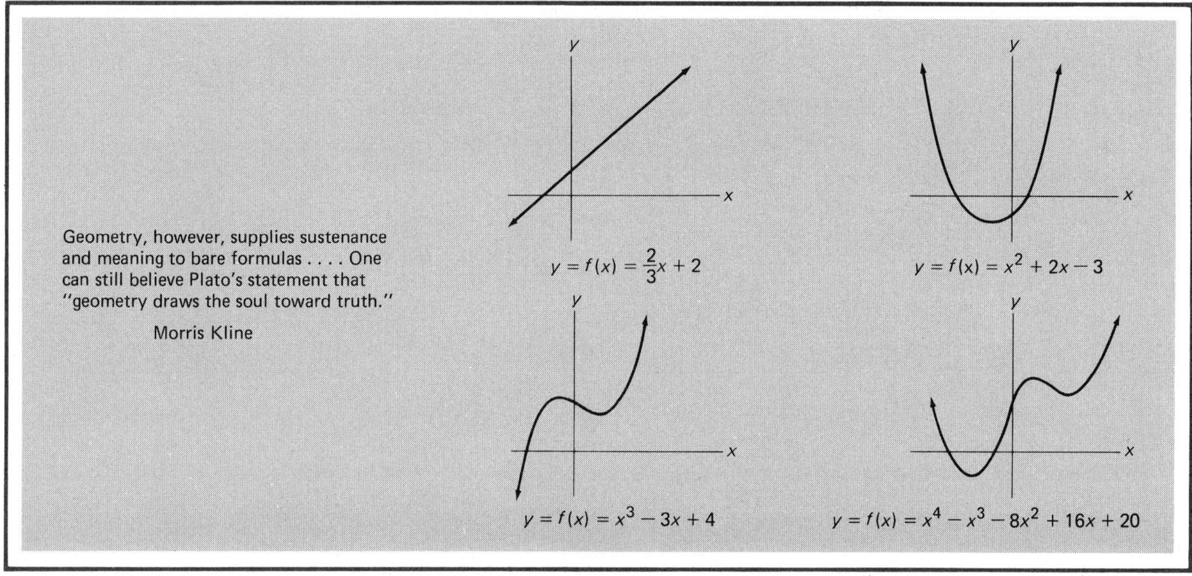

5-2 Graphs of Functions

We have said that functions are usually specified by giving formulas. Formulas are fine for manipulation and almost essential for exact quantitative information, but to grasp the overall qualitative features of a function, we need a picture. The best picture of a function is its graph. And the **graph of a function** f is simply the graph of the equation $y = f(x)$. We learned how to graph equations in the previous chapter.

Polynomial Functions We look first at polynomial functions, that is, functions of the form

$$f(x) = a_n x^n + a_{n-1} x^{n-1} + \cdots + a_1 x + a_0$$

Four typical graphs are shown above. We know from the last chapter that the graph of $f(x) = ax + b$ is always a straight line and that, if $a \neq 0$, the graph of $f(x) = ax^2 + bx + c$ is necessarily a parabola.

The graphs of higher degree polynomial functions are harder to describe, but after we have studied two examples, we can offer some general guidelines. Consider first the cubic function

$$f(x) = x^3 - 3x + 4$$

With the help of a table of values, we sketch its graph.

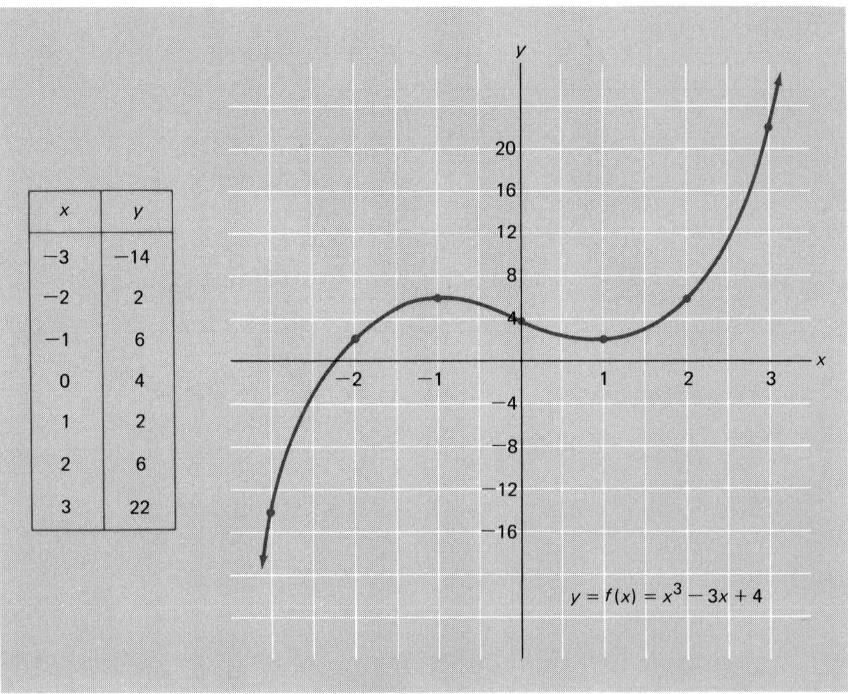

x	y
−3	−14
−2	2
−1	6
0	4
1	2
2	6
3	22

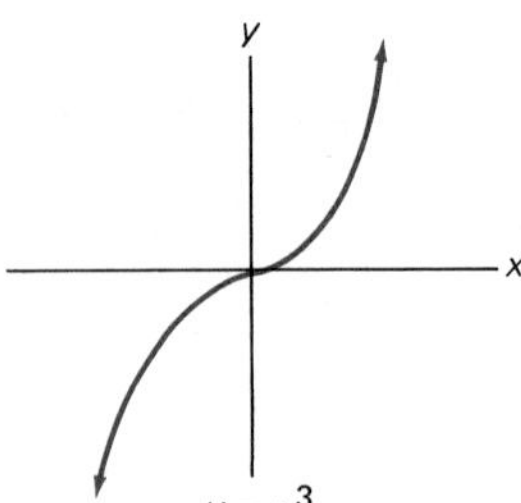

Notice that for large positive values of x, the values of y are large and positive; similarly, for large negative values of x, y is large and negative. This is due to the dominance of the leading term x^3 for large $|x|$. This dominance is responsible for a drooping left arm and a right arm held high on the graph. Notice also that the graph has one hill and one valley. This is typical of the graph of a cubic function, though it is possible for it to have no hills or valleys. The graph of $y = x^3$ illustrates this latter behavior.

Next consider a typical fourth degree polynomial function.

$$f(x) = -x^4 + 4x^3 + 2x^2 - 12x - 3$$

A table of values and the graph are shown at the top of page 183.

The leading term $-x^4$, which is negative for all values of x, determines that the graph has two drooping arms. Note that there are two hills and one valley.

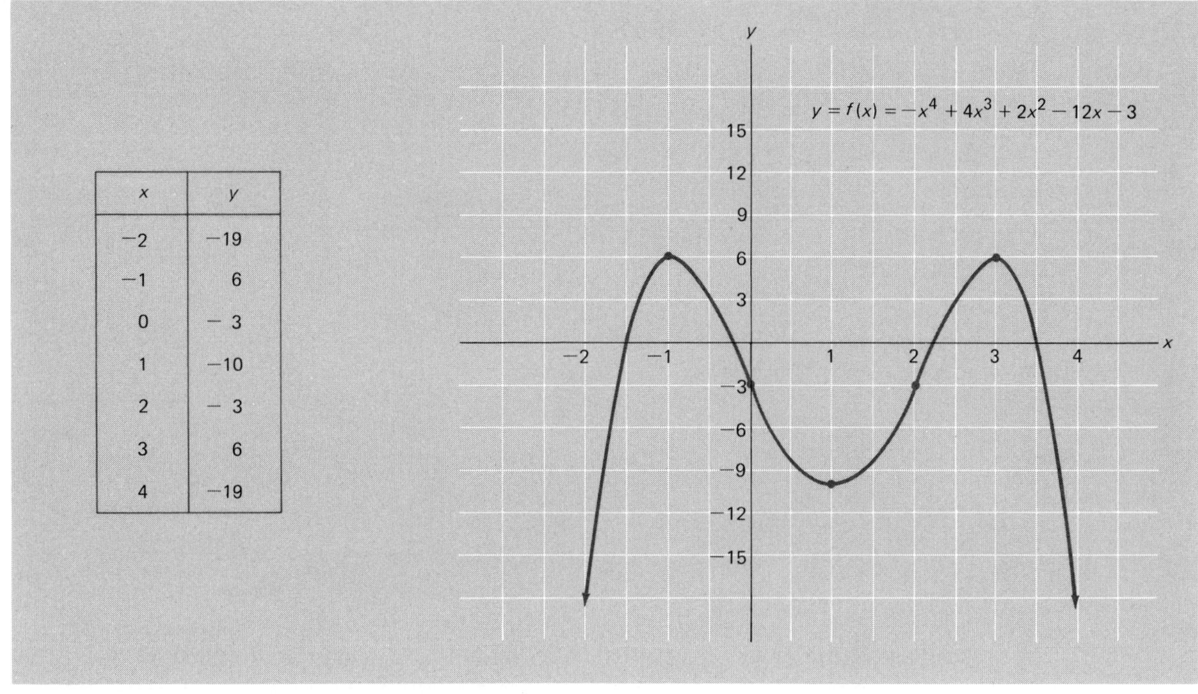

x	y
−2	−19
−1	6
0	− 3
1	−10
2	− 3
3	6
4	−19

In general, we can make the following statements about the graph of

$$f(x) = a_n x^n + a_{n-1}x^{n-1} + \cdots + a_1 x + a_0, \qquad a_n \neq 0$$

1. If n is even and $a_n < 0$, the graph will have two drooping arms; if n is even and $a_n > 0$, it will have both arms raised. This is due to the dominance of $a_n x^n$ for large values of $|x|$.
2. If n is odd, one arm droops and the other points upward. Again, this is dictated by the dominance of $a_n x^n$.
3. The combined number of hills and valleys cannot exceed $n - 1$, although it can be less.

Factored Polynomial Functions The task of graphing can be simplified considerably if our polynomial is factored. The real solutions of $f(x) = 0$ correspond to the x-intercepts of the graph of $y = f(x)$, that is, to the x-coordinates of the points where the graph intersects the x-axis. If the polynomial is factored, these intercepts are easy to find.

Consider as an example

$$y = f(x) = x(x + 3)(x - 1)$$

The solutions of $f(x) = 0$ are 0, -3, and 1; these are the x-intercepts of the graph. With this information and two or three additional points, we can easily sketch the graph. Keep in mind that $f(x)$ cannot change sign between two adjacent x-intercepts (see the discussion on page 116).

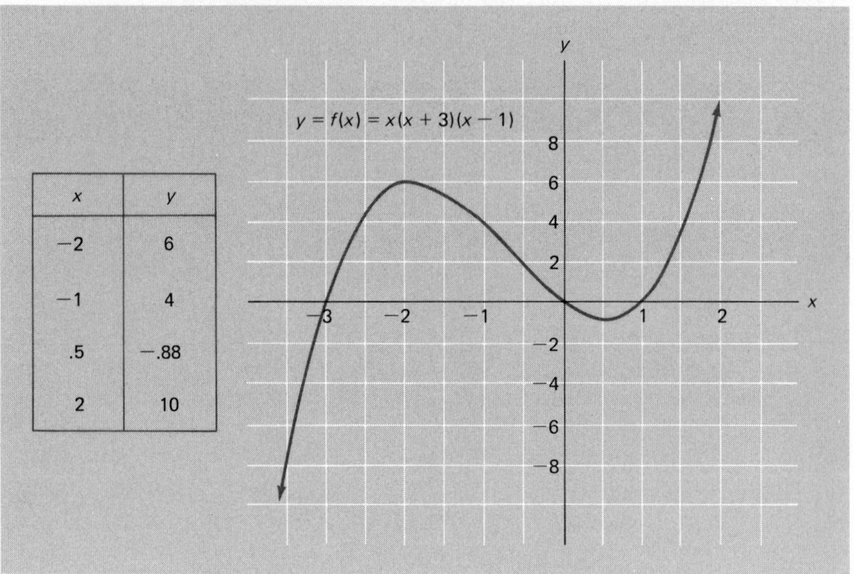

Functions with Multi-Part Rules Sometimes a function has polynomial components even though it is not a polynomial function. Especially notable is the absolute value function $f(x) = |x|$, which has the two-part rule

$$f(x) = \begin{cases} -x & \text{if } x < 0 \\ x & \text{if } x \geq 0 \end{cases}$$

For $x < 0$, the graph of this function coincides with the line $y = -x$; for $x \geq 0$, it coincides with the line $y = x$. Note the sharp corner at the origin.

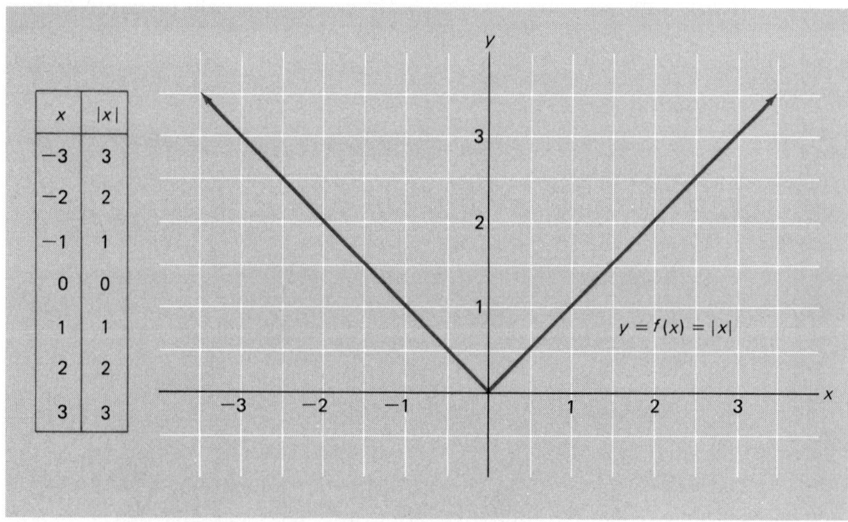

Here is a more complicated example.

$$g(x) = \begin{cases} x + 2 & \text{if} \quad x < 0 \\ x^2 & \text{if} \quad 0 \le x \le 2 \\ 4 & \text{if} \quad x > 2 \end{cases}$$

Though this way of describing a function may seem strange, it is not at all unusual in more advanced courses. The graph of g consists of three pieces.

1. A part of the line $y = x + 2$
2. A part of the parabola $y = x^2$
3. A part of the horizontal line $y = 4$

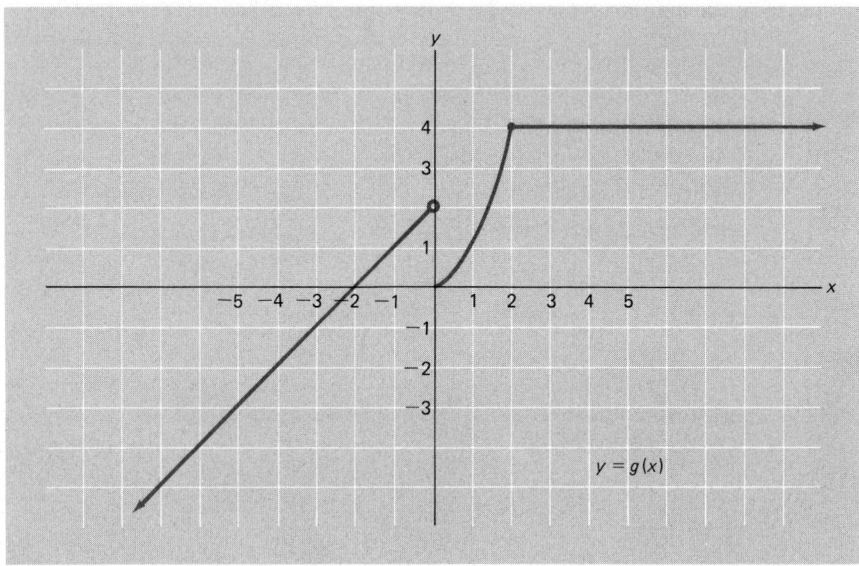

Note the use of the open circle at (0, 2) to indicate that this point is not part of the graph.

Problem Set 5-2

Graph each of the following polynomial functions. The first two are called constant functions.

1. $f(x) = 5$
2. $f(x) = -4$
3. $f(x) = -3x + 5$
4. $f(x) = 4x - 3$
5. $f(x) = x^2 - 5x + 4$
6. $f(x) = x^2 + 2x - 3$
7. $f(x) = x^3 - 9x$
8. $f(x) = x^3 - 16x$
9. [C] $f(x) = 2.12x^3 - 4.13x + 2$
10. [C] $f(x) = -1.2x^3 + 2.3x^2 - 1.4x$

Graph each of the following functions.

11. $f(x) = 2|x|$
12. $f(x) = |x| - 2$
13. $f(x) = |x - 2|$
14. $f(x) = |x| + 2$
15. $f(x) = \begin{cases} x & \text{if } x < 0 \\ 2 & \text{if } x \geq 0 \end{cases}$
16. $f(x) = \begin{cases} -1 & \text{if } x \leq 0 \\ 2x & \text{if } x > 0 \end{cases}$
17. $f(x) = \begin{cases} -5 & \text{if } x \leq -3 \\ 4 - x^2 & \text{if } -3 \leq x \leq 3 \\ -5 & \text{if } x > 3 \end{cases}$
18. $f(x) = \begin{cases} 9 & \text{if } x < 0 \\ 9 - x^2 & \text{if } 0 \leq x \leq 3 \\ x^2 - 9 & \text{if } x > 3 \end{cases}$

Example A (Symmetry properties) A function f is called an **even function** if $f(-x) = f(x)$ for all x in its domain. The graph of an even function is symmetric with respect to the y-axis. A function g is called an **odd function** if $g(-x) = -g(x)$ for all x in its domain; its graph is symmetric with respect to the origin. Graph the following two functions, observing their symmetries.
(a) $f(x) = x^4 + x^2 - 3$ (b) $g(x) = x^3 + 2x$

SOLUTION. Notice that

$$f(-x) = (-x)^4 + (-x)^2 - 3 = x^4 + x^2 - 3 = f(x)$$
$$g(-x) = (-x)^3 + 2(-x) = -x^3 - 2x = -g(x)$$

Thus f is even and g is odd. Their graphs are sketched below.

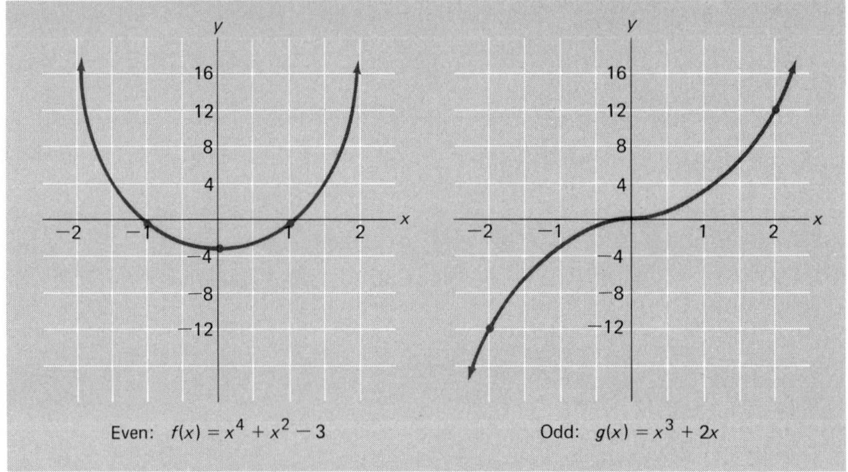

Note that a polynomial function involving only even powers of x is even, while one involving only odd powers of x is odd.

Determine which of the following are even functions, which are odd functions, and which are neither. Then sketch the graphs of those that are even or odd, making use of the symmetry properties.

19. $f(x) = 2x^2 - 5$
20. $f(x) = -3x^2 + 2$
21. $f(x) = x^2 - x + 1$
22. $f(x) = -2x^3$
23. $f(x) = 4x^3 - x$
24. $f(x) = x^3 + x^2$
25. $f(x) = 2x^4 - 5x^2$
26. $f(x) = 3x^4 + x^2$

Example B (More on factored polynomials) Graph

$$f(x) = (x - 1)^2(x - 3)(x + 2)$$

SOLUTION. The x-intercepts are at 1, 3, and -2. The new feature is that $x - 1$ occurs as a square. The factor $(x - 1)^2$ never changes sign, so the graph does not cross the x-axis at $x = 1$; it merely touches the axis there. Note the entries in the table of values corresponding to $x = 0.9$ and $x = 1.1$.

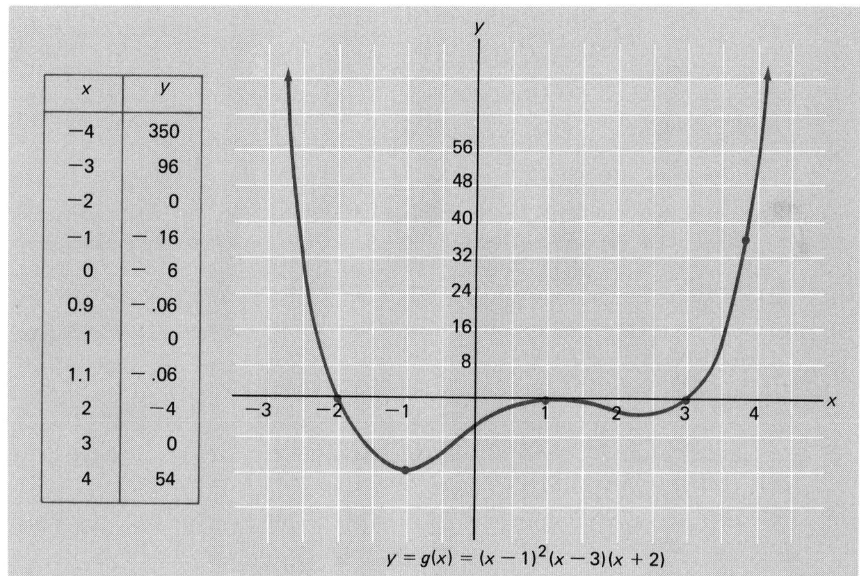

x	y
−4	350
−3	96
−2	0
−1	− 16
0	− 6
0.9	− .06
1	0
1.1	− .06
2	−4
3	0
4	54

$y = g(x) = (x - 1)^2(x - 3)(x + 2)$

Sketch the graph of each of the following.

27. $f(x) = (x + 1)(x - 1)(x - 3)$
28. $f(x) = x(x - 2)(x - 4)$
29. $f(x) = x^2(x - 4)$
30. $f(x) = x(x + 2)^2$
31. $f(x) = (x + 2)^2(x - 2)^2$
32. $f(x) = x(x - 1)^3$

Miscellaneous Problems

Graph each of the functions in Problems 33–42.

33. $f(x) = -3$
34. $f(x) = 2x - 3$
35. $f(x) = x^2 - 1$
36. $f(x) = x^3 - 3x$
37. $f(x) = (x + 2)(x - 1)^2(x - 3)$
38. $f(x) = x(x - 2)^3$
39. $f(x) = \begin{cases} 4 - x^2 & \text{if } -2 \le x < 2 \\ x - 1 & \text{if } \quad 2 \le x \le 4 \end{cases}$
40. $f(x) = \begin{cases} |x| & \text{if } -2 \le x \le 1 \\ x^2 & \text{if } \quad 1 < x \le 2 \\ 4 & \text{if } \quad 2 < x \le 4 \end{cases}$

[C] 41. $f(x) = x^4 + x^2 + 2, \ -2 \le x \le 2$

[C] 42. $f(x) = x^5 + x^3 + x, \ -1 \le x \le 1$

43. The graphs of the three functions $f(x) = x^2$, $g(x) = x^4$, and $h(x) = x^6$ all pass through the points $(-1, 1)$, $(0, 0)$, and $(1, 1)$. Draw careful sketches of these three functions using the same axes. Be sure to show clearly how they differ for $-1 < x < 1$.

[C] 44. Sketch the graph of $f(x) = x^{50}$ for $-1 \le x \le 1$. Be sure to calculate $f(.5)$ and $f(.9)$. What simple figure does the graph resemble?

[C] 45. Notice that $f(x) = 3x^4 + 2x^3 - 3x^2 + x + 1$ can be written as

$$f(x) = [((3x + 2)x - 3)x + 1]x + 1$$

It is now easy to calculate $f(2)$, $f(1.3)$, $f(4.2)$ on a calculator. Do so.

[C] 46. Use the trick described in Problem 45 to evaluate $f(3)$, $f(4.3)$, and $f(-1.6)$ for

$$f(x) = 4x^5 - 3x^4 + 2x^3 - x^2 + 7x - 3$$

47. The function $f(x) = [x]$ is called the *greatest integer function.* It assigns to each real number x the largest integer which is less than or equal to x. For example, $[\frac{5}{2}] = 2$, $[13] = 13$, and $[-14.25] = -15$. Graph this function on the interval $-2 \le x \le 6$.

48. Graph each of the following functions on the interval $-2 \le x \le 8$.
    (a) $f(x) = 2[x]$ (b) $g(x) = 2 + [x]$
    (c) $h(x) = [x - 2]$ (d) $k(x) = x - [x]$

49. A machine purchased for \$8000 is expected to have a scrap value of \$1500 after 12 years. If the machine is depreciated linearly, what will its value $V(x)$ be x years after purchase? Graph this function.

50. It costs the XYZ company $1000 + 10\sqrt{x}$ dollars to make x toy dolls, which sell for \$8 each. Express the total profit $T(x)$ as a function of x and graph $T(x)$.

51. Suppose that the cost of shipping a package is 15 cents for anything less than one ounce and 10 cents for each additional ounce or fraction thereof. Write a formula for the cost $C(x)$ of shipping a package weighing x ounces (using [ ]) and graph it.

52. An open box is made from a piece of 12 inch by 18 inch cardboard by cutting a square of side x inches from each corner and turning up the sides. Express the volume $V(x)$ in terms of x and graph the resulting function. What is the appropriate domain for this function?

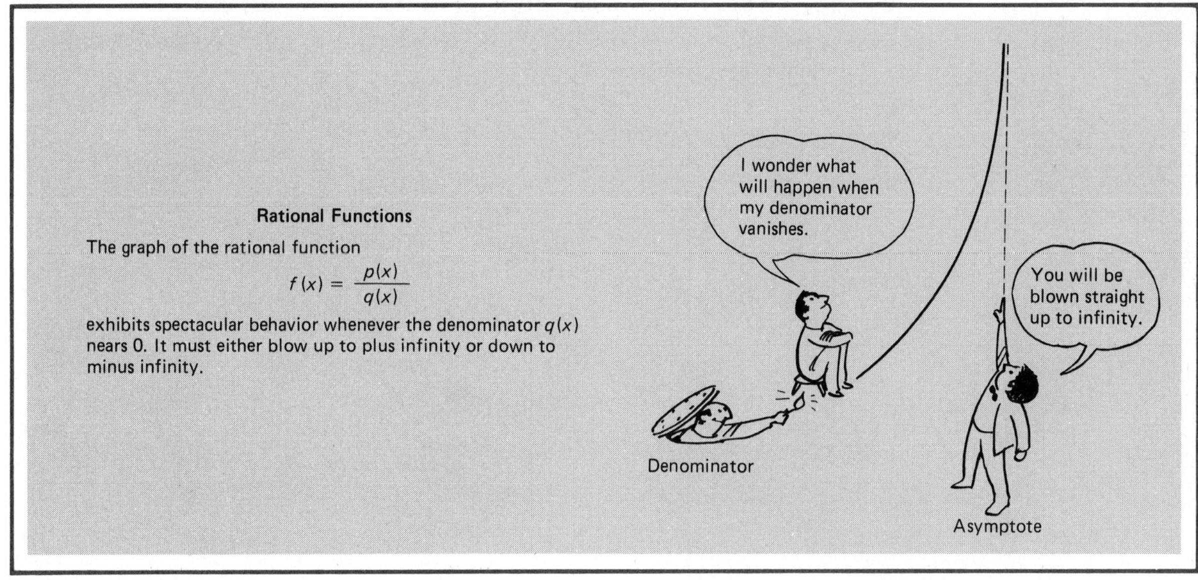

5-3 Graphing Rational Functions

If $f(x)$ is given by

$$f(x) = \frac{p(x)}{q(x)}$$

where $p(x)$ and $q(x)$ are polynomials, then f is called a **rational function**. For simplicity, we shall assume that $f(x)$ is in reduced form, that is, that $p(x)$ and $q(x)$ have no common nontrivial factors. Typical examples of rational functions are

$$f(x) = \frac{x+1}{x^2 - x + 6} = \frac{x+1}{(x-3)(x+2)}$$

$$g(x) = \frac{(x+2)(x-5)}{(x+3)^3}$$

Graphing a rational function can be tricky, primarily because of the denominator $q(x)$. Whenever it is zero, something dramatic is sure to happen to the graph. That is the point of our opening cartoon.

The Graphs of $1/x$ and $1/x^2$ Let us consider two simple cases.

$$f(x) = \frac{1}{x} \qquad g(x) = \frac{1}{x^2}$$

Notice that f is an odd function ($f(-x) = -f(x)$), while g is even ($g(-x) = g(x)$). These facts imply that the graph of f is symmetric with respect to the origin, and that the graph of g is symmetric with respect to the y-axis.

Thus we need to use only positive values of x to calculate y-values. Each calculation yields two points on the graph. Observe particularly the behavior of each graph near $x = 0$.

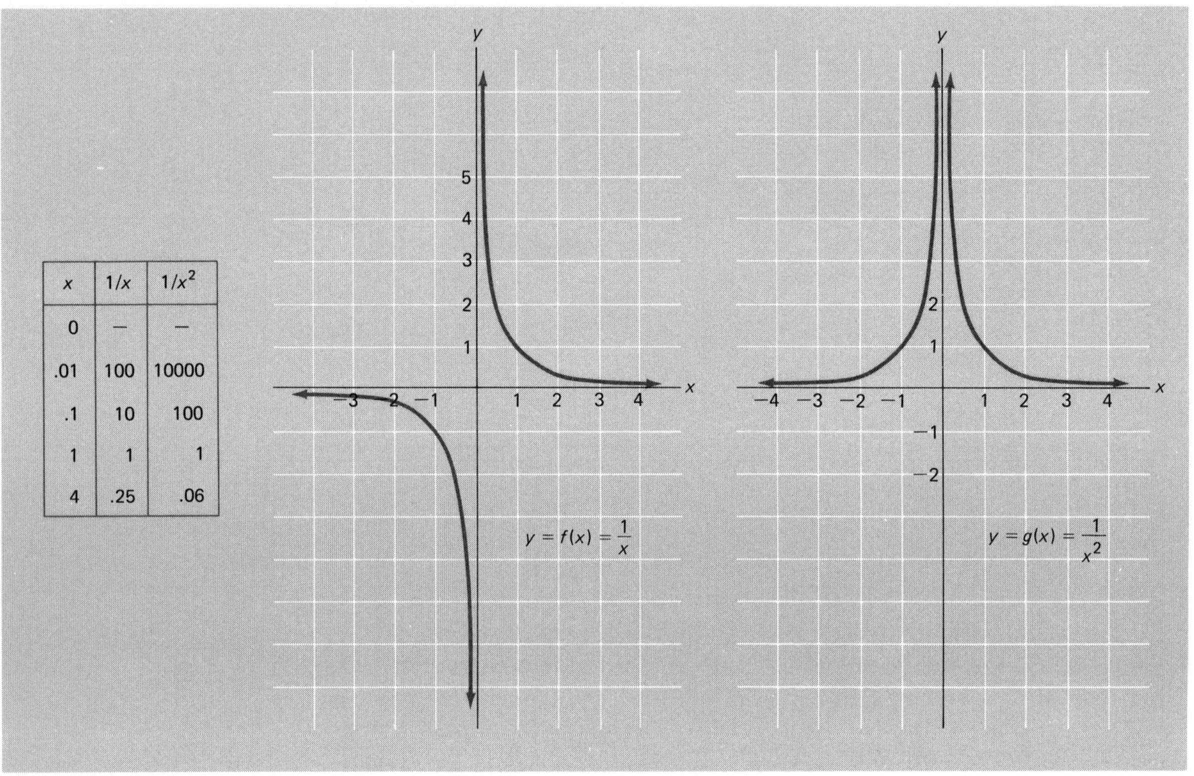

x	$1/x$	$1/x^2$
0	—	—
.01	100	10000
.1	10	100
1	1	1
4	.25	.06

In both cases, the x- and y-axes play special roles; we call them asymptotes for the graphs. If, as a point moves away from the origin along a curve, the distance between it and a line becomes closer and closer to zero, then that line is called an **asymptote** for the curve. Clearly the line $x = 0$ is a vertical asymptote for both of our curves and the line $y = 0$ is a horizontal asymptote for both of them.

The Graphs of $1/(x-2)$ and $1/(x-2)^2$ If we replace x by $x - 2$ in our two functions, we get two new functions.

$$h(x) = \frac{1}{x-2} \qquad k(x) = \frac{1}{(x-2)^2}$$

Their graphs are just like those of f and g except that they are moved two units to the right.

Two observations should be made. The vertical asymptote (dotted line) occurs where the denominator is zero, that is, it is the line $x = 2$. The hori-

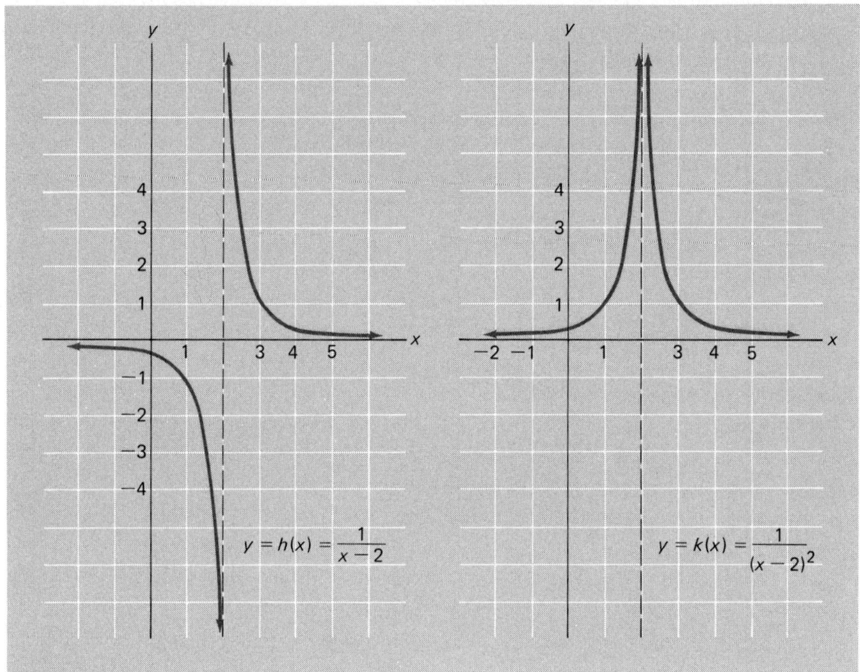

zontal asymptote is again the line $y = 0$. We can find it algebraically by looking at what happens to y as $|x|$ gets larger and larger. It should be clear that y goes to zero.

More Complicated Examples Consider next the rational function determined by

$$y = f(x) = \frac{x}{x^2 + x - 6} = \frac{x}{(x - 2)(x + 3)}$$

We expect its graph to have vertical asymptotes at $x = 2$ and $x = -3$. Again, the line $y = 0$ will be a horizontal asymptote since, as $|x|$ gets large, the term x^2 in the denominator will dominate so that y will behave much like x/x^2 or $1/x$ and will thus approach zero. The graph crosses the x-axis where the numerator is zero, namely, at $x = 0$. Finally, with the help of a table of values, we sketch the graph, shown at the top of page 192.

Lastly we consider

$$y = f(x) = \frac{2x^2 + 2x}{x^2 - 4x + 4} = \frac{2x(x + 1)}{(x - 2)^2}$$

The graph will have one vertical asymptote, at $x = 2$. To check on a horizontal asymptote, we note that for large $|x|$, the numerator behaves like $2x^2$ and the

x	y
−6	−.25
−4	−.67
−2	.50
0	0
1	−.25
3	.50
5	.21

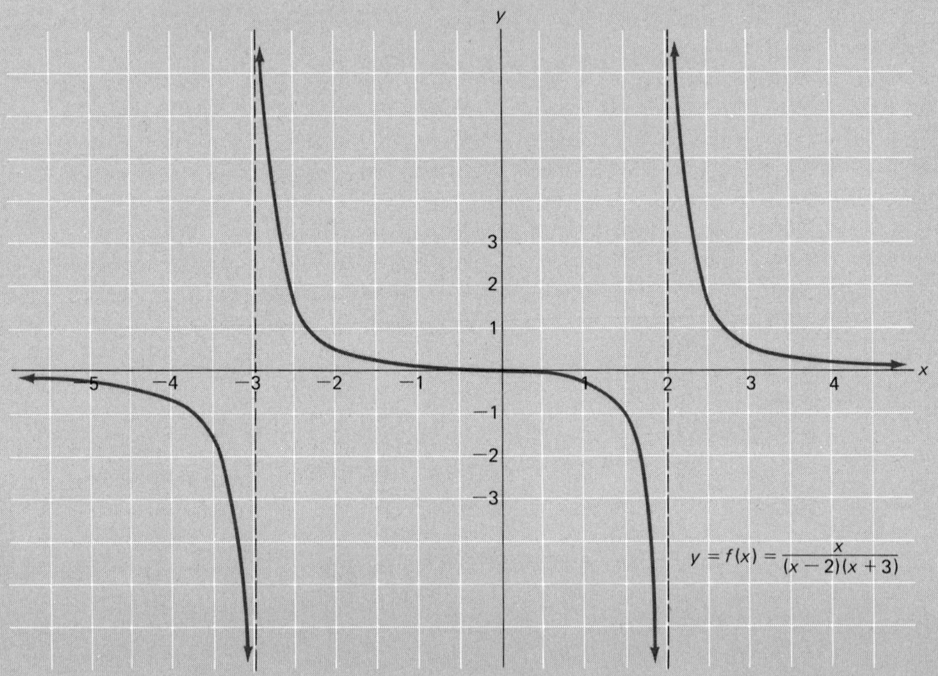

$$y = f(x) = \frac{x}{(x-2)(x+3)}$$

denominator behaves like x^2. It follows that $y = 2$ is a horizontal asymptote. The graph crosses the x-axis where the numerator $2x(x + 1)$ is zero, namely, at $x = 0$ and $x = -1$. A good approximation to the graph is shown below.

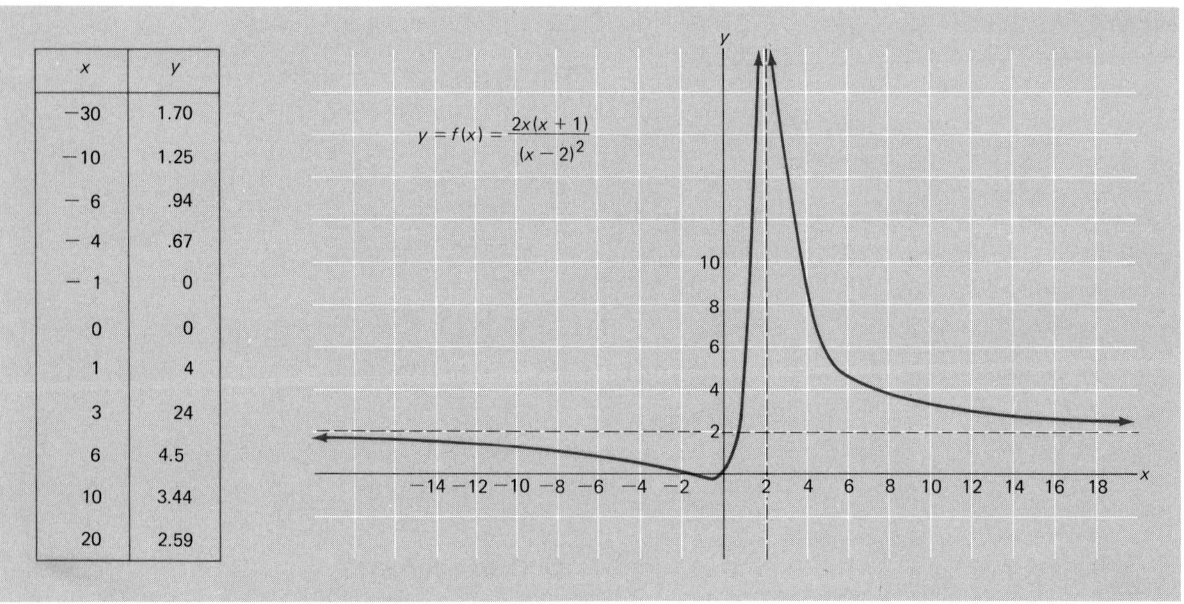

x	y
−30	1.70
−10	1.25
− 6	.94
− 4	.67
− 1	0
0	0
1	4
3	24
6	4.5
10	3.44
20	2.59

A General Procedure Here is an outline of the procedure for graphing a rational function.

1. Check for symmetry with respect to both axes and the origin.
2. Factor the numerator and denominator.
3. Determine the vertical asymptotes (if any) by checking where the denominator is zero. Draw a dotted line for each asymptote.
4. Determine the horizontal asymptotes (if any) by examining the quotient of the leading terms for large $|x|$. Indicate any horizontal asymptote with a dotted line.
5. Determine the x-intercepts (if any). These occur where the numerator is zero.
6. Make a small table of values and plot corresponding points.
7. Sketch the graph.

Problem Set 5-3

Sketch the graph of each of the following functions.

1. $f(x) = \dfrac{2}{x+2}$
2. $f(x) = \dfrac{-1}{x+2}$
3. $f(x) = \dfrac{2}{(x+2)^2}$
4. $f(x) = \dfrac{1}{(x-3)^2}$
5. $f(x) = \dfrac{2x}{x+2}$
6. $f(x) = \dfrac{x+2}{x-3}$
7. $f(x) = \dfrac{1}{(x+2)(x-1)}$
8. $f(x) = \dfrac{3}{x^2-9}$
9. $f(x) = \dfrac{x+1}{(x+2)(x-1)}$
10. $f(x) = \dfrac{3x}{x^2-9}$
11. $f(x) = \dfrac{2x^2}{(x+2)(x-1)}$
12. $f(x) = \dfrac{x^2-4}{x^2-9}$

Example A (No vertical asymptotes) Sketch the graph of

$$f(x) = \frac{x^2-4}{x^2+1} = \frac{(x-2)(x+2)}{x^2+1}$$

SOLUTION. Note that f is an even function, so the graph will be symmetric with respect to the y-axis. The denominator is not zero for any real x, so there are no vertical asymptotes. The line $y = 1$ is a horizontal asymptote, since for large $|x|$, $f(x)$ behaves like x^2/x^2. The x-intercepts are $x = 2$ and $x = -2$. The graph is shown at the top of page 194.

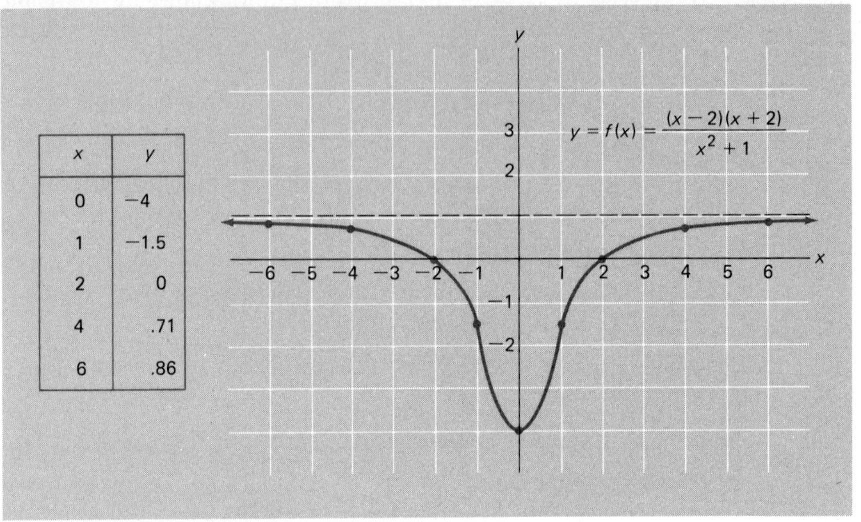

x	y
0	−4
1	−1.5
2	0
4	.71
6	.86

Now sketch the graph of each of the following.

13. $f(x) = \dfrac{1}{x^2 + 2}$

14. $f(x) = \dfrac{x^2 - 2}{x^2 + 2}$

15. $f(x) = \dfrac{x}{x^2 + 2}$

16. $f(x) = \dfrac{x^3}{x^2 + 2}$

Example B (Rational functions that are not in reduced form) Sketch the graph of

$$f(x) = \frac{x^2 + x - 6}{x - 2}$$

SOLUTION. Notice that

$$f(x) = \frac{(x + 3)(x - 2)}{x - 2}$$

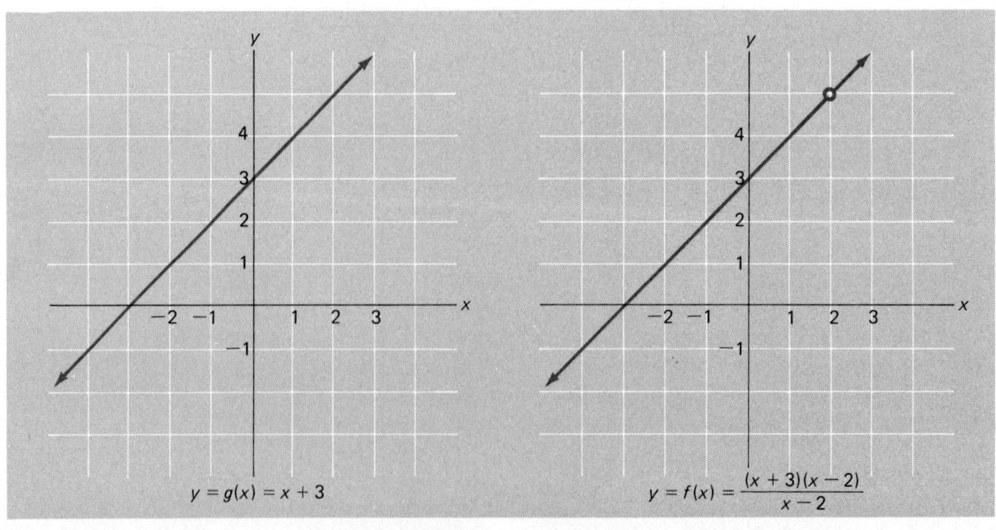

You have every right to expect that we will cancel the factor $x - 2$ from numerator and denominator and graph

$$g(x) = x + 3$$

But note that 2 is in the domain of g but not in the domain of f. Thus f and g and their graphs are exactly alike except at one point, namely, at $x = 2$. Both graphs are shown at the bottom of page 194. You will notice the hole in the graph of $y = f(x)$ at $x = 2$. This technical distinction is occasionally important.

Sketch the graph of each of the following rational functions, which, you will note, are not in reduced form.

17. $f(x) = \dfrac{(x + 2)(x - 4)}{x + 2}$

18. $f(x) = \dfrac{x^2 - 4}{x - 2}$

19. $f(x) = \dfrac{x^3 - x^2 - 12x}{x + 3}$

20. $f(x) = \dfrac{x^3 - 4x}{x^2 - 2x}$

Miscellaneous Problems

Sketch the graphs of the rational functions in Problems 21–26.

21. $f(x) = \dfrac{x}{x + 5}$

22. $f(x) = \dfrac{x - 2}{x + 3}$

23. $f(x) = \dfrac{x^2 - 9}{x^2 - x - 2}$

24. $f(x) = \dfrac{x - 2}{(x + 3)^2}$

25. $f(x) = \dfrac{x^2 - 9}{x^2 - x - 6}$

26. $f(x) = \dfrac{x - 2}{x^2 + 3}$

27. Sketch the graph of $f(x) = x^n/(x^2 + 1)$ for $n = 1$, 2, and 3. Which graph does not have a horizontal asymptote? Which has the x-axis as horizontal asymptote?
28. Let $f(x) = x^4/(x^n + 2x + 3)$.
    (a) Show (by reasoning) that the graph of f has horizontal asymptote $y = 1$ if $n = 4$.
    (b) Show that the graph does not have a horizontal asymptote if $n < 4$.
    (c) Does the graph have a horizontal asymptote if $n > 4$ and, if so, what is it?
29. A manufacturer of gizmos has overhead of \$20,000 per year and direct costs (labor and material) of \$50 per gizmo. Write an expression for $U(x)$, the average cost per unit, if the company makes x gizmos this year. Graph the resulting function.
30. A cylindrical tin can is to contain 10π cubic inches. Write a formula for $S(r)$, the total surface area, in terms of the radius r. Graph the resulting function.

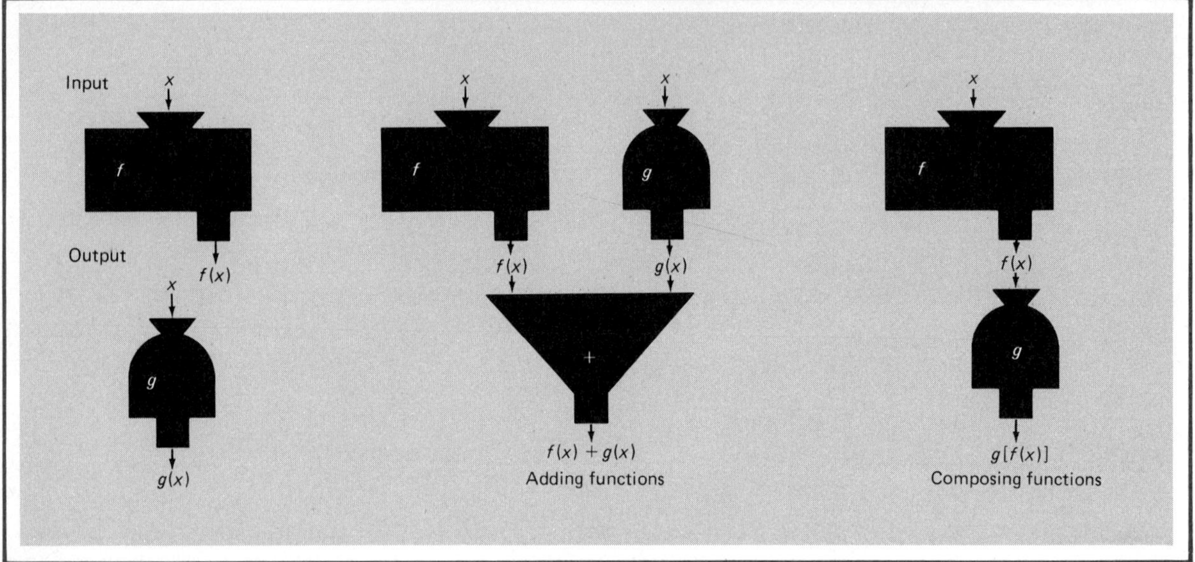

5-4 Putting Functions Together

There is still another way to visualize a function. Think of the function named f as a machine. It accepts a number x as input, operates on it, and then presents the number $f(x)$ as output. Machines can be hooked together to make more complicated machines; similarly, functions can be combined to produce more complicated functions. That is the subject of this section.

Sums, Differences, Products, and Quotients The simplest way to make new functions from old ones is to use the four arithmetic operations on them. Suppose, for example, that the functions f and g have the formulas

$$f(x) = \frac{x - 3}{2} \qquad g(x) = \sqrt{x}$$

We can make a new function $f + g$ by having it assign to x the value $(x - 3)/2 + \sqrt{x}$, that is,

$$(f + g)(x) = f(x) + g(x) = \frac{x - 3}{2} + \sqrt{x}$$

Of course, we must be a little careful about domains. Clearly, x must be a number that both f and g can operate on. In other words, the domain of $f + g$ is the intersection (common part) of the domains of f and g.

The functions $f - g$, $f \cdot g$, and f/g are defined in a completely analogous way. Assuming that f and g have their respective natural domains, namely, all reals and the nonnegative reals, respectively, we have the following.

FORMULA	DOMAIN
$(f + g)(x) = f(x) + g(x) = \dfrac{x-3}{2} + \sqrt{x}$	$x \geq 0$
$(f - g)(x) = f(x) - g(x) = \dfrac{x-3}{2} - \sqrt{x}$	$x \geq 0$
$(f \cdot g)(x) = f(x) \cdot g(x) = \dfrac{x-3}{2}\sqrt{x}$	$x \geq 0$
$(f/g)(x) = f(x)/g(x) = \dfrac{x-3}{2\sqrt{x}}$	$x > 0$

To graph the function $f + g$, it is often best to graph f and g separately in the same coordinate plane and then add the y-coordinates together along vertical lines. We illustrate this method (called **addition of ordinates**) below.

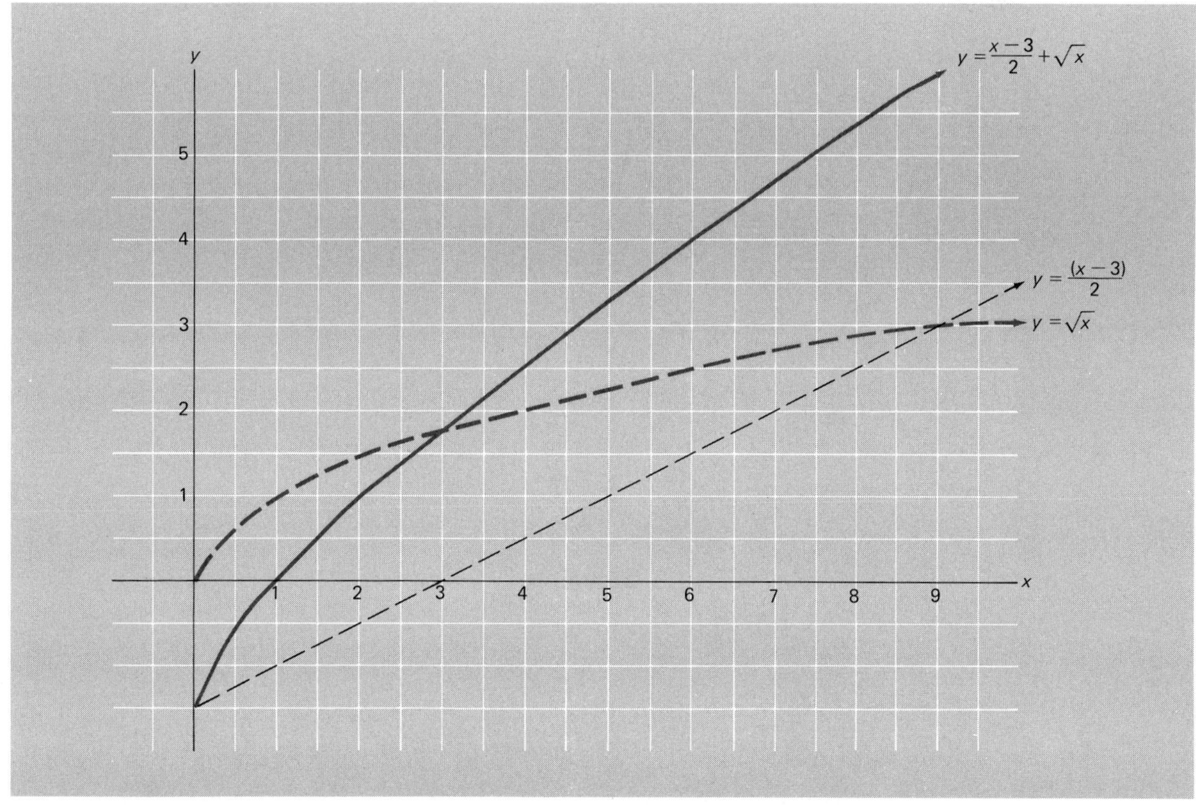

The graph of $f - g$ can be handled similarly; simply graph f and g in the same coordinate plane and subtract ordinates. We can even graph $f \cdot g$ and f/g in the same manner, but that is harder.

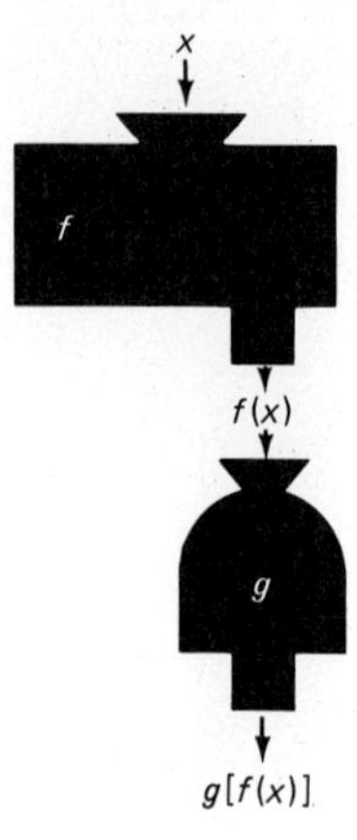

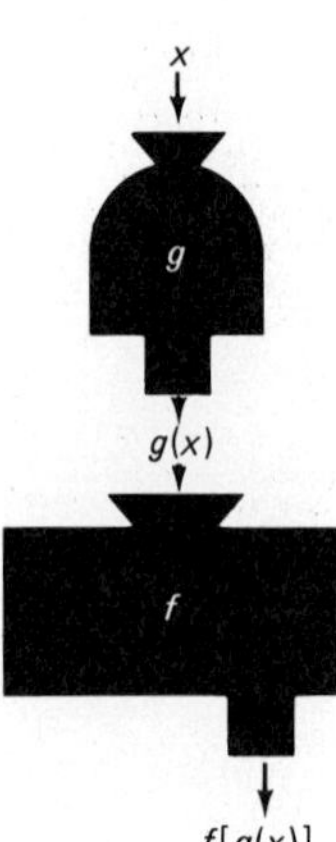

Composition of Functions To compose functions is to string them together in tandem. Part of our opening display (reproduced in the margin) shows how this is done. If f operates on x to produce $f(x)$ and then g operates on $f(x)$ to produce $g(f(x))$, we say that we have composed g and f. The resulting function, called the **composite of** g **with** f, is denoted by $g \circ f$. Thus,

$$(g \circ f)(x) = g(f(x))$$

Recall our earlier examples, $f(x) = (x - 3)/2$ and $g(x) = \sqrt{x}$. We may compose them in two ways.

$$(g \circ f)(x) = g(f(x)) = g\left(\frac{x-3}{2}\right) = \sqrt{\frac{x-3}{2}}$$

$$(f \circ g)(x) = f(g(x)) = f(\sqrt{x}) = \frac{\sqrt{x}-3}{2}$$

We note one thing right away: Composition of functions is not commutative; $g \circ f$ and $f \circ g$ are not the same. We must also be careful in describing the domain of a composite function. The domain of $g \circ f$ is that part of the domain of f for which g can accept $f(x)$ as input. In our example, the domain of $g \circ f$ is $x \geq 3$, not all x or $x \geq 0$ as we might have thought at first glance. The diagram below offers another view of these matters. Note that the shaded portion of the domain of f is not in the domain of $g \circ f$; for x in this portion, $f(x)$ is outside the domain of g.

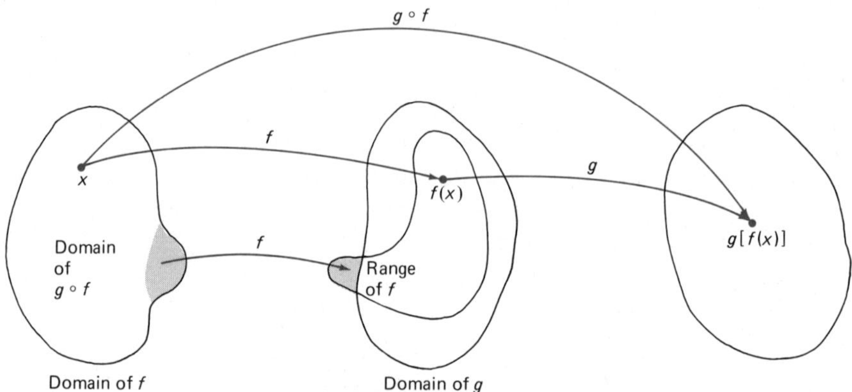

In calculus, we shall often need to take a given function and decompose it, that is, break it into composite pieces. Usually, this can be done in several ways. Take $p(x) = \sqrt{x^2 + 3}$ for example. We may think of it as

$$p(x) = g(f(x)) \quad \text{where} \quad g(x) = \sqrt{x} \quad \text{and} \quad f(x) = x^2 + 3$$

or as

$$p(x) = g(f(x)) \quad \text{where} \quad g(x) = \sqrt{x+3} \quad \text{and} \quad f(x) = x^2$$

Translations Observing how a function is built up from simpler ones can be a big aid in graphing. We may ask this question: How are the graphs of

$$y = f(x) \qquad y = f(x - 3) \qquad y = f(x) + 2 \qquad y = f(x - 3) + 2$$

related to each other? Consider $f(x) = |x|$ as an example. The corresponding four graphs are displayed below.

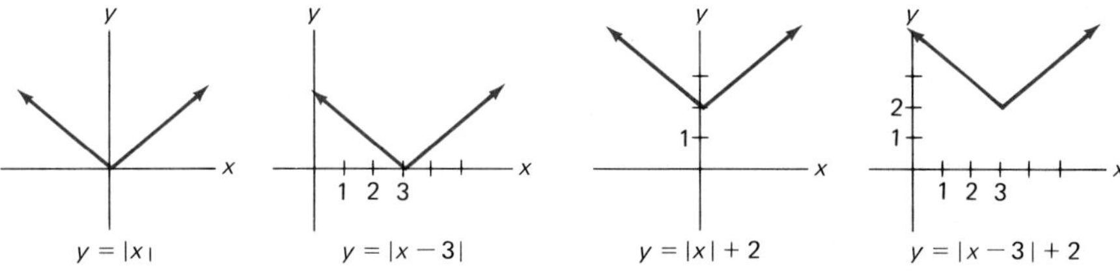

What happened with $f(x) = |x|$ is typical. Notice that all four graphs have the same shape; the last three are just translations of the first. Replacing x by $x - 3$ translates the graph 3 units to the right; adding 2 translates it upward by 2 units.

Here is another illustration of these principles for the function $f(x) = x^3 + x^2$.

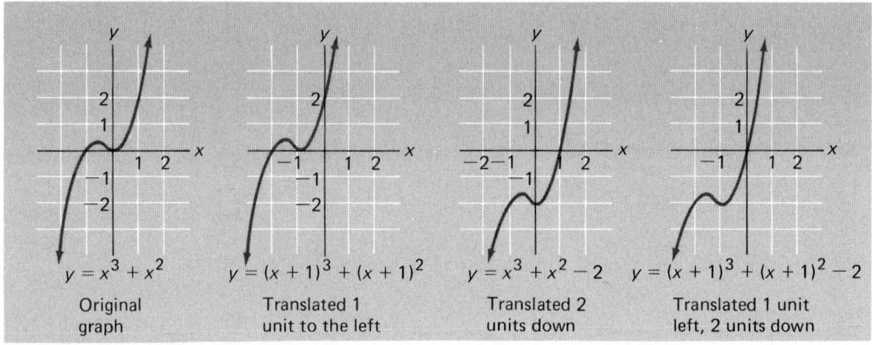

Exactly the same principles apply in the general situation.

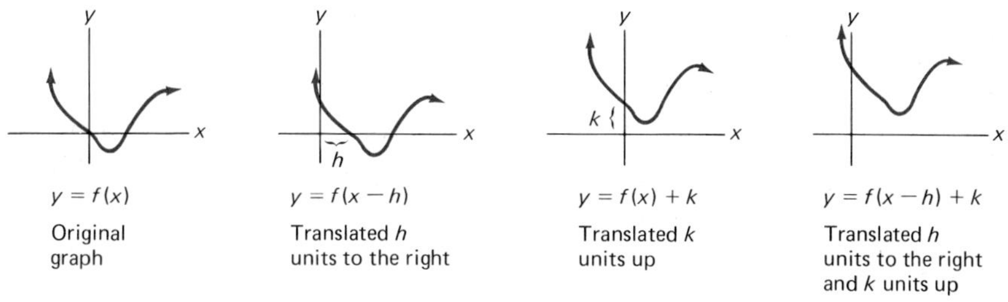

If $h < 0$, the translation is to the left; if $k < 0$, the translation is downward.

Problem Set 5-4

1. Let $f(x) = x^2 - 2x + 2$ and $g(x) = 2/x$. Calculate each of the following.
(a) $(f + g)(2)$ (b) $(f + g)(0)$ (c) $(f - g)(1)$
(d) $(f \cdot g)(-1)$ (e) $(f/g)(2)$ (f) $(g/f)(2)$
(g) $(f \circ g)(-1)$ (h) $(g \circ f)(-1)$ (i) $(g \circ g)(3)$
2. Let $f(x) = 3x + 5$ and $g(x) = |x - 2|$. Perform the calculations in Problem 1 for these functions.

In each of the following, write the formulas for $(f + g)(x)$, $(f - g)(x)$, $(f \cdot g)(x)$, and $(f/g)(x)$ and give the domains of these four functions.

3. $f(x) = x^2$, $g(x) = x - 2$
4. $f(x) = x^3 - 1$, $g(x) = x + 3$
5. $f(x) = x^2$, $g(x) = \sqrt{x}$
6. $f(x) = 2x^2 + 5$, $g(x) = \dfrac{1}{x}$
7. $f(x) = \dfrac{1}{x - 2}$, $g(x) = \dfrac{x}{x - 3}$
8. $f(x) = \dfrac{1}{x^2}$, $g(x) = \dfrac{1}{5 - x}$

For each of the following, write the formulas for $(g \circ f)(x)$ and $(f \circ g)(x)$ and give the domains of these composite functions.

9. $f(x) = x^2$, $g(x) = x - 2$
10. $f(x) = x^3 - 1$, $g(x) = x + 3$
11. $f(x) = \dfrac{1}{x}$, $g(x) = x + 3$
12. $f(x) = 2x^2 + 5$, $g(x) = \dfrac{1}{x}$
13. $f(x) = \sqrt{x - 2}$, $g(x) = x^2 - 2$
14. $f(x) = \sqrt{2x}$, $g(x) = x^2 + 1$
15. $f(x) = 2x - 3$, $g(x) = \frac{1}{2}(x + 3)$
16. $f(x) = x^3 + 1$, $g(x) = \sqrt[3]{x - 1}$

Example (Decomposing functions) In each of the following, H can be thought of as a composite function $g \circ f$. Write formulas for $f(x)$ and $g(x)$.

(a) $H(x) = (2 + 3x)^2$ (b) $H(x) = \dfrac{1}{(x^2 + 4)^3}$

SOLUTION. (a) Think of how you might calculate $H(x)$. You would first calculate $2 + 3x$ and then square the result. That suggests

$$f(x) = 2 + 3x \qquad g(x) = x^2$$

(b) Here there are two obvious ways to proceed. One way would be to let

$$f(x) = x^2 + 4 \qquad g(x) = \frac{1}{x^3}$$

Another selection, which is just as good, is

$$f(x) = \frac{1}{x^2 + 4} \qquad g(x) = x^3$$

We could actually think of H as the composite of four functions. Let

$$f(x) = x^2 \qquad g(x) = x + 4 \qquad h(x) = x^3 \qquad j(x) = \frac{1}{x}$$

Then,

$$H = j \circ h \circ g \circ f$$

You should check this result.

In each of the following, write formulas for $g(x)$ and $f(x)$ so that $H = g \circ f$. The answer is not unique.

17. $H(x) = (x + 4)^3$
18. $H(x) = (2x + 1)^3$
19. $H(x) = \sqrt{x + 2}$
20. $H(x) = \sqrt[3]{2x + 1}$
21. $H(x) = \dfrac{1}{(2x + 5)^3}$
22. $H(x) = \dfrac{6}{(x + 4)^3}$
23. $H(x) = |x^3 - 4|$
24. $H(x) = |4 - x - x^2|$

Use the method of addition or subtraction of ordinates to graph each of the following. That is, graph $y = f(x)$ and $y = g(x)$ in the same coordinate plane and then obtain the graph of $f + g$ or $f - g$ by adding or subtracting ordinates.

25. $f + g$ where $f(x) = x^2$ and $g(x) = x - 2$.
26. $f + g$ where $f(x) = |x|$ and $g(x) = x$.
27. $f - g$ where $f(x) = 1/x$ and $g(x) = x$.
28. $f - g$ where $f(x) = x^3$ and $g(x) = -x + 1$.

In each of the following, graph the function f carefully and then use translations to sketch the graphs of the functions g, h, and j.

29. $f(x) = x^2$, $g(x) = (x - 2)^2$, $h(x) = x^2 - 4$, and $j(x) = (x - 2)^2 + 1$
30. $f(x) = x^3$, $g(x) = (x + 2)^3$, $h(x) = x^3 + 4$, and $j(x) = (x + 2)^3 - 2$
31. $f(x) = \sqrt{x}$, $g(x) = \sqrt{x - 3}$, $h(x) = \sqrt{x + 2}$, and $j(x) = \sqrt{x - 3} - 2$
32. $f(x) = \dfrac{1}{x}$, $g(x) = \dfrac{1}{x - 4}$, $h(x) = \dfrac{1}{x} + 3$, and $j(x) = \dfrac{1}{x - 4} - 5$

Miscellaneous Problems

33. Let $f(x) = 2x + 3$ and $g(x) = x^3$. Write formulas for each of the following.
    (a) $(f + g)(x)$ (b) $(g - f)(x)$
    (c) $(f \cdot g)(x)$ (d) $(f/g)(x)$
    (e) $(f \circ g)(x)$ (f) $(g \circ f)(x)$
    (g) $(f \circ f)(x)$ (h) $(g \circ g \circ g)(x)$
34. If $f(x) = 1/(x - 1)$ and $g(x) = \sqrt{x + 1}$, write formulas for $(f \circ g)(x)$ and $(g \circ f)(x)$ and give the domains of these composite functions.
35. If $f(x) = x^2 - 4$, $g(x) = |x|$, and $h(x) = 1/x$, write a formula for $(h \circ g \circ f)(x)$ and indicate its domain.
36. In general, how many different functions can be obtained by composing three different functions f, g, and h in different orders?
37. In calculus, the *difference quotient*

$$\frac{f(x + h) - f(x)}{h}$$

arises repeatedly. Calculate this expression and simplify it for each of the following.
    (a) $f(x) = x^2$ (b) $f(x) = 2x + 3$
    (c) $f(x) = 1/x$ (d) $f(x) = 2/(x - 2)$

38. Calculate $[g(x - h) - g(x)]/h$ for each of the following. Simplify your answer.
(a) $g(x) = 4x - 9$ (b) $g(x) = x^2 + 2x$
(c) $g(x) = x + 1/x$ (d) $g(x) = x^3$

[C] 39. Let $f(x) = (1 + \sqrt{x})^3/(3x^2 + 1)$. Calculate each of the following.
(a) $f(3.1)$ (b) $f(.03)$

[C] 40. Let $g(x) = (3 + 1/x)^2\sqrt{x^3 + 1}$. Calculate each of the following.
(a) $g(4.2)$ (b) $g(-.91)$

41. The relationship between the price (in cents) for a certain product and the demand (in thousands of units) appears to satisfy

$$P = \sqrt{29 - 3D + D^2}$$

On the other hand, the demand has risen over the past t years according to $D = 2 + \sqrt{t}$.
(a) Express P as a function of t.
[C] (b) Evaluate P when $t = 15$.

42. After being in business x years, a certain tractor manufacturer is making $100 + x + 2x^2$ units each year. The sales price (in dollars) per unit has risen according to the formula $P = 500 + 60x$. Assuming all this is true, write a formula for the manufacturer's yearly revenue $R(x)$ after x years.

5-5 Inverse Functions

Some processes are reversible; most are not. If I take off my shoes, I may put them back on again. The second operation undoes the first one and brings things back to the original state. But if I throw my shoes in the fire, I will have a hard time undoing the damage I have done.

A function f operates on a number x to produce a number $y = f(x)$. It may be that we can find a function g that will operate on y and give back x. For example, if

$$y = f(x) = 2x + 1$$

then

$$g(x) = \tfrac{1}{2}(x - 1)$$

is such a function, since

$$g(y) = g(f(x)) = \tfrac{1}{2}(2x + 1 - 1) = x$$

When we can find such a function g, we call it the *inverse* of f. Not all functions have inverses. Whether they do or not has to do with a concept called one-to-oneness.

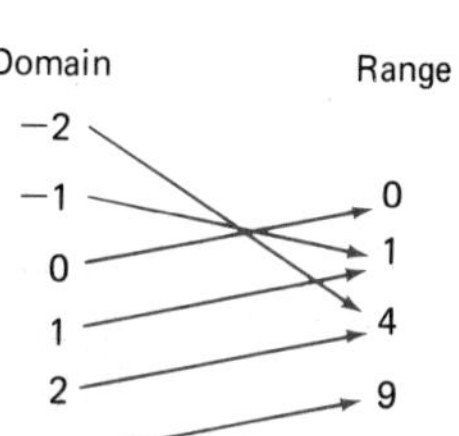

$f(x) = x^2$

One-to-One Functions In the margin, we have reproduced an example we studied earlier, the squaring function with domain $\{-2, -1, 0, 1, 2, 3\}$. It is a perfectly fine function, but it does have one troublesome feature. It may assign the same value to two different x's. In particular, $f(-2) = 4$ and $f(2) = 4$. Such a function cannot possibly have an inverse g. For what would g do with 4? It would not know whether to give back -2 or 2 as the value.

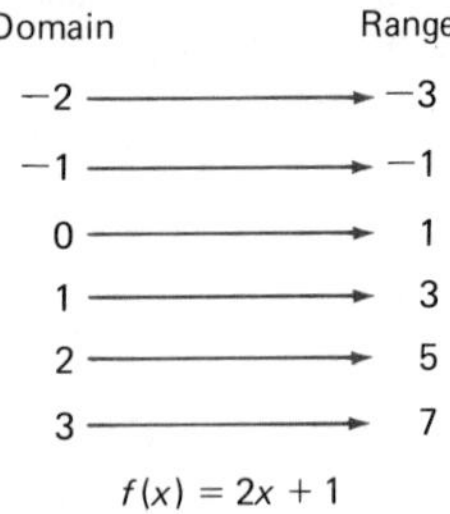

In contrast, consider $f(x) = 2x + 1$, also pictured in the margin. Notice that this function never assigns the same value to two different values of x. Therefore there is an unambiguous way of undoing it.

We say that a function f is **one-to-one** if $x_1 \neq x_2$ implies $f(x_1) \neq f(x_2)$, that is, if different values for x always result in different values for $f(x)$. Some functions are one-to-one; some are not. It would be nice to have a graphical criterion for deciding.

Consider the functions $f(x) = x^2$ and $f(x) = 2x + 1$ again, but now let the domains be the set of all real numbers. Their graphs appear below. In the first case, certain horizontal lines (those which are above the x-axis) meet the graph in two points; in the second case, every horizontal line meets the graph in exactly one point. Notice on the first graph that $f(x_1) = f(x_2)$ even though $x_1 \neq x_2$. On the second graph, this cannot happen. Thus we have the important fact that *if every horizontal line meets the graph of a function f in at most one point, then f is one-to-one.*

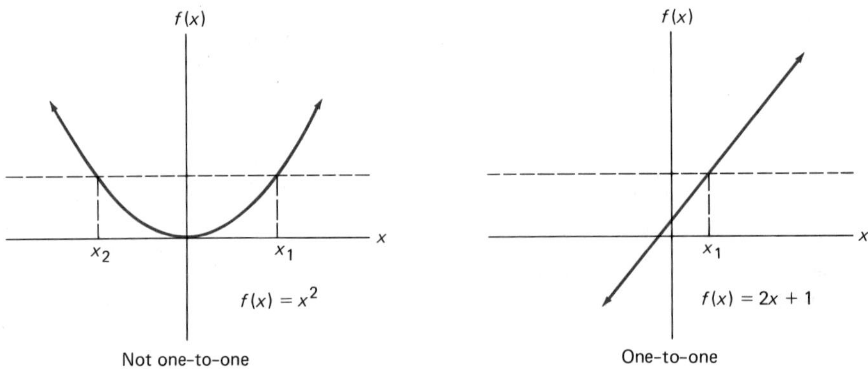

Inverse Functions Now we are ready to give a formal definition of the main idea of this section.

DEFINITION

Let f be a one-to-one function with domain X and range Y. Then the function g with domain Y and range X which satisfies

$$\boxed{g(f(x)) = x}$$

*for all x in X is called the **inverse of f**.*

We make several important observations. First, the boxed formula simply says that g undoes what f did. Second, if g undoes f, then f will undo g, that is,

$$f(g(y)) = y$$

for all y in Y. Third, the function g is usually denoted by the symbol f^{-1}. You are cautioned to remember that f^{-1} does *not* mean $1/f$, as you have the right to expect. Mathematicians decided long ago that f^{-1} should stand for the inverse function (the undoing function). They had a reason; note that

$$(f^{-1} \circ f)(x) = x \quad \text{and} \quad (f \circ f^{-1})(y) = y$$

Under the operation of composition, f and f^{-1} neutralize each other. All of this is summarized in the diagram below.

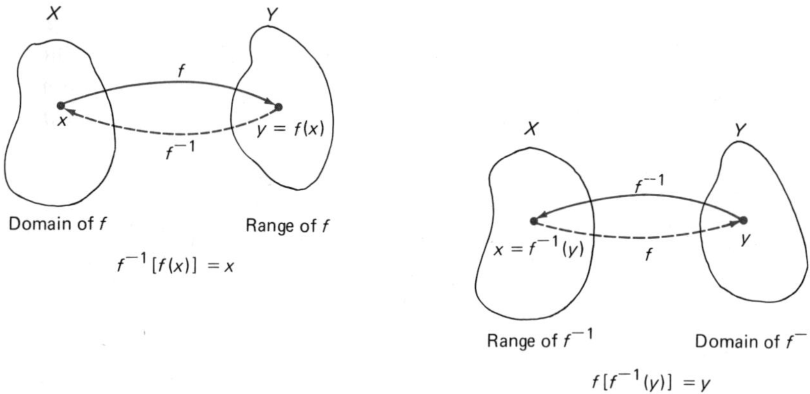

Finding a Formula for f^{-1} If f adds 2, then f^{-1} ought to subtract 2. To say it in symbols, if $f(x) = x + 2$, then we might expect $f^{-1}(y) = y - 2$. And we are right, for

$$f^{-1}[f(x)] = f^{-1}(x + 2) = x + 2 - 2 = x$$

If f divides by 3 and then subtracts 4, we expect f^{-1} to add 4 and multiply by 3. Symbolically, if $f(x) = x/3 - 4$, then we expect $f^{-1}(y) = 3(y + 4)$. Again we are right, for

$$f^{-1}[f(x)] = f^{-1}\left(\frac{x}{3} - 4\right) = 3\left(\frac{x}{3} - 4 + 4\right) = x$$

Note that you must undo things in the reverse order in which you did them (that is, we divided by 3 and then subtracted 4, so to undo this, we first add 4 and then multiply by 3).

When we get to more complicated functions, it is not always easy to find the formula for the inverse function. Here is an important way to look at it.

$$\boxed{x = f^{-1}(y) \quad \text{if and only if} \quad y = f(x)}$$

That means that we can get the formula for f^{-1} by solving the equation $y = f(x)$ for x. Here is an example. Let $y = f(x) = 3/(x - 2)$. Follow the steps below.

$$y = \frac{3}{x - 2}$$

$$(x - 2)y = 3$$

$$xy - 2y = 3$$

$$xy = 3 + 2y$$

$$x = \frac{3 + 2y}{y}$$

Thus

$$f^{-1}(y) = \frac{3 + 2y}{y}$$

In the formula for f^{-1} just derived, there is no need to use y as the variable. We might use u or t or even x. The formulas

$$f^{-1}(u) = \frac{3 + 2u}{u}$$

$$f^{-1}(t) = \frac{3 + 2t}{t}$$

$$f^{-1}(x) = \frac{3 + 2x}{x}$$

all say the same thing in the sense that they give the same rule. It is conventional to give formulas for functions using x as the variable, and so we would write $f^{-1}(x) = (3 + 2x)/x$ as our answer. Let us summarize. To find the formula for $f^{-1}(x)$, use the following steps.

1. Solve $y = f(x)$ for x in terms of y.
2. Use $f^{-1}(y)$ to name the resulting expression in y.
3. Replace y by x to get the formula for $f^{-1}(x)$.

The Graphs of f and f^{-1} Since $y = f(x)$ and $x = f^{-1}(y)$ are equivalent, the graphs of these two equations are the same. Suppose we want to compare the graphs of $y = f(x)$ and $y = f^{-1}(x)$ (where, you will note, we have used x as the domain variable in both cases). To get $y = f^{-1}(x)$ from $x = f^{-1}(y)$, we interchange the roles of x and y. Graphically, this corresponds to folding (reflecting) the graph across the 45° line, that is, across the line $y = x$.

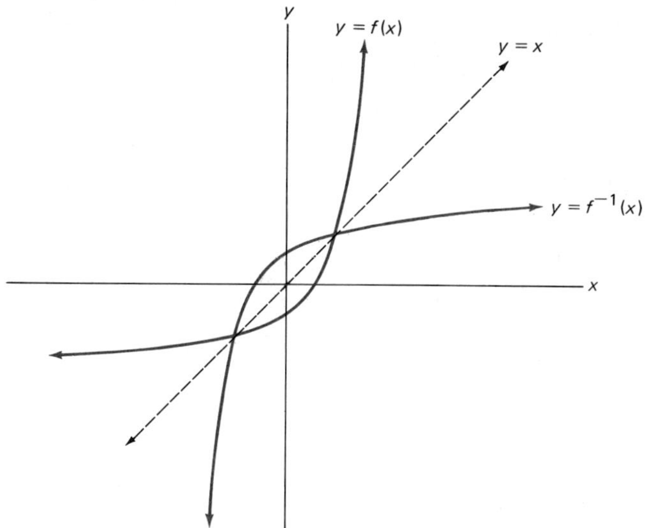

Reflecting a graph across the line $y = x$

Here is a simple example. Let $f(x) = x^3$; then $f^{-1}(x) = \sqrt[3]{x}$, the cube root of x. The graphs of $y = x^3$ and $y = \sqrt[3]{x}$ are shown below, first separately and then on the same coordinate plane. Note that $f(2) = 8$ and $f^{-1}(8) = 2$.

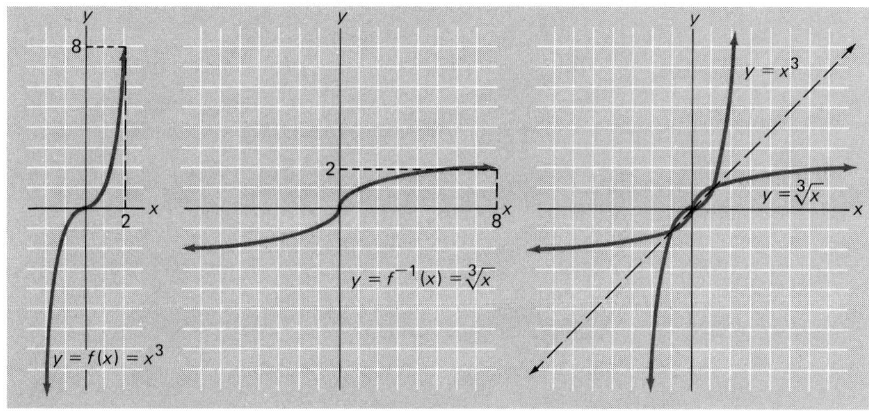

Problem Set 5-5

1. Examine the graphs in the box at the top of page 208.
   (a) Which of these are the graphs of functions with x as domain variable?
   (b) Which of these functions are one-to-one?
   (c) Which of them have inverses?
2. Let each of the following functions have their natural domains. Which of them are one-to-one? (*Hint*: Consider their graphs.)

   (a) $f(x) = x^4$ (b) $f(x) = x^3$ (c) $f(x) = \frac{1}{x}$

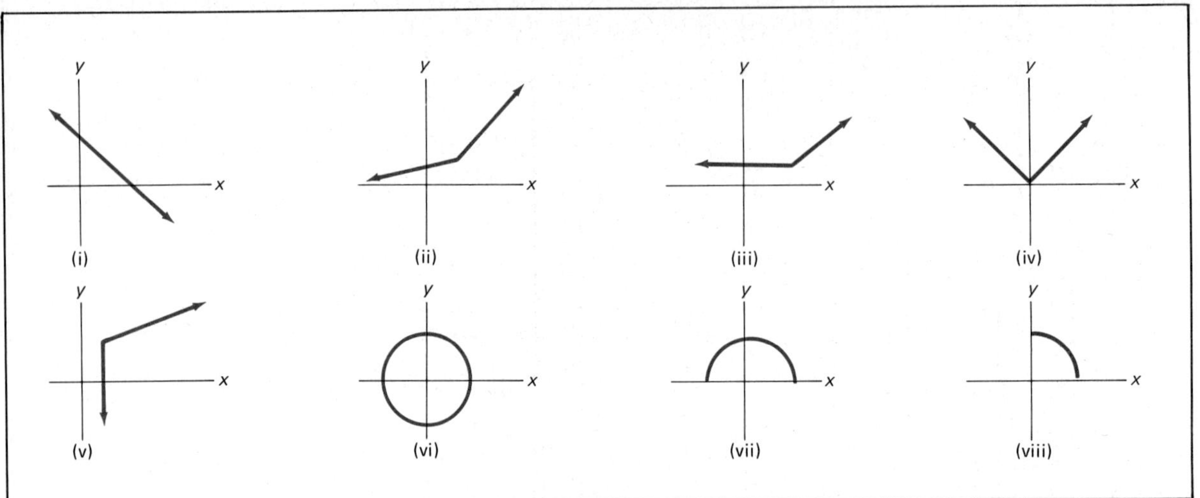

(d) $f(x) = \dfrac{1}{x^2}$ (e) $f(x) = x^2 + 2x + 3$ (f) $f(x) = |x|$

(g) $f(x) = \sqrt{x}$ (h) $f(x) = -3x + 2$

3. Let $f(x) = 3x - 2$. Then $f^{-1}(2) = \frac{4}{3}$ since $f(\frac{4}{3}) = 2$. Find each of the following.
 (a) $f^{-1}(1)$ (b) $f^{-1}(-3)$ (c) $f^{-1}(14)$
4. Let $g(x) = 1/(x - 1)$. Find each of the following.
 (a) $g^{-1}(1)$ (b) $g^{-1}(-1)$ (c) $g^{-1}(14)$
5. Each of the following functions (with their natural domains) has an inverse. Find the formula for $f^{-1}(x)$. Then check your result by calculating $f(f^{-1}(x))$.
 (a) $f(x) = -3x + 2$ (b) $f(x) = 2/(x - 2)$ (c) $f(x) = \sqrt{x} + 2$
6. Follow the directions of Problem 5 for each function.
 (a) $f(x) = \frac{1}{2}x - 3$ (b) $f(x) = x/(x - 3)$ (c) $f(x) = 2\sqrt{x} - 6$
7. In the same coordinate plane, sketch the graphs of $y = f(x)$ and $y = f^{-1}(x)$ for $f(x) = \sqrt{x} + 2$ (see Problem 5c).
8. In the same coordinate plane, sketch the graphs of $y = f(x)$ and $y = f^{-1}(x)$ for $f(x) = x/(x - 3)$ (see Problem 6b).
9. Sketch the graph of $y = f^{-1}(x)$ if the graph of $y = f(x)$ is as shown.

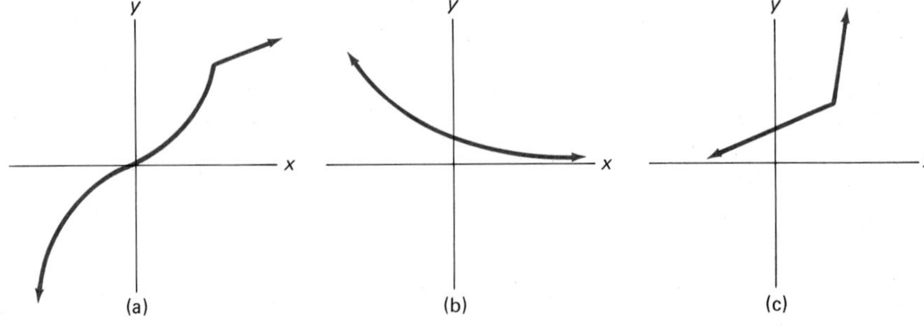

10. Show that $f(x) = 2x/(x - 1)$ and $g(x) = x/(x - 2)$ are inverses of each other by calculating $f(g(x))$ and $g(f(x))$.
11. Show that $f(x) = 3x/(x + 2)$ and $g(x) = 2x/(3 - x)$ are inverses of each other.
12. Sketch the graph of $f(x) = x^3 + 1$ and note that f is one-to-one. Find a formula for $f^{-1}(x)$.

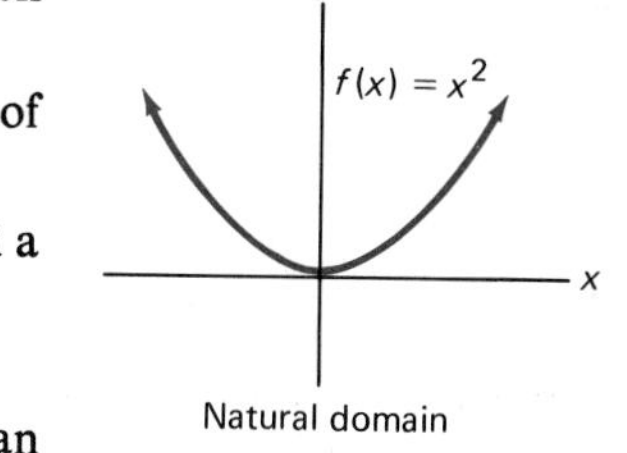

Natural domain

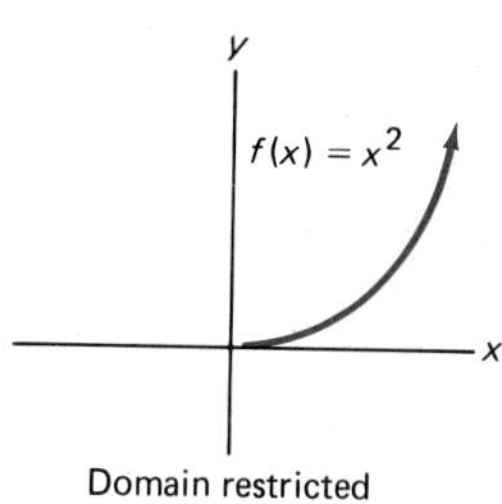

Domain restricted to $x \geq 0$

Example (Restricting the domain) The function $f(x) = x^2$ does not have an inverse if we use its natural domain (all real numbers). However, if we restrict its domain to $x \geq 0$ so that we are considering only its right branch (see graphs in margin), then it has an inverse, $f^{-1}(x) = \sqrt{x}$. Use the same idea to show that $g(x) = x^2 - 2x - 1$ has an inverse when its domain is appropriately restricted. Find $g^{-1}(x)$.

SOLUTION. The graph of $g(x)$ is shown in the margin; it is a parabola with vertex at $x = 1$. Accordingly, we restrict the domain to $x \geq 1$. To find the formula for $g^{-1}(x)$, we first solve $y = x^2 - 2x - 1$ for x using an old trick, completing the square.

$$\begin{aligned} y + 1 &= x^2 - 2x \\ y + 1 + 1 &= x^2 - 2x + 1 \\ y + 2 &= (x - 1)^2 \\ \pm\sqrt{y + 2} &= x - 1 \\ 1 \pm \sqrt{y + 2} &= x \end{aligned}$$

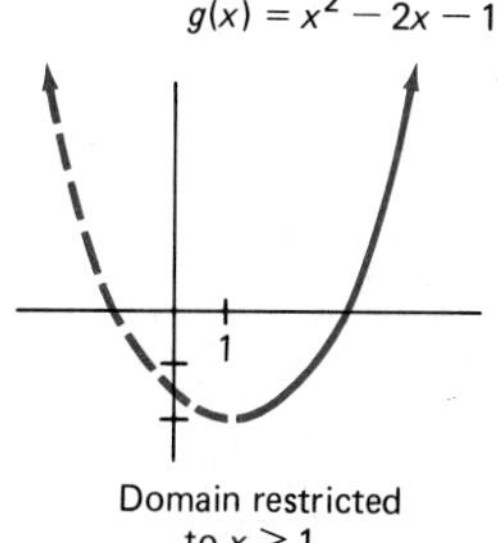

Domain restricted to $x \geq 1$

Notice that there are two expressions for x; they correspond to the two halves of the parabola. We chose to make $x \geq 1$, so $x = 1 + \sqrt{y + 2}$ is the correct expression for $g^{-1}(y)$. Thus

$$g^{-1}(x) = 1 + \sqrt{x + 2}$$

In each of the following, restrict the domain so that f has an inverse. Describe the restricted domain and find a formula for $f^{-1}(x)$.

13. $f(x) = (x - 1)^2$
14. $f(x) = (x + 3)^2$
15. $f(x) = (x + 1)^2 - 4$
16. $f(x) = (x - 2)^2 + 3$
17. $f(x) = x^2 + 6x + 7$
18. $f(x) = x^2 - 4x + 9$
19. $f(x) = |x + 2|$
20. $f(x) = 2|x - 3|$
21. $f(x) = x^2/(2x + 1)$
22. $f(x) = x^2/(4x - 1)$

Miscellaneous Problems

23. Sketch the graph of $f(x) = 1/(x - 1)$. Is f one-to-one? Calculate each of the following.
    (a) $f(3)$ (b) $f^{-1}(\frac{1}{2})$ (c) $f(0)$
    (d) $f^{-1}(-1)$ (e) $f^{-1}(3)$ (f) $f^{-1}(-2)$
24. If $f(x) = x/(x - 2)$, find the formula for $f^{-1}(x)$.
25. Find the formula for $f^{-1}(x)$ if $f(x) = 1/(x - 1)$ and sketch the graph of $y = f^{-1}(x)$. Compare this graph with the graph of $y = f(x)$ that you sketched in Problem 23.

26. The functions $f(x) = 3/(x - 2)$ and $f^{-1}(x) = (3 + 2x)/x$ were shown to be inverses in the text. Sketch the graphs of $y = f(x)$ and $y = f^{-1}(x)$ using the same coordinate axes.
27. Sketch the graph of $f(x) = x^2 - 2x - 3$ and observe that it is not one-to-one. Restrict its domain so it is and then find a formula for $f^{-1}(x)$.
28. Show that $f(x) = x/(x - 1)$ is its own inverse by showing that $f(f(x)) = x$. Can you give other examples of functions with this property? What must be true about their graphs?

In Problems 29–30, find formulas for $f^{-1}(x)$, $g^{-1}(x)$, $(g \circ f)(x)$, $(g \circ f)^{-1}(x)$, *and* $(f^{-1} \circ g^{-1})(x)$.

29. $f(x) = 2x + 5$ and $g(x) = 1/x$
30. $f(x) = x - 4$ and $g(x) = x^3 + 1$
31. From your results in Problems 29–30, what do you conjecture about the relationship between $(g \circ f)^{-1}$ and $f^{-1} \circ g^{-1}$?

CHAPTER SUMMARY

A **function** f is a rule which assigns to each element x in one set (called the **domain**) a value $f(x)$ from another set. The set of all these values is called the **range** of the function. Numerical functions are usually specified by formulas (for example, $g(x) = (x^2 + 1)/(x + 1)$). The **natural domain** for such a function is the largest set of real numbers for which the formula makes sense and gives real values (thus, the natural domain for g is all real numbers except $x = -1$). Related to the notion of function is that of **variation.**

The **graph** of a function is simply the graph of the equation $y = f(x)$. Of special interest are the graphs of **polynomial functions** and **rational functions.** In graphing them, we should show the hills, the valleys, the x**-intercepts**, and, in the case of rational functions, the vertical and horizontal **asymptotes.**

Functions can be combined in many ways. Of these, composition is perhaps the most significant. The **composite** of f with g is defined by $(f \circ g)(x) = f(g(x))$.

Some functions are **one-to-one**; some are not. Those that are one-to-one have undoing functions called inverses. The **inverse** of f, denoted by f^{-1}, satisfies $f^{-1}(f(x)) = x$. Finding a formula for $f^{-1}(x)$ can be tricky; therefore, we described a definite procedure for doing it.

CHAPTER REVIEW PROBLEM SET

1. Let $f(x) = x^2 - 1$ and $g(x) = 2/x$. Calculate if possible.

(a) $f(4)$	(b) $g(\frac{1}{2})$	(c) $g(0)$
(b) $f(1)/g(3)$	(e) $f(g(4))$	(f) $g(f(4))$

2. Find the natural domain of f if $f(x) = \sqrt{x + 1}/(x - 1)$.

3. If y varies directly as the cube of x and $y = 1$ when $x = 2$, find an explicit formula for y in terms of x.

4. If z varies directly as x and inversely as the square of y, and if $z = 1$ when x and y are both 3, find z when $x = 16$ and $y = 2$.

5. Graph each f.
(a) $f(x) = (x - 2)^2$ (b) $f(x) = x^3 + 2x$

(c) $f(x) = \dfrac{1}{x^2 - x - 2}$ (d) $f(x) = \begin{cases} 0 & \text{if } x \le 0 \\ x^2 & \text{if } 0 < x < 1 \\ 1 & \text{if } x \ge 1 \end{cases}$

6. Suppose that g is an even function satisfying $g(x) = \sqrt{x}$ for $x \ge 0$. Sketch its graph on $-4 \le x \le 4$.

7. Sketch the graph of $h(x) = x^2 + 1/x$ by first graphing $y = x^2$ and $y = 1/x$ and then adding ordinates.

8. If $f(x) = x^3 + 1$ and $g(x) = x + 2$, give formulas for each of the following.
(a) $f(g(x))$ (b) $g(f(x))$ (c) $f(f(x))$
(d) $g(g(x))$ (e) $f^{-1}(x)$ (f) $g^{-1}(x)$
(g) $g(x + h)$ (h) $[g(x + h) - g(x)]/h$ (i) $f(3x)$

9. How does the graph of $y = f(x - 2) + 3$ relate to the graph of $y = f(x)$?

10. Which of the following functions are even? Odd? One-to-one?
(a) $f(x) = 1/(x^2 - 1)$ (b) $f(x) = 1/x$
(c) $f(x) = |x|$ (d) $f(x) = 3x + 4$

11. Let $f(x) = x/(x - 2)$. Find a formula for $f^{-1}(x)$. Graph $y = f(x)$ and $y = f^{-1}(x)$ using the same coordinate axes.

12. How could we restrict the domain of $f(x) = (x + 2)^2$ so that f has an inverse?

The method of logarithms, by reducing to a few days the labor of many months, doubles as it were, the life of the astronomer, besides freeing him from the errors and disgust inseparable from long calculation.

P. S. Laplace

CHAPTER SIX

Exponential and Logarithmic Functions

6-1 Radicals

6-2 Exponents and Exponential Functions

6-3 Exponential Growth

6-4 Logarithms and Logarithmic Functions

6-5 Applications of Logarithms

6-6 Common Logarithms (Optional)

6-7 Calculations with Logarithms (Optional)

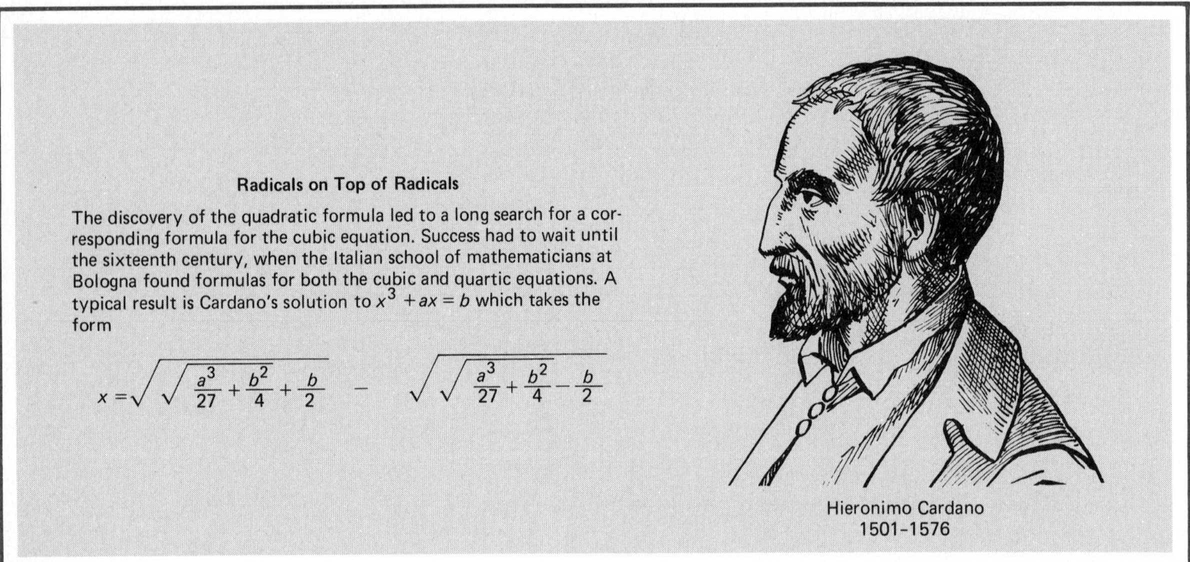

Hieronimo Cardano
1501-1576

6-1 Radicals

Historically, interest in radicals has been associated with the desire to solve equations. Even the general cubic equation leads to very complicated radical expressions. Today, powerful iterative methods make results like Cardano's solution historical curiosities. Yet the need for radicals continues; it is important that we know something about them.

Raising a number to the 3rd power (or cubing it) is a process which can be undone. The inverse process—taking the 3rd root—is denoted by $\sqrt[3]{}$. We call $\sqrt[3]{a}$ a *radical* and read it "the cube root of *a*." Thus $\sqrt[3]{8} = 2$ and $\sqrt[3]{-125} = -5$ since $2^3 = 8$ and $(-5)^3 = -125$.

Our first goal is to give meaning to the symbol $\sqrt[n]{a}$ when n is any positive integer. Naturally, we require that $\sqrt[n]{a}$ be a number which yields a when raised to the nth power, that is,

$$(\sqrt[n]{a})^n = a$$

When n is odd, that is all we need to say, since for any real number a, there is exactly one real number whose nth power is a.

When n is even, we face two serious problems, problems that are already apparent when $n = 2$. We have already discussed square roots, using the symbol $\sqrt{}$ rather than $\sqrt[2]{}$ (see Section 3-4). Recall that if $a < 0$, then $\sqrt{a}$ is not a real number (for example, $\sqrt{-4} = 2i$). Even if $a > 0$, we are in trouble since there are always two real numbers with squares equal to a. For example, both -3 and 3 have squares equal to 9. We agree that in this ambiguous case, $\sqrt{a}$ shall always denote the positive square root of a. Thus $\sqrt{9}$ is equal to 3, not -3.

We make a similar agreement about $\sqrt[n]{a}$ for n an even number greater than 2. First, we shall avoid the case $a < 0$. Second, when $a \geq 0$, $\sqrt[n]{a}$ will always denote the nonnegative number whose nth power is a. Thus $\sqrt[4]{81} = 3$, $\sqrt[4]{16} = 2$, and $\sqrt[4]{0} = 0$; however, the symbol $\sqrt[4]{-16}$ will be assigned no meaning in this book (although there will be a discussion of even roots of negative numbers in Chapter 9). Let us summarize.

> If n is odd, $\sqrt[n]{a}$ is the unique real number satisfying $(\sqrt[n]{a})^n = a$.
> If n is even and $a \geq 0$, $\sqrt[n]{a}$ is the unique nonnegative real number satisfying $(\sqrt[n]{a})^n = a$.

Rules for Radicals Radicals, like exponents, obey certain rules. The most important ones are listed below, where it is assumed that all radicals name real numbers.

RULES FOR RADICALS

1. $(\sqrt[n]{a})^n = a$
2. $\sqrt[n]{a^n} = a \qquad (a \geq 0)$
3. $\sqrt[n]{ab} = \sqrt[n]{a}\sqrt[n]{b}$
4. $\sqrt[n]{\frac{a}{b}} = \frac{\sqrt[n]{a}}{\sqrt[n]{b}}$

These rules can all be proved, but we believe that the following illustrations will be more helpful to you than proofs.

$$(\sqrt[4]{7})^4 = 7$$
$$\sqrt[14]{3^{14}} = 3$$
$$\sqrt{2}\cdot\sqrt{18} = \sqrt{36} = 6$$
$$\frac{\sqrt[3]{750}}{\sqrt[3]{6}} = \sqrt[3]{\frac{750}{6}} = \sqrt[3]{125} = 5$$

Simplifying Radicals One use of the four rules given above is to simplify radicals. We illustrate this by simplifying two radicals.

$$\sqrt[3]{54x^4y^6} \qquad \sqrt[4]{x^8 + x^4y^4}$$

We assume that x and y represent positive numbers.

In the first example, we start by factoring out the largest possible third power.

$$\begin{aligned}\sqrt[3]{54x^4y^6} &= \sqrt[3]{(27x^3y^6)(2x)}\\ &= \sqrt[3]{(3xy^2)^3(2x)}\\ &= \sqrt[3]{(3xy^2)^3}\,\sqrt[3]{2x} \qquad \text{(Rule 3)}\\ &= 3xy^2\,\sqrt[3]{2x} \qquad \text{(Rule 2)}\end{aligned}$$

In the second example, it is tempting to write $\sqrt[4]{x^8 + x^4y^4} = x^2 + xy$, thereby pretending that $\sqrt[4]{a^4 + b^4} = a + b$. This is wrong, because $(a + b)^4 \neq a^4 + b^4$. Here is what we can do.

$$\begin{aligned}\sqrt[4]{x^8 + x^4y^4} &= \sqrt[4]{x^4(x^4 + y^4)}\\ &= \sqrt[4]{x^4}\,\sqrt[4]{x^4 + y^4} \qquad \text{(Rule 3)}\\ &= x\sqrt[4]{x^4 + y^4} \qquad \text{(Rule 2)}\end{aligned}$$

We were able to take x^4 out of the radical because it is a 4th power and a factor of $x^8 + x^4y^4$.

Rationalizing Denominators Fractions with radicals in their denominators are considered to be needlessly complicated. Fortunately, we can usually rewrite a fraction so that its denominator is free of radicals. The process we go through is called **rationalizing the denominator**. Here are two examples.

$$\frac{1}{\sqrt[5]{x}} \qquad \frac{x}{\sqrt{x} + \sqrt{y}}$$

In the first case, we multiply numerator and denominator by $\sqrt[5]{x^4}$, which gives the fifth root of a fifth power in the denominator.

$$\frac{1}{\sqrt[5]{x}} = \frac{1 \cdot \sqrt[5]{x^4}}{\sqrt[5]{x} \cdot \sqrt[5]{x^4}} = \frac{\sqrt[5]{x^4}}{\sqrt[5]{x \cdot x^4}} = \frac{\sqrt[5]{x^4}}{\sqrt[5]{x^5}} = \frac{\sqrt[5]{x^4}}{x}$$

In the second case, we make use of the identity $(a + b)(a - b) = a^2 - b^2$. If we multiply numerator and denominator of the fraction by $\sqrt{x} - \sqrt{y}$, the radicals in the denominator disappear

$$\frac{x}{\sqrt{x} + \sqrt{y}} = \frac{x(\sqrt{x} - \sqrt{y})}{(\sqrt{x} + \sqrt{y})(\sqrt{x} - \sqrt{y})} = \frac{x\sqrt{x} - x\sqrt{y}}{x - y}$$

We should point out that this manipulation is valid provided $x \neq y$.

Problem Set 6-1

Simplify the following. All letters represent positive numbers.

1. $\sqrt{9}$
2. $\sqrt[3]{-8}$
3. $\sqrt[5]{32}$
4. $\sqrt[4]{16}$
5. $(\sqrt[3]{7})^3$
6. $(\sqrt{\pi})^2$

7. $\sqrt[3]{(\frac{3}{2})^3}$

8. $\sqrt[5]{(-2/7)^5}$

9. $(\sqrt{5})^4$

10. $(\sqrt[3]{5})^6$

11. $\sqrt{3}\sqrt{27}$

12. $\sqrt{2}\sqrt{32}$

13. $\sqrt[3]{16}/\sqrt[3]{2}$

14. $\sqrt[4]{48}/\sqrt[4]{3}$

15. $\sqrt[3]{10^{-6}}$

16. $\sqrt[4]{10^8}$

17. $1/\sqrt{2}$

18. $1/\sqrt{3}$

19. $\sqrt{10}/\sqrt{2}$

20. $\sqrt{6}/\sqrt{3}$

21. $\sqrt[3]{54x^4y^5}$

22. $\sqrt[3]{-16x^3y^8}$

23. $\sqrt[4]{(x + 2)^4y^7}$

24. $\sqrt[4]{x^5(y - 1)^8}$

25. $\sqrt{x^2 + x^2y^2}$

26. $\sqrt{25 + 50y^4}$

27. $\sqrt[3]{x^6 - 9x^3y}$

28. $\sqrt[4]{16x^{12} + 64x^8}$

29. $\sqrt[3]{x^4y^{-6}z^6}$

30. $\sqrt[4]{32x^{-4}y^9}$

31. $\dfrac{2}{\sqrt{x} + 3}$

32. $\dfrac{4}{\sqrt{x} - 2}$

33. $\dfrac{2}{\sqrt{x + 3}}$

34. $\dfrac{4}{\sqrt{x - 2}}$

35. $\dfrac{1}{\sqrt[4]{8x^3}}$

36. $\dfrac{1}{\sqrt[3]{5x^2y^4}}$

37. $\sqrt[3]{2x^{-2}y^4}\sqrt[3]{4xy^{-1}}$

38. $\sqrt[4]{125x^5y^3}\sqrt[4]{5x^{-9}y^5}$

39. $\sqrt{50} - 2\sqrt{18} + \sqrt{8}$

40. $\sqrt[3]{24} + \sqrt[3]{375}$

Example (Equations involving radicals) Solve the following equations.

(a) $\sqrt[3]{x - 2} = 3$ (b) $x = \sqrt{2 - x}$

SOLUTION. (a) Raise both sides to the 3rd power and solve for x.

$$\left(\sqrt[3]{x - 2}\right)^3 = 3^3$$
$$x - 2 = 27$$
$$x = 29$$

(b) Square both sides and solve for x.

$$x^2 = 2 - x$$
$$x^2 + x - 2 = 0$$
$$(x - 1)(x + 2) = 0$$
$$x = 1 \qquad x = -2$$

Let us check our answers in part (b) by substituting them in the original equation. When we substitute these numbers for x in $x = \sqrt{2 - x}$, we find that 1 works but -2 does not.

$$1 = \sqrt{2 - 1} \qquad -2 \neq \sqrt{2 - (-2)}$$

In squaring both sides of $x = \sqrt{2 - x}$, we introduced an extraneous solution. That happened because $a = b$ and $a^2 = b^2$ are not equivalent statements. Whenever you square both sides of an equation (or raise both sides of an equation to any even power), be sure to check your answers.

Solve each of the following equations.

41. $\sqrt{x-1} = 5$
42. $\sqrt{x+2} = 3$
43. $\sqrt[3]{2x-1} = 2$
44. $\sqrt[3]{1-5x} = 6$
45. $\sqrt{\dfrac{x}{x+2}} = 4$
46. $\sqrt[3]{\dfrac{x-2}{x+1}} = -2$
47. $\sqrt{x^2+4} = x + 2$
48. $\sqrt{x^2+9} = x - 3$
49. $\sqrt{2x+1} = x - 1$
50. $\sqrt{x} = 12 - x$

Miscellaneous Problems

Simplify the following. Assume that all letters represent positive numbers.

51. $(\sqrt[3]{2})^3$
52. $\sqrt[3]{b^3}$
53. $\sqrt[4]{16y^8}$
54. $\sqrt{\dfrac{9a^2}{b^4}}$
55. $\sqrt[7]{\left(\dfrac{-3x}{y}\right)^7}$
56. $\sqrt{27}\sqrt{3}$
57. $\sqrt{54b^3c^4}$
58. $\sqrt[3]{250x^5y^9}$
59. $\sqrt[4]{x^4 + x^8y^4}$
60. $\dfrac{2}{\sqrt{a} - 1}$
61. $\dfrac{2}{\sqrt{a-1}}$
62. $\dfrac{1}{\sqrt{7bc^3}}$
63. $\sqrt{a}\left(\sqrt{a} + \dfrac{1}{a\sqrt{a}}\right)$
64. $\sqrt{12} + \sqrt{48} - \sqrt{27}$

Solve the equations in Problems 65–70.

65. $\sqrt{3x-2} = 4$
66. $\sqrt{2-3x} = \sqrt{5}$
67. $\sqrt[3]{1-5x} = -4$
68. $\sqrt{x+2} = \sqrt{2x-3}$
69. $\sqrt{4x+1} = x + 1$
70. $2\sqrt{x} = 3x - 1$

[C] 71. Taking a root is the inverse of raising to a power. Thus, on some calculators, you may take roots by using the [INV] and [y^x] keys. For example, to find $\sqrt[4]{31}$, press 31 [INV] [y^x] 4 [=]. You should get 2.3596111. Use your calculator to find each of the following.
(a) $\sqrt[5]{87}$ (b) $\sqrt[3]{213}$ (c) $\sqrt[50]{100}$ (d) $\sqrt[8]{390{,}625}$

[C] 72. Calculate.
(a) $\dfrac{4}{\sqrt[6]{11}}$ (b) $\sqrt[5]{12} - \sqrt[3]{2}$ (c) $\dfrac{\sqrt[4]{29} + \sqrt[3]{6}}{\sqrt[3]{14}}$
(d) $\sqrt[8]{.012}(\sqrt{130} + 4)^2$

73. Show that, for all real numbers x,

$$\boxed{\sqrt{x^2} = |x|}$$

74. Show that, if n is even, $\sqrt[n]{x^n} = |x|$.

75. We know that $f(x) = x^5$ and $g(x) = \sqrt[5]{x}$ are inverse functions. What does this mean about their graphs? Draw both graphs in the same coordinate plane.

76. Draw the graphs of $f(x) = x^4$ and $g(x) = \sqrt[4]{x}$ for $x \geq 0$ in the same coordinate plane.

77. Use algebra to solve $\sqrt{x} \leq \sqrt[3]{x}$. (*Hint*: To what power can you raise both sides to clear the equation of radicals?)

78. Show that there are only two integral values of x between 1 and 1000 for which $\sqrt{x}$ and $\sqrt[3]{x}$ are both integers.

Ⓒ 79. Here is a well-known iterative method for finding the square root of a positive number A to any desired degree of accuracy. Use the formulas

$$x_n = \frac{x_{n-1} + A/x_{n-1}}{2} \qquad n = 2, 3, 4, \ldots$$

where x_1 is a first approximation to $\sqrt{A}$. As an example, let $A = 37$ and take $x_1 = 6$. Then

$$x_2 = \frac{x_1 + 37/x_1}{2} = \frac{6 + 37/6}{2} \approx 6.08$$

$$x_3 = \frac{x_2 + 37/x_2}{2} = \frac{6.08 + 37/6.08}{2} \approx 6.086$$

and so on. Find the second and third approximations, x_2 and x_3, of each of the following square roots using the given first approximation x_1.

(a) $\sqrt{15}$, $x_1 = 4$ (b) $\sqrt{40}$, $x_1 = 6$ (c) $\sqrt{40}$, $x_1 = 7$

80. Show that $\sqrt[3]{2}$ is a solution of the equation

$$2x^9 - 4x^6 + 5x^3 - 10 = 0$$

Some Sense—Some Nonsense

One of the authors once asked a student to write the definition of $2^{4.6}$ on the blackboard. After thinking deeply for a minute, he wrote:

$$2^2 = 2 \cdot 2$$
$$2^3 = 2 \cdot 2 \cdot 2$$
$$2^4 = 2 \cdot 2 \cdot 2 \cdot 2$$
$$2^{4.6} = 2 \cdot 2 \cdot 2 \cdot 2 \cdot \text{ɀ}$$

6-2 Exponents and Exponential Functions

Rules for Exponents

1. $a^m a^n = a^{m+n}$
2. $\dfrac{a^m}{a^n} = a^{m-n}$
3. $(a^m)^n = a^{mn}$

After you have criticized the student mentioned above, ask yourself how you would define $2^{4.6}$. Of course, integral powers of 2 make perfectly good sense, although 2^{-3} and 2^0 became meaningful only after we had *defined* a^{-n} to be $1/a^n$ and a^0 to be 1 (see Section 2-1). Those were good definitions because they were consistent with the familiar rules of exponents. Now we ask what meaning we can give to powers like $2^{1/2}$, $2^{4.6}$, and even 2^{π} so that these familiar rules still hold.

Rational Exponents We assume throughout this section that $a > 0$. If n is any positive integer, we want

$$(a^{1/n})^n = a^{(1/n) \cdot n} = a^1 = a$$

But we know that $(\sqrt[n]{a})^n = a$. Thus we define

$$\boxed{a^{1/n} = \sqrt[n]{a}}$$

For example, $2^{1/2} = \sqrt{2}$, $27^{1/3} = \sqrt[3]{27} = 3$, and $(16)^{1/4} = \sqrt[4]{16} = 2$.

Next, if m and n are positive integers, we want

$$(a^{1/n})^m = a^{m/n} \quad \text{and} \quad (a^m)^{1/n} = a^{m/n}$$

This forces us to define

$$\boxed{a^{m/n} = (\sqrt[n]{a})^m = \sqrt[n]{a^m}}$$

Accordingly,

$$2^{3/2} = (\sqrt{2})^3 = \sqrt{2}\,\sqrt{2}\,\sqrt{2} = 2\sqrt{2}$$

and

$$27^{2/3} = (\sqrt[3]{27})^2 = 3^2 = 9$$

Lastly, we define

$$a^{-m/n} = \frac{1}{a^{m/n}}$$

so that

$$2^{-1/2} = \frac{1}{2^{1/2}} = \frac{1}{\sqrt{2}}$$

and

$$4^{-3/2} = \frac{1}{4^{3/2}} = \frac{1}{(\sqrt{4})^3} = \frac{1}{8}$$

We have just succeeded in defining a^x for all rational numbers x (recall that a rational number is a ratio of two integers). What is more important is that we have done it in such a way that the rules of exponents still hold. Incidentally, we can now answer the question in our opening display.

$$2^{4.6} = 2^4 2^{.6} = 2^4 2^{6/10} = 16(\sqrt[10]{2})^6$$

For simplicity, we have assumed that a is positive in our discussion of $a^{m/n}$. But we should point out that the definition of $a^{m/n}$ given above is also appropriate for the case in which a is negative and n is odd. For example,

$$(-27)^{2/3} = (\sqrt[3]{-27})^2 = (-3)^2 = 9$$

Real Exponents Irrational powers such as 2^π and $3^{\sqrt{2}}$ are intrinsically more difficult to define than are rational powers. Rather than attempt a technical definition, we ask you to consider what 2^π might mean. The decimal expansion of π is 3.14159.... Thus we could look at the sequence of rational powers

$$2^3, \quad 2^{3.1}, \quad 2^{3.14}, \quad 2^{3.141}, \quad 2^{3.1415}, \quad 2^{3.14159}, \ldots$$

As you should suspect, when the exponents get closer and closer to π, the corresponding powers of 2 get closer and closer to a definite number. We shall call that number 2^π.

The process of starting with integral exponents and then extending to rational exponents and finally to real exponents can be clarified by means of three graphs.

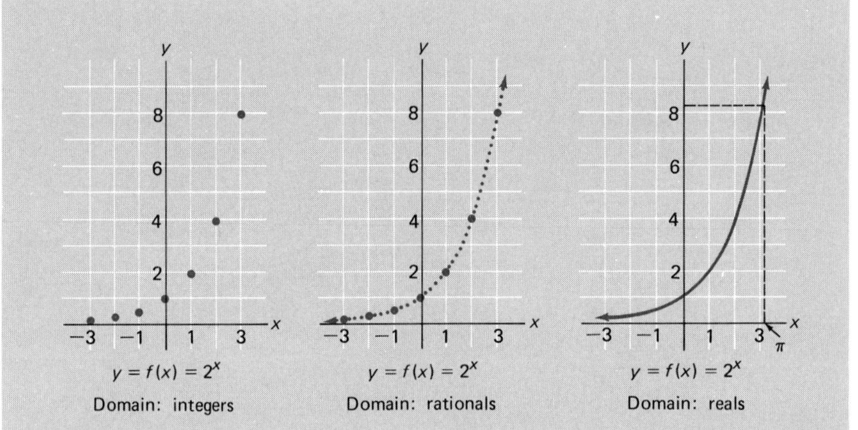

The first graph suggests a curve rising from left to right. The second graph makes the suggestion stronger. The third graph leaves nothing to the imagination; it is a continuous curve and it shows 2^x for all values of x, rational and irrational. As x increases in the positive direction, the values of 2^x increase without bound; in the negative direction, the values of 2^x approach 0. Notice that 2^π is a little less than 9; it is actually given by

$$2^\pi = 8.8249778...$$

Exponential Functions The function $f(x) = 2^x$, graphed above, is one example of an exponential function. But what has been done with 2 can be done with any positive real number a. In general, the formula

$$f(x) = a^x$$

determines a function called an **exponential function with base** a. Its domain is the set of all real numbers and its range is the set of positive numbers.

Let us see what effect the size of a has on the graph of $f(x) = a^x$. We choose $a = 3$ and $a = \frac{1}{3}$, showing both graphs at the top of page 223.

The graph of $f(x) = 3^x$ looks much like the graph of $f(x) = 2^x$, although it rises more rapidly. The graph of $f(x) = 10^x$ would be even steeper. However, in the case of $(\frac{1}{3})^x$, the graph falls from left to right. In fact, you can get the graph of $f(x) = (\frac{1}{3})^x$ by reflecting the graph of $f(x) = 3^x$ about the y-axis. This is because $(\frac{1}{3})^x = 3^{-x}$.

We can summarize what is suggested by our two graphs as follows.

If $a > 1$, $f(x) = a^x$ is an increasing function.
If $0 < a < 1$, $f(x) = a^x$ is a decreasing function.

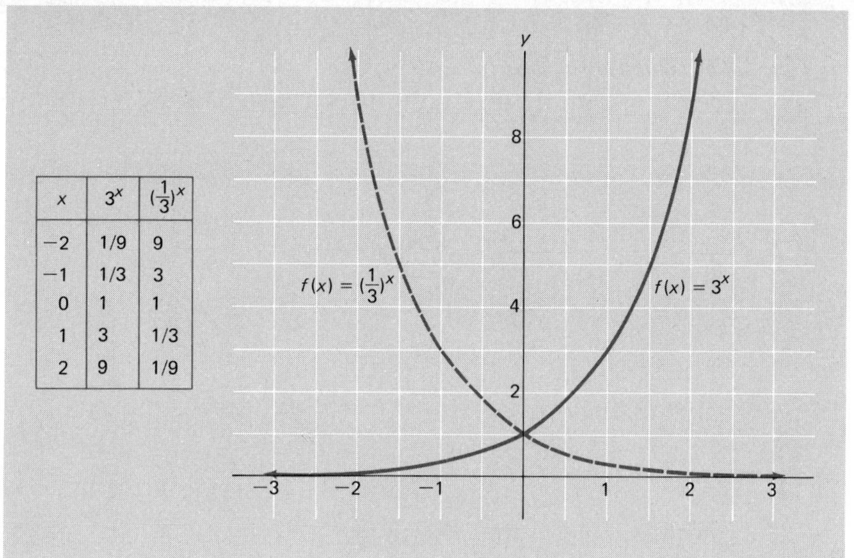

x	3^x	$(\frac{1}{3})^x$
−2	1/9	9
−1	1/3	3
0	1	1
1	3	1/3
2	9	1/9

In both of these cases, the graph of f has the x-axis as an asymptote. The case $a = 1$ is not very interesting since it yields the constant function $f(x) = 1$.

Properties of Exponential Functions It is easy to describe the main properties of exponential functions, since they obey the rules we learned in Section 2-1. Perhaps it is worth repeating them, since we do want to emphasize that they now hold for all *real* exponents x and y (at least for the case where a and b are both positive).

1. $a^x a^y = a^{x+y}$

2. $\dfrac{a^x}{a^y} = a^{x-y}$

3. $(a^x)^y = a^{xy}$

4. $(ab)^x = a^x b^x$

5. $\left(\dfrac{a}{b}\right)^x = \dfrac{a^x}{b^x}$

Here are a number of examples that are worth studying.

$$3^{1/2}3^{3/4} = 3^{1/2+3/4} = 3^{5/4}$$

$$\frac{\pi^4}{\pi^{5/2}} = \pi^{4-5/2} = \pi^{3/2}$$

$$(2^{\sqrt{3}})^4 = 2^{4\sqrt{3}}$$

$$(5^{\sqrt{2}}\, 5^{1-\sqrt{2}})^{-3} = (5^1)^{-3} = 5^{-3}$$

Problem Set 6-2 *Write each of the following as a power of 7.*

1. $\sqrt[3]{7}$
2. $\sqrt[5]{7}$
3. $\sqrt[3]{7^2}$
4. $\sqrt[5]{7^3}$
5. $\dfrac{1}{\sqrt[3]{7}}$
6. $\dfrac{1}{\sqrt[5]{7}}$
7. $\dfrac{1}{\sqrt[3]{7^2}}$
8. $\dfrac{1}{\sqrt[5]{7^3}}$
9. $7\sqrt[3]{7}$
10. $7\sqrt[5]{7}$

Rewrite each of the following using exponents instead of radicals.

11. $\sqrt[3]{x^2}$
12. $\sqrt[4]{x^3}$
13. $x^2\sqrt{x}$
14. $x\sqrt[3]{x}$
15. $\sqrt{(x+y)^3}$
16. $\sqrt[3]{(x+y)^2}$
17. $\sqrt{x^2+y^2}$
18. $\sqrt[3]{x^3+8}$

Rewrite each of the following using radicals instead of fractional exponents.

19. $4^{2/3}$
20. $10^{3/4}$
21. $8^{-3/2}$
22. $12^{-5/6}$
23. $(x^4+y^4)^{1/4}$
24. $(x^2+xy)^{1/2}$
25. $(x^2y^3)^{2/5}$
26. $(3ab^2)^{2/3}$
27. $(x^{1/2}+y^{1/2})^{1/2}$
28. $(x^{1/3}+y^{2/3})^{1/3}$

Simplify each of the following. Give your answer without any negative exponents.

29. $25^{1/2}$
30. $27^{1/3}$
31. $8^{2/3}$
32. $16^{3/2}$
33. $9^{-3/2}$
34. $64^{-2/3}$
35. $(-.008)^{2/3}$
36. $(-.027)^{5/3}$
37. $(.0025)^{3/2}$
38. $(1.44)^{3/2}$
39. $5^{2/3}5^{-5/3}$
40. $4^{3/4}4^{-1/4}$
41. $16^{7/6}16^{-5/6}16^{-4/3}$
42. $9^29^{2/3}9^{-7/6}$
43. $(8^2)^{-2/3}$
44. $(4^{-3})^{2/3}$
45. $(3a^{1/2})(-2a^{3/2})$
46. $(2x^{3/4})(5x^{-3/4})$
47. $(2^{1/2}x^{-2/3})^6$
48. $(\sqrt{3}x^{-1/4}y^{3/4})^4$
49. $(xy^{-2/3})^3(x^{1/2}y)^2$
50. $(a^2b^{-1/4})^2(a^{-1/3}b^{1/2})^3$
51. $\dfrac{(2x^{-1}y^{2/3})^2}{x^2y^{-2/3}}$
52. $\left(\dfrac{a^{1/2}b^{1/3}}{c^{5/6}}\right)^{12}$
53. $\left(\dfrac{x^{-2}y^{3/4}}{x^{1/2}}\right)^{12}$
54. $\dfrac{(x^{1/3}y^{-3/4})}{x^{-2/3}y^{1/2}}$
55. $y^{2/3}(2y^{4/3}-y^{-5/3})$
56. $x^{-3/4}\left(-x^{7/4}+\dfrac{2}{\sqrt[4]{x}}\right)$
57. $(x^{1/2}+y^{1/2})^2$
58. $(a^{3/2}+\pi)^2$

Example (Mixing radicals of different orders) Express $\sqrt{2}\sqrt[3]{5}$ using just one radical.

SOLUTION. Square roots and cube roots mix about as well as oil and water, but exponents can serve as a blender. They allow us to write both $\sqrt{2}$ and $\sqrt[3]{5}$ as sixth roots.

$$\begin{aligned}\sqrt{2}\sqrt[3]{5} &= 2^{1/2} \cdot 5^{1/3} \\ &= 2^{3/6} \cdot 5^{2/6} \\ &= (2^3 \cdot 5^2)^{1/6} \\ &= \sqrt[6]{200}\end{aligned}$$

Express each of the following in terms of at most one radical in simplest form.

59. $\sqrt{2}\sqrt[3]{2}$
60. $\sqrt[3]{2}\sqrt[4]{2}$
61. $\sqrt[4]{2}\sqrt[6]{x}$
62. $\sqrt[3]{5}\sqrt{x}$
63. $\sqrt[3]{x\sqrt{x}}$
64. $\sqrt{x\sqrt[3]{x}}$

Use a calculator to find an approximate value of each of the following.

C 65. $2^{1.34}$
C 66. $2^{-.79}$
C 67. $\pi^{1.34}$
C 68. π^{π}
C 69. $(1.46)^{\sqrt{2}}$
C 70. $\pi^{\sqrt{2}}$
C 71. $(.9)^{50.2}$
C 72. $(1.01)^{50.2}$

Sketch the graph of each of the following functions.

73. $f(x) = 4^x$
74. $f(x) = 4^{-x}$
75. $f(x) = (\frac{2}{3})^x$
76. $f(x) = (\frac{2}{3})^{-x}$
C 77. $f(x) = \pi^x$
C 78. $f(x) = (\sqrt{2})^x$

Miscellaneous Problems

Rewrite each of the following using exponents instead of radicals.

79. $\sqrt[3]{a^2}$
80. $\sqrt[6]{x^3}$
81. $\sqrt[5]{(a + 2b)^3}$
82. $\sqrt{x^2 + 9}$

Rewrite using radicals instead of exponents.

83. $6^{3/4}$
84. $12^{-2/3}$
85. $(4 + x^{1/2})^{-1/2}$
86. $(16a^2b^3)^{3/4}$

Simplify.

87. $8^{1/3}$
88. $8^{2/3}$
89. $8^{-2/3}$
90. $(.04)^{-1/2}$
91. $(.125)^{4/3}$
92. $7^{2/3}7^{1/2}7^{-1/6}$
93. $5^{10}[(25)^3]^{-3/2}$
94. $(y^2z^{-4})^{1/2}$
95. $x^{1/4}(x^{-5/4} + x^{3/4})$
96. $(x^{3/2} + x^{-3/2})^2$

Sketch the graphs of the functions in 97–100.

97. $f(x) = 5^x$

98. $f(x) = (\frac{1}{5})^x$

[C] 99. $f(x) = (.9)^x$

[C] 100. $f(x) = (1.1)^x$

101. Rank the numbers $5^{5/3}$, $5^{7/4}$, and $5^{\sqrt{3}}$ from least to greatest.

102. How would you define $2^{\sqrt{2}}$?

103. Solve $x^{2/3} - 3x^{1/3} + 2 = 0$.

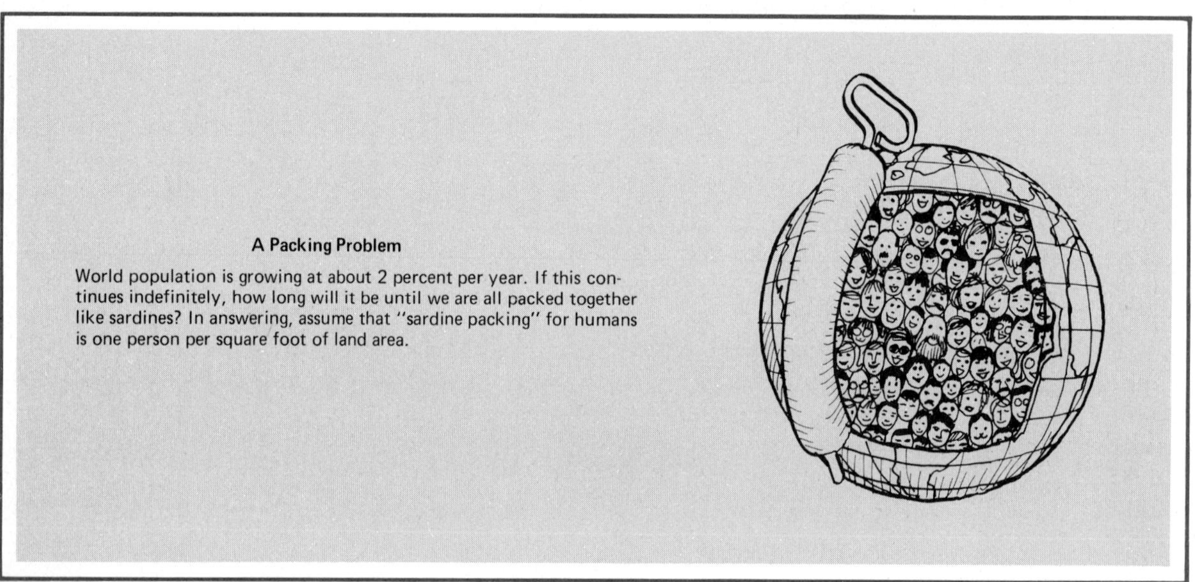

6-3 Exponential Growth

The phrase *exponential growth* is used repeatedly by professors, politicians, and pessimists. Population, energy use, mining of ores, pollution, and the number of books about these things are all said to be growing exponentially. Most people probably do not know what exponential growth means, except that they have heard it guarantees alarming consequences. For students of this book, it is easy to explain its meaning. For y to grow exponentially with time t means that it satisfies the relationship

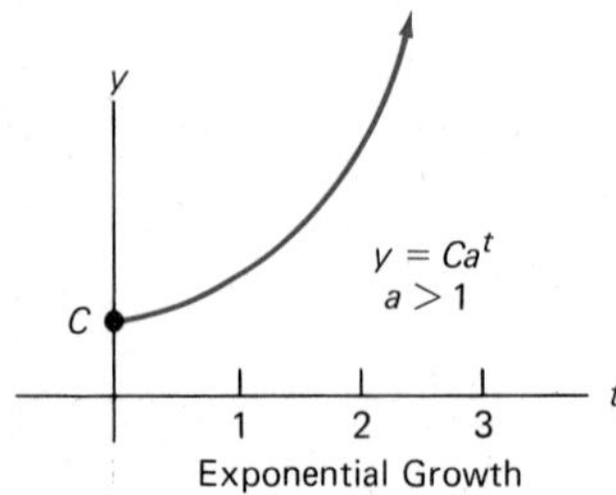

Exponential Growth

$$y = Ca^t$$

for constants C and a, with $C > 0$ and $a > 1$. Why should so many ingredients of modern society behave this way? The basic cause is population growth.

Population Growth Simple organisms reproduce by cell division. If, for example, there is one cell today, that cell may split so that there are two cells tomorrow. Then each of those cells may divide giving four cells the following day. As this process continues, the number of cells on successive days forms the sequence

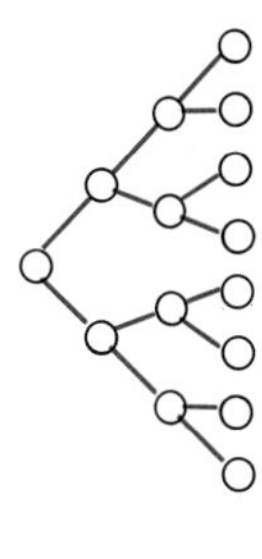

$$1, \quad 2, \quad 4, \quad 8, \quad 16, \quad 32, \ldots$$

If we start with 100 cells and let $f(t)$ denote the number present t days from now, we have the results indicated in the table below.

t	0	1	2	3	4	5
$f(t)$	100	200	400	800	1600	3200

It seems that

$$f(t) = (100)2^t$$

A perceptive reader will ask if this formula is really valid. Does it give the right answer when $t = 5.7$? Is not population growth a discrete process, occurring in unit amounts at distinct times, rather than a continuous process as the formula implies? The answer is that the exponential growth model provides a very good approximation to the growth of simple organisms, provided the initial population is large.

The mechanism of reproduction is different (and more interesting) for people, but the pattern of population growth is similar. World population is presently growing at about 2 percent per year. In 1975, there were about 4 billion people. If present trends continue, there will be $4(1.02)^{10}$ billion people in 1985 (that is, after 10 years) and $4(1.02)^{20}$ billion in 1995. It appears that world population obeys the formula

$$p(t) = 4(1.02)^t$$

where $p(t)$ represents the number of people (in billions) t years after 1975.

t	$(1.02)^t$
5	1.104
10	1.219
15	1.346
20	1.486
25	1.641
30	1.811
35	2.000
40	2.208
45	2.438
50	2.692
55	2.972
60	3.281
65	3.623
70	4.000
75	4.416
80	4.875
85	5.383
90	5.943

Doubling Times One way to get a feeling for the spectacular nature of exponential growth is via the concept of **doubling time**. Suppose that the population given by the formula $f(t) = Ca^t$ doubles in time T. This means that $f(T) = 2f(0)$, or $Ca^T = 2C$. Thus $a^T = 2$. By the time $2T$, the population will have doubled again because $f(2T) = Ca^{2T} = C(a^T)^2 = 4C$. Similarly, $f(3T) = Ca^{3T} = C(a^T)^3 = 8C$. This justifies calling T the doubling time.

The table in the margin indicates the doubling time for world population (growing at 2 percent per year) is 35 years. Thus a population of 4 billion in 1975 will grow to 8 billion in 2010, 16 billion in 2045, and so on.

Now we can answer the question about sardine packing in our opening display. There are slightly more than 1,000,000 billion square feet of land area on the surface of the earth. Sardine packing for humans is about 1 square foot per person. Thus we are asking when $4(1.02)^t$ billion will equal 1,000,000 billion. This leads to the equation

$$(1.02)^t = 250{,}000$$

To solve this exponential equation, we use the following approximations.

$$(1.02)^{35} \approx 2 \qquad 250{,}000 \approx 2^{18}$$

Our equation can then be rewritten as

$$[(1.02)^{35}]^{t/35} = 2^{18}$$

or

$$2^{t/35} = 2^{18}$$

We conclude that

$$\frac{t}{35} = 18$$

$$t = (18)(35) = 630$$

Thus after about 630 years, we will be packed together like sardines.

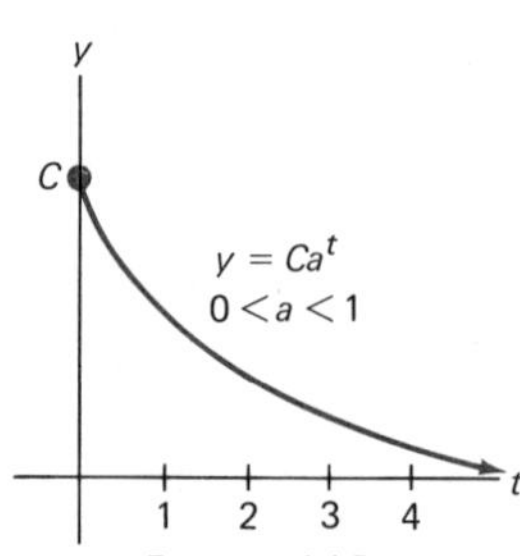

Exponential Decay

Exponential Decay Fortunately, not all things grow; some decline or decay. In fact, some things—notably the radioactive elements—decay exponentially. This means that the amount y present at time t satisfies

$$y = Ca^t$$

for some constants C and a with $0 < a < 1$.

Here an important idea is that of **half-life**, the time required for half of a substance to disappear. For example, radium decays with a half-life of 1690 years. Thus if 1000 grams of radium are present now, 1690 years from now 500 grams will be present, 2(1690) = 3380 years from now only 250 grams will be present, and so on.

The precise nature of radioactive decay is used to date old objects. If an object contains radium and lead (the product to which radium decays) in the ratio 1 to 3, then it is believed that an original amount of pure radium has decayed to $\frac{1}{4}$ its original size. The object must be two half-lives, or 3380 years old. We note that two important assumptions have been made: (a) decay of radium is exactly exponential over long periods of time; and (b) no lead was originally present. Recent research raises some question about the correctness of such assumptions.

Problem Set 6-3

1. In each of the following, indicate whether y grows exponentially or decays exponentially with t.
   (a) $y = 128(\frac{1}{2})^t$ (b) $y = 5(\frac{5}{3})^t$
   (c) $y = 4(10)^9(1.03)^t$ (d) $y = 1000(.99)^t$
2. Find the values of y corresponding to $t = 0$, $t = 1$, and $t = 2$ for each case in Problem 1.

3. For what value of t is $128(\frac{1}{2})^t$ equal to 1? $\frac{1}{16}$? 2^{-10}?
4. For what value of t is $5(\frac{5}{3})^t$ equal to $\frac{3125}{81}$?
5. Suppose that $(1.03)^T = 5$. Find the value of $4(10^9)(1.03)^{2T}$.
6. Suppose that $(.99)^T = \frac{1}{2}$. Find the value of $1000(.99)^{3T}$.
7. Under the assumptions concerning world population used in this section, what will be the approximate number of people on this earth in each year?
   (a) 1990 (that is, 15 years after 1975)
   (b) 2000
   (c) 2065
8. A certain radioactive substance has a half-life of 40 minutes. What fraction of an initial amount of this substance will remain after 1 hour and 20 minutes (that is, after 2 half-lives)? After 2 hours and 40 minutes?

Example A (Compound interest) Depositing money in a bank at compound interest is a practical illustration of exponential growth. Suppose that Amy deposits \$1000 in a bank at 8 percent compound interest with interest compounded annually. How much money will be in her account after 15 years?

SOLUTION. Interest for the first year is (.08)1000 = \$80. At the end of the first year, this is added to Amy's account, giving a principal of \$1080. Notice that this is 1000(1.08). During the second year, \$1080 draws interest. At the end of that year, the bank adds (.08)1080 to the account, bringing the principal to

$$\begin{aligned} 1080 + (.08)1080 &= (1080)(1.08) \\ &= 1000(1.08)(1.08) \\ &= 1000(1.08)^2 \end{aligned}$$

Continuing in this way, we see that Amy's account will have grown to $1000(1.08)^3$ by the end of 3 years, $1000(1.08)^4$ by the end of 4 years, and so on. By the end of 15 years, it will have grown to

$$\begin{aligned} 1000(1.08)^{15} &\approx 1000(3.172169) \\ &= \$3172.17 \end{aligned}$$

To calculate $(1.08)^{15}$, we used the table at the end of the problem set. We could have used a calculator.

9. Use the table at the end of the problem set or a calculator to find each value.
   (a) $(1.08)^{20}$ (b) $(1.12)^{25}$
   (c) $(1.04)^{40}$ (d) $(1.02)^{80}$
10. If you invest \$500 in the bank today, how much will it be worth after 25 years
    (a) at 8 percent compounded annually;
    (b) at 4 percent compounded annually;
    (c) at 12 percent compounded annually?
11. If you put \$3500 in the bank today, how much will it be worth after 40 years
    (a) at 8 percent compounded annually;
    (b) at 12 percent compounded annually?

12. Approximately how long will it take for money to accumulate to twice its value
    (a) if it is invested at 8 percent compounded annually;
    (b) if it is invested at 12 percent compounded annually?
13. Suppose that you invest $\$p$ at r percent compounded annually. Write an expression for the accumulated amount after n years.

Example B (More on compound interest) If \$1000 is invested at 8 percent compounded quarterly, find the accumulated amount after 15 years.

SOLUTION. Now interest calculated at 2 percent ($\frac{1}{4}$ of 8 percent) is converted to principal every 3 months. By the end of the first 3-month period, the account has grown to $1000(1.02) = \$1020$; by the end of the second 3-month period, it has grown to $1000(1.02)^2$; and so on. The accumulated amount after 15 years, or 60 conversion periods, is

$$\begin{aligned} 1000(1.02)^{60} &\approx 1000(3.28103) \\ &= \$3281.03 \end{aligned}$$

If you compare this with \$3172.17, the answer in Example A, you see how much better it is to have interest compounded quarterly than annually.

Find the accumulated amount for the indicated initial principal, compound interest rate, and total time period.

14. \$2000; 8 percent compounded annually; 15 years.
15. \$5000; 8 percent compounded semiannually; 5 years.
16. \$5000; 12 percent compounded monthly; 5 years.
17. \$3000; 4 percent compounded annually; 10 years.
18. \$3000; 4 percent compounded semiannually; 10 years.
19. \$3000; 4 percent compounded quarterly; 10 years.

[C] 20. \$3000; 4 percent compounded monthly; 10 years.

[C] 21. \$1000; 8 percent compounded monthly; 10 years.

[C] 22. \$1000; 8 percent compounded daily; 10 years. (*Hint*: Assume there are 365 days in a year, so that the interest rate per day is .08/365).

Miscellaneous Problems

23. If $y = 1600(\frac{3}{4})^t$, find the values of y corresponding to $t = 0$, $t = 1$, $t = 2$, and $t = 3$.
24. If $f(t) = 2500(\frac{6}{5})^t$, find $f(0)$, $f(1)$, $f(2)$, $f(3)$, and $f(4)$.
25. If $(1.034)^T = 2$, find the value of $3000(1.034)^{3T}$.
26. If $(0.63)^H = \frac{1}{2}$, find the value of $640(0.63)^{4H}$.
27. If \$4000 is invested today, how much will it be worth after 10 years at 8 percent interest if interest is
    (a) compounded annually;
    (b) compounded quarterly;
    [C] (c) compounded monthly;
    [C] (d) compounded daily?
    (*Hint*: In part (d), the interest rate per day is .08/365.)
28. If \$2500 is invested today, how much will it be worth after 5 years at 12 percent interest if interest is

(a) compounded annually;
(b) compounded monthly;
[C] (c) compounded daily?

[C] 29. Find the accumulated amount for the indicated initial principal, compound interest rate, and total time period.
(a) \$2500; 5.25 percent compounded quarterly; 4 years and 6 months.
(b) \$4175; 7.76 percent compounded monthly; 11 years and 9 months.

[C] 30. Suppose that the population of a certain city obeys the formula

$$p(t) = 4600(1.016)^t$$

where $p(t)$ is the population t years after 1980.
(a) What is the population in 2020? In 2080?
(b) What is the doubling time for this population?

31. The number of bacteria in a certain culture triples every hour. Suppose that the count at 12:00 noon is 162,000. What was the count at 11:00 A.M.? at 10:00 A.M.? at 8:00 A.M.?

32. Assuming that the half-life of radium is 1690 years, what fraction of an initial amount of the substance will remain after 3380 years? After 5070 years?

33. Assume that the number of bacteria in a certain culture t hours from now is given by $Q(t) = 5000(2^t)$. Find the number of bacteria
(a) at the present time; (b) after 30 minutes;
(c) after 3 hours; (d) after 90 minutes.

34. Assume in Problem 32 that the quantity of radium in milligrams after t years is given by $q(t) = 30(\frac{1}{2})^{kt}$.
(a) Find the value of k. (Use the half-life.)
[C] (b) How much radium will remain after 2385 years?

35. Carbon 14 has a half-life of 5720 years. If initially there were 10 grams, how much will there be after 17,160 years?

36. About how long does it take a population to double if it is growing at
(a) 1 percent per year;
(b) 4 percent per year;
[C] (c) 1.5 percent per year;
[C] (d) .5 percent per year?

[C] 37. If $f(t) = 1000(.983)^t$ is the formula for an exponential decay, find
(a) $f(10)$; (b) $f(40)$; (c) the approximate half-life.

[C] 38. If \$1 is invested at 10 percent compounded annually, in how many years will it accumulate to \$1,000,000,000?

[C] 39. The amount A (in dollars) of the average life insurance coverage of families in the United States satisfies (approximately)

$$A = 7{,}040(1.074)^t$$

where t is the number of years after 1955. Assume that is valid for the future and find the average coverage of a family in 1995.

[C] 40. A manufacturer of radial tires has found that the percentage of tires P still usable after being driven m miles is given by

$$P = 100(2.71)^{-.000025m}$$

(a) Graph this equation.
(b) What percentage of the tires are still usable at 80,000 miles?

41. One method of depreciation allowed by IRS is the double-declining-balance method. If this is done over N years, the original value C is depreciated each year by $100(2/N)$ percent of its value at the beginning of that year. Thus, the value after n years is

$$V = C\left(1 - \frac{2}{N}\right)^n$$

[C] (a) If a piece of equipment costing \$8000 is depreciated by this method over 10 years, what is its value after 5 years?

(b) Answer the same question if the equipment in part (a) is depreciated to zero linearly over 10 years.

(c) Does the value ever become zero by the double-declining-balance method?

COMPOUND INTEREST TABLE

n	$(1.01)^n$	$(1.02)^n$	$(1.04)^n$	$(1.08)^n$	$(1.12)^n$
1	1.01000000	1.02000000	1.04000000	1.08000000	1.12000000
2	1.02010000	1.04040000	1.08160000	1.16640000	1.25440000
3	1.03030100	1.06120800	1.12486400	1.25971200	1.40492800
4	1.04060401	1.08243216	1.16985856	1.36048896	1.57351936
5	1.05101005	1.10408080	1.21665290	1.46932808	1.76234168
6	1.06152015	1.12616242	1.26531902	1.58687432	1.97382269
7	1.07213535	1.14868567	1.31593178	1.71382427	2.21068141
8	1.08285671	1.17165938	1.36856905	1.85093021	2.47596318
9	1.09368527	1.19509257	1.42331181	1.99900463	2.77307876
10	1.10462213	1.21899442	1.48024428	2.15892500	3.10584821
11	1.11566835	1.24337431	1.53945406	2.33163900	3.47854999
12	1.12682503	1.26824179	1.60103222	2.51817012	3.89597599
15	1.16096896	1.34586834	1.80094351	3.17216911	5.47356576
20	1.22019004	1.48594740	2.19112314	4.66095714	9.64629309
25	1.28243200	1.64060599	2.66583633	6.84847520	17.00006441
30	1.34784892	1.81136158	3.24339751	10.06265689	29.95992212
35	1.41660276	1.99988955	3.94608899	14.78534429	52.79961958
40	1.48886373	2.20803966	4.80102063	21.72452150	93.05097044
45	1.56481075	2.43785421	5.84117568	31.92044939	163.98760387
50	1.64463182	2.69158803	7.10668335	46.90161251	289.00218983
55	1.72852457	2.97173067	8.64636692	68.91385611	509.32060567
60	1.81669670	3.28103079	10.51962741	101.25706367	897.59693349
65	1.90936649	3.62252311	12.79873522	148.77984662	1581.87249060
70	2.00676337	3.99955822	15.57161835	218.60640590	2787.79982770
75	2.10912847	4.41583546	18.94525466	321.20452996	4913.05584077
80	2.21671522	4.87543916	23.04979907	471.95483426	8658.48310008
85	2.32978997	5.38287878	28.04360494	693.45648897	15259.20568055
90	2.44863267	5.94313313	34.11933334	1018.91508928	26891.93422336
95	2.57353755	6.56169920	41.51138594	1497.12054855	47392.77662369
100	2.70481383	7.24464612	50.50494818	2199.76125634	83522.26572652

An active participant in the political and religious battles of his day, the Scot John Napier amused himself by studying mathematics and science. In 1614 he published a book containing the idea that made him famous. He gave it the name *logarithm.*

John Napier
(1550–1617)

6-4 Logarithms and Logarithmic Functions

Napier's approach to logarithms is out of style, but the goal he had in mind is still worth considering. He hoped to replace multiplications by additions. He thought additions were easier to do, and he was right.

Consider the exponential function $f(x) = 2^x$ and recall that

$$2^x \cdot 2^y = 2^{x+y}$$

On the left, we have a multiplication and on the right, an addition. If we are to fulfill Napier's objective, we want logarithms to behave like exponents. That suggests a definition. The logarithm of N to the base 2 is the exponent to which 2 must be raised to yield N. That is,

$$\boxed{\log_2 N = x \quad \text{if and only if} \quad 2^x = N}$$

Thus

$$\begin{aligned} \log_2 4 &= 2 \quad \text{since} \quad 2^2 = 4 \\ \log_2 8 &= 3 \quad \text{since} \quad 2^3 = 8 \\ \log_2 \sqrt{2} &= \tfrac{1}{2} \quad \text{since} \quad 2^{1/2} = \sqrt{2} \end{aligned}$$

and, in general,

$$\log_2(2^x) = x \quad \text{since} \quad 2^x = 2^x$$

Has Napier's goal been achieved? Does the logarithm turn a product into a sum? Yes, for note that

$$\begin{aligned}\log_2(2^x \cdot 2^y) &= \log_2(2^{x+y}) \qquad \text{(property of exponents)}\\ &= x + y \qquad \text{(definition of } \log_2\text{)}\\ &= \log_2(2^x) + \log_2(2^y)\end{aligned}$$

Thus

$$\log_2(2^x \cdot 2^y) = \log_2(2^x) + \log_2(2^y)$$

which has the form

$$\log_2(M \cdot N) = \log_2 M + \log_2 N$$

The General Definition What has been done for 2 can be done for any base $a > 1$. The **logarithm of N to the base a** is the exponent x to which a must be raised to yield N. Thus

$$\log_a N = x \quad \text{if and only if} \quad a^x = N$$

Now we can calculate many kinds of logarithms.

$$\begin{aligned}\log_4 16 &= 2 \quad \text{since} \quad 4^2 = 16\\ \log_{10} 1000 &= 3 \quad \text{since} \quad 10^3 = 1000\\ \log_{10}(.001) &= -3 \quad \text{since} \quad 10^{-3} = \tfrac{1}{1000} = .001\end{aligned}$$

What is $\log_{10} 7$? We are not ready to answer that yet, except to say it is a number x satisfying $10^x = 7$ (see Section 6-6).

We point out that negative numbers and zero do not have logarithms. Suppose -4 and 0 did have logarithms, that is, suppose

$$\log_a(-4) = m \quad \text{and} \quad \log_a 0 = n$$

Then

$$a^m = -4 \quad \text{and} \quad a^n = 0$$

But this is impossible; we learned earlier that a^x is always positive.

Properties of Logarithms There are three main properties of logarithms.

PROPERTIES OF LOGARITHMS

1. $\log_a(M \cdot N) = \log_a M + \log_a N$
2. $\log_a(M/N) = \log_a M - \log_a N$
3. $\log_a(M^p) = p \log_a M$

To establish Property 1, let

$$x = \log_a M \quad \text{and} \quad y = \log_a N$$

Then, by definition,

$$M = a^x \quad \text{and} \quad N = a^y$$

so that

$$M \cdot N = a^x \cdot a^y = a^{x+y}$$

Thus $x + y$ is the exponent to which a must be raised to yield $M \cdot N$, that is,

$$\log_a(M \cdot N) = x + y = \log_a M + \log_a N$$

Properties 2 and 3 are demonstrated in a similar fashion.

The Logarithmic Function The function determined by

$$g(x) = \log_a x$$

is called the **logarithmic function with base *a***. We can get a feeling for the behavior of this function by drawing its graph for $a = 2$.

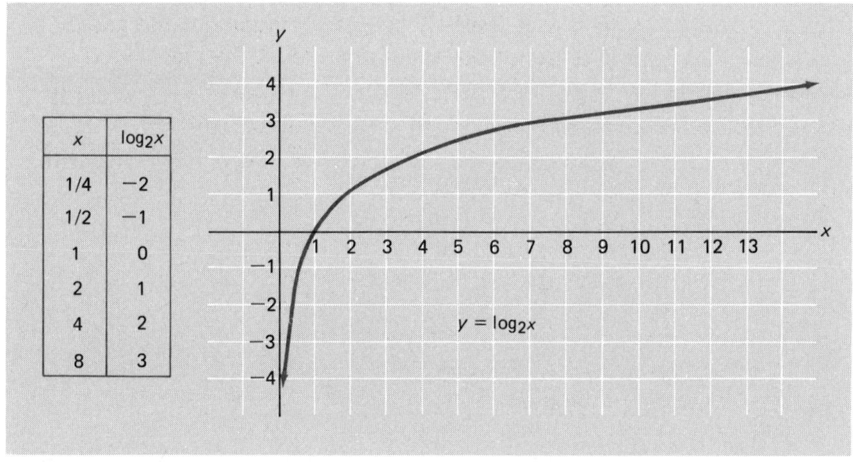

x	$\log_2 x$
1/4	−2
1/2	−1
1	0
2	1
4	2
8	3

Several properties of $y = \log_2 x$ are apparent from this graph. The domain consists of all positive real numbers. If $0 < x < 1$, $\log_2 x$ is negative; if $x > 1$, $\log_2 x$ is positive. The y-axis is a vertical asymptote of the graph since very small positive x values yield large negative y values. Although $\log_2 x$ continues to increase as x increases, even this small part of the complete graph indicates how slowly it grows for large x. In fact, by the time x reaches 1,000,000, $\log_2 x$ is still loafing along at about 20. In this sense, it behaves in a manner opposite to the exponential function 2^x, which grows more and more rapidly as x increases. There is a good reason for this; the two functions are inverses of each other.

Inverse Functions We begin by emphasizing two facts that you must not forget.

$$a^{\log_a x} = x$$
$$\log_a (a^x) = x$$

Both of them are direct consequences of the definition of logarithms; the second is also a special case of Property 3, stated earlier. What these facts tell us is that the logarithmic and exponential functions undo each other.

Let us put it in the language of Section 5-5. If $f(x) = a^x$ and $g(x) = \log_a x$, then

$$f(g(x)) = f(\log_a x) = a^{\log_a x} = x$$

and

$$g(f(x)) = g(a^x) = \log_a(a^x) = x$$

Thus g is really f^{-1}. This fact also tells us something about the graphs of g and f: They are simply reflections of each other about the line $y = x$.

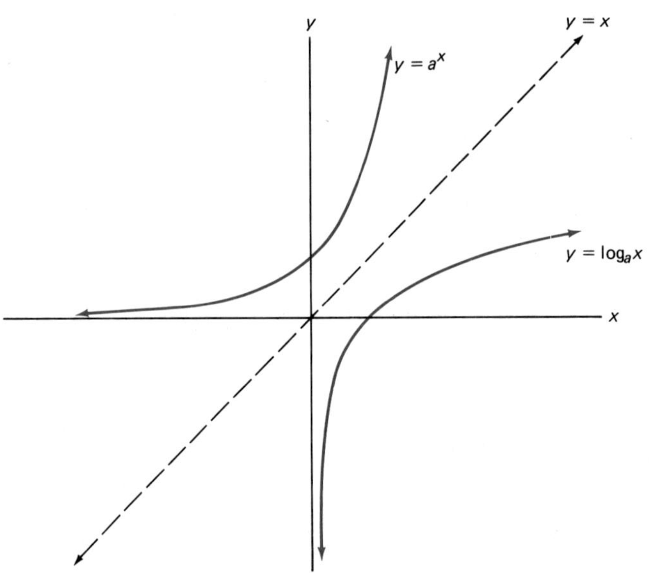

Note finally that $f(x) = a^x$ has the set of all real numbers as its domain and the positive real numbers as its range. Thus its inverse $f^{-1}(x) = \log_a x$ has domain consisting of the positive real numbers and range consisting of all real numbers. We emphasize again a fact you might otherwise forget: *Negative numbers and zero do not have logarithms.*

Problem Set 6-4

Write each of the following in logarithmic form. For example, $3^4 = 81$ *can be written as* $\log_3 81 = 4$.

1. $4^3 = 64$ 2. $7^3 = 343$ 3. $27^{1/3} = 3$
4. $16^{1/4} = 2$ 5. $4^0 = 1$ 6. $81^{-1/2} = \frac{1}{9}$
7. $125^{-2/3} = \frac{1}{25}$ 8. $2^{9/2} = 16\sqrt{2}$ 9. $10^{\sqrt{3}} = a$
10. $5^{\sqrt{2}} = b$ 11. $10^a = \sqrt{3}$ 12. $b^x = y$

Write each of the following in exponential form. For example, $\log_5 125 = 3$ *can be written as* $5^3 = 125$.

13. $\log_5 625 = 4$ 14. $\log_6 216 = 3$ 15. $\log_4 8 = \frac{3}{2}$
16. $\log_{27} 9 = \frac{2}{3}$ 17. $\log_{10}(.01) = -2$ 18. $\log_3(\frac{1}{27}) = -3$
19. $\log_5 25\sqrt{5} = \frac{5}{2}$ 20. $\log_{10} 10\sqrt[3]{10} = \frac{4}{3}$ 21. $\log_b 1 = 0$
22. $\log_c c = 1$ 23. $\log_b N = x$ 24. $\log_c Q = y$

Determine the value of each of the following logarithms.

25. $\log_4 16$ 26. $\log_5 25$ 27. $\log_7 \frac{1}{7}$
28. $\log_3 \frac{1}{3}$ 29. $\log_4 2$ 30. $\log_{27} 3$
31. $\log_{10}(10^{-6})$ 32. $\log_{10}(.0001)$ 33. $\log_8 1$
34. $\log_3 1$ 35. $\log_{100} 1000$ 36. $\log_8 16$

Find the value of c in each of the following.

37. $\log_c 25 = 2$ 38. $\log_c 8 = 3$ 39. $\log_4 c = -\frac{1}{2}$
40. $\log_9 c = -\frac{3}{2}$ 41. $\log_2(2^{5.6}) = c$ 42. $\log_3(3^{-2.9}) = c$

Given $\log_{10} 2 = 0.301$ *and* $\log_{10} 3 = 0.477$, *calculate each of the following without the use of tables. For example, note that* $\log_{10} 6 = \log_{10} 2 + \log_{10} 3$ *in Problem 43.*

43. $\log_{10} 6$ 44. $\log_{10} \frac{3}{2}$ 45. $\log_{10} 16$
46. $\log_{10} 27$ 47. $\log_{10} \frac{1}{4}$ 48. $\log_{10} \frac{1}{27}$
49. $\log_{10} 24$ 50. $\log_{10} 54$ 51. $\log_{10} \frac{8}{9}$
52. $\log_{10} \frac{3}{8}$ 53. $\log_{10} 5$ 54. $\log_{10} \sqrt[3]{3}$

Example A (Combining logarithms) Write the following expression as a single logarithm.

$$2 \log_{10} x + 3 \log_{10}(x + 2) - \log_{10}(x^2 + 5)$$

SOLUTION. We use the properties of logarithms to rewrite this as

$$\log_{10} x^2 + \log_{10}(x + 2)^3 - \log_{10}(x^2 + 5) \qquad \text{(Property 3)}$$
$$= \log_{10} x^2(x + 2)^3 - \log_{10}(x^2 + 5) \qquad \text{(Property 1)}$$
$$= \log_{10}\left[\frac{x^2(x + 2)^3}{x^2 + 5}\right] \qquad \text{(Property 2)}$$

Write each of the following as a single logarithm.

55. $3 \log_{10}(x + 1) + \log_{10}(4x + 7)$
56. $\log_{10}(x^2 + 1) + 5 \log_{10} x$
57. $3 \log_2(x + 2) + \log_2 8x - 2 \log_2(x + 8)$
58. $2 \log_5 x - 3 \log_5(2x + 1) + \log_5(x - 4)$
59. $\frac{1}{2} \log_6 x + \frac{1}{3} \log_6(x^3 + 3)$
60. $-\frac{2}{3} \log_3 x + \frac{5}{2} \log_3(2x^2 + 3)$

Example B (Solving logarithmic equations) Solve the equation

$$\log_2 x + \log_2(x + 2) = 3$$

SOLUTION. First we note that we must have $x > 0$ so that both logarithms exist. Next we rewrite the equation using the first property of logarithms and the fact that $\log_2 8 = 3$.

$$\log_2 x(x + 2) = \log_2 8$$
$$x(x + 2) = 8$$
$$x^2 + 2x - 8 = 0$$
$$(x + 4)(x - 2) = 0$$

We reject $x = -4$ (because $-4 < 0$) and keep $x = 2$. To make sure that 2 is a solution, we substitute 2 for x in the original equation.

$$\log_2 2 + \log_2(2 + 2) \stackrel{?}{=} 3$$
$$1 + 2 = 3$$

Solve each of the following equations.

61. $\log_7(x + 2) = 2$
62. $\log_5(3x + 2) = 1$
63. $\log_2(x + 3) = -2$
64. $\log_4(\frac{1}{64}x + 1) = -3$
65. $\log_2 x - \log_2(x - 2) = 3$
66. $\log_3 x - \log_3(2x + 3) = -2$
67. $\log_2(x - 4) + \log_2(x - 3) = 1$
68. $\log_{10} x + \log_{10}(x - 3) = 1$

Miscellaneous Problems

69. Write in logarithmic form.
(a) $16^{-3/4} = \frac{1}{8}$ (b) $y = 10^{3.24}$ (c) $b = (1.4)^a$
70. Write in exponential form.
(a) $\log_3 \frac{1}{9} = -2$ (b) $\log_8 4 = \frac{2}{3}$
(c) $\log_8 c = b$ (d) $\log_b N = x$
71. Find the value of x in each of the following.
(a) $x = \log_6 36$ (b) $x = \log_4 2$
(c) $\log_{25} x = \frac{3}{2}$ (d) $\log_4 x = \frac{5}{2}$
(e) $\log_x 10\sqrt{10} = \frac{3}{2}$ (f) $\log_x \frac{1}{8} = -\frac{3}{2}$

72. Write each of the following as a single logarithm.
(a) $3 \log_2 5 - 2 \log_2 7$
(b) $\frac{1}{2} \log_5 64 + \frac{1}{3} \log_5 27 - \log_5(x^2 + 4)$
(c) $\frac{2}{3} \log_{10}(x + 5) + 4 \log_{10} x - 2 \log_{10}(x - 3)$

73. What is the domain of the following function?
$$f(x) = \tfrac{2}{3} \log_{10}(x + 5) + 4 \log_{10} x - 2 \log_{10}(x - 3)$$

74. Solve each of the following equations for x.
(a) $\log_5(2x - 1) = 2$
(b) $\log_4\left(\dfrac{x - 2}{2x + 3}\right) = 0$
(c) $\log_4(x - 2) - \log_4(2x + 3) = 0$
(d) $\log_{10} x + \log_{10}(x - 15) = 2$

75. Derive the second property of logarithms, that is, show that $\log_a(M/N) = \log_a M - \log_a N$. (*Hint*: Use the method given in the text to derive Property 1.)

76. Show that $\log_a(M^p) = p \log_a M$.

77. Show that $\log_a b = 1/\log_b a$, where $a, b > 0$. (*Hint*: Let $\log_a b = x$ and $\log_b a = y$.)

78. Show that $\log_2 x = \log_{10} x \cdot \log_2 10$, where $x > 0$. (*Hint*: Let $\log_{10} x = c$. Then $x = 10^c$. Next take $\log_2$ of both sides.)

79. Use the technique outlined in Problem 78 to show that

$$\log_a x = \log_b x \cdot \log_a b$$

This is often called the **change of base formula** for logarithms.

80. Graph the equations $y = 2^x$ and $y = \log_2 x$ on the same coordinate plane.

81. Graph the equations $y = 3^x$ and $y = \log_3 x$ on the same coordinate plane.

82. Give the value(s) of x for which
(a) $2^x = 3^x$; (b) $2^x > 3^x$; (c) $2^x < 3^x$.

83. Give the value(s) of x for which
(a) $\log_2 x = \log_3 x$; (b) $\log_2 x > \log_3 x$; (c) $\log_2 x < \log_3 x$.

84. Give the solution set for each of the following inequalities.
(a) $\log_2 x < 0$
(b) $\log_2 x > 3$
(c) $\log_{10} x < -1$
(d) $\log_{10} x \geq 2$
(e) $2 < \log_3 x < 3$
(f) $0 \leq \log_{10} x \leq 2$
(g) $-3 < \log_2 x < -2$
(h) $-2 \leq \log_{10} x \leq -1$

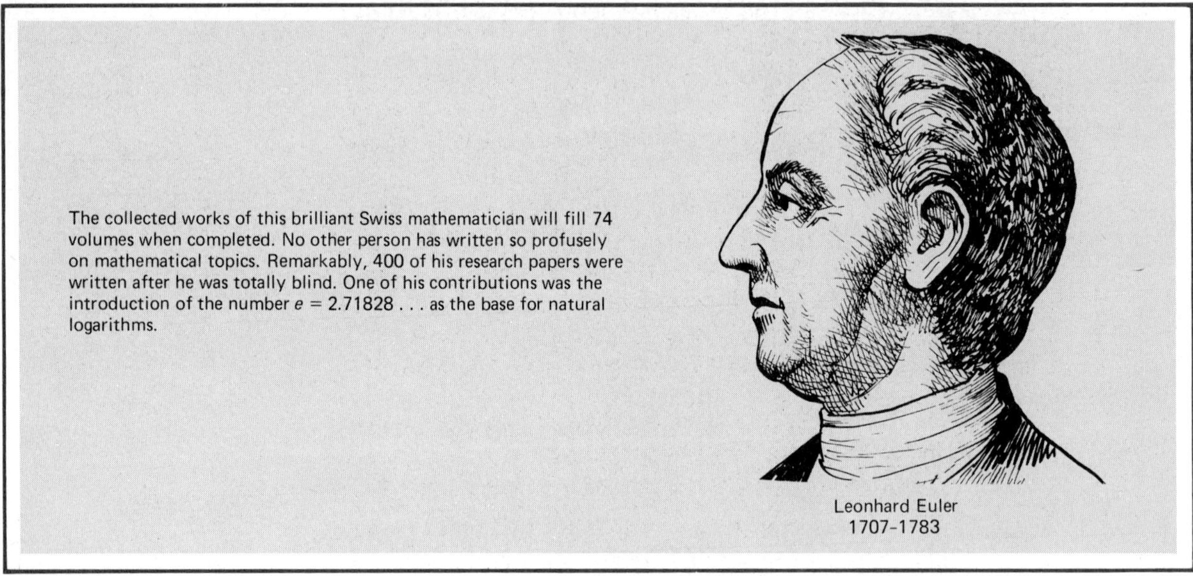

Leonhard Euler
1707–1783

6-5 Applications of Logarithms

Napier invented logarithms to simplify arithmetic calculations. Computers and calculators have reduced that application to minor significance, though we will discuss such a use of logarithms later in this chapter. Here we have in mind deeper applications such as solving exponential equations, defining power functions, and modeling physical phenomena.

In order to make any progress, we will need an easy way to calculate logarithms. Fortunately, this has been done for us as tables of logarithms to several bases are available. For our purposes in this section, one base is as good as another. Base 10 would be an appropriate choice but we would rather defer discussion of logarithms to base 10 (common logarithms) until Section 6-6. We have chosen rather to introduce you to the number e (after Euler), which is used as a base of logarithms in all advanced mathematics courses. You will see the importance of logarithms to this base (**natural logarithms**) when you study calculus. An approximate value of e is

$$e \approx 2.71828$$

and, like π, e is an irrational number. Page 242 shows a table of values of natural logarithms, which we shall denote by ln instead of $\log_e$.

Since ln denotes a genuine logarithm function, we have as in Section 6-4

$$\ln N = x \quad \text{if and only if} \quad e^x = N$$

and consequently

$$\ln e^x = x \quad \text{and} \quad e^{\ln N} = N$$

Moreover the three properties of logarithms hold.

1. $\ln(MN) = \ln M + \ln N$
2. $\ln(M/N) = \ln M - \ln N$
3. $\ln(N^p) = p \ln N$

Solving Exponential Equations Consider first the simple equation

$$5^x = 1.7$$

We call it an *exponential equation* because the unknown is in the exponent. To solve it, we take natural logarithms of both sides.

$$\begin{aligned} 5^x &= 1.7 \\ \ln(5^x) &= \ln 1.7 \\ x \ln 5 &= \ln 1.7 \qquad \text{(Property 3)} \\ x &= \frac{\ln 1.7}{\ln 5} \\ x &= \frac{.531}{1.609} = .330 \qquad \text{(table of ln values)} \end{aligned}$$

We point out that the last step can also be done on a scientific calculator, which has a key for calculating natural logarithms.

Here is a more complicated example.

$$5^{2x-1} = 7^{x+2}$$

Begin by taking natural logarithms of both sides and then solve for x.

$$\begin{aligned} \ln(5^{2x-1}) &= \ln(7^{x+2}) \\ (2x - 1) \ln 5 &= (x + 2) \ln 7 \qquad \text{(Property 3)} \\ 2x \ln 5 - \ln 5 &= x \ln 7 + 2 \ln 7 \\ 2x \ln 5 - x \ln 7 &= \ln 5 + 2 \ln 7 \\ x(2 \ln 5 - \ln 7) &= \ln 5 + 2 \ln 7 \\ x &= \frac{\ln 5 + 2 \ln 7}{2 \ln 5 - \ln 7} \\ &= \frac{1.609 + 2(1.946)}{2(1.609) - 1.946} \qquad \text{(table of ln values or a calculator)} \\ &= 4.325 \end{aligned}$$

Exponential Functions versus Power Functions Look closely at the formulas below.

$$f(x) = 2^x \qquad f(x) = x^2$$

TABLE OF NATURAL LOGARITHMS

x	$\ln x$	x	$\ln x$	x	$\ln x$
		4.0	1.386	8.0	2.079
0.1	−2.303	4.1	1.411	8.1	2.092
0.2	−1.609	4.2	1.435	8.2	2.104
0.3	−1.204	4.3	1.459	8.3	2.116
0.4	−0.916	4.4	1.482	8.4	2.128
0.5	−0.693	4.5	1.504	8.5	2.140
0.6	−0.511	4.6	1.526	8.6	2.152
0.7	−0.357	4.7	1.548	8.7	2.163
0.8	−0.223	4.8	1.569	8.8	2.175
0.9	−0.105	4.9	1.589	8.9	2.186
1.0	0.000	5.0	1.609	9.0	2.197
1.1	0.095	5.1	1.629	9.1	2.208
1.2	0.182	5.2	1.649	9.2	2.219
1.3	0.262	5.3	1.668	9.3	2.230
1.4	0.336	5.4	1.686	9.4	2.241
1.5	0.405	5.5	1.705	9.5	2.251
1.6	0.470	5.6	1.723	9.6	2.262
1.7	0.531	5.7	1.740	9.7	2.272
1.8	0.588	5.8	1.758	9.8	2.282
1.9	0.642	5.9	1.775	9.9	2.293
2.0	0.693	6.0	1.792	10	2.303
2.1	0.742	6.1	1.808	20	2.996
2.2	0.788	6.2	1.825	30	3.401
2.3	0.833	6.3	1.841	40	3.689
2.4	0.875	6.4	1.856	50	3.912
2.5	0.916	6.5	1.872	60	4.094
2.6	0.956	6.6	1.887	70	4.248
2.7	0.993	6.7	1.902	80	4.382
2.8	1.030	6.8	1.917	90	4.500
2.9	1.065	6.9	1.932	100	4.605
3.0	1.099	7.0	1.946		
3.1	1.131	7.1	1.960		
3.2	1.163	7.2	1.974	e	1.000
3.3	1.194	7.3	1.988		
3.4	1.224	7.4	2.001	π	1.145
3.5	1.253	7.5	2.015		
3.6	1.281	7.6	2.028		
3.7	1.308	7.7	2.041		
3.8	1.335	7.8	2.054		
3.9	1.361	7.9	2.067		

To find the natural logarithm of a number N which is either smaller than 0.1 or larger than 10, write N in scientific notation, that is, write $N = c \times 10^k$. Then $\ln N = \ln c + k \ln 10 = \ln c + k(2.303)$. A more complete table of natural logarithms appears as Table A of the Appendix.

They are very different, yet easily confused. The first is an exponential function, while the second is called a power function. Both grow rapidly for large x, but the exponential function ultimately gets far ahead (see the graphs in the margin).

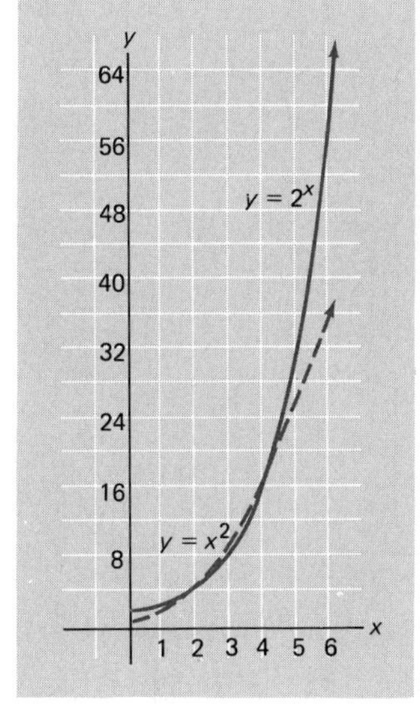

The situation described above is a special instance of two very general classes of functions.

> EXPONENTIAL FUNCTIONS
>
> $$f(x) = ba^x$$
>
> POWER FUNCTIONS
>
> $$f(x) = bx^a$$

While these functions are very different in character, they are related, as you will be asked to show in Problem 54.

Curve Fitting A recurring theme in science is to fit a mathematical curve to a set of experimental data. Suppose that a scientist, studying the relationship between two variables x and y, obtained the data plotted in the margin. In searching for curves to fit these data, the scientist naturally thought of exponential curves and power curves. How did he or she decide if either was appropriate? The scientist took logarithms. Let us see why.

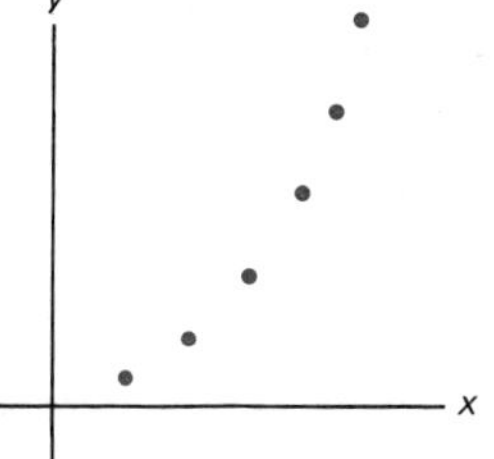

MODEL 1	MODEL 2
$y = ba^x$	$y = bx^a$
$\ln y = \ln b + x \ln a$	$\ln y = \ln b + a \ln x$
$Y = B + Ax$	$Y = B + aX$

Here the scientist made the substitutions $Y = \ln y$, $B = \ln b$, $A = \ln a$, and $X = \ln x$.

In both cases, the final result is a straight line. But note the difference. In the first case, $\ln y$ is a linear function of x, while in the second case, $\ln y$ is a linear function of $\ln x$. These considerations suggest the following procedures. Make two additional plots of the data. In the first, plot $\ln y$ against x and in the second, plot $\ln y$ against $\ln x$. If the first plotting gives data nearly along a straight line, Model 1 is appropriate; if the second does, then Model 2 is appropriate. If neither plot approximates a straight line, our scientist should look for a different and perhaps more complicated model.

Logarithms and Physiology The human body appears to have a built-in logarithmic calculator. What do we mean by this statement?

In 1834, the German physiologist E. Weber noticed an interesting fact. Two heavy objects must differ in weight by considerably more than two light objects if a person is to perceive a difference between them. Other scientists

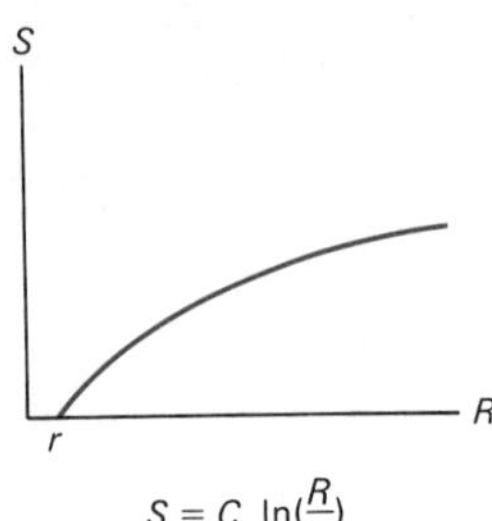

$S = C\ \ln(\frac{R}{r})$

noted the same phenomenon when human subjects tried to differentiate loudness of sounds, pitches of musical tones, brightness of light, and so on. Experiments suggested that people react to stimuli on a logarithmic scale, a result formulated as the Weber-Fechner Law.

$$S = C \ln\left(\frac{R}{r}\right)$$

Here R is the actual intensity of the stimulus, r is the threshold value (smallest value at which the stimulus is observed), C is a constant depending on the type of stimulus, and S is the perceived intensity of the stimulus. Note that a change in R is not as perceptible for large R as for small R because as R increases, the graph of the logarithmic function gets steadily flatter.

Problem Set 6-5

In Problems 1–8, find the value of each natural logarithm.

1. $\ln e$
2. $\ln(e^2)$
3. $\ln 1$
4. $\ln(1/e)$
5. $\ln \sqrt{e}$
6. $\ln(e^{1.1})$
7. $\ln(1/e^3)$
8. $\ln(e^\pi)$

For Problems 9–14, find each value. Assume that $\ln a = 2.5$ *and* $\ln b = -.4$.

9. $\ln(ae)$
10. $\ln(e/b)$
11. $\ln \sqrt{b}$
12. $\ln(a^2b^{10})$
13. $\ln(1/a^3)$
14. $\ln(a^{4/5})$

Use the table of natural logarithms to calculate the values in Problems 15–20.

15. $\ln 120$
    $= \ln 60 + \ln 2$
16. $\ln 150$
17. $\ln 690$
18. $\ln 84$
19. $\ln \frac{6}{5}$
20. $\ln 20{,}000$

In Problems 21–26, use the natural logarithm table to find N.

21. $\ln N = 2.208$
22. $\ln N = 1.808$
23. $\ln N = -0.105$
24. $\ln N = -.916$
25. $\ln N = 4.500$
26. $\ln N = 9.000$

Solve for x in Problems 27–32. Begin by taking natural logarithms of both sides. Then use the natural logarithm table or calculator.

27. $3^x = 20$
28. $5^x = 40$
29. $2^x = .6$
30. $4^x = 3^{2x-1}$
31. $(1.4)^x = 10$
32. $5^x = \frac{1}{2}(4^x)$
33. By finding the natural logarithm of the numbers in each pair, determine which is larger.
    (a) $10^5, 5^{10}$ (b) $10^9, 9^{10}$
    (c) $10^{20}, 20^{10}$ (d) $10^{1000}, 1000^{10}$
34. What do your answers in Problem 33 confirm about the growth of 10^x and x^{10} for large x?
35. On the same coordinate plane, graph $y = 3^x$ and $y = x^3$ for $0 \le x \le 4$.

36. By means of a change of variable(s) (as explained in the text), transform each equation below to a linear equation. Find the slope and Y-intercept of the resulting line.
(a) $y = 3e^{2x}$ (b) $y = 2x^3$ (c) $xy = 12$
(d) $y = x^e$ (e) $y = 5(3^x)$ (f) $y = ex^{1.1}$

Example (Curve fitting) The table in the margin shows the number N of bacteria in a certain culture found after t hours. Which is a better description of these data,

$$N = ba^t \quad \text{or} \quad N = bt^a$$

Find a and b.

t	N
0	100
1	700
2	5000
3	40,000

SOLUTION. Following the discussion of curve fitting in the text, we begin by plotting $\ln N$ against t. If the resulting points lie along a line, we choose $N = ba^t$ as the appropriate model. If not, we will plot $\ln N$ against $\ln t$ to check on the second model.

t	$\ln N$
0	4.6
1	6.6
2	8.5
3	10.6

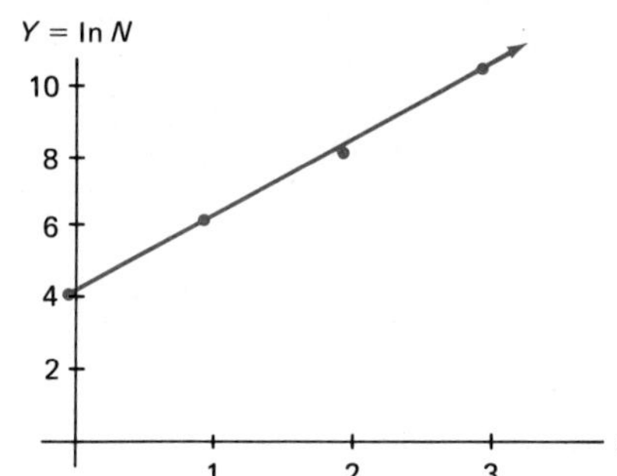

Since the fit to a line is quite good, we accept $N = ba^t$ as our model. To find a and b, we write $N = ba^t$ in the form

$$\ln N = \ln b + t \ln a \quad \text{or} \quad Y = \ln b + (\ln a)t$$

Examination of the line shows that it has a Y-intercept of 4.6 and a slope of about 2; so for its equation we write $Y = 4.6 + 2t$. Comparing this with $Y = \ln b + (\ln a)t$ gives

$$\ln b = 4.6 \qquad \ln a = 2$$

Finally, we use the natural logarithm table to find

$$b \approx 100 \qquad a \approx 7.4$$

Thus the original data are described reasonably well by the equation

$$N = 100(7.4)^t$$

For the data sets below, decide whether $y = ba^x$ or $y = bx^a$ is the best model. Then determine a and b.

37.

x	1	2	3	4
y	96	145	216	325

38.

x	0	1	2	4
y	243	162	108	48

39.

x	1	2	3	5
y	12	190	975	7490

40.

x	1	4	9
y	16	128	432

Miscellaneous Problems

In Problems 41–48, find the value of each expression using the natural logarithm table where necessary.

41. $\ln(e^{3.5})$
42. $\ln(e^{-1.2})$
43. $3\ln(e^{2/3})$
44. $\ln(1/\sqrt{e})$
45. $\ln(10e)$
46. $\ln 4300$
47. $\ln(80^2/\sqrt{6.3})$
48. $\ln(4.1 \times 10^4)$
49. Solve for x.
    (a) $10^{2x+5} = 800$ (b) $e^{2x} = 8^{x-1}$
50. By means of a change of variables, transform each of the following equations to a linear equation. Find the slope and y-intercept of the resulting line.
    (a) $xy^2 = 40$ (b) $y = 9(2^{2x})$
51. The Weber-Fechner Law discussed in the text says that $S = C\ln(R/r)$.
    (a) What is S when $R = r$?
    (b) What is S when $R = r^3$?
    (c) What is S when $R = 2r$?
52. If the perceived intensity S of the stimulus in the Weber-Fechner Law has doubled, what has happened to R/r?
53. A certain substance decays according to the formula

$$y = 100e^{-3t}$$

    where y is the amount present (in grams) after t years. Find its half-life.
54. We know that $N = e^{\ln N}$ for any $N > 0$. By letting x^a take the place of N, show that

$$x^a = e^{a \ln x} \qquad x > 0$$

    This equation is sometimes used as the definition of x^a, especially when a is irrational.
55. Along the lines of Problem 54, how would you define x^x?
56. How would you define $(x^2 + 5)^{2x}$?
57. How would you define $(x - 5)^{1/x}$? How must x be restricted for your definition to make sense?
58. [C] In calculus it is shown that

$$e^x \approx 1 + x + \frac{x^2}{2} + \frac{x^3}{6} + \frac{x^4}{24} + \frac{x^5}{120}$$

    Use this to approximate e and $e^{-.5}$.
59. [C] If $f(x) = [\ln(x^2 + \sqrt{x} + 1/x)]^3$, calculate
    (a) $f(2.1)$; (b) $f(.12)$.

Henry Briggs 1561–1631

When 10 is used as the base for logarithms, some very nice things happen as the display at the right shows. Realizing this, John Napier and his friend Henry Briggs decided to produce a table of these logarithms. The job was accomplished by Briggs after Napier's death and published in 1624 in the famous book *Arithmetica Logarithmica.*

$\log .0853 = .9309 - 2$

$\log .853 = .9309 - 1$

$\log 8.53 = .9309$

$\log 85.3 = .9309 + 1$

$\log 853 = .9309 + 2$

$\log 8530 = .9309 + 3$

6-6 Common Logarithms (Optional)

In Section 6-4, we defined the logarithms of a positive number N to the base a as follows.

$$\log_a N = x \quad \text{if and only if} \quad a^x = N$$

Then in Section 6-5, we introduced base e, calling the result natural logarithms. These are the logarithms that are most important in advanced branches of mathematics such as calculus.

In this section, we shall consider logarithms to the base 10, often called **common logarithms** or Briggsian logarithms. They have been studied by high school and college students for centuries as an aid to computation, a subject we take up in the next section. We shall write $\log N$ instead of $\log_{10} N$ (just as we write $\ln N$ instead of $\log_e N$). Note that

$$\log N = x \quad \text{if and only if} \quad 10^x = N$$

In other words, the common logarithm of any power of 10 is simply its exponent. Thus

$$\begin{aligned} \log 100 &= \log 10^2 = 2 \\ \log 1 &= \log 10^0 = 0 \\ \log .001 &= \log 10^{-3} = -3 \end{aligned}$$

But how do we find common logarithms of numbers that are not integral powers of 10, such as 8.53 or 14,600? That is the next topic.

Finding Common Logarithms We know that $\log 1 = 0$ and $\log 10 = 1$. If $1 < N < 10$, we correctly expect $\log N$ to be between 0 and 1. Table B (in the appendix) gives us four-place approximations of the common logarithms of all three-digit numbers between 1 and 10. For example,

$$\log 8.53 = .9309$$

We find this value by locating 8.5 in the left column and then moving right to the entry with 3 as heading. Similarly,

$$\log 1.08 = .0334$$

and

$$\log 9.69 = .9863$$

You should check these values.

You may have noticed in the opening box that log 8.53, log 8530, and log .0853 all have the same positive fractional part, .9309, called the **mantissa** of the logarithm. They differ only in the integer part, called the **characteristic** of the logarithm. To see why this is so, recall that

$$\log(M \cdot N) = \log M + \log N$$

Thus

$$\begin{aligned}\log 8530 &= \log(8.53 \times 10^3) = \log 8.53 + \log 10^3 \\ &= .9309 + 3 \\ \log .0853 &= \log(8.53 \times 10^{-2}) = \log 8.53 + \log 10^{-2} \\ &= .9309 - 2\end{aligned}$$

Clearly the mantissa .9309 is determined by the sequence of digits 8, 5, 3, while the characteristic is determined by the position of the decimal point. Let us say that the decimal point is in **standard position** when it occurs immediately after the first nonzero digit. The characteristic for the logarithm of a number with the decimal point in standard position is 0. The characteristic is k if the decimal point is k places to the right of standard position; it is $-k$ if it is k places to the left of standard position.

8530000.
6 places right

$$\log 8{,}530{,}000 = .9309 + 6$$

.0000853
5 places left

$$\log .0000853 = .9309 - 5$$

Finding Antilogarithms If we are to make significant use of common logarithms, we must know how to find a number when its logarithm is given. This process is called finding the inverse logarithm or the **antilogarithm**. The process is simple: Use the mantissa to find the sequence of digits and then let the characteristic tell you where to put the decimal point.

Suppose, for example, that you are given

$$\log N = .4031 - 4$$

Locate .4031 in the body of Table B. You will find it across from 2.5 and below 3. Thus the number N must have 2, 5, 3 as its sequence of digits. Since the characteristic is -4, put the decimal point 4 places to the left of standard position. The result is

$$N = .000253$$

As a second example, let us find antilog 5.9547. The mantissa .9547 gives us the digits 9, 0, 1. Since the characteristic is 5,

$$\text{antilog } 5.9547 = 901{,}000$$

Linear Interpolation Suppose that for some function f we know $f(a)$ and $f(b)$, but we want $f(c)$ where c is between a and b (see the diagrams below). As a reasonable approximation, we may pretend that the graph of f is a straight line between a and b. Then

$$f(c) \approx f(a) + d$$

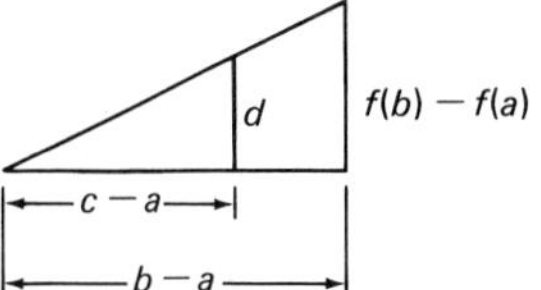

where, by similarity of triangles,

$$\frac{d}{f(b) - f(a)} = \frac{c - a}{b - a}$$

That is,

$$d = \frac{f(b) - f(a)}{b - a}(c - a)$$

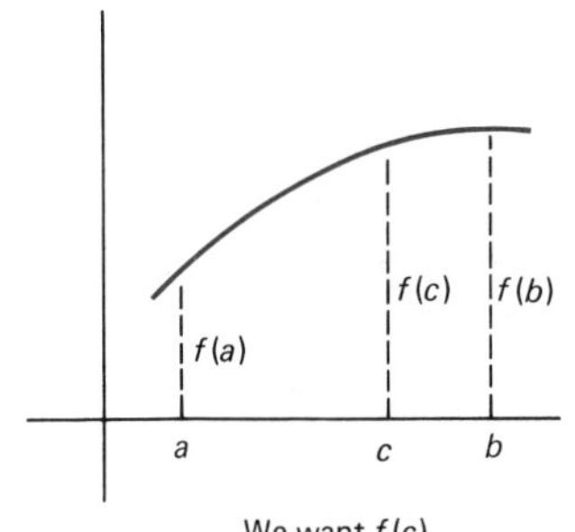

We want $f(c)$

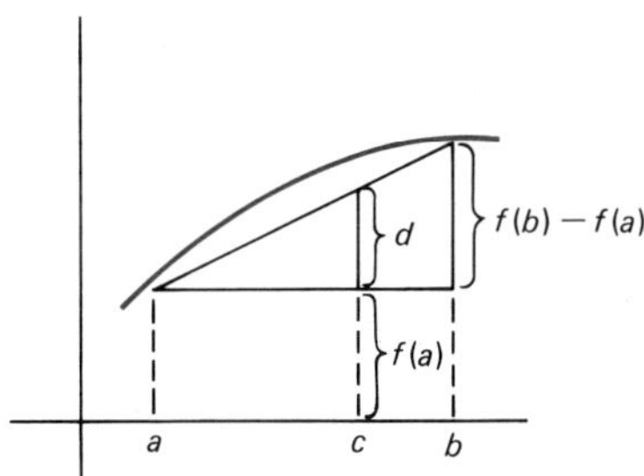

We calculate $f(a) + d$

The process just described is called **linear interpolation**. The process of linear interpolation for the logarithm function is explained in Examples A and B of the problem set, as well as at the beginning of the logarithm tables at the back of the book.

Problem Set 6-6 *Find the common logarithm of each of the following numbers.*

1. 10,000
2. 1,000,000
3. .01
4. .0001
5. $10^4 10^{3/2}$
6. $10^3 10^{4/3}$
7. $(10^3)^{-5}$
8. $(10^5)^{-1/10}$

Find N in each case.

9. $\log N = 4$
10. $\log N = 6$
11. $\log N = -2$
12. $\log N = -5$
13. $\log N = \frac{3}{2}$
14. $\log N = \frac{1}{3}$
15. $\log N = -\frac{3}{4}$
16. $\log N = -\frac{1}{6}$

Use Table B to find the following logarithms.

17. log 4.32
18. log 3.09
19. log 158
20. log 47.3
21. log .0329
22. log .0715
23. log 563,000
24. log 420,000
25. $\log(9.23 \times 10^8)$
26. $\log(2.83 \times 10^{-11})$

Find N in each case.

27. $\log N = 1.5159$
28. $\log N = 3.9015$
29. $\log N = .0043 - 2$
30. $\log N = .8627 - 4$
31. $\log N = 8.5999$
32. $\log N = 4.7427$

Find the antilogarithm of each number.

33. 2.2201
34. 3.8639
35. .9232 − 1
36. .8500 − 5

Example A (Linear interpolation in finding logarithms) Find log 34.67.

SOLUTION. Our table gives the logarithms of 34.6 and 34.7, so we use linear interpolation to get an intermediate value. Here is how we arrange our work.

$$.10\left[\begin{array}{l} .07\left[\begin{array}{l}\log 34.60 = 1.5391 \\ \log 34.67 = \quad ?\end{array}\right]d \\ \quad\ \log 34.70 = 1.5403\end{array}\right].0012$$

$$\frac{d}{.0012} = \frac{.07}{.10} = \frac{7}{10}$$

$$d = \tfrac{7}{10}(.0012) \approx .0008$$

$$\log 34.67 \approx \log 34.60 + d \approx 1.5391 + .0008 = 1.5399$$

Use linear interpolation in Table B to find each value.

37. log 5.237
38. log 9.826
39. log 7234
40. log 68.04
41. log .001234
42. log .09876

Example B (Interpolation in finding antilogarithms) Find antilog 2.5285.

SOLUTION. We find .5285 sandwiched between .5276 and .5289 in the body of Table B.

$$.0013\left[\begin{array}{l} .0009\left[\begin{array}{l}\text{antilog } 2.5276 = 337.0 \\ \text{antilog } 2.5285 = \quad ? \end{array}\right]d \\ \qquad\quad \text{antilog } 2.5289 = 338.0 \end{array}\right]1.0$$

$$\frac{d}{1.0} = \frac{.0009}{.0013} = \frac{9}{13}$$

$$d = \tfrac{9}{13}(1.0) \approx .7$$

$$\text{antilog } 2.5285 \approx 337.0 + .7 = 337.7$$

Find the antilogarithm of each of the following using linear interpolation.

43. 0.8497
44. 0.8516
45. 3.9130
46. 1.9849
47. .6004 − 2
48. .4946 − 4

Miscellaneous Problems

Find the common logarithm of each of the following.

49. $10^{3/2}10^{-1/4}$
50. $(.0001)^{1/3}$
51. $\sqrt{10}\sqrt[3]{10}$
52. $10^{\log .001}$

Find N in each of the following.

53. $\log N = 0$
54. $\log N = -2$
55. $\log N = \frac{2}{3}$
56. $\log N = 10$

Use Table B to find the following logarithms.

57. log 9.83
58. log 46.3
59. log .00219
60. log 492.7
61. log 9.623
62. log .04705

Use Table B to find N to four significant digits in Problems 63–66.

63. $\log N = 2.9325$
64. $\log N = .2695 - 3$
65. $\log N = 0.7300$
66. $\log N = .8619 - 1$
67. Find N if $\log N = -2.4473$. (*Hint*: First write -2.4473 as an integer plus a positive decimal fraction: $-2.4473 = (-2.4473 + 3) - 3 = .5527 - 3$.)
68. Find N.
(a) $\log N = -1.0074$ (b) $\log N = -4.0729$
69. If $b = 100a$ and $\log a = .75$, find $\log b$.
70. If $b = .001a$ and $\log a = 5.5$, find $\log b$.

71. Find antilog(log .78).
72. Find log(antilog(.4275 − 2)).
73. If $.000001 < N < .00001$, what is the characteristic of $\log N$?
74. Find $\log\left(\frac{471 + 328}{471 \times 328}\right)$.

The Laws of Logs

If $M > 0$ and $N > 0$, then

1. $\log M \cdot N = \log M + \log N$
2. $\log \frac{M}{N} = \log M - \log N$
3. $\log M^n = n \log M$

"The miraculous powers of modern calculation are due to three inventions: the Arabic Notation, Decimal Fractions, and Logarithms."

F. Cajori, 1897

6-7 Calculations with Logarithms (Optional)

For 300 years, scientists depended on logarithms to reduce the drudgery associated with long computations. The invention of electronic computers and calculators has diminished the importance of this long-established technique. Still, we think that any student of algebra should know how products, quotients, powers, and roots can be calculated by means of common logarithms.

About all you need are the three laws stated above and Table B of the Appendix. A little common sense and the ability to organize your work will help.

Products Suppose you want to calculate (.00872)(95,300). Call this number x. Then by Law 1 and Table B,

$$\begin{aligned}\log x &= \log .00872 + \log 95{,}300 \\ &= (.9405 - 3) + 4.9791 \\ &= 5.9196 - 3 \\ &= 2.9196\end{aligned}$$

Now use Table B backwards to find that antilog .9196 = 8.31, so x = antilog 2.9196 = 831.

Here is a good way to organize your work in a compact systematic way.

$$\begin{aligned} x &= (.00872)(95{,}300) \\ \log .00872 &= .9405 - 3 \\ (+)\quad \underline{\log 95300} &= \underline{4.9791} \\ \log x &= 5.9196 - 3 \\ x &= 831 \end{aligned}$$

Quotients Suppose we want to calculate $x = .4362/91.84$. Then by Law 2,

$$\begin{aligned} \log x &= \log .4362 - \log 91.84 \\ &= (.6397 - 1) - 1.9630 \\ &= .6397 - 2.9630 \\ &= -2.3233 \end{aligned}$$

What we have done is correct; however, it is poor strategy. The result we found for $\log x$ is not in characteristic-mantissa form and therefore is not usable. Remember that the mantissa must be positive. We can bring this about by adding and subtracting 3.

$$-2.3233 = (-2.3233 + 3) - 3 = .6767 - 3$$

Actually it is better to anticipate the need for doing this and arrange the work as follows.

$$x = \frac{.4362}{91.84}$$

$$\begin{aligned} \log .4362 &= .6397 - 1 = 2.6397 - 3 \\ (-)\quad \underline{\log 91.48} &= \phantom{.6397 - 1 = {}} \underline{1.9630} \\ \log x &= .6767 - 3 \\ x &= .00475 \end{aligned}$$

Powers or Roots Here the main tool is Law 3. We illustrate with two examples.

$$\begin{aligned} x &= (31.4)^{11} \\ \log x &= 11 \log 31.4 \\ \log 31.4 &= 1.4969 \\ 11 \log 31.4 &= 16.4659 \\ \log x &= 16.4659 \\ x &= 29{,}230{,}000{,}000{,}000{,}000 \\ &= 2.923 \times 10^{16} \end{aligned}$$

$$
\begin{aligned}
x &= \sqrt[4]{.427} = (.427)^{1/4} \\
\log x &= \tfrac{1}{4}\log .427 \\
\log .427 &= .6304 - 1 \\
\tfrac{1}{4}\log .427 &= \tfrac{1}{4}(.6304 - 1) = \tfrac{1}{4}(3.6304 - 4) = .9076 - 1 \\
\log x &= .9076 - 1 \\
x &= .8084
\end{aligned}
$$

Notice in the second example that we wrote $3.6304 - 4$ in place of $.6304 - 1$, so that multiplication by $\frac{1}{4}$ gave the logarithm in characteristic-mantissa form.

Problem Set 6-7

Use logarithms and Table B without interpolation to find approximate values for each of the following.

1. $(46.3)(2.76)$
2. $(378)(9.63)$
3. $\dfrac{46.3}{483}$
4. $\dfrac{437}{92300}$
5. $\dfrac{.00912}{.439}$
6. $\dfrac{.0429}{15.7}$
7. $(37.2)^5$
8. $(113)^3$
9. $\sqrt[3]{42.9}$
10. $\sqrt[4]{312}$
11. $\sqrt[5]{.918}$
12. $\sqrt[3]{.0307}$
13. $(14.9)^{2/3}$
14. $(98.6)^{3/4}$

Use logarithms and Table B with interpolation to approximate each of the following.

15. $(31.96)(149)$
16. $(6236)(.00108)$
17. $\dfrac{43.98}{7.16}$
18. $\dfrac{115}{4.623}$
19. $(.1234)^6$
20. $(92.83)^3$

Example A (More complicated calculations) Use logarithms, without interpolation, to calculate

$$\frac{(31.4)^3(.982)}{(.0463)(824)}$$

SOLUTION. Let N denote the entire numerator, D the entire denominator, and x the fraction N/D. Then

$$\log x = \log N - \log D$$

where

$$
\begin{aligned}
\log N &= 3\log 31.4 + \log .982 \\
\log D &= \log .0463 + \log 824
\end{aligned}
$$

Here is a good way to organize the work.

$$\begin{array}{lrl} & \log 31.4 = & 1.4969 \\ & 3\log 31.4 = & 4.4907 \\ (+) & \log .982 = & .9921 - 1 \\ \hline & \log N = & 5.4828 - 1 \\ & = & 4.4828 \end{array} \qquad \begin{array}{lrl} & \log .0463 = & .6656 - 2 \\ (+) & \log 824 = & 2.9159 \\ \hline & \log D = & 3.5815 - 2 \\ & = & 1.5815 \end{array}$$

$$\begin{array}{lrl} & \log N = & 4.4828 \\ (-) & \log D = & 1.5815 \\ \hline & \log x = & 2.9103 \\ & x = & 797 \end{array}$$

Carry out the following calculations using logarithms without interpolation.

21. $\dfrac{(.56)^2(619)}{21.8}$

22. $\dfrac{.413}{(4.9)^2(.724)}$

23. $\dfrac{(14.3)\sqrt{92.3}}{\sqrt[3]{432}}$

24. $\dfrac{(91)(41.3)^{2/3}}{42.6}$

Example B (Solving exponential equations) Solve the equation

$$2^{2x-1} = 13$$

SOLUTION. We begin by taking logarithms of both sides and then solving for x.

$$(2x - 1)\log 2 = \log 13$$

$$2x - 1 = \frac{\log 13}{\log 2} = \frac{1.1139}{.3010}$$

$$\begin{aligned} \log(2x - 1) &= \log 1.1139 - \log .3010 \\ &= (1.0469 - 1) - (.4786 - 1) \\ &= .5683 \\ 2x - 1 &= \text{antilog } .5683 = 3.70 \\ 2x &= 4.70 \\ x &= 2.35 \end{aligned}$$

Notice that we did the division of (log 13)/(log 2) by means of logarithms. We could have done it by long division (or on a calculator) if we preferred.

Use logarithms to solve the following exponential equations. You need not interpolate.

25. $3^x = 300$

26. $5^x = 14$

27. $10^{2-3x} = 6240$

28. $10^{5x-1} = .00425$

29. $2^{3x} = 3^{x+2}$

30. $2^{x^2} = 3^x$

Miscellaneous Problems

Use common logarithms and Table B to approximate the following.

31. $(86.4)(.000139)$

32. $(14.8)^3$

33. $\dfrac{38.9}{6150}$

34. $\sqrt[3]{.0427}$

35. $\dfrac{(42.9)^2(.983)}{\sqrt{323}}$

36. $\dfrac{10^{6.42}}{8^{7.2}}$

Use common logarithms to solve the following equations. You need not interpolate.

37. $4^{2x} = 150$

38. $(.975)^x = .5$

In Problems 39–42, solve for x. (Hint: First combine the logarithms.)

39. $\log(x + 2) - \log x = 1$

40. $\log(2x + 1) - \log(x + 3) = 0$

41. $\log \dfrac{x + 3}{x} + \log 2x^2 = \log 8.$

42. $\log(x + 3) + \log(x - 1) = \log 4x.$

43. Suppose that the amount Q of a radioactive substance (in grams) remaining after t years is

$$Q = (42)2^{-.017t}$$

(a) Use logarithms to calculate the amount remaining after 20 years.
(b) After how many years will the amount remaining be only .42 grams?

44. Suppose that the number of bacteria in a certain culture t hours from now is $(800)3^t$.
(a) What will the bacteria count be 3.12 hours from now?
(b) When will the bacteria count reach 100,000?

45. Assume the 1977 population of the earth was 4.19 billion and that the growth rate is 2 percent per year. Then the population t years after 1977 will be $4.19(10^9)(1.02)^t$.
(a) What will the population of the earth be in the year 2000?
(b) When will the population reach 8.63 billion?

46. Answer the two questions in Problem 45 for a growth rate of 3 percent and then for a growth rate of 1 percent.

47. The volume of a sphere is given by $V = \frac{4}{3}\pi r^3$, where r is the radius of the sphere. Find the radius of a sphere whose volume is 42,900 cubic feet.

CHAPTER SUMMARY

The symbol $\sqrt[n]{a}$, the nth root of a, denotes one of the numbers whose nth power is a. For odd n, that is all that needs to be said. For n even and $a > 0$, we specify that $\sqrt[n]{a}$ signifies the positive nth root. Thus $\sqrt[3]{-8} = -2$ and $\sqrt{16} = \sqrt[2]{16} = 4$. (It is wrong to write $\sqrt{16} = -4$.) These symbols are also called

radicals. These radicals obey four carefully prescribed rules (page 215). These rules allow us to simplify complicated radical expressions, in particular, to **rationalize denominators**.

The key to understanding **rational exponents** is the definition $a^{1/n} = \sqrt[n]{a}$, which implies $a^{m/n} = (\sqrt[n]{a})^m$. Thus $16^{5/4} = (\sqrt[4]{16})^5 = 2^5 = 32$. The meaning of **real exponents** is determined by considering rational approximations. For example, 2^π is the number that the sequence $2^3, 3^{3.1}, 2^{3.14}, \ldots$ approaches. The function $f(x) = a^x$ (and more generally $f(x) = b \cdot a^x$) is called an **exponential function**.

A variable y is **growing exponentially** or **decaying exponentially** according as $y = b \cdot a^x$ for $a > 1$ or for $0 < a < 1$. Typical of the former are biological populations; of the latter, radioactive elements. Corresponding key ideas are **doubling times** and **half-lives**.

Logarithms are exponents. In fact, $\log_a N = x$ means $a^x = N$, that is, $a^{\log_a N} = N$. The functions $f(x) = \log_a x$ and $g(x) = a^x$ are **inverses** of each other. Logarithms have three primary properties (page 234). **Natural logarithms** correspond to the choice of base $a = e = 2.71828\ldots$ and play a fundamental role in advanced courses. **Common logarithms** correspond to base 10 and have historically been used to simplify arithmetic calculations.

CHAPTER REVIEW PROBLEM SET

1. Simplify, rationalizing all denominators. Assume letters represent positive numbers.

   (a) $\sqrt[3]{\dfrac{-8y^6}{z^{14}}}$ (b) $\sqrt[4]{32x^5y^8}$ (c) $\sqrt{4\sqrt[3]{5}}$

   (d) $\dfrac{2}{\sqrt{x} - \sqrt{y}}$ (e) $\sqrt{50 + 25x^2}$ (f) $\sqrt{32} + \sqrt{8}$

2. Solve the equations.
   (a) $\sqrt{x - 3} = 3$ (b) $\sqrt{x} = 6 - x$
3. Simplify, writing your answer in exponential form with all positive exponents.
   (a) $(25a^2)^{3/2}$ (b) $(a^{-1/2}aa^{-3/4})^2$

   (c) $\dfrac{1}{5\sqrt[4]{5^3}}$ (d) $\dfrac{(3x^{-2}y^{3/4})^2}{3x^2y^{-2/3}}$

   (e) $(x^{1/2} - y^{1/2})^2$ (f) $\sqrt[3]{4}\sqrt[4]{4}$
4. Sketch the graph of $y = (\frac{3}{2})^x$ and use your graph to estimate the value of $(\frac{3}{2})^\pi$.
5. A certain radioactive substance has a half-life of 3 days. What fraction of an initial amount will be left after 243 days?
6. A population grows so that its doubling time is 40 years. If this population is 1 million today, what will it be after 160 years?

7. If \$100 is put in the bank at 8 percent interest compounded quarterly, what will it be worth at the end of 10 years?

8. Find x in each of the following.
   (a) $\log_4 64 = x$ (b) $\log_2 x = -3$
   (c) $\log_x 49 = 2$ (d) $\log_4 x = 0$
   (e) $\log_9 27 = x$ (f) $\log_{10} x + \log_{10}(x - 3) = 1$
   (g) $a^{\log_a 10} = x$ (h) $x = \log_a(a^{1.14})$

9. Write as a single logarithm.

$$2 \log_4(3x + 1) - \tfrac{1}{2} \log_4 x + \log_4(x - 1)$$

10. Evaluate.
    (a) $\log_{10} \sqrt{1000}$ (b) $\log_{27} 81$
    (c) $\ln \sqrt{e}$ (d) $\ln(1/2^4)$

11. Use the table of natural logarithms to determine each of the following.
    (a) $\ln(\frac{7}{4})^3$ (b) N if $\ln N = 2.230$
    (c) $\ln[(3.4)(9.9)]$ (d) N if $\ln N = -0.105$

12. By taking ln of both sides, solve $2^{x+1} = 7$.

13. A certain substance decays according to the formula $y = y_0 e^{-.05t}$, where t is measured in years. Find its half-life.

14. A substance initially weighing 100 grams decays exponentially according to the formula $y = 100e^{-kt}$. If 30 grams are left after 10 days, determine k.

15. Sketch the graphs of $y = \log_3 x$ and $y = 3^x$ using the same coordinate axes.

16. Use common logarithms to calculate

$$\frac{(13.2)^4\sqrt{15.2}}{29.6}$$

The great book of Nature lies open before our eyes and true philosophy is written in it. . . . But we cannot read it unless we have first learned the language and characters in which it is written. . . . It is written in mathematical language and the characters are triangles, circles, and other geometrical figures.

Galileo

CHAPTER SEVEN

The Trigonometric Functions

7-1 Right-Triangle Trigonometry

7-2 Angles and Arcs

7-3 The Sine and Cosine Functions

7-4 Four More Trigonometric Functions

7-5 Finding Values of the Trigonometric Functions

7-6 Graphs of the Trigonometric Functions

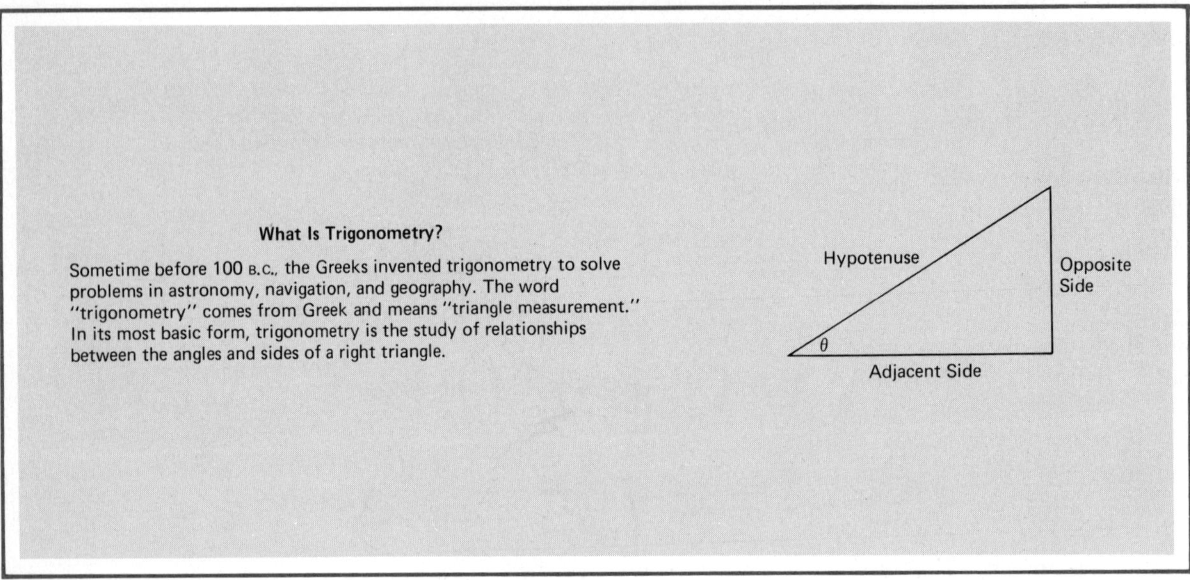

7-1 Right-Triangle Trigonometry

A triangle is called a *right triangle* if one of its angles is a right angle, that is, a 90° angle. The other two angles are necessarily acute angles (less than 90°) since the sum of all three angles in a triangle is 180°. Let θ (the Greek letter theta) denote one of these acute angles. We may label the three sides relative to θ: adjacent side, opposite side, and hypotenuse, as shown in the diagram above. In terms of these sides, we introduce the three fundamental ratios of trigonometry, sine θ, cosine θ, and tangent θ. Using obvious abbreviations, we give the following definitions.

$$\sin\theta = \frac{\text{opp}}{\text{hyp}}$$

$$\cos\theta = \frac{\text{adj}}{\text{hyp}}$$

$$\tan\theta = \frac{\text{opp}}{\text{adj}}$$

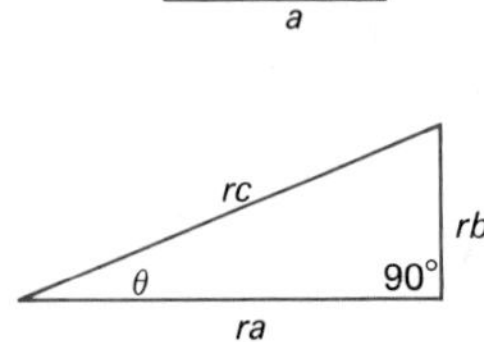

Thus with every acute angle θ, we associate three numbers, $\sin\theta$, $\cos\theta$, and $\tan\theta$. A careful reader might wonder whether these numbers depend only on the size of θ, or if they also depend on the lengths of the sides of the right triangle with which we started. Consider two different right triangles, each with the same angle θ (as in the margin). You may think of the lower triangle as a magnification of the upper one. Each of its sides has length r times that of the corresponding side in the upper triangle. If we calculate $\sin\theta$ from the lower triangle, we get

$$\sin\theta = \frac{\text{opp}}{\text{hyp}} = \frac{rb}{rc} = \frac{b}{c}$$

which is the same result we get using the upper triangle. We conclude that for a given θ, $\sin\theta$ has the same value no matter which right triangle is used to compute it. So do $\cos\theta$ and $\tan\theta$.

Special Angles We can use the Pythagorean Theorem ($a^2 + b^2 = c^2$) to find the values of sine, cosine, and tangent for the special angles 30°, 45°, and 60°. Consider the two right triangles below, which involve these angles.

To see that we have calculated a correctly, note that in the first triangle, $a^2 + a^2 = 2^2$, which gives $a = \sqrt{2}$. In the second, which is half of an equilateral triangle, $a^2 + 1^2 = 2^2$, or $a = \sqrt{3}$.

From these triangles, we obtain the following important facts.

$$\sin 45^\circ = \frac{\sqrt{2}}{2} \qquad \cos 45^\circ = \frac{\sqrt{2}}{2} \qquad \tan 45^\circ = 1$$

$$\sin 30^\circ = \frac{1}{2} \qquad \cos 30^\circ = \frac{\sqrt{3}}{2} \qquad \tan 30^\circ = \frac{1}{\sqrt{3}} = \frac{\sqrt{3}}{3}$$

$$\sin 60^\circ = \frac{\sqrt{3}}{2} \qquad \cos 60^\circ = \frac{1}{2} \qquad \tan 60^\circ = \sqrt{3}$$

Other Angles When you need the sine, cosine, or tangent of an angle other than the special ones just considered, you may do one of two things. If you have a scientific calculator, you may simply push two or three keys and have your answer correct to eight or more significant digits. Otherwise, you will need to use Table C in the appendix.

Several facts about Table C should be noted. First, it gives answers usually to four decimal places. Second, angles are measured in degrees and tenths of degrees. By interpolation (see page 510), it is possible to consider angles measured to the nearest hundredth of a degree. Finally, notice that the left column of the table lists angles from 0° to 45°. For angles from 45° to 90°, use the right column; you must then also use the bottom captions. To make sure that you are reading the table (or your calculator) correctly, check that you get each of the following answers.

$$\tan 33.1^\circ = .6519 \qquad \sin 26.9^\circ = .4524$$
$$\cos 54.3^\circ = .5835 \qquad \tan 82^\circ = 7.115$$

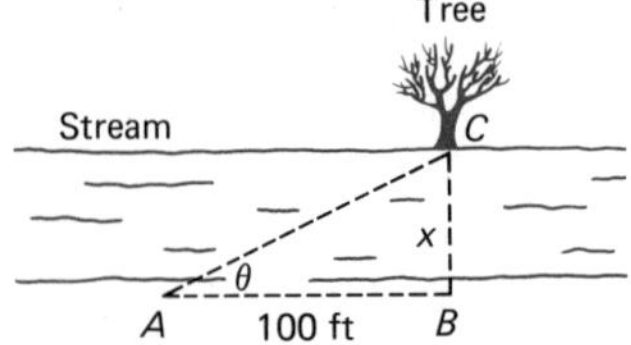

Applications Suppose that you wish to measure the distance across a stream but do not want to get your feet wet. Here is how you might proceed.

Pick out a tree at C on the opposite shore and set a stone at B directly across from it on your shore. Set another stone at A, 100 feet up the shore from B. With an angle measuring device (for example, a protractor or a transit), measure angle θ between AB and AC. Then x, the length of BC, satisfies the following equation.

$$\tan \theta = \frac{\text{opp}}{\text{hyp}} = \frac{x}{100}$$

or

$$x = 100 \tan \theta$$

For example, if θ measures 29°, you find from your scientific calculator or Table C that $\tan 29^\circ = .5543$. Then $x = 100(.5543) = 55.43$ feet. Since you used stones and trees for points, this suggests that you should not give your answer with such accuracy. It would be better to say that the distance x is approximately 55 feet.

As a more difficult example, consider a church with a steeple, as shown in the diagram below. The problem is to calculate the height of the steeple while standing on the ground. To find the height, mark a point B on the ground directly below the steeple and another point A 200 feet away on the ground. At A, measure the angles of elevation α and β to the top and bottom of the steeple. This is all the information you will need, provided you know your trigonometry.

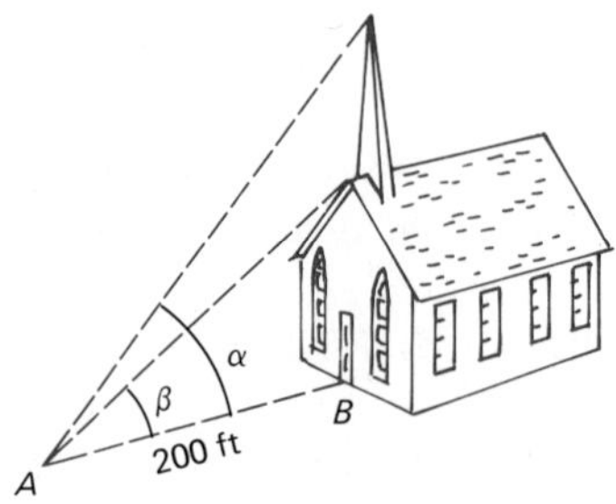

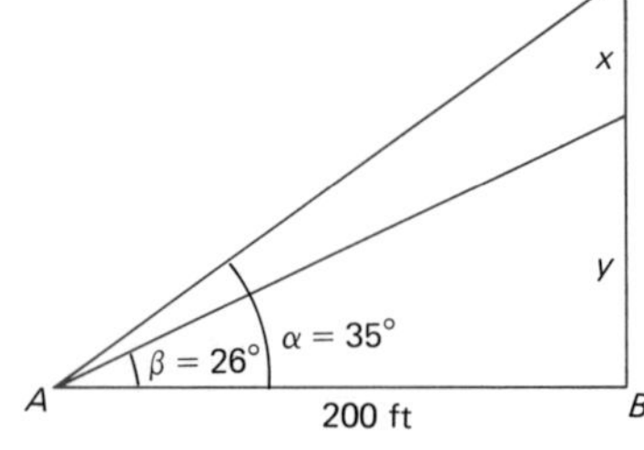

Let x be the height of the steeple and y be the distance from the ground to the bottom of the steeple. Suppose that $\alpha = 35°$ and $\beta = 26°$. Then

$$\tan 35° = \frac{x + y}{200}$$

$$\tan 26° = \frac{y}{200}$$

If you solve for y in the second equation and substitute the value in the first, you will get the following sequence of equations.

$$\tan 35° = \frac{x + 200 \tan 26°}{200}$$

$$\begin{aligned} 200 \tan 35° &= x + 200 \tan 26° \\ x &= 200 \tan 35° - 200 \tan 26° \\ &= 200(.7002 - .4877) \\ x &= 42.5 \text{ feet} \end{aligned}$$

Problem Set 7-1

In Problems 1–6, use Table C to evaluate each expression. If you have a scientific calculator, use it as a check.

1. $\sin 41.3°$
2. $\tan 54.4°$
3. $\cos 49.2°$
4. $\sin 89.3°$
5. $\tan 72.3°$
6. $\cos 38.7°$

In Problems 7–12, use Table C to find θ. We suggest you also do these problems on a calculator. For example, to do Problem 7 on many calculators, press .2164 [INV] [sin].

7. $\sin \theta = .2164$
8. $\tan \theta = .3096$
9. $\tan \theta = 2.311$
10. $\cos \theta = .9354$
11. $\cos \theta = .3535$
12. $\sin \theta = .7302$

Each of the remaining problems in this problem set involves a considerable amount of arithmetic that you can do by hand (using tables) or by using a calculator. (If you use a calculator to find values for the trigonometric functions, be sure that it is in the degree mode.) In Problems 13–18, find x.

13.

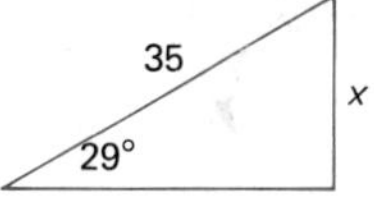

14.

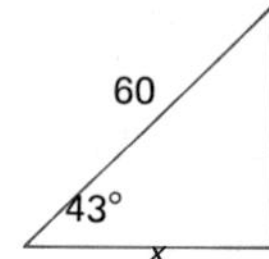

15.

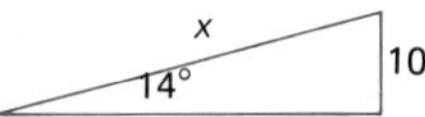

16.

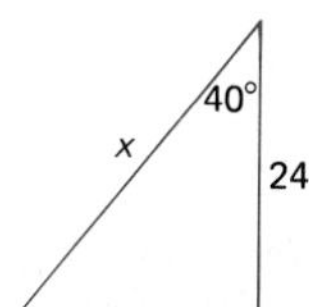

17.

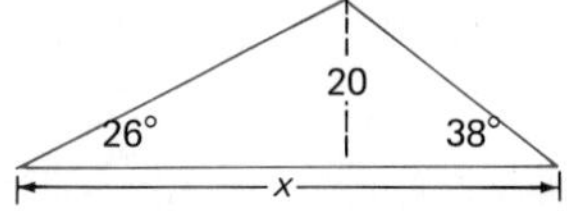

18.

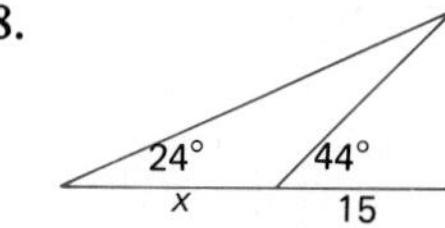

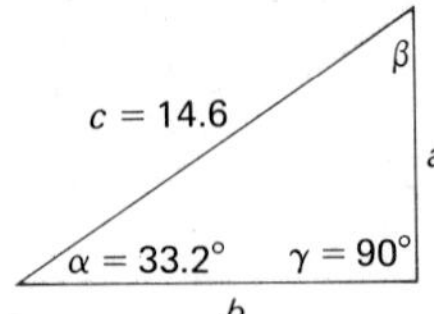

Example A (Solving a right triangle given an angle and a side) To solve a triangle means to determine all its unknown parts. Solve the right triangle which has hypotenuse of length 14.6 and an angle measuring 33.2°.

SOLUTION. First, we draw the triangle labeling the known parts and assigning letters to the unknown parts. Our convention is to use the first three Greek letters, α, β, and γ (alpha, beta, and gamma) for the angles and a, b, and c for the lengths of the respective sides opposite these angles (see diagram). We need to find β, a, and b.

(i) $\beta = 90° - 33.2° = 56.8°$

(ii) $\sin 33.2° = a/14.6$, so

$$a = 14.6 \sin 33.2° = (14.6)(.5476) \approx 7.99$$

(iii) $\cos 33.2° = b/14.6$, so

$$b = 14.6 \cos 33.2° = (14.6)(.8368) \approx 12.2$$

Notice that we gave the answers to three significant digits since the given data have three significant digits.

Solve each of the following triangles. First draw the triangle, labeling it as in the example with $\gamma = 90°$.

19. $\alpha = 42°$, $c = 35$
20. $\beta = 29°$, $c = 50$
21. $\beta = 56.2°$, $c = 91.3$
22. $\alpha = 69.9°$, $c = 10.6$
23. $\alpha = 39.4°$, $a = 120$
24. $\alpha = 40.6°$, $b = 163$

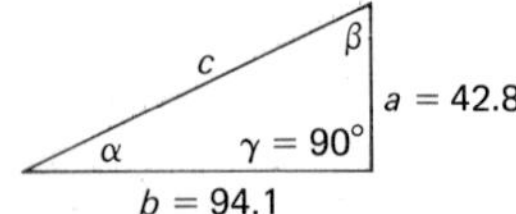

Example B (Solving a right triangle given two sides) Solve the right triangle which has legs $a = 42.8$ and $b = 94.1$.

SOLUTION. First, we draw the triangle and label its parts. We must find α, β, and c.

(i) $$\tan \alpha = \frac{42.8}{94.1} \approx .4548$$

Now we can find α by using Table C backwards, that is, by looking under tangent in the body of the table for .4548 and determining the corresponding angle. Or better, we can use the [INV] [tan] keys on a scientific calculator. On many calculators, press

[(] 42.8 [÷] 94.1 [)] [INV] [tan]

In either case, the result is $\alpha \approx 24.5°$.

(ii) $\beta = 90° - \alpha = 90° - 24.5° = 65.5°$

(iii) We could find c by using $c^2 = a^2 + b^2$. Instead, we use $\sin \alpha$.

$$\sin \alpha = \sin 24.5° = \frac{42.8}{c}$$

$$c = \frac{42.8}{\sin 24.5°} = \frac{42.8}{.4147} \approx 103$$

Solve the right triangles satisfying the given information in Problems 25–32, assuming that c is the hypotenuse. You can do them either with tables or a calculator.

25. $a = 9$, $b = 12$
26. $a = 24$, $b = 10$
27. $a = 40$, $c = 50$
28. $c = 41$, $a = 40$

29. $a = 14.6, c = 32.5$

30. $a = 243, c = 419$

31. $a = 9.52, b = 14.7$

32. $a = .123, b = .456$

33. A straight path leading up a hill rises 26 feet per 100 horizontal feet. What angle does it make with the horizontal?

34. A 20-foot ladder leans against a wall, making an angle of 76° with the level ground. How high up the wall is the upper end of the ladder?

35. Find the angle of elevation of the sun if a woman 5 feet 9 inches tall casts a shadow 46.8 feet long.

36. A guy wire to a pole makes an angle of 69° with the level ground and is 14 feet from the pole at the ground. How high above the ground is the wire attached to the pole?

37. Suppose that the woman in Problem 35 is walking with her daughter Sue, who is 3 feet 10 inches tall. How long is Sue's shadow?

38. Find the length of the supporting wire in Problem 36.

Miscellaneous Problems

39. Find each value.
(a) $\sin 49.3°$ (b) $\cos 31.3°$ (c) $\tan 59.3°$

40. Find θ in each case.
(a) $\sin \theta = .2317$ (b) $\cos \theta = .3387$ (c) $\tan \theta = .3620$

41. Solve the right triangle in which $\alpha = 41.3°$ and $a = 341$.

42. Solve the right triangle in which $a = 44.5$ and $c = 53.6$.

43. A kite is 144 feet high when 645 feet of string are out. Assuming that the string is taut and is held 4 feet above the ground, find the angle of elevation of the string.

44. From the top of a lighthouse 120 feet above sea level, the angle of depression to a boat adrift on the sea is 9.4°. How far from the foot of the lighthouse is the boat?

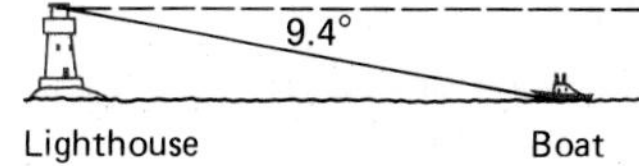

45. Suppose that a manned balloon rises straight up. At a certain instant, an observer 1000 yards away from the launch point on level ground notes its angle of elevation as 22.1°. A minute later the angle is 62.4°. How far did the balloon rise during that minute?

46. Sitting at a window of an office building, I note that in looking directly across the street at an apartment building, the angle of depression of the base of the building is 42.5° and the angle of elevation of the top is 33.5°. Find the height of the apartment building if the distance between the two buildings is 150 feet.

47. Suppose that a regular octagon (8 equal sides) is inscribed in a circle of radius 2. Find the perimeter of the octagon.

48. Find the area of a regular dodecagon (12 sides) inscribed in a circle of radius 8. What percent of the area of the circle is the area of the dodecagon?

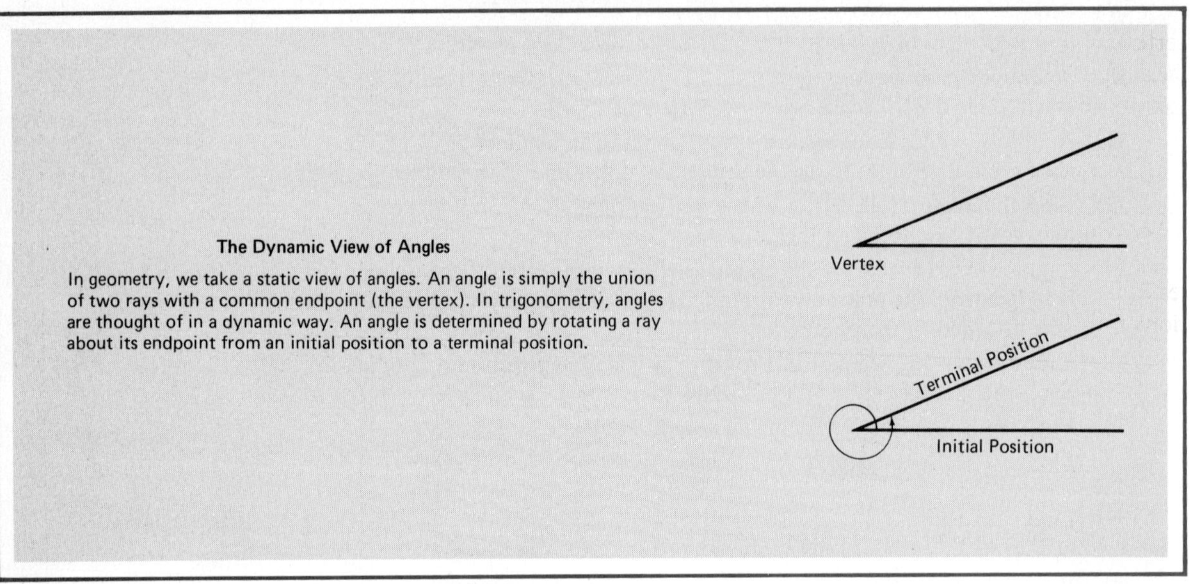

7-2 Angles and Arcs

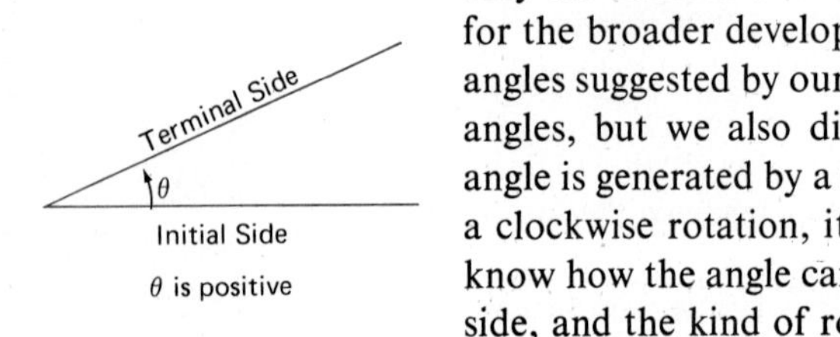

For the solution of right triangles (which involve acute angles), we required only the familiar and simple notion of angle from high-school geometry. But for the broader development of trigonometry, we need the new perspective on angles suggested by our opening display. Not only do we allow arbitrarily large angles, but we also distinguish between positive and negative angles. If an angle is generated by a counterclockwise rotation, it is positive; if generated by a clockwise rotation, it is negative. To know an angle, in trigonometry, is to know how the angle came into being. It is to know the initial side, the terminal side, and the kind of rotation that produced the angle.

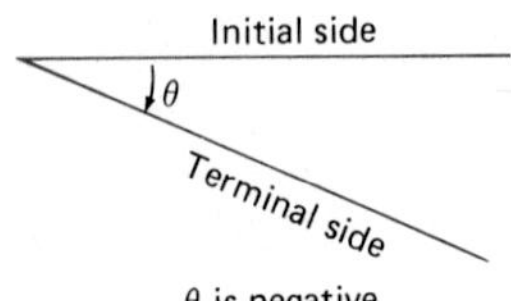

Degree Measurement Take a circle and divide its circumference into 360 equal parts. The angle with vertex at the center determined by one of these parts has measure one **degree** (written 1°). This way of measuring angles is due to the ancient Babylonians and is so familiar that we used it in Section 7-1 without comment. There is a refinement, however, that we avoid. The Babylonians divided each degree into 60 minutes and each minute into 60 seconds; some people still follow this cumbersome practice. If we need to measure angles to finer accuracy than a degree, we will use decimal parts. Thus we write 40.5° rather than $40^\circ 30'$.

It is important that we be familiar with measuring both positive and negative angles, as well as angles resulting from large rotations. Three angles are shown in the margin. Note that all three have the same initial and terminal sides.

Radian Measurement The best way to measure angles is in radians. Take a circle of radius r. The familiar formula $C = 2\pi r$ tells us that the circumference has 2π (about 6.28) arcs of length r around it. The angle with vertex at the center of a circle determined by an arc of length equal to its radius measures one **radian**. Thus an angle of size 360° measures 2π radians or, equivalently,

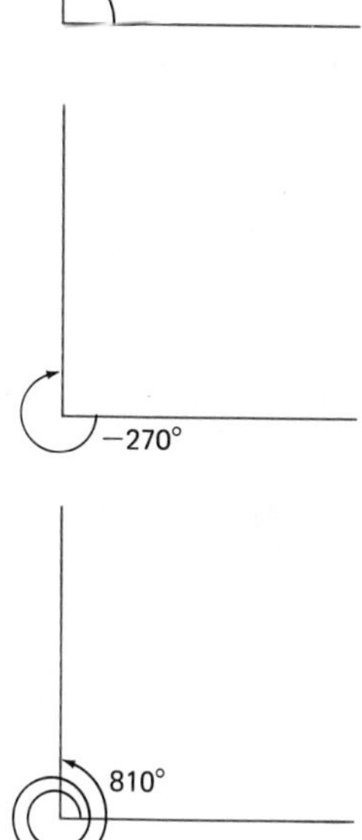

$$\boxed{180^\circ = \pi \text{ radians}}$$

To convert from degrees to radians, all one needs to remember is the result in the box. By dividing by 2, 3, 4, and 6, respectively, we get the conversions for several special angles.

$$90^\circ = \frac{\pi}{2} \text{ radians}$$

$$60^\circ = \frac{\pi}{3} \text{ radians}$$

$$45^\circ = \frac{\pi}{4} \text{ radians}$$

$$30^\circ = \frac{\pi}{6} \text{ radians}$$

Similarly,

$$135^\circ = \tfrac{3}{4}\pi \text{ radians}$$
$$720^\circ = 4\pi \text{ radians}$$
$$-120^\circ = -\tfrac{2}{3}\pi \text{ radians}$$

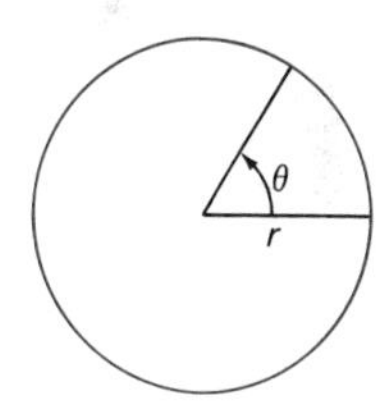

θ measures one radian (about 57.3°)

To change 22° to radians, divide the boxed formula by 180 and then multiply by 22.

$$1^\circ = \frac{\pi}{180} \text{ radians}$$

$$22^\circ = 22\left(\frac{\pi}{180}\right) \text{ radians} \approx .38397 \text{ radians}$$

Conversely, to change 2.3 radians to degrees, divide the boxed formula by π and multiply by 2.3.

$$\frac{180^\circ}{\pi} = 1 \text{ radian}$$

$$2.3 \text{ radians} = (2.3)\,\frac{180^\circ}{\pi} \approx 131.78^\circ$$

If you look at the last steps in these two examples, you see that to change from degree measure to radian measure, we multiply by $\pi/180$; to change from radians to degrees, we multiply by $180/\pi$.

Some scientific calculators have a key that makes these conversions automatically.

Arc Length Radian measure is useful because it is an intrinsic measure. The division of a circle into 360 parts was quite arbitrary; its division into parts of radius length (2π parts) is more natural. Because of this, formulas using radian measure tend to be simple, while those using degree measure are often complicated (see Problems 52 and 53). As an example, consider arc length. Let t be the radian measure of an angle θ with vertex at the center of a circle of radius r. This angle cuts off an arc of length s which satisfies the simple formula

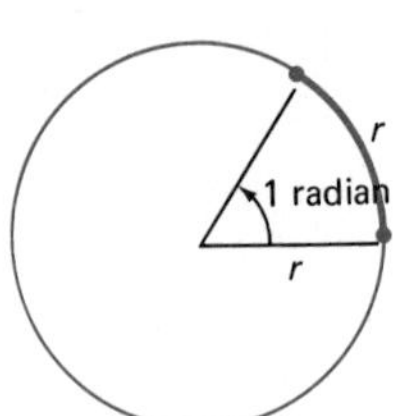

$$s = rt$$

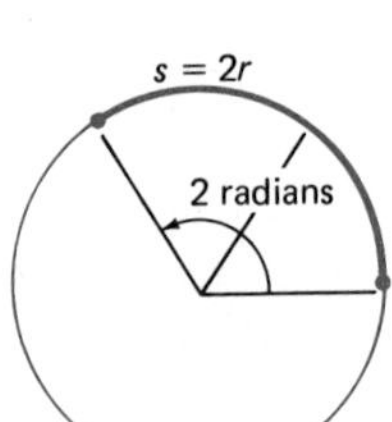

This follows directly from the fact that an angle of one radian ($t = 1$) cuts off an arc of length r.

The Unit Circle The formula for arc length takes a particularly simple form when $r = 1$, namely, $s = t$. We emphasize its meaning. *On the unit circle, the length of an arc is the same as the radian measure of the angle it determines.*

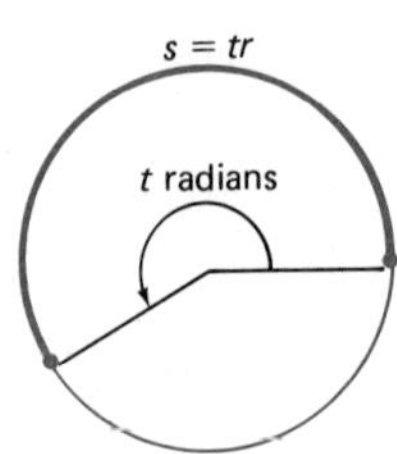

Someone is sure to point out a difficulty in what we have just said. What happens when t is greater than 2π or when t is negative? To understand our meaning, imagine an infinitely long string on which the real number scale has been marked. Think of wrapping this string around the unit circle as shown below.

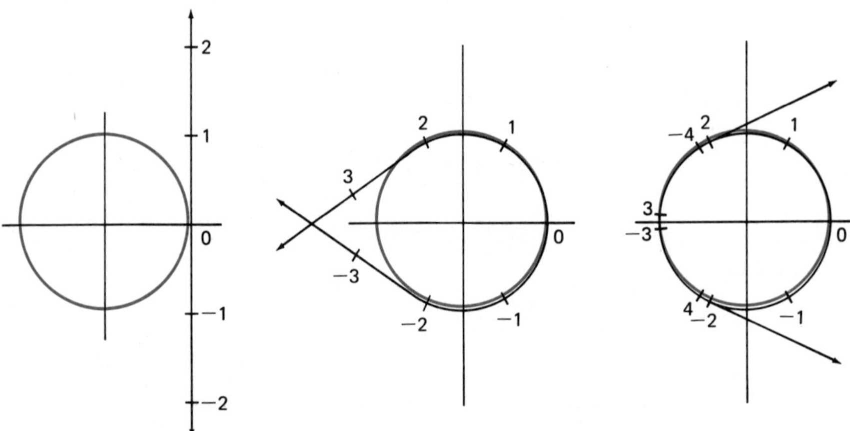

Wrapping the Real Line around the Unit Circle

Now if we think of the directed length (that is, the signed length) of a piece of the string, the formula $s = t$ holds no matter what t is. For example, the length of string corresponding to an angle of 8π radians is 8π. That piece of string wraps counterclockwise around the unit circle exactly 4 times. A piece of string corresponding to an angle of -3π radians would wrap clockwise around the unit circle one and a half times, its directed length being -3π.

Problem Set 7-2

Convert each of the following to radians. You may leave π in your answer.

1. 120° 2. 225° 3. 240°
4. 150° 5. 210° 6. 330°
7. 315° 8. 300° 9. 540°
10. 450° 11. −420° 12. −660°
13. 160° 14. 200° 15. $(20/\pi)°$
16. $(150/\pi)°$

Convert each of the following to degrees. Give your answer correct to the nearest tenth of a degree.

17. $\frac{4}{3}\pi$ radians 18. $\frac{5}{6}\pi$ radians 19. $-\frac{2\pi}{3}$ radians
20. $-\frac{7\pi}{4}$ radians 21. 3π radians 22. 3 radians
[C] 23. 4.52 radians [C] 24. $2\frac{3}{4}$ radians [C] 25. $\frac{1}{\pi}$ radians
[C] 26. $\frac{4}{3\pi}$ radians

27. Find the radian measure of the angle at the center of a circle of radius 6 inches which cuts off an arc
(a) of 12 inches; (b) of 18.84 inches.
28. Find the length of the arc cut off on a circle of radius 3 feet by an angle at the center
(a) of 2 radians; (b) of 5.5 radians;
(c) of $\frac{\pi}{4}$ radians; (d) of $\frac{5\pi}{6}$ radians.
29. Find the radius r for each of the following.

(a)

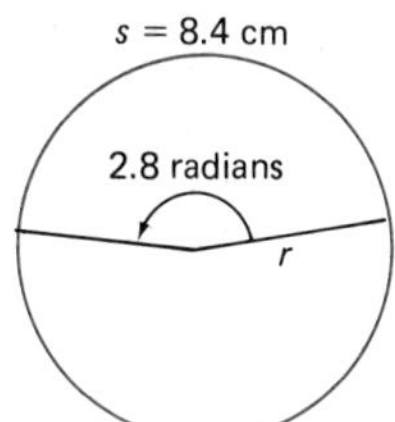

(b)

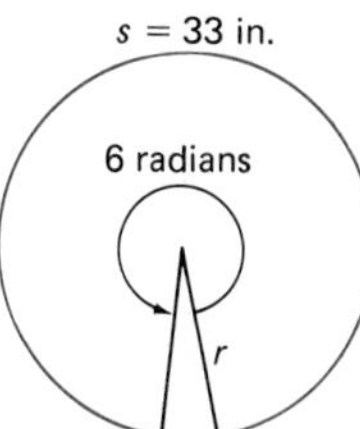

30. Through how many radians does the minute hand of a clock turn in one hour? The hour hand in one hour? The minute hand in five hours?

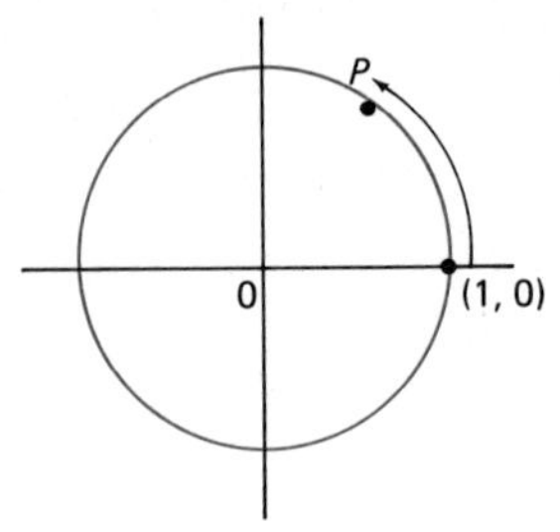

Example (Locating a point on the unit circle) The figure in the margin shows a unit circle with center at the origin. Suppose that a point P moves in a counterclockwise direction around the circle starting at (1, 0). In which quadrant is P when it has traveled a distance of 4 units? Of 40 units?

SOLUTION. Keep in mind that the distance P travels equals the radian measure of the angle through which OP turns. A distance of 4 units puts P in quadrant III since $\pi < 4 < 3\pi/2$. Once around the circle is 2π units. Since $40 = 6(6.28) + 2.32$ ($2\pi \approx 6.28$) and 2.32 is between $\pi/2$ and π, traveling 40 units puts P in quadrant II.

Find the quadrant in which the point P in the example above lies when it has traveled each of the following distances.

31. 3 units
32. 3.2 units
33. 4.7 units
34. 4.8 units
35. $\left(\frac{5\pi}{2} + 1\right)$ units
36. $\left(\frac{9\pi}{2} - 1\right)$ units
37. 100 units
38. 200 units

Miscellaneous Problems

39. Convert to radians.
    (a) $-135°$ (b) $1500°$ (c) $(45/\pi)°$
40. Convert to degrees.
    (a) $\frac{13\pi}{18}$ radians (b) 5.2 radians (c) $\frac{180}{\pi}$ radians
41. Find the length of arc cut off on a circle of radius 2.5 centimeters by each central angle.
    (a) 4 radians (b) 135°
42. Find the radian measure of θ if it cuts off an arc of length 11 meters on a circle of radius 2 meters.
43. A formula that is closely related to the formula $s = rt$ is $v = r\omega$, where ω is the *angular velocity* of a point moving around a circle of radius r and v is the speed of the point. If r is measured in feet and ω in radians per second, then v is in feet per second.
    (a) If r is 2 feet and v is 5 feet per second, find ω.
    (b) If $v = 12$ feet per second and $\omega = 20$ radians per second, find r.
44. How far does a wheel of radius 2 feet roll along level ground in making 300 revolutions?
45. [C] Suppose that a tire on a car has an outer diameter of 2.5 feet. How many revolutions per minute does that tire make when the car is traveling 60 miles per hour? (*Hint*: First find the angular velocity in radians per minute.)

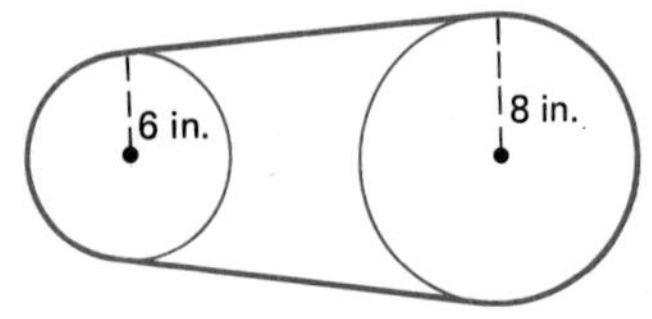

46. [C] A dead fly is stuck to a belt that passes over two pulleys 6 inches and 8 inches in radius. Assuming no slippage, how fast is the fly moving when the large pulley makes 21 revolutions per second? How many revolutions per second does the small pulley make?
47. [C] How long will it take the dead fly in Problem 46 to travel 1 mile?

48. Show that the area of a sector of a circle is given by

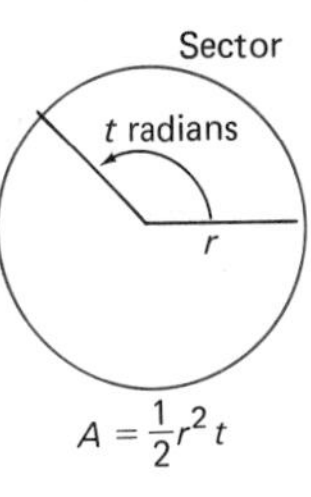

$$A = \tfrac{1}{2}r^2 t$$

where r is the radius of the circle and t is the radian measure of the central angle of the sector. (*Hint*: What fraction of the area of the entire circle (πr^2) is the area of the sector?)

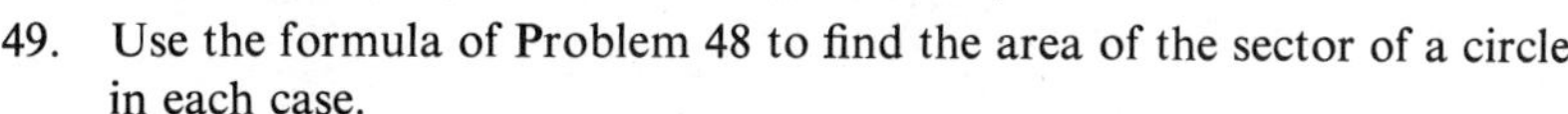

49. Use the formula of Problem 48 to find the area of the sector of a circle in each case.
(a) $r = 1$ and the central angle measures 2 radians.
(b) $r = 3$ and the central angle measures $\frac{7}{18}\pi$ radians.

50. Find the radian measure of the central angle of a sector of a circle if the area of the sector
(a) is $\frac{5}{6}$ of the area of the circle;
(b) is $\frac{17}{18}$ of the area of the circle.

51. Suppose that the area of a sector of a circle is 200 square centimeters and that its central angle measures 2 radians. What is the radius of the circle?

52. Write the formula for arc length on a circle where the angle is measured in degrees.

53. Write the formula for the area of a sector of a circle (see Problem 48) where the angle is measured in degrees.

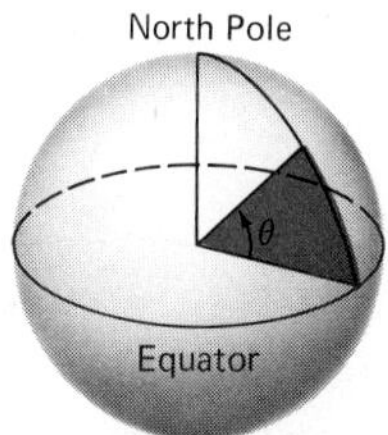

θ measures latitude north

C 54. Assume that the earth is a sphere of radius 3960 miles. One of the authors (Dale Varberg) lives at exactly 45° latitude north. How far does he live from the North Pole?

C 55. Washington, D.C. is at 38.7° latitude north. How far is it from the equator? (See Problem 54.)

C 56. A nautical mile is the length of 1 minute ($\frac{1}{60}$ of a degree) of arc on the equator. How many miles (correct to 2 decimal places) are there in a nautical mile? (See Problem 54.)

C 57. The sun is about 9.3×10^7 miles from the earth. If the angle subtended by the sun on the surface of the earth is 9.3×10^{-3} radians, approximately what is the diameter of the sun?

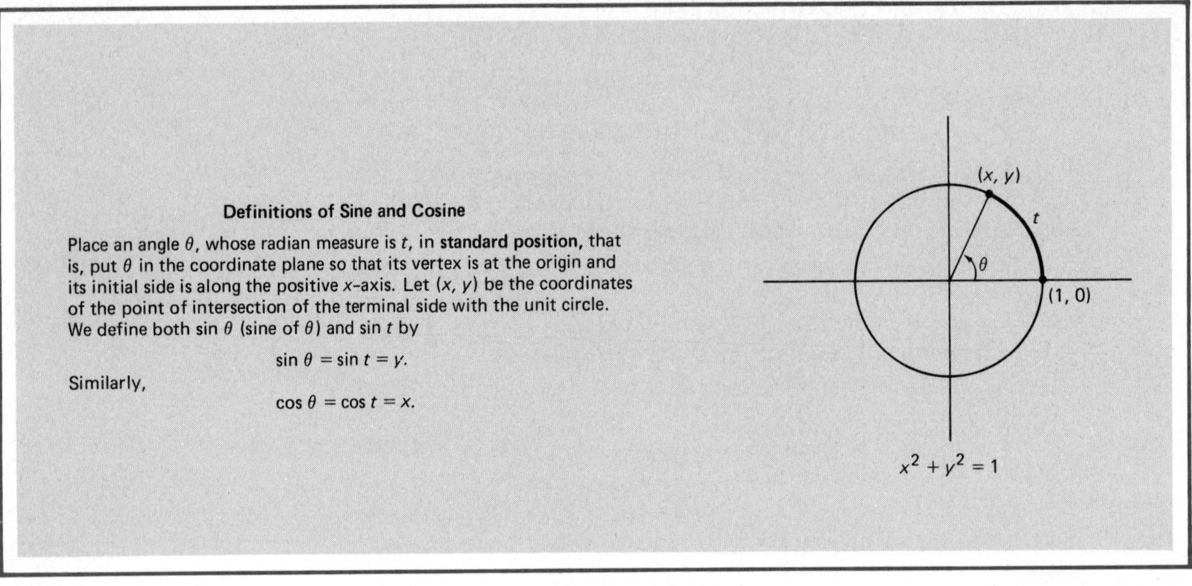

Definitions of Sine and Cosine

Place an angle θ, whose radian measure is t, in **standard position,** that is, put θ in the coordinate plane so that its vertex is at the origin and its initial side is along the positive x-axis. Let (x, y) be the coordinates of the point of intersection of the terminal side with the unit circle. We define both $\sin \theta$ (sine of θ) and $\sin t$ by

$$\sin \theta = \sin t = y.$$

Similarly,

$$\cos \theta = \cos t = x.$$

7-3 The Sine and Cosine Functions

In Section 7-1, we defined the sine and cosine for positive acute angles. The definitions in our opening display are more general and hence more widely applicable. They should be studied carefully. Notice that we have defined the sine and cosine for any angle θ and also for the corresponding number t. Both concepts are important. In geometric situations, angles play a central role; thus we are likely to need sines and cosines of angles. But in most of pure mathematics and in many scientific applications, it is the trigonometric functions of numbers that are important. In this connection, we emphasize that the number t may be positive or negative, large or small. And we may think of it as the radian measure of an angle, as the directed length of an arc on the unit circle, or simply as a number.

Consistency With Earlier Definitions Do the definitions given in Section 7-1 for the sine and cosine of an acute angle harmonize with those given here? Yes. Take a right triangle ABC with an acute angle θ. Place θ in standard position, thus determining a point $B'(x, y)$ on the unit circle and a point $C'(x, 0)$ directly below it on the x-axis (see the middle diagram at the top of page 273). Notice that triangles ABC and $AB'C'$ are similar. It follows that

$$\frac{\text{opp}}{\text{hyp}} = \frac{y}{1} = y$$

$$\frac{\text{adj}}{\text{hyp}} = \frac{x}{1} = x$$

On the left are the old definitions of $\sin \theta$ and $\cos \theta$; on the right are the new ones. They are consistent.

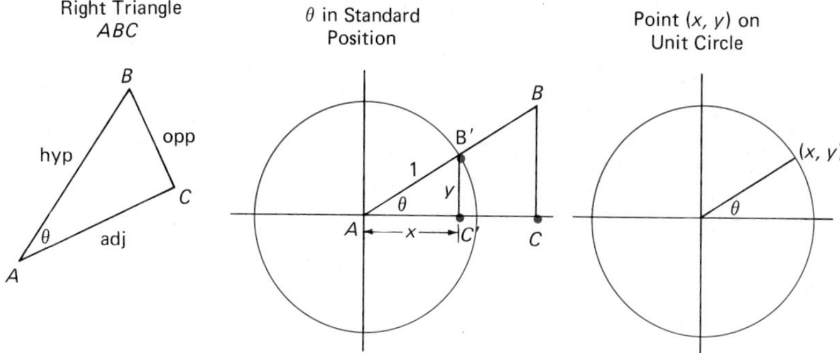

Special Angles In Section 7-1, we learned that

$$\cos 45^\circ = \frac{\sqrt{2}}{2} \qquad \sin 45^\circ = \frac{\sqrt{2}}{2}$$

$$\cos 30^\circ = \frac{\sqrt{3}}{2} \qquad \sin 30^\circ = \frac{1}{2}$$

$$\cos 60^\circ = \frac{1}{2} \qquad \sin 60^\circ = \frac{\sqrt{3}}{2}$$

Making use of the consistency of the old and new definitions of sine and cosine, we conclude that the point on the unit circle corresponding to $\theta = 45^\circ = \pi/4$ radians must have coordinates $(\sqrt{2}/2, \sqrt{2}/2)$. Similarly, the point corresponding to $\theta = 30^\circ = \pi/6$ radians has coordinates $(\sqrt{3}/2, 1/2)$ and the point corresponding to $\theta = 60^\circ = \pi/3$ radians has coordinates $(1/2, \sqrt{3}/2)$.

Now we can make use of obvious symmetries to find the coordinates of many other points on the unit circle. In the figure below, we show a number of these points, noting first the radian measure of the angle and then the coordinates of the corresponding point on the unit circle.

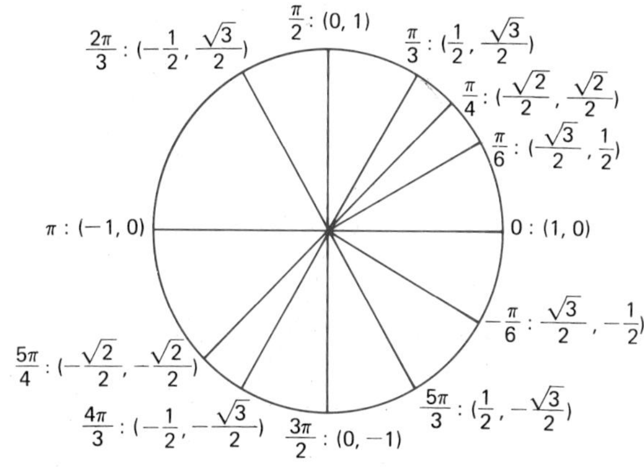

Notice, for example, how the coordinates of the points corresponding to $t = 2\pi/3$, $4\pi/3$, and $5\pi/3$ are related to the point corresponding to $t = \pi/3$. You should have no trouble seeing other relationships.

Once we know the coordinates of a point on the unit circle, we can state the sine and cosine of the corresponding angle. In particular, we get the values in the following table. They are used so often that you should memorize them.

t	0	$\frac{\pi}{6}$	$\frac{\pi}{4}$	$\frac{\pi}{3}$	$\frac{\pi}{2}$	π	$\frac{3\pi}{2}$
$\cos t$	1	$\frac{\sqrt{3}}{2}$	$\frac{\sqrt{2}}{2}$	$\frac{1}{2}$	0	-1	0
$\sin t$	0	$\frac{1}{2}$	$\frac{\sqrt{2}}{2}$	$\frac{\sqrt{3}}{2}$	1	0	-1

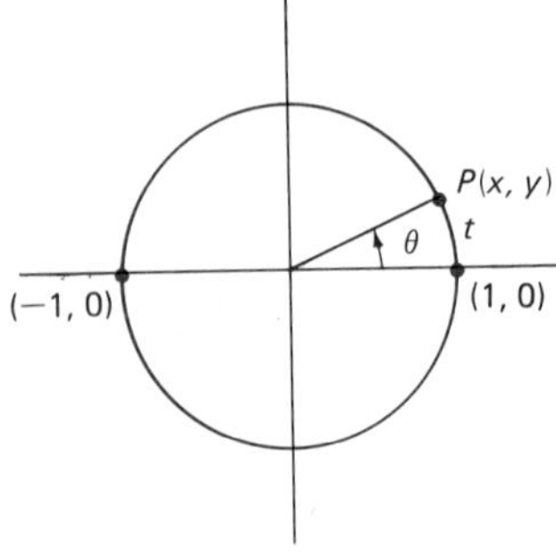

Properties of Sines and Cosines Think of what happens to x and y as t increases from 0 to 2π in the accompanying figure, that is, as P travels all the way around on the unit circle. For example, x steadily decreases until it reaches its smallest value of -1 at $t = \pi$; then it starts to increase until it is back to 1 at $t = 2\pi$. We have just described the behavior of $\cos t$ (or $\cos \theta$) as t increases from 0 to 2π. You should trace the behavior of $\sin t$ in the same way. Notice that both x and y are always between -1 and 1 (inclusive). It follows that

$$-1 \leq \sin t \leq 1$$
$$-1 \leq \cos t \leq 1$$

Since P is on the unit circle, $x^2 + y^2 = 1$, and $x = \cos t$ and $y = \sin t$, it follows that

$$(\sin t)^2 + (\cos t)^2 = 1$$

It is conventional to write $\sin^2 t$ instead of $(\sin t)^2$ and $\cos^2 t$ instead of $(\cos t)^2$. Thus we have

$$\sin^2 t + \cos^2 t = 1$$

This is an identity; it is true for all t. Of course we can just as well write

$$\sin^2 \theta + \cos^2 \theta = 1$$

We have established one basic relationship between the sine and the cosine; here are two others, valid for all t.

$$\sin\left(\frac{\pi}{2} - t\right) = \cos t$$
$$\cos\left(\frac{\pi}{2} - t\right) = \sin t$$

These relationships are easy to see when $0 < t < \pi/2$. Notice that t and $\pi/2 - t$ are measures of complementary angles (two angles with measures totaling 90° or $\pi/2$). That means that t and $\pi/2 - t$ determine points on the unit circle which are reflections of each other about the line $y = x$ (see the diagram in the margin). Thus if one point has coordinates (x, y), the other has coordinates (y, x). The result given above follows from this fact.

Finally, we point out that $t, t \pm 2\pi, t \pm 4\pi, \ldots$ all determine the same point on the unit circle and thus have the same sine and cosine. We express this fact by saying that $\sin t$ and $\cos t$ are **periodic functions** with **period** 2π and write

$$\sin(t + 2\pi) = \sin t$$
$$\cos(t + 2\pi) = \cos t$$

Problem Set 7-3

In this problem set, we will let $P(t)$ denote the point on the unit circle corresponding to the number t. Thus $P(0)$ has coordinates $(1, 0)$, $P(\pi/6)$ has coordinates $(\sqrt{3}/2, 1/2)$, and $P(-\pi/2)$ has coordinates $(0, -1)$.

Example A (Using the unit circle to find sine and cosine values) Find:
(a) $\sin(11\pi/6)$; (b) $\cos(8\pi/3)$; (c) $\cos(27\pi/4)$; (d) $\sin(-27\pi/2)$.
SOLUTION. (a) We locate $P(11\pi/6)$ on the unit circle and see that its y-coordinate is $-\frac{1}{2}$ because of its position relative to $P(\pi/6)$. Thus $\sin(11\pi/6) = -\frac{1}{2}$. (b) Since $8\pi/3 = 2\pi + (2\pi/3)$, the point $P(8\pi/3)$ is the same point as $P(2\pi/3)$, which has coordinates $(-\frac{1}{2}, \sqrt{3}/2)$ (see page 273). Therefore $\cos(8\pi/3) = -\frac{1}{2}$. (c) Since $27\pi/4 = 3(2\pi) + (3\pi/4)$, $P(27\pi/4)$ is the same as $P(3\pi/4)$. The x-coordinate of this point is $-\sqrt{2}/2$, the negative of the x-coordinate of $P(\pi/4)$. Thus $\cos(27\pi/4) = -\sqrt{2}/2$. Note that in both parts (b) and (c), our procedure is to ignore as large a multiple of 2π as possible. (d) $-27\pi/2 = -14\pi + (\pi/2) = -7(2\pi) + (\pi/2)$. Therefore $P(-27\pi/2) = P(\pi/2)$ and $\sin(-27\pi/2) = \sin(\pi/2) = 1$.

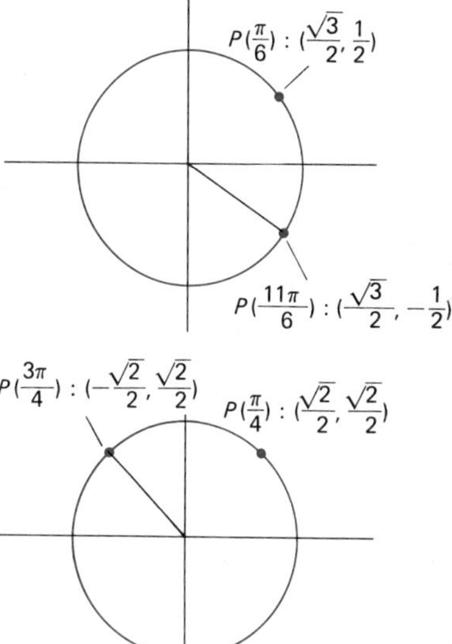

Use the unit circle to find the value of each of the following.

1. $\sin(3\pi/4)$
2. $\sin(5\pi/4)$
3. $\sin(7\pi/4)$
4. $\sin(9\pi/4)$
5. $\cos(3\pi/4)$
6. $\cos(5\pi/4)$
7. $\cos(7\pi/4)$
8. $\cos(9\pi/4)$
9. $\sin(5\pi/2)$

10. $\cos 7\pi$	11. $\sin(-4\pi)$	12. $\cos(7\pi/2)$
13. $\cos(19\pi/6)$	14. $\sin(14\pi/3)$	15. $\cos(-\pi/3)$
16. $\sin(-5\pi/6)$	17. $\cos(125\pi/4)$	18. $\cos(-13\pi/6)$
19. $\sin 510°$	20. $\sin(-390°)$	21. $\cos 840°$
22. $\cos(-720°)$	23. $\cos(-210°)$	24. $\sin 900°$

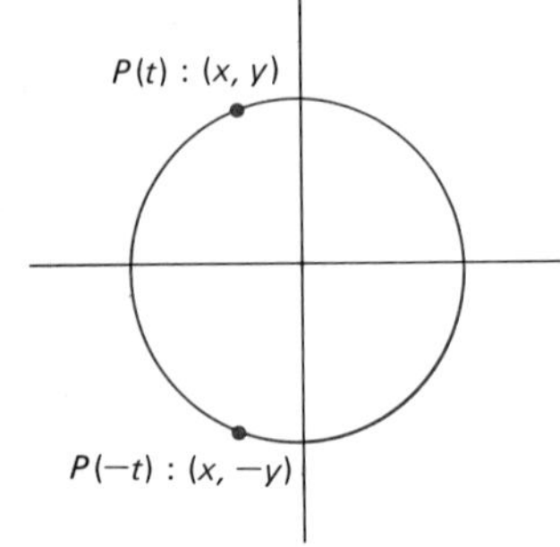

Example B (Sine and cosine of $-t$) Show that for all t

$$\sin(-t) = -\sin t$$
$$\cos(-t) = \cos t$$

that is, sine is an odd function and cosine is an even function.

SOLUTION. The points $P(-t)$ and $P(t)$ are symmetric with respect to the x-axis. Thus if $P(t)$ has coordinates (x, y), $P(-t)$ has coordinates $(x, -y)$ and so

$$\sin(-t) = -y = -\sin t$$
$$\cos(-t) = x = \cos t$$

25. If $\sin 1.87 = .95557$ and $\cos 1.87 = -0.29476$, find $\sin(-1.87)$ and $\cos(-1.87)$.
26. If $\sin 15.2° = 0.2622$ and $\cos 15.2° = 0.9650$, find $\sin(-15.2°)$ and $\cos(-15.2°)$.
27. Given $P(t)$ with coordinates $(1/\sqrt{5}, -2/\sqrt{5})$.
    (a) What are the coordinates of $P(-t)$?
    (b) What are the values of $\sin(-t)$ and $\cos(-t)$?
28. If t is the radian measure of an angle in quadrant III and $\sin t = -\frac{3}{5}$, evaluate each expression.
    (a) $\sin(-t)$
    (b) $\cos(t)$ (*Hint*: Use the fact that $\sin^2 t + \cos^2 t = 1$.)
    (c) $\cos(-t)$

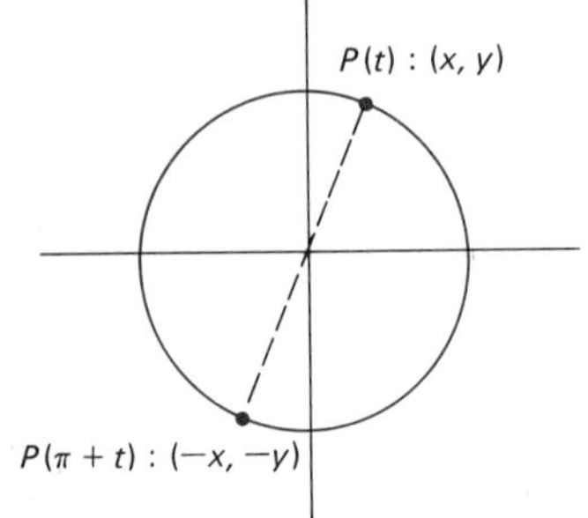

29. Note that $P(t)$ and $P(\pi + t)$ are symmetric with respect to the origin (see the diagram in the margin). Use this to show that
    (a) $\sin(\pi + t) = -\sin t$; (b) $\cos(\pi + t) = -\cos t$.
30. Note that $P(t)$ and $P(\pi - t)$ are symmetric with respect to the y-axis. Use this fact to find identities analogous to those in Problem 29 for $\sin(\pi - t)$ and $\cos(\pi - t)$.

Miscellaneous Problems

31. Find the values of each of the following.
    (a) $\sin(2\pi/3)$ (b) $\cos(3\pi/4)$ (c) $\cos(9\pi/4)$
    (d) $\sin(17\pi/6)$ (e) $\sin(-3\pi/4)$ (f) $\cos 480°$
    (g) $\sin(-675°)$ (h) $\cos 3660°$
32. Let θ denote an angle in standard position. In which quadrants is
    (a) $\sin\theta$ positive?
    (b) $\cos\theta$ negative?
    (c) $\sin\theta\cos\theta$ negative?

33. Suppose $P(t)$ has coordinates $(\frac{3}{5}, -\frac{4}{5})$. Evaluate each expression.
(a) $\cos t$ (b) $\sin t$
(c) $\cos(\pi/2 - t)$ (d) $\sin(\pi/2 - t)$
(e) $\cos(-t)$ (f) $\sin(-t)$

34. Let θ be an angle in quadrant II. Use the identity $\sin^2 \theta + \cos^2 \theta = 1$ to find each value.
(a) $\cos \theta$ if $\sin \theta = \frac{2}{3}$.
(b) $\sin \theta$ if $\cos \theta = -\frac{3}{4}$.

35. Use the unit circle to find three angles θ (measured in degrees) for which
(a) $\sin \theta = 0$; (b) $\cos \theta = 0$.

36. Find four values of t for which each statement is true.
(a) $\sin t = 1$ (b) $\cos t = \frac{1}{2}$
(c) $\sin t = -\sqrt{2}/2$ (d) $\sin t = \cos t$

37. For what values of t between 0 and π is $\sin t$ greater than $\frac{1}{2}$?

38. For what values of t between 0 and π is $\cos t$ greater than $\frac{1}{2}$?

39. Suppose $P(t)$ has coordinates $(-1/\sqrt{5}, 2/\sqrt{5})$. Find the coordinates of each point.
(a) $P(\pi/2 - t)$ (b) $P(\pi - t)$
(c) $P(\pi + t)$ (d) $P(2\pi - t)$

40. Use the unit circle to find identities for $\sin(2\pi - t)$ and $\cos(2\pi - t)$.

"Strange as it may sound, the power of mathematics rests on its evasion of all unnecessary thought and on its wonderful saving of mental operations."

Ernst Mach

New Functions from Old Ones

tangent:	$\tan t = \dfrac{\sin t}{\cos t}$
cotangent:	$\cot t = \dfrac{\cos t}{\sin t}$
secant:	$\sec t = \dfrac{1}{\cos t}$
cosecant:	$\csc t = \dfrac{1}{\sin t}$

7-4 Four More Trigonometric Functions

Without question, the sine and cosine are the most important of the six trigonometric functions. Not only do they occur most frequently in applications, but the other four functions can be defined in terms of them, as our opening box shows. This means that if you learn all you can about sines and cosines, you will automatically know a great deal about tangents, cotangents, secants, and cosecants. Ernst Mach would say that it is a way to evade unnecessary thought.

Look at the definitions in the opening box again. Naturally, we must rule out any values of t for which a denominator is zero. For example, $\tan t$ is not defined for $t = \pm\pi/2, \pm 3\pi/2, \pm 5\pi/2$, and so on. Similarly, $\csc t$ is not defined for such values as $t = 0, \pm\pi$, and $\pm 2\pi$.

Properties of the New Functions The wisdom of the opening paragraph will now be demonstrated. Recall the identity $\sin^2 t + \cos^2 t = 1$. Out of it come two new identities.

$$1 + \tan^2 t = \sec^2 t$$
$$1 + \cot^2 t = \csc^2 t$$

To show that the first identity is correct, we take its left side, express it in terms of sines and cosines, and do a little algebra.

$$
\begin{aligned}
1 + \tan^2 t &= 1 + \left(\frac{\sin t}{\cos t}\right)^2 \\
&= 1 + \frac{\sin^2 t}{\cos^2 t} \\
&= \frac{\cos^2 t + \sin^2 t}{\cos^2 t} \\
&= \frac{1}{\cos^2 t} \\
&= \left(\frac{1}{\cos t}\right)^2 \\
&= \sec^2 t
\end{aligned}
$$

The second identity is verified in a similar fashion.

Suppose we wanted to know whether cotangent is an even or an odd function (or neither). We simply recall that $\sin(-t) = -\sin t$ and $\cos(-t) = \cos t$ and write

$$\cot(-t) = \frac{\cos(-t)}{\sin(-t)} = \frac{\cos t}{-\sin t} = -\frac{\cos t}{\sin t} = -\cot t$$

Thus cotangent is an odd function.

In a similar vein, recall the identities

(i) $$\sin\left(\frac{\pi}{2} - t\right) = \cos t$$

(ii) $$\cos\left(\frac{\pi}{2} - t\right) = \sin t$$

From them, we obtain

(iii) $$\tan\left(\frac{\pi}{2} - t\right) = \frac{\sin(\pi/2 - t)}{\cos(\pi/2 - t)} = \frac{\cos t}{\sin t} = \cot t$$

These three identities are examples of what are called **cofunction identities**. Sine and cosine are cofunctions; so are tangent and cotangent. Notice that identities (i), (ii), and (iii) all have the form

$$\text{function}\left(\frac{\pi}{2} - t\right) = \text{cofunction}(t)$$

Alternative Definitions of the Trigonometric Functions There is another approach to trigonometry favored by some authors. Let θ be an angle

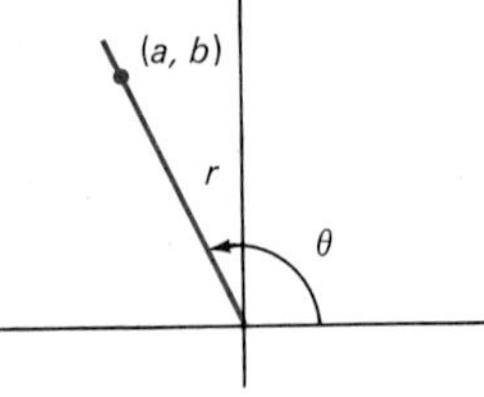

in standard position and suppose that (a, b) is any point on its terminal side at a distance r from the origin. Then

$$\sin\theta = \frac{b}{r} \qquad \cos\theta = \frac{a}{r}$$

$$\tan\theta = \frac{b}{a} \qquad \cot\theta = \frac{a}{b}$$

$$\sec\theta = \frac{r}{a} \qquad \csc\theta = \frac{r}{b}$$

To see that these definitions are equivalent to those we gave earlier, consider first an angle θ with terminal side in quadrant I (see the diagram below).

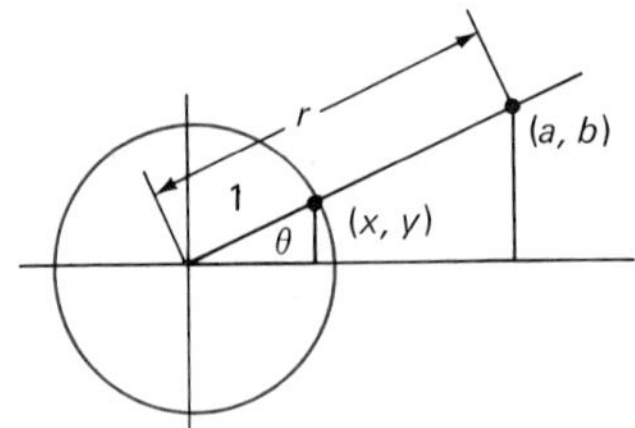

By similar triangles,

$$\frac{b}{r} = \frac{y}{1} \quad \text{and} \quad \frac{a}{r} = \frac{x}{1}$$

Actually these ratios are equal no matter in which quadrant the terminal side of θ is, since b and y always have the same sign, as do a and x. The first two formulas in the box now follow from our original definitions, which say that

$$\sin\theta = y \quad \text{and} \quad \cos\theta = x$$

The others are a consequence of the fact that the remaining four functions can be expressed in terms of sines and cosines.

Problem Set 7-4

1. If $\sin t = \frac{4}{5}$ and $\cos t = -\frac{3}{5}$, evaluate each function.
   (a) $\tan t$ (b) $\cot t$ (c) $\sec t$ (d) $\csc t$
2. If $\sin t = -1/\sqrt{5}$ and $\cos t = 2/\sqrt{5}$, evaluate each function.
   (a) $\tan t$ (b) $\cot t$ (c) $\sec t$ (d) $\csc t$

Keeping in mind what you know about the sines and cosines of special angles, find each of the values in Problems 3–20.

3. $\tan(\pi/6)$
4. $\cot(\pi/6)$
5. $\sec(\pi/6)$
6. $\csc(\pi/6)$
7. $\cot(\pi/4)$
8. $\sec(\pi/4)$
9. $\csc(\pi/3)$
10. $\sec(\pi/3)$
11. $\sin(4\pi/3)$
12. $\cos(4\pi/3)$
13. $\tan(4\pi/3)$
14. $\sec(4\pi/3)$
15. $\tan \pi$
16. $\sec \pi$
17. $\tan 330°$
18. $\cot 120°$
19. $\sec 600°$
20. $\csc(-150°)$
21. For what values of t on $0 \leq t \leq 4\pi$ is each of the following undefined?
    (a) $\sec t$ (b) $\tan t$ (c) $\csc t$ (d) $\cot t$
22. For which values of t on $0 \leq t \leq 4\pi$ is each of the following equal to 1?
    (a) $\sec t$ (b) $\tan t$ (c) $\csc t$ (d) $\cot t$

Example (Using the *a*, *b*, *r* definitions) Suppose that the point $(3, -6)$ is on the terminal side of an angle in standard position. Find $\sin \theta$, $\tan \theta$, and $\sec \theta$.

SOLUTION. First we find r.

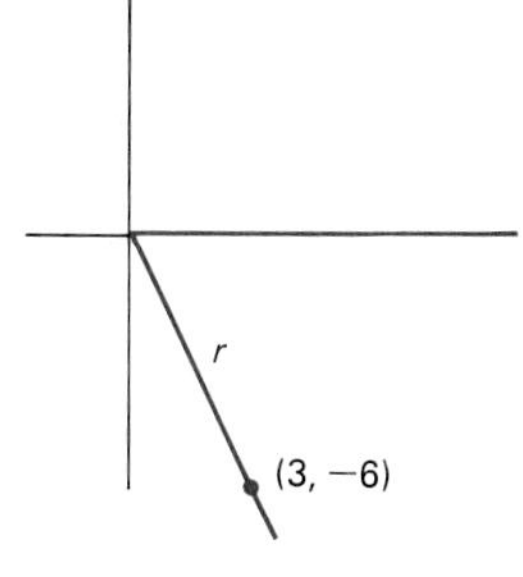

$$r = \sqrt{3^2 + (-6)^2} = \sqrt{45} = 3\sqrt{5}$$

Then

$$\sin \theta = \frac{b}{r} = \frac{-6}{3\sqrt{5}} = -\frac{2}{\sqrt{5}}$$

$$\tan \theta = \frac{b}{a} = \frac{-6}{3} = -2$$

$$\sec \theta = \frac{r}{a} = \frac{3\sqrt{5}}{3} = \sqrt{5}$$

In Problems 23–26, find $\sin \theta$, $\tan \theta$, and $\sec \theta$, assuming that the given point is on the terminal side of θ.

23. $(5, -12)$
24. $(7, 24)$
25. $(-1, -2)$
26. $(-3, 2)$
27. If $\tan \theta = \frac{3}{4}$ and θ is an angle in the first quadrant, find $\sin \theta$ and $\sec \theta$. (*Hint*: The point $(4, 3)$ is on the terminal side of θ.)
28. If $\tan \theta = \frac{3}{4}$ and θ is an angle in the third quadrant, find $\cos \theta$ and $\csc \theta$. (*Hint*: The point $(-4, -3)$ is on the terminal side of θ.)
29. If $\sin \theta = \frac{5}{13}$ and θ is an angle in the second quadrant, find $\cos \theta$ and $\cot \theta$. (*Hint*: A point with y-coordinate 5 and $r = 13$ is on the terminal side of θ. Thus the x-coordinate must be -12.)
30. If $\cos \theta = \frac{4}{5}$ and $\sin \theta < 0$, find $\tan \theta$.
31. Where does the line from the origin to $(5, -12)$ intersect the unit circle?
32. Where does the line from the origin to $(-6, 8)$ intersect the unit circle?

Miscellaneous Problems

33. Evaluate.
(a) $\sin(5\pi/4)$ (b) $\cos(5\pi/4)$ (c) $\tan(5\pi/4)$
(d) $\sec(5\pi/4)$ (e) $\tan(-4\pi/3)$ (f) $\sec 5\pi$
(g) $\cot(5\pi/2)$ (h) $\tan 180°$ (i) $\csc 315°$

34. If $(-1, -2)$ is on the terminal side of an angle θ in the third quadrant, find $\sec \theta$.

35. If $\sin \theta = \frac{2}{3}$ and θ is in quadrant II, find $\tan \theta$ and $\sec \theta$.

36. Verify the identity $1 + \cot^2 t = \csc^2 t$ by expressing the left side in terms of sine and cosine and manipulating the result.

37. Verify each identity.
(a) $\tan(-t) = -\tan t$ (b) $\sec(-t) = \sec t$
(c) $\csc(-t) = \csc t$

38. If $\tan 32° = .6249$, find $\cot 58°$. (*Hint*: Use the cofunction identity.)

39. If $\sin 34° = .5592$, find $\cos 56°$.

40. Show that $|\sec t| \geq 1$ and $|\csc t| \geq 1$ for all values of t for which these functions are defined.

41. In which quadrants is $\sec \theta$ negative? Is $\tan \theta$ positive?

42. Find two values of t for which
(a) $\tan t = -1$; (b) $\sec t = 2$;
(c) $\csc t = -1$; (d) $\cot t = 0$.

43. Show that there is an angle θ between 0° and 90° for which $\tan \theta = 1000$.

44. Is $\cot(.000001)$ small or large?

45. By using the identities of Problem 29 of Section 7.3, namely,

$$\sin(\pi + t) = -\sin t \qquad \cos(\pi + t) = -\cos t$$

show that

$$\tan(t + \pi) = \tan t$$

This means that the tangent is periodic with period π.

Ⓒ 46. Find $\sec \theta$ if θ is in standard position and has $(\sqrt{2} + \ln 3, \pi + \pi^{2/3})$ on its terminal side.

Ⓒ 47. Let $f(t) = (t + \tan t)/\sin t$, which, you will note, is not defined at $t = 0$.
(a) Calculate $f(1)$, $f(.1)$, $f(.01)$, $f(.001)$, and $f(.0001)$.
(b) What number does $f(t)$ appear to be approaching as t approaches 0?

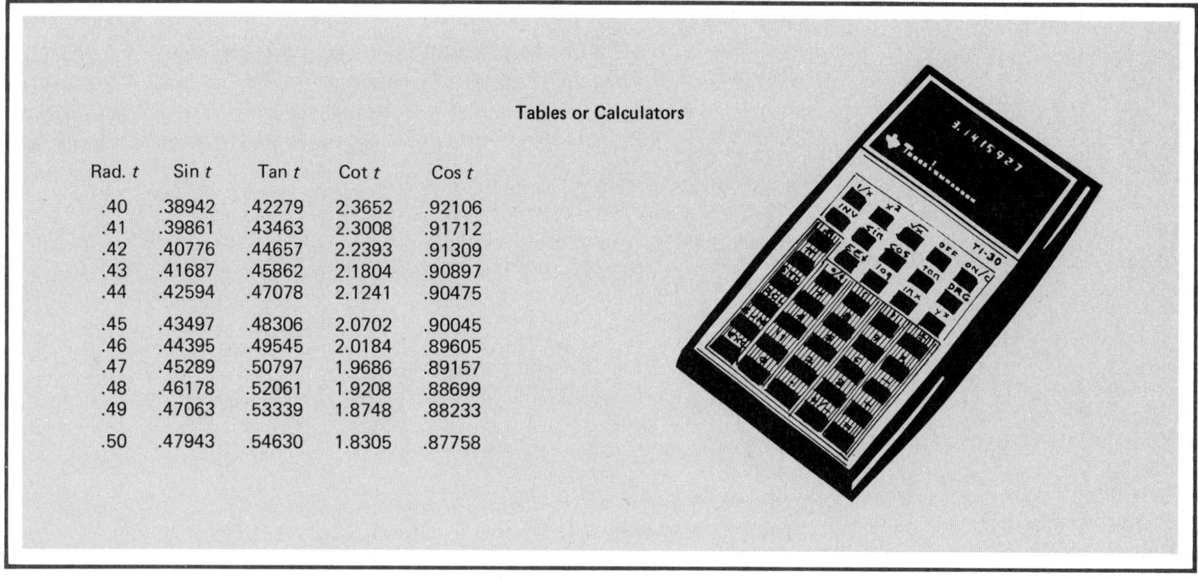

Tables or Calculators

Rad. t	Sin t	Tan t	Cot t	Cos t
.40	.38942	.42279	2.3652	.92106
.41	.39861	.43463	2.3008	.91712
.42	.40776	.44657	2.2393	.91309
.43	.41687	.45862	2.1804	.90897
.44	.42594	.47078	2.1241	.90475
.45	.43497	.48306	2.0702	.90045
.46	.44395	.49545	2.0184	.89605
.47	.45289	.50797	1.9686	.89157
.48	.46178	.52061	1.9208	.88699
.49	.47063	.53339	1.8748	.88233
.50	.47943	.54630	1.8305	.87758

7-5 Finding Values of the Trigonometric Functions

In order to make significant use of the trigonometric functions, we will have to be able to calculate their values for angles other than the special angles we have considered. The simplest procedure is to press the right key on a calculator and read the answer. About the only thing to remember is to make sure the calculator is in the right mode, degree or radian, depending on what we want.

Even though calculators are becoming standard equipment for most mathematics and science students, we think you should also know how to use tables. That is the subject we take up now. We might call it "what to do when your battery goes dead."

The opening display gives a small portion of a five-place table of values for sin t, tan t, cot t, and cos t. (The complete table appears as Table D at the back of the book.) From it we read the following:

$$\sin .44 = .42594 \qquad \tan .44 = .47078$$
$$\cot .44 = 2.1241 \qquad \cos .44 = .90475$$

These results are not exact; they have been rounded off to five significant digits. Keep in mind that you can think of sin .44 in two ways, as the sine of the number .44 or, if you like, as the sine of an angle of radian measure .44.

Table D appears to have two defects. First, t is given only to 2 decimal places. If we need sin .44736, we have to round or perhaps to interpolate (see page 510).

$$\sin .44736 \approx \sin .45 = .43497$$

A more serious defect appears to be the fact that values of t go only to 2.00. This limitation evaporates once we learn about reference angles and reference numbers, our next topic.

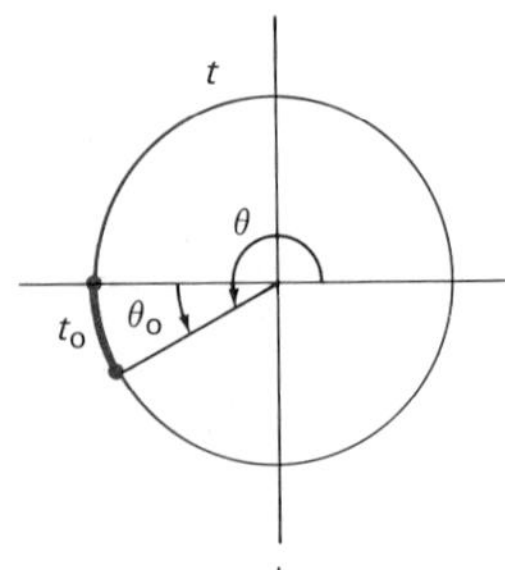

Reference Angles and Reference Numbers Let θ be any angle in standard position and let t be its radian measure. Associated with θ is an acute angle θ_0, called the **reference angle** and defined to be the smallest positive angle between the terminal side of θ and the x-axis (see the diagram in the margin). The radian measure t_0 of θ_0 is called the **reference number** corresponding to t. For example, the reference number for $t = 5\pi/6$ is $t_0 = \pi/6$. Once we know t_0, we can find $\sin t$, $\cos t$, and so on, no matter what t is. Here is how we do it.

Examine the four pictures below.

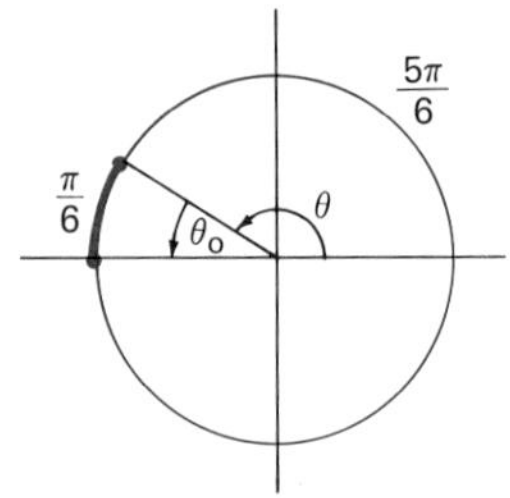

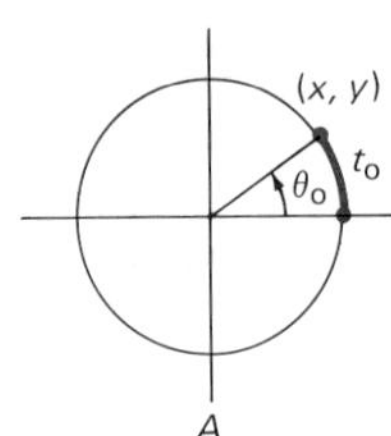

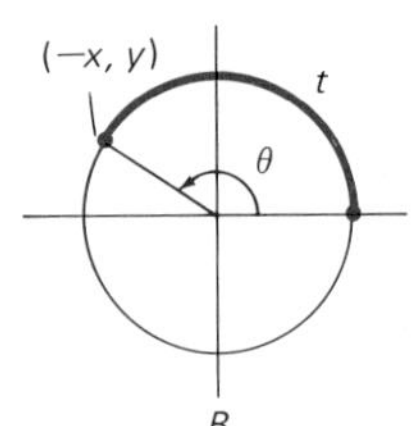

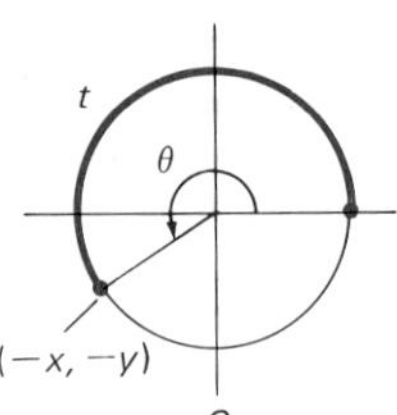

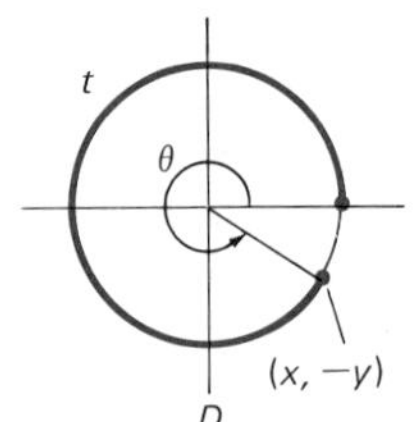

Each angle θ in B, C, and D has θ_0 as its reference angle and, of course, each t has t_0 as its reference number. Now we make a crucial observation. In each case, the point on the unit circle corresponding to t has the same coordinates, except for sign, as the point corresponding to t_0. It follows from this that

$$\sin t = \pm \sin t_0 \qquad \cos t = \pm \cos t_0$$

with the + or − sign being determined by the quadrant in which the terminal side of the angle falls. For example,

$$\sin \frac{5\pi}{6} = \sin \frac{\pi}{6} \qquad \cos \frac{5\pi}{6} = -\cos \frac{\pi}{6}$$

We chose the plus sign for the sine and the minus sign for the cosine because in the second quadrant the sine function is positive, while the cosine function is negative.

What we have just said applies to all six trigonometric functions. If T stands for any one of them, then

$$T(t) = \pm T(t_0) \quad \text{and} \quad T(\theta) = \pm T(\theta_0)$$

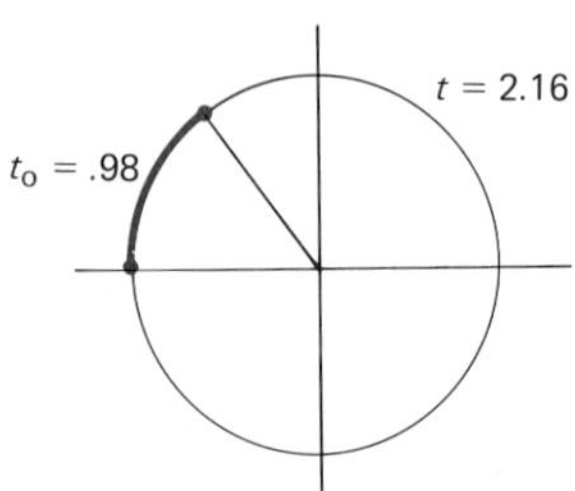

with the plus or minus sign being determined by the quadrant in which the terminal side of θ lies. Of course $T(t_0)$ itself is always nonnegative since $0 \leq t_0 \leq \pi/2$.

Examples If we wish to calculate cos 2.16 using tables, we must first find the reference number for 2.16. Approximating π by 3.14, we find that

$$t_0 = 3.14 - 2.16 = .98$$

and thus, using Table D,

$$\cos 2.16 = -\cos .98 = -.55702$$

Notice that we chose the minus sign because the cosine is negative in quadrant II.

To calculate tan 24.95 is slightly more work. First we remove as large a multiple of 2π as possible from 24.95. Using 6.28 for 2π, we get

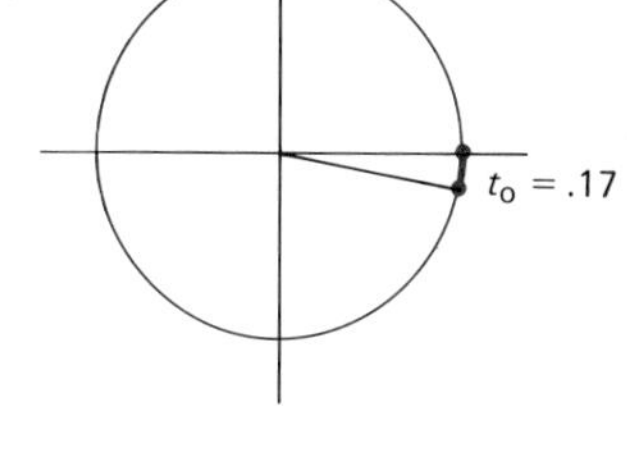

$$24.95 = 3(6.28) + 6.11$$

The reference number for 6.11 is

$$t_0 = 6.28 - 6.11 = .17$$

Thus

$$\tan 24.95 = \tan 6.11 = -\tan .17 = -.17166$$

We choose the minus sign because the tangent is negative in quadrant IV.

Now use your pocket calculator to find tan 24.95 the easy way. Be sure you put it in radian mode. You will get $-.18480$ instead of $-.17166$, a rather large discrepancy. Whom should you believe? We suggest that you trust your calculator. The reason we were so far off is that 6.28 is a rather poor approximation for 2π and multiplying it by 3 made matters worse. Had we used 6.2832 for 2π, we would have obtained $t_0 = .1828$ and $\tan 24.95 = -.18486$.

Problem Set 7-5

Find the reference number t_0 if t has the given value. Use 3.14 for π.

1. 1.84	2. 2.14	3. 3.54
4. 3.74	5. 5.18	6. 6.08
7. 10.48	8. 8.38	9. -1.12
10. -1.86	11. -2.64	12. -4.24

Find the reference number for each of the following. You may leave your answer in terms of π.

13. $13\pi/8$	14. $37\pi/36$	15. $40\pi/3$
16. $-11\pi/5$	17. $3\pi + .24$	18. $3\pi/2 + .17$
19. $3\pi - .24$	20. $3\pi/2 - .17$	21. $11\pi/2$
22. 26π		

Find the value of each of the following using Table D and $\pi = 3.14$. Calculators will give slightly different results because of this crude approximation to π.

23. cos 1.42	24. sin .97	25. tan 1.39
26. cot .08	27. sin 2.14	28. cos 3.08
29. cot 5.62	30. tan 4.11	31. $\cos(-2.54)$
32. $\sin(-4.18)$		

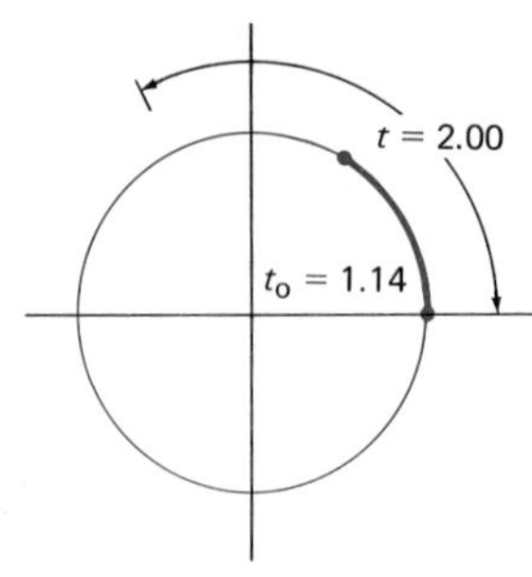

Example A (Finding t when $\sin t$ or $\cos t$ is given) Find 2 values of t between 0 and 2π for which (a) $\sin t = .90863$; (b) $\cos t = -.95824$.

SOLUTION. (a) We get $t = 1.14$ directly from Table D. Since the sine is also positive in quadrant II, we seek a value of t between $\pi/2$ and π for which 1.14 is the reference number. Only one number fits the bill:

$$\pi - 1.14 = 3.14 - 1.14 = 2.00$$

(b) We know that $\cos t_0 = .95824$ and so $t_0 = .29$. Now the cosine is negative in quadrants II and III. Thus we are looking for two numbers between $\pi/2$ and $3\pi/2$ with .29 as reference number. One is $\pi - .29 \approx 3.14 - .29 = 2.85$, and the other is $\pi + .29 \approx 3.14 + .29 = 3.43$.

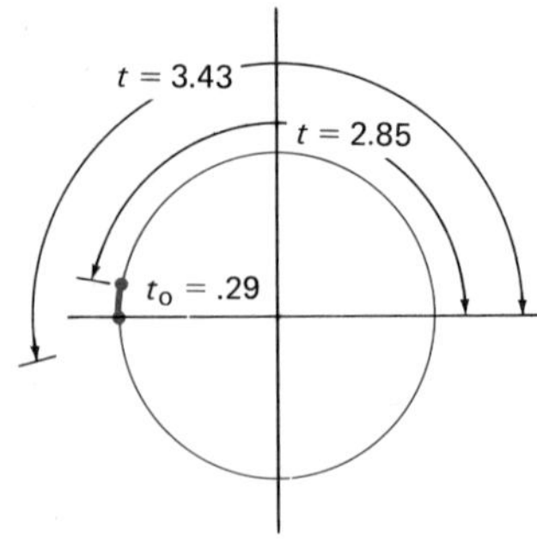

Find 2 values of t between 0 and 2π for which the given equality holds.

33. $\sin t = .94898$
34. $\cos t = .72484$
35. $\cos t = -.08071$
36. $\sin t = -.48818$
37. $\tan t = 4.9131$
38. $\cot t = 1.4007$
39. $\tan t = -3.6021$
40. $\cot t = -.47175$

Example B (Finding $\sin\theta$, $\cos\theta$, and so on, when θ is any angle given in degrees) Find the value of each of the following.

(a) $\cos 214.6°$ (b) $\cot 658°$

SOLUTION. In Section 7-1, we used Table C to find the sine, cosine, and tangent of positive angles measuring less than 90°. Here we do this for angles of arbitrary (degree) measure, and also include the cotangent in our work.

(a) The reference angle is

$$214.6° - 180° = 34.6°$$
$$\cos 214.6° = -\cos 34.6° = -.8231$$

We used the minus sign since cosine is negative in quadrant III.

(b) First we reduce our angle by 360°

$$658° = 360° + 298°$$

The reference angle for 298° is 360° − 298°, or 62°. In the column with cot at the bottom and 62° at the right, we find .5317. Thus $\cot 658° = -.5317$.

Find the value of each of the following.

41. $\sin 156.1°$
42. $\cos 138.7°$
43. $\tan 348.9°$
44. $\cot 224.9°$
45. $\cos(-66.1°)$
46. $\sin 487°$
47. $\cos 441.3°$
48. $\sin 180.2°$
49. $\cot(-134°)$
50. $\tan 311.6°$

Find two different degree values of θ between 0° and 360° for which the given equality holds.

51. $\sin\theta = .3633$
52. $\cos\theta = .9907$
53. $\tan\theta = .4942$
54. $\cot\theta = 1.2799$
55. $\cos\theta = -.9085$
56. $\sin\theta = -.2045$

Miscellaneous Problems

In Problems 57–62, use Table C or Table D to find each value.

57. cos 1.42 58. sin 0.19 59. cos 8.42

60. sin 6.35 61. tan 329° 62. sin 446.9°

63. Use Table D to find the value of t between 0 and $\pi/2$ for which each statement is true.
(a) $\tan t = 2.5722$ (b) $\sin t = .56464$
(c) $\cos t = .29628$ (d) $\cos t = .93233$

[C] 64. Check your answers to Problem 63 using your pocket calculator. For example, to check part (a) on a typical calculator put it in radian mode and press 2.5722 [INV][tan].

65. In Table D, you see that tan 1.57 = 1255.8 and tan 1.58 = −108.65. Explain this tremendous jump.

66. If Table D had given values of t up to 1.57 instead of 2, nothing would have been lost. Explain.

[C] 67. Calculate sin(cos 2.3), ln(sin 1.21), and $\sin^3(2.12)$.

[C] 68. Calculate cos(tan 1.01), cos(cos(cos 1.32)), and $\tan^5(2.31)$.

[C] 69. In calculus, it is shown that for small values of t

$$\sin t \approx t - \frac{t^3}{6} + \frac{t^5}{120} = \left(\left(\frac{t^2}{120} - \frac{1}{6}\right)t^2 + 1\right)t$$

In fact, it is approximations like this that are used to construct Table D. Use this fifth degree polynomial to approximate each value.
(a) sin .1 (b) sin .4 (c) sin 1
Compare your answers with corresponding entries in Table D.

[C] 70. For small values of t,

$$\cos t \approx 1 - \frac{t^2}{2} + \frac{t^4}{24} = \left(\frac{t^2}{24} - \frac{1}{2}\right)t^2 + 1$$

Use this fourth degree polynomial to approximate each value.
(a) cos .1 (b) cos .4 (c) cos 1

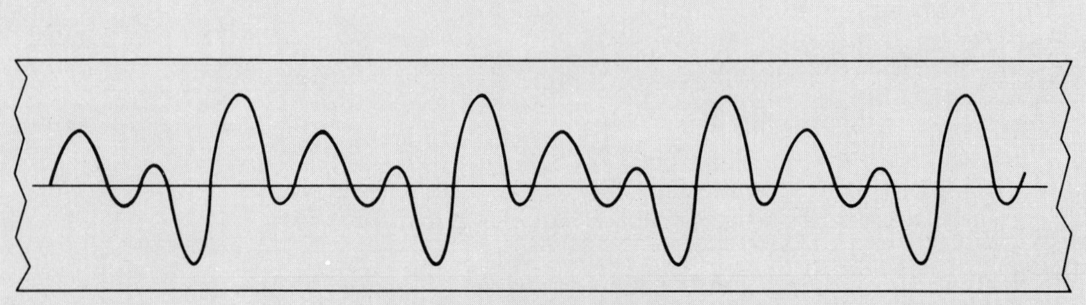

When heart beats, brain activity, or sound waves from a musical instrument are changed into visual images by means of an oscilloscope, they give a regular repetitive pattern which may look something like the diagram above. This repetitive behavior is a characteristic feature of the graphs of the trigonometric functions. In fact, almost any repetitive pattern can be approximated by appropriate combinations of the trigonometric functions.

7-6 Graphs of the Trigonometric Functions

We first discussed graphing of functions in Section 5-2; you might well review that section now. Recall that to graph $y = f(x)$, we first constructed a table of values of ordered pairs (x, y), then plotted the corresponding points, and finally connected those points with a smooth curve. Here we want to graph $y = \sin t$, $y = \cos t$, and so on, and we will follow a similar procedure. Notice that we use t rather than x as the independent variable because we used t as the variable (radian measure of an angle) in our definition of the trigonometric functions.

We begin with the graphs of the sine and cosine functions. You should become so well acquainted with these two graphs that you can sketch them quickly whenever you need them. This will aid you in two ways. First, these graphs will help you remember many of the important properties of the sine and cosine functions. Second, knowing them will help you graph other more complicated trigonometric functions.

The Graph of $y = \sin t$ We begin with a table of values.

t	0	$\frac{\pi}{6}$	$\frac{\pi}{4}$	$\frac{\pi}{3}$	$\frac{\pi}{2}$	$\frac{3\pi}{4}$	π	$\frac{5\pi}{4}$	$\frac{3\pi}{2}$	$\frac{7\pi}{4}$	2π
$y = \sin t$	0	$\frac{1}{2}$	$\frac{\sqrt{2}}{2}$	$\frac{\sqrt{3}}{2}$	1	$\frac{\sqrt{2}}{2}$	0	$-\frac{\sqrt{2}}{2}$	-1	$-\frac{\sqrt{2}}{2}$	0

We have listed values of t between 0 and 2π. That is sufficient to graph one period (shown at the top of page 289 as a heavy curve). From there on we can continue the curve indefinitely in either direction in a repetitive fashion, for we learned earlier that $\sin(t + 2\pi) = \sin t$.

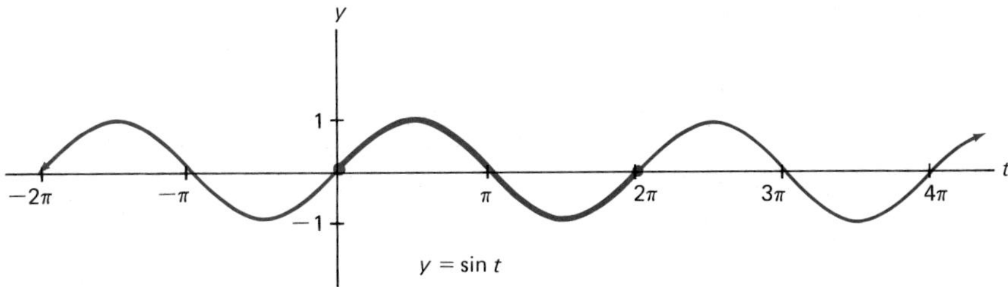

The Graph of $y = \cos t$ The cosine function is a copycat; its graph is just like that of the sine function but pushed $\pi/2$ units to the left.

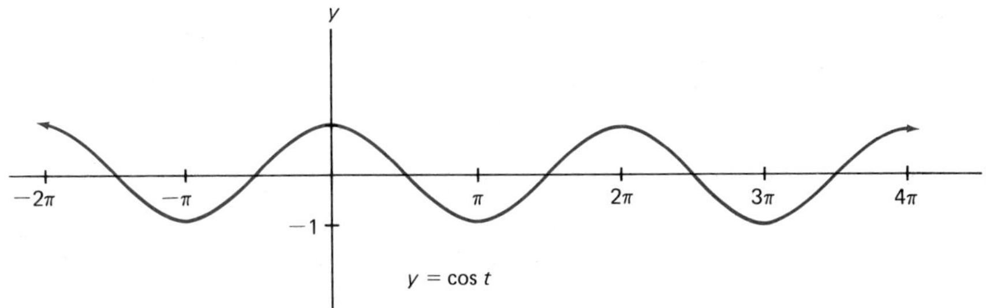

To see that the graph of the cosine function is correct, we might make a table of values and proceed as we did for the sine function. In fact, we ask you to do just that in Problem 1. Alternatively, we can show that

$$\cos t = \sin\left(t + \frac{\pi}{2}\right)$$

This follows directly from identities we have observed earlier.

$$\begin{aligned}\sin\left(t + \frac{\pi}{2}\right) &= \sin\left(\frac{\pi}{2} - (-t)\right) \\ &= \cos(-t) \qquad \text{(cofunction identity)} \\ &= \cos t \qquad \text{(cosine is even)}\end{aligned}$$

Properties Easily Observed from These Graphs

1. Both sine and cosine are periodic with periods of length 2π.
2. $-1 \le \sin t \le 1$ and $-1 \le \cos t \le 1$.
3. $\sin t = 0$ if $t = -\pi, 0, \pi, 2\pi$, and so on.
   $\cos t = 0$ if $t = -\pi/2, \pi/2, 3\pi/2$, and so on.
4. $\sin t > 0$ in quadrants I and II.
   $\cos t > 0$ in quadrants I and IV.
5. $\sin(-t) = -\sin t$ and $\cos(-t) = \cos t$.
   The sine is an odd function; its graph is symmetric with respect to the origin. The cosine is an even function; its graph is symmetric with respect to the y-axis.

6. We can see immediately where the sine and cosine functions are increasing and where they are decreasing. For example, the sine function decreases for $\pi/2 \le t \le 3\pi/2$.

The Graph of $y = \tan t$ Since the tangent function is defined by

$$\tan t = \frac{\sin t}{\cos t}$$

we need to beware of values of t for which $\cos t = 0$: $-\pi/2, \pi/2, 3\pi/2$, and so forth. In fact, from Section 5-3, we know that we should expect vertical asymptotes at these places. Notice also that

$$\tan(-t) = \frac{\sin(-t)}{\cos(-t)} = \frac{-\sin t}{\cos t} = -\tan t$$

t	0	$\frac{\pi}{4}$	$\frac{\pi}{3}$	$\frac{\pi}{2}$	$\frac{2\pi}{3}$	$\frac{3\pi}{4}$	π	$\frac{5\pi}{4}$	$\frac{3\pi}{2}$	$\frac{7\pi}{4}$	2π
$y = \tan t$	0	1	$\sqrt{3}$	undefined	$-\sqrt{3}$	−1	0	1	undefined	−1	0

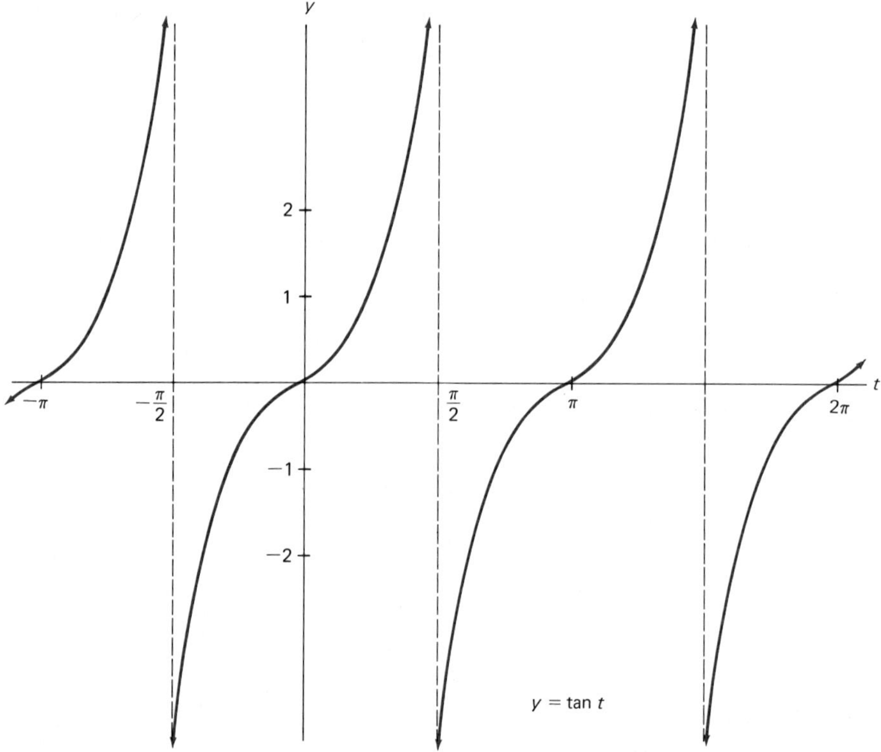

which means that the graph of the tangent will be symmetric with respect to the origin. Using these two pieces of information, a small table of values, and the fact that the tangent is periodic, we obtain the graph on page 290.

To confirm that the graph is correct near $t = \pi/2$, we suggest looking at Table D. Notice that $\tan t$ steadily increases until at $t = 1.57$, we read $\tan t = 1255.8$. But as t takes the short step to 1.58, $\tan t$ takes a tremendous plunge to -108.65. In that short space, t has passed through $\pi/2 \approx 1.5708$ and $\tan t$ has shot up to celestial heights only to fall to a bottomless pit, from which, however, it manages to escape as t moves to the right.

While we knew the tangent would have to repeat itself every 2π units since the sine and cosine do this, we now notice that it actually repeats itself on intervals of length π. We use the word **period** to denote the length of the shortest interval after which a function repeats itself. Thus the tangent function has period π. For an algebraic demonstration, see Problem 45 of Section 7-4.

Be sure to study Example A below. It introduces the graph of $y = A \sin Bt$. This important topic is explored more fully in Section 9.3 in connection with simple harmonic motion.

Problem Set 7-6

1. Make a table of values and then sketch the graph of $y = \cos t$.
2. What real numbers constitute the domain of the cosine? The range?
3. Sketch the graph of $y = \cot t$ for $-2\pi \le t \le 2\pi$, being sure to show the asymptotes.
4. What real numbers constitute the entire domain of the cotangent? The range?
5. Sketch the graph of $y = \sec t$. One way to do this is to make a table of values. Another way is to first graph the cosine and then find reciprocals of ordinates (recall $\sec t = 1/(\cos t)$). Make sure that you show the asymptotes.
6. Sketch the graph of $y = \csc t$.
7. What is the domain of the secant? The range?
8. What is the domain of the cosecant? The range?
9. What is the period of the cotangent? The secant?
10. On the interval $-2\pi \le t \le 2\pi$, where is the cotangent increasing?
11. Which is true: $\cot(-t) = \cot t$ or $\cot(-t) = -\cot t$?
12. Which is true: $\sec(-t) = \sec t$ or $\sec(-t) = -\sec t$?

Example A (Some sine-related graphs) Sketch the graph of each of the following for $-2\pi \le t \le 4\pi$.

(a) $y = 2 \sin t$ (b) $y = \sin 2t$ (c) $y = 3 \sin 4t$

SOLUTION. (a) We could graph $y = 2 \sin t$ from a table of values. It is easier, though to graph $\sin t$ (dotted graph below) and then multiply the ordinates by 2.

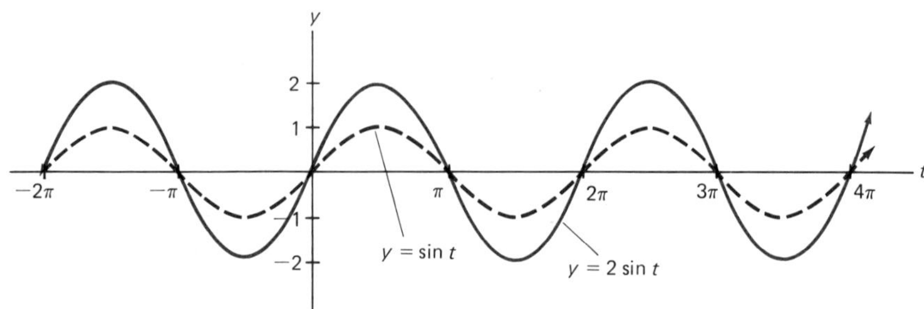

The graph bobs up and down between $y = -2$ and $y = 2$. The period is 2π, the same as for $\sin t$.

(b) Here a table of values is advisable, since this is our first example of this type.

t	$-\pi$	$-\frac{3\pi}{4}$	$-\frac{\pi}{2}$	$-\frac{\pi}{4}$	$-\frac{\pi}{12}$	0	$\frac{\pi}{12}$	$\frac{\pi}{4}$	$\frac{\pi}{2}$	$\frac{3\pi}{4}$	π
$2t$	-2π	$-\frac{3\pi}{2}$	$-\pi$	$-\frac{\pi}{2}$	$-\frac{\pi}{6}$	0	$\frac{\pi}{6}$	$\frac{\pi}{2}$	π	$\frac{3\pi}{2}$	2π
$\sin 2t$	0	1	0	-1	$-\frac{1}{2}$	0	$\frac{1}{2}$	1	0	-1	0

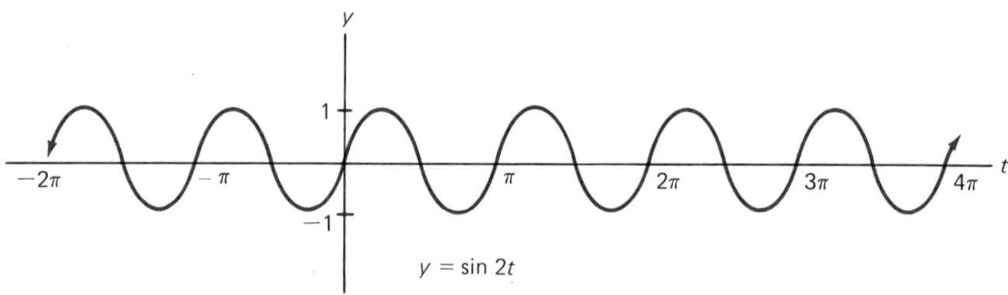

This graph goes through a complete cycle as t increases from 0 to π, that is, the period of $\sin 2t$ is π instead of 2π as it was for $\sin t$. In general, the period of $\sin Bt$ (and of $\cos Bt$) is $2\pi/B$.

(c) We can save a lot of work once we recognize that the character of the graph of $A \sin Bt$ (and $A \cos Bt$) is determined by the numbers A and B. The number A determines how high the graph rises and how far it falls; the number B determines its period. Thus, for a quick sketch we may simply determine the high and low points and connect them with a smooth wavelike graph. For $3 \sin 4t$ the period is $2\pi/4 = \pi/2$ and the maximum height is 3 (see page 293).

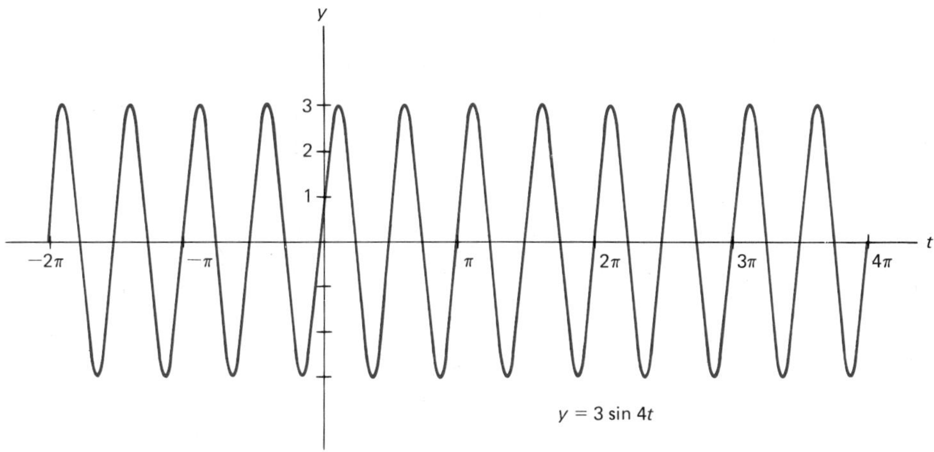

In Problems 13–22, sketch the graph on the indicated interval.

13. $y = 3\cos t,\ -\pi \le t \le \pi$
14. $y = \frac{1}{2}\cos t,\ -\pi \le t \le \pi$
15. $y = -\sin t,\ -\pi \le t \le \pi$
16. $y = -2\cos t,\ -\pi \le t \le \pi$
17. $y = \cos 4t,\ -\pi \le t \le \pi$
18. $y = \cos 3t,\ -\pi/2 \le t \le \pi/2$
19. $y = 2\sin \frac{1}{2}t,\ -2\pi \le t \le 2\pi$
20. $y = 3\sin \frac{1}{3}t,\ -3\pi \le t \le 3\pi$
21. $y = 2\cos 3t,\ -\pi \le t \le \pi$
22. $y = 4\sin 3t,\ -\pi \le t \le \pi$

Example B (Graphing sums of trigonometric functions) Sketch the graph of the equation $y = 2\sin t + \cos 2t$.

SOLUTION. We graph $y = 2\sin t$ and $y = \cos 2t$ on the same coordinate plane (these appear as dotted-line graphs below) and then add ordinates.

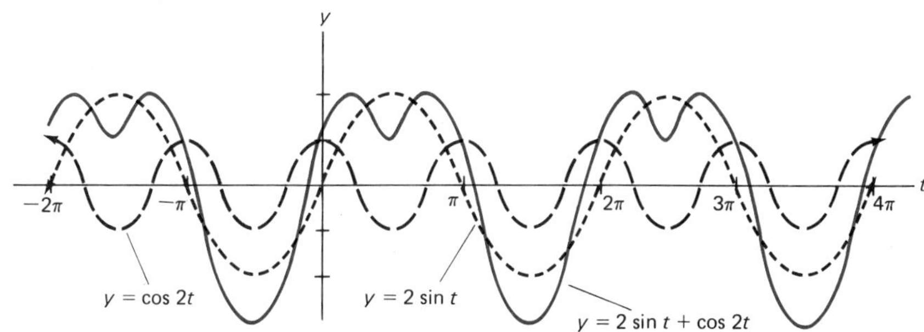

Notice that for any t, the ordinates (y-values) of the dotted curves are added to obtain the desired ordinate. The graph of $y = 2\sin t + \cos 2t$ is quite different from the separate (dotted) graphs but it does repeat itself; it has period 2π.

Sketch each graph by the method of adding ordinates. Show at least one complete period.

23. $y = 2 \sin t + \cos t$
24. $y = \sin t + 2 \cos t$
25. $y = \sin 2t + \cos t$
26. $y = \sin t + \cos 2t$
27. $y = \sin \frac{1}{2}t + \frac{1}{2} \sin t$
28. $y = \cos \frac{1}{2}t + \frac{1}{2} \cos t$

Miscellaneous Problems

In Problems 29–32, sketch each graph on the indicated interval.

29. $y = -\cos t,\ -\pi \le t \le \pi$
30. $y = 3 \sin t,\ -\pi \le t \le \pi$
31. $y = \sin 4t,\ 0 \le t \le \pi$
32. $y = 3 \cos \frac{1}{2}t,\ -2\pi \le t \le 2\pi$
33. What are the periods of the functions in Problems 29 and 31?
34. What are the periods of the functions in Problems 30 and 32?
35. Sketch the graph of $y = \cos 3t + 2 \sin t$ for $-\pi \le t \le \pi$. Use the method of adding ordinates.
36. Sketch the graph of $y = t + \sin t$ on $-4\pi \le t \le 4\pi$. Use the method of adding ordinates.
37. [C] Sketch the graph of $y = t - \cos t$ for $0 \le t \le 6$ by actually calculating the y values corresponding to $t = 0, .5, 1, 1.5, 2, 2.5, \ldots, 6$.
38. By sketching the graphs of $y = t$ and $y = 3 \sin t$ on the same coordinate axes, determine approximately all solutions of $t = 3 \sin t$.
39. The strength I of current (in amperes) in a wire of an alternating current circuit might satisfy

$$I = 30 \sin(120\pi t)$$

where time t is measured in seconds.
(a) What is the period?
(b) How many cycles (periods) are there in one second?
(c) What is the maximum strength of the current?
40. [C] Sketch the graph of $y = (\sin x)/x$ on $-3\pi \le x \le 3\pi$. Be sure to plot several points near $x = 0$.
41. Consider $y = \sin(1/x)$ on the interval $0 < x \le 1$.
(a) Where does its graph cross the x-axis?
(b) Evaluate y when $x = 2/3\pi, 2/5\pi, 2/7\pi, 2/9\pi, \ldots$.
(c) Sketch the graph (as best you can).

CHAPTER SUMMARY

The word **trigonometry** means triangle measurement. In its elementary historical form, it is the study of how to **solve** right triangles when appropriate information is given. The main tools are the three trigonometric ratios $\sin \theta$, $\cos \theta$, and $\tan \theta$, which were first defined only for acute angles θ.

In order to give the subject its modern general form, we first generalized the notion of an angle θ, allowing θ to have arbitrary size and measuring it either in **degrees** or **radians**. Such an angle θ can be placed in **standard position** in a coordinate system, where it will cut off an arc of directed length t (the radian measure of θ) stretching from $(1, 0)$ to (x, y) on the unit circle. This allowed us to make the key definitions

$$\sin \theta = \sin t = y \qquad \cos \theta = \cos t = x$$

on which all of modern trigonometry rests.

From the above definitions, we derived several identities, of which the most important is

$$\sin^2 t + \cos^2 t = 1$$

We also defined four additional functions

$$\tan t = \frac{\sin t}{\cos t} \qquad \cot t = \frac{\cos t}{\sin t}$$

$$\sec t = \frac{1}{\cos t} \qquad \csc t = \frac{1}{\sin t}$$

To evaluate the trigonometric functions, we may use either a scientific calculator or Tables C and D in the Appendix. If the Tables are used, the notions of **reference angle** and **reference number** become important. Finally we graphed several of the trigonometric functions, noting especially their **periodic** behavior.

CHAPTER REVIEW PROBLEM SET

1. Solve the following right triangles ($\gamma = 90°$).
   (a) $\alpha = 47.1°$, $c = 36.9$ (b) $a = 417$, $c = 573$
2. At a distance of 10 feet from a wall, the angle of elevation of the top of a mural with respect to eye level is 18° and the corresponding angle of depression of the bottom is 10°. How high is the mural?
3. Change 33° to radians. Change $9\pi/4$ radians to degrees.
4. How far does a wheel of radius 30 centimeters roll along level ground in making 100 revolutions?
5. Calculate each of the following without use of tables or a calculator.
   (a) $\sin(7\pi/6)$ (b) $\cos(11\pi/6)$
   (c) $\tan(13\pi/4)$ (d) $\sin(41\pi/6)$
6. Evaluate.
   (a) $\sin 411°$ (b) $\cos 1312°$
   (c) $\tan 5.77$ (d) $\sin 13.12$
7. Write in terms of $\sin t$.
   (a) $\sin(-t)$ (b) $\sin(t + 4\pi)$
   (c) $\sin(\pi + t)$ (d) $\cos\left(\frac{\pi}{2} - t\right)$

8. For what values of t between 0 and 2π is
   (a) $\cos t > 0$;   (b) $\cos 2t > 0$?
9. If $(-5, -12)$ is on the terminal side of an angle θ in standard position, find
   (a) $\cot \theta$;   (b) $\sec \theta$.
10. If $\sin \theta = \frac{2}{5}$ and θ is a second quadrant angle, find $\tan \theta$.
11. Sketch the graph of $y = 3 \cos 2t$ for $-\pi \leq t \leq 2\pi$.
12. Sketch the graph of $y = \sin t + \sin 2t$ using the method of adding ordinates.
13. What is the range of the sine function? Of the cosecant function?
14. Using the facts that the sine function is odd and the cosine function is even, show that cotangent is an odd function.
15. Give the general definition of $\cos t$ based on the unit circle.

For just as in nature itself there is no middle ground between truth and falsehood, so in rigorous proofs one must either establish his point beyond doubt, or else beg the question inexcusably. There is no chance of keeping one's feet by invoking limitations, distinctions, verbal distortions, or other mental acrobatics. One must with a few words and at the first assault become Caesar or nothing at all.

Galileo

CHAPTER EIGHT

Trigonometric Identities and Equations

8-1 Identities

8-2 More Identities

8-3 Trigonometric Equations

8-4 Inverse Trigonometric Functions

8-1 Identities

Complicated combinations of the six trigonometric functions occur often in mathematics. It is important that we, like the professor above, be able to write a complicated trigonometric expression in a simpler or more convenient form. To do this requires two things. We must be good at algebra and we must know the fundamental identities of trigonometry.

The Fundamental Identities We list eleven fundamental identities which should be memorized.

1. $\tan t = \dfrac{\sin t}{\cos t}$
2. $\cot t = \dfrac{\cos t}{\sin t} = \dfrac{1}{\tan t}$
3. $\sec t = \dfrac{1}{\cos t}$
4. $\csc t = \dfrac{1}{\sin t}$
5. $\sin^2 t + \cos^2 t = 1$
6. $1 + \tan^2 t = \sec^2 t$
7. $1 + \cot^2 t = \csc^2 t$
8. $\sin\left(\dfrac{\pi}{2} - t\right) = \cos t$

9. $\cos\left(\frac{\pi}{2} - t\right) = \sin t$

10. $\sin(-t) = -\sin t$

11. $\cos(-t) = \cos t$

We have seen all of these identities before. The first four are actually definitions; the others were established either in the text or the problem sets of Sections 7-3 and 7-4.

Proving New Identities The professor's work in our opening cartoon can be viewed in two ways. The more likely way of looking at it is that he wanted to simplify the complicated expression

$$(\sec t + \tan t)(1 - \sin t)$$

But it could be that someone had conjectured that

$$(\sec t + \tan t)(1 - \sin t) = \cos t$$

is an identity and that the professor was trying to prove it. It is this second concept we want to discuss now.

Suppose someone claims that a certain equation is an identity, that is, true for all values of the variable for which both sides make sense. How can you check on such a claim? The procedure used by the professor is one we urge you to follow. Start with the more complicated looking side and try to use a chain of equalities to produce the other side.

Suppose we wish to prove that

$$\sin t + \cos t \cot t = \csc t$$

is an identity. We begin with the left side and rewrite it step by step, using algebra and the fundamental identities, until we get the right side.

$$\begin{aligned}\sin t + \cos t \cot t &= \sin t + \cos t\left(\frac{\cos t}{\sin t}\right)\\ &= \frac{\sin^2 t + \cos^2 t}{\sin t}\\ &= \frac{1}{\sin t}\\ &= \csc t\end{aligned}$$

When proving that an equation is an identity, it pays to look before you leap. Changing the more complicated side to sines and cosines, as in the above example, is often the best thing to do. But not always. Sometimes the simpler side gives us a clue as to how we should reshape the other side. For example, the left side of

$$\tan t = \frac{(\sec t - 1)(\sec t + 1)}{\tan t}$$

suggests that we try to rewrite the right side in terms of $\tan t$. This can be done by multiplying out the numerator and making use of the fundamental identity $\sec^2 t = 1 + \tan^2 t$.

$$\frac{(\sec t - 1)(\sec t + 1)}{\tan t} = \frac{\sec^2 t - 1}{\tan t} = \frac{\tan^2 t}{\tan t} = \tan t$$

A Point of Logic Why all the fuss about working with just one side of a conjectured identity? First of all, it offers good practice in manipulating trigonometric expressions. But there is also a point of logic. If you operate on both sides simultaneously, you are in effect assuming that you already have an identity. That is bad logic and it can be corrected only by carefully checking that each step is reversible. To make this point clear, consider the equation

$$1 - x = x - 1$$

which is certainly not an identity. Yet when we square both sides we get

$$1 - 2x + x^2 = x^2 - 2x - 1$$

which is an identity. The trouble here is that squaring both sides is not a reversible operation.

The situation contrasts sharply with our procedure for solving conditional equations, in which we often perform an operation on both sides. For example, in the case of the equation

$$\sqrt{2x + 1} = 1 - x$$

we even square both sides. We are protected from error here by checking our solutions in the original equation.

Problem Set 8-1

1. Express entirely in terms of $\sin t$.
   (a) $\cos^2 t$
   (b) $\tan t \cos t$
   (c) $\dfrac{3}{\csc^2 t} + 2\cos^2 t - 2$
   (d) $\cot^2 t$
2. Express entirely in terms of $\cos t$.
   (a) $\sin^2 t$
   (b) $\tan^2 t$
   (c) $\csc^2 t$
   (d) $(1 + \sin t)^2 - 2 \sin t$
3. Express entirely in terms of $\tan t$.
   (a) $\cot^3 t$
   (b) $\sec^2 t$
   (c) $\sin t \sec t$
   (d) $2\sec^2 t - 2\tan^2 t + 1$

4. Express entirely in terms of $\sec t$.
   (a) $\cos^4 t$ (b) $\tan^2 t$
   (c) $\tan t \csc t$ (d) $\tan^2 t - 2\sec^2 t + 5$

Prove that each of the following is an identity.

5. $\cos t \sec t = 1$
6. $\sin t \csc t = 1$
7. $\tan x \cot x = 1$
8. $\sin x \sec x = \tan x$
9. $\cos y \csc y = \cot y$
10. $\tan y \cos y = \sin y$
11. $\cot \theta \sin \theta = \cos \theta$
12. $\dfrac{\sec \theta}{\csc \theta} = \tan \theta$
13. $\dfrac{\tan u}{\sin u} = \dfrac{1}{\cos u}$
14. $\dfrac{\sin u}{\csc u} + \dfrac{\cos u}{\sec u} = 1$
15. $(1 + \sin z)(1 - \sin z) = \dfrac{1}{\sec^2 z}$
16. $(\sec z - 1)(\sec z + 1) = \tan^2 z$
17. $(1 - \sin^2 x)(1 + \tan^2 x) = 1$
18. $(1 - \cos^2 x)(1 + \cot^2 x) = 1$
19. $\sec t - \sin t \tan t = \cos t$
20. $\sin t(\csc t - \sin t) = \cos^2 t$
21. $\dfrac{\sec^2 t - 1}{\sec^2 t} = \sin^2 t$
22. $\dfrac{1 - \csc^2 t}{\csc^2 t} = \dfrac{-1}{\sec^2 t}$
23. $\cos t(\tan t + \cot t) = \csc t$
24. $\dfrac{1}{\sin t \cos t} - \dfrac{\cos t}{\sin t} = \tan t$

Example A (Expressing all trigonometric functions in terms of one of them) If $\pi/2 < t < \pi$, express $\cos t$, $\tan t$, $\cot t$, $\sec t$, and $\csc t$ in terms of $\sin t$.

SOLUTION. Since $\cos^2 t = 1 - \sin^2 t$ and cosine is negative in quadrant II,

$$\cos t = -\sqrt{1 - \sin^2 t}$$

Also

$$\tan t = \frac{\sin t}{\cos t} = -\frac{\sin t}{\sqrt{1 - \sin^2 t}}$$

$$\cot t = \frac{1}{\tan t} = -\frac{\sqrt{1 - \sin^2 t}}{\sin t}$$

$$\sec t = \frac{1}{\cos t} = -\frac{1}{\sqrt{1 - \sin^2 t}}$$

$$\csc t = \frac{1}{\sin t}$$

25. If $\pi/2 < t < \pi$, express $\sin t$, $\tan t$, $\cot t$, $\sec t$, and $\csc t$ in terms of $\cos t$.
26. If $\pi < t < 3\pi/2$, express $\sin t$, $\cos t$, $\cot t$, $\sec t$, and $\csc t$ in terms of $\tan t$.
27. If $\pi/2 < t < \pi$ and $\sin t = \frac{4}{5}$, find the values of the other five functions for the same value of t. (*Hint*: Use the results of Example A.)
28. If $\pi < t < 3\pi/2$ and $\tan t = 2$, find $\sin t$, $\cos t$, $\cot t$, $\sec t$, and $\csc t$.

Example B (How to proceed when neither side is simple) Prove that

$$\frac{\sin t}{1 - \cos t} = \frac{1 + \cos t}{\sin t}$$

is an identity.

SOLUTION. Since both sides are equally complicated, it would seem to make no difference which side we choose to manipulate. We will try to transform the left side into the right side. Multiplying by $(1 + \cos t)/(1 + \cos t)$ might bring success, since we want to end up with $1 + \cos t$ in the numerator.

$$\frac{\sin t}{1 - \cos t} = \frac{\sin t}{1 - \cos t} \cdot \frac{1 + \cos t}{1 + \cos t} = \frac{\sin t(1 + \cos t)}{1 - \cos^2 t}$$

$$= \frac{\sin t(1 + \cos t)}{\sin^2 t}$$

$$= \frac{1 + \cos t}{\sin t}$$

Prove that each of the following is an identity.

29. $\dfrac{\sec t - 1}{\tan t} = \dfrac{\tan t}{\sec t + 1}$

30. $\dfrac{1 - \tan \theta}{1 + \tan \theta} = \dfrac{\cot \theta - 1}{\cot \theta + 1}$

(*Hint*: In Problem 30, multiply numerator and denominator of the left side by $\cot \theta$.)

31. $\dfrac{\tan^2 x}{\sec x + 1} = \dfrac{1 - \cos x}{\cos x}$

32. $\dfrac{\cot x}{\csc x + 1} = \dfrac{\csc x - 1}{\cot x}$

33. $\dfrac{\sin t + \cos t}{\tan^2 t - 1} = \dfrac{\cos^2 t}{\sin t - \cos t}$

34. $\dfrac{\sec t - \cos t}{1 + \cos t} = \sec t - 1$

Miscellaneous Problems

35. Express $\sec^2 t + \dfrac{2 \tan t}{\cot t}$

(a) entirely in terms of $\cos t$; (b) entirely in terms of $\tan t$.

36. If $\sin t = .8$ and $\pi/2 < t < \pi$, find the values of $\cos t$ and $\tan t$.

37. If $\tan t = 4$, find the values of $\cot t$ and $\csc^2 t$.

C 38. Use your calculator to find the value of

$$\left(\frac{\cos^3 t}{\sin t}\right)\left(\frac{\tan t}{1 - \sin^2 t}\right)$$

for $t = 1$, 2.15, and 100. Then use fundamental identities to simplify the above expression to see why you got the answers you did.

Prove that the following are identities.

39. $(\tan x + \cot x)(\cos x + \sin x) = \sec x + \csc x$

40. $(\sin^2 x - 1)(\cot^2 x + 1) = 1 - \csc^2 x$

41. $\dfrac{1 + \cos^2 y}{\sin^2 y} = 2 \csc^2 y - 1$

42. $\dfrac{\sin t + \cos t}{\sec t + \csc t} = \dfrac{\sin t}{\sec t}$

43. $\dfrac{\sin t - \cos t}{\sec t - \csc t} = \dfrac{\cos t}{\csc t}$

44. $\dfrac{\cos t}{1 - \sin t} = \sec t + \tan t$

45. $\dfrac{\cos^3 t - \sin^3 t}{\cos t - \sin t} = 1 + \sin t \cos t$

46. $(\csc t - \cot t)^4(\csc t + \cot t)^4 = 1$

47. $\sec^2 t + \tan^2 t = (1 - \sin^4 t) \sec^4 t$

48. $\dfrac{1 - \sin t}{1 + \sin t} = (\sec t - \tan t)^2$

C 49. Calculate

$$\frac{(\tan t)^{\sin^2 t}}{(\cot t)^{\cos^2 t}} - \tan t$$

for $t = 1, .5$, and $.25$. Guess at an identity. Prove it.

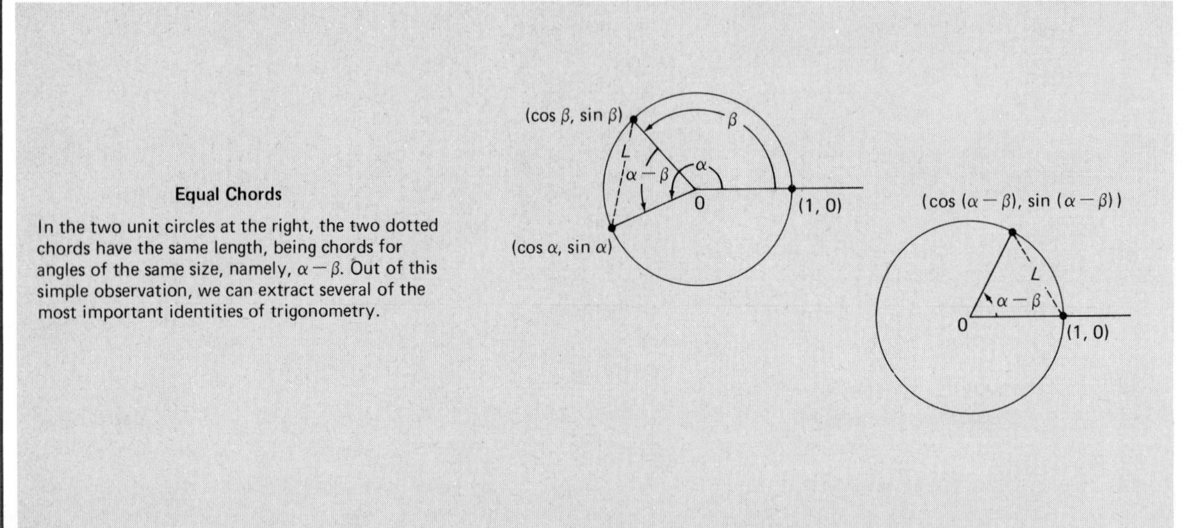

Equal Chords

In the two unit circles at the right, the two dotted chords have the same length, being chords for angles of the same size, namely, $\alpha - \beta$. Out of this simple observation, we can extract several of the most important identities of trigonometry.

8-2 More Identities

When you study calculus, you will meet expressions like $\cos(\alpha + \beta)$ and $\sin(\alpha - \beta)$. It will be very important to rewrite these expressions directly in terms of $\sin \alpha$, $\cos \alpha$, $\sin \beta$, and $\cos \beta$. It might be tempting to replace $\cos(\alpha + \beta)$ by $\cos \alpha + \cos \beta$, but that would be terribly wrong. Try it with $\alpha = \pi/6$ and $\beta = \pi/3$ to convince yourself. To obtain correct expressions is the goal of this section.

> **Distance Formula**
>
> The distance between (x_1, y_1) and (x_2, y_2) is
>
> $\sqrt{(x_2 - x_1)^2 + (y_2 - y_1)^2}$

A Key Identity The opening display shows two chords of equal length L. Using the formula for the distance between two points and the identity $\sin^2 \theta + \cos^2 \theta = 1$, we have the following expression for the square of the chord on the right.

$$\begin{aligned} L^2 &= [\cos(\alpha - \beta) - 1]^2 + \sin^2(\alpha - \beta) \\ &= \cos^2(\alpha - \beta) - 2\cos(\alpha - \beta) + 1 + \sin^2(\alpha - \beta) \\ &= [\cos^2(\alpha - \beta) + \sin^2(\alpha - \beta)] + 1 - 2\cos(\alpha - \beta) \\ &= 2 - 2\cos(\alpha - \beta) \end{aligned}$$

A similar calculation for the square of the chord on the left gives

$$\begin{aligned} L^2 &= (\cos \alpha - \cos \beta)^2 + (\sin \alpha - \sin \beta)^2 \\ &= \cos^2 \alpha - 2\cos \alpha \cos \beta + \cos^2 \beta + \sin^2 \alpha - 2 \sin \alpha \sin \beta + \sin^2 \beta \\ &= 1 - 2\cos \alpha \cos \beta - 2 \sin \alpha \sin \beta + 1 \\ &= 2 - 2(\cos \alpha \cos \beta + \sin \alpha \sin \beta) \end{aligned}$$

When we equate these two expressions for L^2, we get our key identity

$$\cos(\alpha - \beta) = \cos \alpha \cos \beta + \sin \alpha \sin \beta$$

Our derivation is based on a picture in which α and β are positive angles with $\alpha > \beta$. Minor modifications would establish the identity for arbitrary angles α and β and hence also for their radian measures s and t. Thus for all real numbers s and t,

$$\boxed{\cos(s - t) = \cos s \cos t + \sin s \sin t}$$

In words, this identity says: *The cosine of a difference is the cosine of the first times the cosine of the second plus the sine of the first times the sine of the second.* It is important to memorize this identity so you can easily apply it to $\cos(3u - v)$, $\cos[s - (-t)]$, or even $\cos[(\pi/2 - s) - t)]$, as we shall have to do soon.

Related Identities In the boxed identity above, we replace t by $-t$ and use the fundamental identities $\cos(-t) = \cos t$ and $\sin(-t) = -\sin t$ to get

$$\begin{aligned}\cos[s - (-t)] &= \cos s \cos(-t) + \sin s \sin(-t)\\ &= \cos s \cos t + (\sin s)(-\sin t)\end{aligned}$$

This gives us the **addition law for cosines**.

$$\boxed{\cos(s + t) = \cos s \cos t - \sin s \sin t}$$

which also merits memorization.

There is also an identity involving $\sin(s + t)$. To derive this identity, we use the cofunction identity $\sin u = \cos(\pi/2 - u)$ to write

$$\sin(s + t) = \cos\left[\frac{\pi}{2} - (s + t)\right] = \cos\left[\left(\frac{\pi}{2} - s\right) - t\right]$$

Then we use our key identity for the cosine of a difference to obtain

$$\cos\left(\frac{\pi}{2} - s\right)\cos t + \sin\left(\frac{\pi}{2} - s\right)\sin t$$

Two applications of cofunction identities give us the result we want, the **addition law for sines**.

$$\boxed{\sin(s + t) = \sin s \cos t + \cos s \sin t}$$

Finally, replacing t by $-t$ in this last result leads to

$$\boxed{\sin(s - t) = \sin s \cos t - \cos s \sin t}$$

Double-Angle Formulas Out of the addition laws come identities for $\cos 2t$ and $\sin 2t$. We call them **double-angle formulas**, though they might better be called double-number formulas.

$$\cos 2t = \cos^2 t - \sin^2 t$$
$$\sin 2t = 2 \sin t \cos t$$

We derive them by applying the addition laws to $\cos(t + t)$ and $\sin(t + t)$.

$$\cos(t + t) = \cos t \cos t - \sin t \sin t = \cos^2 t - \sin^2 t$$
$$\sin(t + t) = \sin t \cos t + \cos t \sin t = 2 \sin t \cos t$$

There are two other forms of the double-angle formula for the cosine that are often useful. If we replace $\cos^2 t$ by $1 - \sin^2 t$, we obtain

$$\cos 2t = 1 - 2 \sin^2 t$$

and, alternatively, if we replace $\sin^2 t$ by $1 - \cos^2 t$, we have

$$\cos 2t = 2 \cos^2 t - 1$$

Of course, in all that we have done, we may replace the number t by the angle θ; hence the name double-angle formulas.

Half-Angle Formulas In the identity

$$\cos 2t = 1 - 2 \sin^2 t$$

replace t by $t/2$ and then solve for $\sin(t/2)$.

$$\cos t = 1 - 2 \sin^2\left(\frac{t}{2}\right)$$

$$2 \sin^2\left(\frac{t}{2}\right) = 1 - \cos t$$

$$\sin^2\left(\frac{t}{2}\right) = \frac{1 - \cos t}{2}$$

$$\sin\left(\frac{t}{2}\right) = \pm\sqrt{\frac{1 - \cos t}{2}}$$

The corresponding formula for the cosine is

$$\cos\left(\frac{t}{2}\right) = \pm\sqrt{\frac{1 + \cos t}{2}}$$

This formula is derived using the identity $\cos 2t = 2\cos^2 t - 1$, again replacing t by $t/2$. In Example B at the end of this section, we derive a formula for $\tan(t/2)$.

In both of these formulas, the choice of the plus or minus sign is determined by the interval on which $t/2$ lies. For example,

$$\cos\left(\frac{5\pi}{8}\right) = \cos\left(\frac{5\pi/4}{2}\right) = -\sqrt{\frac{1 + \cos(5\pi/4)}{2}}$$

$$= -\sqrt{\frac{1 - \sqrt{2}/2}{2}} = -\frac{\sqrt{2 - \sqrt{2}}}{2}$$

We choose the minus sign because $5\pi/8$ corresponds to an angle in quadrant II where the cosine is negative.

Problem Set 8-2

Find the value of each expression. Note that in each case, the answers to parts (a) and (b) are different.

1. (a) $\sin\frac{\pi}{4} + \sin\frac{\pi}{6}$ (b) $\sin\left(\frac{\pi}{4} + \frac{\pi}{6}\right)$
2. (a) $\cos\frac{\pi}{4} + \cos\frac{\pi}{6}$ (b) $\cos\left(\frac{\pi}{4} + \frac{\pi}{6}\right)$
3. (a) $\cos\frac{\pi}{4} - \cos\frac{\pi}{6}$ (b) $\cos\left(\frac{\pi}{4} - \frac{\pi}{6}\right)$
4. (a) $\sin\frac{\pi}{4} - \sin\frac{\pi}{6}$ (b) $\sin\left(\frac{\pi}{4} - \frac{\pi}{6}\right)$

Write each of the following as a single sine or cosine.

5. $\cos\frac{1}{2}\cos\frac{3}{2} - \sin\frac{1}{2}\sin\frac{3}{2}$
6. $\cos 2\cos 3 + \sin 2\sin 3$
7. $\sin\frac{7\pi}{8}\cos\frac{\pi}{8} + \cos\frac{7\pi}{8}\sin\frac{\pi}{8}$
8. $\sin\frac{5\pi}{16}\cos\frac{\pi}{16} - \cos\frac{5\pi}{16}\sin\frac{\pi}{16}$
9. $\cos 33^\circ \cos 27^\circ - \sin 33^\circ \sin 27^\circ$
10. $\sin 49^\circ \cos 41^\circ + \cos 49^\circ \sin 41^\circ$

Use the identities derived in this section to show that the equalities in Problems 11–18 are identities.

11. $\sin(t + \pi) = -\sin t$

12. $\cos(t + \pi) = -\cos t$

13. $\sin\left(t + \frac{3\pi}{2}\right) = -\cos t$

14. $\cos\left(t + \frac{3\pi}{2}\right) = \sin t$

15. $\sin\left(t - \frac{\pi}{2}\right) = -\cos t$

16. $\cos\left(t - \frac{\pi}{2}\right) = \sin t$

17. $\cos\left(t + \frac{\pi}{3}\right) = \frac{1}{2}\cos t - \frac{\sqrt{3}}{2}\sin t$

18. $\sin\left(t + \frac{\pi}{3}\right) = \frac{1}{2}\sin t + \frac{\sqrt{3}}{2}\cos t$

19. Let α and β be first quadrant angles and suppose $\sin\alpha = \frac{4}{5}$ and $\cos\beta = \frac{5}{13}$. Evaluate each expression.
(a) $\cos\alpha$ (b) $\sin\beta$
(c) $\sin(\alpha + \beta)$ (d) $\cos(\alpha + \beta)$
(e) $\sin(\alpha - \beta)$ (f) $\cos(\alpha - \beta)$
(g) $\tan(\alpha + \beta)$ (h) $\tan(\alpha - \beta)$

20. Suppose that $\sin\alpha = 1/\sqrt{10}$, with α in quadrant I, and that $\cos\beta = -\frac{1}{2}$, with β in quadrant III.
(a) Calculate $\sin(\alpha + \beta)$ and $\cos(\alpha + \beta)$.
(b) In what quadrant is $\alpha + \beta$?
(c) Calculate $\sin(\alpha - \beta)$ and $\cos(\alpha - \beta)$.
(d) In what quadrant is $\alpha - \beta$?

21. Use identities of this section to write a simple expression for each of the following.
(a) $2\sin 32^\circ \cos 32^\circ$ (b) $2\sin\frac{\pi}{12}\cos\frac{\pi}{12}$
(c) $\cos^2\frac{\pi}{12} - \sin^2\frac{\pi}{12}$ (d) $1 - 2\sin^2 41^\circ$
(e) $\frac{1 + \cos 52^\circ}{2}$ (f) $\frac{1 - \cos 64^\circ}{2}$

22. Write a simple expression for each of the following.
(a) $2\sin 15^\circ \cos 15^\circ$ (b) $2\sin(22.5^\circ)\cos(22.5^\circ)$
(c) $2\cos^2 15^\circ - 1$ (d) $\cos^2\frac{\pi}{8} - \sin^2\frac{\pi}{8}$

23. If $\sin t = \frac{3}{5}$ and $\pi/2 < t < \pi$, calculate each of the following.
(a) $\cos t$ (b) $\sin 2t$
(c) $\cos 2t$ (d) $\tan 2t$

24. If $\cos t = -\frac{5}{13}$ and $\pi < t < 3\pi/2$, calculate each of the following.
(a) $\sin 2t$ (b) $\cos 2t$

25. If $\cos\theta = \frac{4}{5}$ and θ is in the first quadrant, calculate each of the following.
(a) $\sin(\theta/2)$ (b) $\cos(\theta/2)$ (c) $\tan(\theta/2)$

26. If $\sin t = -\frac{1}{2}$ and $\pi < t < 3\pi/2$, find each of the following.
(a) $\sin(t/2)$ (b) $\cos(t/2)$ (c) $\tan(t/2)$
(*Note*: $\pi/2 < t/2 < 3\pi/4$.)

27. Write $\sin(t/4)$ in terms of $\cos(t/2)$.

28. Write $\cos(t/6)$ in terms of $\cos(t/3)$.

Example A (Tangent identities) Verify the addition formula for the tangent.

$$\tan(s + t) = \frac{\tan s + \tan t}{1 - \tan s \tan t}$$

SOLUTION.

$$\begin{aligned}
\tan(s + t) &= \frac{\sin(s + t)}{\cos(s + t)} \\
&= \frac{\sin s \cos t + \cos s \sin t}{\cos s \cos t - \sin s \sin t} \\
&= \frac{\dfrac{\sin s \cos t}{\cos s \cos t} + \dfrac{\cos s \sin t}{\cos s \cos t}}{\dfrac{\cos s \cos t}{\cos s \cos t} - \dfrac{\sin s \sin t}{\cos s \cos t}} \\
&= \frac{\tan s + \tan t}{1 - \tan s \tan t}
\end{aligned}$$

The key step was the third one, in which we divided both the numerator and the denominator by $\cos s \cos t$.

Now establish that each of the following are identities.

29. $\tan(s - t) = \dfrac{\tan s - \tan t}{1 + \tan s \tan t}$

30. $\tan(s + \pi) = \tan s$

31. $\tan(2t) = \dfrac{2 \tan t}{1 - \tan^2 t}$

32. $\tan \dfrac{t}{2} = \pm\sqrt{\dfrac{1 - \cos t}{1 + \cos t}}$

Example B (Using double-angle and half-angle formulas) Prove that the following are identities.

(a) $\sin 3t = 3 \sin t - 4 \sin^3 t$

(b) $\tan \dfrac{t}{2} = \dfrac{\sin t}{1 + \cos t}$

SOLUTION. (a) We think of $3t$ as $2t + t$ and use the addition law for sines.

$$\begin{aligned}
\sin 3t &= \sin(2t + t) \\
&= \sin 2t \cos t + \cos 2t \sin t \\
&= (2 \sin t \cos t) \cos t + (1 - 2 \sin^2 t) \sin t \\
&= 2 \sin t(1 - \sin^2 t) + \sin t - 2 \sin^3 t \\
&= 2 \sin t - 2 \sin^3 t + \sin t - 2 \sin^3 t \\
&= 3 \sin t - 4 \sin^3 t
\end{aligned}$$

(b) Thinking of t as $2(t/2)$ and applying double-angle formulas to the right side, we get

$$\begin{aligned}\frac{\sin t}{1+\cos t} &= \frac{\sin(2(t/2))}{1+\cos(2(t/2))}\\ &= \frac{2\sin(t/2)\cos(t/2)}{1+2\cos^2(t/2)-1}\\ &= \frac{\sin(t/2)}{\cos(t/2)}\\ &= \tan\frac{t}{2}\end{aligned}$$

Now prove that each of the following are identities.

33. $\cos 3t = 4\cos^3 t - 3\cos t$

34. $(\sin t + \cos t)^2 = 1 + \sin 2t$

35. $\csc 2t + \cot 2t = \cot t$

36. $\sin^2 t \cos^2 t = \frac{1}{8}(1 - \cos 4t)$

37. $\dfrac{\sin\theta}{1-\cos\theta} = \cot\dfrac{\theta}{2}$

38. $1 - 2\sin^2\theta = 2\cot 2\theta \sin\theta\cos\theta$

39. $\dfrac{2\tan\alpha}{1+\tan^2\alpha} = \sin 2\alpha$

40. $\dfrac{1-\tan^2\alpha}{1+\tan^2\alpha} = \cos 2\alpha$

Miscellaneous Problems

41. Use addition formulas to find:
(a) $\cos 75°$; (b) $\sin 105°$.
(*Hint*: $75° = 45° + 30°$.)

42. Use half-angle formulas to find:
(a) $\tan(\pi/8)$; (b) $\sin 112.5°$.

43. Find $\sin 15°$ by using each fact.
(a) $15° = 45° - 30°$ (b) $15° = 30°/2$

44. Show that $\sin(t + \pi/4) = (\sqrt{2}/2)(\sin t + \cos t)$.

45. Suppose α is in quadrant I with $\sin\alpha = \frac{1}{3}$ and β is in quadrant IV with $\cos\beta = \frac{1}{4}$. Evaluate each expression.
(a) $\cos\alpha$ (b) $\sin\beta$
(c) $\sin(\alpha + \beta)$ (d) $\cos(\alpha - \beta)$
(e) $\sin 2\alpha$ (f) $\cos(\beta/2)$

46. Write $\sin 5u$ in terms of $\cos 10u$.

47. Use the identities

$$\cos(s+t) = \cos s\cos t - \sin s\sin t$$
$$\cos(s-t) = \cos s\cos t + \sin s\sin t$$

to show that
(a) $\cos s\cos t = \frac{1}{2}(\cos(s+t) + \cos(s-t))$;
(b) $\sin s\sin t = -\frac{1}{2}(\cos(s+t) - \cos(s-t))$.
These are called **product formulas**.

48. Use the identities given in Problem 47 to derive the following **sum and difference formulas.**

(a) $\cos A + \cos B = 2 \cos \dfrac{A + B}{2} \cos \dfrac{A - B}{2}$

(*Hint*: Let $A = s + t$ and $B = s - t$.)

(b) $\cos A - \cos B = -2 \sin \dfrac{A + B}{2} \sin \dfrac{A - B}{2}$

49. Using the formulas for $\sin(s + t)$ and $\sin(s - t)$, show that each of the following is true.
(a) $\sin s \cos t = \frac{1}{2}(\sin(s + t) + \sin(s - t))$
(b) $\cos s \sin t = \frac{1}{2}(\sin(s + t) - \sin(s - t))$

50. Let $A = s + t$ and $B = s - t$ in Problem 49 and show that each of the following is true.

(a) $\sin A + \sin B = 2 \sin \dfrac{A + B}{2} \cos \dfrac{A - B}{2}$

(b) $\sin A - \sin B = 2 \cos \dfrac{A + B}{2} \sin \dfrac{A - B}{2}$

51. Use the formulas established in Problems 47 and 49 to write each of the following products as a sum or difference.
(a) $\cos 2x \cos x$ (b) $\sin 2x \sin x$
(c) $\sin 5\alpha \cos 3\alpha$ (d) $\cos 5\alpha \sin 3\alpha$

52. Use the formulas established in Problems 48 and 50 to write each of the following as a product.
(a) $\cos 4x + \cos 2x$ (b) $\cos 4x - \cos 2x$
(c) $\sin 5x + \sin 3x$ (d) $\sin 5x - \sin 3x$

53. Express each of the following as a product.
(a) $\sin(x + h) - \sin x$ (b) $\cos(x + h) - \cos x$

54. Express each of the following as a sum.
(a) $2 \sin(\alpha + \pi/6) \cos(\alpha - \pi/6)$ (b) $2 \cos(\alpha + \pi/2) \cos(\alpha - \pi/2)$

55. Show that

$$\frac{\cos 6\theta + \cos 4\theta}{\sin 6\theta - \sin 4\theta} = \cot \theta$$

56. Show that

$$\frac{\sin y + \sin 3y}{\cos y + \cos 3y} = \tan 2y$$

[C] 57. Calculate $((32 \cos^2 t - 48) \cos^2 t + 18) \cos^2 t - \cos 6t$ for $t = 1, 1.5$, and 2. Guess at an identity and then prove it.

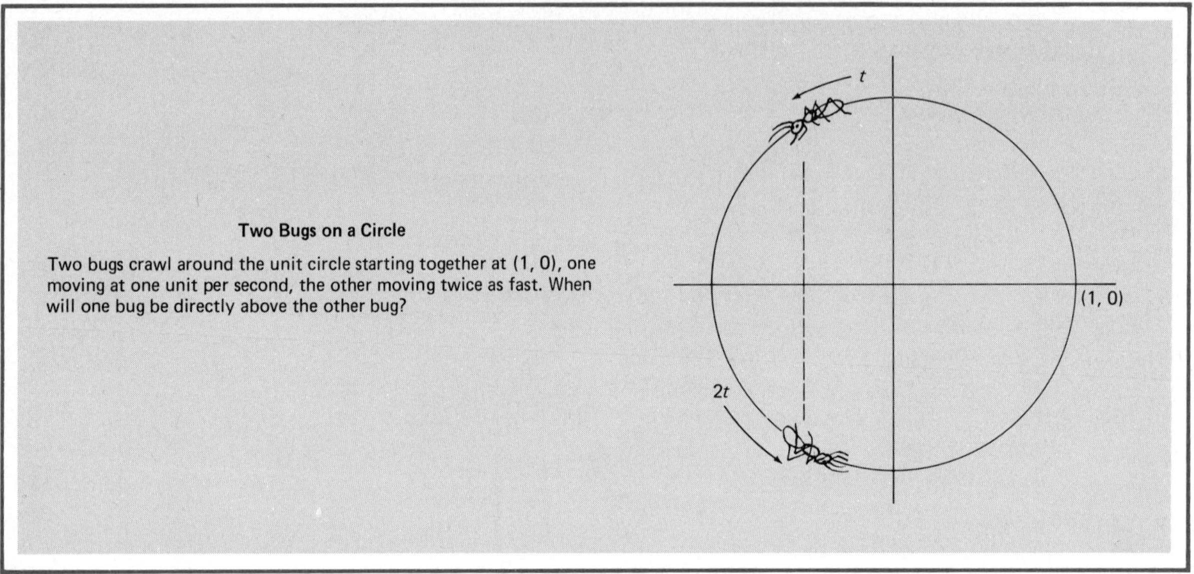

8-3 Trigonometric Equations

What does the bug problem have to do with trigonometric equations? Well, you should agree that after t seconds the slow bug, having traveled t units along the unit circle, is at $(\cos t, \sin t)$. The fast bug is at $(\cos 2t, \sin 2t)$. One bug will be directly above the other bug when their two x-coordinates are equal. This means we must solve the equation

$$\cos 2t = \cos t$$

Specifically, we must find the first $t > 0$ that makes this equality true. We shall solve this equation in due time, but first we ought to solve some simpler trigonometric equations.

Simple Equations Suppose we are asked to solve the equation

$$\sin t = \tfrac{1}{2}$$

for t. The number $t = \pi/6$ occurs to us right away. But that is not the only answer. All numbers that measure angles in the first or second quadrant and have $\pi/6$ as their reference number are solutions. Thus,

$$\ldots, -\frac{11\pi}{6}, -\frac{7\pi}{6}, \frac{\pi}{6}, \frac{5\pi}{6}, \frac{13\pi}{6}, \ldots$$

all work. In fact, one characteristic of trigonometric equations is that, if they have one solution, they have infinitely many solutions.

Let us alter the problem. Suppose we wish to solve $\sin t = \frac{1}{2}$ for $0 \leq t < 2\pi$. Then the answers are $\pi/6$ and $5\pi/6$. In the following pages, we shall assume that, unless otherwise specified, we are to find only those solutions on the interval $0 \leq t < 2\pi$.

For a second example, let us solve

$$\tan t = -1$$

First we should think of the equation $\tan t = 1$, which has $\pi/4$ as a solution. We remember that the tangent is negative in quadrants II and IV. Thus the desired solutions are $3\pi/4$ and $7\pi/4$, which both have $\pi/4$ as their reference number.

Do you remember how we solved the equation $x^2 = 4x$? We rewrote it with 0 on one side, factored the other side, and then set each factor equal to 0 (see margin). We follow exactly the same procedure with our next trigonometric equation. To solve

$$\begin{aligned} x^2 &= 4x \\ x^2 - 4x &= 0 \\ x(x-4) &= 0 \\ x = 0,\ x - 4 &= 0 \\ x &= 0, 4 \end{aligned}$$

$$\cos t \tan t = -\cos t$$

use the steps below.

$$\begin{aligned} \cos t \tan t + \cos t &= 0 \\ \cos t\,(\tan t + 1) &= 0 \\ \cos t = 0,\ \tan t + 1 &= 0 \\ \cos t = 0,\ \tan t &= -1 \end{aligned}$$

Thus our problem is reduced to solving two simple equations. The first has the two solutions $\pi/2$ and $3\pi/2$; the second (solved in the preceding paragraph) has solutions $3\pi/4$ and $7\pi/4$. Thus the set of all solutions of

$$\cos t \tan t = -\cos t$$

on the interval $0 \leq t < 2\pi$ is

$$\left\{\frac{\pi}{2}, \frac{3\pi}{4}, \frac{3\pi}{2}, \frac{7\pi}{4}\right\}$$

Equations of Quadratic Form In Section 3-4, we solved quadratic equations by a number of techniques (factoring, taking square roots, and using the quadratic formula). We use the same techniques here. For example,

$$\cos^2 t = \tfrac{3}{4}$$

is analogous to $x^2 = \frac{3}{4}$. We solve such an equation by taking square roots.

$$\cos t = \pm\frac{\sqrt{3}}{2}$$

The set of solutions on $0 \leq t < 2\pi$ is

$$\left\{\frac{\pi}{6}, \frac{5\pi}{6}, \frac{7\pi}{6}, \frac{11\pi}{6}\right\}$$

As a second example, consider the equation

$$2\sin^2 t - \sin t - 1 = 0$$

Think of it as being like

$$2x^2 - x - 1 = 0$$

Now

$$2x^2 - x - 1 = (2x + 1)(x - 1)$$

and so

$$2\sin^2 t - \sin t - 1 = (2\sin t + 1)(\sin t - 1)$$

When we set each factor equal to zero and solve, we get

$$\begin{aligned} 2\sin t + 1 &= 0 \qquad & \sin t - 1 &= 0 \\ \sin t &= -\tfrac{1}{2} & \sin t &= 1 \\ t &= \frac{7\pi}{6}, \frac{11\pi}{6} & t &= \frac{\pi}{2} \end{aligned}$$

The set of all solutions on $0 \le t < 2\pi$ is

$$\left\{\frac{\pi}{2}, \frac{7\pi}{6}, \frac{11\pi}{6}\right\}$$

Using Identities to Solve Equations Consider the equation

$$\tan^2 x = \sec x + 1$$

The identity $\sec^2 x = \tan^2 x + 1$ suggests writing everything in terms of $\sec x$.

$$\begin{aligned} \sec^2 x - 1 &= \sec x + 1 \\ \sec^2 x - \sec x - 2 &= 0 \\ (\sec x + 1)(\sec x - 2) &= 0 \end{aligned}$$

$$\begin{aligned} \sec x + 1 &= 0 \qquad & \sec x - 2 &= 0 \\ \sec x &= -1 & \sec x &= 2 \\ x &= \pi & x &= \frac{\pi}{3}, \frac{5\pi}{3} \end{aligned}$$

Thus, the set of solutions on $0 \le t < 2\pi$ is $\{\pi/3, \pi, 5\pi/3\}$. Unfamiliarity with the secant may hinder you at the last step. If so, use $\sec x = 1/\cos x$ to write the equations in terms of cosines and solve the equations

$$\cos x = -1 \qquad \cos x = \tfrac{1}{2}$$

Solution to the Two-Bug Problem Our opening display asked when one bug would first be directly above the other. We reduced that problem to solving

$$\cos 2t = \cos t$$

for t. Using a double-angle formula, we may write

$$\begin{aligned} 2\cos^2 t - 1 &= \cos t \\ 2\cos^2 t - \cos t - 1 &= 0 \\ (2\cos t + 1)(\cos t - 1) &= 0 \end{aligned}$$

$$\cos t = -\tfrac{1}{2} \qquad\qquad \cos t = 1$$

$$t = \frac{2\pi}{3}, \frac{4\pi}{3} \qquad\qquad t = 0$$

The smallest positive solution is $t = 2\pi/3$. After a little over 2 seconds, the slow bug will be directly above the fast bug.

Problem Set 8-3

Solve each of the following, finding all solutions on the interval 0 *to* 2π, *excluding* 2π.

1. $\sin t = 0$
2. $\cos t = 1$
3. $\sin t = -1$
4. $\tan t = -\sqrt{3}$
5. $\sin t = 2$
6. $\sec t = \frac{1}{2}$
7. $2\cos x + \sqrt{3} = 0$
8. $2\sin x + 1 = 0$
9. $\tan^2 x = 1$
10. $4\sin^2\theta - 3 = 0$
11. $(2\cos\theta + 1)(2\sin\theta - \sqrt{2}) = 0$
12. $(\sin\theta - 1)(\tan\theta + 1) = 0$
13. $\sin^2 x + \sin x = 0$
14. $2\cos^2 x - \cos x = 0$
15. $\tan^2\theta = \sqrt{3}\tan\theta$
16. $\cot^2\theta = -\cot\theta$
17. $2\sin^2 x = 1 + \cos x$
18. $\sec^2 x = 1 + \tan x$
19. $\sin 2t = -\cos t$
20. $\cos 2t = \sin t$

Example A (Solving by squaring both sides) Solve

$$1 - \cos t = \sqrt{3}\sin t$$

SOLUTION. Since the identity relating sines and cosines involves their squares, we begin by squaring both sides. Then we express everything in terms of $\cos t$ and solve.

$$\begin{aligned} (1 - \cos t)^2 &= 3\sin^2 t \\ 1 - 2\cos t + \cos^2 t &= 3(1 - \cos^2 t) \\ \cos^2 t - 2\cos t + 1 &= 3 - 3\cos^2 t \\ 4\cos^2 t - 2\cos t - 2 &= 0 \\ (4\cos t + 2)(\cos t - 1) &= 0 \end{aligned}$$

$$\cos t = -\tfrac{1}{2} \qquad\qquad \cos t = 1$$

$$t = \frac{2\pi}{3}, \frac{4\pi}{3} \qquad\qquad t = 0$$

Since squaring may introduce extraneous solutions, it is important to check our answers. We find that $4\pi/3$ is extraneous, since substituting $4\pi/3$ for t in the original equation gives us $1 + \frac{1}{2} = -\frac{3}{2}$. However, 0 and $2\pi/3$ are solutions, as you should verify.

Solve each of the following equations on the interval $0 \le t < 2\pi$*; check your answers.*

21. $\sin t + \cos t = 1$
22. $\sin t - \cos t = 1$
23. $\sqrt{3}(1 - \sin t) = \cos t$
24. $1 + \sin t = \sqrt{3}\cos t$
25. $\sec t + \tan t = 1$
26. $\tan t - \sec t = 1$

Example B (Finding all of the solutions) Find the entire set of solutions of the equation $\cos 2t = \cos t$.

SOLUTION. In the text, we found 0, $2\pi/3$, and $4\pi/3$ to be the solutions for $0 \le t < 2\pi$. Clearly we get new solutions by adding 2π again and again to any of these numbers. The same holds true for subtracting 2π. In fact, the entire solution set consists of all those numbers which have the form $2\pi k$, $2\pi/3 + 2\pi k$, or $4\pi/3 + 2\pi k$, where k is any integer.

Find the entire solution set of each of the following equations.

27. $\sin t = \frac{1}{2}$
28. $\cos t = -\frac{1}{2}$
29. $\tan t = 0$
30. $\tan t = -\sqrt{3}$
31. $\sin^2 t = \frac{1}{4}$
32. $\cos^2 t = 1$

Example C (Multiple-angle equations) Find all solutions of $\cos 4t = \frac{1}{2}$ on the interval $0 \le t < 2\pi$.

SOLUTION. There will be more answers than you think. We know that $\cos 4t$ equals $\frac{1}{2}$ when

$$4t = \frac{\pi}{3}, \frac{5\pi}{3}, \frac{7\pi}{3}, \frac{11\pi}{3}, \frac{13\pi}{3}, \frac{17\pi}{3}, \frac{19\pi}{3}, \frac{23\pi}{3}$$

that is, when

$$t = \frac{\pi}{12}, \frac{5\pi}{12}, \frac{7\pi}{12}, \frac{11\pi}{12}, \frac{13\pi}{12}, \frac{17\pi}{12}, \frac{19\pi}{12}, \frac{23\pi}{12}$$

The reason that there are 8 solutions instead of 2 is that $\cos 4t$ completes 4 periods on the interval $0 \le t < 2\pi$.

Solve each of the following equations, finding all solutions on the interval $0 \le t < 2\pi$.

33. $\sin 2t = 0$
34. $\cos 2t = 0$
35. $\sin 4t = 1$
36. $\cos 4t = 1$
37. $\tan 2t = -1$
38. $\tan 3t = 0$

Miscellaneous Problems

Solve each of the equations in Problems 39–48, finding all solutions on the interval 0 to 2π*, excluding* 2π.

39. $\cos t = \dfrac{\sqrt{3}}{2}$
40. $\tan t = -1$

41. $\sin^2 x = 1$

42. $2\cos^2 t - 1 = 0$

43. $\cos x - 2\cos x \sin x = 0$

44. $2\sin x \cos x + \sqrt{3}\sin x = 0$

45. $\sin 2t = \sin t$

46. $\cos 4t = -1$

[C] 47. $\sin^2 t + 2\sin t - 2 = 0$

[C] 48. $\cos^2 t + 3\cos t - 1 = 0$

In Problems 49 and 50, find the entire solution set (see Example B).

49. $(1 + \sin t)(1 - \tan t) = 0$

50. $\cos 2t = 0.5$

In Problems 51–52, find all solutions between 0 and 2π. You will need Table D in the appendix or a calculator.

51. $\sin t = 0.89121$

52. $\cos 3t = 0.87274$

53. A ray of light from the lamp L in the first picture reflects off a mirror to the object O.
(a) Find the distance x.
(b) Write an equation for θ.
(c) Solve this equation.

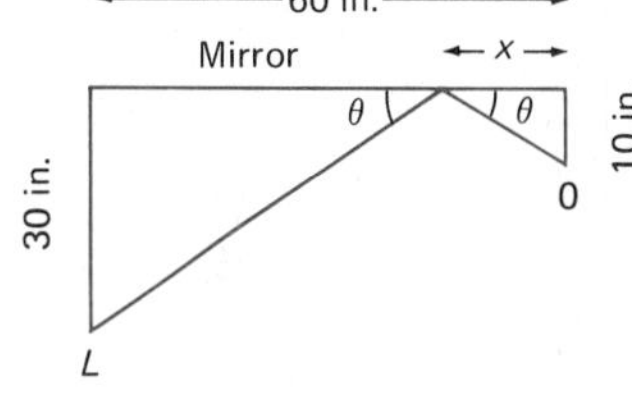

54. Tom and John are lost in a desert 1 mile from a highway, at point A in the second picture. Each strikes out in a different direction to get to the highway. Tom gets to the highway at point B and John arrives at point C, $1 + \sqrt{3}$ miles farther down the road. Write an equation for θ and solve it.

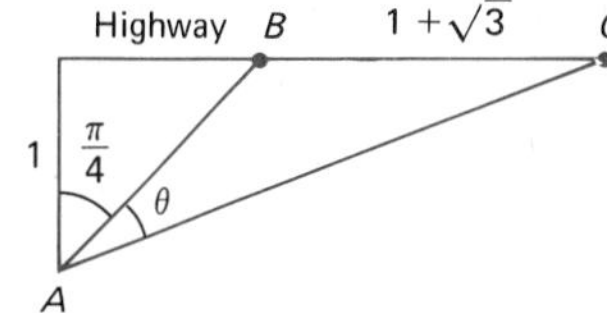

55. Mr. Quincy built a slide with a 10-foot rise and 20-foot base. (a) Find the angle α in degrees. (b) By how much (θ in the third picture) would the angle of the slide increase if he made the rise 15 feet, keeping the base at 20 feet?

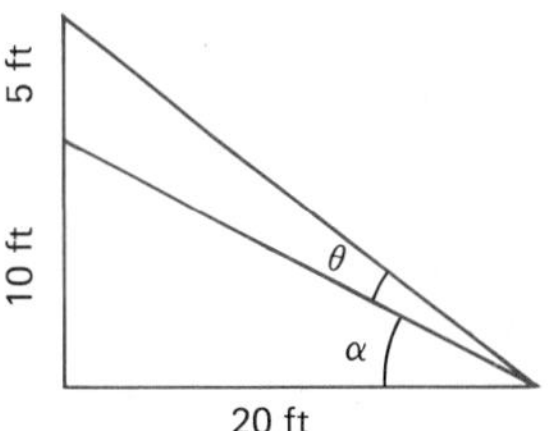

56. Find the angles θ_1, θ_2, and θ_3 shown in the bottom picture. Your answers should convince you that the angle ABC is not trisected.

57. Solve the equation

$$\sin 4t + \sin 3t + \sin 2t = 0$$

Use the identity $\sin u + \sin v = 2\sin((u + v)/2)\cos((u - v)/2)$.

58. Solve the equation

$$\cos 5t + \cos 3t - 2\cos t = 0$$

Use the identity $\cos u + \cos v = 2\cos((u + v)/2)\cos((u - v)/2)$.

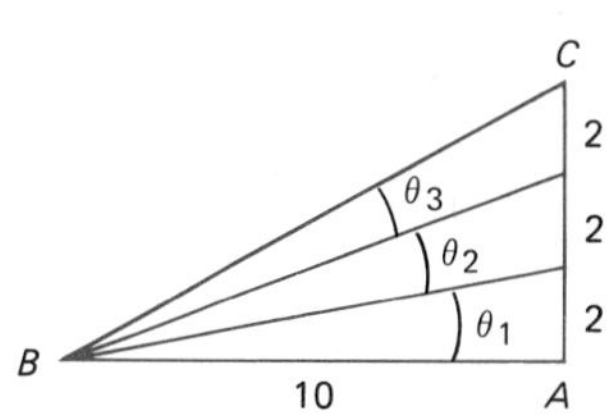

[C] 59. (a) Calculate $5^{\sin^2 t} + 5^{\cos^2 t}$ for $t = 0$, $\pi/4$, and $\pi/2$. (b) Solve

$$5^{\sin^2 t} + 5^{\cos^2 t} = 5$$

for $0 < t < \pi/2$. (*Hint*: Replace $\cos^2 t$ by $1 - \sin^2 t$ and get a quadratic equation in $5^{\sin^2 t}$.)

Find the inverse of $f(x) = 3x$

1. Let $y = 3x$
2. Solve for x

   $x = \frac{y}{3}$

3. Then $x = f^{-1}(y) = \frac{y}{3}$

OK!

Find the inverse of $f(x) = \sin x$

1. Let $y = \sin x$
2. Solve for x

   $x = \frac{y}{\sin}$

3. Then $x = f^{-1}(y) = \frac{y}{\sin}$

WHAT?

8-4 Inverse Trigonometric Functions

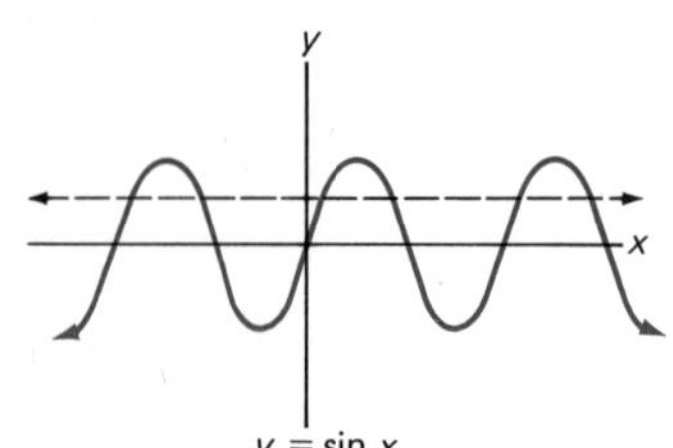

Of course, the derivation at the right above is sheer nonsense. But it does point out an important question. Does the sine function have an inverse? Without some fixing, it does not. We learned in Section 5-5 (a section worth reviewing now) that only one-to-one functions have inverses. The sine function is not one-to-one. In fact, it is about as far from being one-to-one as possible. If the sine function were one-to-one, a horizontal line would meet the graph in at most one point; but, as we see from the graph in the margin, a line may actually intersect it in infinitely many points. To make the sine function have an inverse, we shall have to drastically restrict its domain.

The Inverse Sine Consider the graph of the sine function again. We want to restrict its domain in such a way that the sine assumes its full range of values but takes on each value only once. There are many possible choices, but the one commonly used is $-\pi/2 \leq x \leq \pi/2$. Notice the corresponding part of the sine graph below. From now on, whenever we need an inverse sine function, we always assume the domain of the sine has been restricted to $-\pi/2 \leq x \leq \pi/2$.

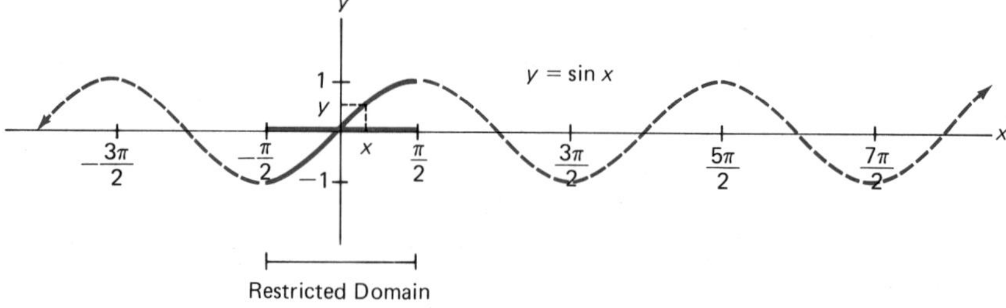

Having done this, we see that each y corresponds to exactly one x. We write $x = \sin^{-1} y$ (x is the inverse sine of y). Thus

$$\sin^{-1}(\tfrac{1}{2}) = \pi/6$$
$$\sin^{-1}(1) = \pi/2$$
$$\sin^{-1}(-1) = -\frac{\pi}{2}$$
$$\sin^{-1}\left(\frac{-\sqrt{2}}{2}\right) = -\frac{\pi}{4}$$

Please note that $\sin^{-1} y$ does not mean $1/(\sin y)$; you should not think of -1 as an exponent when used as a superscript on a function.

An alternate notation for $x = \sin^{-1} y$ is $x = \arcsin y$ (x is the arcsine of y). This is appropriate notation, since $\pi/6 = \arcsin \frac{1}{2}$ could be interpreted as saying that $\pi/6$ is the arc (on the unit circle) whose sine is $\frac{1}{2}$.

Recall from Section 5-5 that if f is a one-to-one function, then

$$x = f^{-1}(y) \quad \text{if and only if} \quad y = f(x)$$

Here the corresponding statement is

$$x = \sin^{-1} y \quad \text{if and only if} \quad y = \sin x \quad \text{and} \quad -\frac{\pi}{2} \le x \le \frac{\pi}{2}$$

Moreover

$$\sin(\sin^{-1} y) = y \quad \text{for} \quad -1 \le y \le 1$$
$$\sin^{-1}(\sin x) = x \quad \text{for} \quad -\frac{\pi}{2} \le x \le \frac{\pi}{2}$$

The inverse sine function plays a significant role in calculus, where we often want to consider $y = \sin^{-1} x$. You will note that we have interchanged the roles of x and y so that x is now the domain variable for $\sin^{-1}$. On the graph, this corresponds to reflecting (folding) the graph of $y = \sin x$ across the line $y = x$ (see the figure in the margin).

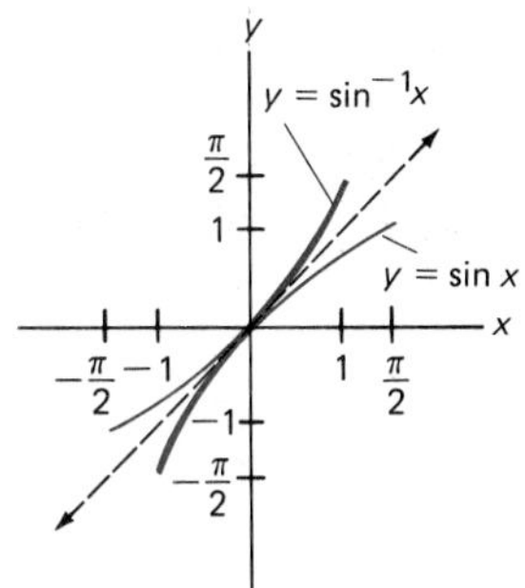

The Inverse Cosine One look at the graph of $y = \cos x$ should convince you that we cannot restrict the domain of the cosine to the same interval as that for the sine. We choose rather to use the interval $0 \le x \le \pi$, in which the cosine is one-to-one.

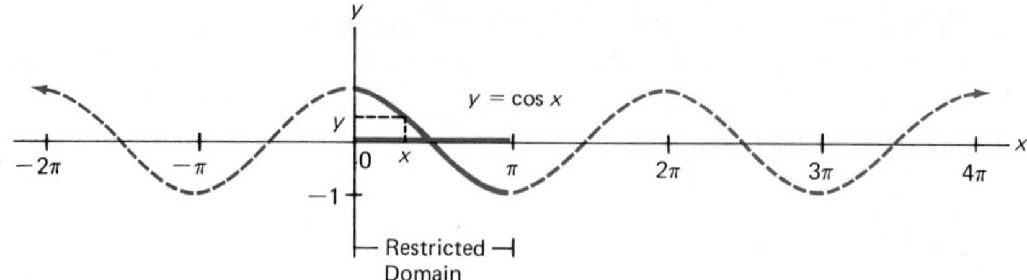

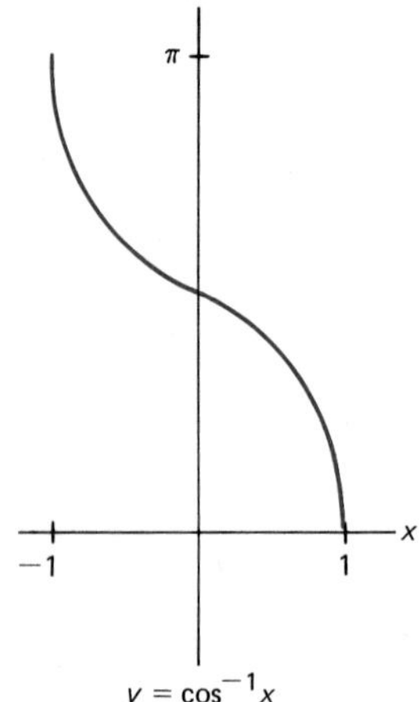

Having made the needed restriction, we may reasonably talk about $\cos^{-1}$. Moreover,

$$x = \cos^{-1} y \quad \text{if and only if} \quad y = \cos x \quad \text{and} \quad 0 \le x \le \pi$$

In particular,

$$\cos^{-1} 1 = 0$$

$$\cos^{-1} \frac{\sqrt{3}}{2} = \frac{\pi}{6}$$

$$\cos^{-1} 0 = \frac{\pi}{2}$$

$$\cos^{-1}(-1) = \pi$$

The graph of $y = \cos^{-1} x$ is shown in the margin. It is the graph of $y = \cos x$ reflected across the line $y = x$.

The Inverse Tangent To make $y = \tan x$ have an inverse, we restrict x to $-\pi/2 < x < \pi/2$. Thus

$$x = \tan^{-1} y \quad \text{if and only if} \quad y = \tan x \quad \text{and} \quad -\frac{\pi}{2} < x < \frac{\pi}{2}$$

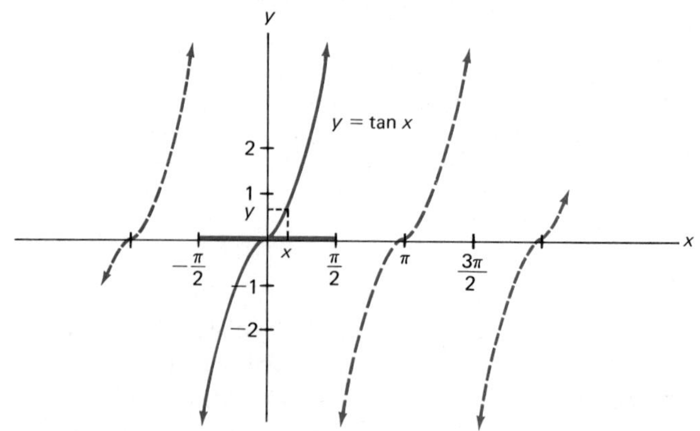

You will be asked to graph $y = \tan^{-1} x$ in Problem 59.

The three other inverse trigonometric functions, $\cot^{-1} x$, $\sec^{-1} x$, and $\csc^{-1} x$, are of less importance. They are introduced in Problem 64.

Three Identities Here are three identities connecting sines, cosines, and their inverses.

(i) $$\cos(\sin^{-1} x) = \sqrt{1 - x^2}$$

(ii) $$\sin(\cos^{-1} x) = \sqrt{1 - x^2}$$

(iii) $$\sin^{-1} x + \cos^{-1} x = \frac{\pi}{2}$$

To prove the first identity, we let $\theta = \sin^{-1} x$. Remember that this means that $x = \sin \theta$, with $-\pi/2 \le \theta \le \pi/2$. Then

$$\cos(\sin^{-1} x) = \cos \theta = \pm\sqrt{1 - \sin^2 \theta} = \pm\sqrt{1 - x^2}$$

Finally, choose the plus sign because $\cos \theta$ is positive for $-\pi/2 \le \theta \le \pi/2$. The second identity is proved in a similar fashion.

To prove the third identity, let

$$\alpha = \sin^{-1} x \qquad \beta = \cos^{-1} x$$

and note that we must show that $\alpha + \beta = \pi/2$. Now

$$\begin{aligned}
\sin(\alpha + \beta) &= \sin \alpha \cos \beta + \cos \alpha \sin \beta \\
&= \sin(\sin^{-1} x) \cos(\cos^{-1} x) + \cos(\sin^{-1} x) \sin(\cos^{-1} x) \\
&= x \cdot x + \sqrt{1 - x^2} \cdot \sqrt{1 - x^2} \\
&= x^2 + 1 - x^2 \\
&= 1
\end{aligned}$$

From this, we conclude that $\alpha + \beta$ is either $\pi/2$ or some number that differs from $\pi/2$ by a multiple of 2π. But since $-\pi/2 \le \alpha \le \pi/2$ and $0 \le \beta \le \pi$, it follows that

$$-\frac{\pi}{2} \le \alpha + \beta \le \frac{3\pi}{2}$$

The only possibility on this interval is $\alpha + \beta = \pi/2$.

Problem Set 8-4

Find the exact value of each of the following.

1. $\sin^{-1}(\sqrt{3}/2)$
2. $\cos^{-1} \frac{1}{2}$
3. $\arcsin(\sqrt{2}/2)$
4. $\arccos(\sqrt{2}/2)$
5. $\tan^{-1} 0$
6. $\tan^{-1} 1$
7. $\tan^{-1} \sqrt{3}$
8. $\tan^{-1}(\sqrt{3}/3)$
9. $\arccos(-\frac{1}{2})$
10. $\arcsin(-\frac{1}{2})$

Use Table D or a calculator to find approximate values of the following. We remark that most scientific calculators are programmed to calculate the inverse trigonometric functions consistently with the definitions given in this section. First put your calculator in the radian mode. Then to do Problem 11, press .21823 [INV] [sin].

11. $\sin^{-1} .21823$
12. $\cos^{-1} .30582$
13. $\sin^{-1}(-0.21823)$
14. $\cos^{-1}(-0.30582)$
15. $\tan^{-1} .20660$
16. $\tan^{-1}(1.2602)$

Solve for t, where $0 \le t \le \pi/2$. *Use a calculator if you have one.*

17. $\sin t = .3416$
18. $\cos t = .9812$
19. $\tan t = 3.345$
20. $\sec t = 1.342$

Find the following without the use of tables or a calculator.

21. $\sin(\sin^{-1} \frac{2}{3})$
22. $\cos(\cos^{-1}(-\frac{1}{4}))$
23. $\tan(\tan^{-1} 10)$
24. $\cos^{-1}(\cos(\pi/2))$
25. $\sin^{-1}(\sin(\pi/3))$
26. $\tan^{-1}(\tan(\pi/4))$
27. $\sin^{-1}(\cos(\pi/4))$
28. $\cos^{-1}(\sin(-\pi/6))$
29. $\cos(\sin^{-1} \frac{4}{5})$
30. $\sin(\cos^{-1} \frac{3}{5})$

(*Hint*: Use the identities established on page 321.)

31. $\cos(\tan^{-1} \frac{1}{2})$
32. $\cos(\tan^{-1}(-\frac{3}{4}))$

Use a calculator to find each value.

[C] 33. $\cos(\sin^{-1}(-.2564))$
[C] 34. $\tan^{-1}(\sin 14.1)$
[C] 35. $\sin^{-1}(\cos 1.12)$
[C] 36. $\cos^{-1}(\cos^{-1} .91)$

Example A (Complicated evaluations involving inverses) Evaluate: (a) $\sin(2 \cos^{-1} \frac{2}{3})$; (b) $\tan(\tan^{-1} 2 + \sin^{-1} \frac{4}{5})$.

SOLUTION. (a) Let $\theta = \cos^{-1}(\frac{2}{3})$ so that $\cos \theta = \frac{2}{3}$ and

$$\sin \theta = \sqrt{1 - \cos^2 \theta} = \sqrt{1 - \tfrac{4}{9}} = \frac{\sqrt{5}}{3}$$

Then apply the double-angle formula for $\sin 2\theta$ as indicated below.

$$\begin{aligned} \sin\left(2 \cos^{-1} \frac{2}{3}\right) &= \sin 2\theta \\ &= 2 \sin \theta \cos \theta \\ &= 2\frac{\sqrt{5}}{3} \cdot \frac{2}{3} \\ &= \tfrac{4}{9}\sqrt{5} \end{aligned}$$

(b) Let $\alpha = \tan^{-1} 2$ and $\beta = \sin^{-1}(\frac{4}{5})$ and apply the identity

$$\tan(\alpha + \beta) = \frac{\tan \alpha + \tan \beta}{1 - \tan \alpha \tan \beta}$$

Now $\tan\alpha = 2$ and

$$\tan\beta = \frac{\sin\beta}{\cos\beta} = \frac{\frac{4}{5}}{\sqrt{1 - (\frac{4}{5})^2}} = \frac{\frac{4}{5}}{\frac{3}{5}} = \frac{4}{3}$$

Therefore

$$\tan(\alpha + \beta) = \frac{2 + \frac{4}{3}}{1 - 2\cdot\frac{4}{3}} = -2$$

Evaluate.

37. $\sin(2\cos^{-1}\frac{3}{5})$

38. $\sin(2\cos^{-1}\frac{1}{2})$

39. $\cos(2\sin^{-1}(-\frac{3}{5}))$

40. $\tan(2\tan^{-1}\frac{1}{3})$

41. $\sin(\cos^{-1}\frac{3}{5} + \cos^{-1}\frac{5}{13})$

42. $\tan(\tan^{-1}\frac{1}{2} + \tan^{-1}(-3))$

Example B (More identities) Show that

$$\cos(2\tan^{-1} x) = \frac{1 - x^2}{1 + x^2}$$

SOLUTION. We will apply the double-angle formula

$$\cos 2\theta = 2\cos^2\theta - 1$$

Here $\theta = \tan^{-1} x$, so that $x = \tan\theta$. Then

$$\begin{aligned}
\cos(2\tan^{-1} x) &= \cos(2\theta) \\
&= 2\cos^2\theta - 1 \\
&= \frac{2}{\sec^2\theta} - 1 \\
&= \frac{2}{1 + \tan^2\theta} - 1 \\
&= \frac{2}{1 + x^2} - 1 \\
&= \frac{1 - x^2}{1 + x^2}
\end{aligned}$$

Show that each of the following is an identity.

43. $\tan(\sin^{-1} x) = \dfrac{x}{\sqrt{1 - x^2}}$

44. $\sin(\tan^{-1} x) = \dfrac{x}{\sqrt{1 + x^2}}$

45. $\tan(2\tan^{-1} x) = \dfrac{2x}{1 - x^2}$

46. $\cos(2\sin^{-1} x) = 1 - 2x^2$

Example C (Finding complicated inverses) If $f(x) = 3 \sin 2x$, find a formula for $f^{-1}(x)$.

SOLUTION. Clearly the domain must be restricted before an inverse exists. We require that $-\pi/2 \le 2x \le \pi/2$ and, hence, $-\pi/4 \le x \le \pi/4$. Then we write $y = 3 \sin 2x$ and solve for x.

$$y = 3 \sin 2x$$

$$\frac{y}{3} = \sin 2x$$

$$2x = \sin^{-1} \frac{y}{3}$$

$$x = f^{-1}(y) = \frac{1}{2} \sin^{-1} \frac{y}{3}$$

Finally, we replace y by x to get the desired formula.

$$f^{-1}(x) = \frac{1}{2} \sin^{-1} \frac{x}{3}$$

In each of the following, indicate the restricted domain for which f has an inverse. Then find a formula for $f^{-1}(x)$.

47. $f(x) = 3 \cos 2x$

48. $f(x) = 2 \sin 3x$

49. $f(x) = \sin \dfrac{1}{x}$

50. $f(x) = \frac{1}{2} \tan x$

Miscellaneous Problems

51. Find the exact value of $\tan^{-1}(\sqrt{3}/3)$.

52. Find the exact value of $\arccos(-1/2)$.

53. Use Table D or a calculator to find the approximate value of each expression.
(a) $\tan^{-1}(0.20660)$ (b) $\cos^{-1}(0.68222)$ (c) $\sin^{-1}(-.88196)$

Find the value of each expression in Problems 54–57.

54. $\cos(\cos^{-1} \frac{4}{5})$

55. $\sin(\sin^{-1}(-0.9))$

56. $\cos(\sin^{-1}(-\frac{5}{13}))$

57. $\sin(\cos^{-1}(\frac{2}{5}) - \tan^{-1} 2)$

58. How should the domain of $f(x) = \tan(x/2)$ be restricted so that it has an inverse? Find $f^{-1}(x)$.

59. Sketch the graph of $y = \tan^{-1} x$, being sure to indicate its horizontal asymptotes. What is the domain of $\tan^{-1}$? The range?

[C] 60. Calculate each value.
(a) $\cos^{-1}(.94)$ (b) $\sin^{-1}(-.49286)$
(c) $\sin(\cos^{-1} .1324)$ (d) $\tan(\tan^{-1} 1.21 + \cos^{-1} .345)$

[C] 61. Calculate each value. Why are they so close to each other?
(a) $\tan^{-1}(8000)$ (b) $\pi/2$

[C] 62. Try to calculate $\ln(\sin^{-1}(-0.2))$. What is wrong?

[C] 63. Calculate

$$\frac{(\sin^{-1}(.41))^2\sqrt{\cos^{-1}(.9)}}{\pi \tan^{-1}(.3)}$$

64. To determine inverses for cotangent, secant, and cosecant, we restrict their domains to $0 < x < \pi$, $0 \le x \le \pi$ $(x \neq \pi/2)$, and $-\pi/2 \le x \le \pi/2$ $(x \neq 0)$, respectively. With these restrictions understood, evaluate each expression.

(a) $\cot^{-1} 1$ (b) $\cot^{-1}(-1)$ (c) $\cot^{-1} 0$
(d) $\sec^{-1} 1$ (e) $\sec^{-1} 2$ (f) $\sec^{-1}(-2)$
(g) $\csc^{-1} 1$ (h) $\csc^{-1} 2$ (i) $\csc^{-1}(-2/\sqrt{3})$

65. It is always true that $\sin(\sin^{-1} x) = x$, but not always true that $\sin^{-1}(\sin x) = x$. For example,

$$\sin^{-1}\left(\sin \frac{3\pi}{2}\right) \neq \frac{3\pi}{2}$$

Instead,

$$\sin^{-1}\left(\sin\frac{3\pi}{2}\right) = \sin^{-1}(-1) = -\frac{\pi}{2}$$

Find each value.

(a) $\sin^{-1}(\sin 4\pi)$ (b) $\sin^{-1}(\sin(2\pi/3))$
(c) $\cos^{-1}(\cos(2\pi/3))$ (d) $\cos^{-1}(\cos(4\pi/3))$

66. Solve for x.

(a) $\cos(\sin^{-1} x) = 1$ (b) $\sin(\cos^{-1} x) = 1$
(c) $\sin^{-1}(2x + 3) = \frac{\pi}{6}$ (d) $\cos^{-1}(x^2 - 1) = \pi$

67. For each of the following right triangles, write θ explicitly in terms of x.

(a)

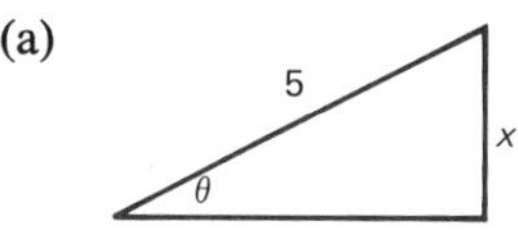

(b)

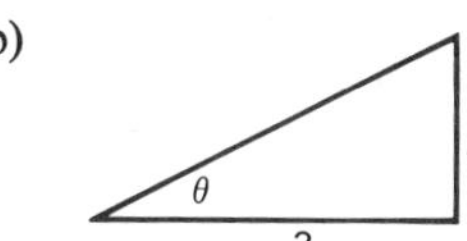

(c)

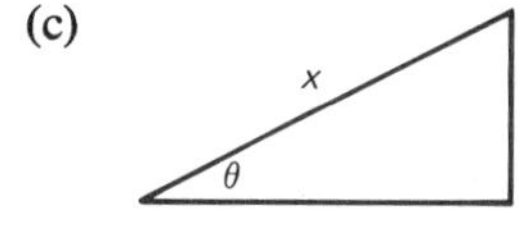

(d)

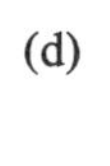
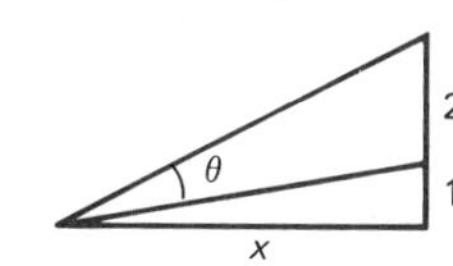

68. A picture 4 feet high is hung on a wall so that its bottom is 9.6 feet from the floor. A viewer whose eye level is 5.2 feet from the floor stands b feet from the wall. Express θ, the vertical angle subtended by the picture at her eye, explicitly in terms of b.

[C] 69. Find θ in Problem 68 when $b = 12.3$ feet.

CHAPTER SUMMARY

An **identity** is an equality that is true for all values of the unknown for which both sides of the equality make sense. Our first task was to establish the fundamental identities of trigonometry, here arranged by category.

Basic Identities

1. $\tan t = \dfrac{\sin t}{\cos t}$

2. $\cot t = \dfrac{\cos t}{\sin t} = \dfrac{1}{\tan t}$

3. $\sec t = \dfrac{1}{\cos t}$

4. $\csc t = \dfrac{1}{\sin t}$

5. $\sin^2 t + \cos^2 t = 1$

6. $1 + \tan^2 t = \sec^2 t$

7. $1 + \cot^2 t = \csc^2 t$

Cofunction Identities

8. $\sin\left(\dfrac{\pi}{2} - t\right) = \cos t$

9. $\cos\left(\dfrac{\pi}{2} - t\right) = \sin t$

Odd-Even Identities

10. $\sin(-t) = -\sin t$

11. $\cos(-t) = \cos t$

Addition Formulas

12. $\sin(s + t) = \sin s \cos t + \cos s \sin t$
13. $\sin(s - t) = \sin s \cos t - \cos s \sin t$
14. $\cos(s + t) = \cos s \cos t - \sin s \sin t$
15. $\cos(s - t) = \cos s \cos t + \sin s \sin t$

Double-Angle Formulas

16. $\sin 2t = 2 \sin t \cos t$
17. $\cos 2t = \cos^2 t - \sin^2 t = 1 - 2 \sin^2 t = 2 \cos^2 t - 1$

Half-Angle Formulas

18. $\sin \dfrac{t}{2} = \pm\sqrt{\dfrac{1 - \cos t}{2}}$

19. $\cos \dfrac{t}{2} = \pm\sqrt{\dfrac{1 + \cos t}{2}}$

Once we have memorized the fundamental identities, we can use them to prove thousands of other identities. The suggested technique is to take one side of a proposed identity and show by a chain of equalities that it is equal to the other.

A **trigonometric equation** is an equality involving trigonometric functions that is true only for some values of the unknown (for example, $\sin 2t = \frac{1}{2}$). Here our job is to solve the equation, that is, to find the values of the unknown that make it true.

With their natural domains, the trigonometric functions are not one-to-one and therefore do not have inverses. However, there are standard ways to restrict the domains so that inverses exist. Here are the results.

$$x = \sin^{-1} y \quad \text{means} \quad y = \sin x \quad \text{and} \quad \frac{-\pi}{2} \le x \le \frac{\pi}{2}$$

$$x = \cos^{-1} y \quad \text{means} \quad y = \cos x \quad \text{and} \quad 0 \le x \le \pi$$

$$x = \tan^{-1} y \quad \text{means} \quad y = \tan x \quad \text{and} \quad \frac{-\pi}{2} < x < \frac{\pi}{2}$$

CHAPTER REVIEW PROBLEM SET

1. Prove that the following are identities.
   (a) $\cot\theta\cos\theta = \csc\theta - \sin\theta$

   (b) $\dfrac{\cos x\tan^2 x}{\sec x + 1} = 1 - \cos x$

2. Express each of the following in terms of $\sin x$ and simplify.

   (a) $\dfrac{(\cos^2 x - 1)(1 + \tan^2 x)}{\csc x}$ (b) $\dfrac{\cos^2 x\csc x}{1 + \csc x}$

3. Use appropriate identities to simplify and then calculate each of the following.
   (a) $2\cos^2 22.5° - 1$
   (b) $\sin 37°\cos 53° + \cos 37°\sin 53°$
   (c) $\cos 108°\cos 63° + \sin 108°\sin 63°$
4. If $\cos t = -\frac{4}{5}$ and $\pi < t < 3\pi/2$, calculate
   (a) $\sin 2t$ (b) $\sin(t/2)$
5. Prove that the following are identities.
   (a) $\sin 2t\cos t - \cos 2t\sin t = \sin t$

   (b) $\sec 2t + \tan 2t = \dfrac{\cos t + \sin t}{\cos t - \sin t}$

   (c) $\dfrac{\cos(\alpha + \beta)}{\cos\alpha\cos\beta} = \tan\alpha(\cot\alpha - \tan\beta)$

6. Solve the following trigonometric equations for t, $0 \le t < 2\pi$.
   (a) $\cos t = -\sqrt{3}/2$
   (b) $(2\sin t + 1)\tan t = 0$
   (c) $\cos^2 t + 2\cos t - 3 = 0$
   (d) $\sin t - \cos t = 1$
   (e) $\sin 3t = 1$
7. What is the standard way to restrict the domain of sine, cosine, and tangent so that they have inverses?
8. Calculate each of the following.
   (a) $\sin^{-1}(-\sqrt{3}/2)$ (b) $\cos^{-1}(-\sqrt{3}/2)$
   (c) $\tan^{-1}(-\sqrt{3})$ (d) $\tan(\tan^{-1} 6)$
   (e) $\cos^{-1}(\cos 3\pi)$ (f) $\sin(\cos^{-1}\frac{2}{3})$
   (g) $\cos(2\cos^{-1} .7)$ (h) $\sin(2\cos^{-1}\frac{5}{13})$
9. Sketch the graph of $y = \tan^{-1} x$.
10. Find an approximate value for $\tan^{-1}(-1000)$.

Thus one sees in the sciences many brilliant theories which have remained unapplied for a long time suddenly becoming the foundation of most important applications, and likewise applications very simple in appearance giving birth to ideas of the most abstract theories.

Marquis de Condorcet

CHAPTER NINE

Applications of Trigonometry

9-1 Oblique Triangles: Law of Sines

9-2 Oblique Triangles: Law of Cosines

9-3 Simple Harmonic Motion

9-4 The Polar Coordinate System

9-5 Polar Representation of Complex Numbers

9-6 Powers and Roots of Complex Numbers

The Law of Sines

Consider an arbitrary triangle with angles α, β, γ, and corresponding opposite sides a, b, c, respectively. Then

$$\frac{\sin \alpha}{a} = \frac{\sin \beta}{b} = \frac{\sin \gamma}{c}$$

Equivalently,

$$\frac{a}{\sin \alpha} = \frac{b}{\sin \beta} = \frac{c}{\sin \gamma}$$

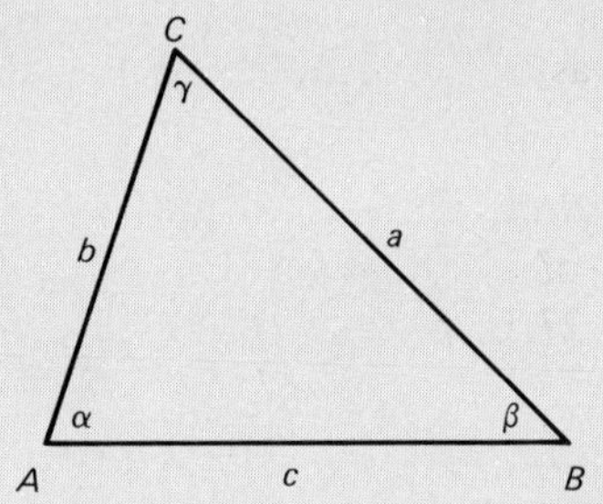

9-1 Oblique Triangles: Law of Sines

We learned in Section 7-1 how to solve a right triangle. But can we solve an oblique triangle, that is, one without a 90° angle? One valuable tool is the **Law of Sines** stated above. It is valid for any triangle whatever, but we initially establish it for the case where all angles are acute.

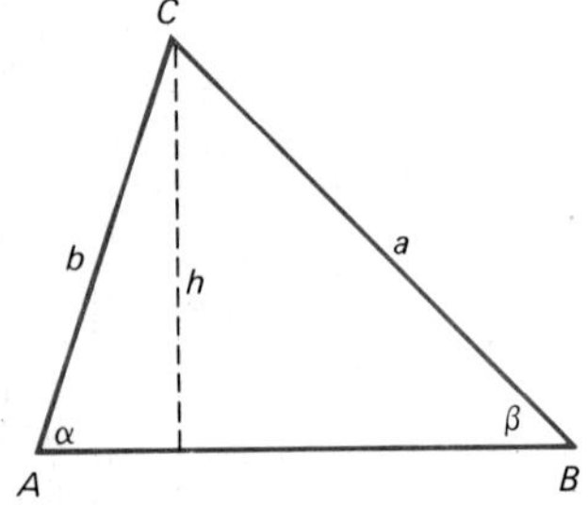

Proof of the Law of Sines Consider a triangle with all acute angles, labeled as in the margin. Drop a perpendicular of length h from vertex C to the opposite side. Then by right-triangle trigonometry,

$$\sin \alpha = \frac{h}{b} \qquad \sin \beta = \frac{h}{a}$$

If we solve for h in these two equations and equate the results, we obtain

$$b \sin \alpha = a \sin \beta$$

Finally, dividing both sides by ab yields

$$\frac{\sin \alpha}{a} = \frac{\sin \beta}{b}$$

Since the roles of β and γ can be interchanged, the same reasoning gives

$$\frac{\sin \alpha}{a} = \frac{\sin \gamma}{c}$$

Next consider a triangle with an obtuse angle α ($90° < \alpha < 180°$). Drop a perpendicular of length h from vertex C to the extension of AB (see the diagram in the margin). Notice that angle α' is the reference angle for α and so $\sin \alpha = \sin \alpha'$. It follows from right-triangle trigonometry that

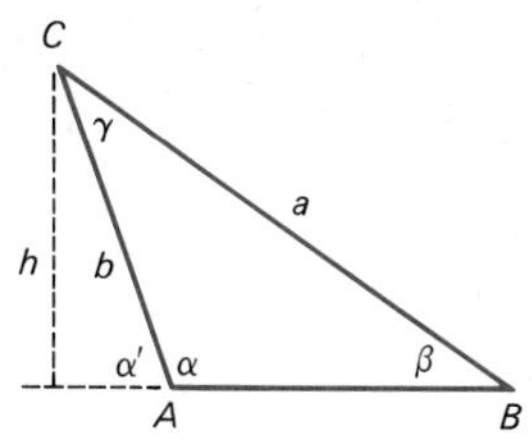

$$\sin \alpha = \sin \alpha' = \frac{h}{b} \qquad \sin \beta = \frac{h}{a}$$

just as in the acute case. The rest of the argument is identical with that case.

Solving a Triangle (AAS) Suppose that we know two angles and any side of a triangle. For example, suppose that in triangle ABC, $\alpha = 103.5°$, $\beta = 27.5°$, and $c = 45.3$. Our task is to find γ, a, and b.

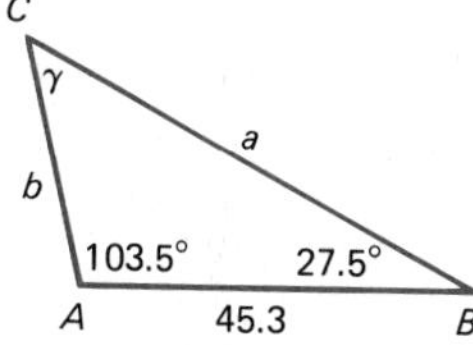

1. Since $\alpha + \beta + \gamma = 180°$, $\gamma = 180° - (103.5° + 27.5°) = 49°$.
2. By the Law of Sines,

$$\frac{a}{\sin 103.5°} = \frac{45.3}{\sin 49°}$$

$$a = \frac{(45.3)(\sin 103.5°)}{\sin 49°}$$

$$= \frac{(45.3)(\sin 76.5°)}{\sin 49°}$$

$$= \frac{(45.3)(.9724)}{.7547}$$

$$\approx 58.4$$

3. Also by the Law of Sines,

$$\frac{b}{\sin 27.5°} = \frac{45.3}{\sin 49°}$$

$$b = \frac{(45.3)(\sin 27.5°)}{\sin 49°}$$

$$= \frac{(45.3)(.4617)}{.7547}$$

$$\approx 27.7$$

Solving a Triangle (SSA) Suppose that two sides and the angle opposite one of them are given. This is called the **ambiguous case** because the given information may not determine a unique triangle.

If α, a, and b are given, we consider trying to construct a triangle fitting these data by first drawing angle α, then marking off b on one of its sides thus determining vertex C. Finally, we attempt to locate vertex B by striking off a circular arc of radius a with center at C. If $a \geq b$, this can always be done in a unique way. The next diagrams illustrate this both for α acute and α obtuse.

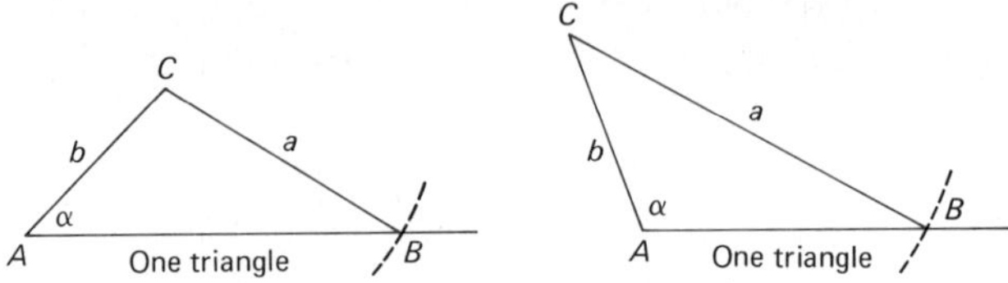

However if $a < b$, there are several possibilities.

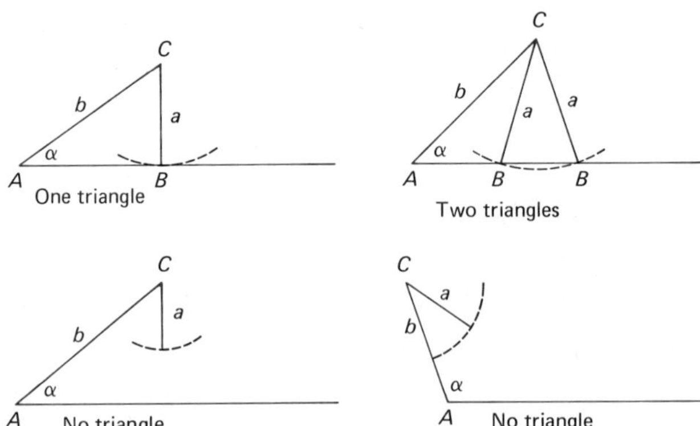

Fortunately, we are able to decide which of these possibilities is the case if we draw an approximate picture and then attempt to apply the Law of Sines. First, note that if $a \geq b$ there is one triangle corresponding to the data and for it β is an acute angle. Application of the Law of Sines will give $\sin \beta$ which allows determination of β.

If $a < b$, we may attempt to apply the Law of Sines. If it yields $\sin \beta = 1$, we have a unique right triangle. If it yields $\sin \beta < 1$, we have two triangles corresponding to the two angles β_1 and β_2 (one acute, the other obtuse) with this sine. If it yields $\sin \beta > 1$, we have an inconsistency in the data; no triangle satisfying the data exists.

Suppose, for example, that we are given $\alpha = 36°$, $a = 9.4$, and $b = 13.1$. We must find β, γ, and c. Since $a < b$, there may be zero, one, or two triangles. We proceed to compute $\sin \beta$.

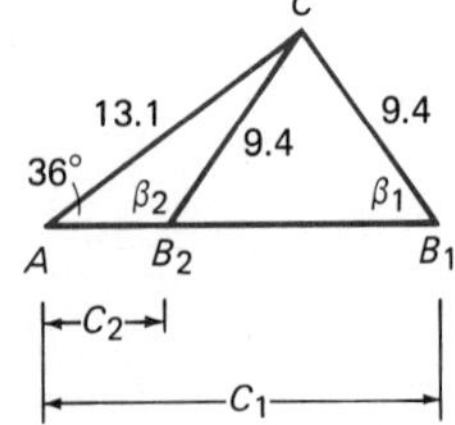

1. $$\frac{\sin \beta}{13.1} = \frac{\sin 36°}{9.4}$$

$$\sin \beta = \frac{(13.1) \sin 36°}{9.4} = \frac{(13.1)(.5878)}{9.4} \approx .8912$$

Since $\sin \beta < 1$, there are two triangles.

$$\beta_1 = 55° \qquad \beta_2 = 125°$$

2. $\gamma_1 = 180° - (36° + 55°) = 89°$
$\gamma_2 = 180° - (36° + 125°) = 19°$

3. $\dfrac{c_1}{\sin 89^\circ} = \dfrac{9.4}{\sin 36^\circ}$

$$c_1 = \frac{9.4}{\sin 36^\circ}(\sin 89^\circ)$$

$$= \frac{9.4}{.5878}(.9998) \approx 16.0$$

$$\frac{c_2}{\sin 19^\circ} = \frac{9.4}{\sin 36^\circ}$$

$$c_2 = \frac{9.4}{\sin 36^\circ}(\sin 19^\circ)$$

$$= \frac{9.4}{.5878}(.3256) \approx 5.2$$

Problem Set 9-1

Solve the triangles of Problems 1–10 using either Table C or a calculator.

1. $\alpha = 42.6^\circ, \beta = 81.9^\circ, a = 14.3$
2. $\beta = 123^\circ, \gamma = 14.2^\circ, a = 295$
3. $\alpha = \gamma = 62^\circ, b = 50$
4. $\alpha = \beta = 14^\circ, c = 30$
5. $\alpha = 115^\circ, a = 46, b = 34$
6. $\beta = 143^\circ, a = 46, b = 84$
7. $\alpha = 30^\circ, a = 8, b = 5$
8. $\beta = 60^\circ, a = 11, b = 12$
9. $\alpha = 30^\circ, a = 5, b = 8$
10. $\beta = 60^\circ, a = 12, b = 11$
11. Two observers stationed 110 meters apart at A and B on the bank of a river are looking at a tower situated at a point C on the opposite bank. They measure angles CAB and CBA to be 43° and 57°, respectively. How far is the first observer from the tower?
12. A telegraph pole leans away from the sun at an angle of 11° to the vertical. The pole casts a shadow 96 feet long on horizontal ground when the angle of elevation of the sun is 23°. Find the length of the pole.
13. A vertical pole 60 feet long is standing by the side of an inclined road. It casts a shadow 138 feet long directly downhill along the road when the angle of elevation of the sun is 58°. Find the angle of inclination of the road.
14. Two forest rangers 15 miles apart at points A and B observe a fire at a point C. The ranger at A measures angle CAB as 43.6° and the one at B measures angle CBA as 79.3°. How far is the fire from each ranger? How far is the fire from a straight road that goes from A to B?

Miscellaneous Problems

15. Solve the triangle having $\alpha = 111^\circ$, $\beta = 34^\circ$, and $a = 360$.
16. Solve the triangle that satisfies $\alpha = 46^\circ$, $a = 23$, and $b = 25$.
17. Suppose that the north side of the roof of a house is 22 feet on the slant and the south side 30 feet. If the north side makes an angle of 75° with the horizontal, how wide is the house from north to south? Assume that the eaves are at the same height on the two sides of the house.

18. A level road leads directly away from the foot of a tower. From the top of the tower the angle of depression of a certain milestone is 58.4° and the angle of depression of the next milestone is 14.9°. How high is the tower?
19. Suppose that a vertical television tower stands on the top of a hill which is inclined 20° from the horizontal. From a point $\frac{1}{2}$ kilometer straight down the hill from the foot of the tower the angle of elevation of the top of the tower is 36°. How high is the tower?
20. A lighthouse stands at a certain distance out from a straight shore. It throws a beam of light that revolves at a constant rate of one revolution per minute. A short time after shining on the nearest point on the shore, the beam reaches a point on the shore that is 2640 feet from the lighthouse, and three seconds later it reaches a point 2000 feet farther along the shore. How far is the lighthouse from the shore?
21. A motorcycle course is in the shape of a triangle determined by points A, B, and C. Suppose that angle ABC measures 46.3° and that it takes a motorcycle half as long to ride from B to A as from A to C. Find the angle ACB.
22. From a point on the ground 100 yards away from the base of a building, the angle of elevation of the top of the building is 47°, and the angle of elevation of the top of a flagpole situated on the nearest edge of the roof of the building is 61°. Find the length of the flagpole and the height of the building.

The Law of Cosines

Consider an arbitrary triangle with angles α, β, γ and corresponding opposite sides $a, b, c,$ respectively. Then

$$a^2 = b^2 + c^2 - 2bc \cos \alpha$$

$$b^2 = a^2 + c^2 - 2ac \cos \beta$$

$$c^2 = a^2 + b^2 - 2ab \cos \gamma$$

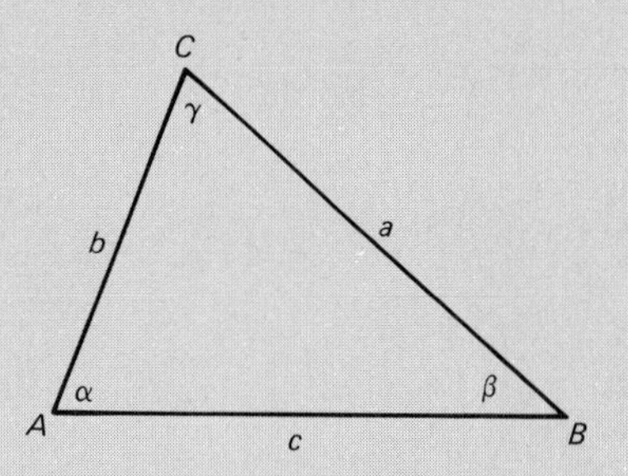

9-2 Oblique Triangles: Law of Cosines

When two sides and the included angle (SAS) or three sides (SSS) of a triangle are given, we cannot apply the Law of Sines to solve the triangle. Rather, we need the Law of Cosines, stated above in symbols. Actually it is wise to learn the law in words.

The square of any side is equal to the sum of the squares of the other two sides minus twice the product of those sides and the cosine of the angle between them.

Notice what happens when $\gamma = 90°$ so that $\cos \gamma = 0$. The Law of Cosines

$$c^2 = a^2 + b^2 - 2ab \cos \gamma$$

becomes

$$c^2 = a^2 + b^2$$

which is just the Pythagorean Theorem. In fact, you should think of the Law of Cosines as a generalization of the Pythagorean Theorem with the term $-2ab \cos \gamma$ acting as a correction term when γ is not 90°.

Proof of the Law of Cosines Assume first that angle α is acute. Drop a perpendicular CD from vertex C to side AB as shown in the diagram at the top of page 336. Label the lengths of CD, AD, and DB by h, x, and $c - x$, respectively.

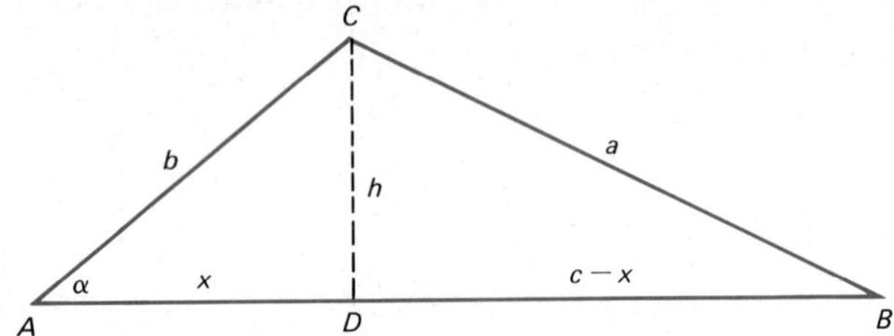

Consider the two right triangles ADC and BDC. By the Pythagorean Theorem

$$h^2 = b^2 - x^2 \quad \text{and} \quad h^2 = a^2 - (c - x)^2$$

Equating these two expressions for h^2 gives

$$\begin{aligned} a^2 - (c - x)^2 &= b^2 - x^2 \\ a^2 &= b^2 - x^2 + (c - x)^2 \\ a^2 &= b^2 - x^2 + c^2 - 2cx + x^2 \\ a^2 &= b^2 + c^2 - 2cx \end{aligned}$$

Now $\cos \alpha = x/b$ and so $x = b \cos \alpha$. Thus

$$a^2 = b^2 + c^2 - 2cb \cos \alpha$$

which is the result we wanted.

Next we give the proof of the Law of Cosines for the obtuse angle case. Again drop a perpendicular from vertex C to side AB extended and label the resulting diagram as shown below.

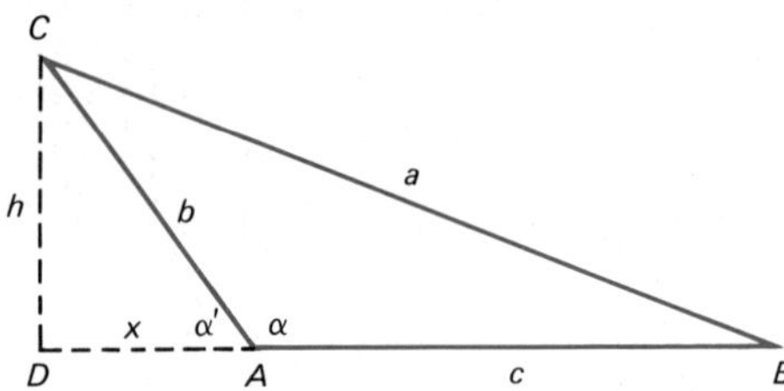

From consideration of triangles ADC and BDC and the Pythagorean Theorem, we obtain

$$h^2 = b^2 - x^2 \quad \text{and} \quad h^2 = a^2 - (c + x)^2$$

Algebra analogous to that used in the acute angle case yields

$$a^2 = b^2 + c^2 + 2cx$$

Now α' is the reference angle for α and so $\cos \alpha = -\cos \alpha'$. Also $\cos \alpha' = x/b$. Therefore

$$x = b \cos \alpha' = -b \cos \alpha$$

When we substitute this expression for x in the equation above, we get

$$a^2 = b^2 + c^2 - 2cb \cos \alpha$$

Solving a Triangle (SAS) Consider a triangle with $b = 18.1$, $c = 12.3$, and $\alpha = 115°$. We want to determine a, β, and γ.

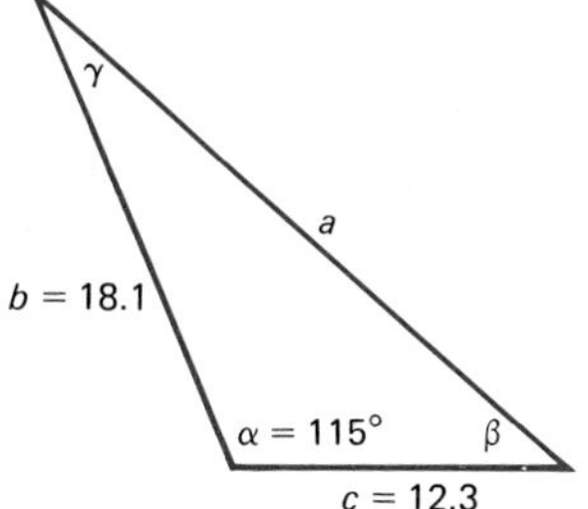

1. By the Law of Cosines,

$$\begin{aligned} a^2 &= (18.1)^2 + (12.3)^2 - 2(18.1)(12.3) \cos 115° \\ &= 327.61 + 151.29 - (445.80)(-\cos 65°) \\ &= 327.61 + 151.29 + (445.80)(.4226) \\ &= 667.30 \\ a &\approx 25.8 \end{aligned}$$

2. Now we can use the Law of Sines.

$$\frac{\sin \beta}{18.1} = \frac{\sin 115°}{25.8}$$

$$\sin \beta = \frac{(18.1) \sin 115°}{25.8} = \frac{(1.81)(\sin 65°)}{25.8}$$

$$= \frac{(18.1)(.9063)}{25.8} = .6358$$

$$\beta \approx 39.5°$$

3. $\gamma \approx 180° - (115° + 39.5°) = 25.5°$.

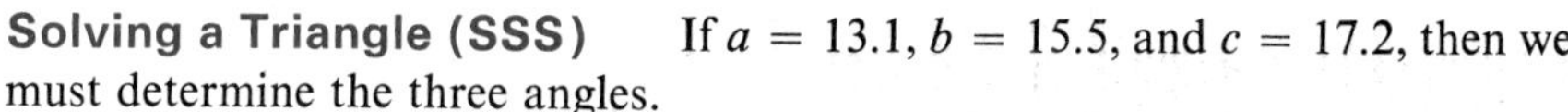

Solving a Triangle (SSS) If $a = 13.1$, $b = 15.5$, and $c = 17.2$, then we must determine the three angles.

1. By the Law of Cosines,

$$a^2 = b^2 + c^2 - 2bc \cos \alpha$$

Thus

$$\cos \alpha = \frac{b^2 + c^2 - a^2}{2bc}$$

$$= \frac{(15.5)^2 + (17.2)^2 - (13.1)^2}{2(15.5)(17.2)} = .6836$$

$$\alpha \approx 46.9°$$

2. By the Law of Sines,

$$\frac{\sin \beta}{15.5} = \frac{\sin 46.9^\circ}{13.1}$$

$$\sin \beta = \frac{(15.5)(\sin 46.9^\circ)}{13.1} = \frac{(15.5)(.7302)}{13.1} = .8640$$

$$\beta \approx 59.8^\circ$$

3. $\gamma \approx 180^\circ - (46.9^\circ + 59.8^\circ) = 73.3^\circ$.

Problem Set 9-2

In Problems 1–8, solve the triangles satisfying the given data. Use either Table C or a calculator.

1. $\alpha = 60^\circ$, $b = 14$, $c = 10$
2. $\beta = 60^\circ$, $a = c = 8$
3. $\gamma = 120^\circ$, $a = 8$, $b = 10$
4. $\alpha = 150^\circ$, $b = 35$, $c = 40$
5. $a = 5$, $b = 6$, $c = 7$
6. $a = 10$, $b = 20$, $c = 25$
7. $a = 12.2$, $b = 19.1$, $c = 23.8$
8. $a = .11$, $b = .21$, $c = .31$
9. At one corner of a triangular field, the angle measures 52.4°. The sides that meet at this corner are 100 meters and 120 meters long. How long is the third side?
10. To approximate the distance between two points A and B on opposite sides of a swamp, a surveyor selects a point C and measures it to be 140 meters from A and 260 meters from B. Then he measures the angle ACB, which turns out to be 49°. What is the calculated distance from A to B?

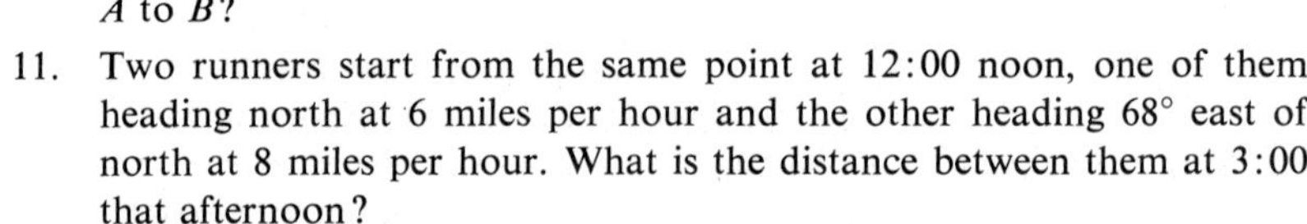

11. Two runners start from the same point at 12:00 noon, one of them heading north at 6 miles per hour and the other heading 68° east of north at 8 miles per hour. What is the distance between them at 3:00 that afternoon?
12. A 50-foot pole stands on top of a hill which slants 20° from the horizontal. How long must a rope be to reach from the top of the pole to a point 88 feet directly downhill (that is, on the slant) from the base of the pole?
13. A triangular garden plot has sides of length 35 meters, 40 meters, and 60 meters. Find the largest angle of the triangle.
14. A piece of wire 60 inches long is bent into the shape of a triangle. Find the angles of the triangle if two of the sides have lengths 24 inches and 20 inches.

Miscellaneous Problems

15. Solve the triangle that satisfies $b = 25$, $c = 37$, and $\alpha = 121^\circ$.
16. Solve the triangle that has sides $a = 39$, $b = 71$, and $c = 65$.
17. A parallelogram has sides of length 30 inches and 10 inches and the longer diagonal has length 36 inches. Find the obtuse angle of the parallelogram and also the length of the shorter diagonal.
18. The larger angle of a parallelogram measures 124°. If its sides have length 20 cm and 24 cm, find the length of the shorter diagonal.

19. A man stands at a point 500 meters from one end of a pond and 700 meters from the other. He reads the bearing of the first end as 43° west of north and that of the other as 56° east of north. Find the length of the pond.

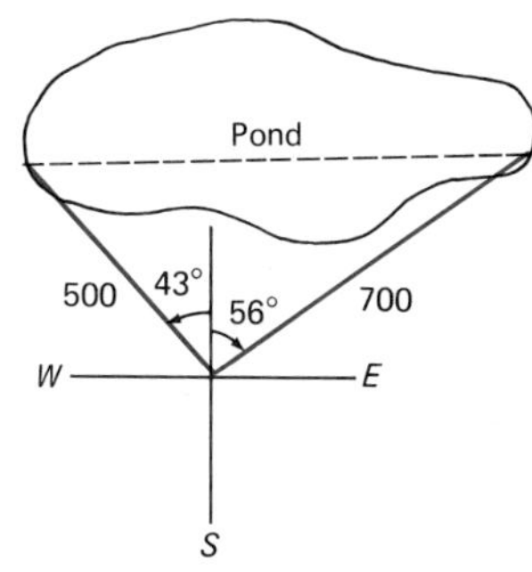

20. Three runners are practicing on a triangular course ABC. Their speeds are in the ratio of 2:3:4. If they start simultaneously from three different corners and also arrive at the next corner, going clockwise, at the same time, what are the three angles of the triangle?

21. Here is an important **area formula**. The area of any triangle is equal to one-half the product of any two sides multiplied by the sine of their included angle. Derive this by showing

$$A = \tfrac{1}{2}bc \sin \alpha$$

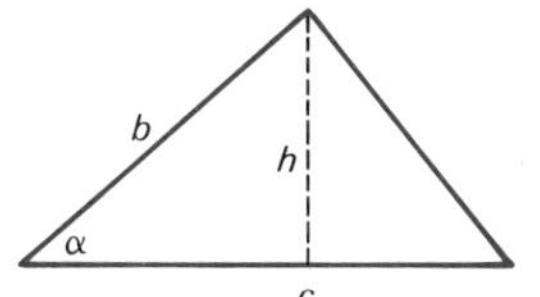

both for the case where α is acute and where α is obtuse. (*Hint*: Consider h in the accompanying diagrams.)

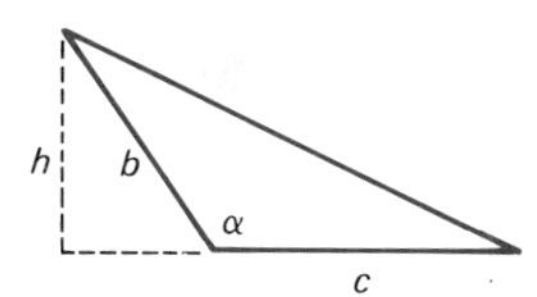

22. Calculate the area of the triangle satisfying the given data $a = 20$, $b = 40$, and $\gamma = 32°$ (see Problem 21).

[C] 23. Calculate the area of the triangle given $a = 14.6$, $b = 32.8$, and $c = 35.4$ (*Hint*: First find any angle and then use Problem 21.)

24. Let r be the radius of the inscribed circle of a triangle ABC. See the diagram in the margin. Show that the area of the triangle is given by rs, where $s = (a + b + c)/2$. (*Hint*: Divide the triangle ABC into three smaller triangles by drawing line segments from A, B, and C to the center of the inscribed circle.)

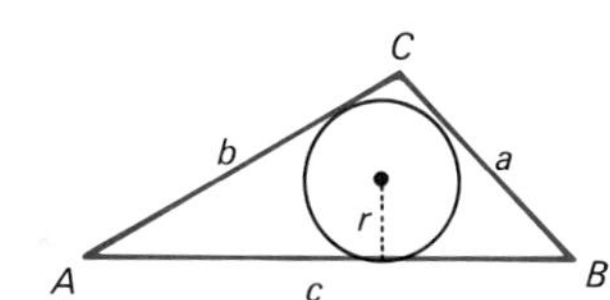

25. Find the radius r of the inscribed circle of the triangle in Problem 23 using your answer to that problem. (See Problem 24.)

26. It can be shown that the area of a triangle ABC is also given by

$$A = \sqrt{s(s - a)(s - b)(s - c)}$$

where $s = \frac{1}{2}(a + b + c)$. This is called Heron's formula. Use the formula to find the area of each of the following triangles.

(a) $a = 14$, $b = 21$, $c = 29$

[C] (b) $a = 13.2$, $b = 49.1$, $c = 55.7$

27. Show that the radius of the inscribed circle of a triangle ABC is given by

$$\sqrt{\frac{(s - a)(s - b)(s - c)}{s}}$$

where $s = (a + b + c)/2$.
(*Hint*: See Problems 24 and 26.)

28. Use the formula in Problem 27 to find the radius of the inscribed circle of the triangle in which $a = 3$, $b = 4$, and $c = 5$.

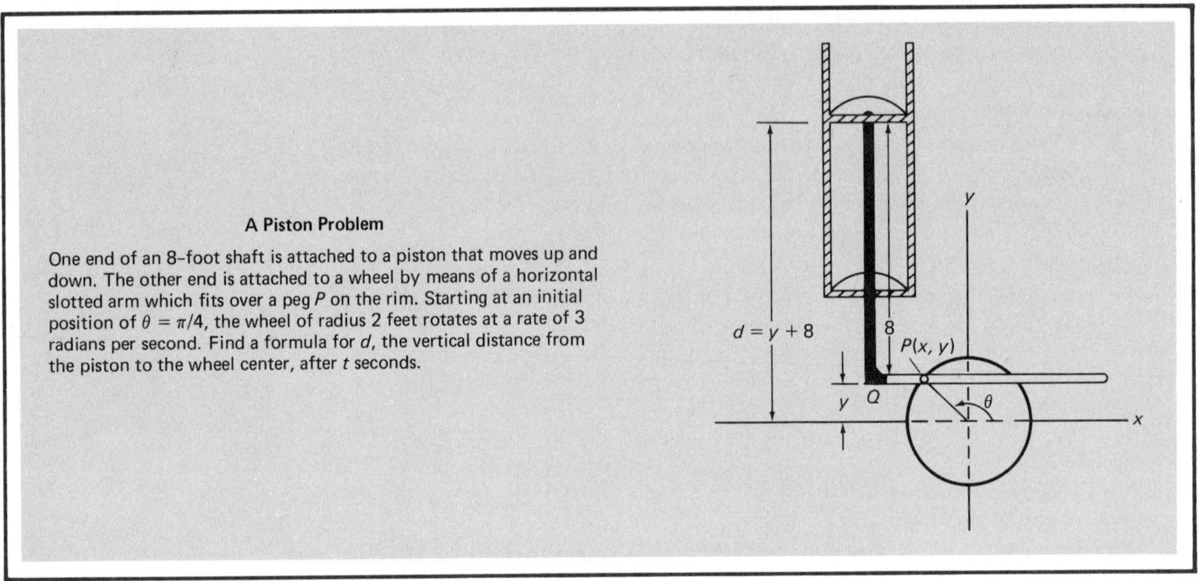

A Piston Problem

One end of an 8-foot shaft is attached to a piston that moves up and down. The other end is attached to a wheel by means of a horizontal slotted arm which fits over a peg P on the rim. Starting at an initial position of $\theta = \pi/4$, the wheel of radius 2 feet rotates at a rate of 3 radians per second. Find a formula for d, the vertical distance from the piston to the wheel center, after t seconds.

9-3 Simple Harmonic Motion

The up and down motion of the piston above is an example of what is called simple harmonic motion. Notice right away that the motion of the piston is essentially the same as that of the point Q. That means we want to find y; and y is just the y-coordinate of the peg P. Thus enters trigonometry, for P is on a circle. It would be nice if it were a unit circle? It is not. In fact, the problem is complicated by a number of scale factors. We think it wise to build up to our problem by considering related examples with nicer numbers.

Case 1. Suppose the wheel had radius 1, that it turned at 1 radian per second, and that it started at $\theta = 0$. Then at time t, θ would measure t radians and P would have y-coordinate

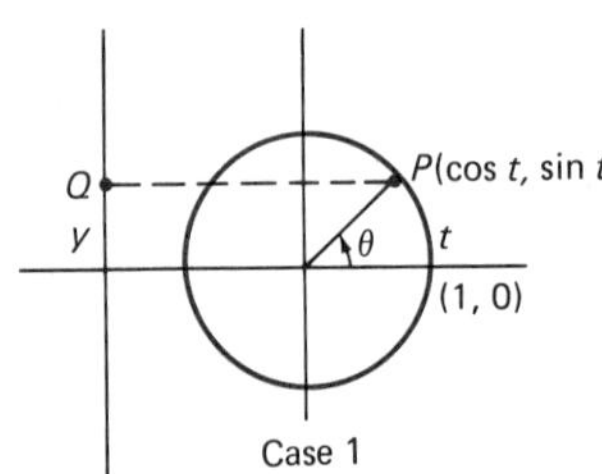

Case 1

$$y = \sin t$$

(see the diagram). Keep in mind that this equation describes the up-and-down motion of Q.

Case 2. Let everything be as in the first case, but now let the wheel turn at 3 radians per second. Then at time t, θ will measure $3t$ radians and both P and Q will have y-coordinate

P(cos 3t, sin 3t)

Q

y

3t

θ

(1, 0)

Case 2

$$y = \sin 3t$$

Case 3. Next increase the radius of the wheel to 2 feet but leave the other information as in Case 2. Now the coordinates of P are $(2 \cos 3t, 2 \sin 3t)$ and

$$\boxed{y = 2 \sin 3t}$$

Case 4. Finally, let the wheel start at $\theta = \pi/4$ rather than $\theta = 0$. With the help of the diagram below, we see that

$$\boxed{y = 2 \sin\left(3t + \frac{\pi}{4}\right)}$$

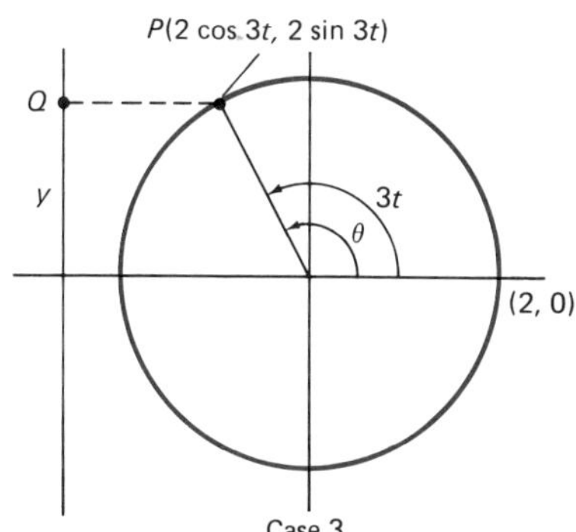

Case 3

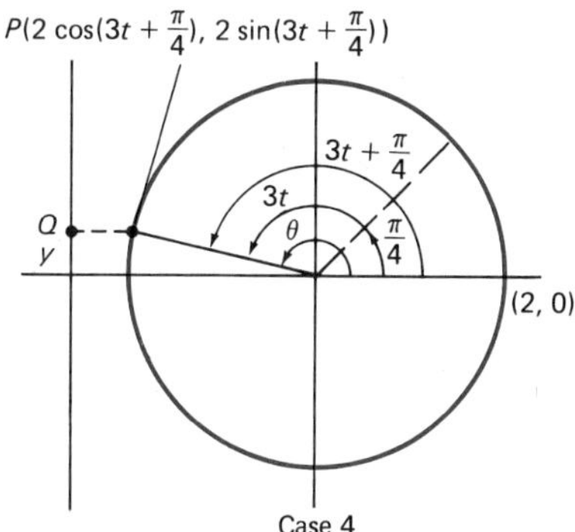

Case 4

Case 4 describes the wheel of the original problem. The number y measures the distance between Q and the x-axis, and $d = y + 8$ is the distance from the piston to the x-axis. Thus the answer to the question first posed is

$$d = 8 + 2 \sin\left(3t + \frac{\pi}{4}\right)$$

The number 8 does not interest us; it is the sine expression that is significant. As a matter of fact, equations of the form

$$y = A \sin(Bt + C) \quad \text{and} \quad y = A \cos(Bt + C)$$

with $B > 0$ arise often in physics. Any straight-line motion which can be described by one of these formulas is called **simple harmonic motion**. Cases 1–4 are examples of this motion. Other examples from physics occur in connection with the motion of a weight attached to a vibrating spring and the motion of a water molecule in an ocean wave. Voltage in an alternating current, although it does not involve motion, is given by the same kind of sine (or cosine) equation.

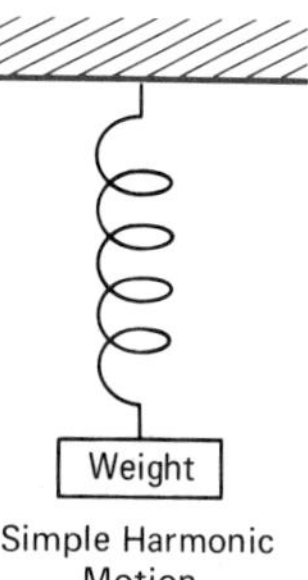

Simple Harmonic Motion

Graphs The graphs of the 4 boxed equations given on pages 340–341 are worthy of study. They are shown below. Note how the graph of $y = \sin t$ is progressively modified as we move from Case 1 to Case 4.

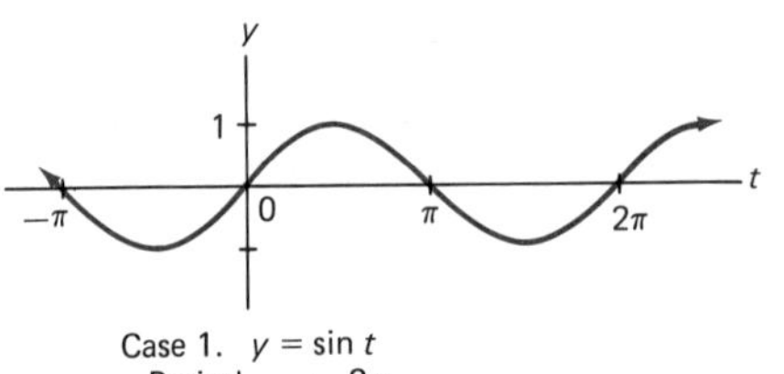

Case 1. $y = \sin t$
Period : 2π
Amplitude: 1
Phase shift: 0

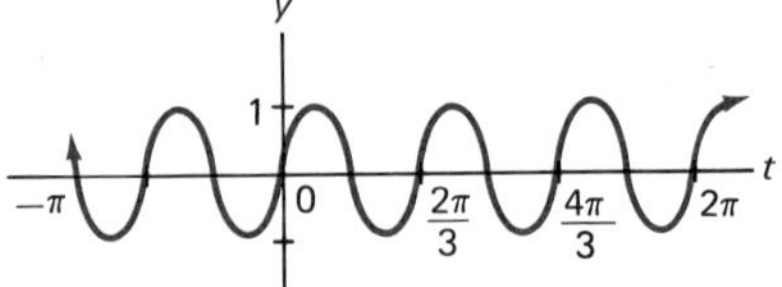

Case 2. $y = \sin 3t$
Period : $\frac{2\pi}{3}$
Amplitude: 1
Phase shift: 0

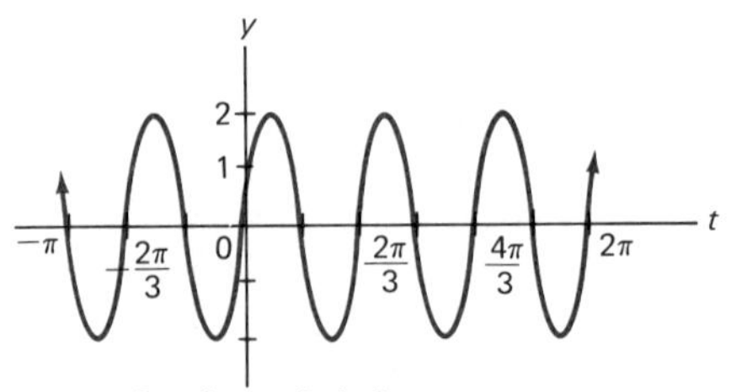

Case 3. $y = 2 \sin 3t$
Period : $\frac{2\pi}{3}$
Amplitude : 2
Phase shift : 0

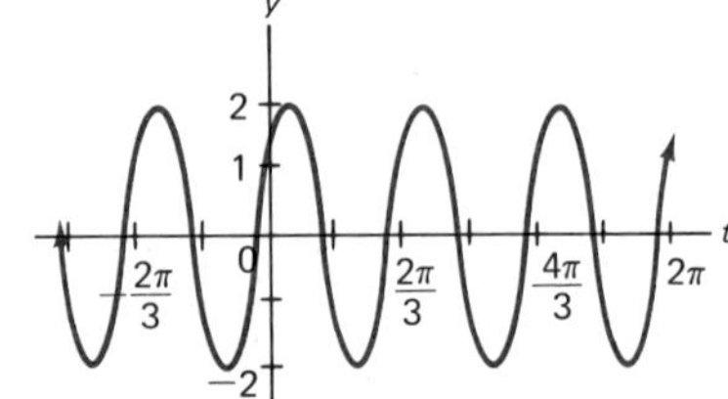

Case 4. $y = 2 \sin(3t + \frac{\pi}{4})$
Period : $\frac{2\pi}{3}$
Amplitude: 2
Phase shift: $-\frac{\pi}{12}$

Under each graph are listed three important numbers, numbers that identify the critical features of the graph. The **period** is the length of the shortest interval after which the graph repeats itself. The **amplitude** is the maximum distance of the graph from its median position (the t-axis). The **phase shift** measures the distance the graph is shifted horizontally from its normal position.

You might have expected a phase shift of $-\pi/4$ in Case 4, since the initial angle of the wheel measured $\pi/4$ radians. But, note that factoring 3 from $3t + \pi/4$ gives

$$y = 2 \sin\left(3t + \frac{\pi}{4}\right) = 2 \sin 3\left(t + \frac{\pi}{12}\right)$$

If you recall our discussion of translations (see Section 5-4), you see why the graph is shifted $\pi/12$ units to the left. Note in particular that y is 0 when $t = -\pi/12$ instead of when $t = 0$.

Graphing in the General Case If

$$y = A \sin(Bt + C) \quad \text{or} \quad y = A \cos(Bt + C)$$

with $B > 0$, all three concepts (period, amplitude, phase shift) make good sense. We have the following formulas.

Period:	$\dfrac{2\pi}{B}$
Amplitude:	$\lvert A \rvert$
Phase shift:	$\dfrac{-C}{B}$

Knowing these three numbers is a great aid in graphing. For example, to graph

$$y = 3\cos\left(4t - \frac{\pi}{4}\right)$$

we recall the graph of $y = \cos t$ and then modify it using the three numbers.

$$\text{Period:} \quad \frac{2\pi}{B} = \frac{2\pi}{4} = \frac{\pi}{2}$$

$$\text{Amplitude:} \quad |A| = |3| = 3$$

$$\text{Phase shift:} \quad -\frac{C}{B} = \frac{\pi/4}{4} = \frac{\pi}{16}$$

The result is shown below.

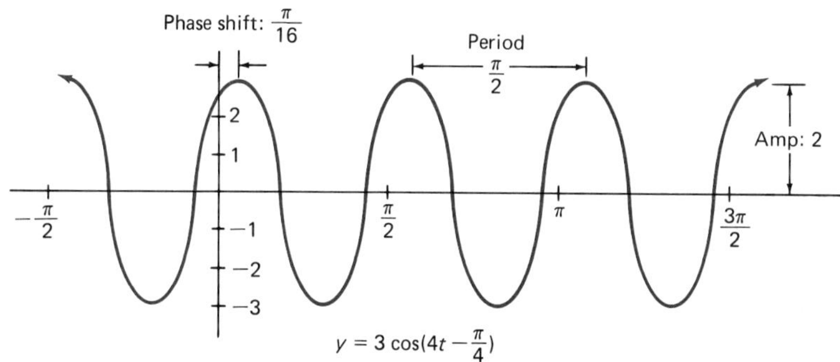

Problem Set 9-3

1. Sketch the graphs of the following equations in the order given. Use the interval $-2\pi \leq t \leq 2\pi$.
   (a) $y = \cos t$ (b) $y = \cos 2t$
   (c) $y = 4\cos 2t$ (d) $y = 4\cos(2t + \pi/3)$

2. Sketch the graphs of the following on $-2\pi \le t \le 4\pi$.

(a) $y = \sin t$ (b) $y = \sin \frac{1}{2} t$

(c) $y = 3 \sin \frac{1}{2} t$ (d) $y = 3 \sin\left(\frac{1}{2} t + \frac{\pi}{2}\right)$

3. Find the period, amplitude, and phase shift for each graph.

(a) $y = 4 \sin 2t$ (b) $y = 3 \cos\left(t + \frac{\pi}{8}\right)$

(c) $y = \sin\left(4t + \frac{\pi}{8}\right)$ (d) $y = 3 \cos\left(3t - \frac{\pi}{2}\right)$

4. Find the period, amplitude, and phase shift for each graph.

(a) $y = \frac{1}{2} \cos 3t$ (b) $y = 3 \sin\left(t - \frac{\pi}{6}\right)$

(c) $y = 2 \sin\left(\frac{1}{2} t + \frac{\pi}{8}\right)$ (d) $y = \frac{1}{2} \sin(2t - 1)$

5. Sketch the graphs of the equations in Problem 3.
6. Sketch the graphs of the equations in Problem 4.

Example (Negative A) Sketch the graph of $y = -3 \cos 2t$.

SOLUTION. We begin by asking how the graph of $y = -3 \cos 2t$ relates to that of $y = 3 \cos 2t$. Clearly, every y value has the opposite sign, which has the effect of reflecting the graph about the t-axis. Then we calculate the three crucial numbers.

$$\text{Period:} \quad \frac{2\pi}{B} = \frac{2\pi}{2} = \pi$$

$$\text{Amplitude:} \quad |A| = |-3| = 3$$

$$\text{Phase shift:} \quad \frac{-C}{B} = \frac{0}{2} = 0$$

Finally we sketch the graph.

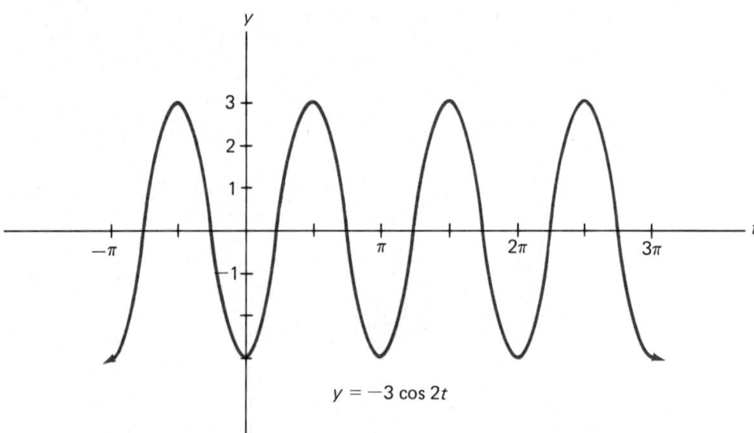

Now sketch the graph of each of the following equations.

7. $y = -2 \sin 3t$
8. $y = -4 \cos \frac{1}{2}t$
9. $y = \sin(2t - \pi/3)$
10. $y = -\cos(3t + \pi)$
11. $y = -2 \cos(t - \frac{1}{6})$
12. $y = -3 \sin(3t + 3)$
13. A wheel with center at the origin is rotating counterclockwise at 4 radians per second. There is a small hole in the wheel 5 centimeters from the center. If that hole has initial coordinates (5, 0), what will its coordinates be after t seconds?

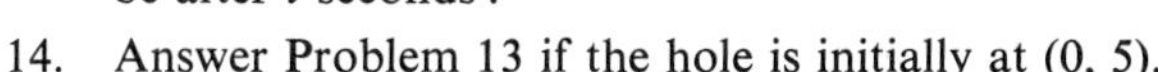

14. Answer Problem 13 if the hole is initially at (0, 5).
15. A free-hanging shaft, 8 centimeters long, is attached to the wheel of Problem 13 by putting a bolt through the hole. What are the coordinates of P, the bottom point of the shaft, at time t?
16. Suppose the wheel of Problem 13 rotates at 3 revolutions per second. What are the coordinates of the hole after t seconds?

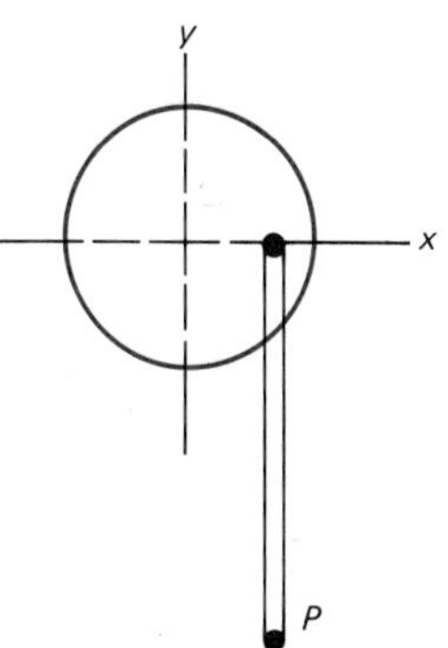

Miscellaneous Problems

17. Find the period, amplitude, and phase shift for each graph.
    (a) $y = \sin 5t$ (b) $y = \frac{3}{2} \cos(\frac{1}{2}t)$
    (c) $y = 2 \cos(4t - \pi)$ (d) $y = -4 \sin(3t + 3\pi/4)$
18. Sketch the graphs of the equations in Problem 17 on the interval $-\pi \leq t \leq 2\pi$.
19. A vibrating spring (see the diagram in the margin) is executing simple harmonic motion according to the formula

$$y = 3 \sin\left(\pi t + \frac{\pi}{2}\right) + 12$$

where y and t are measured in feet and seconds, respectively. How far was the weight from the ceiling at the beginning? What is the closest it gets to the ceiling? How close will it be after 3.5 seconds?

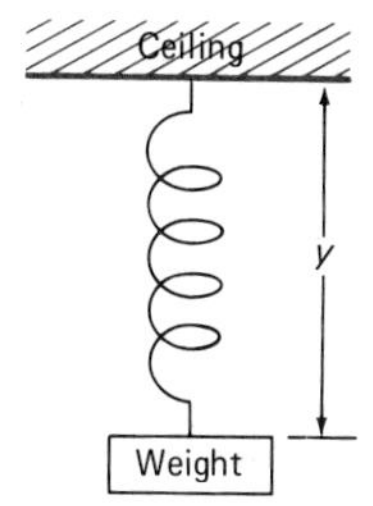

20. Consider the wheel-piston device in the margin. The wheel has radius 1 foot and rotates counterclockwise at 2 radians per second. The connecting rod is 5 feet long. If point P is initially at (1, 0), what is the y-coordinate of Q after t seconds? Assume that the x-coordinate of Q is always zero.
21. Do Problem 20, but assume the wheel is rotating at 60 revolutions per second.
22. Show by use of appropriate identities that

$$3 \cos\left(4t + \frac{\pi}{3}\right) = -3 \sin\left(4t - \frac{\pi}{6}\right)$$

(*Hint*: $\cos(4t + \pi/3) = \sin[\pi/2 - (4t + \pi/3)]$.)

23. Write $2 \sin(3t - \pi/2)$ in the form $A \cos(3t + C)$. (See Problem 22.)
24. Use an addition formula to show that

$$3 \cos\left(4t + \frac{\pi}{3}\right) = \frac{-3\sqrt{3}}{2} \sin 4t + \frac{3}{2} \cos 4t$$

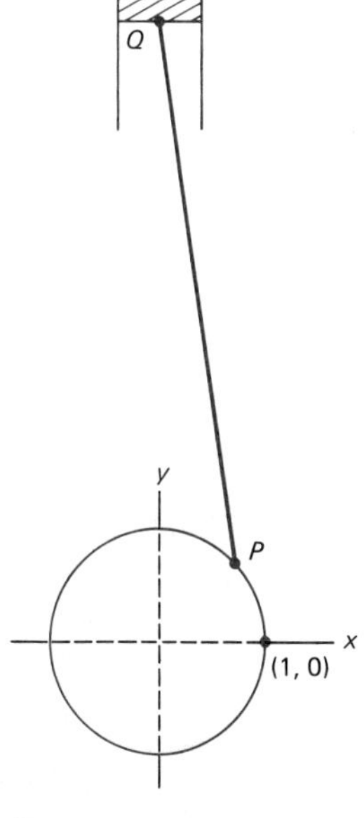

25. Determine A_1 and A_2 so that

$$2 \sin\left(3t - \frac{\pi}{2}\right) = A_1 \sin 3t + A_2 \cos 3t$$

26. Show that

$$3 \sin 2t + 4 \cos 2t = 5 \sin(2t + C)$$

where $C = \sin^{-1} \frac{4}{5}$. (*Hint*: Apply the addition formula to the right side.)

27. Suppose A_1 and A_2 are both positive. Show that

$$A_1 \sin Bt + A_2 \cos Bt = A \sin(Bt + C)$$

where $A = \sqrt{A_1^2 + A_2^2}$ and $C = \sin^{-1}(A_2/A)$.

28. Show that $A_1 \sin Bt + A_2 \cos Bt$ can always be written in the form $A \sin(Bt + C)$. (*Hint*: Choose $A = \sqrt{A_1^2 + A_2^2}$ and let C be the radian measure of an angle that has (A_1, A_2) on its terminal side.)

29. The voltage drop E across the terminals in a certain alternating current circuit is approximately $E = 156 \sin(110\pi t)$, where t is in seconds. What is the maximum voltage drop and what is the **frequency** (number of cycles per second) for this circuit?

30. The carrier wave for the radio wave of a certain FM station has the form $y = A \sin(2\pi \cdot 10^8 t)$, where t is measured in seconds. What is the frequency for this wave?

31. The AM radio wave for a certain station has the form

$$y = 55(1 + .02 \sin(2400\pi t)) \sin(2 \times 10^5 \pi t)$$

(a) Find y when $t = 3$.
(b) Find y when $t = .03216$.
C (c) Find y when $t = .0000321$.

32. In predator-prey systems, the number of predators and the number of prey tend to vary periodically. In a certain region with coyotes as predators and rabbits as prey, the rabbit population R varied according to the formula

$$R = 1000 + 150 \sin 2t$$

where t was measured in years after January 1, 1950.
(a) What was the maximum rabbit population?
(b) When was it first reached?
C (c) What was the population on January 1, 1953?

33. The number of coyotes C in Problem 32 satisfied

$$C = 200 + 50 \sin(2t - .7)$$

Sketch the graphs of C and R using the same coordinate system and attempt to explain the phase shift in C.

C 34. Sketch the graph of $y = 2^{-t} \cos 2t$ for $0 \le t \le 3\pi$. This is an example of damped harmonic motion, which is typical of harmonic motion where there is friction.

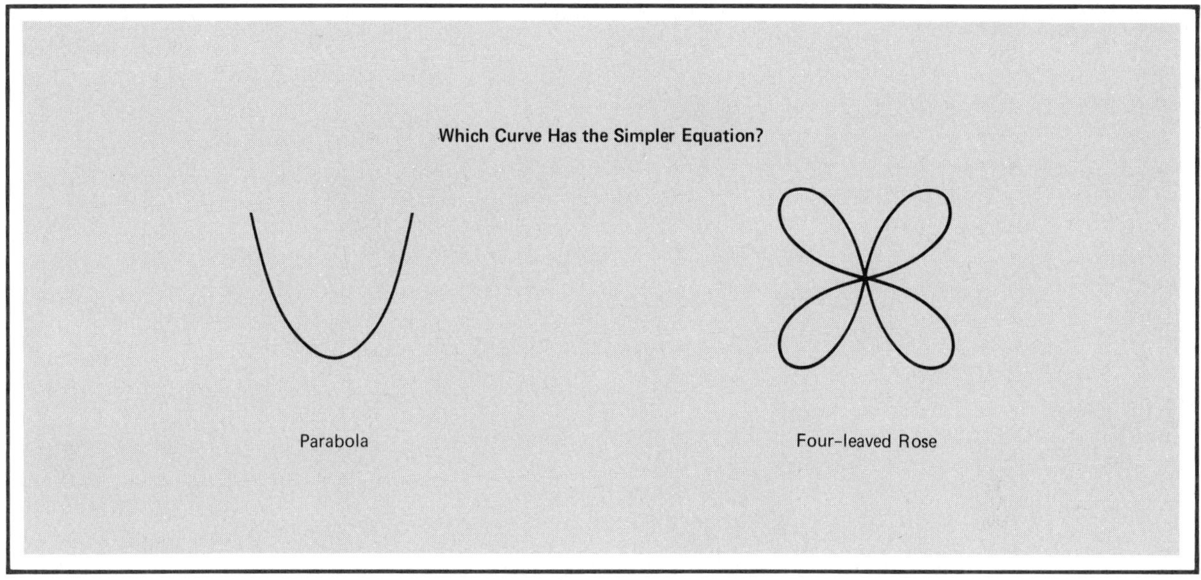

9-4 The Polar Coordinate System

The question we have asked above makes no sense unless coordinate axes are present. Most people would then choose the parabola as having the simpler equation, but the question is still more subtle than one might think. You already know that the complexity of the equation of a curve depends on the placement of the coordinate axes. Placed just right, the equation of the parabola might be as simple as $y = x^2$. Placed less wisely, the equation might be as complicated as $x - 3 = -(y + 7)^2$, or even worse. However, the four-leaved rose has a very messy equation no matter where the x- and y-axes are placed.

But there is another aspect to the question, one that Fermat and Descartes did not think about when they gave us cartesian coordinates. There are many different kinds of coordinate systems, that is, different ways of specifying the position of a point. One of these systems, when placed the best possible way, gives the four-leaved rose a delightfully simple equation (see Example B). This system is called the **polar coordinate system**; it simplifies many problems that arise in calculus.

Polar Coordinates In place of two perpendicular axes as in cartesian coordinates, we introduce in the plane a single horizontal ray, called the **polar axis**, emanating from a fixed point O, called the **pole**. On the polar axis, we mark off the positive half of a number scale with zero at the pole. Any point P other than the pole is the intersection of a unique circle with center at O and a unique ray emanating from O. If r is the radius of the circle and θ is the angle the ray makes with the polar axis, then (r, θ) are the polar coordinates of P.

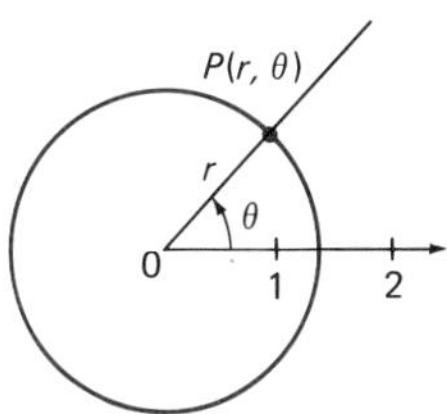

Points specified by polar coordinates are easiest to plot if we use polar graph paper. The grid on this paper consists of concentric circles and rays emanating from their common center. We have reproduced such a grid below and plotted a few points.

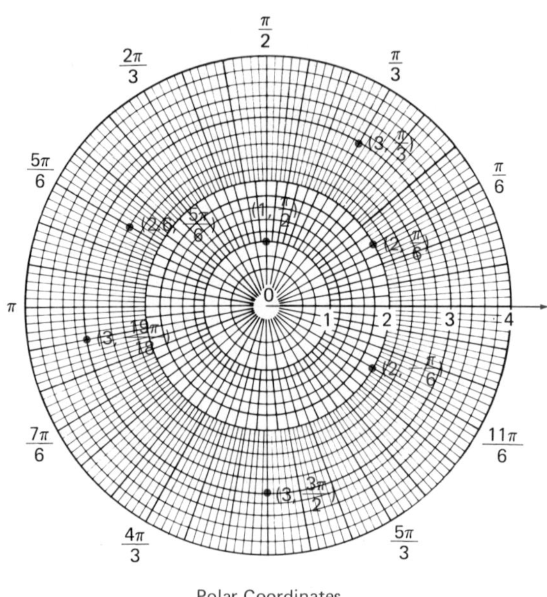

Polar Coordinates

Of course, we can measure the angle θ in degrees as well as radians. More significantly, notice that while a pair of coordinates (r, θ) determine a unique point $P(r, \theta)$, each point has many different pairs of polar coordinates. For example,

$$\left(2, \frac{3\pi}{2}\right) \qquad \left(2, -\frac{\pi}{2}\right) \qquad \left(2, \frac{7\pi}{2}\right)$$

are all coordinates for the same point.

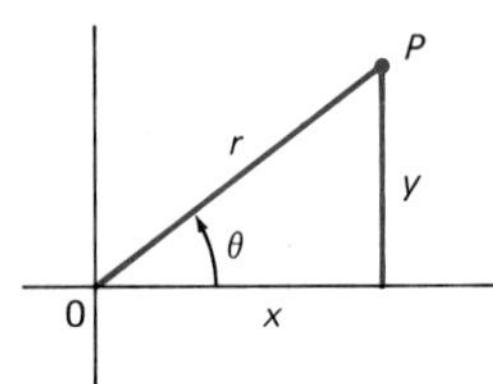

Relation to Cartesian Coordinates Let the positive x-axis of the cartesian coordinate system serve also as the polar axis of a polar coordinate system, the origin coinciding with the pole (see the diagram in the margin). Then cartesian coordinates and polar coordinates are related by two pairs of simple equations.

$$x = r\cos\theta \qquad r = \sqrt{x^2 + y^2}$$

$$y = r\sin\theta \qquad \tan\theta = \frac{y}{x}$$

For example, if $(4, \pi/6)$ are the polar coordinates of a point, then its cartesian coordinates are:

$$x = 4\cos\frac{\pi}{6} = 4\cdot\frac{\sqrt{3}}{2} = 2\sqrt{3}$$

$$y = 4\sin\frac{\pi}{6} = 4\cdot\frac{1}{2} = 2$$

On the other hand, if $(-3, \sqrt{3})$ are the cartesian coordinates of a point, then

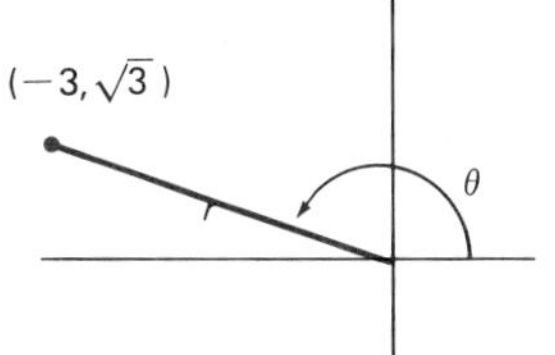

$$r = \sqrt{(-3)^2 + (\sqrt{3})^2} = \sqrt{12} = 2\sqrt{3}$$

$$\tan\theta = \frac{\sqrt{3}}{-3}$$

Since the point is in the second quadrant, we choose $5\pi/6$ as an appropriate value of θ. Thus one choice of polar coordinates for the point in question is $(2\sqrt{3}, 5\pi/6)$.

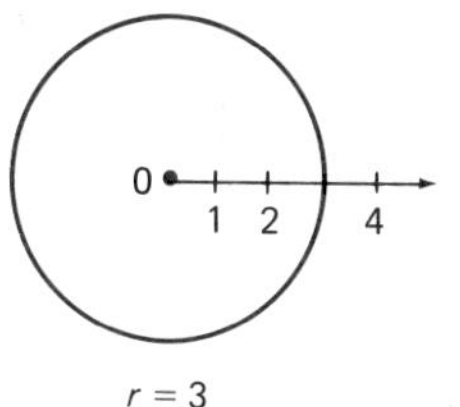

$r = 3$

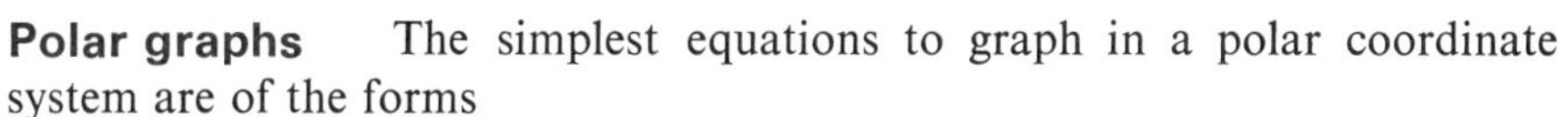

Polar graphs The simplest equations to graph in a polar coordinate system are of the forms

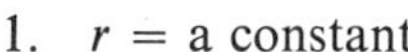

1. $r =$ a constant
2. $\theta =$ a constant

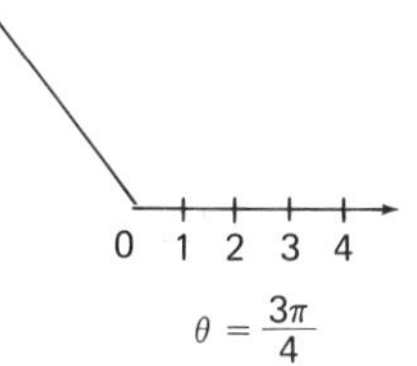

$\theta = \frac{3\pi}{4}$

The first is a circle with center at the pole; the second is a ray emanating from the pole. Examples of both are shown in the margin.

To graph more complicated equations such as

$$r = 2(1 + \cos\theta)$$

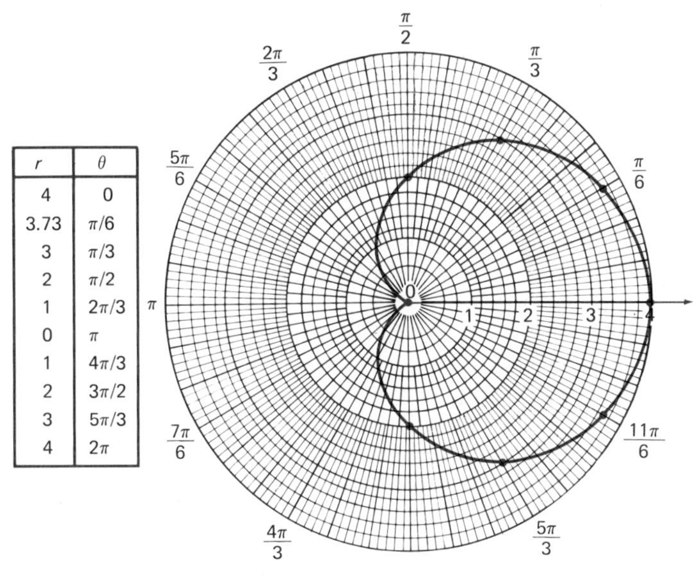

r	θ
4	0
3.73	$\pi/6$
3	$\pi/3$
2	$\pi/2$
1	$2\pi/3$
0	π
1	$4\pi/3$
2	$3\pi/2$
3	$5\pi/3$
4	2π

$r = 2(1 + \cos\theta)$

we follow our usual procedure of making a table of values, plotting the corresponding points, and then connecting them with a smooth curve. The graph (p. 349) of the equation is called a *cardioid* (a heart-like curve). Notice that r approaches 0 as θ approaches π, creating a dimple in the curve.

Problem Set 9-4

Graph each of the following points given in polar coordinates. Polar graph paper will simplify the graphing process.

1. $\left(3, \frac{\pi}{4}\right)$
2. $\left(2, \frac{\pi}{3}\right)$
3. $\left(\frac{3}{2}, \frac{5\pi}{6}\right)$
4. $\left(1, \frac{5\pi}{3}\right)$
5. $(3, \pi)$
6. $\left(2, \frac{\pi}{2}\right)$
7. $(3, -\pi)$
8. $\left(2, -\frac{3\pi}{2}\right)$
9. $(4, 70^\circ)$
10. $(3, 190^\circ)$
11. $\left(\frac{5}{2}, \frac{7\pi}{3}\right)$
12. $\left(\frac{7}{2}, \frac{11\pi}{4}\right)$

Find the cartesian coordinates of the point having the given polar coordinates.

13. $\left(4, \frac{\pi}{4}\right)$
14. $\left(6, \frac{\pi}{6}\right)$
15. $(3, \pi)$
16. $\left(2, \frac{3\pi}{2}\right)$
17. $\left(10, \frac{4\pi}{3}\right)$
18. $\left(8, \frac{11\pi}{6}\right)$
19. $\left(2, -\frac{\pi}{4}\right)$
20. $\left(3, -\frac{2\pi}{3}\right)$

Find polar coordinates for the point with the given cartesian coordinates.

21. $(4, 0)$
22. $(0, 3)$
23. $(-2, 0)$
24. $(0, -5)$
25. $(2, 2)$
26. $(2, -2)$
27. $(-2, 2)$
28. $(-2, -2)$
29. $(1, -\sqrt{3})$
30. $(-2\sqrt{3}, 2)$
31. $(3, -\sqrt{3})$
32. $(-\sqrt{3}, -3)$

Graph each of the following equations. Use polar graph paper if it is available.

33. $r = 2$
34. $r = 5$
35. $\theta = \pi/3$
36. $\theta = -2\pi/3$
37. $r = |\theta|$ (with θ in radians)
38. $r = \theta^2$
39. $r = 2(1 - \cos\theta)$
40. $r = 3(1 + \sin\theta)$
41. $r = 2 + \cos\theta$
42. $r = 2 - \sin\theta$

Example A (Transforming equations) (a) Change the cartesian equation $(x^2 + y^2)^2 = x^2 - y^2$ to a polar equation. (b) Change $r = 2\sin 2\theta$ to a cartesian equation.

SOLUTION. (a) Replacing $x^2 + y^2$ by r^2, x by $r\cos\theta$, and y by $r\sin\theta$, we get

$$\begin{aligned}(r^2)^2 &= r^2\cos^2\theta - r^2\sin^2\theta \\ r^4 &= r^2(\cos^2\theta - \sin^2\theta) \\ r^2 &= \cos 2\theta\end{aligned}$$

Dividing by r^2 at the last step did no harm since the graph of the last equation passes through the pole $r = 0$.

(b)
$$r = 2 \sin 2\theta$$
$$r = 2 \cdot 2 \sin \theta \cos \theta$$

Multiplying both sides by r^2 gives

$$r^3 = 4(r \sin \theta)(r \cos \theta)$$
$$(x^2 + y^2)^{3/2} = 4yx$$

Transform to a polar equation.

43. $x^2 + y^2 = 4$

44. $\sqrt{x^2 + y^2} = 6$

45. $y = x^2$

46. $x^2 + (y - 1)^2 = 1$

Transform to a cartesian equation.

47. $\tan \theta = 2$

48. $r = 3 \cos \theta$

49. $r = \cos 2\theta$

50. $r^2 = \cos \theta$

Example B (Allowing negative values for r) It is sometimes useful to allow r to be negative. By the point $(-3, \pi/4)$, we shall mean the point 3 units from the pole on the ray in the opposite direction from the ray for $\theta = \pi/4$ (see the diagram in margin). Allowing r to be negative, graph

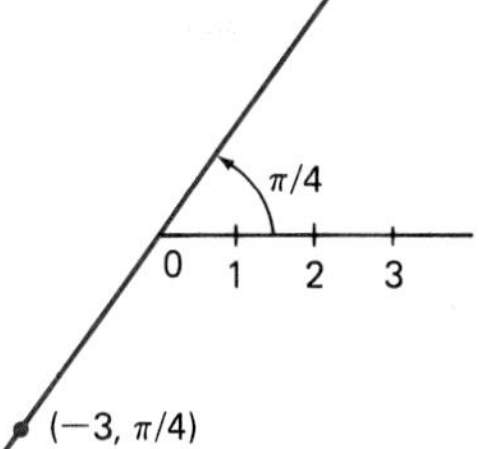

$$r = 2 \sin 2\theta$$

Solution. We begin with a table of values, plot the corresponding points, and then sketch the graph.

θ	0	$\frac{\pi}{12}$	$\frac{\pi}{6}$	$\frac{\pi}{4}$	$\frac{\pi}{3}$	$\frac{5\pi}{12}$	$\frac{\pi}{2}$	$\frac{7\pi}{12}$	$\frac{3\pi}{4}$	$\frac{11\pi}{12}$	π	$\frac{5\pi}{4}$	$\frac{3\pi}{2}$	$\frac{7\pi}{4}$	2π
2θ	0	$\frac{\pi}{6}$	$\frac{\pi}{3}$	$\frac{\pi}{2}$	$\frac{2\pi}{3}$	$\frac{5\pi}{6}$	π	$\frac{7\pi}{6}$	$\frac{3\pi}{2}$	$\frac{11\pi}{6}$	2π	$\frac{5\pi}{2}$	3π	$\frac{7\pi}{2}$	4π
r	0	1	$\sqrt{3}$	2	$\sqrt{3}$	1	0	-1	-2	-1	0	2	0	-2	0
				a					*b*			*c*		*d*	

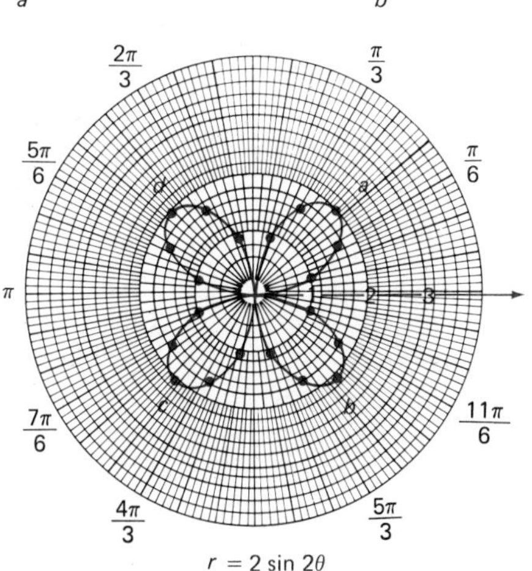

$r = 2 \sin 2\theta$

Note: The four leaves correspond to the four parts (a), (b), (c), and (d) of the table of values. For example, leaf (b) results from values of θ between $\pi/2$ and π where r is negative. This graph is the four-leaved rose of our opening display. Its cartesian equation was obtained in (b) of Example A.

Graph each of the following, allowing r to be negative.

51. $r = 3 \cos 2\theta$
52. $r = \cos 3\theta$
53. $r = \sin 3\theta$
54. $r = 4 \cos \theta$
55. $r = \sin 4\theta$
56. $r = \cos 4\theta$

Miscellaneous Problems

57. Find the cartesian coordinates of the point having the given polar coordinates.

    (a) $\left(3, \frac{4\pi}{3}\right)$ (b) $\left(2, \frac{-3\pi}{4}\right)$

58. Find polar coordinates for the point with given cartesian coordinates.

    (a) $(8, -8)$ (b) $(-2, 2\sqrt{3})$

59. Transform the equation $x^2 + y^2 - 6x + 8y = 0$ to a polar equation.
60. Transform the equation $r = 3/(2 \cos \theta - \sin \theta)$ to a cartesian equation.

Graph each of the polar equations in Problems 61–66.

61. $r = 4$
62. $\theta = -\pi/3$
63. $r = 2(1 - \sin \theta)$
64. $r = 1/\theta, \quad \theta > 0$
65. $r^2 = \sin 2\theta$ (*Caution*: Avoid values of θ which make r^2 negative.)
66. $r = 2^\theta$ (*Note*: Use both negative and positive values for θ.)
67. Solve the following pairs of equations simultaneously for $0 \leq \theta < 2\pi$.
    (a) $r = 4 \sin \theta, r = 4 \cos \theta$ (b) $r = 3 \cos 2\theta, r = 3(-1 + \cos \theta)$
68. Show that a circle with radius a and center at (a, α) has polar equation

$$r = 2a \cos(\theta - \alpha)$$

    (*Hint*: You will need the Law of Cosines (see Section 9-2).)

69. Show that the polar equation of a parabola with focus at the origin and directrix $x = -k$ is

$$r = \frac{k}{1 - \cos \theta}$$

70. The line through the origin perpendicular to a line L intersects L at (p, β). Find the polar equation of line L.

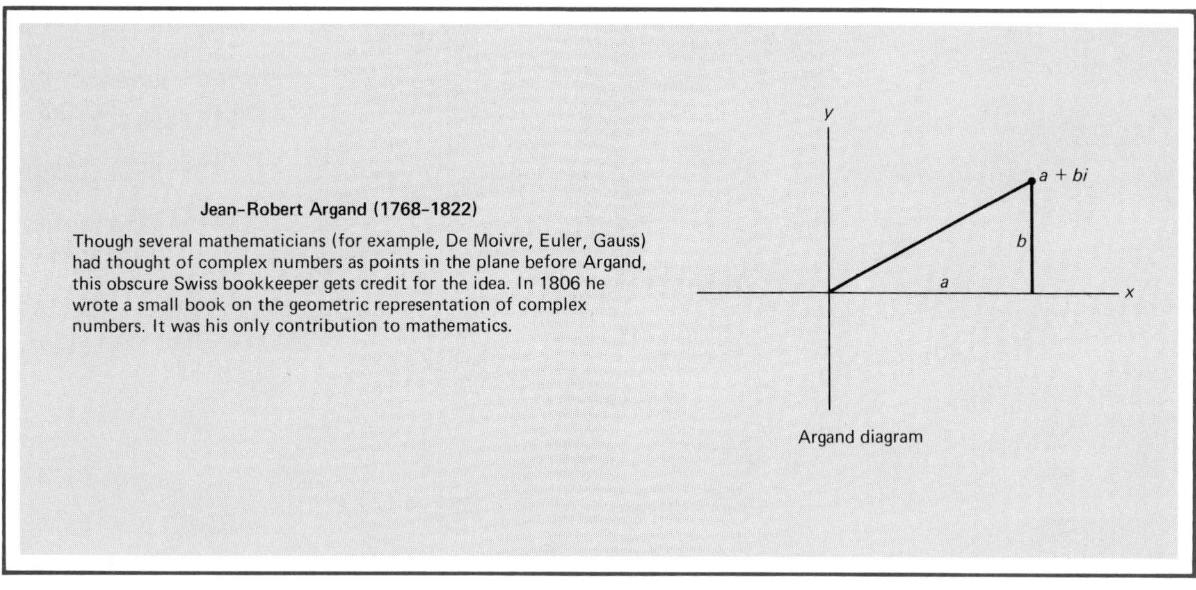

Argand diagram

Jean-Robert Argand (1768–1822)

Though several mathematicians (for example, De Moivre, Euler, Gauss) had thought of complex numbers as points in the plane before Argand, this obscure Swiss bookkeeper gets credit for the idea. In 1806 he wrote a small book on the geometric representation of complex numbers. It was his only contribution to mathematics.

9-5 Polar Representation of Complex Numbers

Throughout this book we have used the fact that a real number can be thought of as a point on a line. Now we are going to learn that a complex number can be represented as a point in the plane. This representation can be accomplished using either cartesian or polar coordinates. The latter leads to the polar form of a complex number, which aids in multiplication and division and greatly facilitates finding powers and roots. Incidentally, if you have forgotten the basic facts about the complex numbers, you should review Section 1-6 before going on.

Complex Numbers as Points in the Plane Consider a complex number $a + bi$. It is determined by the two real numbers a and b, that is, by the ordered pair (a, b). But (a, b), in turn, determines a point in the plane. That point we now label with the complex number $a + bi$. Thus $2 + 4i$, $2 - 4i$, $-3 + 2i$, and all other complex numbers may be used as labels for points in the plane (top diagram on p. 354). The plane with points labeled this way is called the **Argand diagram** or **complex plane**. Note that $3i = 0 + 3i$ labels a point on the y-axis, which we now call the **imaginary axis**, while $4 = 4 + 0i$ corresponds to a point on the x-axis (called the **real axis**).

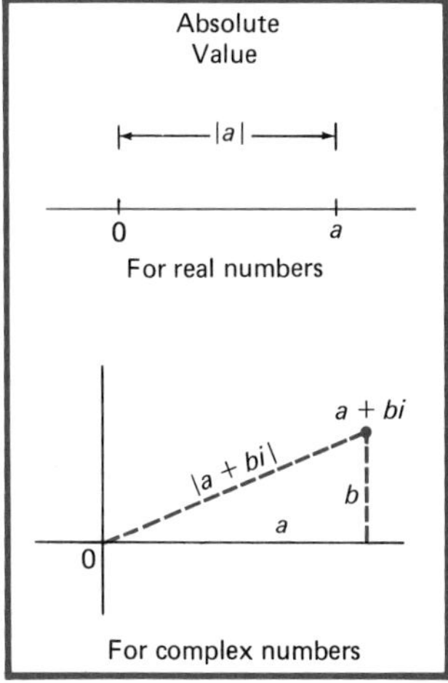

Recall that the absolute value of a real number a (written $|a|$) is its distance from the origin on the real line. The concept of absolute value is extended to a complex number $a + bi$ by defining

$$|a + bi| = \sqrt{a^2 + b^2}$$

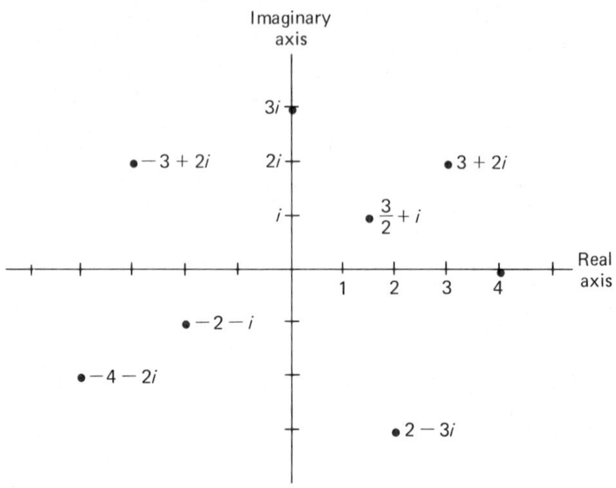

Argand diagram (complex plane)

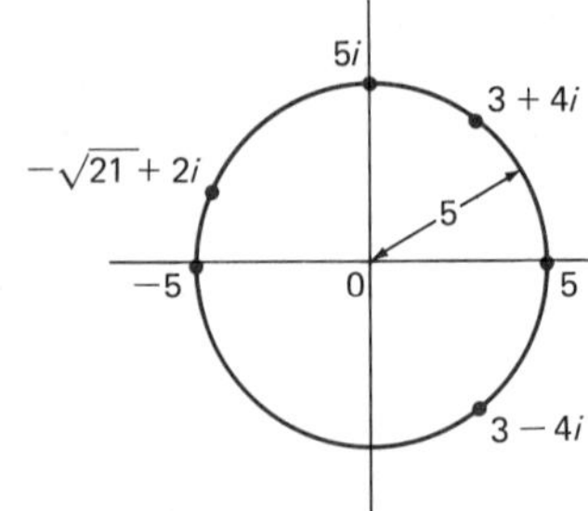

which is also its distance from the origin. Thus while there are only two real numbers with absolute value of 5, namely -5 and 5, there are infinitely many complex numbers with absolute value 5. They include 5, -5, $5i$, $3 + 4i$, $3 - 4i$, $-\sqrt{21} + 2i$, and, in fact, all complex numbers on a circle of radius 5 centered at the origin.

Polar Form Let (r, θ), where $r \geq 0$, be the polar coordinates of the point with cartesian coordinates (a, b) (see the picture in the margin). Then

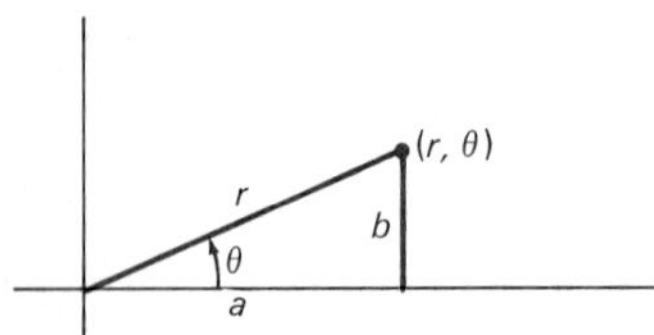

$$a = r \cos \theta \qquad b = r \sin \theta$$

This means that we can write

$$a + bi = r \cos \theta + (r \sin \theta)i$$

or

$$\boxed{a + bi = r(\cos \theta + i \sin \theta)}$$

The boxed expression gives the polar form of $a + bi$. Notice that r is just the absolute value of $a + bi$; we shall refer to θ as its angle.

To put a number $a + bi$ in polar form, we use the formulas

$$r = \sqrt{a^2 + b^2} \qquad \cos \theta = \frac{a}{r}$$

For example, for $2\sqrt{3} - 2i$,

$$r = \sqrt{(2\sqrt{3})^2 + (-2)^2} = \sqrt{12 + 4} = 4$$
$$\cos\theta = \frac{2\sqrt{3}}{4} = \frac{\sqrt{3}}{2}$$

Since $2\sqrt{3} - 2i$ is in Quadrant IV and $\cos\theta = \sqrt{3}/2$, θ can be chosen as an angle of $11\pi/6$ radians or 330°. Thus

$$\begin{aligned} 2\sqrt{3} - 2i &= 4\left(\cos\frac{11\pi}{6} + i\sin\frac{11\pi}{6}\right) \\ &= 4(\cos 330° + i\sin 330°) \end{aligned}$$

For some numbers, finding the polar form is almost trivial. Just picture in your mind (or on paper) where -6 and $4i$ are located in the complex plane and you will know that

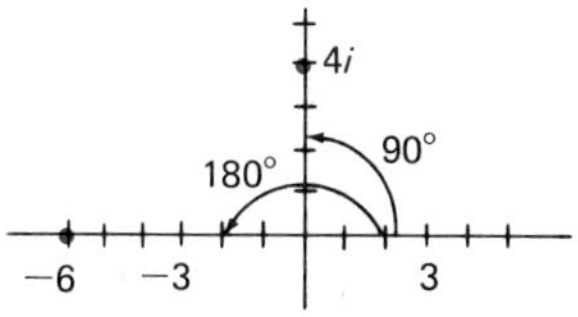

$$\begin{aligned} -6 &= 6(\cos 180° + i\sin 180°) \\ 4i &= 4(\cos 90° + i\sin 90°) \end{aligned}$$

Changing from the cartesian form $a + bi$ to polar form is what we have just illustrated. Going in the opposite direction is much easier. For example, to change the polar form $3(\cos 240° + i\sin 240°)$ to cartesian form, we simply calculate the sine and cosine of 240° and remove the parentheses.

$$\begin{aligned} 3(\cos 240° + i\sin 240°) &= 3\left(-\frac{1}{2} + i\frac{-\sqrt{3}}{2}\right) \\ &= -\frac{3}{2} - \frac{\sqrt{3}}{2}i \end{aligned}$$

Multiplication and Division The polar form is ideally suited for multiplying and dividing complex numbers. Let U and V be complex numbers given in polar form by

$$\begin{aligned} U &= r(\cos\alpha + i\sin\alpha) \\ V &= s(\cos\beta + i\sin\beta) \end{aligned}$$

Then

$$\begin{aligned} U \cdot V &= rs[\cos(\alpha + \beta) + i\sin(\alpha + \beta)] \\ \frac{U}{V} &= \frac{r}{s}[\cos(\alpha - \beta) + i\sin(\alpha - \beta)] \end{aligned}$$

In words, to multiply two complex numbers we multiply their absolute values and add their angles. To divide two complex numbers, we divide their absolute values and subtract their angles (in the correct order). Thus if

$$U = 4(\cos 75° + i \sin 75°)$$
$$V = 3(\cos 60° + i \sin 60°)$$

then

$$U \cdot V = 12(\cos 135° + i \sin 135°)$$

$$\frac{U}{V} = \frac{4}{3}(\cos 15° + i \sin 15°)$$

To establish the multiplication formula we use a bit of trigonometry.

$$\begin{aligned} U \cdot V &= r(\cos \alpha + i \sin \alpha) s(\cos \beta + i \sin \beta) \\ &= rs(\cos \alpha \cos \beta + i \cos \alpha \sin \beta + i \sin \alpha \cos \beta + i^2 \sin \alpha \sin \beta) \\ &= rs[(\cos \alpha \cos \beta - \sin \alpha \sin \beta) + i(\sin \alpha \cos \beta + \cos \alpha \sin \beta)] \\ &= rs[\cos(\alpha + \beta) + i \sin(\alpha + \beta)] \end{aligned}$$

The key step was the last one, where we used the addition laws for the cosine and the sine.

You will be asked to establish the division formula in Problem 53.

Geometric Addition and Multiplication Having learned that the complex numbers can be thought of as points in a plane, we should not be surprised that the operations of addition and multiplication have a geometric interpretation. Let U and V be any two complex numbers, that is, let

$$U = a + bi = r(\cos \alpha + i \sin \alpha)$$
$$V = c + di = s(\cos \beta + i \sin \beta)$$

Addition is accomplished algebraically by adding the real parts and imaginary parts separately.

$$U + V = (a + c) + (b + d)i$$

To accomplish the same thing geometrically, we construct the parallelogram that has O, U, and V as three of its vertices (see the diagram on the left, p. 357). Then $U + V$ corresponds to the vertex opposite the origin, as you should be able to show by finding the coordinates of this vertex.

To multiply algebraically, we use the polar forms of U and V, adding the angles and multiplying the absolute values.

$$U \cdot V = rs[\cos(\alpha + \beta) + i \sin(\alpha + \beta)]$$

To interpret this geometrically (for the case where α and β are between 0° and 180°), first draw triangle OAU, where A is the point $1 + 0i$. Then construct triangle OVW similar to triangle OAU in the manner indicated in the diagram below, on the right. We claim that $W = U \cdot V$. Certainly W has the correct angle, namely, $\alpha + \beta$. Moreover, by similarity of triangles,

$$\frac{\overline{OW}}{\overline{OV}} = \frac{\overline{OU}}{\overline{OA}} = \frac{\overline{OU}}{1}$$

(Here we are using $\overline{OW}$ for the length of the line segment from O to W.) Thus

$$|W| = \overline{OW} = \overline{OU} \cdot \overline{OV} = |U| \cdot |V|$$

so W also has the correct absolute value.

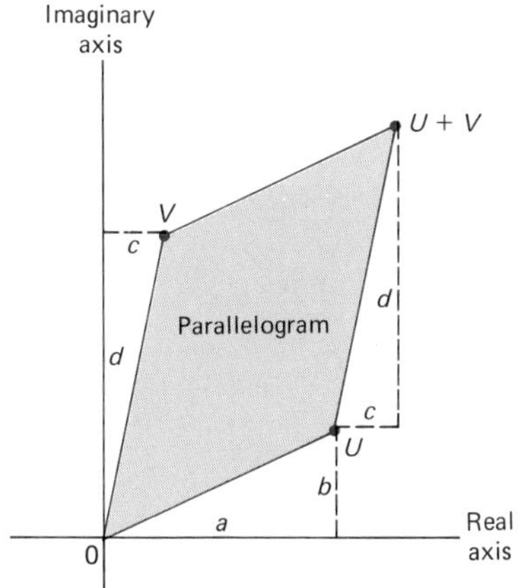

Geometric Addition : $U + V$

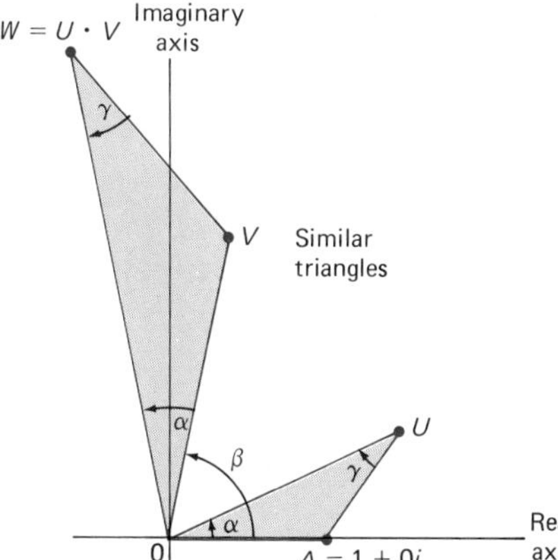

Geometric Multiplication : $U \cdot V$

Problem Set 9-5

In Problems 1–12, plot the given numbers in the complex plane.

1. $2 + 3i$
2. $2 - 3i$
3. $-2 - 3i$
4. $-2 + 3i$
5. 5
6. -6
7. $-4i$
8. $6i$
9. $\frac{3}{5} - \frac{4}{5}i$
10. $-\frac{5}{13} + \frac{12}{13}i$
11. $2\left(\cos \frac{\pi}{4} + i \sin \frac{\pi}{4}\right)$
12. $3\left(\cos \frac{7\pi}{6} + i \sin \frac{7\pi}{6}\right)$
13. Find the absolute values of the numbers in Problems 1, 3, 5, 7, 9, and 11.
14. Find the absolute values of the numbers in Problems 2, 4, 6, 8, and 10.

Express each of the following in the form $a + bi$.

15. $4\left(\cos \frac{3\pi}{2} + i \sin \frac{3\pi}{2}\right)$
16. $5(\cos \pi + i \sin \pi)$
17. $2(\cos 225° + i \sin 225°)$
18. $\frac{3}{2}(\cos 300° + i \sin 300°)$

Express each of the following in polar form.

19. -4
20. 9
21. $-5i$
22. $4i$
23. $2 - 2i$
24. $-5 - 5i$
25. $2\sqrt{3} + 2i$
26. $-4\sqrt{3} + 4i$
27. [C] $5 + 4i$
28. [C] $3 + 2i$

For Problems 29–36 let $u = 2(\cos 140° + i \sin 140°)$, $v = 3(\cos 70° + i \sin 70°)$ *and* $w = \frac{1}{2}(\cos 55° + i \sin 55°)$. *Calculate each product or quotient, leaving your answer in polar form.*

29. uv
30. uw
31. vw
32. uvw
33. u/v
34. uv/w
35. $1/w$
36. $1/v$

Example (Finding products in two ways) Find the product

$$(\sqrt{3} + i)(-4 - 4\sqrt{3}i)$$

directly and then by using the polar form.

SOLUTION. *Method 1* We use the definition of multiplication given in Section 1-6 to get

$$\begin{aligned}(\sqrt{3} + i)(-4 - 4\sqrt{3}i) &= (-4\sqrt{3} + 4\sqrt{3}) + (-4 - 12)i \\ &= -16i\end{aligned}$$

Method 2 We change both numbers to polar form, multiply by the method of this section, and finally change to $a + bi$ form.

$$\begin{aligned}\sqrt{3} + i &= 2(\cos 30° + i \sin 30°) \\ -4 - 4\sqrt{3}i &= 8(\cos 240° + i \sin 240°) \\ (\sqrt{3} + i)(-4 - 4\sqrt{3}i) &= 16(\cos 270° + i \sin 270°) \\ &= 16(0 - i) \\ &= -16i\end{aligned}$$

Find each of the following products in two ways, giving your final answer in $a + bi$ *form.*

37. $(4 - 4i)(2 + 2i)$
38. $(\sqrt{3} + i)(2 - 2\sqrt{3}i)$
39. $(1 + \sqrt{3}i)(1 + \sqrt{3}i)$
40. $(\sqrt{2} + \sqrt{2}i)(\sqrt{2} + \sqrt{2}i)$

Find the following products and quotients, giving your answers in polar form.

41. $4i(2\sqrt{3} - 2i)$
42. $(-2i)(5 + 5i)$
43. $\dfrac{4i}{2\sqrt{3} - 2i}$
44. $\dfrac{-2i}{5 + 5i}$
45. $(2\sqrt{2} - 2\sqrt{2}i)(2\sqrt{2} - 2\sqrt{2}i)$
46. $(1 - \sqrt{3}i)(1 - \sqrt{3}i)$

Miscellaneous Problems

47. Plot the given number in the complex plane and find its absolute value.
(a) $-5 + 12i$ (b) $-4i$ (c) $5(\cos 60° + i \sin 60°)$

48. Express in the form $a + bi$.
(a) $3(\cos 135° + i \sin 135°)$ (b) $8[\cos(-4\pi/3) + i \sin(-4\pi/3)]$

49. Express in polar form
(a) -12 (b) $-3 - 3i$ (c) $-i$
(d) $\sqrt{2} - \sqrt{2}i$ (e) $5 - 5\sqrt{3}i$ (f) $4 - 3i$

Find the products and quotients in Problems 50–51, giving your answer in polar form.

50. $\dfrac{2(\cos 14° + i \sin 14°)6(\cos 131° + i \sin 131°)}{4(\cos 65° + i \sin 65°)}$

51. $(\sqrt{2} - \sqrt{2}i)(5 - 5\sqrt{3}i)$ (See Problems 49d and 49e.)

52. If $u = r(\cos\theta + i \sin\theta)$, show that
(a) $-u = r[\cos(\theta + 180°) + i \sin(\theta + 180°)]$;
(b) $\bar{u} = r[\cos(-\theta) + i \sin(-\theta)]$;
(c) $u^2 = r^2(\cos 2\theta + i \sin 2\theta)$;
(d) $\dfrac{1}{u} = \dfrac{1}{r}[\cos(-\theta) + i \sin(-\theta)]$.

(*Hint*: $1 = 1(\cos 0° + i \sin 0°)$.)

53. Prove that the formula for division

$$\frac{U}{V} = \frac{r}{s}[\cos(\alpha - \beta) + i \sin(\alpha - \beta)]$$

is correct. (*Hint*: Multiply the right side by V and show that the result is U.)

54. Find two complex numbers whose real part is 4 and whose absolute value is 5.

55. Find two complex numbers located on the line $y = 2x$ with absolute value 5.

56. (a) If $z = r(\cos\theta + i \sin\theta)$ and $u = \cos(\pi/4) + i \sin(\pi/4)$, write the polar form of zu.
(b) Describe where zu is located relative to z in the plane.
(c) Show that $zu^4 = -z$.

57. Show that each of the following is correct.

(a) $\left(\cos\dfrac{\pi}{4} + i \sin\dfrac{\pi}{4}\right)^8 = 1$

(b) $\left(\cos\dfrac{5\pi}{4} + i \sin\dfrac{5\pi}{4}\right)^8 = 1$

(c) $\left(\cos\dfrac{k\pi}{4} + i \sin\dfrac{k\pi}{4}\right)^8 = 1, \quad k = 0, 1, 2, \ldots, 7$

58. The eight complex numbers $\cos(k\pi/4) + i \sin(k\pi/4)$, $k = 0, 1, 2, 3, 4, 5, 6, 7$, are all on the unit circle.
(a) Calculate the product of the eight numbers.
(b) Show where the eight points are on the unit circle.
(c) Use a geometric argument to show that the sum of the eight numbers is zero.

Abraham De Moivre (1667–1754)

Though he was a Frenchman, De Moivre spent most of his life in London. There he became an intimate friend of the great Isaac Newton, inventor of calculus. De Moivre made many contributions to mathematics but his reputation rests most securely on the theorem that bears his name.

$$(\cos\theta + i\sin\theta)^n$$
$$= \cos n\theta + i\sin n\theta$$

De Moivre's Theorem

9-6 Powers and Roots of Complex Numbers

De Moivre's Theorem tells us how to raise a complex number of absolute value 1 to an integral power. We can easily extend it to cover the case of any complex number, no matter what its absolute value. Then with a little work, we can use it to find roots of complex numbers. Here we are in for a surprise. Take the number $8i$, for example. After some fumbling around, we find that one of its cube roots is $-2i$, because $(-2i)^3 = -8i^3 = 8i$. We shall find that it has two other cube roots (both nonreal numbers). In fact, we shall see that every number has exactly three cube roots, four 4th roots, five 5th roots, and so on. To put it in a spectacular way, we claim that any number, for example $37 + 3.5i$, has 1,000,000 millionth roots.

Powers of Complex Numbers To raise the complex number $r(\cos\theta + i\sin\theta)$ to the nth power, n a positive integer, we simply find the product of n factors of $r(\cos\theta + i\sin\theta)$. But from Section 9-5, we know that we multiply complex numbers by multiplying their absolute values and adding their angles. Thus,

$$[r(\cos\theta + i\sin\theta)]^n$$

$$= \underbrace{r \cdot r \cdots r}_{n \text{ factors}}[\cos(\underbrace{\theta + \theta + \cdots + \theta}_{n \text{ terms}}) + i\sin(\theta + \theta + \cdots + \theta)]$$

In short,

$$\boxed{[r(\cos\theta + i\sin\theta)]^n = r^n(\cos n\theta + i\sin n\theta)}$$

When $r = 1$, this is De Moivre's Theorem.

As a first illustration, let us find the 6th power of a complex number that is already in polar form.

$$\begin{aligned}\left[2\left(\cos\frac{\pi}{6} + i\sin\frac{\pi}{6}\right)\right]^6 &= 2^6\left[\cos 6\cdot\frac{\pi}{6} + i\sin 6\cdot\frac{\pi}{6}\right]\\ &= 64(\cos\pi + i\sin\pi)\\ &= 64(-1 + i\cdot 0)\\ &= -64\end{aligned}$$

To find $(1 - \sqrt{3}i)^5$, we could use repeated multiplication of $1 - \sqrt{3}i$ by itself. But how much better to change $1 - \sqrt{3}i$ to polar form and use the boxed formula at the bottom of page 360.

$$1 - \sqrt{3}i = 2(\cos 300^\circ + i\sin 300^\circ)$$

Then

$$\begin{aligned}(1 - \sqrt{3}i)^5 &= 2^5(\cos 1500^\circ + i\sin 1500^\circ)\\ &= 32(\cos 60^\circ + i\sin 60^\circ)\\ &= 32\left(\frac{1}{2} + i\frac{\sqrt{3}}{2}\right)\\ &= 16 + 16\sqrt{3}i\end{aligned}$$

The Three Cube Roots of $8i$ Because finding roots is tricky, we begin with an example before attempting the general case. We have already noted that $-2i$ is one cube root of $8i$, but now we claim there are two others. How shall we find them? We begin by writing $8i$ in polar form.

$$8i = 8(\cos 90^\circ + i\sin 90^\circ)$$

Finding cube roots is the opposite of cubing. That suggests that we take the real cube root (rather than the cube) of 8 and divide (rather than multiply) the angle 90° by 3. This would give us one cube root

$$2(\cos 30^\circ + i\sin 30^\circ)$$

which reduces to

$$2\left(\frac{\sqrt{3}}{2} + \frac{1}{2}i\right) = \sqrt{3} + i$$

Is this really a cube root of $8i$? For fear that you might be suspicious of the polar form, we will cube it the old-fashioned way and check.

$$\begin{aligned}(\sqrt{3} + i)^3 &= (\sqrt{3} + i)(\sqrt{3} + i)(\sqrt{3} + i)\\ &= [(3 - 1) + 2\sqrt{3}i](\sqrt{3} + i)\\ &= 2(1 + \sqrt{3}i)(\sqrt{3} + i)\\ &= 2(0 + 4i)\\ &= 8i\end{aligned}$$

Of course, the check using polar form is more direct.

$$\begin{aligned}[2(\cos 30^\circ + i \sin 30^\circ)]^3 &= 2^3(\cos 90^\circ + i \sin 90^\circ)\\ &= 8(0 + i)\\ &= 8i\end{aligned}$$

The process described above yielded one cube root of $8i$ (namely, $\sqrt{3} + i$); there are two others. Let us go back to our representation of $8i$ in polar form. We used the angle 90°; we could as well have used $90^\circ + 360^\circ = 450^\circ$.

$$8i = 8(\cos 450^\circ + i \sin 450^\circ)$$

Now if we take the real cube root of 8 and divide 450° by 3 we get

$$2(\cos 150^\circ + i \sin 150^\circ) = 2\left(-\frac{\sqrt{3}}{2} + \frac{1}{2}i\right) = -\sqrt{3} + i$$

We could again check that this is indeed a cube root of $8i$.

What worked once might work twice. Let us write $8i$ in polar form in a third way, this time adding $2(360^\circ)$ to its angle of 90°.

$$8i = 8(\cos 810^\circ + i \sin 810^\circ)$$

The corresponding cube root is

$$2(\cos 270^\circ + i \sin 270^\circ) = 2(0 - i) = -2i$$

This does not come as a surprise, since we knew that $-2i$ was one of the cube roots of $8i$.

If we add $3(360^\circ)$ (that is, 1080°) to 90°, do we get still another cube root of $8i$? No, for if we write

$$8i = 8(\cos 1170^\circ + i \sin 1170^\circ)$$

the corresponding cube root of $8i$ would be

$$2(\cos 390^\circ + i \sin 390^\circ) = 2(\cos 30^\circ + i \sin 30^\circ)$$

But this is the same as the first cube root we found. The truth is that we have found all the cube roots of $8i$, namely, $\sqrt{3} + i$, $-\sqrt{3} + i$, and $-2i$.

Let us summarize. The number $8i$ has three cube roots given by

$$2\left[\cos\left(\frac{90^\circ}{3}\right) + i\sin\left(\frac{90^\circ}{3}\right)\right]$$

$$2\left[\cos\left(\frac{90^\circ + 360^\circ}{3}\right) + i\sin\left(\frac{90^\circ + 360^\circ}{3}\right)\right]$$

$$2\left[\cos\left(\frac{90^\circ + 720^\circ}{3}\right) + i\sin\left(\frac{90^\circ + 720^\circ}{3}\right)\right]$$

We can say the same thing in a shorter way by writing

$$2\left[\cos\left(\frac{90^\circ + k\cdot 360^\circ}{3}\right) + i\sin\left(\frac{90^\circ + k\cdot 360^\circ}{3}\right)\right] \qquad k = 0, 1, 2$$

Roots of Complex Numbers We are ready to generalize. If

$$u = r(\cos\theta + i\sin\theta)$$

then u has n distinct nth roots $u_0, u_1, \ldots, u_{n-1}$ given by

$$u_k = \sqrt[n]{r}\left[\cos\left(\frac{\theta + k\cdot 360^\circ}{n}\right) + i\sin\left(\frac{\theta + k\cdot 360^\circ}{n}\right)\right]$$
$$k = 0, 1, 2, \ldots, n-1$$

Recall that $\sqrt[n]{r}$ denotes the positive real nth root of $r = |u|$. In our example, it was $\sqrt[3]{|8i|} = \sqrt[3]{8} = 2$. To see that each value of u_k is an nth root, simply raise it to the nth power. In each case, you should get u.

The boxed formula assumes that θ is given in degrees. If θ is in radians, the formula takes the following form.

$$u_k = \sqrt[n]{r}\left[\cos\left(\frac{\theta + 2k\pi}{n}\right) + i\sin\left(\frac{\theta + 2k\pi}{n}\right)\right]$$
$$k = 0, 1, 2, \ldots, n-1$$

A Real Example Let us use the boxed formula to find the six 6th roots of 64. (Keep in mind that a real number is a special kind of complex number.) Changing to polar form, we write

$$64 = 64(\cos 0^\circ + i\sin 0^\circ)$$

Applying the formula with $r = |64| = 64$, $\theta = 0°$, and $n = 6$ gives

$$u_0 = 2(\cos 0° + i \sin 0°) = 2$$
$$u_1 = 2(\cos 60° + i \sin 60°) = 1 + \sqrt{3}i$$
$$u_2 = 2(\cos 120° + i \sin 120°) = -1 + \sqrt{3}i$$
$$u_3 = 2(\cos 180° + i \sin 180°) = -2$$
$$u_4 = 2(\cos 240° + i \sin 240°) = -1 - \sqrt{3}i$$
$$u_5 = 2(\cos 300° + i \sin 300°) = 1 - \sqrt{3}i$$

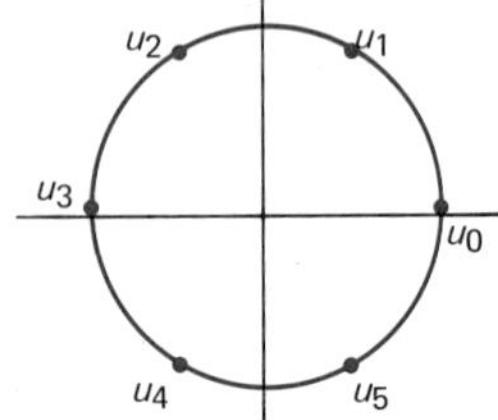

Notice that two of the roots, 2 and -2, are real; the other four are not real.

If you plot these six numbers (see the figure in the margin), you will find that they lie on a circle of radius 2 centered at the origin and that they are equally spaced around the circle. This is typical of what happens in general.

Problem Set 9-6

Find each of the following, leaving your answer in polar form.

1. $\left[2\left(\cos \frac{\pi}{4} + i \sin \frac{\pi}{4}\right)\right]^3$
2. $\left[3\left(\cos \frac{5\pi}{6} + i \sin \frac{5\pi}{6}\right)\right]^2$
3. $[\sqrt{5}(\cos 11° + i \sin 11°)]^6$
4. $[\frac{1}{3}(\cos 12.5° + i \sin 12.5°)]^4$
5. $(1 + i)^8$
6. $(1 - i)^4$

Find each of the following powers. Write your answer in a + bi form.

7. $(\cos 36° + i \sin 36°)^{10}$
8. $(\cos 27° + i \sin 27°)^{10}$
9. $(\sqrt{3} + i)^5$
10. $(2 - 2\sqrt{3}i)^4$

Find the nth roots of u for the given u and n, leaving your answers in polar form. Plot these roots in the complex plane.

11. $u = 125(\cos 45° + i \sin 45°)$; $n = 3$
12. $u = 81(\cos 80° + i \sin 80°)$; $n = 4$
13. $u = 64\left(\cos \frac{\pi}{2} + i \sin \frac{\pi}{2}\right)$; $n = 6$
14. $u = 3^8\left(\cos \frac{2\pi}{3} + i \sin \frac{2\pi}{3}\right)$; $n = 8$
15. $u = 4(\cos 112° + i \sin 112°)$; $n = 4$
16. $u = 7(\cos 200° + i \sin 200°)$; $n = 5$

Find the nth roots of u for the given u and n. Write your answers in the a + bi form.

17. $u = 16$, $n = 4$
18. $u = -16$, $n = 4$
19. $u = 4i$, $n = 2$
20. $u = -27i$, $n = 3$
21. $u = -4 + 4\sqrt{3}i$, $n = 2$
22. $u = -2 - 2\sqrt{3}i$, $n = 4$

Example (Roots of unity) The nth roots of 1, called the ***n*th roots of unity**, play an important role in advanced algebra. Find the five 5th roots of unity, plot them, and show that four of the roots are powers of the 5th root.

SOLUTION. First we represent 1 in polar form.

$$1 = 1(\cos 0^\circ + i \sin 0^\circ)$$

The five 5th roots are (according to a formula developed in this section)

$$\begin{aligned} u_0 &= \cos 0^\circ + i \sin 0^\circ = 1 \\ u_1 &= \cos 72^\circ + i \sin 72^\circ \\ u_2 &= \cos 144^\circ + i \sin 144^\circ \\ u_3 &= \cos 216^\circ + i \sin 216^\circ \\ u_4 &= \cos 288^\circ + i \sin 288^\circ \end{aligned}$$

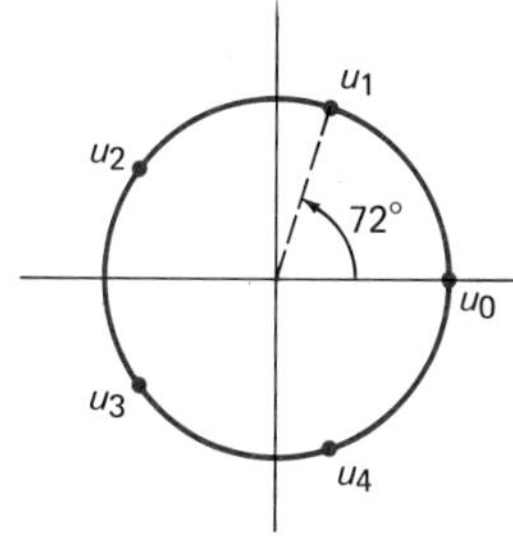

These roots are plotted in the margin. They lie on the unit circle and are equally spaced around it. Finally notice that

$$\begin{aligned} u_1 &= u_1 \\ u_2 &= u_1^2 \\ u_3 &= u_1^3 \\ u_4 &= u_1^4 \\ u_0 &= u_1^5 \end{aligned}$$

Thus all the roots are powers of u_1. These powers of u_1 repeat in cycles of 5. For example, note that

$$\begin{aligned} u_1^6 &= u_1^5 u_1 = u_1 \\ u_1^7 &= u_1^5 u_1^2 = u_1^2 \end{aligned}$$

In each of the following, find all the nth *roots of unity for the given* n *and plot them in the complex plane.*

23. $n = 4$
24. $n = 6$
25. $n = 10$
26. $n = 12$

Miscellaneous Problems

Find the indicated power in Problems 27–30, leaving your answer in polar form.

27. $[\sqrt{3}(\cos 13^\circ + i \sin 13^\circ)]^8$
28. Ⓒ $[.132(\cos 2.12 + i \sin 2.12)]^7$
29. $(2 - 2\sqrt{3}i)^5$
30. $[(\sqrt{3} + i)(2 + 2i)]^6$
31. Write your answer to Problem 29 in the form $a + bi$.
32. Ⓒ Write your answer to Problem 28 in the form $a + bi$.
33. Find the four 4th roots of $16(\cos 220^\circ + i \sin 220^\circ)$, leaving your answers in polar form.
34. Find the five 5th roots of -32, leaving your answers in polar form.
35. Find the four 4th roots of -1. Then find their sum and their product. Write your answers in the form $a + bi$.
36. Solve the equation $x^3 - 8 = 0$
    (a) by finding the 3 cube roots of 8;
    (b) by writing $x^3 - 8 = (x - 2)(x^2 + 2x + 4)$ and using the quadratic formula.

37. Show that $\cos(\pi/4) + i\sin(\pi/4)$ is a solution of $x^4 + x^2 + 1 = i$.
38. Show that $\cos(\pi/6) + i\sin(\pi/6)$ is a solution of $x^9 + x^6 + x^3 + 1 = 0$.
39. Show that De Moivre's Theorem is valid when n is a negative integer.
40. Use De Moivre's Theorem to establish the following familiar identities.

$$\cos 2\theta = \cos^2\theta - \sin^2\theta$$
$$\sin 2\theta = 2\sin\theta\cos\theta$$

(*Hint*: Multiply out the left side of

$$(\cos\theta + i\sin\theta)^2 = \cos 2\theta + i\sin 2\theta)$$

and then equate real and imaginary parts.)

41. Use the method of Problem 40 to find corresponding identities for $\cos 3\theta$ and $\sin 3\theta$.

CHAPTER SUMMARY

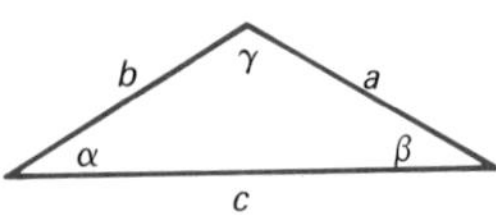

A triangle like the one in the margin that has no right angle is called an oblique triangle. If any three of the six parts α, β, γ, a, b, and c—including at least one side—are given, we can find the remaining parts by using the **Law of Sines**

$$\frac{a}{\sin\alpha} = \frac{b}{\sin\beta} = \frac{c}{\sin\gamma}$$

and the **Law of Cosines**

$$a^2 = b^2 + c^2 - 2bc\cos\alpha \qquad \text{(one of three forms)}$$

There is, however, one case (given two sides and an angle opposite one of them) in which there might be no solution or two solutions. We call it the ambiguous case.

The equations $y = A\sin(Bt + C)$ and $y = A\cos(Bt + C)$ describe a common phenomenon known as **simple harmonic motion**. We can quickly draw the graphs of these equations by making use of three key numbers: the **amplitude**, $|A|$, the **period**, $2\pi/B$, and the **phase shift**, $-C/B$.

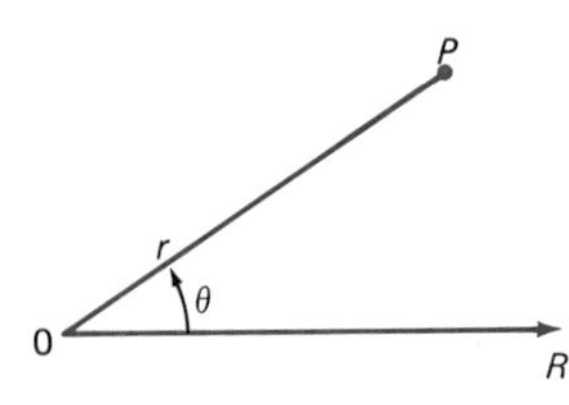

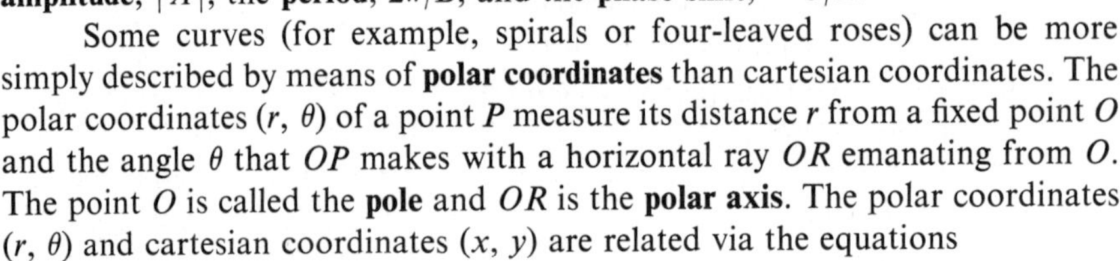

Some curves (for example, spirals or four-leaved roses) can be more simply described by means of **polar coordinates** than cartesian coordinates. The polar coordinates (r, θ) of a point P measure its distance r from a fixed point O and the angle θ that OP makes with a horizontal ray OR emanating from O. The point O is called the **pole** and OR is the **polar axis**. The polar coordinates (r, θ) and cartesian coordinates (x, y) are related via the equations

$$x = r\cos\theta \qquad y = r\sin\theta \qquad r^2 = x^2 + y^2$$

A complex number $a + bi$ can be represented geometrically as a point (a, b) in a plane called the **complex plane**, or **Argand diagram**. The horizontal and vertical axes are known as the **real axis** and **imaginary axis**, respectively. The distance from the origin to (a, b) is $\sqrt{a^2 + b^2}$; it is also the absolute value of $a + bi$, denoted as with real numbers by $|a + bi|$. If polar coordinates (r, θ) are used in place of cartesian coordinates (a, b), we get the **polar form** of $a + bi$, namely

$$r(\cos\theta + i\sin\theta)$$

This form facilitates multiplication and division and is especially helpful in finding powers and roots of a number. An important result is the formula

$$[r(\cos\theta + i\sin\theta)]^n = r^n(\cos n\theta + i\sin n\theta)]$$

CHAPTER REVIEW PROBLEM SET

1. Solve each of the following triangles using Table C or a calculator.
   (a) $\alpha = 104.9°$, $\gamma = 36°$, $b = 149$
   (b) $a = 14.6$, $b = 89.2$, $c = 75.8$
   (c) $\gamma = 35°$, $a = 14$, $b = 22$
   (d) $\beta = 48.6°$, $c = 39.2$, $b = 57.6$
2. For the triangle in the margin, find x and the area of the triangle.

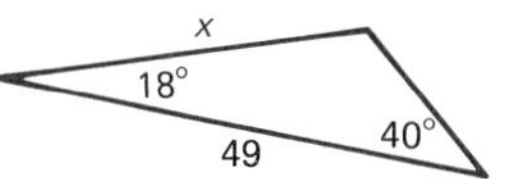

3. Find the period, amplitude, and phase shift for the graph of each equation.
   (a) $y = \cos 2t$ (b) $y = 3\cos 4t$
   (c) $y = 2\sin(3t - \pi/2)$ (d) $y = -2\sin(\frac{1}{2}t + \pi)$
4. Sketch the graphs of the equations in Problem 3 on $-\pi \le t \le \pi$.
5. A wheel of radius 4 feet with center at the origin is rotating counterclockwise at $3\pi/4$ radians per second. If a paint speck P has coordinates $(-4, 0)$ initially, what will its coordinates be after t seconds?

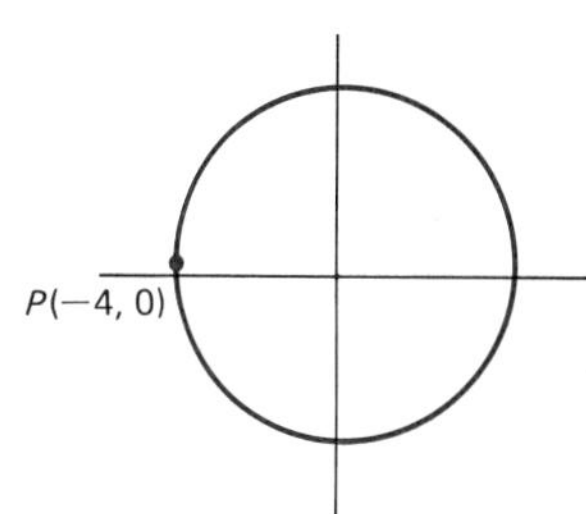

6. A point is moving in a straight line according to the equation $x = 3\cos(5t + 3\pi)$, where x is in feet and t in seconds.
   (a) What is the period of the motion?
   (b) Find the initial position of the point relative to the point $x = 0$.
   (c) When is $x = 3$ for the first time?
7. Plot the points with the following polar coordinates.
   (a) $(3, 2\pi/3)$ (b) $(2, -3\pi/2)$ (c) $(6, 210°)$
8. Find the cartesian coordinates of the points in Problem 7.
9. Find the polar coordinates of the point having the given cartesian coordinates.
   (a) $(5, 0)$ (b) $(2\sqrt{2}, -2\sqrt{2})$ (c) $(-2\sqrt{3}, 2)$
10. Graph each of the following polar equations.
    (a) $r = 4$ (b) $r = 4\sin\theta$ (c) $r = 4\cos 3\theta$
11. Transform $xy = 4$ to a polar equation. Transform $r = \sin 2\theta$ to a cartesian equation.

12. Plot the following numbers in the complex plane.
(a) $3 - 4i$ (b) -6
(c) $5i$ (d) $3\left(\cos\frac{3\pi}{4} + i\sin\frac{3\pi}{4}\right)$
(e) $4(\cos 300° + i\sin 300°)$
13. Find the absolute value of each number in Problem 12.
14. Express $4(\cos 150° + i\sin 150°)$ in the form $a + bi$.
15. Express in polar form.
(a) $3i$ (b) -6
(c) $-1 - i$ (d) $2\sqrt{3} - 2i$
16. Let

$$u = 8(\cos 105° + i\sin 105°)$$
$$v = 4(\cos 40° + i\sin 40°)$$

Calculate each of the following, leaving your answer in polar form.
(a) uv (b) u/v
(c) u^3 (d) u^2v^3
17. Find all the 6th roots of

$$2^6(\cos 120° + i\sin 120°)$$

leaving your answers in polar form.
18. Find the five solutions to $x^5 - 1 = 0$.

He who loves practice without theory is like the sailor who boards ship without a rudder and compass and never knows where he may cast.

Leonardo da Vinci

CHAPTER TEN

Theory of Polynomial Equations

10-1 Division of Polynomials

10-2 Factorization Theory for Polynomials

10-3 Polynomial Equations With Real Coefficients

10-4 The Method of Successive Approximations

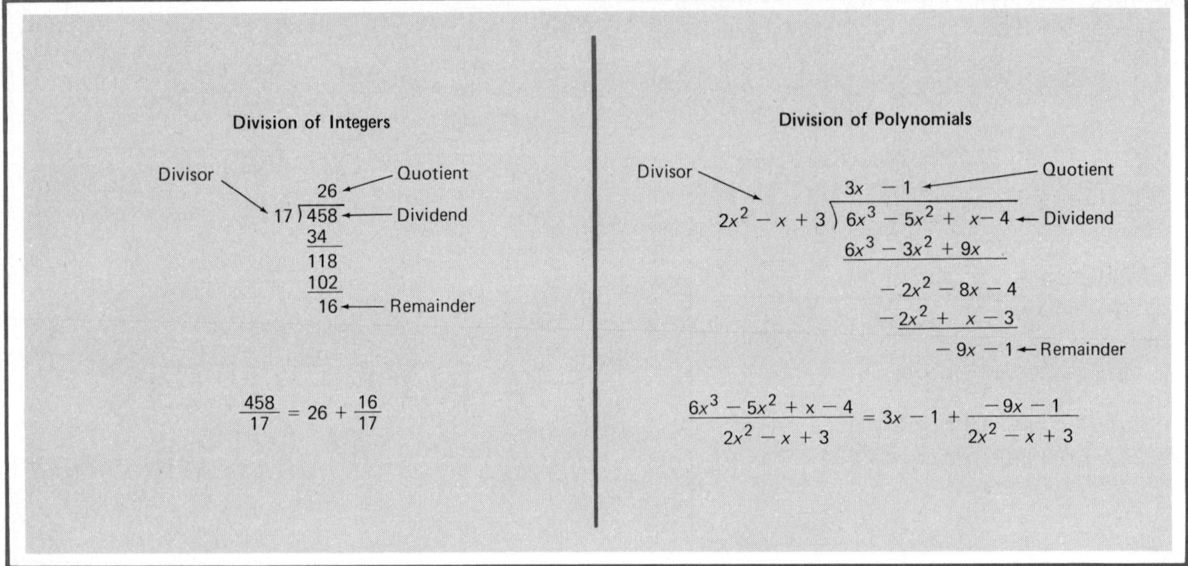

10-1 Division of Polynomials

In Section 2-4, we learned that polynomials in x can always be added, subtracted, and multiplied; the result in every case is a polynomial. For example,

$$
\begin{aligned}
(3x^2 + x - 2) + (3x - 2) &= 3x^2 + 4x - 4 \\
(3x^2 + x - 2) - (3x - 2) &= 3x^2 - 2x \\
(3x^2 + x - 2)\cdot(3x - 2) &= 9x^3 - 3x^2 - 8x + 4
\end{aligned}
$$

Now we are going to study division of polynomials. Occasionally the division is exact.

$$(3x^2 + x - 2) \div (3x - 2) = \frac{(3x - 2)(x + 1)}{3x - 2} = x + 1$$

We say $3x - 2$ is an exact divisor, or factor, of $3x^2 + x - 2$. More often than not, the division is inexact and there is a nonzero remainder.

The Division Algorithm The opening display shows that the process of division for polynomials is much the same as for integers. Both processes (we call them *algorithms*) involve subtraction. The first one shows how many times 17 can be subtracted from 458 before obtaining a remainder less than 17. The answer is 26 times, with a remainder of 16. The second algorithm shows how many times $2x^2 - x + 3$ can be subtracted from $6x^3 - 5x^2 + x - 4$ before obtaining a remainder of lower degree than $2x^2 - x + 3$. The answer is $3x - 1$ times, with a remainder of $-9x - 1$. Thus in these two examples we have

$$
\begin{gathered}
458 - 26(17) = 16 \\
(6x^3 - 5x^2 + x - 4) - (3x - 1)(2x^2 - x + 3) = -9x - 1
\end{gathered}
$$

Of course, we can also write these equalities as

$$458 = (17)(26) + 16$$
$$6x^3 - 5x^2 + x - 4 = (2x^2 - x + 3)(3x - 1) + (-9x - 1)$$

Notice that both of them can be summarized in the words

$$\text{dividend} = (\text{divisor})\cdot(\text{quotient}) + \text{remainder}$$

This statement is very important and is worth stating again for polynomials in very precise language.

THE DIVISION LAW FOR POLYNOMIALS

If $P(x)$ and $D(x)$ are any two nonconstant polynomials, then there are unique polynomials $Q(x)$ and $R(x)$ such that

$$\boxed{P(x) = D(x)Q(x) + R(x)}$$

where $R(x)$ is either zero or it is of lower degree than $D(x)$.

Here you should think of $P(x)$ as the dividend, $D(x)$ as the divisor, $Q(x)$ as the quotient, and $R(x)$ as the remainder.

The algorithm we use to find $Q(x)$ and $R(x)$ was illustrated in the opening display. Here is another illustration, this time for polynomials with some nonreal coefficients. Notice that we always arrange the divisor and dividend in descending powers of x before we start the division process.

$$\begin{array}{r|l} & x - i \\ \hline 3x + i & 3x^2 - 2ix + 7 \\ & \underline{3x^2 + ix} \\ & \quad -3ix + 7 \\ & \quad \underline{-3ix + 1} \\ & \qquad\quad 6 \end{array} \qquad \begin{array}{l} Q(x) = x - i \\ R(x) = 6 \end{array}$$

Synthetic Division It is often necessary to divide by polynomials of the form $x - c$. For such a division, there is a shortcut called **synthetic division.** We illustrate how it works for

$$(2x^3 - x^2 + 5) \div (x - 2)$$

Certainly the result depends on the coefficients; the powers of x serve mainly to determine the placement of the coefficients. Below, we show the division in its usual form and then in a skeletal form with the x's omitted. Note that we leave a blank space for the missing first degree term in the long division but indicate it with a 0 in the skeletal form.

LONG DIVISION

$$\begin{array}{r|l}
 & 2x^2 + 3x \;+ 6 \\
\hline
x - 2 & 2x^3 - \;x^2 \qquad\quad + 5 \\
 & 2x^3 - 4x^2 \\
\hline
 & \quad\;\; 3x^2 \\
 & \quad\;\; 3x^2 - 6x \\
\hline
 & \qquad\qquad 6x + 5 \\
 & \qquad\qquad 6x - 12 \\
\hline
 & \qquad\qquad\quad\;\; 17
\end{array}$$

FIRST CONDENSATION

$$\begin{array}{r|rrrr}
 & ② & ③ & ⑥ & \\
\hline
① - 2 & 2 & -1 & 0 & 5 \\
 & 2 & -4 & & \\
\hline
 & & 3 & ⓪ & \\
 & & ③ & -6 & \\
\hline
 & & & 6 & ⑤ \\
 & & & ⑥ & -12 \\
\hline
 & & & & 17
\end{array}$$

We can condense things still more by discarding all of the circled digits. The coefficients of the quotient, 2, 3, and 6, the remainder, 17, and the numbers from which they were calculated remain. All of the important numbers appear in the diagram below, on the left. On the right, we show the final modification. There we have changed the divisor from -2 to 2 to allow us to do addition rather than subtraction at each stage.

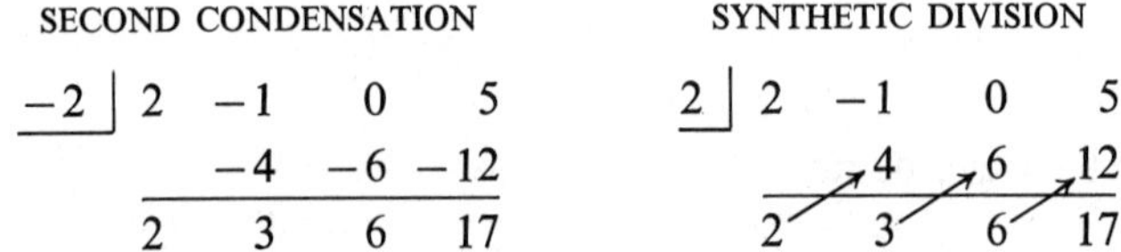

The process shown in the final format is called **synthetic division**. We can describe it as a series of steps to follow.

1. To divide by $x - 2$, use 2 as the synthetic divisor.
2. Write down the coefficients of the dividend. Be sure to write zeros for missing powers.
3. Bring down the first coefficient.
4. Follow the arrows, first multiplying by 2 (the divisor), then adding, multiplying the sum by 2, adding, and so on.
5. The last number in the third row is the remainder and the others are the coefficients of the quotient.

Here is another example. We use synthetic division to divide $3x^3 - x^2 + 15x - 5$ by $x - \frac{1}{3}$.

$$\begin{array}{r|rrrr}
\frac{1}{3} & 3 & -1 & 15 & -5 \\
 & & 1 & 0 & 5 \\
\hline
 & 3 & 0 & 15 & 0
\end{array}$$

Since the remainder is 0, the division is exact. We conclude that

$$3x^3 - x^2 + 15x - 5 = (x - \tfrac{1}{3})(3x^2 + 15)$$

Proper and Improper Rational Expressions A rational expression (ratio of two polynomials) is said to be **proper** if the degree of its numerator is smaller than that of its denominator. Thus

$$\frac{x + 1}{x^2 - 3x + 2}$$

is a proper rational expression but

$$\frac{2x^3}{x^2 - 3}$$

is improper. The Division Law, $P(x) = D(x)Q(x) + R(x)$, can be written as

$$\frac{P(x)}{D(x)} = Q(x) + \frac{R(x)}{D(x)}$$

It implies that any improper rational expression can be rewritten as the sum of a polynomial and a proper rational expression. For example,

$$\frac{2x^3}{x^2 - 3} = 2x + \frac{6x}{x^2 - 3}$$

a result we obtained by dividing $x^2 - 3$ into $2x^3$.

Problem Set 10-1

In Problems 1–8 find the quotient and the remainder if the first polynomial is divided by the second.

1. $x^3 - x^2 + x + 3;\ x^2 - 2x + 3$
2. $x^3 - 2x^2 - 7x - 4;\ x^2 + 2x + 1$
3. $6x^3 + 7x^2 - 18x + 15;\ 2x^2 + 3x - 5$
4. $10x^3 + 13x^2 + 5x + 12;\ 5x^2 - x + 4$
5. $4x^4 - x^2 - 6x - 9;\ 2x^2 - x - 3$
6. $25x^4 - 20x^3 + 4x^2 - 4;\ 5x^2 - 2x + 2$
7. $2x^5 - 2x^4 + 9x^3 - 12x^2 + 4x - 16;\ 2x^3 - 2x^2 + x - 4$
8. $3x^5 - x^4 - 8x^3 - x^2 - 3x + 12;\ 3x^3 - x^2 + x - 4$

In Problems 9–12, write each rational expression as the sum of a polynomial and a proper rational expression.

9. $\dfrac{x^3 + 2x^2 + 5}{x^2}$
10. $\dfrac{2x^3 - 4x^2 - 3}{x^2 + 1}$
11. $\dfrac{x^3 - 4x + 5}{x^2 + x - 2}$
12. $\dfrac{2x^3 + x - 8}{x^2 - x + 4}$

In Problems 13–18, use synthetic division to find the quotient and remainder if the first polynomial is divided by the second.

13. $2x^3 - x^2 + x - 4;\ x - 1$
14. $3x^3 + 2x^2 - 4x + 5;\ x - 2$
15. $3x^3 + 5x^2 + 2x - 10;\ x - 1$
16. $2x^3 - 5x^2 + 4x - 4;\ x - 2$
17. $x^4 - 2x^2 - 1;\ x - 3$
18. $x^4 + 3x^2 - 340;\ x - 4$

Example A (Synthetic division by $x + c$) Find the quotient and remainder when $2x^3 - 2x^2 + 5$ is divided by $x + 3$.

SOLUTION. Since $x + 3 = x - (-3)$, we may use synthetic division with -3 as the synthetic divisor.

$$\begin{array}{r|rrrr} -3 & 2 & -2 & 0 & 5 \\ & & -6 & 24 & -72 \\ \hline & 2 & -8 & 24 & -67 \end{array}$$

The quotient is $2x^2 - 8x + 24$ and the remainder is -67.

Find the quotient and the remainder when the first polynomial is divided by the second.

19. $x^3 + 2x^2 - 3x + 2;\ x + 1$
20. $x^3 - x^2 + 11x - 1;\ x + 1$
21. $x^4 + 2x^3 + 4x^2 + 7x + 4;\ x + 1$
22. $2x^4 + x^3 + x^2 + 10x - 8;\ x + 2$

Example B (Division by $x - (a + bi)$) Find the quotient and remainder when $x^3 - 2x^2 - 6ix + 18$ is divided by $x - 2 - 3i$.

SOLUTION. Synthetic division works just fine even when some or all of the coefficients are nonreal. Since $x - 2 - 3i = x - (2 + 3i)$, we use $2 + 3i$ as the synthetic divisor.

$$\begin{array}{r|rrrr} 2+3i & 1 & -2 & -6i & 18 \\ & & 2+3i & -9+6i & -18-27i \\ \hline & 1 & 3i & -9 & -27i \end{array}$$

Quotient: $x^2 + 3ix - 9$; Remainder: $-27i$.

Use synthetic division to find the quotient and remainder when the first polynomial is divided by the second.

23. $x^3 - 2x^2 + 5x + 30;\ x - 2 + 3i$
24. $2x^3 - 11x^2 + 44x + 35;\ x - 3 - 4i$
25. $x^4 - 17;\ x - 2i$
26. $x^4 + 18x^2 + 90;\ x - 3i$

Miscellaneous Problems

27. Find the quotient and remainder when $x^4 + 6x^3 - 2x^2 + 4x - 15$ is divided by $x^2 - 2x + 3$.
28. Express $(3x^2 + 4x)/(x - 2)$ as the sum of a polynomial and a proper rational function.

In Problems 29–32, find the quotient and remainder when the first polynomial is divided by the second.

29. $2x^3 + 5x - 2$; $x + 2$
30. $x^4 - 4x^3 + 27$; $x - 3$
31. $x^5 - 32$; $x - 2$
32. $x^3 + (2i - 2)x^2 - (9 + 10i)x + 20 + 12i$; $x - 3$

In Problems 33–36, use division to show that the second polynomial is a factor of the first. You may use synthetic division where it applies.

33. $x^5 + x^4 - 16x - 16$; $x - 2$
34. $x^5 + x^4 - 16x - 16$; $x - 2i$
35. $x^5 + x^4 - 16x - 16$; $x + 2i$
36. $x^5 - 32$; $x^4 + 2x^3 + 4x^2 + 8x^3 + 16$

In Problems 37–40, determine by inspection the remainder when the first polynomial is divided by the second.

37. $(x - 2)^3 - 3(x - 2)^2 + 5(x - 2) + 11$; $x - 2$
    (*Hint*: The first polynomial can be written as $(x - 2)[(x - 2)^2 - 3(x - 2) + 5] + 11$.)
38. $2(x + 3)^2 + 118(x + 3) - 14$; $x + 3$
39. $(x - 4)^6 + 14(x - 4)^3 + 25$; $(x - 4)^3$
40. $(x + b)^3 - 10(x + b)^2 + 4(x + b) - 11$; $x + b$

Young Scholar: And does $x^4 + 99x^3 + 21$ have a zero?
Carl Gauss: Yes.
Young Scholar: How about $\pi x^{67} - \sqrt{3}\,x^{19} + 4i$?
Carl Gauss: It does.
Young Scholar: How can you be sure?
Carl Gauss: When I was young, 22 I think, I proved that every non-constant polynomial, no matter how complicated, has at least one zero. Many people call it the *Fundamental Theorem of Algebra.*

Carl F. Gauss (1777–1855)
"The Prince of Mathematicians"

10-2 Factorization Theory for Polynomials

Our young scholar could have asked a harder question. How do you find the zeros of a polynomial? Even the eminent Gauss would have had trouble with that question. You see, it is one thing to know a polynomial has zeros; it is quite another thing to find them.

Even though it is a difficult task and one at which we will have only limited success, our goal for this and the next two sections is to develop methods for finding zeros of polynomials. Remember that a polynomial is an expression of the form

$$P(x) = a_n x^n + a_{n-1}x^{n-1} + \cdots + a_1 x + a_0$$

Unless otherwise specified, the coefficients (the a_i's) are allowed to be *complex* numbers. And by a **zero** of $P(x)$, we mean any complex number c (real or nonreal) such that $P(c) = 0$.

The Remainder and Factor Theorems Recall the Division Law from Section 10-1, which had as its conclusion

$$P(x) = D(x)Q(x) + R(x)$$

If $D(x)$ has the form $x - c$, this becomes

$$P(x) = (x - c)Q(x) + R$$

where R, which is of lower degree than $x - c$, must be a constant. This last equation is an identity; it is true for all values of x, including $x = c$. Thus

$$P(c) = (c - c)Q(c) + R = 0 + R$$

We have just proved an important result.

REMAINDER THEOREM

If a polynomial $P(x)$ is divided by $x - c$, then the constant remainder R is given by $R = P(c)$.

Here is a nice example. Suppose we want to know the remainder R when $P(x) = x^{1000} + x^{22} - 15$ is divided by $x - 1$. We could, of course, divide it out—but that would be like going through purgatory, especially since we know the Remainder Theorem. From it we learn that

$$R = P(1) = 1^{1000} + 1^{22} - 15 = -13$$

Much more important than the mere calculation of remainders is a consequence called the Factor Theorem. Since $R = P(c)$, as we have just seen, we may rewrite the Division Law as

$$P(x) = (x - c)Q(x) + P(c)$$

Now it is plain to see that $P(c) = 0$ if and only if the division of $P(x)$ by $x - c$ is exact, that is, if and only if $x - c$ is a factor of $P(x)$.

FACTOR THEOREM

A polynomial $P(x)$ has c as a zero if and only if it has $x - c$ as a factor.

Sometimes it is easy to spot one zero of a polynomial. If so, the Factor Theorem may help us find the other zeros. Consider the polynomial

$$P(x) = 3x^3 - 8x^2 + 3x + 2$$

Notice that

$$P(1) = 3 - 8 + 3 + 2 = 0$$

so 1 is a zero. By the Factor Theorem, $x - 1$ is a factor of $P(x)$. We can use synthetic division to find the other factor.

$$\begin{array}{r|rrrr} 1 & 3 & -8 & 3 & 2 \\ & & 3 & -5 & -2 \\ \hline & 3 & -5 & -2 & 0 \end{array}$$

The remainder is 0 as we expected, and

$$P(x) = (x - 1)(3x^2 - 5x - 2)$$

Using the quadratic formula, we find the zeros of $3x^2 - 5x - 2$ to be $(5 \pm \sqrt{49})/6$, which simplify to 2 and $-\frac{1}{3}$. Thus $P(x)$ has 1, 2, and $-\frac{1}{3}$ as its three zeros.

Complete Factorization of Polynomials In the example above, we did not really need the quadratic formula. If we had been clever, we would have factored $3x^2 - 5x - 2$.

$$\begin{aligned} 3x^2 - 5x - 2 &= (3x + 1)(x - 2) \\ &= 3(x + \tfrac{1}{3})(x - 2) \end{aligned}$$

Thus $P(x)$, our original polynomial, may be written as

$$P(x) = 3(x - 1)(x + \tfrac{1}{3})(x - 2)$$

from which all three of the zeros are immediately evident.

But now we make another key observation. Notice that $P(x)$ factored as a product of its leading coefficient and three factors of the form $(x - c)$, where the c's are the zeros of $P(x)$. That holds true in general.

COMPLETE FACTORIZATION THEOREM

If

$$P(x) = a_n x^n + a_{n-1}x^{n-1} + \cdots + a_1 x + a_0, \qquad n > 0$$

is an nth *degree polynomial, then there are n numbers* $c_1, c_2, \ldots, c_n$, *not necessarily distinct, such that*

$$P(x) = a_n(x - c_1)(x - c_2)\cdots(x - c_n)$$

The c*'s are the zeros of* $P(x)$.

To prove the Complete Factorization Theorem, we must go back to Carl Gauss and our opening display. In his doctoral dissertation in 1799, Gauss proved a fundamental result that we need.

FUNDAMENTAL THEOREM OF ALGEBRA

Every nonconstant polynomial has at least one zero.

Taking this for granted, consider a polynomial $P(x)$ of degree $n > 0$. By the Fundamental Theorem, it has a zero which we may call c_1. By the Factor Theorem, $x - c_1$ is a factor of $P(x)$; that is,

$$P(x) = (x - c_1)P_1(x)$$

where $P_1(x)$ is a polynomial of degree $n - 1$ and with the same leading coefficient as $P(x)$, namely, a_n.

If $n - 1 > 0$, we may repeat the argument on $P_1(x)$. It has a zero c_2 and hence a factor $x - c_2$, that is,

$$P_1(x) = (x - c_2)P_2(x)$$

where $P_2(x)$ has degree $n - 2$. For our original polynomial $P(x)$, we may now write

$$P(x) = (x - c_1)(x - c_2)P_2(x)$$

Continuing in the pattern now established, we eventually get

$$P(x) = (x - c_1)(x - c_2)\cdots(x - c_n)P_n$$

where P_n has degree zero; that is, P_n is a constant. In fact, $P_n = a_n$ since the leading coefficient stayed the same at each step. This establishes the Complete Factorization Theorem.

About the Number of Zeros Each of the numbers c_i in

$$P(x) = a_n(x - c_1)(x - c_2)\cdots(x - c_n)$$

is a zero of $P(x)$. Are there any other zeros? No, for if d is any number different from each of the c_i's, then

$$P(d) = a_n(d - c_1)(d - c_2)\cdots(d - c_n) \neq 0$$

All of this tempts us to say that a polynomial of degree n has exactly n zeros. But hold on! The numbers $c_1, c_2, \ldots, c_n$ need not all be different. For example, the sixth degree polynomial

$$P(x) = 4(x - 2)^3(x + 1)(x - 4)^2$$

has only three distinct zeros, 2, -1, and 4. We have to settle for the following statement.

An nth *degree polynomial has at most n distinct zeros.*

There is a way in which we can say that there are exactly n zeros. Call c a **zero of multiplicity** k of $P(x)$ if $x - c$ appears k times in its complete factorization. For example, in

$$P(x) = 4(x - 2)^3(x + 1)(x - 4)^2$$

the zeros 2, -1, and 4 have multiplicities 3, 1, and 2, respectively. A zero of multiplicity 1 is called a **simple zero**. Notice in our example that the multiplicities add to 6, the degree of the polynomial. In general we may say that an nth degree polynomial has exactly n zeros if we count a zero of multiplicity k as k zeros.

Problem Set 10-2

Use the Remainder Theorem to find P(c). Check your answer by substituting c for x.

1. $P(x) = 2x^3 - 5x^2 + 3x - 4$; $c = 2$
2. $P(x) = x^3 + 4x^2 - 11x - 5$; $c = 3$

3. $P(x) = 8x^4 - 3x^2 - 2$; $c = \frac{1}{2}$
4. $P(x) = 2x^4 + \frac{3}{4}x + \frac{3}{2}$; $c = -\frac{1}{2}$

Find the remainder if the first polynomial is divided by the second. Do it without actually dividing.

5. $x^{10} - 15x + 8$; $x - 1$
6. $2x^{20} + 5$; $x + 1$
7. $64x^6 + 13$; $x + \frac{1}{2}$
8. $81x^3 + 9x^2 - 2$; $x - \frac{1}{3}$

Find all of the zeros of the given polynomial and give their multiplicities.

9. $(x - 1)(x + 2)(x - 3)$
10. $(x + 2)(x + 5)(x - 7)$
11. $(2x - 1)(x - 2)^2x^3$
12. $(3x + 1)(x + 1)^3x^2$
13. $3(x - 1 - 2i)(x + \frac{2}{3})$
14. $5(x - 2 + \sqrt{5})(x - \frac{4}{5})$

In Problems 15–18, show that $x - c$ is a factor of $P(x)$.

15. $P(x) = 2x^3 - 7x^2 + 9x - 4$; $c = 1$
16. $P(x) = 3x^3 + 4x^2 - 6x - 1$; $c = 1$
17. $P(x) = x^3 - 7x^2 + 16x - 12$; $c = 3$
18. $P(x) = x^3 - 8x^2 + 13x + 10$; $c = 5$
19. In Problem 15, you know that $P(x)$ has 1 as a zero. Find the other zeros.
20. Find all of the zeros of $P(x)$ in Problem 16.
21. Find all of the zeros of $P(x)$ in Problem 17.
22. Find all of the zeros of $P(x)$ in Problem 18.

In Problems 23–26, factor the given polynomial into linear factors. You should be able to do it by inspection.

23. $x^2 - 5x + 6$
24. $2x^2 - 14x + 24$
25. $x^4 - 5x^2 + 4$
26. $x^4 - 13x^2 + 36$

In Problems 27–30, factor $P(x)$ into linear factors given that c is a zero of $P(x)$.

27. $P(x) = x^3 - 3x^2 - 28x + 60$; $c = 2$
28. $P(x) = x^3 - 2x^2 - 29x - 42$; $c = -2$
29. $P(x) = x^3 + 3x^2 - 10x - 12$; $c = -1$
30. $P(x) = x^3 + 11x^2 - 5x - 55$; $c = -11$

Example A (Factoring a polynomial with a given nonreal zero) Show that $1 + 2i$ is a zero of $P(x) = x^3 - (3 + 2i)x^2 + (4i - 13)x + 30i + 15$. Then factor $P(x)$ into linear factors.

SOLUTION. We start by using synthetic division.

$$\begin{array}{r|rrrr} 1+2i & 1 & -3-2i & -13+4i & 15+30i \\ & & 1+2i & -2-4i & -15-30i \\ \hline & 1 & -2 & -15 & 0 \end{array}$$

Therefore $1 + 2i$ is a zero and $x - 1 - 2i$ is a factor of $P(x)$, and

$$\begin{aligned} P(x) &= (x - 1 - 2i)(x^2 - 2x - 15) \\ &= (x - 1 - 2i)(x + 3)(x - 5) \end{aligned}$$

In each of the following, factor $P(x)$ into linear factors given that c is a zero of $P(x)$.

31. $P(x) = x^3 - (3 + 2i)x^2 + (2 + 6i)x - 4i$; $c = 2i$
32. $P(x) = x^3 + 3ix^2 - 9x - 27i$; $c = -3i$
33. $P(x) = x^3 + (1 - i)x^2 - (1 + 2i)x - 1 - i$; $c = 1 + i$
34. $P(x) = x^3 - 3ix^2 + (3i - 3)x - 2 + 6i$; $c = -1 + 3i$

Example B (Finding a polynomial from its zeros)
(a) Find a cubic polynomial having simple zeros 3, -2, and $2i$.
(b) Find a polynomial $P(x)$ with integral coefficients and having $\frac{1}{2}$ and $-\frac{2}{3}$ as simple zeros and 1 as a zero of multiplicity 2.

SOLUTION. (a) Let us call the required polynomial $P(x)$. Then $P(x) = a(x - 3)(x + 2)(x - 2i)$, where a can be any complex number. Choosing $a = 1$ and multiplying, we have

$$P(x) = (x^2 - x - 6)(x - 2i) = x^3 - (1 + 2i)x^2 + (2i - 6)x + 12i$$

(b) $P(x) = a(x - \frac{1}{2})(x + \frac{2}{3})(x - 1)^2$

We choose $a = 6$ to eliminate fractions.

$$\begin{aligned} P(x) &= 6(x - \tfrac{1}{2})(x + \tfrac{2}{3})(x - 1)^2 \\ &= 2(x - \tfrac{1}{2})3(x + \tfrac{2}{3})(x - 1)^2 \\ &= (2x - 1)(3x + 2)(x^2 - 2x + 1) \\ &= 6x^4 - 11x^3 + 2x^2 + 5x - 2 \end{aligned}$$

In Problems 35–42, find a polynomial $P(x)$ with integral coefficients having the given zeros. Assume each zero to be simple (multiplicity 1) unless otherwise indicated.

35. 2, 1, and -4
36. 3, -2, and 5
37. $\frac{1}{2}, -\frac{5}{6}$
38. $\frac{3}{7}, \frac{3}{4}$
39. 2, $\sqrt{5}$, $-\sqrt{5}$
40. -3, $\sqrt{7}$, $-\sqrt{7}$
41. $\frac{1}{2}$(multiplicity 2), -2(multiplicity 3)
42. 0, -2, $\frac{3}{4}$(multiplicity 3)

In Problems 43 and 44, find a polynomial $P(x)$ having only the indicated simple zeros.

43. 2, -5, $2 + 3i$
44. -3, 2, $1 - 4i$

Example C (More on zeros of polynomials) Show that 2 is a zero of multiplicity 3 of the polynomial

$$P(x) = 2x^5 - 17x^4 + 51x^3 - 58x^2 + 4x + 24$$

and find the remaining zeros.

SOLUTION. We must show that $x - 2$ appears as a factor 3 times in the factored form of $P(x)$. Synthetic division can be used successively as shown below.

2	2	-17	51	-58	4	24	
		4	-26	50	-16	-24	
2	2	-13	25	-8	-12	(0)	This shows that $x - 2$ is a factor;
		4	-18	14	12		
2	2	-9	7	6	(0)		again;
		4	-10	-6			
	2	-5	-3	(0)			and again.

The final quotient $2x^2 - 5x - 3$ factors as $(2x + 1)(x - 3)$. Therefore the remaining two zeros are $-\frac{1}{2}$ and 3. The factored form of $P(x)$ is

$$2(x - 2)^3(x + \tfrac{1}{2})(x - 3)$$

45. Show that 1 is a zero of multiplicity 3 of the polynomial $x^5 + 2x^4 - 6x^3 - 4x^2 + 13x - 6$ and find the remaining zeros.
46. Show that the polynomial $x^5 - 11x^4 + 46x^3 - 90x^2 + 81x - 27$ has 3 as a zero of multiplicity 3 and find the remaining zeros.
47. Show that the polynomial

$$x^6 - 8x^5 + 7x^4 + 32x^3 + 31x^2 + 40x + 25$$

has -1 and 5 as zeros of multiplicity 2 and find the remaining zeros.

48. Show that the polynomial

$$x^6 + 3x^5 - 9x^4 - 50x^3 - 84x^2 - 72x - 32$$

has 4 as a simple zero and -2 as a zero of multiplicity 3. Find the remaining zeros.

Miscellaneous Problems

49. Without actually dividing, find the remainder when $x^5 - 10$ is divided by $x - 2$.

In Problems 50–52 find all the zeros of the given polynomials and give their multiplicities.

50. $(x^2 - 4)^3$
51. $(x^2 - 3x + 2)^2$
52. $(x^2 + 2x + 4)(x + 2)^2$
53. Find a polynomial $P(x)$ with integral coefficients that has the given zeros.
(a) $\frac{3}{4}, 2, -\frac{2}{3}$ (b) $3, -2$(multiplicity 2)

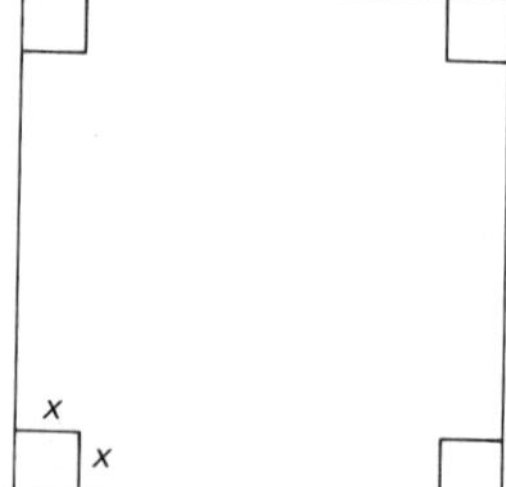

54. Find the polynomial $P(x) = ax^2 + bx + c$ which has 1 and -2 as zeros and which satisfies $P(3) = 40$. (*Hint*: First find a from the factored form of $P(x)$.)
55. Sketch the graph of $y = x^3 + 4x^2 - 2x$. Now find the three values of x for which $y = 8$. (*Hint*: One of these is an integer.)
56. Refer to the graph of Problem 55 and note that there is only one real x (namely, $x = 2$) for which $y = 20$. What does this tell you about the other two solutions of $x^3 + 4x^2 - 2x = 20$? Find the other two solutions.
57. A tray is to be constructed from a piece of sheet metal 16 inches square by cutting equal squares from the corners and then folding up the flaps (see the diagrams in the margin). How large must these squares be if the volume is to be 300 cubic inches? (*Hint*: There are two possible answers, one of which is an integer.)
58. Take a number, add 2, multiply the result by the original number, and then add 9. Now multiply this result by the original number, subtract 2, again multiply by the original number, and finally subtract 10. What must your original number be if the final outcome is 0?

59. Find $P(x) = a_4x^4 + a_3x^3 + a_2x^2 + a_1x + a_0$ if $P(x)$ has $\frac{1}{2}$ as a zero of multiplicity 4 and $P(0) = 1$.

60. Use the Factor Theorem to show that
(a) $x^n - a^n$ has $x - a$ as a factor for any positive integer n, where a is any complex number;
(b) $x^n + a^n$ has $x + a$ as a factor if n is odd;
(c) $x^n + a^n$ does not have $x + a$ as a factor if n is even.

61. Show that the polynomial

$$P(x) = x^{40} + 10x^{22} + 4x^8 + 16x^2 + 15$$

has no real zeros.

62. Find the zeros of $(x^2 + 1)^4$ and their multiplicities.

63. Show that if

$$P(x) = a_nx^n + a_{n-1}x^{n-1} + \cdots + a_1x + a_0$$

has $n + 1$ distinct zeros, then all of the coefficients have to be zero.

64. Let

$$P_1(x) = x^n + a_{n-1}x^{n-1} + a_{n-2}x^{n-2} + \cdots + a_1x + a_0$$

and

$$P_2(x) = x^n + b_{n-1}x^{n-1} + b_{n-2}x^{n-2} + \cdots + b_1x + b_0$$

Show that if $P_1(x) = P_2(x)$ for n distinct values of x, then the two polynomials are identical; that is, $a_i = b_i$ for each i. (*Hint*: Let $P(x) = P_1(x) - P_2(x)$ and apply Problem 63.)

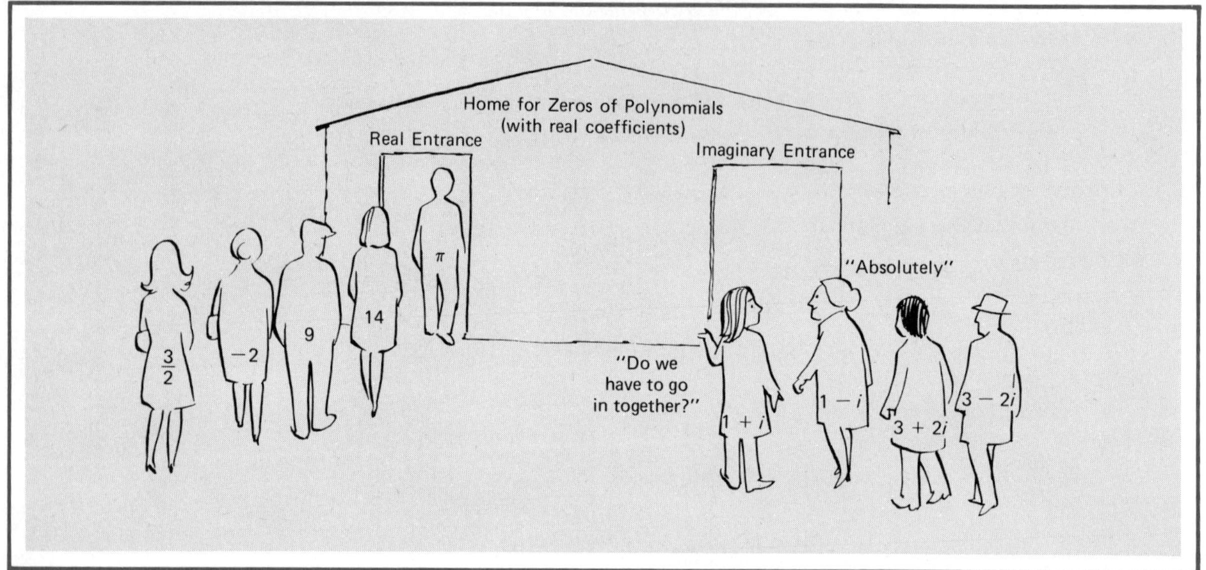

10-3 Polynomial Equations with Real Coefficients

Finding the zeros of a polynomial $P(x)$ is the same as finding the solutions of the equation $P(x) = 0$. We have solved equations before, especially in Chapter 3. It would be a good idea to review Section 3-4 on the quadratic equation now. In particular, recall that the two solutions of a quadratic equation with real coefficients are either both real or both nonreal. For example, the equation $x^2 - 7x + 12 = 0$ has the real solutions 3 and 4. On the other hand, $x^2 - 8x + 17 = 0$ has two nonreal solutions $4 + i$ and $4 - i$.

In this section, we shall see that if a polynomial equation has real coefficients, then its nonreal solutions (if any) must occur in pairs—conjugate pairs. In the opening display $1 + i$ and $1 - i$ must enter together or not at all; so must $3 + 2i$ and $3 - 2i$.

Properties of Conjugates We indicate the conjugate of a complex number by putting a bar over it. If $u = a + bi$, then $\bar{u} = a - bi$. For example,

$$\overline{1 + i} = 1 - i$$
$$\overline{3 - 2i} = 3 + 2i$$
$$\bar{4} = 4$$

The operation of taking the conjugate behaves nicely in both addition and multiplication. There are two pertinent properties, stated first in words.

1. The conjugate of a sum is the sum of the conjugates.
2. The conjugate of a product is the product of the conjugates.

In symbols these properties become

1. $\overline{u_1 + u_2 + \cdots + u_n} = \bar{u}_1 + \bar{u}_2 + \bar{u}_3 + \cdots + \bar{u}_n$
2. $\overline{u_1 u_2 u_3 \cdots u_n} = \bar{u}_1 \cdot \bar{u}_2 \cdot \bar{u}_3 \cdots \bar{u}_n$

A third property follows from the second property, if we set all u_i's equal to u.

3. $\overline{u^n} = (\bar{u})^n$

Rather than prove these properties, we shall illustrate them.

1. $\overline{(2 + 3i) + (1 - 4i)} = \overline{2 + 3i} + \overline{1 - 4i}$
2. $\overline{(2 + 3i)(1 - 4i)} = \overline{(2 + 3i)}\,\overline{(1 - 4i)}$
3. $\overline{(2 + 3i)^3} = (\overline{2 + 3i})^3$

Let us check that the second statement is correct. We will do it by computing both sides independently.

$$\overline{(2 + 3i)(1 - 4i)} = \overline{2 + 12 + (-8 + 3)i} = \overline{14 - 5i} = 14 + 5i$$
$$\overline{(2 + 3i)}\,\overline{(1 - 4i)} = (2 - 3i)(1 + 4i) = 2 + 12 + (8 - 3)i = 14 + 5i$$

Now let us use the three properties in more and more complicated situations. First, note that

$$\begin{aligned} \overline{2(5 + 2i)^2} &= \bar{2} \cdot \overline{(5 + 2i)^2} && \text{(Property 2)} \\ &= \bar{2} \cdot (\overline{5 + 2i})^2 && \text{(Property 3)} \\ &= 2(5 - 2i)^2 \end{aligned}$$

Next using Property 1 and the result just obtained, we get

$$\begin{aligned} \overline{2(5 + 2i)^2 - 3(5 + 2i) + 4} &= \overline{2(5 + 2i)^2} + \overline{(-3)(5 + 2i)} + \bar{4} \\ &= 2(5 - 2i)^2 - 3(5 - 2i) + 4 \end{aligned}$$

What we just did for $5 + 2i$ could be done for any complex number u, that is,

$$\overline{2u^2 - 3u + 4} = 2\bar{u}^2 - 3\bar{u} + 4$$

In a similar fashion we could show that if a, b, c, and d are any real numbers, then

$$\overline{au^3 + bu^2 + cu + d} = a\bar{u}^3 + b\bar{u}^2 + c\bar{u} + d$$

In particular, if the left side happens to be zero, so is the right side.

Of what use is this? Well, it tells us that if u is a solution of the polynomial equation

$$ax^3 + bx^2 + cx + d = 0$$

then $\bar{u}$ is also a solution. The reason is clear. For u to be a solution means

$$au^3 + bu^2 + cu + d = 0$$

But then

$$\overline{au^3 + bu^2 + cu + d} = \bar{0} = 0$$

that is,

$$a\bar{u}^3 + b\bar{u}^2 + c\bar{u} + d = 0$$

Nonreal Solutions Occur in Pairs We are ready to state the main theorem of this section.

CONJUGATE PAIR THEOREM

Let

$$a_nx^n + a_{n-1}x^{n-1} + \cdots + a_1x + a_0 = 0$$

be a polynomial equation with real coefficients. If u is a solution, its conjugate $\bar{u}$ is also a solution.

We feel confident that you will be willing to accept the truth of this theorem without further argument. The formal proof would mimic the proof given above for the cubic equation.

As an illustration of one use of this theorem, suppose that we know that $3 + 4i$ is a solution of the equation

$$x^3 - 8x^2 + 37x - 50 = 0$$

Then we know that $3 - 4i$ is also a solution. We can easily find the third solution, which incidentally must be real. (Why?) Here is how we do it.

$$\begin{array}{r|rrrr}
3 + 4i & 1 & -8 & 37 & -50 \\
 & & 3 + 4i & -31 - 8i & 50 \\
\hline
3 - 4i & 1 & -5 + 4i & 6 - 8i & 0 \\
 & & 3 - 4i & -6 + 8i & \\
\hline
 & 1 & -2 & 0 &
\end{array}$$

From this synthetic division, we conclude that $x - 2$ is a factor of the left side of our equation. It follows that 2 is the third solution.

Rational Solutions How does one get started on solving an equation of high degree? So far, all we can suggest is to guess. If you are lucky and find a solution, you can use synthetic division to reduce the degree of the equation to be solved. Eventually you may get it down to a quadratic equation, for which we have the quadratic formula.

Guessing would not be so bad if there were not so many possibilities to consider. Is there an intelligent way to guess? There is, but unfortunately it works only if the coefficients are integers, and then it only helps us find rational solutions.

Consider

$$3x^3 + 13x^2 - x - 6 = 0$$

which, as you will note, has integral coefficients. Suppose it has a rational solution c/d which is in reduced form (that is, c and d are integers without common divisors greater than 1 and $d > 0$). Then

$$3 \cdot \frac{c^3}{d^3} + 13 \cdot \frac{c^2}{d^2} - \frac{c}{d} - 6 = 0$$

or, after multiplying by d^3,

$$3c^3 + 13c^2d - cd^2 - 6d^3 = 0$$

We can rewrite this as

$$c(3c^2 + 13cd - d^2) = 6d^3$$

and also as

$$d(13c^2 - cd - 6d^2) = -3c^3$$

The first of these tells us that c divides $6d^3$ and the second that d divides $-3c^3$. But c and d have no common divisors. Therefore c must divide 6 and d must divide 3.

The only possibilities for c are ± 1, ± 2, ± 3, and ± 6; for d, the only possibilities are 1 and 3. Thus the possible rational solutions must come from the list below.

$$\frac{c}{d}: \quad \pm 1, \pm 2, \pm 3, \pm 6, \pm \tfrac{1}{3}, \pm \tfrac{2}{3}$$

Upon checking all 12 numbers (which takes time, but a bit less time than checking *all* numbers would take!) we find that only $\frac{2}{3}$ works.

$$\begin{array}{r|rrrr} \frac{2}{3} & 3 & 13 & -1 & -6 \\ & & 2 & 10 & 6 \\ \hline & 3 & 15 & 9 & 0 \end{array}$$

We could prove the following theorem by using similar reasoning.

RATIONAL SOLUTION THEOREM

Let

$$a_nx^n + a_{n-1}x^{n-1} + \cdots + a_1x + a_0 = 0$$

have integral coefficients. If c/d is a rational solution in reduced form, then c divides a_0 and d divides a_n.

Problem Set 10-3 *In Problems 1–10, write the conjugate of the number.*

1. $2 + 3i$
2. $3 - 5i$
3. $4i$
4. $-6i$
5. $4 + \sqrt{6}$
6. $3 - \sqrt{5}$
7. $(2 - 3i)^8$
8. $(3 + 4i)^{12}$
9. $2(1 + 2i)^3 - 3(1 + 2i)^2 + 5$
10. $4(6 - i)^4 + 11(6 - i) - 23$
11. If $P(x)$ is a cubic polynomial with real coefficients and has -3 and $5 - i$ as zeros, what other zero does it have?
12. If $P(x)$ is a cubic polynomial with real coefficients and has 0 and $\sqrt{2} + 3i$ as zeros, what other zero does it have?
13. Suppose that $P(x)$ has real coefficients and is of the fourth degree. If it has $3 - 2i$ and $5 + 4i$ as two of its zeros, what other zeros does it have?
14. If $P(x)$ is a fourth degree polynomial with real coefficients and has $5 + 6i$ as a zero of multiplicity 2, what are its other zeros?

Example A (Solving an equation given some solutions) Given that -1 and $1 + 2i$ are solutions of the equation

$$2x^4 - 5x^3 + 9x^2 + x - 15 = 0$$

find the other solutions.

SOLUTION. Since the coefficients are real, $1 - 2i$ is a solution. The fourth solution is found by progressively using synthetic division.

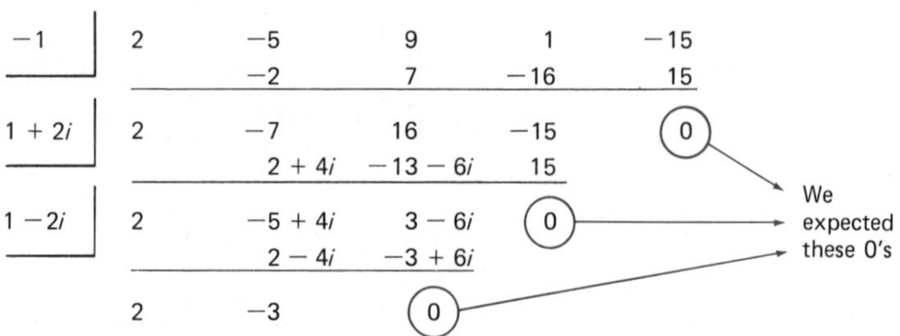

Last quotient: $2x - 3$

The 4th solution: $\frac{3}{2}$

In Problems 15–18, one or more solutions of the specified equations are given. Find the other solutions.

15. $2x^3 - x^2 + 2x - 1 = 0$; i
16. $x^3 - 3x^2 + 4x - 12 = 0$; $2i$
17. $x^4 + x^3 + 6x^2 + 26x + 20 = 0$; $1 + 3i$
18. $x^6 + 2x^5 + 4x^4 + 4x^3 - 4x^2 - 16x - 16 = 0$; $2i$, $-1 + i$

Example B (Obtaining a polynomial given some of its zeros) Find a cubic polynomial with real coefficients that has 3 and $2 + 3i$ as zeros. Make the leading coefficient 1.

SOLUTION. The third zero has to be $2 - 3i$. In factored form, our polynomial is

$$(x - 3)(x - 2 - 3i)(x - 2 + 3i)$$

We multiply this out in stages.

$$\begin{gathered}(x - 3)[(x - 2)^2 + 9] \\ (x - 3)(x^2 - 4x + 13) \\ x^3 - 7x^2 + 25x - 39\end{gathered}$$

Find a polynomial with real coefficients that has the indicated degree and the given zero(s). Make the leading coefficient 1.

19. Degree 2; zero: $2 + 5i$.
20. Degree 2; zero: $\sqrt{6}i$.
21. Degree 3; zeros: -3, $2i$.
22. Degree 3; zeros: 5, $-3i$.
23. Degree 5; zeros: 2, $3i$ of multiplicity 2.
24. Degree 5; zeros: 1, $1 - i$ of multiplicity 2.

Example C (Finding rational solutions) Find the rational solutions of the equation

$$3x^4 + 2x^3 + 2x^2 + 2x - 1 = 0$$

Then find the remaining solutions.

SOLUTION. The only way that c/d (in reduced form) can be a solution is for c to be 1 or -1 and for d to be 1 or 3. This means that the possibilities for c/d are ± 1 and $\pm\frac{1}{3}$. Synthetic division shows that -1 and $\frac{1}{3}$ work.

$$\begin{array}{r|rrrrr}
-1 & 3 & 2 & 2 & 2 & -1 \\
 & & -3 & 1 & -3 & 1 \\ \hline
\frac{1}{3} & 3 & -1 & 3 & -1 & 0 \\
 & & 1 & 0 & 1 & \\ \hline
 & 3 & 0 & 3 & 0 &
\end{array}$$

Setting the final quotient, $3x^2 + 3$, equal to zero and solving, we get

$$\begin{aligned}3x^2 &= -3 \\ x^2 &= -1 \\ x &= \pm i\end{aligned}$$

The complete solution set is $\{-1, \frac{1}{3}, i, -i\}$.

In Problems 25–30, find the rational solutions of each equation. If possible, find the other solutions.

25. $x^3 - 3x^2 - x + 3 = 0$
26. $x^3 + 3x^2 - 4x - 12 = 0$
27. $2x^3 + 3x^2 - 4x + 1 = 0$

28. $5x^3 - x^2 + 5x - 1 = 0$

29. $\frac{1}{3}x^3 - \frac{1}{2}x^2 - \frac{1}{6}x + \frac{1}{6} = 0$ (*Hint*: Clear the equation of fractions.)

30. $\frac{2}{3}x^3 - \frac{1}{2}x^2 + \frac{2}{3}x - \frac{1}{2} = 0$

Miscellaneous Problems

31. The equation $2x^3 - 11x^2 + 20x + 13 = 0$ has $3 - 2i$ as a solution. Find the other solutions.

32. Find the fourth degree polynomial with real coefficients and leading coefficient 1 that has 2, -4, and $1 + 2i$ as three of its zeros.

33. Find all the solutions of

$$x^4 - 3x^3 - 20x^2 - 24x - 8 = 0$$

Start by looking for rational solutions.

34. Find all the solutions of

$$2x^4 - x^3 + x^2 - x - 1 = 0$$

35. Answer *true* or *false*.
    (a) Every nth degree polynomial has n distinct zeros.
    (b) Every quadratic polynomial with real coefficients has either 2 distinct real zeros or 2 distinct nonreal zeros.
    (c) If $2 + 3i$ is a solution of a polynomial equation, so is $2 - 3i$.
    (d) It is possible for a sixth degree polynomial to have only 3 distinct zeros.
    (e) There is exactly one fourth degree polynomial which has 3, -2, and $2 \pm 5i$ as zeros.

36. Show that a polynomial equation with real coefficients and of odd degree has at least one real solution.

37. Show by means of an example that the statement in Problem 36 is not necessarily true if the coefficients of the equation are complex.

38. A cubic equation with real coefficients has either 3 real solutions (not necessarily distinct), or 1 real solution and a pair of nonreal solutions. State all of the possibilities for:
    (a) A fourth degree equation with real coefficients;
    (b) A fifth degree equation with real coefficients.

39. Suppose that a polynomial has integral coefficients and a leading coefficient of 1. Show that if this equation has a rational solution, it must be an integer.

40. Show that the following equation has no rational solution.

$$x^4 + x^3 - 4x^2 - 5x - 5 = 0$$

41. Show that the following equation has 1 as a solution of multiplicity 4. Find the remaining solution.

$$x^5 + 6x^4 - 34x^3 + 56x^2 - 39x + 10 = 0$$

42. (a) Show that if u is a complex number, then $u + \bar{u}$ and $u \cdot \bar{u}$ are real numbers. (*Hint*: Let $u = a + bi$.)
    (b) Show that $(x - u)(x - \bar{u})$, when multiplied out, has real coefficients.

(c) Suppose $P(x)$ is a fifth degree polynomial with real coefficients, having one real zero c and two pairs of nonreal zeros u and $\bar{u}$, v and $\bar{v}$. Show that $P(x)$ can be expressed as a product of linear and quadratic polynomials with real coefficients.

(d) Now let $P(x)$ be any polynomial with real coefficients. Prove that $P(x)$ can be expressed as a product of linear and quadratic polynomials with real coefficients.

43. (a) Solve the equation $x^4 + 1 = 0$. (*Hint*: This is the same as finding the four 4th roots of -1. See the example in Problem Set 9-6.)

(b) Find the rational solution(s) of $x^5 - 2x^4 + x - 2 = 0$. Then find the remaining solutions.

10-4 The Method of Successive Approximations

Even with the theory developed so far, we are often unable to get started on solving an equation of high degree. Imagine being given a fifth degree equation whose true solutions are $\sqrt{2} + 5i$, $\sqrt{2} - 5i$, $\sqrt[3]{19}$, 1.597, and 3π. How would you ever find them? Nothing you have learned until now would be of much help.

There is a general method of solving problems known to all resourceful people. We call it "muddling through" or "trial and error." Given a cup of tea, we add sugar a bit at a time until it tastes just right. Given a stopper too large for a hole, we whittle it down until it fits. We change the solution a step at a time, continually improving the accuracy until we are satisfied. Mathematicians call it the **method of successive approximations**.

We explore two such methods in this section. The first is a graphical method; the second is Newton's algebraic method. Both are designed to find the real solutions of polynomial equations with real coefficients. Both require many computations. We suggest that you keep your pocket calculator handy.

Method of Successive Enlargements Consider the equation

$$x^3 - 3x - 5 = 0$$

We begin by graphing

$$y = x^3 - 3x - 5$$

looking for the point (or points) where the graph crosses the x-axis. These points correspond to the real solutions.

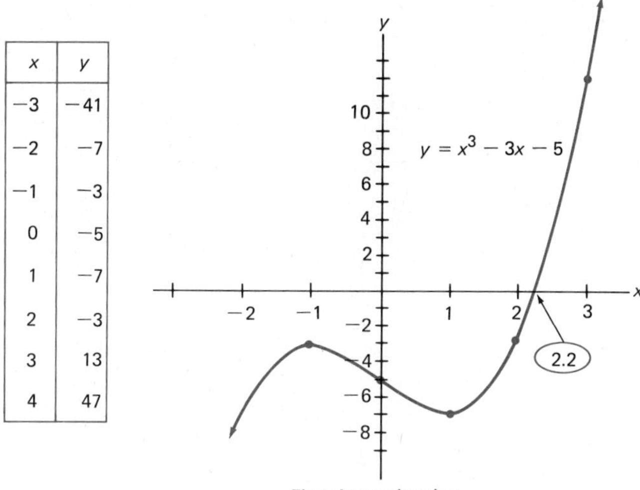

x	y
−3	−41
−2	−7
−1	−3
0	−5
1	−7
2	−3
3	13
4	47

First Approximation

Clearly there is only one real solution and it is between 2 and 3; a good first guess might be 2.2. Now we calculate y for values of x near 2.2 (for instance, 2.1, 2.2, and 2.3) until we find an interval of length .1 on which y changes sign. The interval is $2.2 \leq x \leq 2.3$. On this interval, we pretend the graph is a straight line. The point at which this line crosses the x-axis gives us our next approximation. It is about 2.28.

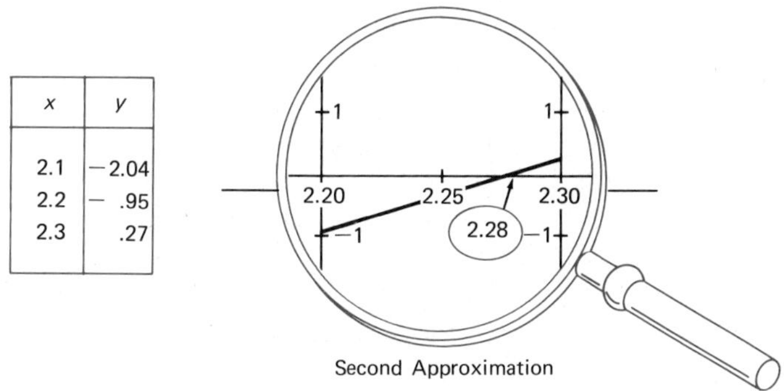

x	y
2.1	−2.04
2.2	− .95
2.3	.27

Second Approximation

Now we calculate y for values of x near 2.28 until we find an interval of length .01 on which y changes sign. This occurs on the interval $2.27 \leq x \leq 2.28$. Using a straight-line graph for this interval, we read our next approximation as 2.279.

In effect, we are enlarging the graph (using a more and more powerful magnifying glass) at each stage, increasing the accuracy by one digit each time. We can continue this process as long as we have the patience to do the necessary calculations.

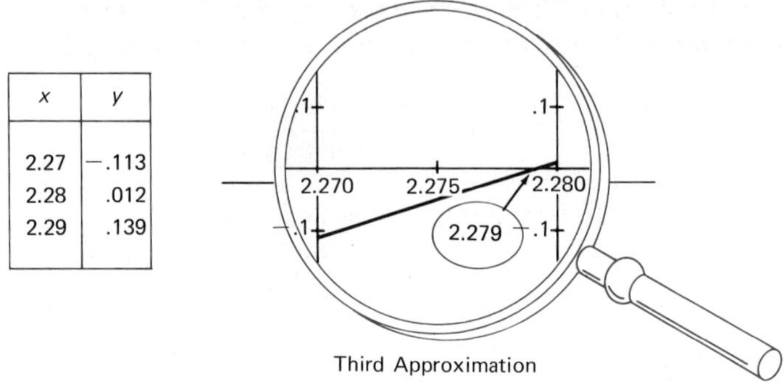

x	y
2.27	−.113
2.28	.012
2.29	.139

Third Approximation

Newton's Method Let $P(x) = 0$ be a polynomial equation with real coefficients. Suppose that by some means (perhaps graphing), we discover that it has a real solution r which we guess to be about x_1. Then, as the accompanying diagram suggests, a better approximation to r is x_2, the point at which the tangent line to the curve at x_1 crosses the x-axis.

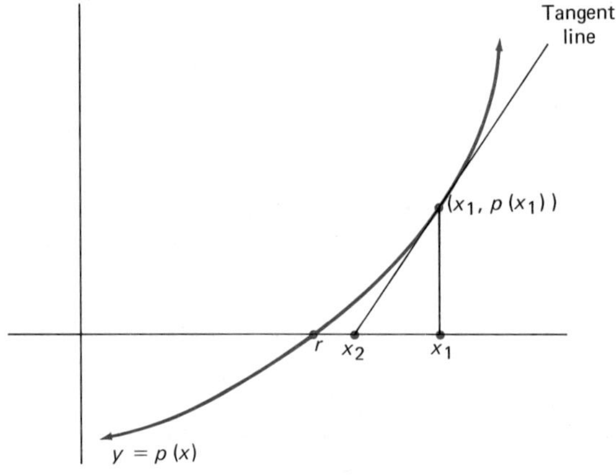

If the slope of the tangent line is m_1, then

$$m_1 = \frac{\text{rise}}{\text{run}} = \frac{P(x_1)}{x_1 - x_2}$$

Solving this for x_2 yields

$$x_2 = x_1 - \frac{P(x_1)}{m_1}$$

What has been done once can be repeated. A still better approximation to r would be x_3, obtained from x_2 as follows:

$$x_3 = x_2 - \frac{P(x_2)}{m_2}$$

where m_2 is the slope of the tangent line at $(x_2, P(x_2))$. In general, we find the $(k + 1)$st approximation from the kth by using Newton's formula

$$x_{k+1} = x_k - \frac{P(x_k)}{m_k}$$

where m_k is the slope of the tangent line at $(x_k, P(x_k))$.

The rub, of course, is that we do not know how to calculate m_k. That is precisely where Newton made his biggest contribution. He showed how to find the slope of the tangent line to any curve. This is done in every calculus course; we state one result without proof.

SLOPE THEOREM

If

$$P(x) = a_n x^n + a_{n-1} x^{n-1} + \cdots + a_2 x^2 + a_1 x + a_0$$

is a polynomial with real coefficients, then the slope of the tangent line to the graph of $y = P(x)$ at x is $P'(x)$, where $P'(x)$ is the polynomial

$$P'(x) = n a_n x^{n-1} + (n - 1) a_{n-1} x^{n-2} + \cdots + 2 a_2 x + a_1$$

For example, if

$$P(x) = 2x^4 + 4x^3 - 6x^2 + 2x + 15$$

then

$$\begin{aligned} P'(x) &= 4 \cdot 2 \cdot x^3 + 3 \cdot 4 \cdot x^2 - 2 \cdot 6 \cdot x + 2 \\ &= 8x^3 + 12x^2 - 12x + 2 \end{aligned}$$

In particular, the slope of the tangent line at $x = 2$ is $P'(2) = 8 \cdot 2^3 + 12 \cdot 2^2 - 12 \cdot 2 + 2 = 90$.

Taking the Slope Theorem for granted, we may write Newton's formula in the useful form.

$$\boxed{x_{k+1} = x_k - \frac{P(x_k)}{P'(x_k)}}$$

Using Newton's Method Consider the equation

$$x^3 - 3x - 5 = 0$$

again. If we let

$$P(x) = x^3 - 3x - 5$$

then, by the Slope Theorem,

$$P'(x) = 3x^2 - 3$$

and Newton's formula becomes

$$x_{k+1} = x_k - \frac{x_k^3 - 3x_k - 5}{3x_k^2 - 3}$$

If we take $x_1 = 3$ as our initial guess, then

$$\begin{aligned} x_2 &= x_1 - \frac{x_1^3 - 3x_1 - 5}{3x_1^2 - 3} \\ &= 3 - \frac{3^3 - 3\cdot 3 - 5}{3\cdot 3^2 - 3} \\ &\approx \boxed{2.5} \\ x_3 &= x_2 - \frac{x_2^3 - 3x_2 - 5}{3x_2^2 - 3} \\ &= 2.5 - \frac{(2.5)^3 - 3(2.5) - 5}{3(2.5)^2 - 3} \\ &\approx \boxed{2.30} \\ x_4 &= x_3 - \frac{x_3^3 - 3x_3 - 5}{3x_3^2 - 3} \\ &= 2.30 - \frac{(2.30)^3 - 3(2.30) - 5}{3(2.30)^2 - 3} \\ &\approx \boxed{2.279} \end{aligned}$$

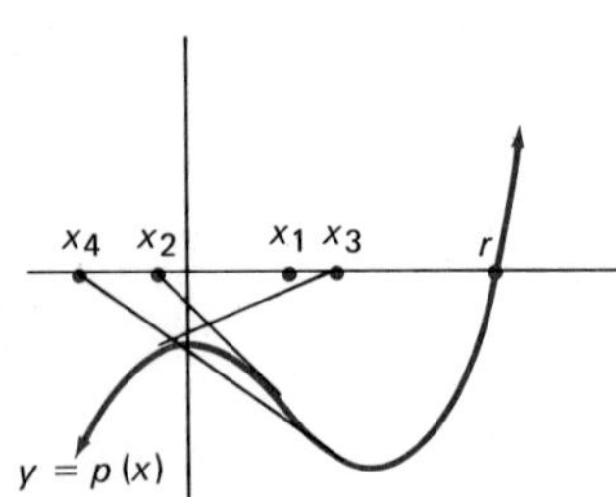

We can continue this repetitive process until we have the accuracy we desire.

When you use Newton's method, it is important to make your initial guess reasonably good. The sketch in the margin shows how badly the method can lead you astray if you choose x_1 too far off the mark.

Problem Set 10-4

In Problems 1–6, use the Slope Theorem to find the slope polynomial $P'(x)$ if $P(x)$ is the given polynomial. Then find the slope of the tangent line at $x = 1$.

1. $2x^2 - 5x + 6$
2. $3x^2 + 4x - 9$
3. $2x^2 + x - 2$
4. $x^3 - 5x + 8$
5. $2x^5 + x^4 - 2x^3 + 8x - 4$
6. $x^6 - 3x^4 + 7x^3 - 4x^2 + 5x - 4$

Each of the equations in Problems 7–10 has exactly one real solution. By means of a graph, make an initial guess at the solution. Then use the method of successive enlargements to find a second and a third approximation.

7. [C] $x^3 + 2x - 5 = 0$
8. [C] $x^3 + x - 32 = 0$
9. [C] $x^3 - 3x - 10 = 0$
10. [C] $2x^3 - 6x - 15 = 0$

Ⓒ 11. Use Newton's method to approximate the real solution of the equation $x^3 + 2x - 5 = 0$. Take as x_1 the initial guess you made in Problem 7, and then find x_2, x_3, and x_4.

Ⓒ 12–14. Follow the instructions of Problem 11 for the equations in Problems 8–10.

Example (Finding all real solutions by Newton's method) Find all real solutions of the following equation by Newton's method.

$$P(x) = x^4 - 8x^3 + 22x^2 - 24x + 6 = 0$$

SOLUTION. First we sketch the graph of $P(x)$.

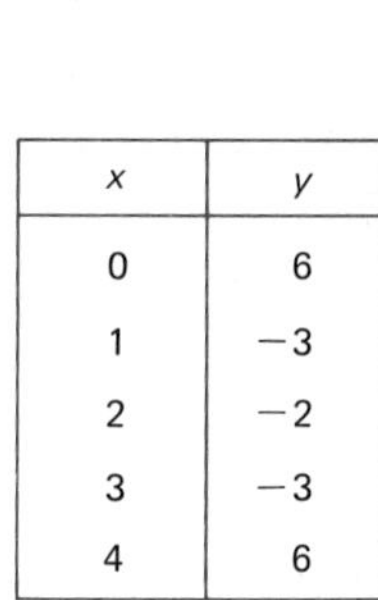

x	y
0	6
1	−3
2	−2
3	−3
4	6

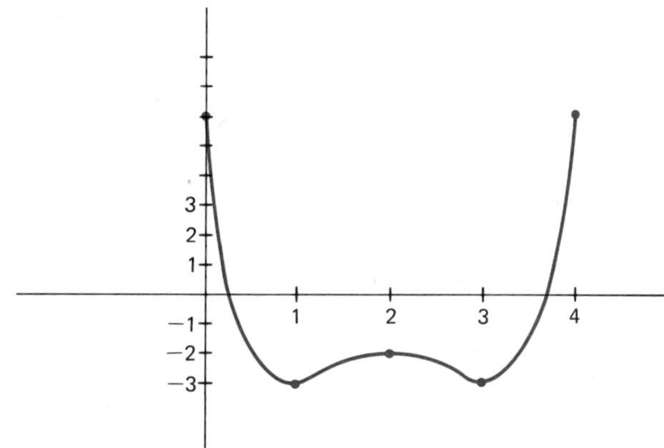

The graph crosses the x-axis at approximately .4 and 3.6. The slope polynomial is

$$P'(x) = 4x^3 - 24x^2 + 44x - 24$$

Take $x_1 = .4$,

$$x_2 = .4 - \frac{P(.4)}{P'(.4)}$$

$$= .4 - \frac{(-.57)}{(-9.98)} \approx .34$$

$$x_3 = .34 - \frac{P(.34)}{P'(.34)}$$

$$= .34 - \frac{.0821}{(-4.657)} \approx .347$$

Take $x_1 = 3.6$,

$$x_2 = 3.6 - \frac{P(3.6)}{P'(3.6)}$$

$$= 3.6 - \frac{(-.566)}{9.98} \approx 3.66$$

$$x_3 = 3.66 - \frac{P(3.66)}{P'(3.66)}$$

$$= 3.66 - \frac{.0821}{11.66} \approx 3.653$$

Each equation in Problems 15–18 has several real solutions. Draw a graph to get your first estimates. Then use Newton's method to find these solutions to three decimal places.

Ⓒ 15. $x^4 + x^3 - 3x^2 + 4x - 28 = 0$

Ⓒ 16. $x^4 + x^3 - 6x^2 - 7x - 7 = 0$

Ⓒ 17. $x^3 - 3x + 1 = 0$

Ⓒ 18. $x^3 - 12x + 1 = 0$

Miscellaneous Problems

Ⓒ 19. The line $y = x + 3$ intersects the curve $y = x^3 - 3x + 4$ at three points. Use Newton's method to find the x-coordinates of these points correct to 2 decimal places.

Ⓒ 20. A spherical shell has a thickness of 1 centimeter. What is the outer radius r of the shell if the volume of the shell is equal to the volume of the space inside? First find an equation for r and then solve it by Newton's method.

Ⓒ 21. Solve Problem 20 if the volume of the hollow space is twice the volume of the shell.

Ⓒ 22. The dimensions of a rectangular box are 6, 8, and 10 feet. Suppose that the volume of the box is increased by 300 cubic feet by equal elongations of the three dimensions. Find this elongation correct to two decimal places.

Ⓒ 23. What rate of interest compounded annually is implied in an offer to sell a house for \$50,000 cash, or in annual installments of \$20,000 each payable 1, 2, and 3 years from now? (*Hint*: The amount of \$50,000 with interest for 3 years should equal the sum of the first payment accumulated for 2 years, the second accumulated for 1 year, and the third payment. Hence, if i is the interest rate, $50{,}000(1 + i)^3 = 20{,}000(1 + i)^2 + 20{,}000(1 + i) + 20{,}000$. Dividing by 10,000 and writing x for $1 + i$ gives the equation $5x^3 = 2x^2 + 2x + 2$.)

Ⓒ 24. Find the rate of interest implied if a house is offered for sale at \$80,000 cash or in 4 annual installments of \$23,000, the first payable now. (See Problem 23.)

25. Find the equation of the tangent line to the graph of the equation $y = 3x^2$ at the point (2, 12). (*Hint*: Remember that the line through (2, 12) with slope m has equation $y - 12 = m(x - 2)$. Find m by evaluating the slope polynomial at $x = 2$.)

26. Find the equation of the tangent line to the curve
(a) $y = x^2 + x$ at the point (2, 6);
(b) $y = 2x^3 - 4x + 5$ at the point $(-1, 7)$;
(c) $y = \frac{1}{5}x^5$ at the point $(2, \frac{32}{5})$.
(See the hint for Problem 25.)

27. Draw a careful graph of the equation $y = x^2$ for $-3 \le x \le 3$. Then draw tangent lines to the curve at $x = -2$, $x = 0$, $x = 1$, $x = 2$, and $x = 3$, and estimate their slopes. How well do your answers compare with the corresponding values of the slope polynomial for x^2, that is, with the values of $2x$?

28. The point $P(2, 12)$ is on the graph of the equation $y = 3x^2$. Find the slope of the line segment PQ, where Q is also on the graph and
(a) has x-coordinate 2.5;
(b) has x-coordinate 2.1;
(c) has x-coordinate $2 + h$.
What number does your answer to part (c) approach as h approaches zero? How does this compare with the value of the slope polynomial of $3x^2$ for $x = 2$?

CHAPTER SUMMARY

The **division law for polynomials** asserts that if $P(x)$ and $D(x)$ are any given nonconstant polynomials, then there are unique polynomials $Q(x)$ and $R(x)$ such that

$$P(x) = D(x)Q(x) + R(x)$$

where $R(x)$ is either 0 or of lower degree than $D(x)$. In fact, we can find $Q(x)$ and $R(x)$ by the **division algorithm**, which is just a fancy name for ordinary long division. When $D(x)$ has the form $x - c$, $R(x)$ will have to be a constant R, since it is of lower degree than $D(x)$. The substitution $x = c$ then gives

$$P(c) = R$$

a result known as the **Remainder Theorem**. An immediate consequence is the **Factor Theorem**, which says that c is a zero of $P(x)$ if and only if $x - c$ is a factor of $P(x)$. Division of a polynomial by $x - c$ can be greatly simplified by use of **synthetic division**.

That every nonconstant polynomial has at least one zero is guaranteed by Gauss's **Fundamental Theorem of Algebra**. But we can say much more than that. For any nonconstant polynomial

$$P(x) = a_n x^n + a_{n-1}x^{n-1} + \cdots + a_1 x + a_0$$

there are n numbers $c_1, c_2, \ldots, c_n$ (not necessarily all different) such that

$$P(x) = a_n(x - c_1)(x - c_2)\cdots(x - c_n)$$

We call the latter result the **Complete Factorization Theorem**.

If the polynomial equation

$$P(x) = a_n x^n + a_{n-1}x^{n-1} + \cdots + a_1 x + a_0 = 0$$

has real coefficients, then its nonreal solutions (if any) must occur in conjugate pairs $a + bi$ and $a - bi$. If the coefficients are integers and if c/d is a rational solution in reduced form, then c divides a_0 and d divides a_n.

To find exact solutions to a polynomial equation may be very difficult; often we are more than happy to find good approximations. Two good methods for doing this are the **method of successive enlargements** and **Newton's method**. Both require plenty of calculating power.

CHAPTER REVIEW PROBLEM SET

1. Find the quotient and remainder if the first polynomial is divided by the second.
   (a) $2x^3 - x^2 + 4x - 5;\ x^2 + 2x - 3$
   (b) $x^4 - 8x^2 + 5;\ x^2 + 3x$

2. Use synthetic division to find the quotient and remainder if the first polynomial is divided by the second.
(a) $x^3 - 2x^2 - 4x + 7$; $x - 2$
(b) $2x^4 - 15x^2 + 4x - 3$; $x + 3$
(c) $x^3 + (3 - 3i)x^2 - (9 + 15i)x - 3 - 3i$; $x - 2 - 3i$
3. Without dividing, find the remainder if $2x^4 - 6x^3 + 17$ is divided by $x - 2$; if it is divided by $x + 2$.
4. Find the zeros of the given polynomial and give their multiplicities.
(a) $(x^2 - 1)^2(x^2 + 1)$
(b) $x(x^2 - 2x + 4)(x + \pi)^3$

In Problems 5 and 6, use synthetic division to show that $x - c$ is a factor of $P(x)$. Then factor $P(x)$ completely into linear factors.

5. $P(x) = 2x^3 - x^2 - 18x + 9$; $c = 3$
6. $P(x) = x^3 + 4x^2 - 7x - 28$; $c = -4$

In Problems 7 and 8, find a polynomial $P(x)$ with integral coefficients that has the given zeros. Assume each zero to be simple unless otherwise indicated.

7. 3, −2, 4 (multiplicity 2)
8. $3 + \sqrt{7}, 3 - \sqrt{7}, 2 - i, 2 + i$
9. Show that 1 is a zero of multiplicity 2 of the polynomial $x^4 - 4x^3 - 3x^2 + 14x - 8$ and find the remaining zeros.
10. Find the polynomial $P(x) = a_3x^3 + a_2x^2 + a_1x + a_0$ which has zeros $\frac{1}{2}, -\frac{1}{3}$, and 4 and for which $P(2) = -42$.
11. Find the value of k so that $\sqrt{2}$ is a zero of $x^3 + 3x^2 - 2x + k$.
12. Find a cubic polynomial with real coefficients that has $4 + 3i$ and -2 as two of its zeros.
13. Solve the equation $x^4 - 4x^3 + 24x^2 + 20x - 145 = 0$, given that $2 + 5i$ is one of its solutions.
14. The equation $2x^3 - 15x^2 + 20x - 3 = 0$ has a rational solution. Find it and then find the other solutions.
15. The equation $x^3 - x^2 - x - 7 = 0$ has at least one real solution. Why? Show that it has no rational solution.

[C] 16. The equation $x^3 - 6x + 6 = 0$ has exactly one real solution. By means of a graph, make an initial guess at the solution. Then use the method of successive enlargements to find a second and a third approximation.

[C] 17. Use Newton's method to approximate the real solution of the equation in Problem 16. Take as x_1 the initial guess you made in Problem 16, and find x_2 and x_3.

Geometry may sometimes appear to take the lead over analysis but in fact precedes it only as a servant goes before the master to clear the path and light him on his way.

James Joseph Sylvester

CHAPTER ELEVEN

Systems of Equations and Inequalities

11-1 Equivalent Systems of Equations

11-2 Matrix Methods

11-3 The Algebra of Matrices

11-4 Multiplicative Inverses

11-5 Second- and Third-Order Determinants

11-6 Higher-Order Determinants

11-7 Systems of Inequalities

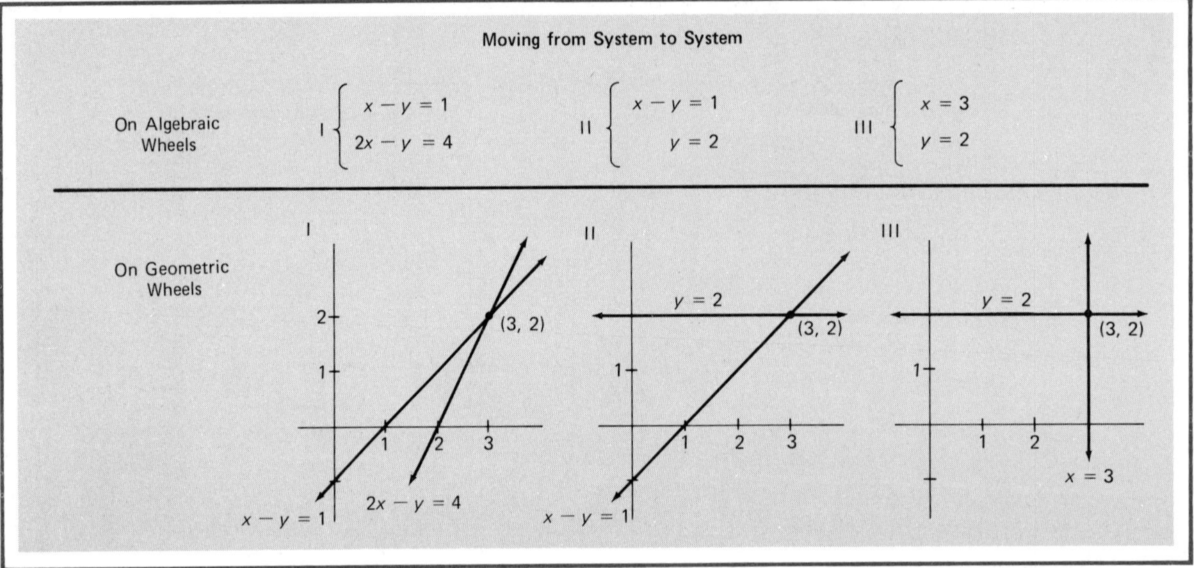

11-1 Equivalent Systems of Equations

In Section 3-3, you learned how to solve a system of two equations in two unknowns, but you probably did not think of the process as one of replacing the given system by another having the same solutions. It is this point of view that we now want to explore.

In the display above, system I is replaced by system II, which is simpler; system II is in turn replaced by system III, which is simpler yet. To go from system I to system II, we eliminated x from the second equation; we then substituted $y = 2$ in the first equation to get system III. What happened geometrically is shown in the bottom half of our display. Notice that the three pairs of lines have the same point of intersection (3, 2).

Because the notion of changing from one system of equations to another having the same solutions is so important, we make a formal definition. We say that two systems of equations are **equivalent** if they have the same solutions.

Operations That Lead to Equivalent Systems Now we face a big question. What operations can we perform on a system without changing its solutions?

Operation 1. *We can interchange the position of two equations.*
Operation 2. *We can multiply an equation by a nonzero constant, that is, we can replace an equation by a nonzero multiple of itself.*
Operation 3. *We can add a multiple of one equation to another, that is, we can replace an equation by the sum of that equation and a multiple of another.*

Operation 3 is the workhorse of the set. We show how it is used in the example of the opening display.

$$\text{I}\quad \begin{cases} x - y = 1 \\ 2x - y = 4 \end{cases}$$

If we add -2 times the first equation to the second, we obtain

$$\text{II}\quad \begin{cases} x - y = 1 \\ \phantom{x -{}} y = 2 \end{cases}$$

We then add the second equation to the first. This gives

$$\text{III}\quad \begin{cases} x = 3 \\ y = 2 \end{cases}$$

This is one way to write the solution. Alternatively, we say that the solution is the ordered pair (3, 2).

The Three Possibilities for a Linear System We are mainly interested in linear systems, that is, systems of linear equations, and we shall restrict our discussion to the case where there is the same number of equations as unknowns. There are three possibilities for the set of solutions: The set may be empty, it may have just one point, or it may have infinitely many points. These three cases are illustrated below.

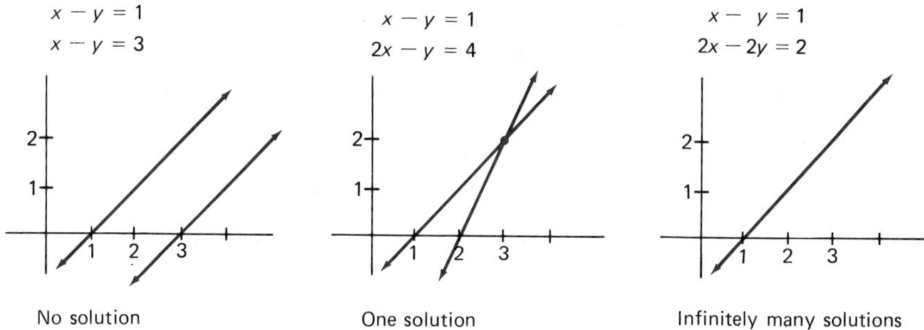

Someone is sure to object and ask if we cannot have a linear system with exactly two solutions or exactly three solutions. The answer is no. If a linear system has two solutions, it has infinitely many. This is obvious in the case of two equations in two unknowns, since two points determine a line, but it is also true for n equations in n unknowns.

Linear Systems in More Than Two Unknowns When we consider large systems, it is a good idea to be very systematic about our method of attack. Our method is to reduce the system to **triangular form** and then use **back substitution**. Let us explain. Consider

$$\begin{aligned} x - 2y + z &= -4 \\ 5y - 3z &= 18 \\ 2z &= -2 \end{aligned}$$

which is already in triangular form. (The name arises from the fact that the terms within the dotted triangle have zero coefficients.) This system is easy to solve. Solve the third equation first ($z = -1$). Then substitute that value in the second equation

$$5y - 3(-1) = 18$$

which gives $y = 3$. Finally substitute these two values in the first equation

$$x - 2(3) + (-1) = -4$$

which gives $x = 3$. Thus the solution of the system is $(3, 3, -1)$. This process, called **back substitution**, works on any linear system that is in triangular form. Just start at the bottom and work your way up.

If the system is not in triangular form initially, we try to operate on it until it is. Suppose we start with

$$\begin{aligned} 2x - 4y + 2z &= -8 \\ 2x + y - z &= 10 \\ 3x - y + 2z &= 4 \end{aligned}$$

Begin by multiplying the first equation by $\frac{1}{2}$ so that its leading coefficient is 1.

$$\begin{aligned} x - 2y + z &= -4 \\ 2x + y - z &= 10 \\ 3x - y + 2z &= 4 \end{aligned}$$

Next add -2 times the first equation to the second equation. Also add -3 times the first equation to the third.

$$\begin{aligned} x - 2y + z &= -4 \\ 5y - 3z &= 18 \\ 5y - z &= 16 \end{aligned}$$

Finally, add -1 times the second equation to the third.

$$\begin{aligned} x - 2y + z &= -4 \\ 5y - 3z &= 18 \\ 2z &= -2 \end{aligned}$$

The system is now in triangular form and can be solved by back substitution. It is, in fact, the triangular system we discussed earlier. The solution is $(3, 3, -1)$.

In Examples A and B of the problem set, we show how this process works when there are infinitely many solutions and when there are no solutions.

Problem Set 11-1

Solve each of the following systems of equations.

1. $\begin{aligned} 2x - 3y &= 7 \\ y &= -1 \end{aligned}$

2. $\begin{aligned} 5x - 3y &= -25 \\ y &= 5 \end{aligned}$

3. $\begin{aligned} x &= -2 \\ 2x + 7y &= 24 \end{aligned}$

4. $\begin{aligned} x &= 5 \\ 3x + 4y &= 3 \end{aligned}$

5. $\begin{aligned} x - 3y &= 7 \\ 4x + y &= 2 \end{aligned}$

6. $\begin{aligned} 5x + 6y &= 27 \\ x - y &= 1 \end{aligned}$

7. $\begin{aligned} 2x - y + 3z &= -6 \\ 2y - z &= 2 \\ z &= -2 \end{aligned}$

8. $\begin{aligned} x + 2y - z &= -4 \\ 3y + z &= 2 \\ z &= 5 \end{aligned}$

9. $\begin{aligned} 3x - 2y + 5z &= -10 \\ y - 4z &= 8 \\ 2y + z &= 7 \end{aligned}$

10. $\begin{aligned} 4x + 5y - 6z &= 31 \\ y - 2z &= 7 \\ 5y + z &= 2 \end{aligned}$

11. $\begin{aligned} x + 2y + z &= 8 \\ 2x - y + 3z &= 15 \\ -x + 3y - 3z &= -11 \end{aligned}$

12. $\begin{aligned} x + y + z &= 5 \\ -4x + 2y - 3z &= -9 \\ 2x - 3y + 2z &= 5 \end{aligned}$

13. $\begin{aligned} x - 2y + 3z &= 0 \\ 2x - 3y - 4z &= 0 \\ x + y - 4z &= 0 \end{aligned}$

14. $\begin{aligned} x + 4y - z &= 0 \\ -x - 3y + 5z &= 0 \\ 3x + y - 2z &= 0 \end{aligned}$

15. $\begin{aligned} x + y + z + w &= 10 \\ y + 3z - w &= 7 \\ x + y + 2z &= 11 \\ x - 3y + w &= -14 \end{aligned}$

16. $\begin{aligned} 2x + y + z &= 3 \\ y + z + w &= 5 \\ 4x + z + w &= 0 \\ 3y - z + 2w &= 0 \end{aligned}$

Example A (Infinitely many solutions) Solve

$$\begin{aligned} x - 2y + 3z &= 10 \\ 2x - 3y - z &= 8 \\ 4x - 7y + 5z &= 28 \end{aligned}$$

SOLUTION. Using Operation 3, we may eliminate x from the last two equations. First add -2 times the first equation to the second equation. Then add -4 times the first equation to the third equation. We obtain

$$\begin{aligned} x - 2y + 3z &= 10 \\ y - 7z &= -12 \\ y - 7z &= -12 \end{aligned}$$

Next add -1 times the second equation to the third equation.

$$\begin{aligned} x - 2y + 3z &= 10 \\ y - 7z &= -12 \\ 0 &= 0 \end{aligned}$$

Finally, we solve the second equation for y in terms of z and substitute that result in the first equation.

$$
\begin{aligned}
y &= 7z - 12 \\
x &= 2y - 3z + 10 = 2(7z - 12) - 3z + 10 = 11z - 14
\end{aligned}
$$

Notice that there are infinitely many solutions; we can give any value we like to z, calculate the corresponding x and y values, and come up with a solution. Here is the format we use to list all the solutions to the system.

$$
\begin{aligned}
x &= 11z - 14 \\
y &= 7z - 12 \\
z &\ \ \text{arbitrary}
\end{aligned}
$$

We could also say that the set of solutions consists of all ordered triples of the form $(11z - 14, 7z - 12, z)$. Thus if $z = 0$, we get the solution $(-14, -12, 0)$; if $z = 2$, we get $(8, 2, 2)$. Of course, it does not have to be z that is arbitrary; in our example, it could just as well be x or y. For example, if we had arranged things so y is treated as the arbitrary variable, we would have obtained.

$$
\begin{aligned}
x &= \tfrac{11}{7}y + \tfrac{34}{7} \\
z &= \tfrac{1}{7}y + \tfrac{12}{7} \\
y &\ \ \text{arbitrary}
\end{aligned}
$$

Note that the solution corresponding to $y = 2$ is $(8, 2, 2)$, which agrees with one found above.

Solve each of the following systems. Some, but not all, have infinitely many solutions.

17. $\begin{aligned} x - 4y + z &= 18 \\ 2x - 7y - 2z &= 4 \\ 3x - 11y - z &= 22 \end{aligned}$

18. $\begin{aligned} x + y - 3z &= 10 \\ 2x + 5y + z &= 18 \\ 5x + 8y - 8z &= 48 \end{aligned}$

19. $\begin{aligned} x - 2y + 3z &= -2 \\ 3x - 6y + 9z &= -6 \\ -2x + 4y - 6z &= 4 \end{aligned}$

20. $\begin{aligned} -4x + y - z &= 5 \\ 4x - y + z &= -5 \\ -24x + 6y - 6z &= 30 \end{aligned}$

21. $\begin{aligned} 2x - y + 4z &= 0 \\ 3x + 2y - z &= 0 \\ 9x - y + 11z &= 0 \end{aligned}$

22. $\begin{aligned} x + 3y - 2z &= 0 \\ 2x + y + z &= 0 \\ y - z &= 0 \end{aligned}$

Example B (No solution) Solve

$$
\begin{aligned}
x - 2y + 3z &= 10 \\
2x - 3y - z &= 8 \\
5x - 9y + 8z &= 20
\end{aligned}
$$

SOLUTION. Using Operation 3, we eliminate x from the last two equations to obtain

$$
\begin{aligned}
x - 2y + 3z &= 10 \\
y - 7z &= -12 \\
y - 7z &= -30
\end{aligned}
$$

It is already apparent that the last two equations cannot get along with each other. Let us continue anyway, putting the system in triangular form by adding -1 times the second equation to the third.

$$\begin{aligned} x - 2y + 3z &= 10 \\ y - 7z &= -12 \\ 0 &= -18 \end{aligned}$$

This system has no solution; we say it is **inconsistent.**

Solve the following systems or show that they are inconsistent.

23. $\begin{aligned} x - 4y + z &= 18 \\ 2x - 7y - 2z &= 4 \\ 3x - 11y - z &= 10 \end{aligned}$

24. $\begin{aligned} x + y - 3z &= 10 \\ 2x + 5y + z &= 18 \\ 5x + 8y - 8z &= 50 \end{aligned}$

25. $\begin{aligned} x + 3y - 2z &= 10 \\ 2x + y + z &= 4 \\ 5y - 5z &= 16 \end{aligned}$

26. $\begin{aligned} x - 2y + 3z &= -2 \\ 3x - 6y + 9z &= -6 \\ -2x + 4y - 6z &= 0 \end{aligned}$

Example C (Nonlinear systems) Solve the following system of equations.

$$\begin{aligned} x^2 + y^2 &= 25 \\ x^2 + y^2 - 2x - 4y &= 5 \end{aligned}$$

SOLUTION. We can use the same operations on this system as we did on linear systems. Adding (-1) times the first equation to the second, we get

$$\begin{aligned} x^2 + y^2 &= 25 \\ -2x - 4y &= -20 \end{aligned}$$

We solve the second equation for x in terms of y, substitute in the first equation, and then solve the resulting quadratic equation in y.

$$\begin{aligned} x &= 10 - 2y \\ (10 - 2y)^2 + y^2 &= 25 \\ 5y^2 - 40y + 75 &= 0 \\ y^2 - 8y + 15 &= 0 \\ (y - 3)(y - 5) &= 0 \end{aligned}$$

From this we get $y = 3$ or 5. Substituting these values in the equation $x = 10 - 2y$ yields two solutions to the original system, (4, 3) and (0, 5).

Solve each of the following systems.

27. $\begin{aligned} x + 2y &= 10 \\ x^2 + y^2 - 10x &= 0 \end{aligned}$

28. $\begin{aligned} x + y &= 10 \\ x^2 + y^2 - 10x - 10y &= 0 \end{aligned}$

29. $\begin{aligned} x^2 + y^2 - 4x + 6y &= 12 \\ x^2 + y^2 + 10x + 4y &= 96 \end{aligned}$

30. $\begin{aligned} x^2 + y^2 - 16y &= 45 \\ x^2 + y^2 + 4x - 20y &= 65 \end{aligned}$

31. $\begin{aligned} y &= 4x^2 - 2 \\ y &= x^2 + 1 \end{aligned}$

32. $\begin{aligned} x &= 3y^2 - 5 \\ x &= y^2 + 3 \end{aligned}$

Miscellaneous Problems

Solve each system in Problems 33–38 or show that it is inconsistent.

33. $3x + 4y = 3$
$2x - y = 5$

34. $2x - y = 6$
$-4x + 2y = 8$

35. $3x - y = 6$
$-4x + 2y = -12$

36. $-4x + y - z = 5$
$4x - y + z = -5$
$-24x + 6y - 6z = 10$

37. $x - y + 3z = 1$
$3x - 2y + 4z = 0$
$4x + 2y - z = 3$

C 38. $.43x - .79y + 4.24z = .67$
$3.16y - 9.74z = 2$
$y + 1.22z = 1.67$

In Problems 39 and 40, find the values of a and b for which the system has infinitely many solutions.

39. $2x - 3y = 5$
$ax + by = -15$

40. $x + 2y = 4$
$ax + 3y = b$

41. Tommy claims that he has \$4.10 in nickels, dimes, and quarters, that he has twice as many quarters as dimes, and that he has 28 coins in all. Is this possible? If so, find how many coins of each kind he has.

42. Suppose that Tommy makes the same claim as in Problem 41, except that instead of having twice as many quarters as dimes, he says that the number of dimes added to four times the number of quarters is 40. Is that possible? If so, find how many coins of each kind he has.

43. A three-digit number equals 25 times the sum of its digits. If the digits are reversed, the resulting number is greater than the given number by 198. The sum of the hundreds digit and the ones digit is 8. Find the number.

44. Find the dimensions of a rectangle whose diagonal and perimeter measure 25 and 62 inches, respectively.

45. A certain rectangle has an area of 130 square inches. Increasing the width by 3 inches and decreasing the length by 2 inches increases the area by 13 square inches. Find the dimensions of the original rectangle.

46. Suppose that the distances from the origin to the points (x, y) and $(x + 2, y + 16)$ are 5 and 13, respectively. Find x and y.

47. A chemist mixes three different nitric acid solutions with concentrations of 25 percent, 40 percent, and 50 percent to form 100 liters of a 32 percent solution. If she uses twice as much of the 25 percent solution as the 40 percent solution, how many liters of each kind does she use?

Arthur Cayley, lawyer, painter, mountaineer, Cambridge professor, but most of all creative mathematician, made his biggest contributions in the field of algebra. To him we owe the idea of replacing a linear system by its matrix.

$$\begin{aligned} 2x + 3y - z &= 1 \\ x + 4y - z &= 4 \\ 3x + y + 2z &= 5 \end{aligned} \qquad \begin{bmatrix} 2 & 3 & -1 & 1 \\ 1 & 4 & -1 & 4 \\ 3 & 1 & 2 & 5 \end{bmatrix}$$

Arthur Cayley (1821-1895)

11-2 Matrix Methods

Contrary to what many people think, mathematicians do not enjoy long, involved calculations. What they do enjoy is looking for shortcuts, for labor saving devices, and for elegant ways of doing things. Consider the problem of solving a system of linear equations, which as you know can become very complicated. Is there any way to simplify and systematize this process? There is. It is the method of matrices (plural of matrix).

A **matrix** is just a rectangular array of numbers. One example is shown above. It has 3 rows and 4 columns and is referred to as a 3×4 matrix. We follow the standard practice of enclosing a matrix in brackets.

An Example with Three Equations Look at our opening display again. Notice how we obtained the matrix from the system of equations. We just suppressed all the unknowns, the plus signs, and the equal signs, and supplied some 1's. We call this matrix the **matrix of the system**. We are going to solve this system, keeping track of what happens to the matrix as we move from step to step.

$$\begin{aligned} 2x + 3y - z &= 1 \\ x + 4y - z &= 4 \\ 3x + y + 2z &= 5 \end{aligned} \qquad \begin{bmatrix} 2 & 3 & -1 & 1 \\ 1 & 4 & -1 & 4 \\ 3 & 1 & 2 & 5 \end{bmatrix}$$

Interchange the first and second equations.

$$\begin{aligned} x + 4y - z &= 4 \\ 2x + 3y - z &= 1 \\ 3x + y + 2z &= 5 \end{aligned} \qquad \begin{bmatrix} 1 & 4 & -1 & 4 \\ 2 & 3 & -1 & 1 \\ 3 & 1 & 2 & 5 \end{bmatrix}$$

Add -2 times the first equation to the second; then add -3 times the first equation to the third.

$$\begin{aligned} x + 4y - z &= 4 \\ -5y + z &= -7 \\ -11y + 5z &= -7 \end{aligned} \qquad \begin{bmatrix} 1 & 4 & -1 & 4 \\ 0 & -5 & 1 & -7 \\ 0 & -11 & 5 & -7 \end{bmatrix}$$

Multiply the second equation by $-\frac{1}{5}$.

$$\begin{aligned} x + 4y - z &= 4 \\ y - \tfrac{1}{5}z &= \tfrac{7}{5} \\ -11y + 5z &= -7 \end{aligned} \qquad \begin{bmatrix} 1 & 4 & -1 & 4 \\ 0 & 1 & -\frac{1}{5} & \frac{7}{5} \\ 0 & -11 & 5 & -7 \end{bmatrix}$$

Add 11 times the second equation to the third.

$$\begin{aligned} x + 4y - z &= 4 \\ y - \tfrac{1}{5}z &= \tfrac{7}{5} \\ \tfrac{14}{5}z &= \tfrac{42}{5} \end{aligned} \qquad \begin{bmatrix} 1 & 4 & -1 & 4 \\ 0 & 1 & -\frac{1}{5} & \frac{7}{5} \\ 0 & 0 & \frac{14}{5} & \frac{42}{5} \end{bmatrix}$$

Now the system is in triangular form and can be solved by backward substitution. The result is $z = 3$, $y = 2$, and $x = -1$; we say the solution is $(-1, 2, 3)$.

We make two points about what we have just done. First, the process is not unique. We happen to prefer having a leading coefficient of 1; that was the reason for our first step. One could have started by multiplying the first equation by $-\frac{1}{2}$ and adding to the second, then multiplying the first equation by $-\frac{3}{2}$ and adding to the third. Any process that ultimately puts the system in triangular form is fine.

The second and main point is this. It is unnecessary to carry along all the x's and y's. Why not work with just the numbers? Why not do all the operations on the matrix of the system? Well, why not?

Equivalent Matrices Guided by our knowledge of systems of equations, we say that matrices **A** and **B** are **equivalent** if **B** can be obtained from **A** by applying the operations below (a finite number of times).

Operation 1. *Interchanging two rows.*
Operation 2. *Multiplying a row by a nonzero number.*
Operation 3. *Replacing a row by the sum of that row and a multiple of another row.*

When **A** and **B** are equivalent, we write $\mathbf{A} \sim \mathbf{B}$. If $\mathbf{A} \sim \mathbf{B}$, then $\mathbf{B} \sim \mathbf{A}$. If $\mathbf{A} \sim \mathbf{B}$ and $\mathbf{B} \sim \mathbf{C}$, then $\mathbf{A} \sim \mathbf{C}$.

An Example with Four Equations Consider

$$\begin{aligned} x + 3y + z &= 1 \\ 2x + 7y + z - w &= -1 \\ 3x - 2y + 4w &= 8 \\ -x + y - 3z - w &= -6 \end{aligned}$$

To solve this system, we take its matrix and transform it to triangular form using the operations above. Here is one possible sequence of steps.

$$\begin{bmatrix} 1 & 3 & 1 & 0 & 1 \\ 2 & 7 & 1 & -1 & -1 \\ 3 & -2 & 0 & 4 & 8 \\ -1 & 1 & -3 & -1 & -6 \end{bmatrix}$$

Add -2 times the first row to the second; -3 times the first row to the third row; 1 times the first row to the fourth row.

$$\begin{bmatrix} 1 & 3 & 1 & 0 & 1 \\ 0 & 1 & -1 & -1 & -3 \\ 0 & -11 & -3 & 4 & 5 \\ 0 & 4 & -2 & -1 & -5 \end{bmatrix}$$

Add 11 times the second row to the third and -4 times the second row to the fourth.

$$\begin{bmatrix} 1 & 3 & 1 & 0 & 1 \\ 0 & 1 & -1 & -1 & -3 \\ 0 & 0 & -14 & -7 & -28 \\ 0 & 0 & 2 & 3 & 7 \end{bmatrix}$$

Multiply the third row by $-\frac{1}{14}$.

$$\begin{bmatrix} 1 & 3 & 1 & 0 & 1 \\ 0 & 1 & -1 & -1 & -3 \\ 0 & 0 & 1 & \frac{1}{2} & 2 \\ 0 & 0 & 2 & 3 & 7 \end{bmatrix}$$

Add -2 times the third row to the fourth row.

$$\begin{bmatrix} 1 & 3 & 1 & 0 & 1 \\ 0 & 1 & -1 & -1 & -3 \\ 0 & 0 & 1 & \frac{1}{2} & 2 \\ 0 & 0 & 0 & 2 & 3 \end{bmatrix}$$

This last matrix represents the system

$$\begin{aligned} x + 3y + z \quad\quad &= 1 \\ y - z - \quad w &= -3 \\ z + \tfrac{1}{2}w &= 2 \\ 2w &= 3 \end{aligned}$$

If we use back substitution, we get $w = \frac{3}{2}$, $z = \frac{5}{4}$, $y = -\frac{1}{4}$, and $x = \frac{1}{2}$. We say the solution is $(\frac{1}{2}, -\frac{1}{4}, \frac{5}{4}, \frac{3}{2})$.

The Cases with Many Solutions and No Solution A system of equations need not have a unique solution; it may have infinitely many solutions or none at all. We need to be able to analyze the latter two cases by our matrix method. Fortunately, this is easy to do. Consider Example A of Section 11-1 first. Here is how we handle it using matrices.

$$\begin{bmatrix} 1 & -2 & 3 & 10 \\ 2 & -3 & -1 & 8 \\ 4 & -7 & 5 & 28 \end{bmatrix} \sim \begin{bmatrix} 1 & -2 & 3 & 10 \\ 0 & 1 & -7 & -12 \\ 0 & 1 & -7 & -12 \end{bmatrix}$$

$$\sim \begin{bmatrix} 1 & -2 & 3 & 10 \\ 0 & 1 & -7 & -12 \\ 0 & 0 & 0 & 0 \end{bmatrix}$$

The appearance of a row of zeros tells us that we have infinitely many solutions. The set of solutions is obtained by considering the equations corresponding to the first two rows.

$$\begin{aligned} x - 2y + 3z &= 10 \\ y - 7z &= -12 \end{aligned}$$

When we solve for y in the second equation and substitute in the first, we obtain

$$\begin{aligned} x &= 11z - 14 \\ y &= \ \ 7z - 12 \\ z &\ \ \text{arbitrary} \end{aligned}$$

Next consider the inconsistent example treated in Section 11-1 (Example B). Here is what happens when the matrix method is applied to this example.

$$\begin{bmatrix} 1 & -2 & 3 & 10 \\ 2 & -3 & -1 & 8 \\ 5 & -9 & 8 & 20 \end{bmatrix} \sim \begin{bmatrix} 1 & -2 & 3 & 10 \\ 0 & 1 & -7 & -12 \\ 0 & 1 & -7 & -30 \end{bmatrix}$$

$$\sim \begin{bmatrix} 1 & -2 & 3 & 10 \\ 0 & 1 & -7 & -12 \\ 0 & 0 & 0 & -18 \end{bmatrix}$$

We are tipped off to the inconsistency of the system by the third row of the matrix. It corresponds to the equation

$$0x + 0y + 0z = -18$$

which has no solution. Consequently, the system as a whole has no solution.

We may summarize our discussion as follows. If the process of transforming the matrix of a system of n equations in n unknowns to triangular form leads to a row in which all elements but the last one are zero, then the system is inconsistent; that is, it has no solution. If we are led to a matrix with one or more rows consisting entirely of zeros, then the system has infinitely many solutions.

Problem Set 11-2

Write the matrix of each system in Problems 1–8.

1. $\begin{aligned} 2x - y &= 4 \\ x - 3y &= -2 \end{aligned}$

2. $\begin{aligned} x + 2y &= 13 \\ 11x - y &= 0 \end{aligned}$

3. $\begin{aligned} x - 2y + z &= 3 \\ 2x + y &= 5 \\ x + y + 3z &= -4 \end{aligned}$

4. $\begin{aligned} x + 4z &= 10 \\ 2y - z &= 0 \\ 3x - y &= 20 \end{aligned}$

5. $\begin{aligned} 2x &= 3y - 4 \\ 3x + 2 &= -y \end{aligned}$

6. $\begin{aligned} x &= 4y + 3 \\ y &= -2x + 5 \end{aligned}$

7. $\begin{aligned} x &= 5 \\ 2y + x - z &= 4 \\ 3x - y + 13 &= 5z \end{aligned}$

8. $\begin{aligned} z &= 2 \\ 2x - z &= -4 \\ x + 2y + 4z &= -8 \end{aligned}$

Regard each matrix in Problems 9–18 as a matrix of a linear system of equations. Tell whether the system has a unique solution, infinitely many solutions, or no solution. You need not solve any of the systems.

9. $\begin{bmatrix} 1 & -2 & 3 \\ 0 & 1 & -4 \end{bmatrix}$

10. $\begin{bmatrix} 2 & 5 & 0 \\ 0 & -3 & 5 \end{bmatrix}$

11. $\begin{bmatrix} 1 & -3 & 5 \\ 2 & -6 & -10 \end{bmatrix}$

12. $\begin{bmatrix} 2 & 1 & -4 \\ -6 & -3 & 12 \end{bmatrix}$

13. $\begin{bmatrix} 1 & -2 & 4 & -2 \\ 0 & 3 & 1 & 4 \\ 0 & 0 & 1 & -3 \end{bmatrix}$

14. $\begin{bmatrix} 5 & 4 & 0 & -11 \\ 0 & 1 & -4 & 0 \\ 0 & 0 & 2 & -4 \end{bmatrix}$

15. $\begin{bmatrix} 2 & 1 & 5 & 4 \\ 0 & 3 & -2 & 10 \\ 0 & 3 & -2 & 10 \end{bmatrix}$

16. $\begin{bmatrix} 4 & 1 & -3 & 5 \\ 0 & 0 & 1 & -4 \\ 0 & 0 & 1 & -4 \end{bmatrix}$

17. $\begin{bmatrix} 3 & 2 & -1 & 0 \\ 0 & 1 & 0 & -4 \\ 0 & 1 & 0 & 5 \end{bmatrix}$

18. $\begin{bmatrix} -1 & 5 & 6 & -3 \\ 0 & 0 & 0 & 0 \\ 0 & 0 & 0 & 4 \end{bmatrix}$

In Problems 19–30, use matrices to solve each system or to show that it has no solution.

19. $\begin{aligned} x + 2y &= 5 \\ 2x - 5y &= -8 \end{aligned}$

20. $\begin{aligned} 2x + 4y &= 16 \\ 3x - y &= 10 \end{aligned}$

21. $\begin{aligned} 3x - 2y &= 1 \\ -6x + 4y &= -2 \end{aligned}$

22. $\begin{aligned} x + 3y &= 12 \\ 5x + 15y &= 12 \end{aligned}$

23. $\begin{aligned} 3x - 2y + 5z &= -10 \\ y - 4z &= 8 \\ 2y + z &= 7 \end{aligned}$

24. $\begin{aligned} 4x + 5y + 2z &= 25 \\ y - 2z &= 7 \\ 5y + z &= 2 \end{aligned}$

25. $\begin{aligned} x + y - 3z &= 10 \\ 2x + 5y + z &= 18 \\ 5x + 8y - 8z &= 48 \end{aligned}$

26. $\begin{aligned} x - 4y + z &= 18 \\ 2x - 7y - 2z &= 4 \\ 3x - 11y - z &= 22 \end{aligned}$

27. $\begin{aligned} 2x + 5y + 2z &= 6 \\ x + 2y - z &= 3 \\ 3x - y + 2z &= 9 \end{aligned}$

28. $\begin{aligned} x - 2y + 3z &= -2 \\ 3x - 6y + 9z &= -6 \\ -2x + 4y - 6z &= 0 \end{aligned}$

[C] 29. $\begin{aligned} x + 1.2y - 2.3z &= 8.1 \\ 1.3x + .7y + .4z &= 6.2 \\ .5x + 1.2y + .5z &= 3.2 \end{aligned}$

30. $\begin{aligned} 3x + 2y \quad &= 4 \\ 3x - 4y + 6z &= 16 \\ 3x - y + z &= 6 \end{aligned}$

Miscellaneous Problems

31. Write the matrix which represents the following system.

$$\begin{aligned} z &= 2x - y + 1 \\ y &= 4x - 3 \\ 2x - y &= 7 \end{aligned}$$

Regard each matrix in Problems 32–35 as the matrix of a linear system of equations. Without solving the system, tell whether it has a unique solution, infinitely many solutions, or no solution.

32. $\begin{bmatrix} 3 & -2 & 5 \\ 0 & 1 & -3 \end{bmatrix}$

33. $\begin{bmatrix} 2 & -1 & 5 \\ -4 & 2 & 8 \end{bmatrix}$

34. $\begin{bmatrix} 2 & -1 & 4 & 6 \\ 0 & 4 & -1 & 5 \\ 0 & 0 & 2 & 1 \end{bmatrix}$

35. $\begin{bmatrix} 3 & 3 & 0 & -4 \\ 0 & 1 & -3 & 2 \\ 0 & 0 & 0 & 0 \end{bmatrix}$

In Problems 36–38, use matrices to solve each system or to show that it has no solution.

36. $\begin{aligned} 3x - 2y + 4z &= 0 \\ x - y + 3z &= 1 \\ 4x + 2y - z &= 3 \end{aligned}$

37. $\begin{aligned} -4x + y - z &= 5 \\ 4x - y + z &= -5 \\ -24x + 6y - 6z &= 10 \end{aligned}$

38. $\begin{aligned} 2x + 4y - z &= 8 \\ 4x + 9y + 3z &= 42 \\ 8x + 17y + z &= 58 \end{aligned}$

39. Find the values of a, b, and c for which the parabola $y = ax^2 + bx + c$ passes through the points $(-2, -32)$, $(1, 4)$, and $(3, -12)$.

40. Show that there is no parabola $y = ax^2 + bx + c$ which passes through the given set of points.
    (a) $(1, 2)$, $(4, 8)$, and $(5, 10)$
    (b) $(1, 2)$, $(4, 8)$, and $(1, -4)$

41. Find the values of D, E, and F for which the circle $x^2 + y^2 + Dx + Ey + F = 0$ passes through the points $(3, -3)$, $(8, 2)$, and $(6, 6)$.

Cayley's Weapons

$$\begin{bmatrix} a & b \\ c & d \end{bmatrix} + \begin{bmatrix} A & B \\ C & D \end{bmatrix} = \begin{bmatrix} a+A & b+B \\ c+C & d+D \end{bmatrix}$$

$$\begin{bmatrix} a & b \\ c & d \end{bmatrix} \cdot \begin{bmatrix} A & B \\ C & D \end{bmatrix} = \begin{bmatrix} aA+bC & aB+bD \\ cA+dC & cB+dD \end{bmatrix}$$

"Cayley is forging the weapons for future generations of physicists."

P. G. Tait

11-3 The Algebra of Matrices

When Arthur Cayley introduced matrices, he had much more in mind than the application described in the previous section. There, matrices served as a device to simplify solving systems of equations. Cayley saw that these number boxes could be studied independently of equations, that they could be thought of as a new type of mathematical object. He realized that if he could give appropriate definitions of addition and multiplication, he would create a mathematical system that might stand with the real numbers and the complex numbers as a potential model for many applications. Cayley did all of this in a major paper in 1858. Some of his contemporaries saw little of significance in this new abstraction. But one of them, P. G. Tait, uttered the prophetic words quoted in the opening box. Tait was right. During the 1920's, Werner Heisenberg found that matrices were just the tool he needed to formulate his quantum mechanics. And by 1950, it was generally recognized that matrix theory provides the best model for many problems in economics and the social sciences.

To simplify our discussion, we initially consider only 2×2 matrices, that is, matrices with two rows and two columns. Examples are

$$\begin{bmatrix} -1 & 3 \\ 4 & 0 \end{bmatrix} \quad \begin{bmatrix} \log .1 & \frac{6}{2} \\ \frac{12}{3} & \log 1 \end{bmatrix} \quad \begin{bmatrix} a & b \\ c & d \end{bmatrix}$$

The first two of these matrices are said to be equal. In fact, two matrices are **equal** if and only if the entries in corresponding positions are equal. Be sure to distinguish the notion of equality (written =) from that of equivalence (written ~) introduced in Section 11-2. For example,

$$\begin{bmatrix} 2 & 1 \\ -3 & 4 \end{bmatrix} \text{ and } \begin{bmatrix} -3 & 4 \\ 2 & 1 \end{bmatrix}$$

are equivalent matrices; however, they are not equal.

Addition and Subtraction Cayley's definition of addition is straightforward. To add two matrices, add the entries in corresponding positions. Thus

$$\begin{bmatrix} 1 & 3 \\ -1 & 4 \end{bmatrix} + \begin{bmatrix} 6 & -2 \\ 5 & 1 \end{bmatrix} = \begin{bmatrix} 1+6 & 3+(-2) \\ -1+5 & 4+1 \end{bmatrix} = \begin{bmatrix} 7 & 1 \\ 4 & 5 \end{bmatrix}$$

and in general

$$\begin{bmatrix} a & b \\ c & d \end{bmatrix} + \begin{bmatrix} A & B \\ C & D \end{bmatrix} = \begin{bmatrix} a+A & b+B \\ c+C & d+D \end{bmatrix}$$

It is easy to check that the commutative and associative properties for addition are valid. If $\mathbf{U}$, $\mathbf{V}$, and $\mathbf{W}$ are any three matrices,

1. **(Commutativity +).** $\mathbf{U} + \mathbf{V} = \mathbf{V} + \mathbf{U}$
2. **(Associativity +).** $\mathbf{U} + (\mathbf{V} + \mathbf{W}) = (\mathbf{U} + \mathbf{V}) + \mathbf{W}$

The matrix

$$\mathbf{O} = \begin{bmatrix} 0 & 0 \\ 0 & 0 \end{bmatrix}$$

behaves as the "zero" for matrices. And the additive inverse of the matrix

$$\mathbf{U} = \begin{bmatrix} a & b \\ c & d \end{bmatrix}$$

is given by

$$-\mathbf{U} = \begin{bmatrix} -a & -b \\ -c & -d \end{bmatrix}$$

We may summarize these statements as follows.

3. **(Neutral element +).** There is a matrix $\mathbf{O}$ satisfying $\mathbf{O} + \mathbf{U} = \mathbf{U} + \mathbf{O} = \mathbf{U}$.
4. **(Additive inverses).** For each matrix $\mathbf{U}$, there is a matrix $-\mathbf{U}$ satisfying

$$\mathbf{U} + (-\mathbf{U}) = (-\mathbf{U}) + \mathbf{U} = \mathbf{O}$$

With the existence of an additive inverse settled, we can define subtraction by $\mathbf{U} - \mathbf{V} = \mathbf{U} + (-\mathbf{V})$. This amounts to subtracting the entries of **V** from the corresponding entries of **U**. Thus

$$\begin{bmatrix} 1 & 3 \\ -1 & 4 \end{bmatrix} - \begin{bmatrix} 6 & -2 \\ 5 & 1 \end{bmatrix} = \begin{bmatrix} -5 & 5 \\ -6 & 3 \end{bmatrix}$$

So far, all has been straightforward and nice. But with multiplication, Cayley hit a snag.

Multiplication Cayley's definition of multiplication may seem odd at first glance. He was led to it by consideration of a special problem that we do not have time to describe. It is enough to say that Cayley's definition is the one that proves useful in modern applications (as you will see).

Here it is in symbols.

$$\begin{bmatrix} a & b \\ c & d \end{bmatrix} \cdot \begin{bmatrix} A & B \\ C & D \end{bmatrix} = \begin{bmatrix} aA + bC & aB + bD \\ cA + dC & cB + dD \end{bmatrix}$$

Stated in words, we multiply two matrices by multiplying the rows of the left matrix by the columns of the right matrix in pairwise entry fashion, adding the results. For example, the entry in the second row and first column of the product is obtained by multiplying the entries of the second row of the left matrix by the corresponding entries of the first column of the right matrix, adding the results. Until you get used to it, it may help to use your fingers as shown in the diagram below.

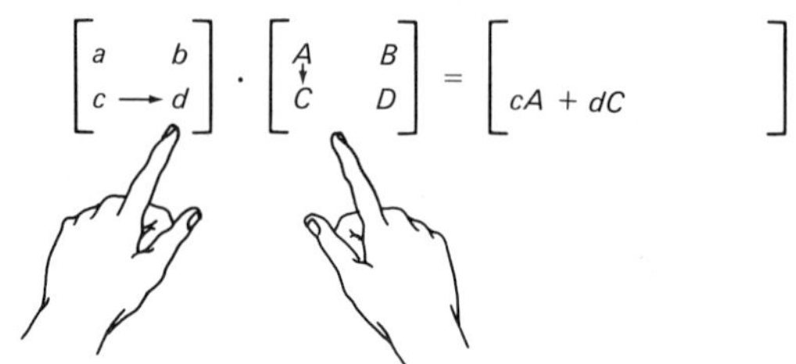

Here is an example worked out in detail.

$$\begin{bmatrix} 1 & 3 \\ -1 & 4 \end{bmatrix} \begin{bmatrix} 6 & -2 \\ 5 & 1 \end{bmatrix} = \begin{bmatrix} (1)(6) + (3)(5) & (1)(-2) + (3)(1) \\ (-1)(6) + (4)(5) & (-1)(-2) + (4)(1) \end{bmatrix}$$

$$= \begin{bmatrix} 21 & 1 \\ 14 & 6 \end{bmatrix}$$

Here is the same problem, but with the matrices multiplied in the opposite order.

$$\begin{bmatrix} 6 & -2 \\ 5 & 1 \end{bmatrix}\begin{bmatrix} 1 & 3 \\ -1 & 4 \end{bmatrix} = \begin{bmatrix} (6)(1) + (-2)(-1) & (6)(3) + (-2)(4) \\ (5)(1) + (1)(-1) & (5)(3) + (1)(4) \end{bmatrix}$$

$$= \begin{bmatrix} 8 & 10 \\ 4 & 19 \end{bmatrix}$$

Now you see the snag about which we warned you. The commutative property for multiplication fails. That is troublesome, but not fatal. We manage to get along in a world that is largely noncommutative (try removing your clothes and taking a shower in the opposite order). We just have to remember never to commute matrices under multiplication. Fortunately two other nice properties do hold.

5. **(Associativity ·).** $\mathbf{U}\cdot(\mathbf{V}\cdot\mathbf{W}) = (\mathbf{U}\cdot\mathbf{V})\cdot\mathbf{W}$
6. **(Distributivity).** $\mathbf{U}\cdot(\mathbf{V} + \mathbf{W}) = \mathbf{U}\cdot\mathbf{V} + \mathbf{U}\cdot\mathbf{W}$
   $(\mathbf{V} + \mathbf{W})\cdot\mathbf{U} = \mathbf{V}\cdot\mathbf{U} + \mathbf{W}\cdot\mathbf{U}$

We have not said anything about multiplicative inverses; that comes in the next section. We do, however, want to mention a special operation called **scalar multiplication**, that is, multiplication of a matrix by a scalar (number). To multiply a matrix by a number, multiply each entry by that number. That is,

$$k\begin{bmatrix} a & b \\ c & d \end{bmatrix} = \begin{bmatrix} ka & kb \\ kc & kd \end{bmatrix}$$

Scalar multiplication satisfies the expected properties.

7. $k(\mathbf{U} + \mathbf{V}) = k\mathbf{U} + k\mathbf{V}$
8. $(k + m)\mathbf{U} = k\mathbf{U} + m\mathbf{U}$
9. $(km)\mathbf{U} = k(m\mathbf{U})$

Larger Matrices and Compatibility So far we have considered only 2×2 matrices. That is an unnecessary restriction; however, to perform operations on arbitrary matrices, we must make sure they are **compatible**. For addition, this simply means that the matrices must be of the same size. Thus

$$\begin{bmatrix} 1 & -1 & 3 \\ 4 & -5 & 2 \end{bmatrix} + \begin{bmatrix} 2 & 6 & 0 \\ -3 & 2 & 4 \end{bmatrix} = \begin{bmatrix} 3 & 5 & 3 \\ 1 & -3 & 6 \end{bmatrix}$$

but

$$\begin{bmatrix} 1 & -1 & 3 \\ 4 & -5 & 2 \end{bmatrix} + \begin{bmatrix} 2 & 6 \\ -3 & 2 \end{bmatrix}$$

makes no sense.

Two matrices are compatible for multiplication if the left matrix has the same number of columns as the right matrix has rows. For example

$$\begin{bmatrix} 1 & -1 & 3 \\ 4 & -5 & 2 \end{bmatrix} \begin{bmatrix} 2 & 1 & 5 & 0 \\ -1 & 3 & 2 & 1 \\ 1 & -2 & 0 & 2 \end{bmatrix} = \begin{bmatrix} 6 & -8 & 3 & 5 \\ 15 & -15 & 10 & -1 \end{bmatrix}$$

The left matrix is 2×3, the right one is 3×4, and the result is 2×4. In general, we can multiply an $m \times n$ matrix by an $n \times p$ matrix, the result being an $m \times p$ matrix. All of the properties mentioned earlier are valid, provided we work with compatible matrices.

A Business Application The ABC Company sells precut lumber for two types of summer cottages, standard and deluxe. The standard model requires 30,000 board feet of lumber and 100 worker-hours of cutting; the deluxe model takes 40,000 board feet of lumber and 110 worker-hours of cutting. This year, the ABC Company buys its lumber at \$.20 per board foot and pays its laborers \$9.00 per hour. Next year it expects these costs to be \$.25 and \$10.00, respectively. This information can be displayed in matrix form as follows.

REQUIREMENTS **A**

	lumber	labor
standard	30,000	100
deluxe	40,000	110

UNITS COSTS **B**

	this year	next year
lumber	\$.20	\$.25
labor	\$9.00	\$10.00

Now we ask whether the product matrix **AB** has economic significance. It does: It gives the total dollar cost of standard and deluxe cottages both for this year and next. You can see this from the following calculation.

$$\mathbf{AB} = \begin{bmatrix} (30{,}000)(.20) + (100)(9) & (30{,}000)(.25) + (100)(10) \\ (40{,}000)(.20) + (110)(9) & (40{,}000)(.25) + (110)(10) \end{bmatrix}$$

	this year	next year	
=	\$6900	\$8500	standard
	\$8990	\$11,100	deluxe

Problem Set 11-3

Calculate **A** + **B**, **A** − **B**, *and* 3**A** *in Problems 1–4.*

1. $\mathbf{A} = \begin{bmatrix} 2 & -1 \\ 3 & 7 \end{bmatrix}, \mathbf{B} = \begin{bmatrix} 6 & 5 \\ -2 & 3 \end{bmatrix}$

2. $\mathbf{A} = \begin{bmatrix} -1 & 0 \\ 5 & 4 \end{bmatrix}, \mathbf{B} = \begin{bmatrix} 2 & -2 \\ 3 & 7 \end{bmatrix}$

3. $\mathbf{A} = \begin{bmatrix} 3 & -2 & 5 \\ 4 & 0 & -3 \end{bmatrix}$, $\mathbf{B} = \begin{bmatrix} 2 & 6 & -1 \\ 4 & 3 & -3 \end{bmatrix}$

4. $\mathbf{A} = \begin{bmatrix} 1 & 2 & 3 \\ 4 & 5 & 6 \\ 7 & 8 & 9 \end{bmatrix}$, $\mathbf{B} = \begin{bmatrix} -1 & -2 & -2 \\ -4 & -5 & -6 \\ -7 & -8 & -9 \end{bmatrix}$

Calculate **AB** *and* **BA** *if possible in Problems 5–12.*

5. $\mathbf{A} = \begin{bmatrix} 2 & -1 \\ 3 & 7 \end{bmatrix}$, $\mathbf{B} = \begin{bmatrix} 6 & 5 \\ -2 & 3 \end{bmatrix}$

6. $\mathbf{A} = \begin{bmatrix} -1 & 0 \\ 5 & 4 \end{bmatrix}$, $\mathbf{B} = \begin{bmatrix} 2 & -2 \\ 3 & 7 \end{bmatrix}$

7. $\mathbf{A} = \begin{bmatrix} 1 & -1 & 2 \\ 3 & 4 & -4 \\ 2 & 1 & 3 \end{bmatrix}$, $\mathbf{B} = \begin{bmatrix} 0 & 2 & -3 \\ 1 & 2 & 3 \\ -1 & -2 & 4 \end{bmatrix}$

8. $\mathbf{A} = \begin{bmatrix} -2 & 5 & 1 \\ 0 & -2 & 3 \\ 1 & 2 & -1 \end{bmatrix}$, $\mathbf{B} = \begin{bmatrix} -3 & 4 & 1 \\ 2 & 5 & 1 \\ 1 & 2 & 3 \end{bmatrix}$

9. $\mathbf{A} = \begin{bmatrix} 1 & -2 & 3 & 4 \\ 3 & 2 & -5 & 1 \end{bmatrix}$, $\mathbf{B} = \begin{bmatrix} 1 & 2 \\ 3 & 4 \end{bmatrix}$

10. $\mathbf{A} = \begin{bmatrix} -1 & 3 \\ 4 & 2 \\ 1 & 5 \end{bmatrix}$, $\mathbf{B} = \begin{bmatrix} -1 & 2 & 3 & 4 \\ 0 & -3 & 2 & 1 \end{bmatrix}$

11. $\mathbf{A} = \begin{bmatrix} 3 & 1 & -1 \\ 2 & 4 & 2 \\ -3 & 2 & -1 \end{bmatrix}$, $\mathbf{B} = \begin{bmatrix} 1 \\ 2 \\ 3 \end{bmatrix}$

12. $\mathbf{A} = [1 \quad 2 \quad -1]$, $\mathbf{B} = \begin{bmatrix} 4 & 3 \\ 0 & 2 \\ -1 & 4 \end{bmatrix}$

13. Calculate **AB** and **BA** for

$$\mathbf{A} = \begin{bmatrix} 0 & 0 \\ 0 & 0 \end{bmatrix} \qquad \mathbf{B} = \begin{bmatrix} 2 & -1 \\ 3 & 4 \end{bmatrix}$$

14. State the general property illustrated by Problem 13.

15. Find **X** if

$$\begin{bmatrix} 2 & 1 & -3 \\ 1 & 5 & 0 \end{bmatrix} + \mathbf{X} = 2\begin{bmatrix} -1 & 4 & 3 \\ -2 & 0 & 4 \end{bmatrix}$$

16. Solve for **X**.

$$-3\mathbf{X} + 2\begin{bmatrix} 1 & -2 \\ 5 & 6 \end{bmatrix} = -\begin{bmatrix} 5 & -14 \\ 8 & 15 \end{bmatrix}$$

17. Calculate $\mathbf{A}(\mathbf{B} + \mathbf{C})$ and $\mathbf{AB} + \mathbf{AC}$ for

$$\mathbf{A} = \begin{bmatrix} 2 & -1 \\ 3 & 4 \end{bmatrix} \quad \mathbf{B} = \begin{bmatrix} 2 & 4 \\ 6 & 1 \end{bmatrix} \quad \mathbf{C} = \begin{bmatrix} -1 & -2 \\ 3 & 6 \end{bmatrix}$$

What property does this illustrate?

18. Calculate $(\mathbf{A} + \mathbf{B})(\mathbf{A} - \mathbf{B})$ and $\mathbf{A}^2 - \mathbf{B}^2$ for

$$\mathbf{A} = \begin{bmatrix} 3 & -2 \\ 1 & 4 \end{bmatrix} \quad \mathbf{B} = \begin{bmatrix} 6 & -3 \\ 2 & 5 \end{bmatrix}$$

Why are your two answers different?

C 19. Find the entry in the third row and second column of the product

$$\begin{bmatrix} 1.39 & 4.13 & -2.78 \\ 4.72 & -3.69 & 5.41 \\ 8.09 & -6.73 & 5.03 \end{bmatrix} \begin{bmatrix} 5.45 & 6.31 \\ 7.24 & -5.32 \\ 6.06 & 1.34 \end{bmatrix}$$

C 20. Find the entry in the second row and first column of the product in Problem 19.

Miscellaneous Problems

21. Compute $\mathbf{A} - 2\mathbf{B}$, $\mathbf{AB}$, and $\mathbf{A}^2$ for

$$\mathbf{A} = \begin{bmatrix} 4 & -1 & 3 \\ 2 & 5 & 3 \\ 6 & 2 & 1 \end{bmatrix} \quad \mathbf{B} = \begin{bmatrix} 1 & -3 & 2 \\ 5 & 0 & 3 \\ -5 & 2 & 1 \end{bmatrix}$$

22. Let $\mathbf{A}$ and $\mathbf{B}$ be 3×4 matrices, $\mathbf{C}$ a 4×5 matrix, and $\mathbf{D}$ a 3×5 matrix. Which of the following satisfy the compatibility conditions?
(a) $\mathbf{AB}$ (b) $\mathbf{AC}$ (c) $\mathbf{AC} + \mathbf{D}$
(d) $(\mathbf{A} - \mathbf{B})\mathbf{C}$ (e) $(\mathbf{AC})\mathbf{D}$ (f) $\mathbf{A}(\mathbf{CD})$
(g) $\mathbf{A}^2$

23. Expand $(\mathbf{A} + \mathbf{B})^2$. Does it equal $\mathbf{A}^2 + 2\mathbf{AB} + \mathbf{B}^2$?

24. Calculate

$$\begin{bmatrix} 1 & 5 \\ 3 & -4 \end{bmatrix} \begin{bmatrix} 3 & 0 \\ -2 & 0 \end{bmatrix}$$

25. If the second row of $\mathbf{A}$ consists of zeros, what can you say about the second row of $\mathbf{AB}$?

26. Calculate $\mathbf{AB}$ and $\mathbf{BA}$ for

$$\mathbf{A} = \begin{bmatrix} 1 & 2 & 3 & 4 \end{bmatrix} \quad \mathbf{B} = \begin{bmatrix} 2 \\ 1 \\ -1 \\ -2 \end{bmatrix}$$

27. Show that if both $\mathbf{AB}$ and $\mathbf{BA}$ make sense ($\mathbf{B}$ and $\mathbf{A}$ are compatible for multiplication in either order), then $\mathbf{AB}$ and $\mathbf{BA}$ are both square matrices.

28. Show that if **AB** and **BC** both satisfy the compatibility condition, then (**AB**)**C** and **A**(**BC**) do also.

29. Art, Bob, and Curt work for a company that makes Flukes, Gizmos, and Horks. They are paid for their labor on a piecework basis, receiving \$2 for each Fluke, \$3 for each Gizmo, and \$4 for each Hork. Below are matrices **U** and **V** representing their outputs on Monday and Tuesday. Matrix **X** is the wage/unit matrix.

MONDAY'S OUTPUT **U**

	F	G	H
Art	4	3	2
Bob	5	1	2
Curt	3	4	1

TUESDAY'S OUTPUT **V**

	F	G	H
Art	3	6	1
Bob	4	2	2
Curt	5	1	3

WAGE/UNIT **X**

F	1
G	2
H	3

Compute the following matrices and decide what they represent.
(a) **UX** (b) **VX** (c) **U** + **V** (d) (**U** + **V**)**X**

30. Four friends A, B, C, and D have unlisted telephone numbers. Whether or not one person knows another's number is indicated by the matrix **U** below, where 1 indicates knowing and 0 indicates not knowing. For example, the 1 in row 3 and column 1 means that C knows A's number.

$$\mathbf{U} = \begin{array}{c} \\ A \\ B \\ C \\ D \end{array}\begin{array}{c} \begin{array}{cccc} A & B & C & D \end{array} \\ \begin{bmatrix} 1 & 0 & 1 & 0 \\ 0 & 1 & 1 & 0 \\ 1 & 0 & 1 & 1 \\ 0 & 1 & 0 & 1 \end{bmatrix} \end{array}$$

(a) Calculate $\mathbf{U}^2$.
(b) Interpret $\mathbf{U}^2$ in terms of the possibility of each person being able to get a telephone message to another.
(c) Can D get a message to A via one other person?
(d) Interpret $\mathbf{U}^3$.

31. Let

$$\mathbf{A} = \begin{bmatrix} 3 & 0 & 0 \\ 0 & 4 & 0 \\ 0 & 0 & 5 \end{bmatrix}$$

If **B** is any 3×3 matrix, what does multiplication on the left by **A** do to **B**? Multiplication on the right by **A**?

32. Calculate $\mathbf{A}^2$ and $\mathbf{A}^3$ for the matrix of Problem 31. State a general result for raising diagonal matrices to powers.

"What was that?" inquired Alice. "Reeling and Writhing, of course, to begin with," the mock turtle replied, "and then the different branches of Arithmetic-Ambition, Distraction, Uglification, and Derision."

from *Alice's Adventures in Wonderland* by Lewis Carroll

Is this Derision?

$$\frac{\begin{bmatrix} 2 & 3 \\ -4 & 1 \end{bmatrix}}{\begin{bmatrix} 6 & 7 \\ 1 & 2 \end{bmatrix}} = \begin{bmatrix} ? & ? \\ ? & ? \end{bmatrix}$$

11-4 Multiplicative Inverses

Even for ordinary numbers, the notion of division seems more difficult than that of addition, subtraction, and multiplication. Certainly this is true for division of matrices. Look at the example displayed above. It could tempt more than a Mock Turtle to derision. However, Arthur Cayley saw no need to sneer. He noted that in the case of numbers,

$$\frac{U}{V} = U \cdot \frac{1}{V} = U \cdot V^{-1}$$

What is needed is a concept of "one" for matrices; then we need the concept of multiplicative inverse. The first is easy.

The Multiplicative Identity for Matrices Let

$$\mathbf{I} = \begin{bmatrix} 1 & 0 \\ 0 & 1 \end{bmatrix}$$

Then for any 2×2 matrix **U**,

$$\mathbf{UI} = \mathbf{U} = \mathbf{IU}$$

This can be checked by noting that

$$\begin{bmatrix} a & b \\ c & d \end{bmatrix}\begin{bmatrix} 1 & 0 \\ 0 & 1 \end{bmatrix} = \begin{bmatrix} a & b \\ c & d \end{bmatrix} = \begin{bmatrix} 1 & 0 \\ 0 & 1 \end{bmatrix}\begin{bmatrix} a & b \\ c & d \end{bmatrix}$$

The symbol **I** is chosen because it is often called the **multiplicative identity**. In accordance with Section 1-4, it is also called the neutral element for multiplication.

For 3×3 matrices, the multiplicative identity has the form

$$\begin{bmatrix} 1 & 0 & 0 \\ 0 & 1 & 0 \\ 0 & 0 & 1 \end{bmatrix}$$

You should be able to guess its form for 4×4 and higher order matrices.

Inverses of 2 × 2 Matrices Suppose we want to find the multiplicative inverse of

$$\mathbf{V} = \begin{bmatrix} 6 & 7 \\ 1 & 2 \end{bmatrix}$$

We are looking for a matrix

$$\mathbf{W} = \begin{bmatrix} a & b \\ c & d \end{bmatrix}$$

that satisfies $\mathbf{VW} = \mathbf{I}$ and $\mathbf{WV} = \mathbf{I}$. Taking $\mathbf{VW} = \mathbf{I}$ first, we want

$$\begin{bmatrix} 6 & 7 \\ 1 & 2 \end{bmatrix}\begin{bmatrix} a & b \\ c & d \end{bmatrix} = \begin{bmatrix} 1 & 0 \\ 0 & 1 \end{bmatrix}$$

which means

$$\begin{bmatrix} 6a + 7c & 6b + 7d \\ a + 2c & b + 2d \end{bmatrix} = \begin{bmatrix} 1 & 0 \\ 0 & 1 \end{bmatrix}$$

or

$$\begin{aligned} 6a + 7c &= 1 \qquad & 6b + 7d &= 0 \\ a + 2c &= 0 \qquad & b + 2d &= 1 \end{aligned}$$

When these four equations are solved for a, b, c, d, we have

$$\mathbf{W} = \begin{bmatrix} \frac{2}{5} & -\frac{7}{5} \\ -\frac{1}{5} & \frac{6}{5} \end{bmatrix}$$

as a tentative solution to our problem. We say tentative, because so far we know only that $\mathbf{VW} = \mathbf{I}$. Happily, **W** works on the other side of **V** too, as we can check. (In this exceptional case, we do have commutativity.)

$$\mathbf{WV} = \begin{bmatrix} \frac{2}{5} & -\frac{7}{5} \\ -\frac{1}{5} & \frac{6}{5} \end{bmatrix}\begin{bmatrix} 6 & 7 \\ 1 & 2 \end{bmatrix} = \begin{bmatrix} 1 & 0 \\ 0 & 1 \end{bmatrix}$$

Success! **W** is the inverse of **V**; we denote it by the symbol $\mathbf{V}^{-1}$.

The process just described can be carried out for any specific 2 × 2 matrix, or better, it can be carried out for a general 2 × 2 matrix. But before we give the result, we make an important comment. There is no reason to think that every 2 × 2 matrix has a multiplicative inverse. Remember that the number 0 does not have such an inverse; neither does the matrix **O**. But here is a mild surprise. Many other 2 × 2 matrices do not have inverses. The following theorem identifies in a very precise way those that do, and then gives a formula for their inverses.

THEOREM (MULTIPLICATIVE INVERSES)

The matrix

$$\mathbf{V} = \begin{bmatrix} a & b \\ c & d \end{bmatrix}$$

has a multiplicative inverse if and only if $D = ad - bc$ *is nonzero. If* $D \neq 0$, *then*

$$\mathbf{V}^{-1} = \begin{bmatrix} d/D & -b/D \\ -c/D & a/D \end{bmatrix}$$

Thus the number D determines whether a matrix has an inverse. This number, which we shall call a *determinant*, will be studied in detail in the next section. Each 2 × 2 matrix has such a number associated with it. Let us look at two examples.

$$\mathbf{X} = \begin{bmatrix} 2 & -3 \\ -4 & 6 \end{bmatrix} \qquad \mathbf{Y} = \begin{bmatrix} 5 & -3 \\ -4 & 3 \end{bmatrix}$$

$$D = (2)(6) - (-3)(-4) = 0 \qquad D = (5)(3) - (-3)(-4) = 3$$

$$\mathbf{X}^{-1} \text{ does not exist} \qquad \mathbf{Y}^{-1} = \begin{bmatrix} \frac{3}{3} & \frac{3}{3} \\ \frac{4}{3} & \frac{5}{3} \end{bmatrix}$$

Inverses for Higher-Order Matrices There is a theorem like the one above for square matrices of any size, which Cayley found in 1858. It is complicated and, rather than try to state it, we are going to illustrate a process which yields the inverse of a matrix whenever it exists. Briefly described, it is this. Take any square matrix **V** and write the corresponding identity matrix **I** next to it on the right. By using the three row operations of Section 11-2, attempt to reduce **V** to the identity matrix while simultaneously performing the same operations on **I**. If you can reduce **V** to **I**, you will simultaneously turn **I** into $\mathbf{V}^{-1}$. If you cannot reduce **V** to **I**, **V** has no inverse.

Here is an illustration for the 2×2 matrix $\mathbf{V}$ that we used earlier.

$$\left[\begin{array}{cc|cc} 6 & 7 & 1 & 0 \\ 1 & 2 & 0 & 1 \end{array}\right]$$

$$\sim \left[\begin{array}{cc|cc} 1 & 2 & 0 & 1 \\ 6 & 7 & 1 & 0 \end{array}\right] \quad \text{(interchange rows)}$$

$$\sim \left[\begin{array}{cc|cc} 1 & 2 & 0 & 1 \\ 0 & -5 & 1 & -6 \end{array}\right] \quad \text{(add } -6 \text{ times row 1 to row 2)}$$

$$\sim \left[\begin{array}{cc|cc} 1 & 2 & 0 & 1 \\ 0 & 1 & -\frac{1}{5} & \frac{6}{5} \end{array}\right] \quad \text{(divide row 2 by } -5\text{)}$$

$$\sim \left[\begin{array}{cc|cc} 1 & 0 & \frac{2}{5} & -\frac{7}{5} \\ 0 & 1 & -\frac{1}{5} & \frac{6}{5} \end{array}\right] \quad \text{(add } -2 \text{ times row 2 to row 1)}$$

Notice that the matrix $\mathbf{V}^{-1}$ that we obtained earlier appears on the right. We illustrate the same process for a 3×3 matrix in Example A of the problem set.

An Application Consider the system of equations

$$\begin{aligned} 2x + 6y + 6z &= 8 \\ 2x + 7y + 6z &= 10 \\ 2x + 7y + 7z &= 9 \end{aligned}$$

If we introduce matrices

$$\mathbf{A} = \begin{bmatrix} 2 & 6 & 6 \\ 2 & 7 & 6 \\ 2 & 7 & 7 \end{bmatrix} \qquad \mathbf{X} = \begin{bmatrix} x \\ y \\ z \end{bmatrix} \qquad \mathbf{B} = \begin{bmatrix} 8 \\ 10 \\ 9 \end{bmatrix}$$

this system can be written in the form

$$\mathbf{AX} = \mathbf{B}$$

Now divide both sides by $\mathbf{A}$, by which we mean, of course, multiply both sides by $\mathbf{A}^{-1}$. We must be more precise. Multiply both sides on the left by $\mathbf{A}^{-1}$ (do not forget the lack of commutativity).

$$\begin{aligned} \mathbf{A}^{-1}\mathbf{AX} &= \mathbf{A}^{-1}\mathbf{B} \\ \mathbf{IX} &= \mathbf{A}^{-1}\mathbf{B} \\ \mathbf{X} &= \mathbf{A}^{-1}\mathbf{B} \end{aligned}$$

In Example A on page 428, $\mathbf{A}^{-1}$ is found to be

$$\mathbf{A}^{-1} = \begin{bmatrix} \frac{7}{2} & 0 & -3 \\ -1 & 1 & 0 \\ 0 & -1 & 1 \end{bmatrix}$$

Thus,

$$\mathbf{X} = \begin{bmatrix} \frac{7}{2} & 0 & -3 \\ -1 & 1 & 0 \\ 0 & -1 & 1 \end{bmatrix} \begin{bmatrix} 8 \\ 10 \\ 9 \end{bmatrix} = \begin{bmatrix} 1 \\ 2 \\ -1 \end{bmatrix}$$

and therefore $(1, 2, -1)$ is the solution to our system.

This method of solution is particularly useful when many systems with the same coefficient matrix **A** are under consideration. Once we have $\mathbf{A}^{-1}$, we can obtain any solution simply by doing an easy matrix multiplication. If only one system is being studied, the method of Section 11-2 is best.

Problem Set 11-4

Find the multiplicative inverse of each matrix.

1. $\begin{bmatrix} 2 & 3 \\ -1 & -1 \end{bmatrix}$

2. $\begin{bmatrix} 4 & 3 \\ 1 & 2 \end{bmatrix}$

3. $\begin{bmatrix} 6 & -14 \\ 0 & 2 \end{bmatrix}$

4. $\begin{bmatrix} 0 & 3 \\ 2 & 4 \end{bmatrix}$

5. $\begin{bmatrix} 1 & 0 \\ 0 & 1 \end{bmatrix}$

6. $\begin{bmatrix} 4 & 0 \\ 0 & 5 \end{bmatrix}$

7. $\begin{bmatrix} a & 0 \\ 0 & b \end{bmatrix}$

8. $\begin{bmatrix} 3 & 0 & 0 \\ 0 & 4 & 0 \\ 0 & 0 & 5 \end{bmatrix}$

Example A (Inverses of large matrices) Find the multiplicative inverse of

$$\begin{bmatrix} 2 & 6 & 6 \\ 2 & 7 & 6 \\ 2 & 7 & 7 \end{bmatrix}$$

SOLUTION. We use the reduction method described in the text.

$$\left[\begin{array}{ccc|ccc} 2 & 6 & 6 & 1 & 0 & 0 \\ 2 & 7 & 6 & 0 & 1 & 0 \\ 2 & 7 & 7 & 0 & 0 & 1 \end{array}\right]$$

$$\sim \left[\begin{array}{ccc|ccc} 1 & 3 & 3 & \frac{1}{2} & 0 & 0 \\ 2 & 7 & 6 & 0 & 1 & 0 \\ 2 & 7 & 7 & 0 & 0 & 1 \end{array}\right] \quad \text{(divide row 1 by 2)}$$

$$\sim \left[\begin{array}{ccc|ccc} 1 & 3 & 3 & \frac{1}{2} & 0 & 0 \\ 0 & 1 & 0 & -1 & 1 & 0 \\ 0 & 1 & 1 & -1 & 0 & 1 \end{array}\right] \quad \text{(add } -2 \text{ times row 1 to row 2 and to row 3)}$$

$$\sim \left[\begin{array}{ccc|ccc} 1 & 3 & 3 & \frac{1}{2} & 0 & 0 \\ 0 & 1 & 0 & -1 & 1 & 0 \\ 0 & 0 & 1 & 0 & -1 & 1 \end{array}\right] \quad \text{(add } -1 \text{ times row 2 to row 3)}$$

$$\sim \left[\begin{array}{ccc|ccc} 1 & 0 & 3 & \frac{7}{2} & -3 & 0 \\ 0 & 1 & 0 & -1 & 1 & 0 \\ 0 & 0 & 1 & 0 & -1 & 1 \end{array}\right] \quad \text{(add } -3 \text{ times row 2 to row 1)}$$

$$\sim \left[\begin{array}{ccc|ccc} 1 & 0 & 0 & \frac{7}{2} & 0 & -3 \\ 0 & 1 & 0 & -1 & 1 & 0 \\ 0 & 0 & 1 & 0 & -1 & 1 \end{array}\right] \quad \text{(add } -3 \text{ times row 3 to row 1)}$$

Thus the desired inverse is

$$\begin{bmatrix} \frac{7}{2} & 0 & -3 \\ -1 & 1 & 0 \\ 0 & -1 & 1 \end{bmatrix}$$

Use the method illustrated above to find the multiplicative inverse of each of the following.

9. $\begin{bmatrix} 1 & 3 \\ 2 & 4 \end{bmatrix}$

10. $\begin{bmatrix} 2 & 6 \\ 3 & 1 \end{bmatrix}$

11. $\begin{bmatrix} 1 & 1 & 1 \\ 1 & -1 & 2 \\ 3 & 2 & 0 \end{bmatrix}$

12. $\begin{bmatrix} 2 & 1 & 1 \\ 1 & 3 & 1 \\ -1 & 4 & 0 \end{bmatrix}$

13. $\begin{bmatrix} 3 & 1 & 2 \\ 4 & 1 & -6 \\ 1 & 0 & 1 \end{bmatrix}$

14. $\begin{bmatrix} 2 & 4 & 6 \\ 3 & 2 & -5 \\ 2 & 3 & 1 \end{bmatrix}$

15. $\begin{bmatrix} 1 & 2 & 1 & 1 \\ 0 & 2 & 3 & 2 \\ 0 & 0 & 1 & 3 \\ 0 & 0 & 0 & 4 \end{bmatrix}$

16. $\begin{bmatrix} 1 & 1 & 1 & 1 \\ 1 & 1 & 1 & -1 \\ 1 & 1 & -1 & 1 \\ 1 & -1 & 1 & 1 \end{bmatrix}$

Solve the following systems by making use of the inverses you found in Problems 11–14. Begin by writing the system in the matrix form **AX** = **B**.

17. $\begin{aligned} x + y + z &= 2 \\ x - y + 2z &= -1 \\ 3x + 2y \quad &= 5 \end{aligned}$

18. $\begin{aligned} 2x + y + z &= 4 \\ x + 3y + z &= 5 \\ -x + 4y \quad &= 0 \end{aligned}$

19. $\begin{aligned} 3x + y + 2z &= 3 \\ 4x + y - 6z &= 2 \\ x \quad + z &= 6 \end{aligned}$

20. $\begin{aligned} 2x + 4y + 6z &= 9 \\ 3x + 2y - 5z &= 2 \\ 2x + 3y + z &= 4 \end{aligned}$

Example B (Matrices without inverses) Try to find the multiplicative inverse of

$$\mathbf{U} = \begin{bmatrix} 1 & 4 & 2 \\ 0 & 2 & 4 \\ 0 & -3 & -6 \end{bmatrix}$$

SOLUTION.

$$\left[\begin{array}{ccc|ccc} 1 & 4 & 2 & 1 & 0 & 0 \\ 0 & 2 & 4 & 0 & 1 & 0 \\ 0 & -3 & -6 & 0 & 0 & 1 \end{array}\right] \sim \left[\begin{array}{ccc|ccc} 1 & 4 & 2 & 1 & 0 & 0 \\ 0 & 1 & 2 & 0 & \frac{1}{2} & 0 \\ 0 & -3 & -6 & 0 & 0 & 1 \end{array}\right]$$

$$\sim \left[\begin{array}{ccc|ccc} 1 & 4 & 2 & 1 & 0 & 0 \\ 0 & 1 & 2 & 0 & \frac{1}{2} & 0 \\ 0 & 0 & 0 & 0 & \frac{3}{2} & 1 \end{array}\right]$$

Since we got a row of zeros in the left half above, we know we can never reduce it to the identity matrix **I**. The matrix **U** does not have an inverse.

Show that neither of the following matrices has a multiplicative inverse.

21. $\begin{bmatrix} 1 & 3 & 4 \\ 2 & 1 & -1 \\ 4 & 7 & 7 \end{bmatrix}$

22. $\begin{bmatrix} 2 & -2 & 4 \\ 5 & 3 & 2 \\ 3 & 5 & -2 \end{bmatrix}$

Miscellaneous Problems

In Problems 23–26, find the multiplicative inverse or indicate that it does not exist.

23. $\begin{bmatrix} -2 & 5 \\ 1 & -\frac{5}{2} \end{bmatrix}$

24. $\begin{bmatrix} 3 & -1 \\ 4 & 2 \end{bmatrix}$

25. $\begin{bmatrix} 1 & -3 & 4 \\ 2 & 3 & 5 \\ -1 & 4 & 2 \end{bmatrix}$

26. $\begin{bmatrix} -2 & 4 & 2 \\ 3 & 5 & 6 \\ 1 & 9 & 8 \end{bmatrix}$

27. Find the multiplicative inverse of

$$\begin{bmatrix} 2 & 0 & 0 \\ 0 & 3 & 0 \\ 0 & 0 & -4 \end{bmatrix}$$

28. Give a formula for $\mathbf{U}^{-1}$ if

$$\mathbf{U} = \begin{bmatrix} a & 0 & 0 \\ 0 & b & 0 \\ 0 & 0 & c \end{bmatrix}$$

When does the matrix **U** fail to have an inverse?

29. Use your result from Problem 25 to solve the system

$$\begin{aligned} x - 3y + 4z &= a \\ 2x + 3y + 5z &= b \\ -x + 4y + 2z &= c \end{aligned}$$

30. Let **A** and **B** be 3×3 matrices with inverses $\mathbf{A}^{-1}$ and $\mathbf{B}^{-1}$. Show that **AB** has an inverse given by $\mathbf{B}^{-1}\mathbf{A}^{-1}$. (*Hint*: The product in either order must be **I**.)

31. Show that

$$\begin{bmatrix} 0 & 0 \\ 2 & 3 \end{bmatrix}\begin{bmatrix} -6 & 0 \\ 4 & 0 \end{bmatrix} = \begin{bmatrix} 0 & 0 \\ 0 & 0 \end{bmatrix}$$

Thus **AB** = **O** but neither **A** nor **B** is **O**. This is another way in which matrices differ from ordinary numbers.

32. Suppose **AB** = **O** and **A** has a multiplicative inverse. Show that **B** = **O**. (See Problem 31.)

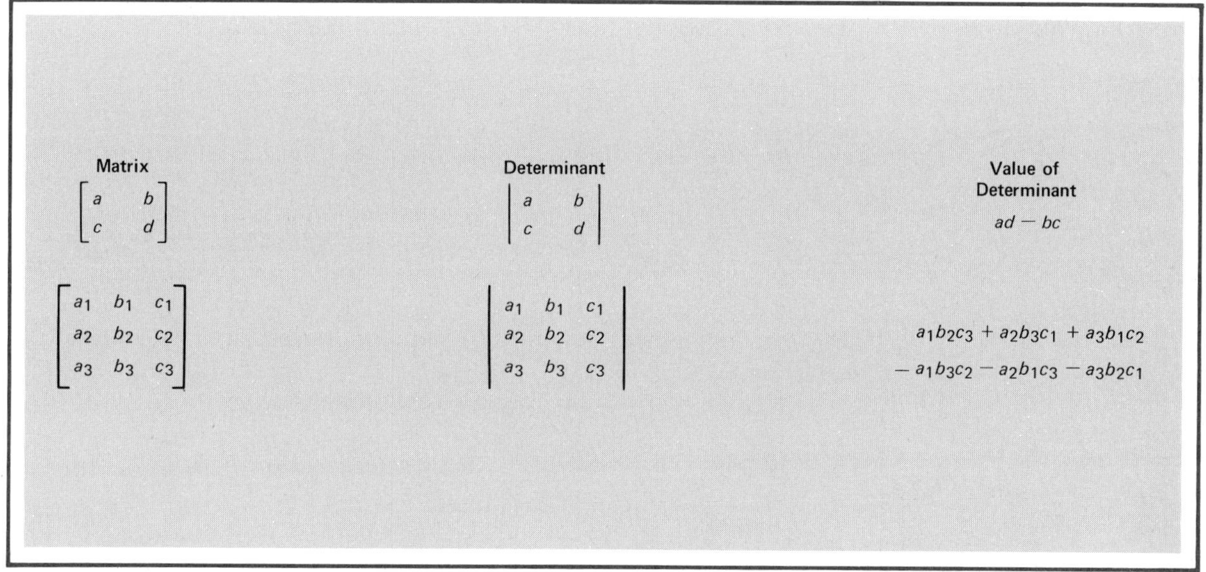

11-5 Second- and Third-Order Determinants

The notion of a determinant is usually attributed to the German mathematician Gottfried Wilhelm Leibniz (1646–1716), but it seems that Seki Kōwa of Japan had the idea somewhat earlier. It grew out of the study of systems of equations.

Second-Order Determinants Consider the general system of two equations in two unknowns

$$\begin{aligned} ax + by &= r \\ cx + dy &= s \end{aligned}$$

If we multiply the second equation by a and then add $-c$ times the first equation to it, we obtain the equivalent triangular system.

$$\begin{aligned} ax + \qquad by &= r \\ (ad - bc)y &= as - cr \end{aligned}$$

If $ad - bc \neq 0$, we can solve by backward substitution.

$$x = \frac{rd - bs}{ad - bc}$$

$$y = \frac{as - rc}{ad - bc}$$

These formulas are hard to remember unless we associate special symbols with the numbers $ad - bc$, $rd - bs$, and $as - rc$. For the first of these, we propose

$$\begin{vmatrix} a & b \\ c & d \end{vmatrix} = ad - bc$$

The symbol on the left is called a **second-order determinant** and we say that $ad - bc$ is its value. Thus

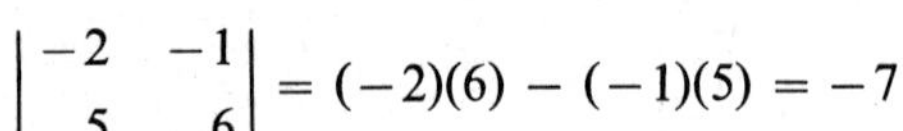

$$\begin{vmatrix} -2 & -1 \\ 5 & 6 \end{vmatrix} = (-2)(6) - (-1)(5) = -7$$

Memory Aid

+ −

a b

c d

$ad - bc$

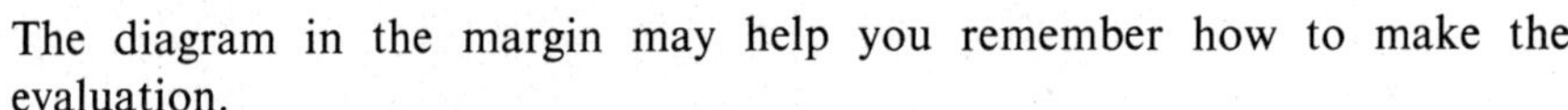

The diagram in the margin may help you remember how to make the evaluation.

With this new symbol, we can write the solution to

$$\begin{aligned} ax + by &= r \\ cx + dy &= s \end{aligned}$$

as

$$x = \frac{rd - bs}{ad - bc} = \frac{\begin{vmatrix} r & b \\ s & d \end{vmatrix}}{\begin{vmatrix} a & b \\ c & d \end{vmatrix}}$$

$$y = \frac{as - rc}{ad - bc} = \frac{\begin{vmatrix} a & r \\ c & s \end{vmatrix}}{\begin{vmatrix} a & b \\ c & d \end{vmatrix}}$$

These results are easy to remember when we notice that the denominator is the determinant of the coefficient matrix, and that the numerator is the same except that the coefficients of the unknown we are seeking are replaced by the constants from the right side of the system.

Here is an example.

$$3x - 2y = 7$$
$$4x + 5y = 2$$

$$x = \frac{\begin{vmatrix} 7 & -2 \\ 2 & 5 \end{vmatrix}}{\begin{vmatrix} 3 & -2 \\ 4 & 5 \end{vmatrix}} = \frac{(7)(5) - (-2)(2)}{(3)(5) - (-2)(4)} = \frac{39}{23}$$

$$y = \frac{\begin{vmatrix} 3 & 7 \\ 4 & 2 \end{vmatrix}}{\begin{vmatrix} 3 & -2 \\ 4 & 5 \end{vmatrix}} = \frac{(3)(2) - (7)(4)}{23} = -\frac{22}{23}$$

The choice of the name *determinant* is appropriate, for the determinants of a system completely *determine* its character.

1. If $ad - bc \neq 0$, the system has a unique solution, the one given on page 432.
2. If $ad - bc = 0$, $as - rc = 0$, and $rd - bs = 0$, then a, b, and r are proportional to c, d, and s and the system has infinitely many solutions. Here is an example.

$$\begin{aligned} 3x - 2y &= 7 \\ 6x - 4y &= 14 \end{aligned} \qquad \frac{3}{6} = \frac{-2}{-4} = \frac{7}{14}$$

3. If $ad - bc = 0$ and $as - rc \neq 0$ or $rd - bs \neq 0$, then a and b are proportional to c and d, but this proportionality does not extend to r and s; the system has no solution. This is illustrated by the following.

$$\begin{aligned} 3x - 2y &= 7 \\ 6x - 4y &= 10 \end{aligned} \qquad \frac{3}{6} = \frac{-2}{-4} \neq \frac{7}{10}$$

Third-Order Determinants When we consider the general system of three equations in three unknowns

$$\begin{aligned} a_1x + b_1y + c_1z &= d_1 \\ a_2x + b_2y + c_2z &= d_2 \\ a_3x + b_3y + c_3z &= d_3 \end{aligned}$$

things get more complicated, but the results are similar. The appropriate determinant symbol and its corresponding value are

$$\begin{vmatrix} a_1 & b_1 & c_1 \\ a_2 & b_2 & c_2 \\ a_3 & b_3 & c_3 \end{vmatrix} = a_1b_2c_3 + b_1c_2a_3 + c_1b_3a_2 - c_1b_2a_3 - b_1a_2c_3 - a_1b_3c_2$$

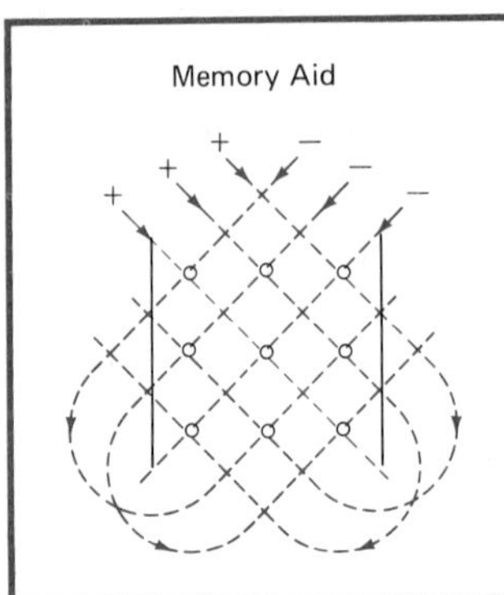

There are six terms in the sum on the right, three with a plus sign and three with a minus sign. The diagram in the margin will help you remember the products that enter each term. Just follow the arrows. Here is an example.

$$\begin{vmatrix} 3 & 2 & 4 \\ 4 & -2 & 6 \\ 8 & 3 & 5 \end{vmatrix} = (3)(-2)(5)+(2)(6)(8)+(4)(3)(4)-(4)(-2)(8)-(2)(4)(5)-(3)(3)(6)$$

$$=84$$

Cramer's Rule We saw that the solutions for x and y in a second-order system could be written as the quotients of two determinants. That fact generalizes to the third-order case. We present it without proof. Consider

$$\begin{aligned} a_1x + b_1y + c_1z &= d_1 \\ a_2x + b_2y + c_2z &= d_2 \\ a_3x + b_3y + c_3z &= d_3 \end{aligned}$$

If

$$D = \begin{vmatrix} a_1 & b_1 & c_1 \\ a_2 & b_2 & c_2 \\ a_3 & b_3 & c_3 \end{vmatrix} \neq 0$$

then the system above has a unique solution given by

$$x = \frac{1}{D}\begin{vmatrix} d_1 & b_1 & c_1 \\ d_2 & b_2 & c_2 \\ d_3 & b_3 & c_3 \end{vmatrix} \qquad y = \frac{1}{D}\begin{vmatrix} a_1 & d_1 & c_1 \\ a_2 & d_2 & c_2 \\ a_3 & d_3 & c_3 \end{vmatrix} \qquad z = \frac{1}{D}\begin{vmatrix} a_1 & b_1 & d_1 \\ a_2 & b_2 & d_2 \\ a_3 & b_3 & d_3 \end{vmatrix}$$

The pattern is the same as in the second-order situation. The denominator D is the determinant of the coefficient matrix. The numerator in each case is obtained from D by replacing the coefficients of the unknown by the constants from the right side of the system.

This method of solving a system of equations is named after one of its discoverers, Gabriel Cramer (1704–1752). Historically, it has been a popular method. However, note that even for a system of three equations in three unknowns, it requires the evaluation of four determinants. The method of matrices (Section 11-2) is considerably more efficient, both for hand and computer calculation. Consequently, Cramer's rule is now primarily of theoretical rather than practical interest.

Properties of Determinants We are interested in how the matrix operations considered in Section 11-2 affect the values of the corresponding determinants.

1. *Interchanging two rows changes the sign of the determinant; for example,*

$$\begin{vmatrix} a & b \\ c & d \end{vmatrix} = -\begin{vmatrix} c & d \\ a & b \end{vmatrix}$$

2. *Multiplying a row by a constant k multiplies the value of the determinant by k; for example,*

$$\begin{vmatrix} ka & kb \\ c & d \end{vmatrix} = k\begin{vmatrix} a & b \\ c & d \end{vmatrix}$$

3. *Adding a multiple of one row to another does not affect the value of the determinant; for example,*

$$\begin{vmatrix} a & b \\ c + ka & d + kb \end{vmatrix} = \begin{vmatrix} a & b \\ c & d \end{vmatrix}$$

We mention also the effect of a new operation.

4. *Interchanging the rows and columns (pairwise) does not affect the value of the determinant; for example,*

$$\begin{vmatrix} a & c \\ b & d \end{vmatrix} = \begin{vmatrix} a & b \\ c & d \end{vmatrix}$$

Though they are harder to prove in the third-order case, we emphasize that all four properties hold for both second- and third-order determinants. We offer only one proof, a proof of Property 3 in the second-order case.

$$\begin{aligned} \begin{vmatrix} a & b \\ c + ka & d + kb \end{vmatrix} &= a(d + kb) - b(c + ka) \\ &= ad + akb - bc - bka \\ &= ad - bc \\ &= \begin{vmatrix} a & b \\ c & d \end{vmatrix} \end{aligned}$$

Property 3 can be a great aid in evaluating a determinant. Using it, we can transform a matrix to triangular form without changing the value of its determinant. But the determinant of a triangular matrix is just the product of the elements on the main diagonal, since this is the only nonzero term in the determinant formula. Here is an example.

$$\begin{vmatrix} 1 & 3 & 4 \\ -1 & -2 & 3 \\ 2 & -6 & 11 \end{vmatrix} = \begin{vmatrix} 1 & 3 & 4 \\ 0 & 1 & 7 \\ 0 & -12 & 3 \end{vmatrix} = \begin{vmatrix} 1 & 3 & 4 \\ 0 & 1 & 7 \\ 0 & 0 & 87 \end{vmatrix} = 87$$

Problem Set 11-5 *Evaluate each of the determinants in Problems 1–8 by inspection.*

1. $\begin{vmatrix} 4 & 0 \\ 0 & -2 \end{vmatrix}$

2. $\begin{vmatrix} 8 & 0 \\ 5 & 0 \end{vmatrix}$

3. $\begin{vmatrix} 11 & 4 \\ 0 & 2 \end{vmatrix}$

4. $\begin{vmatrix} 2 & -1 & 5 \\ 0 & 4 & 2 \\ 0 & 0 & -1 \end{vmatrix}$

5. $\begin{vmatrix} -1 & -7 & 9 \\ 0 & 5 & 4 \\ 0 & 0 & 10 \end{vmatrix}$

6. $\begin{vmatrix} 3 & -2 & 1 \\ 0 & 0 & 0 \\ 1 & 5 & -8 \end{vmatrix}$

7. $\begin{vmatrix} 3 & 0 & 8 \\ 10 & 0 & 2 \\ -1 & 0 & -9 \end{vmatrix}$

8. $\begin{vmatrix} 9 & 0 & 0 \\ 0 & 0 & -2 \\ 0 & 4 & 0 \end{vmatrix}$

9. If

$$\begin{vmatrix} a_1 & b_1 & c_1 \\ a_2 & b_2 & c_2 \\ a_3 & b_3 & c_3 \end{vmatrix} = 12$$

find the value of each of the following determinants.

(a) $\begin{vmatrix} a_1 & a_2 & a_3 \\ b_1 & b_2 & b_3 \\ c_1 & c_2 & c_3 \end{vmatrix}$

(b) $\begin{vmatrix} a_3 & b_3 & c_3 \\ a_2 & b_2 & c_2 \\ a_1 & b_1 & c_1 \end{vmatrix}$

(c) $\begin{vmatrix} a_1 & b_1 & c_1 \\ a_2 & b_2 & c_2 \\ 3a_3 & 3b_3 & 3c_3 \end{vmatrix}$

(d) $\begin{vmatrix} a_1 + 3a_3 & b_1 + 3b_3 & c_1 + 3c_3 \\ a_2 & b_2 & c_2 \\ a_3 & b_3 & c_3 \end{vmatrix}$

Evaluate each of the determinants in Problems 10–17.

10. $\begin{vmatrix} 3 & 2 \\ 5 & 6 \end{vmatrix}$

11. $\begin{vmatrix} 5 & 3 \\ 5 & -3 \end{vmatrix}$

12. $\begin{vmatrix} 3 & 0 & 0 \\ -2 & 5 & 4 \\ 1 & 2 & -9 \end{vmatrix}$

13. $\begin{vmatrix} 4 & 8 & -2 \\ 1 & -2 & 0 \\ 2 & 4 & 0 \end{vmatrix}$

14. $\begin{vmatrix} 3 & 2 & -4 \\ 1 & 0 & 5 \\ 4 & -2 & 3 \end{vmatrix}$

15. $\begin{vmatrix} 2 & 4 & 1 \\ 1 & 3 & 6 \\ 2 & 3 & -1 \end{vmatrix}$

Ⓒ 16. $\begin{vmatrix} 5.1 & -3.2 & 2.6 \\ 1.3 & 4.5 & 2.3 \\ 3.4 & -2.2 & 1.9 \end{vmatrix}$

Ⓒ 17. $\begin{vmatrix} 2.03 & 5.41 & -3.14 \\ 0 & 6.22 & 0 \\ -1.93 & 7.13 & 6.34 \end{vmatrix}$

Use Cramer's Rule to solve the systems of equations in Problems 18–21.

18. $2x - 3y = -11$
$x + 2y = -2$

19. $5x + y = 7$
$3x - 4y = 18$

20. $2x + 4y + z = 15$
$x + 3y + 6z = 15$
$2x + 3y - z = 11$

21. $5x - 3y + 2z = 18$
$x + 4y + 2z = -4$
$3x - 2y + z = 11$

Evaluate the determinants in Problems 22–26.

Miscellaneous Problems

22. $\begin{vmatrix} 11 & 4 \\ 0 & 2 \end{vmatrix}$

23. $\begin{vmatrix} 3 & -2 & 5 \\ 0 & 0 & 0 \\ 1 & -4 & 6 \end{vmatrix}$

24. $\begin{vmatrix} -2 & 5 & 100 \\ 0 & 4 & 96 \\ 0 & 0 & -1 \end{vmatrix}$

25. $\begin{vmatrix} 2 & 3 & -4 \\ 1 & 2 & 5 \\ 5 & 8 & -3 \end{vmatrix}$

26. $\begin{vmatrix} 4 & 3 & -2 \\ -1 & 12 & 13 \\ 1 & 3 & 5 \end{vmatrix}$

27. Solve by Cramer's Rule.

$$\begin{aligned} 3x - y + 2z &= 7 \\ 2x \quad + z &= 5 \\ y - 2z &= -4 \end{aligned}$$

28. Show that interchanging two rows of a second-order determinant changes the sign of the determinant; that is,

$$\begin{vmatrix} a & b \\ c & d \end{vmatrix} = -\begin{vmatrix} c & d \\ a & b \end{vmatrix}$$

29. Show that

$$\begin{vmatrix} ka & kb \\ c & d \end{vmatrix} = k\begin{vmatrix} a & b \\ c & d \end{vmatrix}$$

30. Show that

$$\begin{vmatrix} a & c \\ b & d \end{vmatrix} = \begin{vmatrix} a & b \\ c & d \end{vmatrix}$$

31. Let the determinant

$$\begin{vmatrix} a_1 & b_1 & c_1 \\ a_2 & b_2 & c_2 \\ a_1 & b_1 & c_1 \end{vmatrix}$$

have value D. Interchanging rows 1 and 3 leaves the determinant unchanged, so its value is still D. But Property 1 tells us that the value should now be $-D$. What must be true about D?

32. If we apply Property 3 to the determinant in Problem 31 we see that

$$D = \begin{vmatrix} a_1 & b_1 & c_1 \\ a_2 & b_2 & c_2 \\ a_1 & b_1 & c_1 \end{vmatrix} = \begin{vmatrix} 0 & 0 & 0 \\ a_2 & b_2 & c_2 \\ a_1 & b_1 & c_1 \end{vmatrix}$$

What is the value of the determinant on the right? Does your answer agree with what you got in Problem 31?

33. If you have worked Problems 31 and 32 you have probably reached the verdict that if two rows (or two columns) of a determinant are identical, the determinant has a value of zero. This is true also if two rows (or two columns) are proportional. Convince yourself of this by showing that

$$\begin{vmatrix} a_1 & b_1 & c_1 \\ a_2 & b_2 & c_2 \\ ka_1 & kb_1 & kc_1 \end{vmatrix} = 0$$

34. (a) Show that the determinant equation

$$\begin{vmatrix} x & y & 1 \\ 2 & 4 & 1 \\ -3 & 5 & 1 \end{vmatrix} = 0$$

is the equation of a line.

(b) How can you tell without expanding the determinant that the points (2, 4) and (−3, 5) are on the line?

(c) Write a determinant equation for the line that passes through the points (5, −1) and (4, 11).

A Mathematician and a Poet

One of the most consistent workers on the theory of determinants over a period of 50 years was the Englishman, James Joseph Sylvester. Known as a poet, a wit, and a mathematician, he taught in England and in America. During his stay at Johns Hopkins University in Baltimore (1877–1883), he helped establish one of the first graduate programs in mathematics in America. Under his tutelage, mathematics began to flourish in the United States.

James Joseph Sylvester
1814–1897

11-6 Higher-Order Determinants

Having defined determinants for 2×2 and 3×3 matrices, we expect to do it for 4×4 matrices, 5×5 matrices, and so on. Our problem is to do it in such a way that Cramer's Rule and the determinant properties of Section 11-5 still hold. This will take some work.

Minors We begin by introducing the standard notation for a general $n \times n$ matrix.

$$\begin{bmatrix} a_{11} & a_{12} & a_{13} & \cdots & a_{1n} \\ a_{21} & a_{22} & a_{23} & \cdots & a_{2n} \\ a_{31} & a_{32} & a_{33} & \cdots & a_{3n} \\ \vdots & \vdots & \vdots & & \vdots \\ a_{n1} & a_{n2} & a_{n3} & \cdots & a_{nn} \end{bmatrix}$$

Note the use of the double subscript on each entry: the first subscript gives the row in which a_{ij} is and the second gives the column. For example, a_{32} is the entry in the third row and second column.

Associated with each entry a_{ij} in an $n \times n$ matrix is a determinant M_{ij} of order $n - 1$ called the **minor** of a_{ij}. It is obtained by taking the determinant of the submatrix that results when we blot out the row and column in which a_{ij} stands. For example, the minor M_{13} of a_{13} in the 4×4 matrix

$$\begin{bmatrix} a_{11} & a_{12} & a_{13} & a_{14} \\ a_{21} & a_{22} & a_{23} & a_{24} \\ a_{31} & a_{32} & a_{33} & a_{34} \\ a_{41} & a_{42} & a_{43} & a_{44} \end{bmatrix}$$

is the third-order determinant

$$\begin{vmatrix} a_{21} & a_{22} & a_{24} \\ a_{31} & a_{32} & a_{34} \\ a_{41} & a_{42} & a_{44} \end{vmatrix}$$

The General *n*th-Order Determinant Here is the definition that we have been leading up to.

$$\begin{vmatrix} a_{11} & a_{12} & \cdots & a_{1n} \\ a_{21} & a_{22} & \cdots & a_{2n} \\ \vdots & \vdots & & \vdots \\ a_{n1} & a_{n2} & \cdots & a_{nn} \end{vmatrix} = a_{11}M_{11} - a_{12}M_{12} + a_{13}M_{13} \cdots + (-1)^{n+1}a_{1n}M_{1n}$$

There are three important questions to answer regarding this definition.

Does this definition really define? Only if the minors M_{ij} can be evaluated. They are themselves determinants, but here is the key point: They are of order $n - 1$, one less than the order of the determinant we started with. They can, in turn, be expressed in terms of determinants of order $n - 2$, and so on, using the same definition. Thus, for example, a fifth-order determinant can be expressed in terms of fourth-order determinants, and these fourth-order determinants can be expressed in terms of third-order determinants. But we know how to evaluate third-order determinants from Section 11-5.

Is this definition consistent with the earlier definition when applied to third-order determinants? Yes, for if we apply it to a general third-order determinant, we get

$$\begin{vmatrix} a_1 & b_1 & c_1 \\ a_2 & b_2 & c_2 \\ a_3 & b_3 & c_3 \end{vmatrix} = a_1\begin{vmatrix} b_2 & c_2 \\ b_3 & c_3 \end{vmatrix} - b_1\begin{vmatrix} a_2 & c_2 \\ a_3 & c_3 \end{vmatrix} + c_1\begin{vmatrix} a_2 & b_2 \\ a_3 & b_3 \end{vmatrix}$$

$$= a_1b_2c_3 - a_1c_2b_3 - b_1a_2c_3 + b_1c_2a_3 + c_1a_2b_3 - c_1b_2a_3$$

This is the same value we gave in Section 11-5.

Does this definition preserve Cramer's rule and the properties of Section 11-5? Yes, it does. We shall not prove this because the proofs are lengthy and difficult.

Expansion According to Any Row or Column Our definition expressed the value of a determinant in terms of the entries and minors of the first row; we call it an expansion according to the first row. It is a remarkable fact that we can expand a determinant according to any row or column (and always get the same answer).

Before we can show what we mean, we must explain a sign convention. We associate a plus or minus sign with every position in a matrix. To the *ij* position, we assign a plus sign if $i + j$ is even and a minus sign otherwise. Thus for a 4×4 matrix, we have this pattern of signs.

$$\begin{bmatrix} + & - & + & - \\ - & + & - & + \\ + & - & + & - \\ - & + & - & + \end{bmatrix}$$

There is always a + in the upper left position and then the signs alternate.

With this understanding about signs, we may expand according to any row or column. For example, to evaluate a fourth-order determinant, we can expand according to the second column if we wish. We multiply each entry in that column by its minor, prefixing each product with a plus or minus sign according to the pattern above. Then we add the results.

$$\begin{vmatrix} a_{11} & a_{12} & a_{13} & a_{14} \\ a_{21} & a_{22} & a_{23} & a_{24} \\ a_{31} & a_{32} & a_{33} & a_{34} \\ a_{41} & a_{42} & a_{43} & a_{45} \end{vmatrix} = -a_{12}M_{12} + a_{22}M_{22} - a_{32}M_{32} + a_{42}M_{42}$$

Example To evaluate

$$\begin{vmatrix} 6 & 0 & 4 & -1 \\ 2 & 0 & -1 & 4 \\ -2 & 4 & -2 & 3 \\ 4 & 0 & 5 & -4 \end{vmatrix}$$

it is obviously best to expand according to the second column, since three of the four resulting terms are zero. The single nonzero term is just $(-1)(4)$ times the minor M_{32}, that is,

$$-4\begin{vmatrix} 6 & 4 & -1 \\ 2 & -1 & 4 \\ 4 & 5 & -4 \end{vmatrix}$$

We could now evaluate this third-order determinant as in Section 11-5. But having seen the usefulness of zeros, let us take a different tack. It is easy to get two zeros in the first column. Simply add -3 times the second row to the first row and -2 times the second row to the third. We get

$$-4\begin{vmatrix} 0 & 7 & -13 \\ 2 & -1 & 4 \\ 0 & 7 & -12 \end{vmatrix}$$

Finally, expand according to the first column.

$$(-4)(-1)(2)\begin{vmatrix} 7 & -13 \\ 7 & -12 \end{vmatrix} = 8(-84 + 91) = 56$$

The reason for the factor of -1 is that the entry 2 is in a minus position in the 3×3 pattern of signs.

Problem Set 11-6

Evaluate each of the determinants in Problems 1–6 according to a row or column of your choice. Make a good choice or suffer the consequences!

1. $\begin{vmatrix} 3 & -2 & 4 \\ 1 & 5 & 0 \\ 3 & 10 & 0 \end{vmatrix}$

2. $\begin{vmatrix} 4 & 0 & -6 \\ -2 & 3 & 5 \\ 1 & 0 & 8 \end{vmatrix}$

3. $\begin{vmatrix} 1 & 2 & 3 \\ 0 & 2 & 3 \\ 1 & 3 & 4 \end{vmatrix}$

4. $\begin{vmatrix} 2 & -1 & -1 \\ 3 & 4 & 2 \\ 0 & -1 & -1 \end{vmatrix}$

5. $\begin{vmatrix} 3 & 0 & 0 & 0 \\ -1 & 1 & 4 & 2 \\ 2 & 0 & 2 & -3 \\ -4 & 0 & 1 & 5 \end{vmatrix}$

6. $\begin{vmatrix} 0 & 5 & 0 & 0 \\ 1 & -3 & 0 & 2 \\ 4 & 1 & 2 & 8 \\ -3 & 2 & 0 & 5 \end{vmatrix}$

Evaluate each of the determinants in Problems 7–10 by first getting some zeros in a row or column and then expanding according to that row or column.

7. $\begin{vmatrix} 3 & 5 & -10 \\ 2 & 4 & 6 \\ -3 & -5 & 12 \end{vmatrix}$

8. $\begin{vmatrix} 2 & -1 & 2 \\ 4 & 3 & 4 \\ 7 & -5 & 10 \end{vmatrix}$

9. $\begin{vmatrix} 1 & -2 & 1 & 4 \\ -2 & 5 & -3 & 1 \\ 0 & 7 & -4 & 2 \\ 3 & -2 & 2 & 6 \end{vmatrix}$

10. $\begin{vmatrix} 1 & -2 & 0 & -4 \\ 3 & -4 & 3 & -10 \\ 2 & 1 & -2 & 1 \\ 4 & -5 & 1 & 4 \end{vmatrix}$

11. Solve the following system for x only.

$$\begin{aligned} x - 2y + z + 4w &= 1 \\ -2x + 5y - 3z + w &= -2 \\ 7y - 4z + 2w &= 3 \\ 3x - 2y + 2z + 6w &= 6 \end{aligned}$$

(Make use of your answer to Problem 9.)

12. Solve the following system for z only.

$$\begin{aligned} x - 2y - 4w &= -14 \\ 3x - 4y + 3z - 10w &= -28 \\ 2x + y - 2z + w &= 0 \\ 4x - 5y + z + 4w &= 9 \end{aligned}$$

(Make use of your answer to Problem 10.)

Miscellaneous Problems

Evaluate the determinants in Problems 13–18.

13. $\begin{vmatrix} 2 & -3 & 2 \\ 1 & 0 & -4 \\ -1 & 0 & 6 \end{vmatrix}$

14. $\begin{vmatrix} 3 & 1 & -5 \\ 2 & -2 & 7 \\ 1 & 0 & -1 \end{vmatrix}$

15. $\begin{vmatrix} 2 & -3 & 4 & 5 \\ 2 & -3 & 4 & 7 \\ 1 & 6 & 4 & 5 \\ 2 & 6 & 4 & -8 \end{vmatrix}$

16. $\begin{vmatrix} 2 & 2 & 3 & 7 \\ 1 & 2 & 3 & -2 \\ 4 & -3 & 9 & 6 \\ 1 & 2 & 3 & -1 \end{vmatrix}$

17. $\begin{vmatrix} 1 & 2 & -3 & 1 & 2 \\ -1 & 0 & 2 & 5 & -3 \\ 5 & 0 & 0 & -2 & 4 \\ 0 & 0 & 0 & 6 & 3 \\ 0 & 0 & 0 & 2 & -7 \end{vmatrix}$

18. $\begin{vmatrix} 1 & 2 & 3 & 4 & 5 \\ 2 & 1 & 1 & 1 & 1 \\ 3 & 1 & 1 & 1 & 1 \\ 4 & 1 & 1 & 1 & 1 \\ 5 & 1 & 1 & 1 & 1 \end{vmatrix}$

(*Hint*: Subtract row 2 from row 3.)

19. Evaluate the following determinant.

$$\begin{vmatrix} a & b & c & d \\ 0 & e & f & g \\ 0 & 0 & h & i \\ 0 & 0 & 0 & j \end{vmatrix}$$

Conjecture a general result about the determinant of a triangular matrix.

[C] 20. Use the result of Problem 19 to evaluate

$$\begin{vmatrix} 2.12 & 3.14 & -1.61 & 1.72 \\ 0 & -2.36 & 5.91 & 7.82 \\ 0 & 0 & 1.46 & 3.34 \\ 0 & 0 & 0 & 3.31 \end{vmatrix}$$

[C] 21. Evaluate by reducing to triangular form and using Problem 19.

$$\begin{vmatrix} 1 & 2 & 2.6 & 1.5 \\ 2.3 & 5.6 & -1.3 & 9.8 \\ 2.7 & 1.3 & 4.2 & -1.9 \\ 5.5 & 6.2 & 3.0 & 1.4 \end{vmatrix}$$

22. Show that

$$\begin{vmatrix} a_1 + d_1 & b_1 & c_1 \\ a_2 + d_2 & b_2 & c_2 \\ a_3 + d_3 & b_3 & c_3 \end{vmatrix} = \begin{vmatrix} a_1 & b_1 & c_1 \\ a_2 & b_2 & c_2 \\ a_3 & b_3 & c_3 \end{vmatrix} + \begin{vmatrix} d_1 & b_1 & c_1 \\ d_2 & b_2 & c_2 \\ d_3 & b_3 & c_3 \end{vmatrix}$$

23. Show that the determinant

$$\begin{vmatrix} a_1 & b_1 & c_1 \\ a_2 & b_2 & c_2 \\ ra_1 + sa_2 & rb_1 + sb_2 & rc_1 + sc_2 \end{vmatrix}$$

is zero for any values of r and s.

24. Show that for any value of n,

$$\begin{vmatrix} n+1 & n+2 & n+3 \\ n+4 & n+5 & n+6 \\ n+7 & n+8 & n+9 \end{vmatrix} = 0$$

25. Show that

$$\begin{vmatrix} 1 & 1 & 1 \\ p & q & r \\ p^2 & q^2 & r^2 \end{vmatrix} = (p - q)(q - r)(r - p)$$

(*Hint*: First get two zeros in the first row; then look for common factors in two of the columns.)

26. Show that if all the entries in a determinant are integers, its value is an integer.

27. From Problem 26 and Cramer's Rule, conclude something about the kind of numbers that can arise as solutions to n linear equations in n unknowns if the coefficients are all integers.

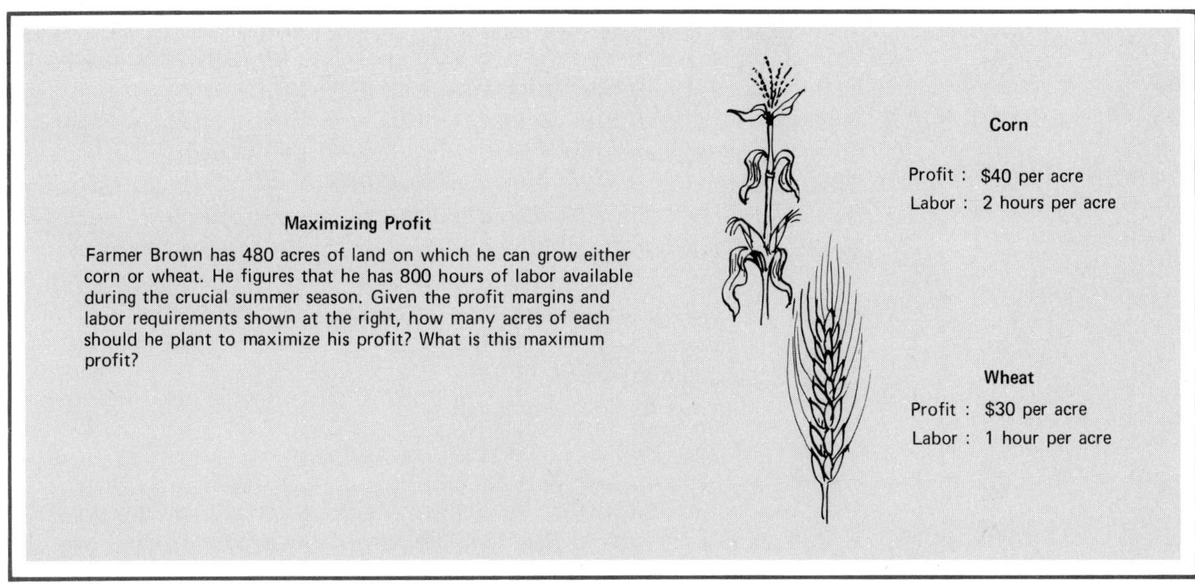

11-7 Systems of Inequalities

At first glance you might think that Farmer Brown should put all of his land into corn. Unfortunately, however, that requires 960 hours of labor and he has only 800 available. Well, maybe he should plant 400 acres of corn, using his allotted 800 hours of labor on them, and let the remaining 80 acres lie idle. Or would it be wise to at least plant enough wheat so all his land is in use? This problem is complicated enough so that no one is likely to find the best solution without a lot of work. And would not a method be better than blind experimenting? That is our subject—a method for handling Farmer Brown's problem and others of the same type.

Like all individuals and businesses, Farmer Brown must operate within certain limitations; we call them **constraints**. Suppose he plants x acres of corn and y acres of wheat. His constraints can be translated into inequalities.

Land constraint:	$x + y \leq 480$
Labor constraint:	$2x + y \leq 800$
Nonnegativity constraints:	$x \geq 0 \qquad y \geq 0$

His task is to maximize the profit $P = 40x + 30y$ subject to these constraints. Before we can solve his problem, we will need to know more about inequalities.

The Graph of a Linear Inequality The best way to visualize an inequality is by means of its graph. Consider, for example,

$$2x + y \leq 6$$

which can be rewritten as

$$y \leq -2x + 6$$

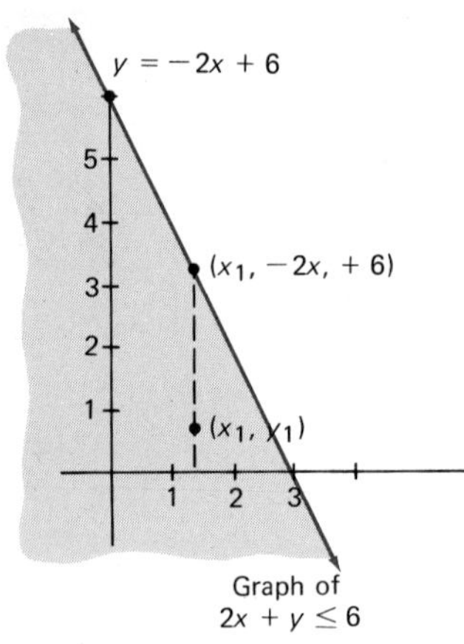

Graph of
$2x + y \leq 6$

The complete graph consists of those points which satisfy $y = -2x + 6$ (a line), together with those that satisfy $y < -2x + 6$ (the points below the line). To see that this description is correct, note that for any abscissa x_1, the point $(x_1, -2x_1 + 6)$ is on the line $y = -2x + 6$. The point (x_1, y_1) is directly below that point if and only if $y_1 < -2x_1 + 6$ (see the diagram in the margin). Thus the graph of $y \leq -2x + 6$ is the **closed half-plane** that we have shaded on the diagram. We refer to it as *closed* because the edge $y = -2x + 6$ is included. Correspondingly, the graph of $y < -2x + 6$ is called an **open half-plane**.

The graph of any linear inequality in x and y is a half-plane, open or closed. To sketch the graph, first draw the corresponding edge. Then determine the correct half-plane by taking a sample point, not on the edge, and checking to see if it satisfies the inequality.

To illustrate this procedure, consider

$$2x - 3y < -6$$

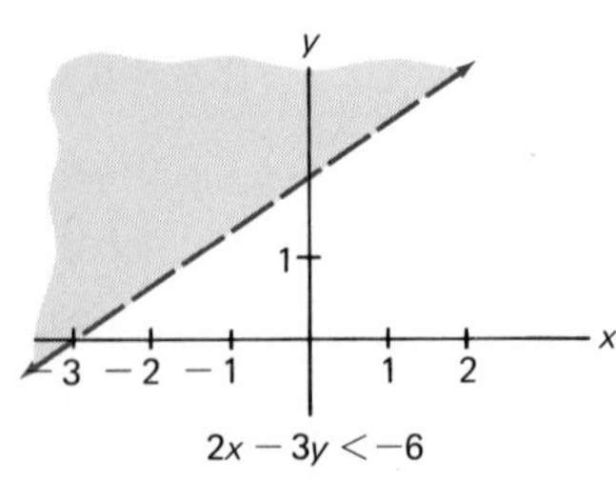

$2x - 3y < -6$

Its graph does not include the line $2x - 3y = -6$, although that line is crucial in determining the graph. We therefore show it as a dotted line. Since the sample point (0, 0) does not satisfy the inequality, we choose the half-plane on the opposite side of the line from it. The complete graph is the shaded open half-plane shown in the margin.

Graphing a System of Linear Inequalities The graph of a system of inequalities like Farmer Brown's constraints

$$\begin{aligned} x + y &\leq 480 \\ 2x + y &\leq 800 \\ x \geq 0 \quad & y \geq 0 \end{aligned}$$

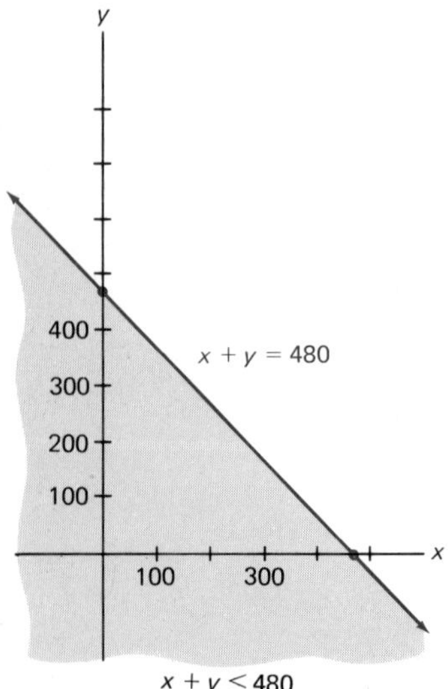

$x + y \leq 480$

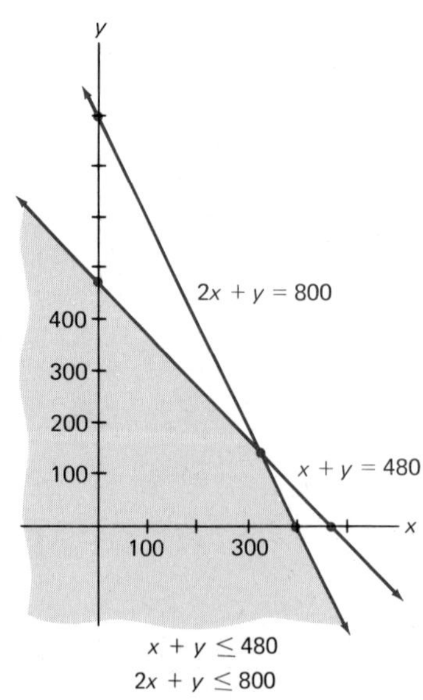

$x + y \leq 480$
$2x + y \leq 800$

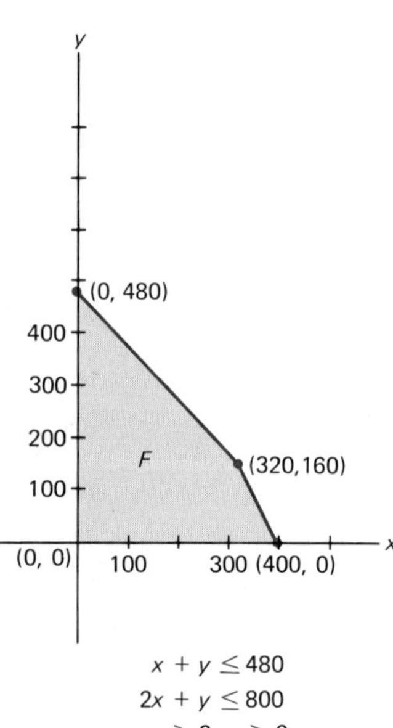

$x + y \leq 480$
$2x + y \leq 800$
$x \geq 0 \quad y \geq 0$

is simply the intersection of the graphs of the individual inequalities. We can construct the graph in stages as we did on page 446, though we are confident that you will quickly learn to do it in one operation.

The diagram on the right (page 446) is the one we want. All the points in the shaded region F satisfy the four inequalities simultaneously. The points (0, 0), (400, 0), (320, 160), and (0, 480) are called the **vertices** (or corner points) of F. Incidentally, the point (320, 160) was obtained by solving the two equations $2x + y = 800$ and $x + y = 480$ simultaneously.

The region F has three important properties.

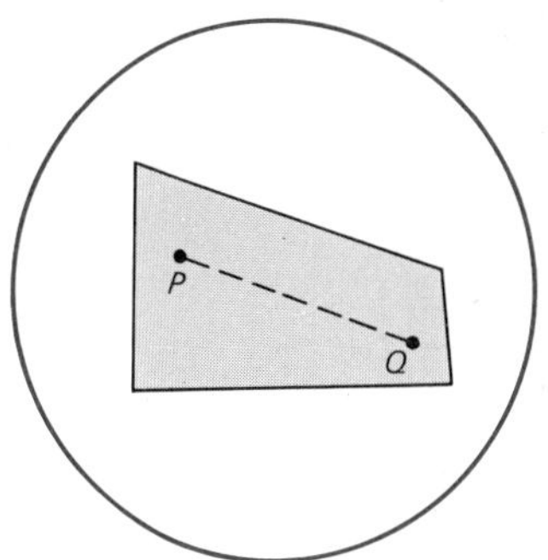

Polygonal, convex, and bounded

1. It is polygonal (its boundary consists of line segments).
2. It is convex (if points P and Q are in the region, then the line segment PQ lies entirely within the region).
3. It is bounded (it can be enclosed in a circle).

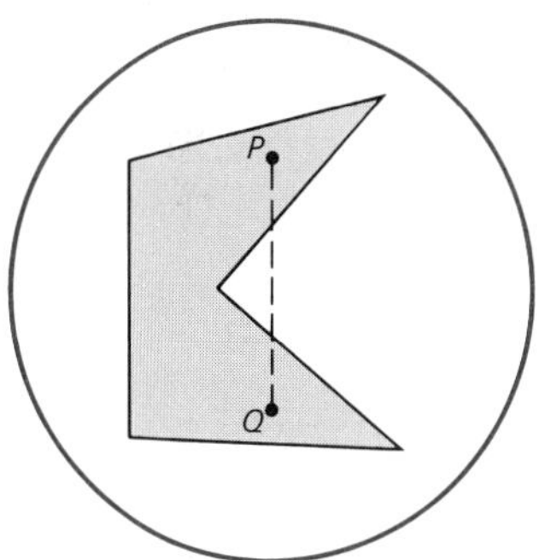

Polygonal and bounded but not convex

As a matter of fact, every region that arises as the solution set of a system of linear inequalities is polygonal and convex, though it need not be bounded. The shaded region in the lower diagram in the margin could not be the solution set for a system of linear inequalities because it is not convex.

Linear Programming Problems It is time that we solved Farmer Brown's problem.

Maximize

$$P = 40x + 30y$$

subject to

$$\begin{cases} x + y \leq 480 \\ 2x + y \leq 800 \\ x \geq 0 \qquad y \geq 0 \end{cases}$$

Any problem that asks us to find the maximum (or minimum) of a linear function subject to linear inequality constraints is called a **linear programming problem**. Here is a method for solving such problems.

1. Graph the solution set corresponding to the inequality constraints.
2. Find the coordinates of the vertices of the solution set.
3. Evaluate the linear function that you want to maximize (or minimize) at each of these vertices. The largest of these gives the maximum, while the smallest gives the minimum.

To see why this method works, consider the diagram for Farmer Brown's problem at the top of page 448.

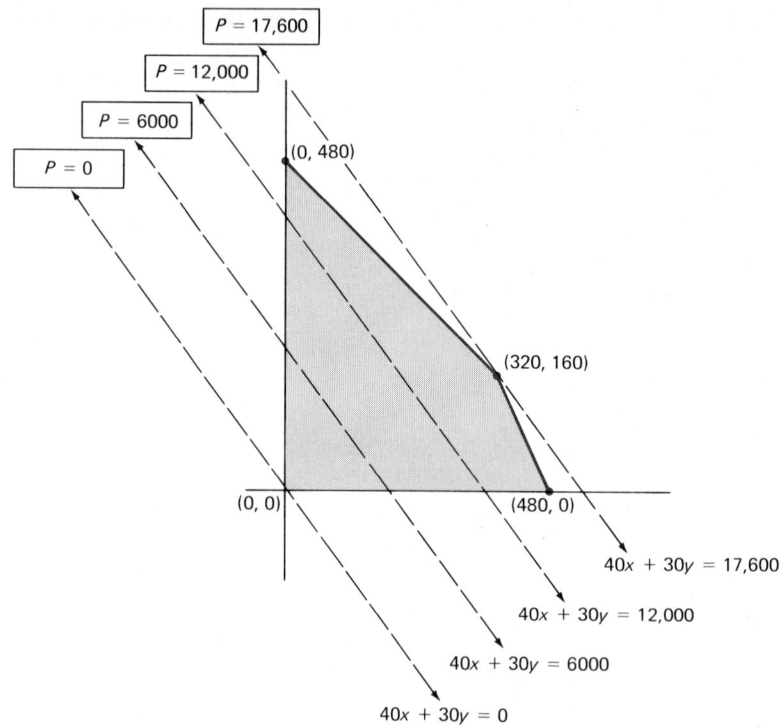

Vertex	$P = 40x + 30y$
(0, 0)	0
(0, 480)	14,400
(320, 160)	17,600
(400, 0)	16,000

Vertex	$P = 80x + 30y$
(0, 0)	0
(0, 480)	14,400
(320, 160)	30,400
(400, 0)	32,000

Vertex	$P = 40x + 40y$
(0, 0)	0
(0, 480)	19,200
(320, 160)	19,200
(400, 0)	16,000

The dotted lines are profit lines, each with slope $-\frac{4}{3}$; they are the graphs of $40x + 30y = P$ for various values of P. All the points on a dotted line give the same total profit. Imagine a profit line moving from left to right across the shaded region with the slope constant, as indicated in the diagram. During this motion, the profit is zero at (0, 0) and increases to its maximum of 17,600 at (320, 160). It should be clear that such a moving line (no matter what its slope) will always enter the shaded set at a vertex and leave it at another vertex. The particular vertex depends upon the slope of the profit lines. In Farmer Brown's problem, the minimum profit of \$0 occurs at (0, 0); the maximum profit of \$17,600 occurs at (320, 160). In the margin, we show the total profit for each of the four vertices. Clearly Farmer Brown should plant 320 acres of corn and 160 acres of wheat.

Now suppose the price of corn goes up so that Farmer Brown can expect a profit of \$80 per acre on corn but still only \$30 per acre on wheat. Would this change his strategy? In the second table in the margin, we show his total profit $P = 80x + 30y$ at each of the four vertices. Evidently he should plant 400 acres of corn and no wheat to achieve maximum profit. Note that this means he should leave 80 acres of his land idle.

Finally, suppose that the profit per acre is \$40 both for wheat and for corn. The table for this case shows the same total profit at the vertices (0, 480) and (320, 160). This means that the moving profit line leaves the shaded region along the side determined by those two vertices, so that every point on that side gives a maximum profit. It is still true, however, that the maximum profit occurs at a vertex.

The situation with an unbounded constraint set is slightly more complicated. It is discussed in Example A. Here we simply point out that in the unbounded case, there may not be a maximum (or minimum), but if there is one, it will still occur at a vertex.

Problem Set 11-7

In Problems 1–6, graph the solution set of each inequality in the xy-plane.

1. $4x + y \leq 8$
2. $2x + 5y \leq 20$
3. $x \leq 3$
4. $y \leq -2$
5. $4x - y \geq 8$
6. $2x - 5y \geq -20$

In Problems 7–10, graph the solution set of the given system. On the graph, label the coordinates of the vertices.

7. $4x + y \leq 8$
   $2x + 3y \leq 14$
   $x \geq 0 \quad y \geq 0$
8. $2x + 5y \leq 20$
   $4x + y \leq 22$
   $x \geq 0 \quad y \leq 0$
9. $4x + y \leq 8$
   $x - y \leq -2$
   $x \geq 0$
10. $2x + 5y \leq 20$
    $x - 2y \geq 1$
    $y \geq 0$

In Problems 11–14, find the maximum and minimum value of the given linear function P subject to the given inequalities.

11. $P = 2x + y$; the inequalities of Problem 7.
12. $P = 3x + 2y$; the inequalities of Problem 8.
13. $P = 2x - y$; the inequalities of Problem 9.
14. $P = 3x - 2y$; the inequalities of Problem 10.

Example A (Unbounded region) Find the maximum and minimum values of the function $3x + 4y$ subject to the constraints

$$\begin{cases} 3x + 2y \geq 13 \\ x + y \geq 5 \\ x \geq 1 \\ y \geq 0 \end{cases}$$

SOLUTION. We proceed to graph the solution set of our system, noting that the region must lie above the lines $3x + 2y = 13$ and $x + y = 5$ and to the right of the line $x = 1$. It is shown at the top of page 450.
Notice that the region is unbounded and has (5, 0), (3, 2), and (1, 5) as its vertices. The point (1, 4) at which the lines $x + y = 5$ and $x = 1$ intersect is not a vertex. It should be clear right away that $3x + 4y$ does not assume a maximum value in our region; its values can be made as large as we please

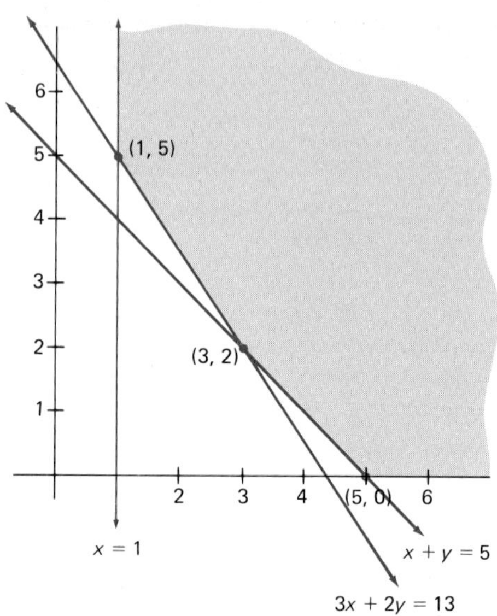

by increasing x and y. To find the minimum value, we calculate $3x + 4y$ at the three vertices.

$$\begin{aligned}
(5, 0)&: \quad 3 \cdot 5 + 4 \cdot 0 = 15 \\
(3, 2)&: \quad 3 \cdot 3 + 4 \cdot 2 = 17 \\
(1, 5)&: \quad 3 \cdot 1 + 4 \cdot 5 = 23
\end{aligned}$$

The minimum value of $3x + 4y$ is 15.

Solve each of the following problems.

15. Minimize $5x + 2y$ subject to

$$\begin{cases} x + y \geq 4 \\ x \geq 2 \\ y \geq 0 \end{cases}$$

16. Minimize $2x - y$ subject to

$$\begin{cases} x - 2y \geq 2 \\ y \geq 2 \end{cases}$$

17. Minimize $2x + y$ subject to

$$\begin{cases} 4x + y \geq 7 \\ 2x + 3y \geq 6 \\ x \geq 1 \\ y \geq 0 \end{cases}$$

18. Minimize $3x + 2y$ subject to

$$\begin{cases} x - 2y \leq 2 \\ x - 2y \geq -2 \\ 3x - 2y \geq 10 \end{cases}$$

Example B (Systems with nonlinear inequalities) Graph the solution set of the following system of inequalities.

$$\begin{cases} y \geq 2x^2 \\ y \leq 2x + 4 \end{cases}$$

SOLUTION. It helps to find the points at which the parabola $y = 2x^2$ intersects the line $y = 2x + 4$. Eliminating y between the two equations and then solving for x, we get

$$2x^2 = 2x + 4$$
$$x^2 - x - 2 = 0$$
$$(x - 2)(x + 1) = 0$$
$$x = 2 \qquad x = -1$$

The corresponding values of y are 8 and 2, respectively; so the points of intersection are $(-1, 2)$ and $(2, 8)$. Making use of these points, we draw the parabola and the line. Since the point $(0, 2)$ satisfies the inequality $y \geq 2x^2$, the desired region is above and including the parabola. The graph of the linear inequality $y \leq 2x + 4$ is to the right of and including the line. The shaded region in the diagram below is the graph we want.

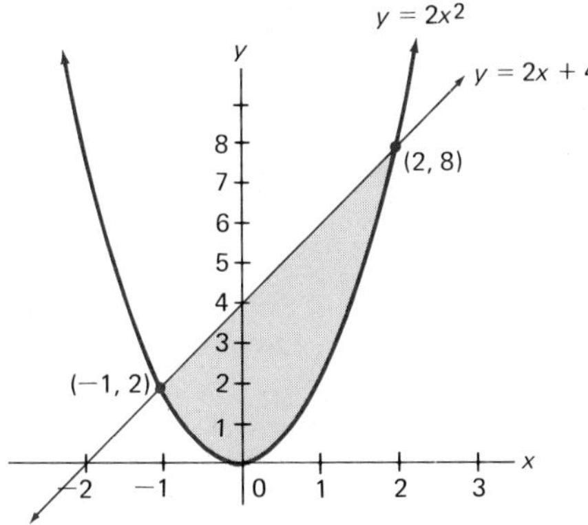

In Problems 19–22, graph the solution set of each system of inequalities.

19. $y \leq 4x - x^2$
    $y \leq x$
    $x \geq 0 \qquad y \geq 0$

20. $y \geq 2^x$
    $y \leq 8$
    $x \geq 0$

21. $y \leq \log_2 x$
    $x \leq 8$
    $y \geq 0$

22. $x^2 + y^2 \geq 9$
    $0 \leq x \leq 3$
    $0 \leq y \leq 3$

Miscellaneous Problems

In Problems 23 and 24, graph the solution set and label the coordinates of the vertices.

23. $4x + y \leq 8$
    $x - y \geq -2$
    $x \geq 0 \qquad y \geq 0$

24. $2x + 5y \leq 20$
    $x - 2y \leq 1$
    $x \geq 0$

25. Find the maximum and minimum values of $P = 2x - y$ subject to the inequalities of Problem 23.

26. Find the maximum and minimum values of $P = 3x - 2y$ subject to the inequalities of Problem 24.

27. Find the maximum and minimum values of $x + y$ (if they exist) subject to the following inequalities.

$$\begin{aligned} x - y &\geq -1 \\ x - 2y &\leq 5 \\ 3x + y &\leq 10 \\ x \geq 0 \qquad y &\geq 0 \end{aligned}$$

28. Find the maximum and minimum values of $x - 2y$ (if they exist) subject to the inequalities of Problem 27.

29. An oil refinery has a maximum production of 2000 barrels of oil per day. It produces two types of oil: type A, which is used for gasoline and type B, which is used for heating oil. There is a requirement that at least 300 barrels of type B be produced each day. If the profit is \$3 a barrel for type A and \$2 a barrel for type B, find the maximum profit per day.

30. A company makes a single product on two production lines A and B. A labor force of 900 hours per week is available, and weekly running costs shall not exceed \$1500. It takes 4 hours to produce one item on production line A and 3 hours on production line B. The cost per item is \$5 on line A and \$6 on line B. Find the largest number of items that can be produced in one day.

31. A shoemaker has a supply of 100 square feet of type A leather which is used for soles and 600 square feet of type B leather used for the rest of the shoe. The average shoe uses $\frac{1}{4}$ square feet of type A leather and 1 square foot of type B leather. The average boot uses $\frac{1}{4}$ square feet and 3 square feet of types A and B leather, respectively. If shoes sell at \$40 a pair and boots at \$60 a pair, find the maximum income.

32. A manufacturer of trailers wishes to determine how many camper units and how many house trailers she should produce in order to make the best possible use of her resources. She has 42 units of wood, 56 worker-weeks of labor and 16 units of aluminum. (Assume that all other needed resources are available and have no effect on her decision.) The amount of each resource needed to produce each camper and each trailer is given below.

	Wood	Worker-weeks	Aluminum
Per camper	3	7	3
Per trailer	6	7	1

If the manufacturer realizes a profit of \$600 on a camper and \$800 on a trailer, what should be her production in order to maximize her profit?

33. A grain farmer has 100 acres available for sowing oats and wheat. The seed oats costs \$5 per acre and the seed wheat costs \$8 per acre. The labor costs are \$20 per acre for oats and \$12 per acre for wheat. The farmer expects an income from oats of \$220 per acre and from wheat of \$250 per acre. How many acres of each crop should he sow to maximize his profit, if he does not wish to spend more than \$620 for seed and \$1800 for labor?

34. Suppose that the minimum monthly requirements for one person are 60 units of carbohydrates, 40 units of protein, and 35 units of fat. Two

foods A and B contain the following numbers of units of the three diet components per pound.

	Carbohydrates	Protein	Fat
A	5	3	5
B	2	2	1

If food A costs \$3.00 a pound and food B costs \$1.40 a pound, how many pounds of each should a person purchase per month to minimize the cost?

CHAPTER SUMMARY

Two systems of equations are **equivalent** if they have the same solutions. Three elementary operations (multiplying a row by a nonzero constant, interchanging two rows, and adding a multiple of one row to another) lead to equivalent systems. Use of these operations allows us to transform a system of linear equations to **triangular form** and then to solve the system by **back substitution** or to show it is **inconsistent**.

A **matrix** is a rectangular array of numbers. In solving a system of linear equations, it is efficient to work with just the **matrix of the system**. We solve the system by transforming its matrix to triangular form using the three operations mentioned above.

Addition and multiplication are defined for matrices with resulting algebraic rules. Matrices behave much like numbers, with the exception that the commutative law for multiplication fails. Even the notion of **multiplicative inverse** has meaning, though the process for finding such an inverse is lengthy.

Associated with every square matrix is a number called its **determinant**. For 2×2 and 3×3 matrices, the value of the determinant can be found by using certain arrow diagrams. For higher order cases, the value of a determinant is found by expanding it in terms of the elements of a row (or column) and their **minors** (determinants whose order is lower than the given determinant by 1). In doing this, it is helpful to know what happens to the determinant of a matrix when any of the three elementary operations are applied to it. **Cramer's Rule** provides a direct way of solving a system of n equations in n unknowns using determinants.

The graph of a **linear inequality** in x and y is a **half-plane** (closed or open according as the inequality sign does or does not include the equal sign). The graph of a **system of linear inequalities** is the intersection of the half-planes corresponding to the separate inequalities. Such a graph is always **polygonal** and **convex** but may be **bounded** or **unbounded**. A **linear programming problem** asks us to find the maximum (or minimum) of a linear function (such as $2x + 5y$) subject to a system of linear inequalities called **constraints**. The maximum (or minimum) always occurs at a **vertex**, that is, at a corner point of the graph of the inequality constraints.

CHAPTER REVIEW PROBLEM SET

In Problems 1–5, solve each system or show that there is no solution.

1. $$\begin{aligned} 3x + y &= 12 \\ 2x - y &= -2 \end{aligned}$$

2. $$\begin{aligned} x - 3y + 2z &= 5 \\ 2y - z &= 1 \\ z &= 3 \end{aligned}$$

3. $$\begin{aligned} 2x - y + 3z &= 10 \\ y - 2z &= 4 \\ -2y + 4z &= 8 \end{aligned}$$

4. $$\begin{aligned} x - 3y + 2z &= -5 \\ 4x + y - z &= 4 \\ 5x + 11y - 8z &= 20 \end{aligned}$$

5. $$\begin{aligned} x + 2y + 3z + 4w &= 20 \\ y - 4z + w &= 6 \\ z + 2w &= 4 \\ 2z + 4w &= 8 \end{aligned}$$

6. Evaluate each expression for

$$\mathbf{A} = \begin{bmatrix} 1 & 1 & 1 \\ 3 & 1 & -1 \\ 2 & 2 & -1 \end{bmatrix} \quad \text{and} \quad \mathbf{B} = \begin{bmatrix} -3 & 6 & -4 \\ 2 & -3 & 5 \\ 1 & 9 & 2 \end{bmatrix}$$

(a) $2\mathbf{A} + \mathbf{B}$ (b) $\mathbf{A} - 2\mathbf{B}$ (c) $\mathbf{AB}$ (d) $\mathbf{BA}$

7. Find $\mathbf{A}^{-1}$ for the matrix $\mathbf{A}$ of Problem 6.

8. Consider the system

$$\begin{aligned} x + y + z &= 4 \\ 3x + y - z &= -4 \\ 2x + 2y - z &= -1 \end{aligned}$$

Write this system in matrix form and then use the result of Problem 7 to solve it.

Evaluate the determinants in Problems 9–13.

9. $$\begin{vmatrix} -2 & 5 \\ 2 & -6 \end{vmatrix}$$

10. $$\begin{vmatrix} 2 & 3 \\ -4 & -6 \end{vmatrix}$$

11. $$\begin{vmatrix} -2 & 1 & 4 \\ 0 & 5 & -1 \\ 4 & 0 & 3 \end{vmatrix}$$

12. $$\begin{vmatrix} 1 & -2 & 3 \\ 4 & 1 & 5 \\ 7 & -5 & 14 \end{vmatrix}$$

13. $$\begin{vmatrix} 6 & 0 & 0 & 0 \\ -1 & 3 & 1 & 0 \\ 3 & 2 & 3 & 2 \\ 4 & 5 & 1 & -4 \end{vmatrix}$$

14. Use Cramer's Rule to solve the following system.

$$\begin{aligned} 2x + y - z &= -4 \\ x - 3y - 2z &= -1 \\ 3x + 2y + 3z &= 11 \end{aligned}$$

In Problems 15 and 16, graph the solution set of the given system. On the graph, label the coordinates of the vertices.

15. $x + y \leq 7$
    $3x + y \leq 15$
    $x \geq 0 \quad y \geq 0$

16. $x - 2y + 4 \geq 0$
    $x + y - 11 \geq 0$
    $x \geq 0 \quad y \geq 0$

17. Find the maximum value of the function $P = x + 2y$ subject to the inequalities of Problem 15.

18. Find the minimum value of the function $P = x + 2y$ subject to the inequalities of Problem 16.

19. A certain company has 100 employees, some of whom get \$4 an hour, others \$5, and the rest \$8. Half as many make \$8 an hour as \$5 an hour. If the total paid out in hourly wages is \$544, find the number of employees who make \$8 an hour.

20. A tailor has 110 yards of cotton material and 160 yards of woolen material. It takes $1\frac{1}{2}$ yards of cotton and 1 yard of wool to make a suit, while a dress requires 1 yard of cotton and 2 yards of wool. If a suit sells for \$100 and a dress for \$80, how many of each should the tailor make to maximize the total income?

Method consists entirely in properly ordering and arranging things to which we should pay attention.

René Descartes

CHAPTER TWELVE

Sequences and Counting Problems

12-1 Number Sequences

12-2 Arithmetic Sequences

12-3 Geometric Sequences

12-4 Mathematical Induction

12-5 Counting Ordered Arrangements

12-6 Counting Unordered Collections

12-7 The Binomial Formula

What Comes Next?

(a) 1, 4, 7, 10, 13, 16, □, □, . . .

(b) 2, 4, 6, 8, 10, 12, □, □, . . .

(c) 1, 4, 9, 16, 25, 36, □, □, . . .

(d) 1, 4, 9, 16, 27, 40, □, □, . . .

(e) 1, 1, 2, 3, 5, 8, □, □, . . .

12-1 Number Sequences

Try filling in the boxes of our opening display. You will have little trouble with *a*, *b*, and *c* but *d* and *e* may offer quite a challenge. We will give the answers we had in mind later. Right now, we merely point out that each sequence has a pattern; we used a definite rule in writing the first six terms of each of them.

The word *sequence* is commonly used in ordinary language. For example, your history teacher may talk about a sequence of events that led to World War II (for instance, the Versailles Treaty, world depression, Hitler's ascendancy, Munich Agreement). What characterizes this sequence is the notion of one event following another in a definite order. There is a first event, a second event, a third event, and so on. We might even give them labels.

E_1: Versailles Treaty
E_2: World depression
E_3: Hitler's ascendancy
E_4: Munich Agreement

We use a similar notation for number sequences. Thus

$$a_1, a_2, a_3, a_4, \ldots$$

could denote sequence (a) of our opening display. Then

$$\begin{aligned} a_1 &= 1 \\ a_2 &= 4 \\ a_3 &= 7 \\ a_4 &= 10 \\ &\vdots \end{aligned}$$

Note that a_3 stands for the 3rd term; a_{10} would represent the 10th term. The subscript indicates the position of the term in the sequence. For the general term, that is, the nth term, we use the symbol a_n. The three dots indicate that the sequence continues indefinitely.

There is another way to describe a number sequence. A **number sequence** is a function whose domain is the set of positive integers. That means it is a rule that associates with each positive integer n a definite number a_n. In conformity with Chapter 5, we could use the notation $a(n)$, but tradition dictates that we use a_n instead. We usually specify functions by giving formulas; this is true of sequences also.

Explicit Formulas Rather than give the first few terms of a sequence and hope that our readers see the pattern intended (different people sometimes see different patterns in the first few terms of a sequence), it is better to give a formula. Take sequence (a) for example. The formula

$$a_n = 3n - 2$$

tells all there is to know about that sequence. In particular,

$$\begin{aligned} a_1 &= 3 \cdot 1 - 2 = 1 \\ a_2 &= 3 \cdot 2 - 2 = 4 \\ a_3 &= 3 \cdot 3 - 2 = 7 \\ a_{100} &= 3 \cdot 100 - 2 = 298 \end{aligned}$$

How does one find the formula for a sequence? Look at sequence (b). Suppose we let b_n stand for the nth term. Then $b_1 = 2$, $b_2 = 4$, $b_3 = 6$, $b_4 = 8$, and so on. Our job is to relate the value of b_n to the subscript n. Clearly, it is just twice the subscript, that is,

$$b_n = 2n$$

Knowing this formula, we can calculate the value of any term. For example,

$$\begin{aligned} b_{10} &= 2 \cdot 10 = 20 \\ b_{281} &= 2 \cdot 281 = 562 \end{aligned}$$

If we follow the same procedure for sequence (c), we have

$$c_1 = 1 \qquad c_2 = 4 \qquad c_3 = 9 \qquad c_4 = 16$$

from which we infer the formula

$$c_n = n^2$$

Now look at sequence (d).

$$1, 4, 9, 16, 27, 40, \ldots$$

The fact that it starts out just like sequence (c) suggests that the pattern is subtle and incidentally warns us that we may have to look at many terms of a sequence before we can discover its rule of construction. Here, as in many sequences, it is a good idea to observe how each term relates to the previous one. Let us write the sequence again, indicating below it the numbers to be added as we progress from term to term.

$$\begin{array}{ccccccccccc} 1 & & 4 & & 9 & & 16 & & 27 & & 40 \\ & 3 & & 5 & & 7 & & 11 & & 13 & \end{array}$$

You may recognize the second row of numbers as consecutive primes (starting with 3). Thus the next two terms in sequence (d) are

$$40 + 17 = 57$$
$$57 + 19 = 76$$

But observing a pattern does not necessarily mean we can write an explicit formula. Though many have tried, no one has found a formula for the nth prime and, thus, no one is likely to find a formula for sequence (d).

Sequence (e) is a famous one. It was introduced by Leonardo Fibonacci around 1200 A.D. in connection with rabbit reproduction (see Problem 27). If you are a keen observer, you have noticed that any term (after the second) is the sum of the preceding two. It was not until 1724 that mathematician Daniel Bernoulli found the explicit formula for this sequence. You will agree that it is complicated, but at least you can check it for $n = 1, 2, 3$.

$$e_k = \frac{1}{\sqrt{5}}\left[\left(\frac{1 + \sqrt{5}}{2}\right)^n - \left(\frac{1 - \sqrt{5}}{2}\right)^n\right]$$

If it took 500 years to discover this formula, you should not be surprised when we say that explicit formulas are often difficult to find (the problem set will give more evidence). There is another type of formula that is usually easier to discover.

Recursion Formulas An explicit formula relates the value of a_n to its subscript n (for example, $a_n = 3n - 2$). Often the pattern we first observe relates a term to the preceding term (or terms). If so, we may be able to describe this pattern by a recursion formula. Look at sequence (a) again. To get a term from the preceding one, we always add 3, that is,

$$a_n = a_{n-1} + 3$$

Or look at sequence (b). There we add 2 each time.

$$b_n = b_{n-1} + 2$$

Sequence (e) is more interesting. There we add together the two previous terms, that is,

$$e_n = e_{n-1} + e_{n-2}$$

We summarize our knowledge about the five sequences in the following chart.

SEQUENCE	EXPLICIT FORMULA	RECURSION FORMULA
(a) 1, 4, 7, 10, 13, 16, . . .	$a_n = 3n - 2$	$a_n = a_{n-1} + 3$
(b) 2, 4, 6, 8, 10, 12, . . .	$b_n = 2n$	$b_n = b_{n-1} + 2$
(c) 1, 4, 9, 16, 25, 36, . . .	$c_n = n^2$	$c_n = c_{n-1} + 2n - 1$
(d) 1, 4, 9, 16, 27, 40, . . .	?	$d_n = d_{n-1} + n$th prime
(e) 1, 1, 2, 3, 5, 8, . . .	$e_n = \frac{1}{\sqrt{5}}\left[\left(\frac{1+\sqrt{5}}{2}\right)^n - \left(\frac{1-\sqrt{5}}{2}\right)^n\right]$	$e_n = e_{n-1} + e_{n-2}$

Recursion formulas are themselves not quite enough to determine a sequence. For example, the recursion formula

$$f_n = 3f_{n-1}$$

does not determine a sequence until we specify the first term. But with the additional information that $f_1 = 2$, we can find any term. Thus

$$\begin{aligned} f_1 &= 2 \\ f_2 &= 3f_1 = 3\cdot 2 = 6 \\ f_3 &= 3f_2 = 3\cdot 6 = 18 \\ f_4 &= 3f_3 = 3\cdot 18 = 54 \\ &\vdots \end{aligned}$$

The disadvantage of a recursion formula is apparent. To find the 100th term, we must first calculate the 99 previous terms. But if it is hard work, it is at least possible. Programmable calculators are particularly adept at calculating sequences by means of recursion formulas.

Problem Set 12-1

1. Discover a pattern and use it to fill in the boxes.
   (a) 1, 3, 5, 7, □, □, . . .
   (b) 17, 14, 11, 8, □, □, . . .
   (c) 1, $\frac{1}{2}$, $\frac{1}{4}$, $\frac{1}{8}$, □, □, . . .
   (d) 1, 9, 25, 49, □, □, . . .
2. Fill in the boxes.
   (a) 1, 3, 9, 27, □, □, . . .
   (b) 2, 2.5, 3, 3.5, □, □, . . .
   (c) 1, 8, 27, 64, □, □, . . .
   (d) $\frac{1}{2}$, $\frac{2}{3}$, $\frac{3}{4}$, $\frac{4}{5}$, □, □, . . .
3. In each case an explicit formula is given. Find the indicated terms.
   (a) $a_n = 2n + 3$; $a_4 = \square$; $a_{20} = \square$
   (b) $a_n = \dfrac{n}{n+1}$; $a_5 = \square$; $a_9 = \square$
   (c) $a_n = (2n - 1)^2$; $a_4 = \square$; $a_5 = \square$
   (d) $a_n = (-3)^n$; $a_3 = \square$; $a_4 = \square$

4. Find the indicated terms.
   (a) $a_n = 2n - 5$; $a_4 = \square$; $a_{20} = \square$
   (b) $a_n = 1/n$; $a_5 = \square$; $a_{50} = \square$
   (c) $a_n = (2n)^2$; $a_5 = \square$; $a_{10} = \square$
   (d) $a_n = 4 - \frac{1}{2}n$; $a_5 = \square$; $a_{10} = \square$
5. Give an explicit formula for each sequence in Problem 1. Recall that you must relate the value of a term to its subscript (see the examples on page 459).
6. Give an explicit formula for each sequence in Problem 2.
7. In each case below, an initial term and a recursion formula are given. Find a_5. (*Hint*: First find a_2, a_3, and a_4, and then find a_5.)
   (a) $a_1 = 2$; $a_n = a_{n-1} + 3$
   (b) $a_1 = 2$; $a_n = 3a_{n-1}$
   (c) $a_1 = 8$; $a_n = \frac{1}{2}a_{n-1}$
   (d) $a_1 = 1$; $a_n = a_{n-1} + 8(n - 1)$
8. Find a_4 for each of the following sequences.
   (a) $a_1 = 2$; $a_n = 2a_{n-1} + 1$
   (b) $a_1 = 2$; $a_n = a_{n-1} + 3$
   (c) $a_1 = 1$; $a_n = a_{n-1} + 3n^2 - 3n + 1$
   (d) $a_1 = 3$; $a_n = a_{n-1} + .5$
9. Try to find a recursion formula for each of the sequences in Problem 1.
10. Try to find a recursion formula for each of the sequences in Problem 2.

Example (Sum sequences) Corresponding to a sequence $a_1, a_2, a_3, \ldots$, we introduce another sequence A_n, called the sum sequence, by

$$A_n = a_1 + a_2 + a_3 + \cdots + a_n$$

Thus

$$\begin{aligned} A_1 &= a_1 \\ A_2 &= a_1 + a_2 \\ A_3 &= a_1 + a_2 + a_3 \\ &\vdots \end{aligned}$$

Find A_5 for the sequence given by $a_n = 3n - 2$.
SOLUTION. We begin by finding the first five terms of sequence a_n.

$$\begin{aligned} a_1 &= 1 \\ a_2 &= 4 \\ a_3 &= 7 \\ a_4 &= 10 \\ a_5 &= 13 \end{aligned}$$

Then

$$\begin{aligned} A_5 &= a_1 + a_2 + a_3 + a_4 + a_5 \\ &= 1 + 4 + 7 + 10 + 13 \\ &= 35 \end{aligned}$$

In each of the following problems, find A_6.

11. $a_n = 2n + 1$
12. $a_n = 2^n$
13. $a_n = (-2)^n$
14. $a_n = n^2$
15. $a_n = n^2 - 2$
16. $a_n = 3n - 4$
17. $a_1 = 4; a_n = a_{n-1} + 3$
18. $a_1 = 1; a_2 = 1;$ $a_n = a_{n-1} + 2a_{n-2}$

Miscellaneous Problems

19. Find a_4 and a_{20} for each sequence.
(a) $a_n = 3n - 1$ (b) $a_n = 2^n$
20. Find recursion formulas for the sequences of Problem 19.
21. Find a_4 for each of the following.
(a) $a_1 = 8; a_n = \frac{3}{2}a_{n-1}$
(b) $a_1 = 6; a_2 = 4; a_n = \dfrac{a_{n-1}}{a_{n-2}}$
22. Give an explicit formula for the first sequence of Problem 21.
23. For each sequence in Problem 19, find A_5, the sum of the first five terms.
24. Find a pattern in each of the following sequences and use it to fill in the boxes.
(a) 2, 6, 18, 54, □, □, . . .
(b) 2, 6, 10, 14, □, □, . . .
(c) 2, 4, 8, 14, □, □, . . .
(d) 2, 4, 6, 10, 16, □, □, . . .
(e) 2, 1, $\frac{1}{2}$, $\frac{1}{4}$, □, □, . . .
(f) 2, 5, 10, 17, □, □, . . .
25. Try to find an explicit formula for each sequence in Problem 24.
26. Find a recursion formula for each sequence in Problem 24.
27. Suppose that a pair of rabbits consisting of a male and a female matures so that it reproduces another male-female pair after two months and continues to do so each month thereafter. If each new male-female pair of rabbits has the same reproductive habits as its parents, how many rabbit pairs will there be after 1 month? 2 months? 3 months? 4 months? n months?
28. Let f_n denote the Fibonacci sequence determined by $f_1 = f_2 = 1$ and $f_n = f_{n-1} + f_{n-2}$. Let $F_n = f_1 + f_2 + \cdots + f_n$. Calculate F_1, F_2, F_3, F_4, F_5, and F_6.
29. With regard to Problem 28, see if you can find a nice formula that connects F_n and f_{n+2}.
30. Let a_n be the remainder when n is divided by 5.
(a) Find a_{11} and a_{400}.
(b) If $a_m = 3$, find a_{m+4}.
31. Let a_n be the number of primes that are less than n. (The smallest prime is 2.)
(a) Find a_{10} and a_{20}.
(b) If m is a number such that $a_{m+20} = a_m$, what can you conclude?

Square Numbers

$s_1 = 1$

$s_2 = 4$

$s_3 = 9$

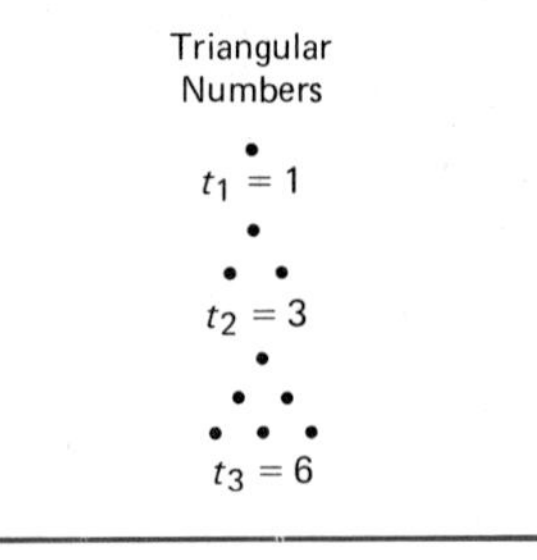

32. Let a_n be the sum of the proper divisors of n. For example, $a_{12} = 1 + 2 + 3 + 4 + 5 + 6 = 16$ and $a_{25} = 1 + 5 = 6$.
    (a) Find a_{10}, a_{16}, and a_{40}.
    (b) If k is a prime, what value does a_k have?
33. The Greeks were enchanted with sequences that arose in a geometric way (see the diagrams in the margin).
    (a) Find an explicit formula for s_n.
    (b) Find a recursion formula for s_n.
    (c) Find a recursion formula for t_n.
    (d) Find an explicit formula for t_n.
34. The numbers 1, 5, 12, 22, . . . are called *pentagonal numbers*. See if you can figure out why and then guess at an explicit formula for p_n, the nth pentagonal number. Use diagrams.
35. Let a_n be the nth digit in the decimal expansion of $\frac{1}{7} = .1428\ldots$. Thus $a_1 = 1$, $a_2 = 4$, $a_3 = 2$, and so on. Find a pattern and use it to determine a_8, a_{27}, and a_{53}.
36. Suppose that January 1 occurs on Wednesday. Let a_n be the day of the week corresponding to the nth day of the year. Thus $a_1 =$ Wednesday, $a_2 =$ Thursday, and so forth. Find a_{39}, a_{57}, and a_{84}.
37. Try to find a pattern in the following array of numbers:

$$3, 3, 5, 4, 4, 3, 5, 5, 4, 3, 6, 6, \ldots$$

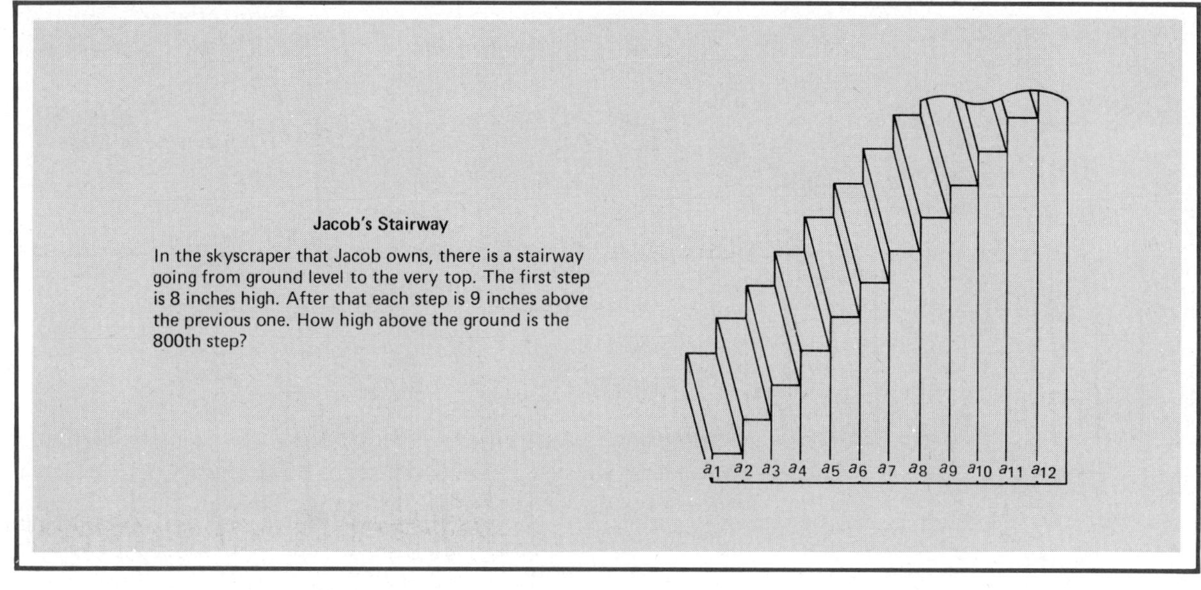

12-2 Arithmetic Sequences

We are going to answer the question above and others like it. Notice that if a_n denotes the height of the nth step, then

$$
\begin{aligned}
a_1 &= 8 \\
a_2 &= 8 + 1(9) = 17 \\
a_3 &= 8 + 2(9) = 26 \\
a_4 &= 8 + 3(9) = 35 \\
&\vdots \\
a_{800} &= 8 + 799(9) = 7199
\end{aligned}
$$

The 800th step of Jacob's stairway is 7199 inches (almost 600 feet) above the ground.

Formulas Now consider the following number sequences. When you see a pattern, fill in the boxes.

(a) 5, 9, 13, 17, □, □, . . .
(b) 2, 2.5, 3, 3.5, □, □, . . .
(c) 8, 5, 2, −1, □, □, . . .

What is it that these three sequences have in common? Simply this: In each case, you can get a term by adding a fixed number to the preceding term. In (a), you add 4 each time, in (b) you add 0.5, and in (c), you add −3. Such sequences are called **arithmetic sequences**. If we denote such a sequence by $a_1, a_2, a_3, \ldots,$ it satisfies the recursion formula

$$\boxed{a_n = a_{n-1} + d}$$

where d is a fixed number called the **common difference**.

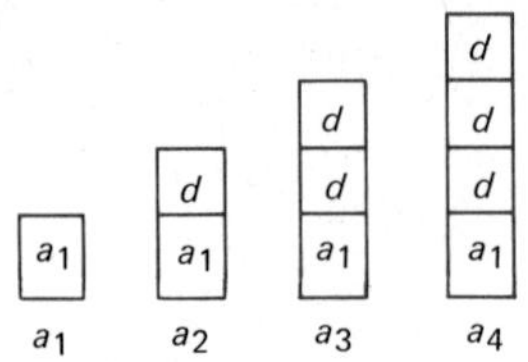

Can we also obtain an explicit formula? Yes. The diagram in the margin should help. Notice that the number of d's to be added to a_1 is one less than the subscript. This means that

$$a_n = a_1 + (n - 1)d$$

Now we can give explicit formulas for each of the sequences (a), (b), and (c) above.

$$\begin{aligned} a_n &= 5 + (n - 1)4 = 1 + 4n \\ b_n &= 2 + (n - 1)(.5) = 1.5 + .5n \\ c_n &= 8 + (n - 1)(-3) = 11 - 3n \end{aligned}$$

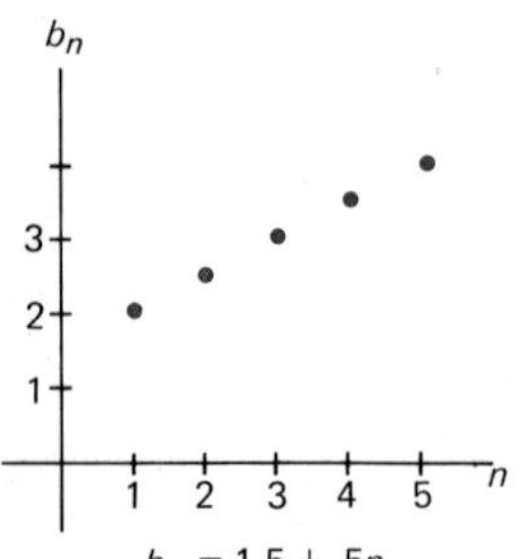

$b_n = 1.5 + .5n$

Arithmetic Sequences and Linear Functions We have said that a sequence is a function whose domain is the set of positive integers. Functions are best visualized by drawing their graphs. Consider the sequence b_n discussed above; its explicit formula is

$$b_n = 1.5 + .5n$$

Its graph is shown in the margin.

Even a cursory look at this graph suggests that the points lie along a straight line. Consider now the function

$$b(x) = 1.5 + 0.5x = 0.5x + 1.5$$

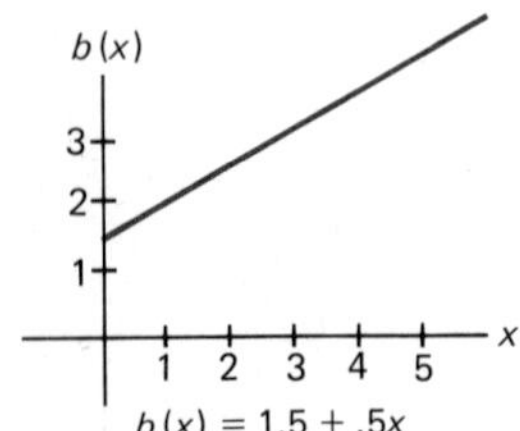

$b(x) = 1.5 + .5x$

where x is allowed to be any real number. This is a linear function, being of the form $mx + b$ (see Section 4-3). Its graph is a straight line (see the second graph in the margin), and its values at $x = 1, 2, 3, \ldots$ are equal to $b_1, b_2, b_3, \ldots$.

The relationship illustrated above between an arithmetic sequence and a linear function holds in general. An arithmetic sequence is just a linear function whose domain has been restricted to the positive integers.

Sums of Arithmetic Sequences There is an old story about Karl Gauss that aptly illustrates the next idea. We are not sure if the story is true, but if not, it should be.

When Gauss was about ten years old, he was admitted to an arithmetic class. To keep the class busy, the teacher often assigned long addition problems. One day he asked his students to add the numbers from 1 to 100. Hardly had he made the assignment when young Gauss laid his slate on the teacher's desk with the answer 5050 written on it.

Here is how Gauss probably thought about the problem.

$$1 + 2 + \ldots + 49 + 50 + 51 + 52 + \ldots + 99 + 100$$

Each of the indicated pairs has 101 as its sum and there are 50 such pairs. Thus the answer is $50(101) = 5050$. For a ten-year-old boy, that is good thinking.

Gauss's trick works perfectly well on any arithmetic sequence where we want to add an even number of terms. And there is a slight modification that works whether the number of terms to be added is even or odd.

Suppose $a_1, a_2, a_3, \ldots$ is an arithmetic sequence and let

$$A_n = a_1 + a_2 + a_3 + \cdots + a_{n-1} + a_n$$

Write this sum twice, once forwards and once backwards, and then add.

$$\begin{array}{rcccccccc} A_n = & a_1 & + & a_2 & + \cdots + & a_{n-1} & + & a_n \\ \underline{A_n} = & a_n & + & a_{n-1} & + \cdots + & a_2 & + & a_1 \\ \hline 2A_n = & \multicolumn{7}{l}{(a_1 + a_n) + (a_2 + a_{n-1}) + \cdots + (a_2 + a_{n-1}) + (a_1 + a_n)} \end{array}$$

Each group on the right has the same sum, namely, $a_1 + a_n$. For example,

$$a_2 + a_{n-1} = a_1 + d + a_n - d = a_1 + a_n$$

There are n such groups and so

$$2A_n = n(a_1 + a_n)$$

$$\boxed{A_n = \frac{n}{2}(a_1 + a_n)}$$

We call this the **sum formula** for an arithmetic sequence. You can remember this sum as being n times the average term $(a_1 + a_n)/2$.

Here is how we apply this formula to Gauss's problem. We want the sum of 100 terms of the sequence $1, 2, 3, \ldots$, that is, we want A_{100}. Here $n = 100$, $a_1 = 1$, and $a_n = 100$. Therefore

$$A_{100} = \tfrac{100}{2}(1 + 100) = 50(101) = 5050$$

For a second example, suppose we want to add the first 350 odd numbers, that is, the first 350 terms of the sequence $1, 3, 5, \ldots$. We can calculate the 350th odd number from the formula $a_n = a_1 + (n - 1)d$.

$$a_{350} = 1 + (349)2 = 699$$

Then we use the sum formula with $n = 350$.

$$\begin{aligned} A_{350} &= 1 + 3 + 5 + \cdots + 699 \\ &= \tfrac{350}{2}(1 + 699) = 122{,}500 \end{aligned}$$

Sigma Notation There is a convenient shorthand that is frequently employed in connection with sums. The first letter of the word *sum* is s; the Greek letter for S is Σ (sigma). We use $\sum$ in mathematics to stand for the operation of summation. In particular,

$$\sum_{i=1}^{n} a_i = a_1 + a_2 + a_3 + \cdots + a_n$$

The symbol $i = 1$ underneath the sigma tells where to start adding the terms a_i and the n at the top tells where to stop. Thus,

$$\sum_{i=1}^{4} a_i = a_1 + a_2 + a_3 + a_4$$

$$\sum_{i=3}^{7} b_i = b_3 + b_4 + b_5 + b_6 + b_7$$

$$\sum_{i=1}^{5} i^2 = 1^2 + 2^2 + 3^2 + 4^2 + 5^2$$

$$\sum_{i=1}^{30} 3i = 3 + 6 + 9 + \cdots + 90$$

If $a_1, a_2, a_3, \ldots$ is an *arithmetic sequence*, then the sum formula previously derived may be written

$$\sum_{i=1}^{n} a_i = \frac{n}{2}(a_1 + a_n)$$

Problem Set 12-2

1. Fill in the boxes below.
   (a) 1, 4, 7, 10, □, □, . . .
   (b) 2, 2.3, 2.6, 2.9, □, □, . . .
   (c) 28, 24, 20, 16, □, □, . . .
2. Fill in the boxes.
   (a) 4, 6, 8, 10, □, □, . . .
   (b) 4, 4.2, 4.4, 4.6, □, □, . . .
   (c) 4, 3.8, 3.6, 3.4, □, □, . . .
3. Determine d and the 30th term of each sequence in Problem 1.
4. Determine d and the 101st term of each sequence in Problem 2.
5. Determine $A_{30} = a_1 + a_2 + \cdots + a_{30}$ for the sequence in part (a) of Problem 1. Similarly, find B_{30} and C_{30} for the sequences in parts (b) and (c).
6. Determine A_{100}, B_{100}, and C_{100} for the sequences of Problem 2.
7. If $a_1 = 5$ and $a_{40} = 24.5$ in an arithmetic sequence, determine d.
8. If $b_1 = 6$ and $b_{30} = -52$ in an arithmetic sequence, determine d.

9. Calculate each sum.
   (a) $2 + 4 + 6 + \cdots + 200$
   (b) $1 + 3 + 5 + \cdots + 199$
   (c) $3 + 6 + 9 + \cdots + 198$
   (*Hint*: Before using the sum formula, you have to determine n. In part (a), n is 100 since we are adding the doubles of the integers from 1 to 100.)

10. Calculate each sum.
    (a) $4 + 8 + 12 + \cdots + 100$
    (b) $10 + 15 + 20 + \cdots + 200$
    (c) $6 + 9 + 12 + \cdots + 72$

11. The bottom rung of a tapered ladder is 30 centimeters long and the top rung is 15 centimeters long. If there are 17 rungs, how many centimeters of rung material are needed to make the ladder, assuming no waste?

12. A clock strikes once at 1:00, twice at 2:00, and so on. How many times does it strike between 10:30 A.M. on Monday and 10:30 P.M. on Tuesday?

13. If 3, a, b, c, d, 7, . . . is an arithmetic sequence, find a, b, c, and d.

14. If 8, a, b, c, 5 is an arithmetic sequence, find a, b, and c.

15. How many multiples of 9 are there between 200 and 300? Find their sum.

16. If Johnny is paid \$10 on January 1, \$20 on January 2, \$30 on January 3, and so on, how much does he earn during January?

17. Calculate each sum.

    (a) $\sum_{i=2}^{6} i^2$ (b) $\sum_{i=1}^{4} \frac{2}{i}$

    (c) $\sum_{i=1}^{100} (3i + 2)$ (d) $\sum_{i=2}^{100} (2i - 3)$

18. Calculate each sum.

    (a) $\sum_{i=1}^{6} 2^i$ (b) $\sum_{i=1}^{5} (i^2 - 2i)$

    (c) $\sum_{i=1}^{101} (2i - 6)$ (d) $\sum_{i=3}^{102} (3i + 5)$

19. Write in sigma notation.
    (a) $b_3 + b_4 + \cdots + b_{20}$
    (b) $1^2 + 2^2 + \cdots + 19^2$
    (c) $1 + \frac{1}{2} + \frac{1}{3} + \cdots + \frac{1}{n}$

20. Write in sigma notation.
    (a) $a_6 + a_7 + a_8 + \cdots + a_{70}$
    (b) $2^3 + 3^3 + 4^3 + \cdots + 100^3$
    (c) $1 + 3 + 5 + 7 + \cdots + 99$

Miscellaneous Problems

21. For the arithmetic sequence 5, 7.5, 10, 12.5, . . .
    (a) find d;
    (b) find the 51st term;
    (c) find the sum of the first 51 terms.
22. If $a_1 = 12$ and $a_{21} = 38$ in an arithmetic sequence,
    (a) find d;
    (b) find a_{56};
    (c) find m if $a_m = 61.4$.
23. Calculate the sum $4.25 + 4.5 + 4.75 + 5 + \cdots + 21.75$.
24. If 10, a, b, c, d, e, 30, . . . is an arithmetic sequence, find a, b, c, d, and e.
25. Find the sum of all multiples of 8 between 150 and 450.
26. Calculate each sum.

    (a) $\sum_{i=1}^{4} i^3$

    (b) $\sum_{i=1}^{100} \left(\frac{i}{i+1} - \frac{i-1}{i}\right)$

    (*Hint*: Write the first three or four terms of this sum without simplifying. You will see a pattern.)
27. Write in sigma notation.
    (a) $b_1 + b_2 + b_3 + \cdots + b_{112}$
    (b) $19 + 26 + 33 + \cdots + 719$
28. At a club meeting with 300 people present, everyone shook hands with every other person exactly once. How many handshakes were there? (*Hint*: Number the people from 1 to 300. Person 1 shakes hands with each of the other 299 people, person 2 shakes hands with 298 people (the handshake with person 1 should not be counted again), and so on.)
29. A pile of logs has 70 logs in the bottom layer, 69 in the second layer, 68 in the third layer, and so on. If there are 59 layers, how many logs are there in the pile?
30. Consider the sequence for which $a_n = n^2$, starting with $n = 0$.
    (a) If $b_n = a_n - a_{n-1}$, show that $b_1, b_2, b_3, \ldots$ is an arithmetic sequence.
    (b) Find b_{1001}.
    (c) Use your answer to part (b) to calculate $a_{1001} = 1001^2$.
31. Consider the sequence for which $a_n = n^3$, starting with $n = 0$.
    (a) If $b_n = a_n - a_{n-1}$, find b_1, b_2, b_3, b_4, and b_5.
    (b) Let $c_n = b_n - b_{n-1}$, find c_2, c_3, c_4, and c_5. The sequence $c_2, c_3, c_4, \ldots$ seems to be what kind of sequence?

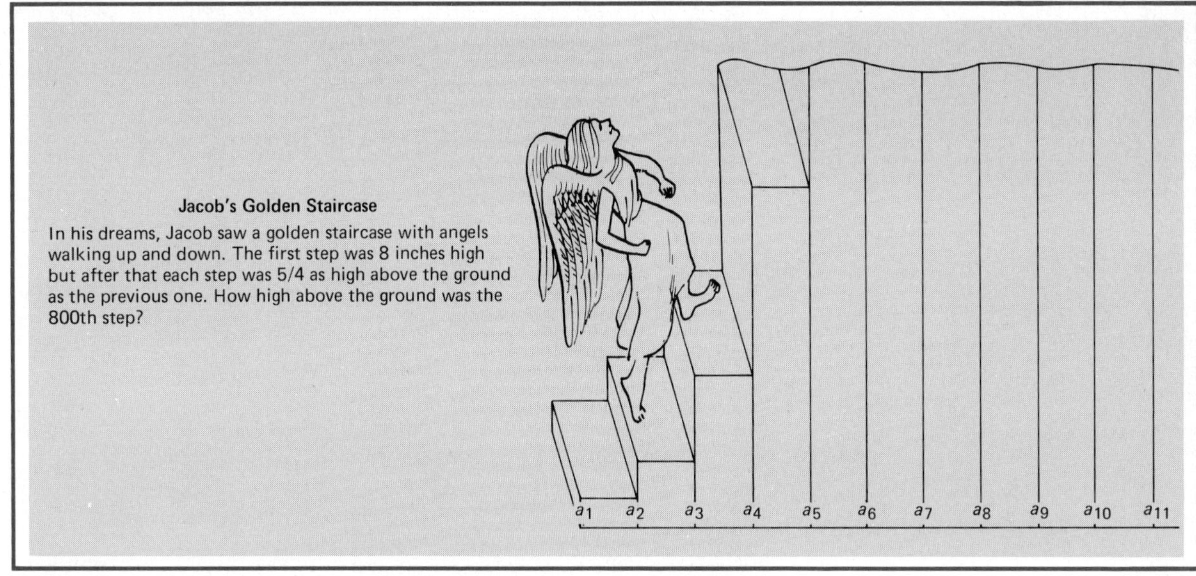

12-3 Geometric Sequences

The staircase of Jacob's dream is most certainly one for angels, not for people. The 800th step actually stands 3.4×10^{73} miles high. By way of comparison, it is 9.3×10^6 miles to the sun and 2.5×10^{13} miles to Alpha Centauri, our nearest star beyond the sun. You might say the golden staircase reaches to heaven.

To see how to calculate the height of the 800th step, notice the pattern of heights for the first few steps and then generalize.

$$\begin{aligned} a_1 &= 8 \\ a_2 &= 8(\tfrac{5}{4}) \\ a_3 &= 8(\tfrac{5}{4})^2 \\ a_4 &= 8(\tfrac{5}{4})^3 \\ &\vdots \\ a_{800} &= 8(\tfrac{5}{4})^{799} \end{aligned}$$

With a pocket calculator, it is easy to calculate $8(\frac{5}{4})^{799}$ and then change this number of inches to miles; the result is the figure given above.

Formulas In the sequence above, each term was $\frac{5}{4}$ times the preceding one. You should be able to find a similar pattern in each of the following sequences. When you do, fill in the boxes.

(a) 3, 6, 12, 24, □, □, . . .
(b) 12, 4, $\frac{4}{3}$, $\frac{4}{9}$, □, □, . . .
(c) .6, 6, 60, 600, □, □, . . .

The common feature of these three sequences is that in each case, you can get a term by multiplying the preceding term by a fixed number. In sequence (a), you multiply by 2; in (b), by $\frac{1}{3}$; and in (c), by 10. We call such sequences **geometric sequences**. Thus a geometric sequence $a_1, a_2, a_3, \ldots$ satisfies the recursion formula

$$a_n = ra_{n-1}$$

where r is a fixed number called the **common ratio**.

To obtain the corresponding explicit formula, note that

$$\begin{aligned} a_2 &= ra_1 \\ a_3 &= ra_2 = r(ra_1) = r^2a_1 \\ a_4 &= ra_3 = r(r^2a_1) = r^3a_1 \end{aligned}$$

In each case, the exponent on r is one less than the subscript on a. Thus,

$$a_n = a_1r^{n-1}$$

From this, we can get explicit formulas for each of the sequences (a), (b), and (c) on page 471.

$$\begin{aligned} a_n &= 3\cdot 2^{n-1} \\ b_n &= 12(\tfrac{1}{3})^{n-1} \\ c_n &= (.6)(10)^{n-1} \end{aligned}$$

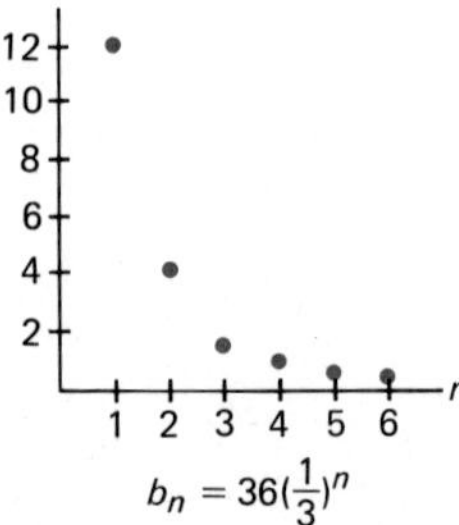

$b_n = 36(\frac{1}{3})^n$

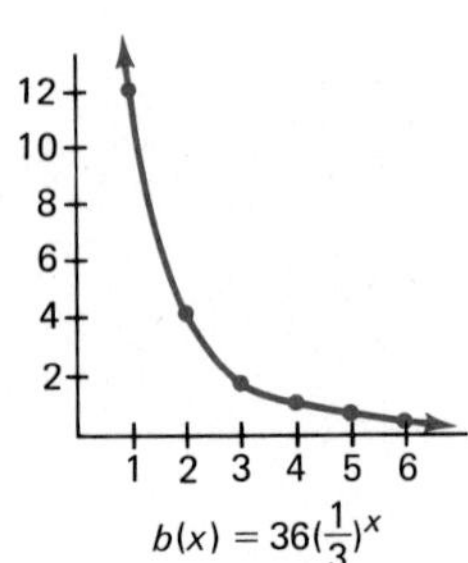

$b(x) = 36(\frac{1}{3})^x$

Geometric Sequences and Exponential Functions Let us consider sequence (b) once more; its explicit formula is

$$b_n = 12(\tfrac{1}{3})^{n-1} = 36(\tfrac{1}{3})^n$$

We have graphed this sequence and also the exponential function

$$b(x) = 36(\tfrac{1}{3})^x$$

in the margin. (See Section 6-2 for a discussion of exponential functions.) It should be clear that the sequence b_n is the function $b(x)$ with its domain restricted to the positive integers.

What we have observed in this example is true in general. A geometric sequence is simply an exponential function with its domain restricted to the positive integers.

Sums of Geometric Sequences There is an old legend about geometric sequences and chessboards. When the king of Persia learned to play chess, he was so enchanted with the game that he determined to reward the inventor, a man named Sessa. Calling Sessa to the palace, the king promised to fulfill any request he might make. With an air of modesty, wily Sessa asked for

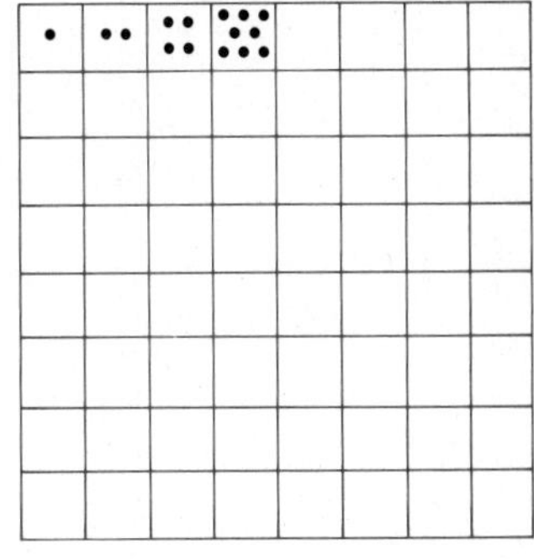

one grain of wheat for the first square of the chessboard, two for the second, four for the third, and so on. The king was amused at such an odd request; nevertheless, he called a servant, told him to get a bag of wheat, and start counting. To the king's surprise, it soon became apparent that Sessa's request could never be fulfilled. The world's total production of wheat for a whole century would not be sufficient.

Sessa was really asking for

$$1 + 2 + 2^2 + 2^3 + \cdots + 2^{63}$$

grains of wheat, the sum of the first 64 terms of the geometric sequence $1, 2, 4, 8, \ldots$. We are going to develop a formula for this sum and for all others that arise from adding the terms of a geometric sequence.

Let $a_1, a_2, a_3, \ldots$ be a geometric sequence with ratio $r \neq 1$. As usual, let

$$A_n = a_1 + a_2 + a_3 + \cdots + a_n$$

which can be written

$$A_n = a_1 + a_1r + a_1r^2 + \cdots + a_1r^{n-1}$$

Now multiply A_n by r, subtract the result from A_n, and use a little algebra to solve for A_n. We obtain

$$\begin{aligned} A_n &= a_1 + a_1r + a_1r^2 + \cdots + a_1r^{n-1} \\ rA_n &= \phantom{a_1 + {}} a_1r + a_1r^2 + \cdots + a_1r^{n-1} + a_1r^n \\ A_n - rA_n &= a_1 + 0 + 0 + \cdots + 0 - a_1r^n \\ A_n(1 - r) &= a_1(1 - r^n) \end{aligned}$$

$$\boxed{A_n = \frac{a_1(1 - r^n)}{1 - r} \qquad r \neq 1}$$

In sigma notation, this is

$$\sum_{i=1}^{n} a_i = \frac{a_1(1 - r^n)}{1 - r} \qquad r \neq 1$$

In the case where $r = 1$,

$$\sum_{i=1}^{n} a_i = a_1 + a_1 + \cdots + a_1 = na_1$$

Applying the sum formula to Sessa's problem (using $n = 64$, $a_1 = 1$, and $r = 2$), we get

$$A_{64} = \frac{1(1 - 2^{64})}{1 - 2} = 2^{64} - 1$$

Ignoring the -1 and using the approximation $2^{10} \approx 1000$ gives

$$A_{64} \approx 2^{64} \approx 2^4(1000)^6 = 1.6 \times 10^{19}$$

If you do this problem on a calculator, you will get $A_{64} \approx 1.845 \times 10^{19}$.

The Sum of the Whole Sequence Is it possible to add infinitely many numbers? Do the following sums make sense?

$$\tfrac{1}{2} + \tfrac{1}{4} + \tfrac{1}{8} + \tfrac{1}{16} + \cdots$$
$$1 + 3 + 9 + 27 + \cdots$$

Questions like this have intrigued great thinkers since Zeno first introduced his famous paradoxes of the infinite over 2000 years ago. We now show that we can make sense out of the first of these two sums but not the second.

Consider a string of length 1 kilometer. We may imagine cutting it into infinitely many pieces, as indicated in the figure below.

$\frac{1}{2}$ $\frac{1}{4}$ $\frac{1}{8}$ $\frac{1}{16}$ $\frac{1}{32}$

Since these pieces together make a string of length 1, it seems natural to say

$$\tfrac{1}{2} + \tfrac{1}{4} + \tfrac{1}{8} + \tfrac{1}{16} + \cdots = 1$$

Let us look at it another way. The sum of the first n terms of the geometric sequence $\frac{1}{2}, \frac{1}{4}, \frac{1}{8}, \frac{1}{16}, \ldots$ is given by

$$A_n = \frac{\frac{1}{2}[1 - (\frac{1}{2})^n]}{1 - \frac{1}{2}} = 1 - (\tfrac{1}{2})^n$$

As n gets larger and larger (goes to infinity), $(\frac{1}{2})^n$ gets smaller and smaller (approaches 0). Thus A_n goes to 1 as n goes to infinity. We therefore say that 1 is the sum of *all* the terms of this sequence.

Now consider any geometric sequence with ratio r satisfying $|r| < 1$. We claim that when n gets large, r^n approaches 0. (As evidence, try calculating $(0.99)^{100}$, $(0.99)^{1000}$, and $(0.99)^{10,000}$ on your calculator.) Thus, as n gets large,

$$A_n = \frac{a_1(1 - r^n)}{1 - r}$$

approaches $a_1/(1 - r)$. We write

$$\sum_{i=1}^{\infty} a_i = \frac{a_1}{1 - r} \qquad |r| < 1$$

For an important use of this formula in a familiar context, see the example in the problem set.

We emphasize that what we have just done is valid if $|r| < 1$. There is no way to make sense out of adding all the terms of a geometric sequence if $|r| \geq 1$.

Problem Set 12-3

1. Fill in the boxes.
   (a) $\frac{1}{2}$, 1, 2, 4, □, □, . . .
   (b) 8, 4, 2, 1, □, □, . . .
   (c) .3, 03, .003, .0003, □, □, . . .
2. Fill in the boxes.
   (a) 1, 3, 9, 27, □, □, . . .
   (b) 27, 9, 3, 1, □, □, . . .
   (c) .2, .02, .002, .0002, □, □, . . .
3. Determine r for each of the sequences in Problem 1 and write an explicit formula for the 30th term.
4. Write a formula for the 100th term of each sequence in Problem 2.
5. Use the sum formula to find the sum of the first five terms of each sequence in Problem 1.
6. Use the sum formula to find the sum of the first five terms of each sequence in Problem 2.
7. A certain culture of bacteria doubles every week. If there are 100 bacteria now, how many will there be after 10 full weeks?
8. A water lily grows so rapidly that each day it covers twice the area it covered the day before. At the end of 20 days, it completely covers a pond. If we start with two lilies, how long will it take to cover the same pond?
9. Johnny is paid \$1 on January 1, \$2 on January 2, \$4 on January 3, and so on. Approximately how much will he earn during January?
10. If you were offered 1¢ today, 2¢ tomorrow, 4¢ the third day, and so on for 20 days or a lump sum of \$10,000, which would you choose? Show why.
11. Calculate:

    (a) $\sum_{t=1}^{\infty} (\frac{1}{3})^t$ (b) $\sum_{t=2}^{\infty} (\frac{2}{5})^t$
12. Calculate:

    (a) $\sum_{t=1}^{\infty} (\frac{2}{3})^t$ (b) $\sum_{t=3}^{\infty} (\frac{1}{6})^t$
13. A ball is dropped from a height of 10 feet. At each bounce, it rises to a height of $\frac{1}{2}$ the previous height. How far will it travel altogether (up and down) by the time it comes to rest? (*Hint*: Think of the total distance as being the sum of the "down" distances $(10 + 5 + \frac{5}{2} + \cdots)$ and the "up" distances $(5 + \frac{5}{2} + \frac{5}{4} + \cdots)$.)
14. Do Problem 13 assuming the ball rises to $\frac{2}{3}$ its previous height at each bounce.

Example (Repeating decimals) Show that $.333\overline{3}\ldots$ and $.2323\overline{23}\ldots$ are rational numbers by using the methods of this section.

SOLUTION.

$$.333\overline{3} = \frac{3}{10} + \frac{3}{100} + \frac{3}{1000} + \cdots$$

Thus we must add all the terms of an infinite geometric sequence with ratio $\frac{1}{10}$. Using the formula $a_1/(1 - r)$, we get

$$\frac{\frac{3}{10}}{1 - \frac{1}{10}} = \frac{\frac{3}{10}}{\frac{9}{10}} = \frac{1}{3}$$

Similarly,

$$.2323\overline{23} = \frac{23}{100} + \frac{23}{10000} + \frac{23}{1000000} + \cdots$$

$$= \frac{\frac{23}{100}}{1 - \frac{1}{100}} = \frac{\frac{23}{100}}{\frac{99}{100}} = \frac{23}{99}$$

Use this method to express each of the following as the ratio of two integers.

15. $.11\overline{1}$
16. $77\overline{7}$
17. $.2525\overline{25}$
18. $.99\overline{9}$
19. $1.234\overline{34}$
20. $.341\overline{41}$

Miscellaneous Problems

21. If $a_n = 625(0.2)^{n-1}$, find a_1, a_2, a_3, a_4, and a_5.
22. Which of the following sequences are geometric, which are arithmetic, and which are neither?
    (a) 120, 24, 4.8, 0.96, . . .
    (b) .1, .02, .003, .0004, . . .
    (c) $4\pi, 2\pi, 0, -2\pi, \ldots$
    (d) 120, 60, −30, −15, 7.5, . . .
    (e) $100(1.08), 100(1.08)^2, 100(1.08)^3, \ldots$
    (f) 100(1.08), 100(1.10), 100(1.12), 100(1.14), . . .
23. Write an explicit formula for each geometric or arithmetic sequence in Problem 22.
24. Express $0.441\overline{441}$ as a ratio of positive integers in reduced form.
25. [C] Calculate each sum to three decimal places.
    (a) $\pi + \pi^2 + \pi^3 + \cdots + \pi^7$
    (b) $\sum_{k=1}^{\infty} \left(\frac{1}{\pi}\right)^k$
    (c) $1 + .982 + (.982)^2 + \cdots + (.982)^{99}$
    (d) $\sum_{k=0}^{\infty} \left(\frac{\sqrt{2}}{\sqrt{2} + 1}\right)^k$

[C] *In Problems 26–29, write a formula for the answer and then use a calculator to evaluate.*

26. If \$1 is put in the bank at 8 percent interest compounded annually, it will be worth $(1.08)^n$ dollars after n years. How much will \$100 be worth after 10 years?
27. If \$1 is put in the bank at 8 percent interest compounded quarterly, it will be worth $(1.02)^n$ dollars after n quarters. How much will \$100 be worth after 10 years (40 quarters)?
28. Suppose Karen puts \$100 in the bank today and \$100 at the beginning of each of the following 9 years. If this money earns interest at 8 percent compounded annually, what will it be worth at the end of 10 years?
29. José makes 40 deposits of \$25 each in a bank at intervals of three months, making the first deposit today. If money earns interest at 8 percent compounded quarterly, what will it all be worth at the end of 10 years (40 quarters)?
30. Suppose the government pumps an extra billion dollars into the economy. Assume that each business and individual saves 25 percent of its income and spends the rest, so that of the initial one billion dollars, 75 percent is re-spent by individuals and businesses. Of that amount, 75 percent is spent, and so forth. What is the total increase in spending due to the government action? (This is called the *multiplier effect* in economics.)
31. Try to explain why it does not make sense to talk about the total sum of a geometric sequence when $r \geq 1$ or $r \leq -1$.
32. Show that $.9\overline{9} = 1$.

The Principle of Mathematical Induction

Let $P_1, P_2, P_3, \ldots$ be a sequence of statements with the following two properties:

1. P_1 is true.
2. The truth of P_k implies the truth of P_{k+1} $(P_k \Rightarrow P_{k+1})$.

Then the statement P_n is true for every positive integer n.

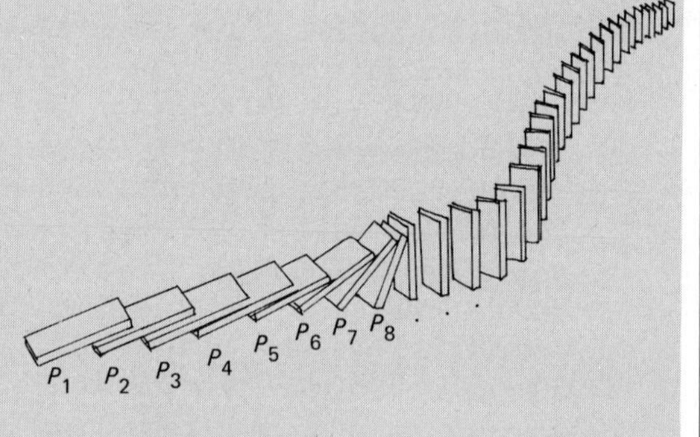

12-4 Mathematical Induction

The principle of mathematical induction deals with a sequence of statements. A **statement** is a sentence which is either true or false. In a sequence of statements, there is a statement corresponding to each positive integer. Here are four examples.

$$P_n: \quad \frac{1}{1\cdot 2} + \frac{1}{2\cdot 3} + \frac{1}{3\cdot 4} + \cdots + \frac{1}{n(n+1)} = \frac{n}{n+1}$$

$$Q_n: \quad n^2 - n + 41 \text{ is a prime number}$$

$$R_n: \quad (a+b)^n = a^n + b^n$$

$$S_n: \quad 1 + 2 + 3 + \cdots + n = \frac{n^2 + n - 6}{2}$$

To be sure we understand the notation, let us write each of these statements for the case $n = 3$.

$$P_3: \quad \frac{1}{1\cdot 2} + \frac{1}{2\cdot 3} + \frac{1}{3\cdot 4} = \frac{3}{4}$$

$$Q_3: \quad 3^2 - 3 + 41 \text{ is a prime number}$$

$$R_3: \quad (a+b)^3 = a^3 + b^3$$

$$S_3: \quad 1 + 2 + 3 = \frac{3^2 + 3 - 6}{2}$$

Of these, P_3 and Q_3 are true, while R_3 and S_3 are false; you should verify this fact. A careful study of these four sequences will indicate the wide range of behavior that sequences of statements can display.

While it certainly is not obvious, we claim that P_n is true for every positive integer n; we are going to prove it soon. Q_n is a well-known sequence. It was thought by some to be true for all n and, in fact, it is true for $n = 1, 2, 3, \ldots, 40$ (see the accompanying table). However, it fails for $n = 41$, a fact that allows us to make an important point. Establishing the truth of Q_n for a finite number of cases, no matter how many, does not prove its truth for *all* n. Sequences R_n and S_n are rather hopeless cases, since R_n is true only for $n = 1$ and S_n is never true.

n	$n^2 - n + 41$
1	41
2	43
3	47
4	53
5	61
6	71
7	83
.	.
.	.
.	.
40	1601
41	$1681 = 41^2$

Proof by Mathematical Induction How does one prove that something is true for all n? The tool uniquely designed for this purpose is the **principle of mathematical induction**; it was stated in our opening display. Let us use mathematical induction to show that

$$P_n: \quad \frac{1}{1\cdot 2} + \frac{1}{2\cdot 3} + \frac{1}{3\cdot 4} + \cdots + \frac{1}{(n-1)n} + \frac{1}{n(n+1)} = \frac{n}{n+1}$$

is true for every positive integer n. There are two steps to the proof. We must show that

1. P_1 is true;
2. $P_k \Rightarrow P_{k+1}$; that is, the truth of P_k implies the truth of P_{k+1}.

The first step is easy. P_1 is just the statement

$$\frac{1}{1\cdot 2} = \frac{1}{1+1}$$

which is clearly true.

To handle the second step ($P_k \Rightarrow P_{k+1}$), it is a good idea to write down the statements corresponding to P_k and P_{k+1} (at least on scratch paper). We get them by substituting k and $k + 1$ for n in the statement for P_n.

$$P_k: \quad \frac{1}{1\cdot 2} + \frac{1}{2\cdot 3} + \cdots + \frac{1}{(k-1)k} + \frac{1}{k(k+1)} = \frac{k}{k+1}$$

$$P_{k+1}: \quad \frac{1}{1\cdot 2} + \frac{1}{2\cdot 3} + \cdots + \frac{1}{k(k+1)} + \frac{1}{(k+1)(k+2)} = \frac{k+1}{k+2}$$

Notice that the left side of P_{k+1} is the same as that of P_k except for the addition of one more term, $1/(k + 1)(k + 2)$.

Suppose for the moment that P_k is true and consider how this assumption allows us to simplify the left side of P_{k+1}.

$$\left[\frac{1}{1\cdot 2} + \frac{1}{2\cdot 3} + \cdots + \frac{1}{k(k+1)}\right] + \frac{1}{(k+1)(k+2)} = \frac{k}{k+1} + \frac{1}{(k+1)(k+2)}$$

$$= \frac{k(k+2)+1}{(k+1)(k+2)}$$

$$= \frac{(k+1)^2}{(k+1)(k+2)}$$

$$= \frac{k+1}{k+2}$$

If you read this chain of equalities from top to bottom, you will see that we have established the truth of P_{k+1}, but under the *assumption that* P_k *is true.* That is, we have established that the truth of P_k implies the truth of P_{k+1}.

Some Comments About Mathematical Induction Students never have any trouble with the verification step (showing that P_1 is true). The inductive step (showing that $P_k \Rightarrow P_{k+1}$) is harder and more subtle. In that step, we do *not* prove that P_k or P_{k+1} is true, but rather that the truth of P_k implies the truth of P_{k+1}. For a vivid illustration of the difference, we point out that in the fourth example of our opening paragraph, the truth of S_k does imply the truth of S_{k+1} ($S_k \Rightarrow S_{k+1}$) and yet not a single statement in that sequence is true (see Problems 31 and 32). To put it another way, what $S_k \Rightarrow S_{k+1}$ means is that *if* S_k were true, then S_{k+1} would be true also. It is like saying that if spinach were ice cream, then kids would want two helpings at every meal.

Perhaps the dominoes in the opening display can help illuminate the idea. For all the dominoes to fall it is sufficient that

1. The first domino is pushed over;
2. If any domino falls (say the kth one), it pushes over the next one (the $(k + 1)$st one).

The diagram on page 481 illustrates what happens to the dominoes in the four examples of our opening paragraph. Study them carefully.

Another Example Consider the statement

$$P_n:\quad 1^2 + 2^2 + 3^2 + \cdots + n^2 = \frac{n(n+1)(2n+1)}{6}$$

We are going to prove that P_n is true for all n by mathematical induction. For $n = 1$, k, and $k + 1$, the statements P_n are:

$$P_1:\quad 1^2 = \frac{1(2)(3)}{6}$$

$$P_k:\quad 1^2 + 2^2 + 3^2 + \cdots + k^2 = \frac{k(k + 1)(2k + 1)}{6}$$

$$P_{k+1}:\quad 1^2 + 2^2 + 3^2 + \cdots + k^2 + (k + 1)^2 = \frac{(k + 1)(k + 2)(2k + 3)}{6}$$

Clearly P_1 is true.

Assuming that P_k is true, we can write the left side of P_{k+1} as shown at the top of page 482.

Why They Fall and Why They Don't

$P_n : \frac{1}{1 \cdot 2} + \frac{1}{2 \cdot 3} + \ldots + \frac{1}{n(n + 1)} = \frac{n}{n + 1}$ P_1 is true $P_k \Rightarrow P_k + 1$	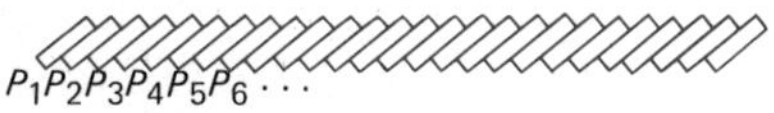 $P_1 P_2 P_3 P_4 P_5 P_6 \cdots$ First domino is pushed over. Each falling domino pushes over the next one.
$Q_n : n^2 - n + 41$ is prime. $Q_1, Q_2, \ldots, Q_{40}$ are true. $Q_k \not\Rightarrow Q_k + 1$	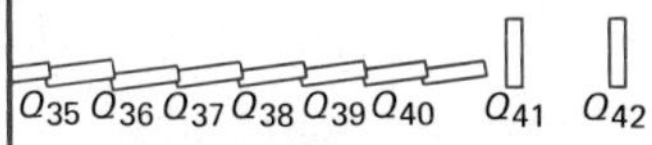 $Q_{35}\ Q_{36}\ Q_{37}\ Q_{38}\ Q_{39}\ Q_{40}\quad Q_{41}\quad Q_{42}$ First 40 dominoes are pushed over. 41st domino remains standing.
$R_n : (a + b)^n = a^n + b^n$ R_1 is true. $R_k \not\Rightarrow R_k + 1$	 $R_1\quad R_2\quad R_3\quad R_4\quad R_5\quad R_6\quad R_7$ First domino is pushed over but dominoes are spaced too far apart to push each other over.
$S_n : 1 + 2 + 3 + \ldots + n = \frac{n^2 + n - 6}{2}$ S_1 is false. $S_k \Rightarrow S_k + 1$	 $S_1 S_2 S_3 S_4 S_5$ Spacing is just right but no one can push over the first domino.

$$1^2 + 2^2 + 3^2 + \cdots + k^2 + (k+1)^2 = \frac{k(k+1)(2k+1)}{6} + (k+1)^2$$

$$= \frac{(k+1)[k(2k+1) + 6(k+1)]}{6}$$

$$= \frac{(k+1)(2k^2 + 7k + 6)}{6}$$

$$= \frac{(k+1)(k+2)(2k+3)}{6}$$

Thus the truth of P_k does imply the truth of P_{k+1}. We conclude by mathematical induction that P_n is true for every positive integer n. Incidentally, the result just proved will be used in calculus.

Problem Set 12-4

In Problems 1–8, prove by mathematical induction that P_n is true for every positive integer n.

1. P_n: $1 + 2 + 3 + \cdots + n = \dfrac{n(n+1)}{2}$
2. P_n: $1 + 3 + 5 + \cdots + (2n-1) = n^2$
3. P_n: $3 + 7 + 11 + \cdots + (4n-1) = n(2n+1)$
4. P_n: $2 + 9 + 16 + \cdots + (7n-5) = \dfrac{n(7n-3)}{2}$
5. P_n: $1\cdot 2 + 2\cdot 3 + 3\cdot 4 + \cdots + n(n+1) = \frac{1}{3}n(n+1)(n+2)$
6. P_n: $\dfrac{1}{1\cdot 3} + \dfrac{1}{3\cdot 5} + \dfrac{1}{5\cdot 7} + \cdots + \dfrac{1}{(2n-1)(2n+1)} = \dfrac{n}{2n+1}$
7. P_n: $2 + 2^2 + 2^3 + \cdots + 2^n = 2(2^n - 1)$
8. P_n: $1^2 + 3^2 + 5^2 + \cdots + (2n-1)^2 = \dfrac{n(2n-1)(2n+1)}{3}$

In Problems 9–18, tell what you can conclude from the information given about the sequence of statements. For example, if you are given that P_4 is true and that $P_k \Rightarrow P_{k+1}$ for any k, then you can conclude that P_n is true for every integer $n \geq 4$.

9. P_8 is true and $P_k \Rightarrow P_{k+1}$.
10. P_8 is not true and $P_k \Rightarrow P_{k+1}$.
11. P_1 is true but P_k does not imply P_{k+1}.
12. $P_1, P_2, \ldots, P_{1000}$ are all true.
13. P_1 is true and $P_k \Rightarrow P_{k+2}$.
14. P_{40} is true and $P_k \Rightarrow P_{k-1}$.
15. P_1 and P_2 are true; P_k and P_{k+1} together imply P_{k+2}.
16. P_1 and P_2 are true and $P_k \Rightarrow P_{k+2}$.
17. P_1 is true and $P_k \Rightarrow P_{4k}$.
18. P_1 is true, $P_k \Rightarrow P_{4k}$, and $P_k \Rightarrow P_{k-1}$.

Example A (Mathematical induction applied to inequalities) Show that the following statement is true for every integer $n \geq 4$.

$$3^n > 2^n + 20$$

SOLUTION. Let P_n represent the given statement. You might check that P_1, P_2, and P_3 are false. However, that does not matter to us. What we need to do is to show that P_4 is true and that $P_k \Rightarrow P_{k+1}$ for any $k \geq 4$.

$$\begin{aligned} P_4&: \quad 3^4 > 2^4 + 20 \\ P_k&: \quad 3^k > 2^k + 20 \\ P_{k+1}&: \quad 3^{k+1} > 2^{k+1} + 20 \end{aligned}$$

Clearly P_4 is true (81 is greater than 36). Next we assume P_k to be true (for $k \geq 4$) and seek to show that this would force P_{k+1} to be true. Working with the left side of P_{k+1} and using the assumption that $3^k > 2^k + 20$, we get

$$3^{k+1} = 3 \cdot 3^k > 3(2^k + 20) > 2(2^k + 20) = 2^{k+1} + 40 > 2^{k+1} + 20$$

Therefore P_{k+1} is true, provided P_k is true. We conclude that P_n is true for every integer $n \geq 4$.

In Problems 19–24, find the smallest positive integer n for which the given statement is true. Then prove that the statement is true for all integers greater than that smallest value.

19. $n + 5 < 2^n$
20. $3n \leq 3^n$
21. $\log_{10} n < n$ (*Hint*: $k + 1 < 10k$)
22. $n^2 \leq 2^n$ (*Hint*: $k^2 + 2k + 1 = k(k + 2 + 1/k) < k(k + k)$)
23. $(1 + x)^n \geq 1 + nx$, where $x \geq -1$
24. $|\sin nx| \leq |\sin x| \cdot n$ for all x

Example B (Mathematical induction and divisibility) Prove that the statement

$$P_n: \quad x - y \text{ is a factor of } x^n - y^n$$

is true for every positive integer n.

SOLUTION. Trivially, $x - y$ is a factor of $x - y$ since $x - y = 1(x - y)$; so P_1 is true. Now suppose that P_k is true, that is, that

$$x - y \text{ is a factor of } x^k - y^k$$

This means that there is a polynomial $Q(x, y)$ such that

$$x^k - y^k = Q(x, y)(x - y)$$

Using this assumption, we may write

$$\begin{aligned} x^{k+1} - y^{k+1} &= x^{k+1} - x^k y + x^k y - y^{k+1} \\ &= x^k(x - y) + y(x^k - y^k) \\ &= x^k(x - y) + yQ(x, y)(x - y) \\ &= [x^k + yQ(x, y)](x - y) \end{aligned}$$

Thus $x - y$ is a factor of $x^{k+1} - y^{k+1}$. We have shown that $P_k \Rightarrow P_{k+1}$ and that P_1 is true; we therefore conclude that P_n is true for all n.

Use mathematical induction to prove that each of the following is true for every positive integer n.

25. $x + y$ is a factor of $x^{2n} - y^{2n}$. (*Hint*: $x^{2k+2} - y^{2k+2} = x^{2k+2} - x^{2k}y^2 + x^{2k}y^2 - y^{2k+2}$.)
26. $x + y$ is a factor of $x^{2n-1} + y^{2n-1}$.
27. $n^2 - n$ is even (that is, has 2 as a factor).
28. $n^3 - n$ is divisible by 6.

Miscellaneous Problems

In Problems 29–30, use mathematical induction to prove that P_n is true for every positive integer n.

29. P_n: $5 + 15 + 25 + \cdots + (10n - 5) = 5n^2$
30. P_n: $(2n + 1)^2 - 1$ is divisible by 8.
31. Consider the statement

$$S_n: \quad 1 + 2 + 3 + \cdots + n = \frac{n^2 + n - 6}{2}$$

Show that $S_k \Rightarrow S_{k+1}$. Does this mean that S_n is true for any n? (See Problem 32.)

32. You know (Problem 1 or Section 12-2) that

$$1 + 2 + 3 + \cdots + n = \frac{n(n + 1)}{2}$$

Show that this means that S_n of Problem 31 is not true for any positive integer.

33. Use mathematical induction to prove

$$1^3 + 2^3 + 3^3 + \cdots + n^3 = \left[\frac{n(n + 1)}{2}\right]^2$$

34. Note that Problem 33 and the formula of Problem 32 imply that

$$1^3 + 2^3 + 3^3 + \cdots + n^3 = (1 + 2 + 3 + \cdots + n)^2$$

Check to see that this strange formula holds for $n = 1, 2, 3$, and 4.

35. Prove by mathematical induction that the sum formulas for arithmetic and geometric sequences are correct, that is, prove each of the following.

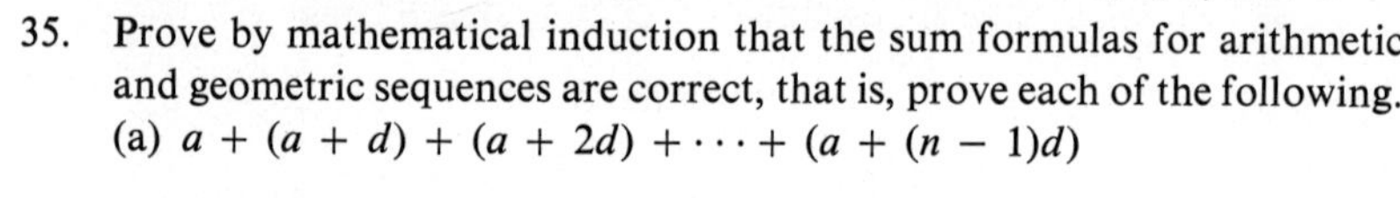

(a) $a + (a + d) + (a + 2d) + \cdots + (a + (n - 1)d)$

$$= \frac{n}{2}[2a + (n - 1)d]$$

$n = 4$

(b) $a + ar + ar^2 + \cdots + ar^{n-1} = \dfrac{a(1 - r^n)}{1 - r} \qquad (r \neq 1)$

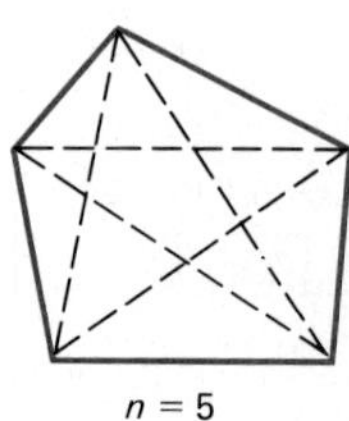

$n = 5$

36. Prove that the number of diagonals of an n-sided convex (that is, no holes or dents) polygon is $n(n - 3)/2$ for $n \geq 3$. The diagrams in the margin show the situation for $n = 4$ and $n = 5$.
37. Prove that the sum of the measures of the angles in an n-sided convex polygon is $(n - 2)180°$ for $n \geq 3$.

38. Prove that, for $n \geq 3$,

$$\frac{1}{n+1} + \frac{1}{n+2} + \cdots + \frac{1}{2n} > \frac{3}{5}$$

39. Prove that, for $n \geq 2$,

$$\left(1 - \frac{1}{4}\right)\left(1 - \frac{1}{9}\right)\left(1 - \frac{1}{16}\right)\cdots\left(1 - \frac{1}{n^2}\right) = \frac{n+1}{2n}$$

40. Let $f_1 = 1, f_2 = 1$, and $f_n = f_{n-1} + f_{n-2}$ for $n \geq 2$ (this is the Fibonacci sequence of Section 12-1). Show by mathematical induction that

$$f_n = \frac{1}{\sqrt{5}}\left[\left(\frac{1+\sqrt{5}}{2}\right)^n - \left(\frac{1-\sqrt{5}}{2}\right)^n\right]$$

(*Hint*: In the inductive step, show that P_k and P_{k+1} together imply P_{k+2}.)

41. Let $F_n = f_1 + f_2 + \cdots + f_n$, where f_n is as in Problem 40. Show by mathematical induction that $F_n = f_{n+2} - 1$.

42. Let $a_0 = 0$, $a_1 = 1$, and $a_{n+2} = 2a_{n+1} - a_n$ for $n \geq 0$. Prove that $a_n = n$ for $n \geq 0$. (See the hint of Problem 40.)

43. Let $a_0 = 0$, $a_1 = 1$, and $a_{n+2} = (a_{n+1} + a_n)/2$ for $n \geq 0$. Prove that

$$a_n = \tfrac{2}{3}[1 - (-\tfrac{1}{2})^n]$$

for $n \geq 0$.

44. What is wrong with the following argument?
Theorem. All horses in the world have the same color.
Proof. Let P_n be the statement: All the horses in any set of n horses are identically colored. Certainly P_1 is true. Suppose that P_k is true, that is, that all the horses in any set of k horses are identically colored. Let W be any set of $k + 1$ horses. Now we may think of W as the union of two overlapping sets X and Y, each with k horses. (The situation for $k = 4$ is shown in the margin.) By assumption, the horses in X are identically colored and the horses in Y are identically colored. Since X and Y overlap, all the horses in $X \cup Y$ must be identically colored. We conclude that P_n is true for all n. Thus the set of all horses in the world (some finite number) have the same color.

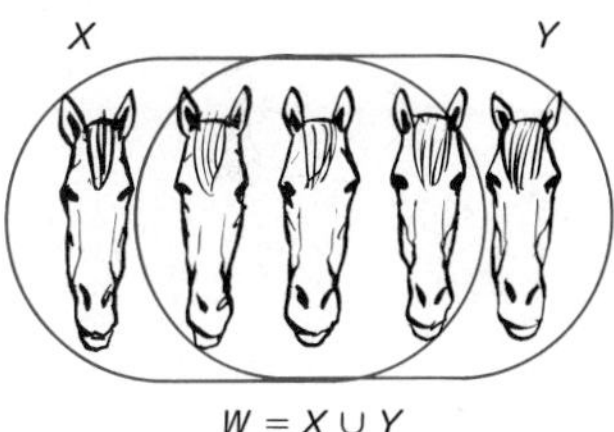

$W = X \cup Y$

12-5 Counting Ordered Arrangements

The Senior Birdwatchers' Club, consisting of 4 women and 2 men, is about to hold its annual meeting. In addition to having a group picture taken, they plan to elect a president, a vice president, and a secretary. Here are some questions that they (and we) might consider.

1. In how many ways can they line up for their group picture?
2. In how many ways can they elect their three officers if there are no restrictions as to sex?
3. In how many ways can they elect their three officers if the president is required to be female and the vice president male?
4. In how many ways can they elect their three officers if the president is to be of one sex and the vice president and secretary of the other?

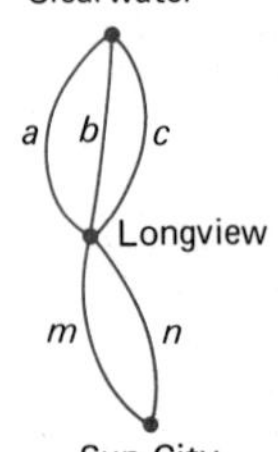

Possible Routes

am	*an*
bm	*bn*
cm	*cn*

In order to answer these questions and others of a similar nature, we need two counting principles. One involves multiplication and the other involves addition.

Multiplication Principle in Counting Suppose that there are three roads a, b, and c leading from Clearwater to Longview and two roads m and n from Longview to Sun City. How many different routes can you choose from Clearwater to Sun City going through Longview? The diagram in the margin clarifies the situation. For each of the 3 choices from Clearwater to Longview, you have 2 choices from Longview to Sun City. Thus, you have $3 \cdot 2$ routes from Clearwater to Sun City. Here is the general principle.

MULTIPLICATION PRINCIPLE

Suppose that an event H can occur in h ways and, after it has occurred, event K can occur in k ways. Then the number of ways in which both H ***and*** *K can occur is hk.*

This principle extends in an obvious way to three or more events.

Consider now the Birdwatchers' third question. It involves three events.

P: Elect a female president.
V: Elect a male president.
S: Elect a secretary of either sex.

We understand that the election will take place in the order indicated and that no person can fill more than one position. Thus

P can occur in 4 ways (there are 4 women).
V can occur in 2 ways (there are 2 men).
S can occur in 4 ways (after P and V occur, there are 4 people left from whom to select a secretary).

The entire selection process can be accomplished in $4 \cdot 2 \cdot 4 = 32$ ways.

Permutations To permute a set of objects is to rearrange them. Thus a **permutation** of a set of objects is an ordered arrangement of those objects. Take the set of letters in the word *FACTOR* as an example. Imagine that these six letters are printed on small cards so they can be arranged at will. Then we may form words like *COTARF*, *TRAFOC*, and *FRACTO*, none of which are in a dictionary but all of which are perfectly good words from our point of view. Let us call them code words. How many six-letter code words can be made from the letters of the word *FACTOR*; that is, how many permutations of 6 objects are there?

Permutations of *ART*

ART
ATR
RAT
RTA
TAR
TRA

Think of this as the problem of filling six slots.

☐ ☐ ☐ ☐ ☐ ☐

We may fill the first slot in 6 ways. Having done that, we may fill the second slot in 5 ways, the third in 4 ways, and so on. By the multiplication principle, we can fill all six slots in

$$6 \cdot 5 \cdot 4 \cdot 3 \cdot 2 \cdot 1 = 720$$

ways.

Do you see that this is also the answer to the first question about the Birdwatchers, which asked in how many ways the six members could be arranged for a group picture? Let us identify each person by a letter; the letters of *FACTOR* will do just fine. Then to arrange the Birdwatchers is to make a six-letter code word out of *FACTOR*. It can be done in 720 ways.

What if we want to make three-letter code words from the letters of the word *FACTOR*, words like *ACT*, *COF*, and *TAC*? How many such words can be made? This is the problem of filling three slots with six letters available. We can fill the first slot in 6 ways, then the second in 5 ways, and then the third in 4 ways. Therefore we can make $6 \cdot 5 \cdot 4 = 120$ three-letter code words from the word *FACTOR*.

Pres		VP		Sec
6	·	5	·	4

The number 120 is also the answer to the second question about the Birdwatchers. If there are no restrictions as to sex, they can elect a president, vice president, and secretary in $6 \cdot 5 \cdot 4 = 120$ ways.

Consider the corresponding general problem. Suppose that from n distinguishable objects, we select r of them and arrange them in a row. The resulting arrangement is called a **permutation of n things taken r at a time**. The number of such permutations is denoted by the symbol ${}_nP_r$. Thus

$$
\begin{aligned}
{}_6P_3 &= 6 \cdot 5 \cdot 4 = 120 \\
{}_6P_6 &= 6 \cdot 5 \cdot 4 \cdot 3 \cdot 2 \cdot 1 = 720 \\
{}_8P_2 &= 8 \cdot 7 = 56
\end{aligned}
$$

and in general

$$
{}_nP_r = n(n-1)(n-2)\cdots(n-r+2)(n-r+1)
$$

Notice that ${}_nP_r$ is the product of r consecutive positive integers starting with n and going down. In particular, ${}_nP_n$ is the product of n positive integers starting with n and going all the way down to 1, that is,

$$
{}_nP_n = n(n-1)(n-2)\cdots 3 \cdot 2 \cdot 1
$$

The symbol $n!$ (read n **factorial**) is also used for this product. Thus

$$
\begin{aligned}
{}_5P_5 &= 5! = 5 \cdot 4 \cdot 3 \cdot 2 \cdot 1 = 120 \\
{}_4P_4 &= 4! = 4 \cdot 3 \cdot 2 \cdot 1 = 24
\end{aligned}
$$

Addition Principle in Counting We still have not answered the fourth Birdwatchers' question. In how many ways can they elect their three officers if the president is to be of one sex and the vice president and secretary of the other? This means that the president should be female and the other two officers male, *or* the president should be male and the other two female. To answer a question like this we need another principle.

ADDITION PRINCIPLE

Let H and K be disjoint events, that is, events that cannot happen simultaneously. If H can occur in h ways and K in k ways, then H **or** *K can occur in $h + k$ ways.*

This principle generalizes to three or more disjoint events.

Applying this principle to the question at hand, we define H and K as follows.

H: Elect a female president, male vice president, and male secretary.
K: Elect a male president, female vice president, and female secretary.

Clearly H and K are disjoint. From the multiplication principle,

H can occur in $4 \cdot 2 \cdot 1 = 8$ ways;
K can occur in $2 \cdot 4 \cdot 3 = 24$ ways.

Then by the addition principle, H or K can occur in $8 + 24 = 32$ ways.

Here is another question that requires the addition principle. Consider again the letters of *FACTOR*, which we supposed were printed on six cards. How many code words of any length can we make using these six letters? We immediately translate this into six disjoint events: make six-letter words, or five-letter words, or four-letter words, or three-letter words, or two-letter words, or one-letter words. We can do this in the following number of ways.

$$\begin{aligned}
&{}_6P_6 + {}_6P_5 + {}_6P_4 + {}_6P_3 + {}_6P_2 + {}_6P_1 \\
&= 6 \cdot 5 \cdot 4 \cdot 3 \cdot 2 \cdot 1 + 6 \cdot 5 \cdot 4 \cdot 3 \cdot 2 + 6 \cdot 5 \cdot 4 \cdot 3 + 6 \cdot 5 \cdot 4 + 6 \cdot 5 + 6 \\
&= 720 + 720 + 360 + 120 + 30 + 6 \\
&= 1956
\end{aligned}$$

Students sometimes find it hard to decide whether to multiply or to add in a counting problem. Notice that the words **and** and **or** are in boldface type in the statements of the multiplication principle and of the addition principle. They are the key words: **and** goes with multiplication; **or** goes with addition.

Problem Set 12-5

1. Calculate.
   (a) $3!$ (b) $(3!)(2!)$ (c) $10!/8!$
2. Calculate.
   (a) $7!$ (b) $7! + 5!$ (c) $12!/9!$
3. Calculate.
   (a) ${}_5P_2$ (b) ${}_9P_4$ (c) ${}_{10}P_3$
4. Calculate.
   (a) ${}_4P_3$ (b) ${}_8P_4$ (c) ${}_{20}P_3$
5. In how many ways can a president and a secretary be chosen from a group of 6 people?
6. Suppose that a club consists of 3 women and 2 men. In how many ways can a president and a secretary be chosen
   (a) if the president is to be female and the secretary male;
   (b) if the president is to be male and the secretary female;
   (c) if the president and secretary are to be of opposite sex?

7. A box contains 12 cards numbered 1 through 12. Suppose one card is drawn from the box. Find the number of ways each of the following can occur.
   (a) The number drawn is even.
   (b) The number is greater than 9 or less than 3.
8. Suppose that two cards are drawn in succession from the box in Problem 7. Assume that the first card is not replaced before the second one is drawn. In how many ways can each of the following occur?
   (a) Both numbers are even.
   (b) The two numbers are both even or both odd?
   (c) The first number is greater than 9 and the second one less than 3.
9. Do Problem 8 with the assumption that the first card is replaced before the second one is drawn.
10. In how many ways can a president, a vice president, and a secretary be chosen from a group of 10 people?
11. How many four-letter code words can be made from the letters of the word *EQUATION*? (Letters are not to be repeated.)
12. How many three-letter code words can be made from the letters of the word *PROBLEM* if
   (a) Letters cannot be repeated;
   (b) Letters can be repeated?
13. Five roads connect Cheer City and Glumville. Starting at Cheer City, how many different ways can Smith drive to Glumville and return, that is, how many different round trips can he make? How many different round trips can he make if he wishes to return by a different road than he took to Glumville?
14. Filipe has 4 ties, 6 shirts, and 3 pairs of trousers. How many different outfits can he wear? Assume that he wears one of each kind of article.
15. Papa's Pizza Place offers 3 choices of salad, 20 kinds of pizza, and 4 different desserts. How many different three-course meals can one order?
16. Minnesota license plate numbers consist of 3 letters followed by 3 digits (for example, AFF033). How many different plates could be issued? (You need not multiply out your answer.)
17. The letters of the word *CREAM* are printed on 5 cards. How many three-, four-, or five-letter code words can be formed?

Example A (Arrangements with side conditions) Suppose that the letters of the word *COMPLEX* are printed on 7 cards. How many three-letter code words can be formed from these letters if
(a) The first and last letters must be consonants (that is, C, M, P, L, or X);
(b) All vowels used (if any) must occur in the right-hand portion of a word (that is, a vowel cannot be followed by a consonant)?

SOLUTION. (a) Let c denote consonant, v vowel, and a any letter. We must fill the three slots below.

$$\boxed{}\ \boxed{}\ \boxed{}$$
$$c \quad a \quad c$$

We begin by filling the two restricted slots, which can be done in $5 \cdot 4 = 20$ ways. Then we fill the unrestricted slot using one of the 5 remaining letters. It can be done in 5 ways. There are $20 \cdot 5 = 100$ code words of the required type. The diagram below summarizes the procedure.

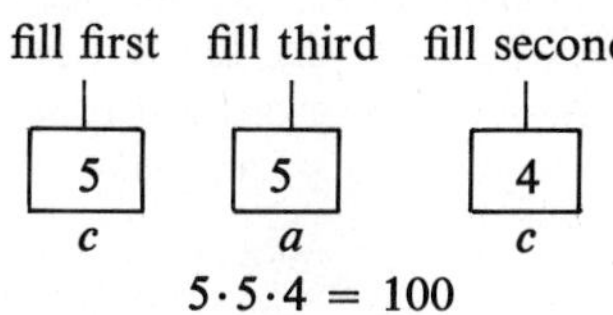

$$5 \cdot 5 \cdot 4 = 100$$

(b) We want to count words of the form *cvv*, *ccv*, or *ccc*. Note the use of the addition principle (as well as the multiplication principle) in the following solution.

5	2	1	or	5	4	2	or	5	4	3
c	*v*	*v*		*c*	*c*	*v*		*c*	*c*	*c*

$$\begin{aligned} & 5 \cdot 2 \cdot 1 + 5 \cdot 4 \cdot 2 + 5 \cdot 4 \cdot 3 \\ &= 10 + 40 + 60 \\ &= 110 \end{aligned}$$

18. Using the letters of the word *FACTOR* (without repetition), how many four-letter code words can be formed
    (a) Starting with *R*;
    (b) With vowels in the two middle positions;
    (c) With only consonants;
    (d) With vowels and consonants alternating;
    (e) With all the vowels (if any) in the left-hand portion of a word (that is, a vowel cannot be preceded by a consonant)?

19. Using the letters of the word *EQUATION* (without repetition), how many four-letter code words can be formed
    (a) Starting with *T* and ending with *N*;
    (b) Starting and ending with a consonant;
    (c) With vowels only;
    (d) With three consonants;
    (e) With all the vowels (if any) in the right-hand portion of the word?

20. Three brothers and 3 sisters are lining up to be photographed. How many arrangements are there
    (a) Altogether;
    (b) With brothers and sisters in alternating positions;
    (c) With the 3 sisters standing together?

21. A baseball team is to be formed from a squad of 12 men. Two teams made up of the same 9 men are different if at least some of the men are assigned different positions. In how many ways can a team be formed if
    (a) There are no restrictions;
    (b) Only two of the men can pitch and these two cannot play any other position;
    (c) Only two of the men can pitch but they can also play any other position?

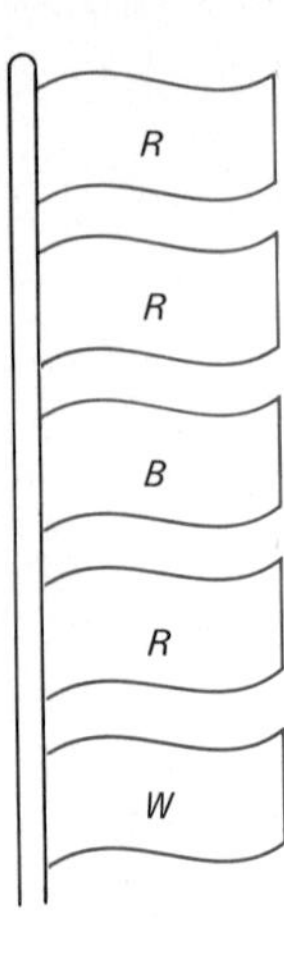

Example B (Permutations with some indistinguishable objects) Lucy has 3 identical red flags, 1 white flag, and 1 blue flag. How many different five-flag signals could she display from the flagpole of her small boat?

SOLUTION. If the 3 red flags were distinguishable, the answer would be ${}_5P_5 = 5! = 120$. Pretending they are distinguishable leads to counting an arrangement such as *RRBRW* six times, corresponding to the 3! ways of arranging the 3 red flags (see margin). For this reason, we must divide by 3!. Thus the number of signals Lucy can make is

$$\frac{5!}{3!} = \frac{5\cdot 4\cdot \not{3}\cdot \not{2}\cdot \not{1}}{\not{3}\cdot \not{2}\cdot \not{1}} = 20$$

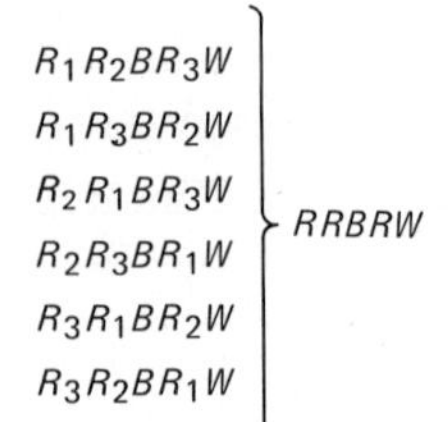

This result can be generalized. For example, given a set of n objects in which j are of one kind, k of a second kind, and m of a third kind, then the number of distinguishable permutations is

$$\frac{n!}{j!\,k!\,m!}$$

22. How many different signals consisting of 8 flags can be made using 4 white flags, 3 red flags, and 1 blue flag?
23. How many different signals consisting of 7 flags can be made using 3 white, 2 red, and 2 blue flags?
24. How many different five-letter code words can be made from the five letters of the word *MIAMI*?
25. How many different eleven-letter code words can be made from the eleven letters of the word *MISSISSIPPI*?
26. In how many different ways can a^4b^6 be written without using exponents? (*Hint*: One way is *aaaabbbbbb*.)
27. In how many different ways can a^3bc^6 be written without using exponents?
28. Consider the part of a city map shown in the margin. How many different shortest routes (no backtracking, no cutting across blocks) are there from A to C? Note that the route shown might be given the designation *EENENNNE*, with E denoting *East* and N denoting *North*.
29. How many different shortest routes are there from A to B (see Problem 28)?

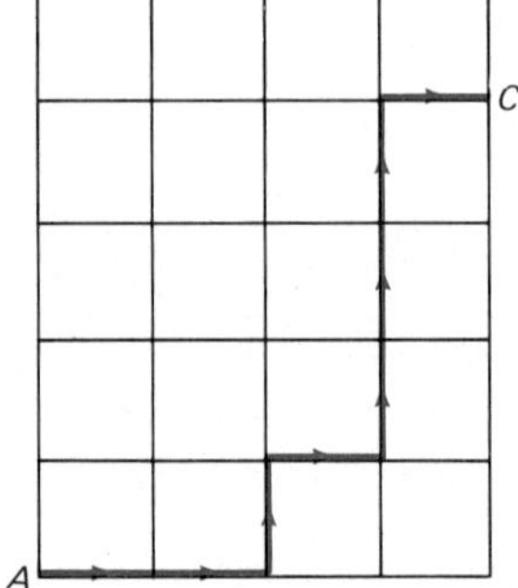

Miscellaneous Problems

30. Calculate.

(a) $\dfrac{7!}{4!}$ (b) $\dfrac{7!}{4!\,3!}$ (c) $7! - 4!$

31. The permutation symbol ${}_7P_3$ can be expressed in factorial notation as follows.

$${}_7P_3 = \frac{{}_7P_3\cdot 4!}{4!} = \frac{(7\cdot 6\cdot 5)(4\cdot 3\cdot 2\cdot 1)}{4!} = \frac{7!}{4!}$$

In the same manner, express each of the following in terms of factorials.

(a) ${}_5P_3$ (b) ${}_{10}P_8$ (c) ${}_8P_2$

32. Write ${}_nP_r$ in factorial notation. Assume $r < n$. (See Problem 31.)

33. Six horses run in a race.
    (a) How many different orders of finishing are there?
    (b) How many possibilities are there for the first three places?

34. In how many ways can first and second prizes be awarded in a baking competition in which there are 8 entries?

35. (a) In how many different ways can the letters of the word *CYCLIC* be arranged?
    (b) In how many of these arrangements are the three *C*'s in consecutive positions?

36. In how many different ways can John select a dinner roll, a salad, a main entrée, and a dessert if there are 2 kinds of dinner rolls, 3 salads, 4 main entrées, and 3 desserts?

37. In how many different ways can a true-false test of 10 questions be answered?

38. The letters of the word *WRONG* are written on five cards. How many code words can be made? (Count one-, two-, three-, four-, and five-letter words.)

39. How many seven-digit phone numbers are possible assuming the first digit cannot be 0?

40. How many four-digit numbers are there that use only the digits 1, 2, 3, 4, 5? How many of these are even? (*Note*: The question is worded in a way that allows repetition of digits.)

41. How many three-digit numbers are there that use only the digits 0, 1, 2, 3, 4? (Be careful; A number cannot start with 0.)

42. In how many ways can 6 people be seated at a round table? (*Note*: Moving each person one place to the right (or left) does not constitute a new arrangement.)

43. A man and his wife invite 4 couples to dinner. The dinner table is rectangular. They decide on a seating arrangement in which the hostess will sit at the end nearest the kitchen, the host at the opposite end, and 4 guests on each side. Furthermore, no man shall sit next to another man, nor shall he sit next to his own wife. In how many ways can this be done?

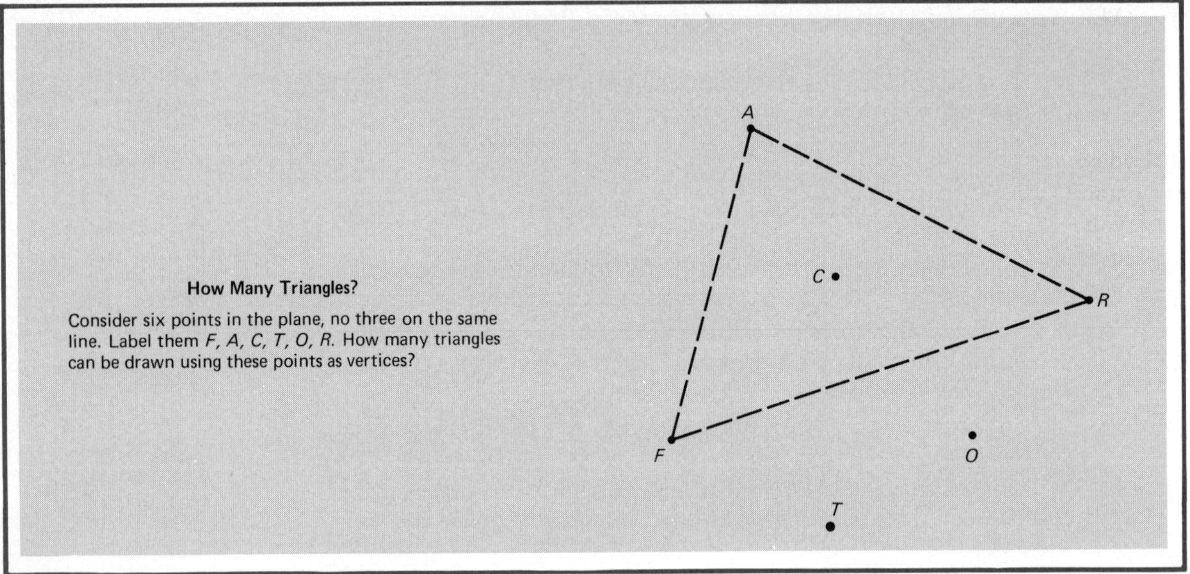

12-6 Counting Unordered Collections

For each choice of three points in the opening display, we can draw a triangle. But notice that the order in which we choose the three points does not matter. For example, *FAR*, *FRA*, *AFR*, *ARF*, *RAF*, and *RFA* all determine the same triangle, namely, the one that is shown by dotted lines in the picture. The question about triangles is very different from the question about three-letter code words raised in the last section; yet there is a connection.

We learned that we can make

$$_6P_3 = 6 \cdot 5 \cdot 4 = 120$$

three-letter code words out of the letters of *FACTOR*. However, every triangle can be labeled with $3! = 6$ different code words. To find the number of triangles, we should therefore divide the number of code words by $3!$. We conclude that the number of triangles that can be drawn is

$$\frac{_6P_3}{3!} = \frac{6 \cdot 5 \cdot 4}{3 \cdot 2 \cdot 1} = 20$$

Combinations An unordered collection of objects is called a **combination** of those objects. If we select r objects from a set of n distinguishable objects, the resulting subset is called a **combination of n things taken r at a time**. The number of such combinations is denoted by $_nC_r$. Thus $_6C_3$ is the number of combinations of 6 things taken 3 at a time. We calculated it in connection with the triangle problem.

$$_6C_3 = \frac{_6P_3}{3!} = \frac{6 \cdot 5 \cdot 4}{3 \cdot 2 \cdot 1} = 20$$

More generally, if $1 \le r \le n$, the combination symbol ${}_nC_r$ is given by

$$ {}_nC_r = \frac{{}_nP_r}{r!} = \frac{n(n-1)(n-2)\cdots(n-r+2)(n-r+1)}{r(r-1)(r-2)\cdots 2\cdot 1} $$

A good way to remember this is that you want r factors in both numerator and denominator. In the numerator, you start with n and go down; in the denominator you start with r and go down. Incidentally, the answer must be an integer. That means the denominator has to divide the numerator. Here are some more examples.

$$ {}_5C_2 = \frac{5\cdot \overset{2}{\cancel{4}}}{\cancel{2}\cdot 1} = 10 $$

$$ {}_{12}C_4 = \frac{\cancel{12}\cdot 11\cdot \overset{5}{\cancel{10}}\cdot 9}{\cancel{4}\cdot\cancel{3}\cdot\cancel{2}\cdot 1} = 495 $$

$$ {}_{10}C_8 = \frac{\overset{5}{\cancel{10}}\cdot 9\cdot\cancel{8}\cdot\cancel{7}\cdot\cancel{6}\cdot\cancel{5}\cdot\cancel{4}\cdot\cancel{3}}{\cancel{8}\cdot\cancel{7}\cdot\cancel{6}\cdot\cancel{5}\cdot\cancel{4}\cdot\cancel{3}\cdot\cancel{2}\cdot 1} = 45 $$

$$ {}_{10}C_2 = \frac{\overset{5}{\cancel{10}}\cdot 9}{\cancel{2}\cdot 1} = 45 $$

Notice that ${}_{10}C_8 = {}_{10}C_2$. This is not surprising, since in selecting a subset of 8 objects out of 10, you automatically select 2 to leave behind (you might call it selection by omission). The general fact that follows by the same reasoning is

$$ \boxed{{}_nC_r = {}_nC_{n-r}} $$

We can express ${}_nC_r$ entirely in terms of factorials. Recall that

$$ {}_nC_r = \frac{n(n-1)(n-2)\cdots(n-r+1)}{r!} $$

If we multiply both numerator and denominator by $(n-r)!$, we get

$$ \boxed{{}_nC_r = \frac{n!}{r!\,(n-r)!}} $$

For this formula to hold when $r = 0$ and $r = n$, it is necessary to define ${}_nC_0 = 1$ and $0! = 1$. Even in mathematics, there is truth to that old proverb "Necessity is the mother of invention."

Combinations versus Permutations Whenever we are faced with the problem of counting the number of ways of selecting r objects from n objects, we are faced with this question. Is the notion of order significant? If the answer is yes, it is a permutation problem; if no, it is a combination problem.

Consider the Birdwatchers' Club of Section 12-5 again. Suppose the club members want to select a president, a vice president, and a secretary. Is order significant? Yes. The selection can be done in

$$_6P_3 = 6 \cdot 5 \cdot 4 = 120$$

ways.

But suppose they decide simply to choose an executive committee consisting of 3 members. Is order relevant? No. A committee consisting of Filipe, Celia, and Amanda is the same as a committee consisting of Celia, Amanda, and Filipe. A three-member committee can be chosen from 6 people in

$$_6C_3 = \frac{6 \cdot 5 \cdot 4}{3 \cdot 2 \cdot 1} = 20$$

ways.

The words *arrangement*, *lineup*, and *signal* all suggest order. The words *set*, *committee*, *group*, and *collection* do not.

Problem Set 12-6

1. Calculate each of the following.
   (a) $_{10}P_3$ (b) $_{10}C_3$
   (c) $_5P_5$ (d) $_5C_5$
   (e) $_6P_1$ (f) $_6C_1$
2. Calculate each of the following.
   (a) $_{12}P_2$ (b) $_{12}C_2$
   (c) $_4P_4$ (d) $_4C_4$
   (e) $_{10}P_1$ (f) $_{10}C_1$
3. Use the fact that $_nC_r = {}_nC_{n-r}$ to calculate each of the following.
   (a) $_{20}C_{17}$ (b) $_{100}C_{97}$
4. Calculate each of the following.
   (a) $_{41}C_{39}$ (b) $_{1000}C_{998}$
   (*Hint*: See Problem 3.)
5. In how many ways can a committee of 3 be selected from a class of 8 students?
6. In how many ways can a committee of 5 be selected from a class of 8 students?
7. A political science professor must select 4 students from her class of 12 students for a field trip to the state legislature. In how many ways can she do it?
8. The professor of Problem 7 was asked to rank the top 4 students in her class of 12. In how many ways could that be done?
9. A police chief needs to pick 3 officers from the 10 available to control traffic at junctions A, B, and C. In how many ways can he do it?

10. If 12 horses are entered in a race, in how many ways can the first 3 places (win, place, show) be taken?

11. From a class of 6 members, in how many ways can a committee of any size be selected (including a committee of one)?

12. From a penny, a nickel, a dime, a quarter, and a half dollar, how many different sums can be made?

Example A (More on committees) A committee of 4 is to be selected from a group of 3 seniors, 4 juniors, and 5 sophomores. In how many ways can it be done if

(a) There are no restrictions on the selection;
(b) The committee must have 2 sophomores, 1 junior, and 1 senior;
(c) The committee must have at least 3 sophomores;
(d) The committee must have at least 1 senior?

SOLUTION.

(a) $${}_{12}C_4 = \frac{\overset{5}{\cancel{12}} \cdot 11 \cdot \cancel{10} \cdot 9}{\cancel{4} \cdot \cancel{3} \cdot \cancel{2} \cdot 1} = 495$$

(b) Two sophomores can be chosen in ${}_5C_2$ ways, 1 junior in ${}_4C_1$ ways, and 1 senior in ${}_3C_1$ ways. By the multiplication principle of counting, the committee can be chosen in

$${}_5C_2 \cdot {}_4C_1 \cdot {}_3C_1 = 10 \cdot 4 \cdot 3 = 120$$

ways. We used the multiplication principle because we choose 2 sophomores *and* 1 junior *and* 1 senior.

(c) At least 3 sophomores means 3 sophomores and 1 nonsophomore *or* 4 sophomores. The word *or* tells us to use the addition principle of counting. We get

$${}_5C_3 \cdot {}_7C_1 + {}_5C_4 = 10 \cdot 7 + 5 = 75$$

(d) Let x be the number of selections with at least one senior and let y be the number of selections with no seniors. Then $x + y$ is the total number of selections, that is, $x + y = 495$ (see part (a)). We calculate y rather than x because it is easier.

$$y = {}_9C_4 = \frac{9 \cdot 8 \cdot 7 \cdot 6}{4 \cdot 3 \cdot 2 \cdot 1} = 126$$

$$x = 495 - 126 = 369$$

13. An investment club has a membership of 4 women and 6 men. A research committee of 3 is to be formed. In how many ways can this be done if
    (a) There are to be 2 women and 1 man on the committee;
    (b) There is to be at least 1 woman on the committee;
    (c) All 3 are to be of the same sex?

14. A senate committee of 4 is to be formed from a group consisting of 5 Republicans and 6 Democrats. In how many ways can this be done if
    (a) There are to be 2 Republicans and 2 Democrats on the committee;
    (b) There are to be no Republicans on the committee;
    (c) There is to be at most one Republican on the committee?

15. Suppose that a bag contains 4 black and 7 white balls. In how many ways can a group of 3 balls be drawn from the bag consisting of
 (a) 1 black and 2 white balls;
 (b) Balls of just one color;
 (c) At least 1 black ball?
 (*Note*: Assume the balls are distinguishable; for example, they may be numbered.)
16. John is going on a vacation trip and wants to take 5 books with him from his personal library, which consists of 6 science books and 10 novels. In how many ways can he make his selection if he wants to take
 (a) 2 science books and 3 novels;
 (b) At least 1 science book;
 (c) 1 book of one kind and 4 books of the other kind?

Example B (Bridge card problems) A standard deck consists of 52 cards. There are 4 suits (spades, clubs, hearts, diamonds), each with 13 cards (2, 3, 4, . . ., 10, jack, queen, king, ace). A bridge hand consists of 13 cards.
(a) How many different possible bridge hands are there?
(b) How many of them have exactly 3 aces?
(c) How many of them have no aces?
(d) How many of them have cards from just 3 suits?

SOLUTION. (a) The order of the cards in a hand is irrelevant; it is a combination problem. We can select 13 cards out of 52 in ${}_{52}C_{13}$ ways, a number so large we will not bother to calculate it.
(b) The three aces can be selected in ${}_4C_3$ ways, the 10 remaining cards in ${}_{48}C_{10}$ ways. The answer (using the multiplication principle) is ${}_4C_3 \cdot {}_{48}C_{10}$.
(c) From 48 nonaces, we select 13 cards; the answer is ${}_{48}C_{13}$.
(d) We think of this as no clubs, or no spades, or no hearts, or no diamonds and use the addition principle.

$${}_{39}C_{13} + {}_{39}C_{13} + {}_{39}C_{13} + {}_{39}C_{13} = 4 \cdot {}_{39}C_{13}$$

Problems 17–22 deal with bridge hands. Leave your answers in terms of combination symbols.

17. How many of the hands have only red cards? (*Note*: Half of the cards are red.)
18. How many of the hands have only honor cards (aces, kings, queens, and jacks)?
19. How many of the hands have one card of each kind (one ace, one king, one queen, and so on)?
20. How many of the hands have exactly 2 kings?
21. How many of the hands have 2 or more kings?
22. How many of the hands have exactly 2 aces and 2 kings?

Problems 23–26 deal with poker hands, which consist of 5 cards.

23. How many different poker hands are possible?
24. How many of them have exactly 2 hearts and 2 diamonds?

25. How many have 2 pairs of different kinds (for example, 2 aces and 2 fives)?

26. How many are 5-card straights (for example, 7, 8, 9, 10, jack)? An ace may count either as the highest or the lowest card, that is, as 1 or 13.

Miscellaneous Problems

27. From a committee of 7 members, in how many ways can a subcommittee of 3 be chosen?

28. If 5 distinguishable coins are tossed, in how many ways can they fall?

29. In how many ways can a student select 4 college courses from a set of 9 courses?

30. A man drives from town A to town B and then returns by a different route. If there are 5 routes between the two towns, how many choices of round trips does he have?

31. From a cent, a nickel, a dime, a quarter, and a half dollar, how many sums can be formed consisting of
(a) 3 coins each;
(b) 2 coins each;
(c) At least 3 coins each?

32. From a group of 5 representatives of labor, 4 representatives of business, and 7 representatives of the general public, how many mediation committees can be formed consisting of 2 people from each group?

33. In how many ways can 9 presents be distributed to 3 children, if each is to receive 3 presents?

34. A committee of six is to be formed from a group of 5 freshmen, 6 sophomores, and 8 seniors. In how many ways can this be done if
(a) There shall be 2 members of each class on the committee;
(b) There shall be no freshmen on the committee;
(c) There shall be exactly 4 seniors on the committee;
(d) There shall be 3 members of each of two classes on the committee?

35. Calculate each of the following.
(a) ${}_2C_0 + {}_2C_1 + {}_2C_2$
(b) ${}_3C_0 + {}_3C_1 + {}_3C_2 + {}_3C_3$
(c) ${}_4C_0 + {}_4C_1 + {}_4C_2 + {}_4C_3 + {}_4C_4$
Now conjecture a formula for

$${}_nC_0 + {}_nC_1 + {}_nC_2 + \cdots + {}_nC_n$$

36. Use the factorial formula for ${}_nC_r$ to prove that

$${}_{n+1}C_r = {}_nC_{r-1} + {}_nC_r$$

37. A class of 10 will elect a president, a secretary, and a social committee of 3 with no overlapping of positions. In how many ways can it be done?

38. A test consists of 10 true-false items.
(a) How many different sets of answers are possible?
(b) How many of them have exactly 4 answers right?

39. An ice cream parlor has 10 different flavors. How many different double dip cones can be made if
    (a) The two dips must be of different flavors but the order of putting them on the cone does not matter;
    (b) The two dips must be different and order does matter;
    (c) The two dips need not be different but order matters;
    (d) The two dips need not be different and order does not matter?
40. How many positive integers less than 1,000,000 involve only the digits 1 and 2?
41. Two distinguishable dice are tossed.
    (a) In how many ways can they fall?
    (b) How many of them give a total of 7?
    (c) How many give a total less than 7?
42. Answer the questions of Problem 41 if 3 distinguishable dice are tossed.
43. In how many ways can a class of 6 girls and 6 boys be seated in a room with 12 chairs if the boys must take the odd-numbered seats?

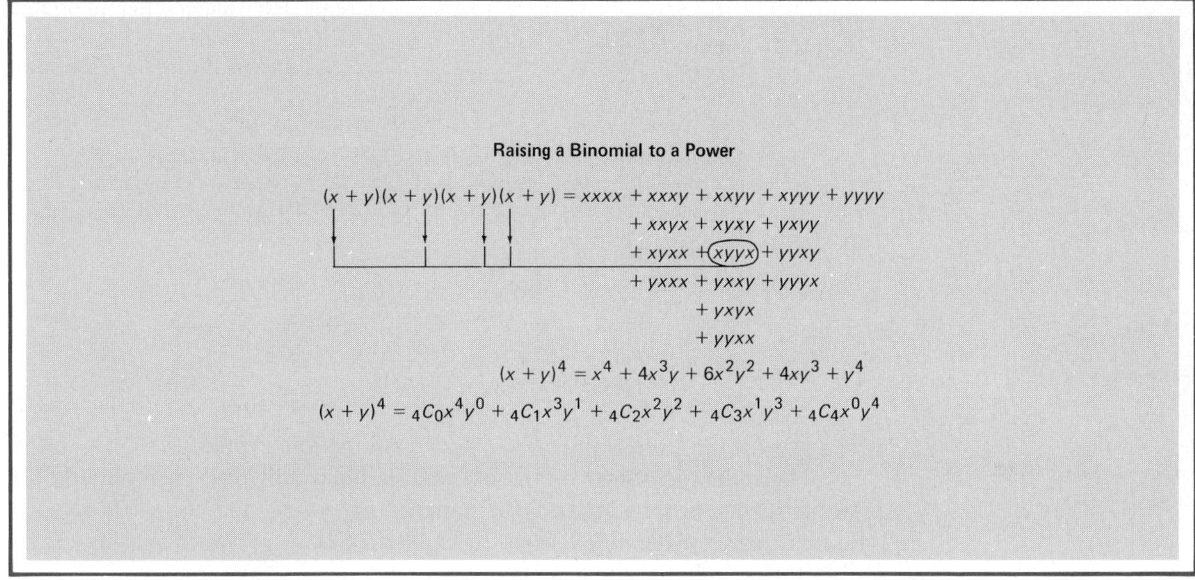

12-7 The Binomial Formula

In the opening box, we have shown how to expand $(x + y)^4$. Admittedly, it looks complicated; however, it leads to the remarkable formula at the bottom of the display. That formula generalizes to handle $(x + y)^n$, where n is any positive integer. It is worth a careful investigation.

To produce any given term in the expansion of $(x + y)^4$, each of the four factors $x + y$ contributes either an x or a y. There are $2 \cdot 2 \cdot 2 \cdot 2 = 16$ ways in which they can make this contribution, hence the 16 terms in the long expanded form. But many of these terms are alike; in fact, only five different types occur, namely, x^4, x^3y, x^2y^2, xy^3, and y^4. The number of times each occurs is ${}_4C_0$, ${}_4C_1$, ${}_4C_2$, ${}_4C_3$, and ${}_4C_4$, respectively. (Remember we defined ${}_4C_0$ to be 1.)

Why do the combination symbols arise in this expansion? For example, why is ${}_4C_2$ the coefficient of x^2y^2? If you follow the arrows in the opening display, you see how the term $xyyx$ comes about. It gets its two y's from the second and third $x + y$ factors (the x's then must come from the first and fourth $x + y$ factors). Thus, the number of terms of the form x^2y^2 is the number of ways of selecting two factors out of four from which to take y's (the x's must come from the remaining two factors). We can select two objects out of four in ${}_4C_2$ ways; hence the coefficient of x^2y^2 is ${}_4C_2$.

The Binomial Formula What we have just done for $(x + y)^4$ can be carried out for $(x + y)^n$, where n is any positive integer. The result is called the **Binomial Formula**.

$$(x + y)^n = {}_nC_0x^ny^0 + {}_nC_1x^{n-1}y^1 + \cdots + {}_nC_{n-1}x^1y^{n-1} + {}_nC_nx^0y^n = \sum_{k=0}^{n} {}_nC_kx^{n-k}y^k$$

Notice that the k in ${}_nC_k$ is the exponent on y and that the two exponents in each term sum to n.

Let us apply the Binomial Formula with $n = 6$.

$$\begin{aligned}(x + y)^6 &= {}_6C_0x^6 + {}_6C_1x^5y + {}_6C_2x^4y^2 + \cdots + {}_6C_6y^6 \\ &= x^6 + 6x^5y + 15x^4y^2 + 20x^3y^3 + 15x^2y^4 + 6xy^5 + y^6\end{aligned}$$

This same result applies to the expansion of $(2a - b^2)^6$. We simply think of $2a$ as x and $-b^2$ as y. Thus

$$\begin{aligned}[2a + (-b^2)]^6 &= (2a)^6 + 6(2a)^5(-b^2) + 15(2a)^4(-b^2)^2 \\ &\quad + 20(2a)^3(-b^2)^3 + 15(2a)^2(-b^2)^4 + 6(2a)(-b^2)^5 + (-b^2)^6 \\ &= 64a^6 - 192a^5b^2 + 240a^4b^4 - 160a^3b^6 \\ &\quad + 60a^2b^8 - 12ab^{10} + b^{12}\end{aligned}$$

The Binomial Coefficients The combination symbols ${}_nC_k$ are often called **binomial coefficients** for reasons that should be obvious. Their remarkable properties have been studied for hundreds of years. Let us see if we can discover some of them. We begin by expanding $(x + y)^k$ for increasing values of k, listing only the coefficients.

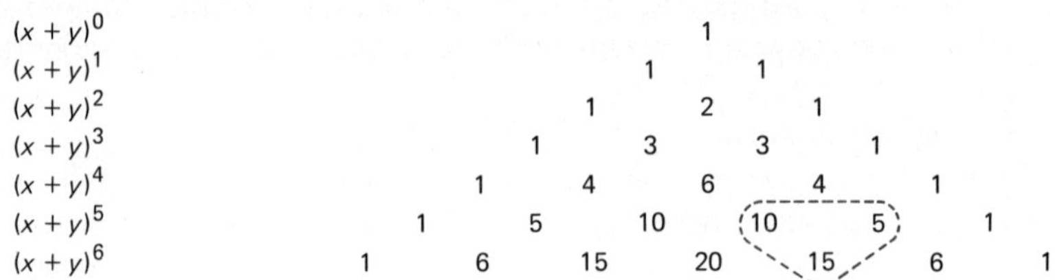

The triangle of numbers composed of the binomial coefficients is called **Pascal's Triangle** after the gifted French philosopher and mathematician, Blaise Pascal (1622–1662). Notice its symmetry. If folded across a vertical center line, the numbers match. This corresponds to an algebraic fact you learned earlier.

$$\boxed{{}_nC_k = {}_nC_{n-k}}$$

Next notice that any number in the body of the triangle is the sum of the two numbers closest to it in the line above the number. For example, $15 = 10 + 5$, as the dotted triangle

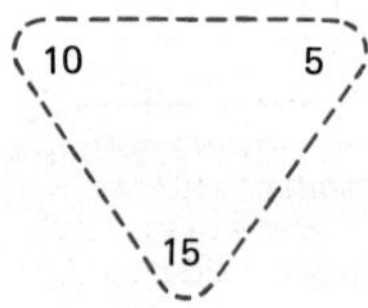

was meant to suggest. In symbols

$$\boxed{{}_{n+1}C_k = {}_nC_{k-1} + {}_nC_k}$$

a fact that can be proved rigorously by using the factorial formula for ${}_nC_k$ (see Problem 36 of Section 12-6).

Now add the numbers in each row of Pascal's triangle.

$$\begin{aligned} 1 &= 1 = 2^0 \\ 1 + 1 &= 2 = 2^1 \\ 1 + 2 + 1 &= 4 = 2^2 \\ 1 + 3 + 3 + 1 &= 8 = 2^3 \end{aligned}$$

This suggests the formula

$$\boxed{{}_nC_0 + {}_nC_1 + {}_nC_2 + \cdots + {}_nC_n = 2^n}$$

Its truth can be demonstrated by substituting $x = 1$ and $y = 1$ in the Binomial Formula. It has an important interpretation. Take a set with n elements. This set has ${}_nC_n$ subsets of size n, ${}_nC_{n-1}$ subsets of size $n - 1$, and so on. The left side of the boxed formula is the total number of subsets (including the empty set) of a set with n elements; remarkably, it is just 2^n. For example, $\{a, b, c\}$ has $2^3 = 8$ subsets.

$$\{a, b, c\} \quad \{a, b\} \quad \{a, c\} \quad \{b, c\} \quad \{a\} \quad \{b\} \quad \{c\} \quad \varnothing$$

A related formula is

$${}_nC_0 - {}_nC_1 + {}_nC_2 - \cdots + (-1)^n{}_nC_n = 0$$

which can be obtained by setting $x = 1$ and $y = -1$ in the Binomial Formula. When rewritten as

$${}_nC_0 + {}_nC_2 + {}_nC_4 + \cdots = {}_nC_1 + {}_nC_3 + {}_nC_5 + \cdots,$$

it says that the number of subsets with an even number of elements is equal to the number of subsets with an odd number of elements.

Problem Set 12-7

In Problems 1–6, expand and simplify.

1. $(x + y)^3$
2. $(x - y)^3$
3. $(x - 2y)^3$
4. $(3x + b)^3$
5. $(c^2 - 3d^3)^4$
6. $(xy - 2z^2)^4$

Write the first 3 terms of each expansion in Problems 7–10 in simplified form.

7. $(x + y)^{20}$

8. $(x + y)^{30}$

9. $\left(x + \dfrac{1}{x^5}\right)^{20}$

10. $\left(xy^2 + \dfrac{1}{y}\right)^{14}$

11. Find the number of subsets of each of the following sets.
(a) $\{a, b, c, d\}$ (b) $\{1, 2, 3, 4, 5\}$
(c) $\{x_1, x_2, x_3, x_4, x_5, x_6\}$

Example A (Finding a specific term of a binomial expansion) Find the term in the expansion of $(2x + y^2)^{10}$ that involves y^{12}.

SOLUTION. This term will arise from raising y^2 to the 6th power. It is therefore

$$_{10}C_6(2x)^4(y^2)^6 = 210 \cdot 16x^4y^{12} = 3360x^4y^{12}$$

12. Find the term in the expansion of $(y^2 - z^3)^{10}$ that involves z^9.
13. Find the term in the expansion of $(3x - y^3)^{10}$ that involves y^{24}.
14. Find the term in the expansion of $(2a - b)^{12}$ that involves a^3.
15. Find the term in the expansion of $(x^2 - 2/x)^5$ that involves x^4.

Example B (An application to compound interest) If \$100 is invested at 12 percent compounded monthly, it will accumulate to $100(1.01)^{12}$ dollars by the end of one year. Use the binomial formula to approximate this amount.

SOLUTION.

$$\begin{aligned} 100(1.01)^{12} &= 100(1 + .01)^{12} \\ &= 100\left[1 + 12(.01) + \frac{12 \cdot 11}{2}(.01)^2 + \frac{12 \cdot 11 \cdot 10}{6}(.01)^3 + \cdots\right] \\ &= 100[1 + .12 + .0066 + .00022 + \cdots] \\ &\approx 100(1.12682) \approx 112.68 \end{aligned}$$

This answer of \$112.68 is accurate to the nearest penny since the last 9 terms of the expansion do not add up to as much as a penny.

In Problems 16–19 use the first 3 terms of a binomial expansion to find an approximate value of the given expression.

16. $20(1.02)^8$
17. $100(1.002)^{20}$
18. $500(1.005)^{20}$
19. $200(1.04)^{10}$
20. Bacteria multiply in a certain medium so that by the end of k hours their number N is $N = 100(1.02)^k$. Approximate the number of bacteria after 20 hours.
21. Do Problem 20 assuming $N = 1000(1.01)^k$.

Miscellaneous Problems

In Problems 22–23, expand and simplify.

22. $(x + h)^4$

23. $\left(2x^3 - \dfrac{1}{x}\right)^5$

In Problems 24 and 25, write and simplify the first 3 terms of the expansion.

24. $(a^2 + 2b)^{15}$ 25. $(b^3 + 3c)^{12}$

26. Simplify the following expression.

$$[2(x + h)^4 - 3(x + h)^2] - [2x^4 - 3x^2]$$

27. Let $A = \{a_1, a_2, \ldots, a_n\}$ be a set with n elements. To make a subset B is to decide for each element a_k whether or not it goes into B. Show why this approach also leads to the conclusion that A has 2^n subsets.

28. Calculate $(0.99)^{10}$ accurate to six decimal places using the Binomial Theorem.

29. Show that $(1.0003)^{10} > 1.003004$.

30. How many committees consisting of 2 or more members can be selected from a group of 10 people? (*Hint*: Think of subsets.)

31. How many selections of 1 or more books can be made from a set of 8 books?

32. Use the Binomial Formula to expand $(x^2 + x + 1)^3$. (*Hint*: Write $x^2 + x + 1$ as $x^2 + (x + 1)$.)

33. In the expansion of the trinomial $(x + y + z)^n$, the coefficient of $x^r y^s z^t$, where $r + s + t = n$, is

$$\frac{n!}{r!\, s!\, t!}$$

What is the coefficient of the term x^2y^4z in the expansion of $(x + y + z)^7$?

34. Using the information given in Problem 33, write the term of the expansion of $(x + y + z)^{10}$ that involves y^2z^3.

CHAPTER SUMMARY

A **number sequence** $a_1, a_2, a_3, \ldots$ is a function that associates with each positive integer n a number a_n. Such a sequence may be described by an **explicit formula** (for instance, $a_n = 2n + 1$), by a **recursion formula** (for instance, $a_n = 3a_{n-1}$), or by giving enough terms so a pattern is evident (for instance, 1, 11, 21, 31, 41, . . .).

If any term in a sequence can be obtained by adding a fixed number d to the preceding term, we call it an **arithmetic sequence**. There are three key formulas associated with this type of sequence.

Recursion formula: $a_n = a_{n-1} + d$

Explicit formula: $a_n = a_1 + (n - 1)d$

Sum formula: $A_n = \frac{n}{2}(a_1 + a_n)$

In the last formula, A_n represents

$$A_n = a_1 + a_2 + \cdots + a_n = \sum_{i=1}^{n} a_i$$

A **geometric sequence** is one in which any term results from multiplying the previous term by a fixed number r. The corresponding key formulas are

$$\begin{aligned} \text{Recursion formula:}\quad a_n &= ra_{n-1} \\ \text{Explicit formula:}\quad a_n &= a_1r^{n-1} \\ \text{Sum formula:}\quad A_n &= \frac{a_1(1-r^n)}{1-r}, r \neq 1 \end{aligned}$$

In the last formula, we may ask what happens as n grows larger and larger. If $|r| < 1$, the value of A_n gets closer and closer to $a_1/(1 - r)$, which we regard as the sum of *all* the terms of the sequence.

Often in mathematics, we wish to demonstrate that a whole **sequence of statements P_n** is true. For this, a powerful tool is the **Principle of Mathematical Induction**, which asserts that if P_1 is true and if the truth of P_k implies the truth of P_{k+1}, then all the statements of the sequence are true.

Given enough time, anyone can count the number of elements in a set. But if the set consists of arrangements of objects (for example, the letters in a word), the work can be greatly simplified by using two principles, the **multiplication principle** and the **addition principle**. Of special interest are the number of **permutations** (ordered arrangements) of n things taken r at a time and the number of **combinations** (unordered collections) of n things taken r at a time. They can be calculated from the formulas

$$_nP_r = n(n-1)(n-2)\cdots(n-r+1)$$

$$_nC_r = \frac{_nP_r}{r!} = \frac{n(n-1)(n-2)\cdots(n-r+1)}{r(r-1)(r-2)\cdots 1}$$

One important use of the symbol $_nC_r$ is in the **Binomial Formula**

$$(x+y)^n = {}_nC_0x^ny^0 + {}_nC_1x^{n-1}y^1 + \cdots + {}_nC_{n-1}x^1y^{n-1} + {}_nC_nx^0y^n$$

CHAPTER REVIEW PROBLEM SET

Problems 1–6 refer to the sequences below.

(a) 2, 5, 8, 11, 14, . . .
(b) 2, 6, 18, 54, . . .
(c) 2, 1.5, 1, 0.5, 0, . . .
(d) 2, 4, 6, 10, 16, . . .
(e) 2, $\frac{2}{3}$, $\frac{2}{9}$, $\frac{2}{27}$, $\frac{2}{81}$, . . .

1. Which of these sequences are arithmetic? Which are geometric?
2. Give a recursion formula for each of the sequences (a)–(e).
3. Give an explicit formula for sequences (a) and (b).
4. Find the sum of the first 67 terms of sequence (a).
5. Write a formula for the sum of the first 100 terms of sequence (b).
6. Find the sum of *all* the terms of sequence (e).

7. If $a_n = 3a_{n-1} - a_{n-2}$, $a_1 = 1$, and $a_2 = 2$, find a_6.
8. If $a_n = n^2 - n$, find $A_5 = \sum_{n=1}^{5} a_n$.
9. Calculate $2 + 4 + 6 + 8 + \cdots + 1000$.
10. Write $.555\bar{5}$ as a ratio of two integers.
11. If \$100 is put in the bank today earning 8 percent interest compounded quarterly, write a formula for its value at the end of 12 years.
12. Show by mathematical induction that
    (a) $5 + 9 + \cdots + (4n + 1) = 2n^2 + 3n$;
    (b) $n! > 2^n$ when $n \geq 4$.
13. Suppose P_3 is true and $P_k \Rightarrow P_{k+3}$. What can we conclude about the sequence P_n?
14. John has 4 sport coats, 3 pairs of trousers, and 5 shirts. How many different outfits could he wear?
15. How many code words of all lengths can be made from the letters of the word *SNOW*, assuming letters cannot be repeated?
16. Evaluate.
    (a) ${}_{10}P_4$ (b) ${}_{10}C_4$ (c) $6!/4!$
    (d) ${}_{50}C_{48}$ (e) $8!/(2!\,6!)$ (f) $10!/(2!\,3!\,5!)$
17. In how many ways can a class of 5 girls and 4 boys select
    (a) A president, vice president, and secretary;
    (b) A president, vice president, and secretary if the secretary must be a boy;
    (c) A social committee of 3 people;
    (d) A social committee of 3 people consisting of 2 girls and 1 boy?
18. The letters of the word *BARBARIAN* are written on 9 cards. How many nine-letter code words can be formed?
19. Consider 8 points in the plane, no 3 on the same line. How many triangles can be formed using these points as vertices?
20. How many subsets does a set of 8 elements have?
21. Find the first 4 terms in simplified form in the expansion of $(x + 2y)^{10}$.
22. Find the term involving a^3b^6 in the expansion of $(a - b^2)^6$.
23. Find $(1.002)^{20}$ accurate to 4 decimal places.

Does the pursuit of truth give you as much pleasure as before? Surely it is not the knowing but the learning, not the possessing but the acquiring, not the being-there but the getting-there, that afford the greatest satisfaction. If I have clarified and exhausted something, I leave it in order to go again into the dark. Thus is that insatiable man so strange: when he has completed a structure it is not in order to dwell in it comfortably, but to start another.

Karl Friedrich Gauss

Appendix

Use of Tables

Table A. **Natural Logarithms**

Table B. **Common Logarithms**

Table C. **Trigonometric Functions (degrees)**

Table D. **Trigonometric Functions (radians)**

Use of Tables

Table A. Natural logarithms: See Section 6-5.
Table B. Common logarithms: See Section 6-6 and 6-7.
Table C. Trigonometric functions (degrees): See Sections 7-1 and 7-5.
Table D. Trigonometry functions (radians): See Sections 7-2 and 7-5.

To find values between those given in any of these tables, we suggest a process called **linear interpolation**. If we know $f(a)$ and $f(b)$ and want $f(c)$, where c is between a and b, we may write

$$f(c) \approx f(a) + d$$

where d is obtained by pretending that the graph of $y = f(x)$ is a straight line on the interval $a \leq x \leq b$. A complete description is given in Section 6-6; the following examples illustrate this process.

Example A (Natural logarithms) Find ln 2.133.

SOLUTION.

$$.010\left[\begin{array}{l} .003\left[\begin{array}{l}\ln 2.130 = .7561 \\ \ln 2.133 = \quad ?\end{array}\right]d \\ \qquad \ln 2.240 = .7608\end{array}\right].0047$$

$$\frac{d}{.0047} = \frac{.003}{.010} = .3$$

$$d = .3(.0047) \approx .0014$$

$$\ln 2.134 \approx \ln 2.130 + d = .7561 + .0014 = .7575$$

Example B (Common logarithms) Find log 63.26.

SOLUTION.

$$.10\left[\begin{array}{l} .06\left[\begin{array}{l}\log 63.20 = 1.8007 \\ \log 63.26 = \quad ?\end{array}\right]d \\ \qquad \log 63.30 = 1.8014\end{array}\right].0007$$

$$\frac{d}{.0007} = \frac{.06}{.10} = .6$$

$$d = .6(.0007) \approx .0004$$

$$\log 63.26 \approx \log 63.20 + d = 1.8007 + .0004 = 1.8011$$

Example C (Trigonometric functions (degrees)) Find sin 57.44°.

SOLUTION. Note that we must use the degree column on the right and the bottom caption in Table C.

$$.10\left[\begin{array}{l} .04\left[\begin{array}{l}\sin 57.40^\circ = .8425 \\ \sin 57.44^\circ = \quad ?\end{array}\right]d \\ \qquad \sin 57.50^\circ = .8434\end{array}\right].0009$$

$$\frac{d}{.0009} = \frac{.04}{.10} = .4$$

$$d = .4(.0009) \approx .0004$$

$$\sin 57.44^\circ \approx \sin 57.40^\circ + d = .8425 + .0004 = .8429$$

Example D (Trigonometric functions (radians)) Find cos(1.436).

SOLUTION. The cosine is a decreasing function on the interval $0 \le t \le \pi/2$. This causes d to be negative.

$$.010\left[\begin{array}{l} .006\left[\begin{array}{l}\cos 1.430 = .14033 \\ \cos 1.436 = \quad ?\end{array}\right]d \\ \qquad \cos 1.440 = .13042\end{array}\right] -.00991$$

$$\frac{d}{-.00991} = \frac{.006}{.010} = .6$$

$$d = .6(-.00991) \approx -.00595$$

$$\cos(1.436) \approx \cos(1.430) + d = .14033 - .00595 = .13438$$

Example E (Angle, given a trigonometric function) Find θ if $\tan\theta = .43600$. Give the answer in radians.

SOLUTION. We use Table D and find that θ is between .41 and .42.

$$.01\left[\begin{array}{l} d\left[\begin{array}{l}\tan .41 = .43463 \\ \tan\ \theta = .43600\end{array}\right].00137 \\ \quad \tan .42 = .44657\end{array}\right].01194$$

$$\frac{d}{.01} = \frac{.00137}{.01194}$$

$$d \approx (.01)(.115) \approx .001$$

$$\theta \approx .41 + .001 = .411$$

TABLE A. NATURAL LOGARITHMS

	.00	.01	.02	.03	.04	.05	.06	.07	.08	.09
1.0	0.0000	0.0100	0.0198	0.0296	0.0392	0.0488	0.0583	0.0677	0.0770	0.0862
1.1	0.0953	0.1044	0.1133	0.1222	0.1310	0.1398	0.1484	0.1570	0.1655	0.1740
1.2	0.1823	0.1906	0.1989	0.2070	0.2151	0.2231	0.2311	0.2390	0.2469	0.2546
1.3	0.2624	0.2700	0.2776	0.2852	0.2927	0.3001	0.3075	0.3148	0.3221	0.3293
1.4	0.3365	0.3436	0.3507	0.3577	0.3646	0.3716	0.3784	0.3853	0.3920	0.3988
1.5	0.4055	0.4121	0.4187	0.4253	0.4318	0.4383	0.4447	0.4511	0.4574	0.4637
1.6	0.4700	0.4762	0.4824	0.4886	0.4947	0.5008	0.5068	0.5128	0.5188	0.5247
1.7	0.5306	0.5365	0.5423	0.5481	0.5539	0.5596	0.5653	0.5710	0.5766	0.5822
1.8	0.5878	0.5933	0.5988	0.6043	0.6098	0.6152	0.6206	0.6259	0.6313	0.6366
1.9	0.6419	0.6471	0.6523	0.6575	0.6627	0.6678	0.6729	0.6780	0.6831	0.6881
2.0	0.6931	0.6981	0.7031	0.7080	0.7130	0.7178	0.7227	0.7275	0.7324	0.7372
2.1	0.7419	0.7467	0.7514	0.7561	0.7608	0.7655	0.7701	0.7747	0.7793	0.7839
2.2	0.7885	0.7930	0.7975	0.8020	0.8065	0.8109	0.8154	0.8198	0.8242	0.8286
2.3	0.8329	0.8372	0.8416	0.8459	0.8502	0.8544	0.8587	0.8629	0.8671	0.8713
2.4	0.8755	0.8796	0.8838	0.8879	0.8920	0.8961	0.9002	0.9042	0.9083	0.9123
2.5	0.9163	0.9203	0.9243	0.9282	0.9322	0.9361	0.9400	0.9439	0.9478	0.9517
2.6	0.9555	0.9594	0.9632	0.9670	0.9708	0.9746	0.9783	0.9821	0.9858	0.9895
2.7	0.9933	0.9969	1.0006	1.0043	1.0080	1.0116	1.0152	1.0188	1.0225	1.0260
2.8	1.0296	1.0332	1.0367	1.0403	1.0438	1.0473	1.0508	1.0543	1.0578	1.0613
2.9	1.0647	1.0682	1.0716	1.0750	1.0784	1.0818	1.0852	1.0886	1.0919	1.0953
3.0	1.0986	1.1019	1.1053	1.1086	1.1119	1.1151	1.1184	1.1217	1.1249	1.1282
3.1	1.1314	1.1346	1.1378	1.1410	1.1442	1.1474	1.1506	1.1537	1.1569	1.1600
3.2	1.1632	1.1663	1.1694	1.1725	1.1756	1.1787	1.1817	1.1848	1.1878	1.1909
3.3	1.1939	1.1970	1.2000	1.2030	1.2060	1.2090	1.2119	1.2149	1.2179	1.2208
3.4	1.2238	1.2267	1.2296	1.2326	1.2355	1.2384	1.2413	1.2442	1.2470	1.2499
3.5	1.2528	1.2556	1.2585	1.2613	1.2641	1.2669	1.2698	1.2726	1.2754	1.2782
3.6	1.2809	1.2837	1.2865	1.2892	1.2920	1.2947	1.2975	1.3002	1.3029	1.3056
3.7	1.3083	1.3110	1.3137	1.3164	1.3191	1.3218	1.3244	1.3271	1.3297	1.3324
3.8	1.3350	1.3376	1.3403	1.3429	1.3455	1.3481	1.3507	1.3533	1.3558	1.3584
3.9	1.3610	1.3635	1.3661	1.3686	1.3712	1.3737	1.3762	1.3788	1.3813	1.3838
4.0	1.3863	1.3888	1.3913	1.3938	1.3962	1.3987	1.4012	1.4036	1.4061	1.4085
4.1	1.4110	1.4134	1.4159	1.4183	1.4207	1.4231	1.4255	1.4279	1.4303	1.4327
4.2	1.4351	1.4375	1.4398	1.4422	1.4446	1.4469	1.4493	1.4516	1.4540	1.4563
4.3	1.4586	1.4609	1.4633	1.4656	1.4679	1.4702	1.4725	1.4748	1.4770	1.4793
4.4	1.4816	1.4839	1.4861	1.4884	1.4907	1.4929	1.4952	1.4974	1.4996	1.5019
4.5	1.5041	1.5063	1.5085	1.5107	1.5129	1.5151	1.5173	1.5195	1.5217	1.5239
4.6	1.5261	1.5282	1.5304	1.5326	1.5347	1.5369	1.5390	1.5412	1.5433	1.5454
4.7	1.5476	1.5497	1.5518	1.5539	1.5560	1.5581	1.5602	1.5623	1.5644	1.5665
4.8	1.5686	1.5707	1.5728	1.5748	1.5769	1.5790	1.5810	1.5831	1.5851	1.5872
4.9	1.5892	1.5913	1.5933	1.5953	1.5974	1.5994	1.6014	1.6034	1.6054	1.6074
5.0	1.6094	1.6114	1.6134	1.6154	1.6174	1.6194	1.6214	1.6233	1.6253	1.6273
5.1	1.6292	1.6312	1.6332	1.6351	1.6371	1.6390	1.6409	1.6429	1.6448	1.6467
5.2	1.6487	1.6506	1.6525	1.6544	1.6563	1.6582	1.6601	1.6620	1.6639	1.6658
5.3	1.6677	1.6696	1.6715	1.6734	1.6753	1.6771	1.6790	1.6808	1.6827	1.6845
5.4	1.6864	1.6882	1.6901	1.6919	1.6938	1.6956	1.6974	1.6993	1.7011	1.7029

$\ln(N \cdot 10^m) = \ln N + m \ln 10, \qquad \ln 10 = 2.3026$

TABLE A. NATURAL LOGARITHMS

	.00	.01	.02	.03	.04	.05	.06	.07	.08	.09
5.5	1.7047	1.7066	1.7084	1.7102	1.7120	1.7138	1.7156	1.7174	1.7192	1.7210
5.6	1.7228	1.7246	1.7263	1.7281	1.7299	1.7317	1.7334	1.7352	1.7370	1.7387
5.7	1.7405	1.7422	1.7440	1.7457	1.7475	1.7492	1.7509	1.7527	1.7544	1.7561
5.8	1.7579	1.7596	1.7613	1.7630	1.7647	1.7664	1.7682	1.7699	1.7716	1.7733
5.9	1.7750	1.7766	1.7783	1.7800	1.7817	1.7834	1.7851	1.7867	1.7884	1.7901
6.0	1.7918	1.7934	1.7951	1.7967	1.7984	1.8001	1.8017	1.8034	1.8050	1.8066
6.1	1.8083	1.8099	1.8116	1.8132	1.8148	1.8165	1.8181	1.8197	1.8213	1.8229
6.2	1.8245	1.8262	1.8278	1.8294	1.8310	1.8326	1.8342	1.8358	1.8374	1.8390
6.3	1.8406	1.8421	1.8437	1.8453	1.8469	1.8485	1.8500	1.8516	1.8532	1.8547
6.4	1.8563	1.8579	1.8594	1.8610	1.8625	1.8641	1.8656	1.8672	1.8687	1.8703
6.5	1.8718	1.8733	1.8749	1.8764	1.8779	1.8795	1.8810	1.8825	1.8840	1.8856
6.6	1.8871	1.8886	1.8901	1.8916	1.8931	1.8946	1.8961	1.8976	1.8991	1.9006
6.7	1.9021	1.9036	1.9051	1.9066	1.9081	1.9095	1.9110	1.9125	1.9140	1.9155
6.8	1.9169	1.9184	1.9199	1.9213	1.9228	1.9242	1.9257	1.9272	1.9286	1.9301
6.9	1.9315	1.9330	1.9344	1.9359	1.9373	1.9387	1.9402	1.9416	1.9430	1.9445
7.0	1.9459	1.9473	1.9488	1.9502	1.9516	1.9530	1.9544	1.9559	1.9573	1.9587
7.1	1.9601	1.9615	1.9629	1.9643	1.9657	1.9671	1.9685	1.9699	1.9713	1.9727
7.2	1.9741	1.9755	1.9769	1.9782	1.9796	1.9810	1.9824	1.9838	1.9851	1.9865
7.3	1.9879	1.9892	1.9906	1.9920	1.9933	1.9947	1.9961	1.9974	1.9988	2.0001
7.4	2.0015	2.0028	2.0042	2.0055	2.0069	2.0082	2.0096	2.0109	2.0122	2.0136
7.5	2.0149	2.0162	2.0176	2.0189	2.0202	2.0215	2.0229	2.0242	2.0255	2.0268
7.6	2.0282	2.0295	2.0308	2.0321	2.0334	2.0347	2.0360	2.0373	2.0386	2.0399
7.7	2.0412	2.0425	2.0438	2.0451	2.0464	2.0477	2.0490	2.0503	2.0516	2.0528
7.8	2.0541	2.0554	2.0567	2.0580	2.0592	2.0605	2.0618	2.0631	2.0643	2.0656
7.9	2.0669	2.0681	2.0694	2.0707	2.0719	2.0732	2.0744	2.0757	2.0769	2.0782
8.0	2.0794	2.0807	2.0819	2.0832	2.0844	2.0857	2.0869	2.0882	2.0894	2.0906
8.1	2.0919	2.0931	2.0943	2.0956	2.0968	2.0980	2.0992	2.1005	2.1017	2.1029
8.2	2.1041	2.1054	2.1066	2.1078	2.1090	2.1102	2.1114	2.1126	2.1138	2.1150
8.3	2.1163	2.1175	2.1187	2.1199	2.1211	2.1223	2.1235	2.1247	2.1258	2.1270
8.4	2.1282	2.1294	2.1306	2.1318	2.1330	2.1342	2.1353	2.1365	2.1377	2.1389
8.5	2.1401	2.1412	2.1424	2.1436	2.1448	2.1459	2.1471	2.1483	2.1494	2.1506
8.6	2.1518	2.1529	2.1541	2.1552	2.1564	2.1576	2.1587	2.1599	2.1610	2.1622
8.7	2.1633	2.1645	2.1656	2.1668	2.1679	2.1691	2.1702	2.1713	2.1725	2.1736
8.8	2.1748	2.1759	2.1770	2.1782	2.1793	2.1804	2.1815	2.1827	2.1838	2.1849
8.9	2.1861	2.1872	2.1883	2.1894	2.1905	2.1917	2.1928	2.1939	2.1950	2.1961
9.0	2.1972	2.1983	2.1994	2.2006	2.2017	2.2028	2.2039	2.2050	2.2061	2.2072
9.1	2.2083	2.2094	2.2105	2.2116	2.2127	2.2138	2.2148	2.2159	2.2170	2.2181
9.2	2.2192	2.2203	2.2214	2.2225	2.2235	2.2246	2.2257	2.2268	2.2279	2.2289
9.3	2.2300	2.2311	2.2322	2.2332	2.2343	2.2354	2.2364	2.2375	2.2386	2.2396
9.4	2.2407	2.2418	2.2428	2.2439	2.2450	2.2460	2.2471	2.2481	2.2492	2.2502
9.5	2.2513	2.2523	2.2534	2.2544	2.2555	2.2565	2.2576	2.2586	2.2597	2.2607
9.6	2.2618	2.2628	2.2638	2.2649	2.2659	2.2670	2.2680	2.2690	2.2701	2.2711
9.7	2.2721	2.2732	2.2742	2.2752	2.2762	2.2773	2.2783	2.2793	2.2803	2.2814
9.8	2.2824	2.2834	2.2844	2.2854	2.2865	2.2875	2.2885	2.2895	2.2905	2.2915
9.9	2.2925	2.2935	2.2946	2.2956	2.2966	2.2976	2.2986	2.2996	2.3006	2.3016

TABLE B. COMMON LOGARITHMS

n	0	1	2	3	4	5	6	7	8	9
1.0	.0000	.0043	.0086	.0128	.0170	.0212	.0253	.0294	.0334	.0374
1.1	.0414	.0453	.0492	.0531	.0569	.0607	.0645	.0682	.0719	.0755
1.2	.0792	.0828	.0864	.0899	.0934	.0969	.1004	.1038	.1072	.1106
1.3	.1139	.1173	.1206	.1239	.1271	.1303	.1335	.1367	.1399	.1430
1.4	.1461	.1492	.1523	.1553	.1584	.1614	.1644	.1673	.1703	.1732
1.5	.1761	.1790	.1818	.1847	.1875	.1903	.1931	.1959	.1987	.2014
1.6	.2041	.2068	.2095	.2122	.2148	.2175	.2201	.2227	.2253	.2279
1.7	.2304	.2330	.2355	.2380	.2405	.2430	.2455	.2480	.2504	.2529
1.8	.2553	.2577	.2601	.2625	.2648	.2672	.2695	.2718	.2742	.2765
1.9	.2788	.2810	.2833	.2856	.2878	.2900	.2923	.2945	.2967	.2989
2.0	.3010	.3032	.3054	.3075	.3096	.3118	.3139	.3160	.3181	.3201
2.1	.3222	.3243	.3263	.3284	.3304	.3324	.3345	.3365	.3385	.3404
2.2	.3424	.3444	.3464	.3483	.3502	.3522	.3541	.3560	.3579	.3598
2.3	.3617	.3636	.3655	.3674	.3692	.3711	.3729	.3747	.3766	.3784
2.4	.3802	.3820	.3838	.3856	.3874	.3892	.3909	.3927	.3945	.3962
2.5	.3979	.3997	.4014	.4031	.4048	.4065	.4082	.4099	.4116	.4133
2.6	.4150	.4166	.4183	.4200	.4216	.4232	.4249	.4265	.4281	.4298
2.7	.4314	.4330	.4346	.4362	.4378	.4393	.4409	.4425	.4440	.4456
2.8	.4472	.4487	.4502	.4518	.4533	.4548	.4564	.4579	.4594	.4609
2.9	.4624	.4639	.4654	.4669	.4683	.4698	.4713	.4728	.4742	.4757
3.0	.4771	.4786	.4800	.4814	.4829	.4843	.4857	.4871	.4886	.4900
3.1	.4914	.4928	.4942	.4955	.4969	.4983	.4997	.5011	.5024	.5038
3.2	.5051	.5065	.5079	.5092	.5105	.5119	.5132	.5145	.5159	.5172
3.3	.5185	.5198	.5211	.5224	.5237	.5250	.5263	.5276	.5289	.5302
3.4	.5315	.5328	.5340	.5353	.5366	.5378	.5391	.5403	.5416	.5428
3.5	.5441	.5453	.5465	.5478	.5490	.5502	.5514	.5527	.5539	.5551
3.6	.5563	.5575	.5587	.5599	.5611	.5623	.5635	.5647	.5658	.5670
3.7	.5682	.5694	.5705	.5717	.5729	.5740	.5752	.5763	.5775	.5786
3.8	.5798	.5809	.5821	.5832	.5843	.5855	.5866	.5877	.5888	.5899
3.9	.5911	.5922	.5933	.5944	.5955	.5966	.5977	.5988	.5999	.6010
4.0	.6021	.6031	.6042	.6053	.6064	.6075	.6085	.6096	.6107	.6117
4.1	.6128	.6138	.6149	.6160	.6170	.6180	.6191	.6201	.6212	.6222
4.2	.6232	.6243	.6253	.6263	.6274	.6284	.6294	.6304	.6314	.6325
4.3	.6335	.6345	.6355	.6365	.6375	.6385	.6395	.6405	.6415	.6425
4.4	.6435	.6444	.6454	.6464	.6474	.6484	.6493	.6503	.6513	.6522
4.5	.6532	.6542	.6551	.6561	.6571	.6580	.6590	.6599	.6609	.6618
4.6	.6628	.6637	.6646	.6656	.6665	.6675	.6684	.6693	.6702	.6712
4.7	.6721	.6730	.6739	.6749	.6758	.6767	.6776	.6785	.6794	.6803
4.8	.6812	.6821	.6830	.6839	.6848	.6857	.6866	.6875	.6884	.6893
4.9	.6902	.6911	.6920	.6928	.6937	.6946	.6955	.6964	.6972	.6981
5.0	.6990	.6998	.7007	.7016	.7024	.7033	.7042	.7050	.7059	.7067
5.1	.7076	.7084	.7093	.7101	.7110	.7118	.7126	.7135	.7143	.7152
5.2	.7160	.7168	.7177	.7185	.7193	.7202	.7210	.7218	.7226	.7235
5.3	.7243	.7251	.7259	.7267	.7275	.7284	.7292	.7300	.7308	.7316
5.4	.7324	.7332	.7340	.7348	.7356	.7364	.7372	.7380	.7388	.7396

TABLE B. COMMON LOGARITHMS

n	0	1	2	3	4	5	6	7	8	9
5.5	.7404	.7412	.7419	.7427	.7435	.7443	.7451	.7459	.7466	.7474
5.6	.7482	.7490	.7497	.7505	.7513	.7520	.7528	.7536	.7543	.7551
5.7	.7559	.7566	.7574	.7582	.7589	.7597	.7604	.7612	.7619	.7627
5.8	.7634	.7642	.7649	.7657	.7664	.7672	.7679	.7686	.7694	.7701
5.9	.7709	.7716	.7723	.7731	.7738	.7745	.7752	.7760	.7767	.7774
6.0	.7782	.7789	.7796	.7803	.7810	.7818	.7825	.7832	.7839	.7846
6.1	.7853	.7860	.7868	.7875	.7882	.7889	.7896	.7903	.7910	.7917
6.2	.7924	.7931	.7938	.7945	.7952	.7959	.7966	.7973	.7980	.7987
6.3	.7993	.8000	.8007	.8014	.8021	.8028	.8035	.8041	.8048	.8055
6.4	.8062	.8069	.8075	.8082	.8089	.8096	.8102	.8109	.8116	.8122
6.5	.8129	.8136	.8142	.8149	.8156	.8162	.8169	.8176	.8182	.8189
6.6	.8195	.8202	.8209	.8215	.8222	.8228	.8235	.8241	.8248	.8254
6.7	.8261	.8267	.8274	.8280	.8287	.8293	.8299	.8306	.8312	.8319
6.8	.8325	.8331	.8338	.8344	.8351	.8357	.8363	.8370	.8376	.8382
6.9	.8388	.8395	.8401	.8407	.8414	.8420	.8426	.8432	.8439	.8445
7.0	.8451	.8457	.8463	.8470	.8476	.8482	.8488	.8494	.8500	.8506
7.1	.8513	.8519	.8525	.8531	.8537	.8543	.8549	.8555	.8561	.8567
7.2	.8573	.8579	.8585	.8591	.8597	.8603	.8609	.8615	.8621	.8627
7.3	.8633	.8639	.8645	.8651	.8657	.8663	.8669	.8675	.8681	.8686
7.4	.8692	.8698	.8704	.8710	.8716	.8722	.8727	.8733	.8739	.8745
7.5	.8751	.8756	.8762	.8768	.8774	.8779	.8785	.8791	.8797	.8802
7.6	.8808	.8814	.8820	.8825	.8831	.8837	.8842	.8848	.8854	.8859
7.7	.8865	.8871	.8876	.8882	.8887	.8893	.8899	.8904	.8910	.8915
7.8	.8921	.8927	.8932	.8938	.8943	.8949	.8954	.8960	.8965	.8971
7.9	.8976	.8982	.8987	.8993	.8998	.9004	.9009	.9015	.9020	.9025
8.0	.9031	.9036	.9042	.9047	.9053	.9058	.9063	.9069	.9074	.9079
8.1	.9085	.9090	.9096	.9101	.9106	.9112	.9117	.9122	.9128	.9133
8.2	.9138	.9143	.9149	.9154	.9159	.9165	.9170	.9175	.9180	.9186
8.3	.9191	.9196	.9201	.9206	.9212	.9217	.9222	.9227	.9232	.9238
8.4	.9243	.9248	.9253	.9258	.9263	.9269	.9274	.9279	.9284	.9289
8.5	.9294	.9299	.9304	.9309	.9315	.9320	.9325	.9330	.9335	.9340
8.6	.9345	.9350	.9355	.9360	.9365	.9370	.9375	.9380	.9385	.9390
8.7	.9395	.9400	.9405	.9410	.9415	.9420	.9425	.9430	.9435	.9440
8.8	.9445	.9450	.9455	.9460	.9465	.9469	.9474	.9479	.9484	.9489
8.9	.9494	.9499	.9504	.9509	.9513	.9518	.9523	.9528	.9533	.9538
9.0	.9542	.9547	.9552	.9557	.9562	.9566	.9571	.9576	.9581	.9586
9.1	.9590	.9595	.9600	.9605	.9609	.9614	.9619	.9624	.9628	.9633
9.2	.9638	.9643	.9647	.9652	.9657	.9661	.9666	.9671	.9675	.9680
9.3	.9685	.9689	.9694	.9699	.9703	.9708	.9713	.9717	.9722	.9727
9.4	.9731	.9736	.9741	.9745	.9750	.9754	.9759	.9763	.9768	.9773
9.5	.9777	.9782	.9786	.9791	.9795	.9800	.9805	.9809	.9814	.9818
9.6	.9823	.9827	.9832	.9836	.9841	.9845	.9850	.9854	.9859	.9863
9.7	.9868	.9872	.9877	.9881	.9886	.9890	.9894	.9899	.9903	.9908
9.8	.9912	.9917	.9921	.9926	.9930	.9934	.9939	.9943	.9948	.9952
9.9	.9956	.9961	.9965	.9969	.9974	.9978	.9983	.9987	.9991	.9996

TABLE C. TRIGONOMETRIC FUNCTIONS (degrees)

Deg.	Sin	Tan	Cot	Cos		Deg.	Sin	Tan	Cot	Cos	
0.0	0.00000	0.00000	∞	1.0000	**90.0**	**6.0**	0.10453	0.10510	9.514	0.9945	**84.0**
.1	.00175	.00175	573.0	1.0000	89.9	.1	.10626	.10687	9.357	.9943	83.9
.2	.00349	.00349	286.5	1.0000	.8	.2	.10800	.10863	9.205	.9942	.8
.3	.00524	.00524	191.0	1.0000	.7	.3	.10973	.11040	9.058	.9940	.7
.4	.00698	.00698	143.24	1.0000	.6	.4	.11147	.11217	8.915	.9938	.6
.5	.00873	.00873	114.59	1.0000	.5	.5	.11320	.11394	8.777	.9936	.5
.6	.01047	.01047	95.49	0.9999	.4	.6	.11494	.11570	8.643	.9934	.4
.7	.01222	.01222	81.85	.9999	.3	.7	.11667	.11747	8.513	.9932	.3
.8	.01396	.01396	71.62	.9999	.2	.8	.11840	.11924	8.386	.9930	.8
.9	.01571	.01571	63.66	.9999	89.1	.9	.12014	.12101	8.264	.9928	83.1
1.0	0.01745	0.01746	57.29	0.9998	**89.0**	**7.0**	0.12187	0.12278	8.144	0.9925	**83.0**
.1	.01920	.01920	52.08	.9998	88.9	.1	.12360	.12456	8.028	.9923	82.9
.2	.02094	.02095	47.74	.9998	.8	.2	.12533	.12633	7.916	.9921	.8
.3	.02269	.02269	44.07	.9997	.7	.3	.12706	.12810	7.806	.9919	.7
.4	.02443	.02444	40.92	.9997	.6	.4	.12880	.12988	7.700	.9917	.6
.5	.02618	.02619	38.19	.9997	.5	.5	.13053	.13165	7.596	.9914	.5
.6	.02792	.02793	35.80	.9996	.4	.6	.13226	.13343	7.495	.9912	.4
.7	.02967	.02968	33.69	.9996	.3	.7	.13399	.13521	7.396	.9910	.3
.8	.03141	.03143	31.82	.9995	.2	.8	.13572	.13698	7.300	.9907	.2
.9	.03316	.03317	30.14	.9995	88.1	.9	.13744	.13876	7.207	.9905	82.1
2.0	0.03490	0.03492	28.64	0.9994	**88.0**	**8.0**	0.13917	0.14054	7.115	0.9903	**82.0**
.1	.03664	.03667	27.27	.9993	87.9	.1	.14090	.14232	7.026	.9900	81.9
.2	.03839	.03842	26.03	.9993	.8	.2	.14263	.14410	6.940	.9898	.8
.3	.04013	.04016	24.90	.9992	.7	.3	.14436	.14588	6.855	.9895	.7
.4	.04188	.04191	23.86	.9991	.6	.4	.14608	.14767	6.772	.9893	.6
.5	.04362	.04366	22.90	.9990	.5	.5	.14781	.14945	6.691	.9890	.5
.6	.04536	.04541	22.02	.9990	.4	.6	.14954	.15124	6.612	.9888	.4
.7	.04711	.04716	21.20	.9989	.3	.7	.15126	.15302	6.535	.9885	.3
.8	.04885	.04891	20.45	.9988	.2	.8	.15299	.15481	6.460	.9882	.2
.9	.05059	.05066	19.74	.9987	87.1	.9	.15471	.15660	6.386	.9880	81.1
3.0	0.05234	0.05241	19.081	0.9986	**87.0**	**9.0**	0.15643	0.15838	6.314	0.9877	**81.0**
.1	.05408	.05416	18.464	.9985	86.9	.1	.15816	.16017	6.243	.9874	80.9
.2	.05582	.05591	17.886	.9984	.8	.2	.15988	.16196	6.174	.9871	.8
.3	.05756	.05766	17.343	.9983	.7	.3	.16160	.16376	6.107	.9869	.7
.4	.05931	.05941	16.832	.9982	.6	.4	.16333	.16555	6.041	.9866	.6
.5	.06105	.06116	16.350	.9981	.5	.5	.16505	.16734	5.976	.9863	.5
.6	.06279	.06291	15.895	.9980	.4	.6	.16677	.16914	5.912	.9860	.4
.7	.06453	.06467	15.464	.9979	.3	.7	.16849	.17093	5.850	.9857	.3
.8	.06627	.06642	15.056	.9978	.2	.8	.17021	.17273	5.789	.9854	.2
.9	.06802	.06817	14.669	.9977	86.1	.9	.17193	.17453	5.730	.9851	80.1
4.0	0.06976	0.06993	14.301	0.9976	**86.0**	**10.0**	0.1736	0.1763	5.671	0.9848	**80.0**
.1	.07150	.07168	13.951	.9974	85.9	.1	.1754	.1781	5.614	.9845	79.9
.2	.07324	.07344	13.617	.9973	.8	.2	.1771	.1799	5.558	.9842	.8
.3	.07498	.07519	13.300	.9972	.7	.3	.1788	.1817	5.503	.9839	.7
.4	.07672	.07695	12.996	.9971	.6	.4	.1805	.1835	5.449	.9836	.6
.5	.07846	.07870	12.706	.9969	.5	.5	.1822	.1853	5.396	.9833	.5
.6	.08020	.08046	12.429	.9968	.4	.6	.1840	.1871	5.343	.9829	.4
.7	.08194	.08221	12.163	.9966	.3	.7	.1857	.1890	5.292	.9826	.3
.8	.08368	.08397	11.909	.9965	.2	.8	.1874	.1908	5.242	.9823	.2
.9	.08542	.08573	11.664	.9963	85.1	.9	.1891	.1926	5.193	.9820	79.1
5.0	0.08716	0.08749	11.430	0.9962	**85.0**	**11.0**	0.1908	0.1944	5.145	0.9816	**79.0**
.1	.08889	.08925	11.205	.9960	84.9	.1	.1925	.1962	5.079	.9813	78.9
.2	.09063	.09101	10.988	.9959	.8	.2	.1942	.1980	5.050	.9810	.8
.3	.09237	.09277	10.780	.9957	.7	.3	.1959	.1998	5.005	.9806	.7
.4	.09411	.09453	10.579	.9956	.6	.4	.1977	.2016	4.959	.9803	.6
.5	.09585	.09629	10.385	.9954	.5	.5	.1994	.2035	4.915	.9799	.5
.6	.09758	.09805	10.199	.9952	.4	.6	.2011	.2053	4.872	.9796	.4
.7	.09932	.09981	10.019	.9951	.3	.7	.2028	.2071	4.829	.9792	.3
.8	.10106	.10158	9.845	.9949	.2	.8	.2045	.2089	4.787	.9789	.2
.9	.10279	.10334	9.677	.9947	84.1	.9	.2062	.2107	4.745	.9785	78.1
6.0	0.10453	0.10510	9.514	0.9945	**84.0**	**12.0**	0.2079	0.2126	4.705	0.9781	**78.0**
	Cos	Cot	Tan	Sin	Deg.		Cos	Cot	Tan	Sin	Deg.

TABLE C. TRIGONOMETRIC FUNCTIONS (degrees)

Deg.	Sin	Tan	Cot	Cos		Deg.	Sin	Tan	Cot	Cos	
12.0	0.2079	0.2126	4.705	0.9781	**78.0**	**18.0**	0.3090	0.3249	3.078	0.9511	**72.0**
.1	.2096	.2144	4.665	.9778	77.9	.1	.3107	.3269	3.060	.9505	71.9
.2	.2113	.2162	4.625	.9774	.8	.2	.3123	.3288	3.042	.9500	.8
.3	.2130	.2180	4.586	.9770	.7	.3	.3140	.3307	3.024	.9494	.7
.4	.2147	.2199	4.548	.9767	.6	.4	.3156	.3327	3.006	.9489	.6
.5	.2164	.2217	4.511	.9763	.5	.5	.3173	.3346	2.989	.9483	.5
.6	.2181	.2235	4.474	.9759	.4	.6	.3190	.3365	2.971	.9478	.4
.7	.2198	.2254	4.437	.9755	.3	.7	.3206	.3385	2.954	.9472	.3
.8	.2215	.2272	4.402	.9751	.2	.8	.3223	.3404	2.937	.9466	.2
.9	.2233	.2290	4.366	.9748	77.1	.9	.3239	.3424	2.921	.9461	71.1
13.0	0.2250	0.2309	4.331	0.9744	**77.0**	**19.0**	0.3256	0.3443	2.904	0.9455	**71.0**
.1	.2267	.2327	4.297	.9740	76.9	.1	.3272	.3463	2.888	.9449	70.9
.2	.2284	.2345	4.264	.9736	.8	.2	.3289	.3482	2.872	.9444	.8
.3	.2300	.2364	4.230	.9732	.7	.3	.3305	.3502	2.856	.9438	.7
.4	.2317	.2382	4.198	.9728	.6	.4	.3322	.3522	2.840	.9432	.6
.5	.2334	.2401	4.165	.9724	.5	.5	.3338	.3541	2.824	.9426	.5
.6	.2351	.2419	4.134	.9720	.4	.6	.3355	.3561	2.808	.9421	.4
.7	.2368	.2438	4.102	.9715	.3	.7	.3371	.3581	2.793	.9415	.3
.8	.2385	.2456	4.071	.9711	.2	.8	.3387	.3600	2.778	.9409	.2
.9	.2402	.2475	4.041	.9707	76.1	.9	.3404	.3620	2.762	.9403	70.1
14.0	0.2419	0.2493	4.011	0.9703	**76.0**	**20.0**	0.3420	0.3640	2.747	0.9397	**70.0**
.1	.2436	.2512	3.981	.9699	75.9	.1	.3437	.3659	2.733	.9391	69.9
.2	.2453	.2530	3.952	.9694	.8	.2	.3453	.3679	2.718	.9385	.8
.3	.2470	.2549	3.923	.9690	.7	.3	.3469	.3699	2.703	.9379	.7
.4	.2487	.2568	3.895	.9686	.6	.4	.3486	.3719	2.689	.9373	.6
.5	.2504	.2586	3.867	.9681	.5	.5	.3502	.3739	2.675	.9367	.5
.6	.2521	.2605	3.839	.9677	.4	.6	.3518	.3759	2.660	.9361	.4
.7	.2538	.2623	3.812	.9673	.3	.7	.3535	.3779	2.646	.9354	.3
.8	.2554	.2642	3.785	.9668	.2	.8	.3551	.3799	2.633	.9348	.2
.9	.2571	.2661	3.758	.9664	75.1	.9	.3567	.3819	2.619	.9342	69.1
15.0	0.2588	0.2679	3.732	0.9659	**75.0**	**21.0**	0.3584	0.3839	2.605	0.9336	**69.0**
.1	.2605	.2698	3.706	.9655	74.9	.1	.3600	.3859	2.592	.9330	68.9
.2	.2622	.2717	3.681	.9650	.8	.2	.3616	.3879	2.578	.9323	.8
.3	.2639	.2736	3.655	.9646	.7	.3	.3633	.3899	2.565	.9317	.7
.4	.2656	.2754	3.630	.9641	.6	.4	.3649	.3919	2.552	.9311	.6
.5	.2672	.2773	3.606	.9636	.5	.5	.3665	.3939	2.539	.9304	.5
.6	.2689	.2792	3.582	.9632	.4	.6	.3681	.3959	2.526	.9298	.4
.7	.2706	.2811	3.558	.9627	.3	.7	.3697	.3979	2.513	.9291	.3
.8	.2723	.2830	3.534	.9622	.2	.8	.3714	.4000	2.500	.9285	.2
.9	.2740	.2849	3.511	.9617	74.1	.9	.3730	.4020	2.488	.9278	68.1
16.0	0.2756	0.2867	3.487	0.9613	**74.0**	**22.0**	0.3746	0.4040	2.475	0.9272	**68.0**
.1	.2773	.2886	3.465	.9608	73.9	.1	.3762	.4061	2.463	.9265	67.9
.2	.2790	.2905	3.442	.9603	.8	.2	.3778	.4081	2.450	.9259	.8
.3	.2807	.2924	3.420	.9598	.7	.3	.3795	.4101	2.438	.9252	.7
.4	.2823	.2943	3.398	.9593	.6	.4	.3811	.4122	2.426	.9245	.6
.5	.2840	.2962	3.376	.9588	.5	.5	.3827	.4142	2.414	.9239	.5
.6	.2857	.2981	3.354	.9583	.4	.6	.3843	.4163	2.402	.9232	.4
.7	.2874	.3000	3.333	.9578	.3	.7	.3859	.4183	2.391	.9225	.3
.8	.2890	.3019	3.312	.9573	.2	.8	.3875	.4204	2.379	.9219	.2
.9	.2907	.3038	3.291	.9568	73.1	.9	.3891	.4224	2.367	.9212	67.1
17.0	0.2924	0.3057	3.271	0.9563	**73.0**	**23.0**	0.3907	0.4245	2.356	0.9205	**67.0**
.1	.2940	.3076	3.251	.9558	72.9	.1	.3923	.4265	2.344	.9198	66.9
.2	.2957	.3096	3.230	.9553	.8	.2	.3939	.4286	2.333	.9191	.8
.3	.2974	.3115	3.211	.9548	.7	.3	.3955	.4307	2.322	.9184	.7
.4	.2990	.3134	3.191	.9542	.6	.4	.3971	.4327	2.311	.9178	.6
.5	.3007	.3153	3.172	.9537	.5	.5	.3987	.4348	2.300	.9171	.5
.6	.3024	.3172	3.152	.9532	.4	.6	.4003	.4369	2.289	.9164	.4
.7	.3040	.3191	3.133	.9527	.3	.7	.4019	.4390	2.278	.9157	.3
.8	.3057	.3211	3.115	.9521	.2	.8	.4035	.4411	2.267	.9150	.2
.9	.3074	.3230	3.096	.9516	72.1	.9	.4051	.4431	2.257	.9143	66.1
18.0	0.3090	0.3249	3.078	0.9511	**72.0**	**24.0**	0.4067	0.4452	2.246	0.9135	**66.0**
	Cos	Cot	Tan	Sin	Deg.		Cos	Cot	Tan	Sin	Deg.

TABLE C. TRIGONOMETRIC FUNCTIONS (degrees)

Deg.	Sin.	Tan	Cot	Cos		Deg.	Sin	Tan	Cot	Cos	
24.0	0.4067	0.4452	2.246	0.9135	**66.0**	**30.0**	0.5000	0.5774	1.7321	0.8660	**60.0**
.1	.4083	.4473	2.236	.9128	65.9	.1	.5015	.5797	1.7251	.8652	59.9
.2	.4099	.4494	2.225	.9121	.8	.2	.5030	.5820	1.7182	.8643	.8
.3	.4115	.4515	2.215	.9114	.7	.3	.5045	.5844	1.7113	.8634	.7
.4	.4131	.4536	2.204	.9107	.6	.4	.5060	.5867	1.7045	.8625	.6
.5	.4147	.4557	2.194	.9100	.5	.5	.5075	.5890	1.6977	.8616	.5
.6	.4163	.4578	2.184	.9092	.4	.6	.5090	.5914	1.6909	.8607	.4
.7	.4179	.4599	2.174	.9085	.3	.7	.5105	.5938	1.6842	.8599	.3
.8	.4195	.4621	2.164	.9078	.2	.8	.5120	.5961	1.6775	.8590	.2
.9	.4210	.4642	2.154	.9070	65.1	.9	.5135	.5985	1.6709	.8581	59.1
25.0	0.4226	0.4663	2.145	0.9063	**65.0**	**31.0**	0.5150	0.6009	1.6643	0.8572	**59.0**
.1	.4242	.4684	2.135	.9056	64.9	.1	.5165	.6032	1.6577	.8563	58.9
.2	.4258	.4706	2.125	.9048	.8	.2	.5180	.6056	1.6512	.8554	.8
.3	.4274	.4727	2.116	.9041	.7	.3	.5195	.6080	1.6447	.8545	.7
.4	.4289	.4748	2.106	.9033	.6	.4	.5210	.6104	1.6383	.8536	.6
.5	.4305	.4770	2.097	.9026	.5	.5	.5225	.6128	1.6319	.8526	.5
.6	.4321	.4791	2.087	.9018	.4	.6	.5240	.6152	1.6255	.8517	.4
.7	.4337	.4813	2.078	.9011	.3	.7	.5255	.6176	1.6191	.8508	.3
.8	.4352	.4834	2.069	.9003	.2	.8	.5270	.6200	1.6128	.8499	.2
.9	.4368	.4856	2.059	.8996	64.1	.9	.5284	.6224	1.6066	.8490	58.1
26.0	0.4384	0.4887	2.050	0.8988	**64.0**	**32.0**	0.5299	0.6249	1.6003	0.8480	**58.0**
.1	.4399	.4899	2.041	.8980	63.9	.1	.5314	.6273	1.5941	.8471	57.9
.2	.4415	.4921	2.032	.8973	.8	.2	.5329	.6297	1.5880	.8462	.8
.3	.4431	.4942	2.023	.8965	.7	.3	.5344	.6322	1.5818	.8453	.7
.4	.4446	.4964	2.014	.8957	.6	.4	.5358	.6346	1.5757	.8443	.6
.5	.4462	.4986	2.006	.8949	.5	.5	.5373	.6371	1.5697	.8434	.5
.6	.4478	.5008	1.997	.8942	.4	.6	.5388	.6395	1.5637	.8425	.4
.7	.4493	.5029	1.988	.8934	.3	.7	.5402	.6420	1.5577	.8415	.3
.8	.4509	.5051	1.980	.8926	.2	.8	.5417	.6445	1.5517	.8406	.2
.9	.4524	.5073	1.971	.8918	63.1	.9	.5432	.6469	1.5458	.8396	57.1
27.0	0.4540	0.5095	1.963	0.8910	**63.0**	**33.0**	0.5446	0.6494	1.5399	0.8387	**57.0**
.1	.4555	.5117	1.954	.8902	62.9	.1	.5461	.6519	1.5340	.8377	56.9
.2	.4571	.5139	1.946	.8894	.8	.2	.5476	.6544	1.5282	.8368	.8
.3	.4586	.5161	1.937	.8886	.7	.3	.5490	.6569	1.5224	.8358	.7
.4	.4602	.5184	1.929	.8878	.6	.4	.5505	.6594	1.5166	.8348	.6
.5	.4617	.5206	1.921	.8870	.5	.5	.5519	.6619	1.5108	.8339	.5
.6	.4633	.5228	1.913	.8862	.4	.6	.5534	.6644	1.5051	.8329	.4
.7	.4648	.5250	1.905	.8854	.3	.7	.5548	.6669	1.4994	.8320	.3
.8	.4664	.5272	1.897	.8846	.2	.8	.5563	.6694	1.4938	.8310	.2
.9	.4679	.5295	1.889	.8838	62.1	.9	.5577	.6720	1.4882	.8300	56.1
28.0	0.4695	0.5317	1.881	0.8829	**62.0**	**34.0**	0.5592	0.6745	1.4826	0.8290	**56.0**
.1	.4710	.5340	1.873	.8821	61.9	.1	.5606	.6771	1.4770	.8281	55.9
.2	.4726	.5362	1.865	.8813	.8	.2	.5621	.6796	1.4715	.8271	.8
.3	.4741	.5384	1.857	.8805	.7	.3	.5635	.6822	1.4659	.8261	.7
.4	.4756	.5407	1.849	.8796	.6	.4	.5650	.6847	1.4605	.8251	.6
.5	.4772	.5430	1.842	.8788	.5	.5	.5664	.6873	1.4550	.8241	.5
.6	.4787	.5452	1.834	.8780	.4	.6	.5678	.6899	1.4496	.8231	.4
.7	.4802	.5475	1.827	.8771	.3	.7	.5693	.6924	1.4442	.8221	.3
.8	.4818	.5498	1.819	.8763	.2	.8	.5707	.6950	1.4388	.8211	.2
.9	.4833	.5520	1.811	.8755	61.1	.9	.5721	.6976	1.4335	.8202	55.1
29.0	0.4848	0.5543	1.804	0.8746	**61.0**	**35.0**	0.5736	0.7002	1.4281	0.8192	**55.0**
.1	.4863	.5566	1.797	.8738	60.9	.1	.5750	.7028	1.4229	.8181	54.9
.2	.4879	.5589	1.789	.8729	.8	.2	.5764	.7054	1.4176	.8171	.8
.3	.4894	.5612	1.782	.8721	.7	.3	.5779	.7080	1.4124	.8161	.7
.4	.4909	.5635	1.775	.8712	.6	.4	.5793	.7107	1.4071	.8151	.6
.5	.4924	.5658	1.767	.8704	.5	.5	.5807	.7133	1.4019	.8141	.5
.6	.4939	.5681	1.760	.8695	.4	.6	.5821	.7159	1.3968	.8131	.4
.7	.4955	.5704	1.753	.8686	.3	.7	.5835	.7186	1.3916	.8121	.3
.8	.4970	.5727	1.746	.8678	.2	.8	.5850	.7212	1.3865	.8111	.2
.9	.4985	.5750	1.739	.8669	60.1	.9	.5864	.7239	1.3814	.8100	54.1
30.0	0.5000	0.5774	1.732	0.8660	**60.0**	**36.0**	0.5878	0.7265	1.3764	0.8090	**54.0**
	Cos	Cot	Tan	Sin	Deg.		Cos	Cot	Tan	Sin	Deg.

TABLE C. TRIGONOMETRIC FUNCTIONS (degrees)

Deg.	Sin	Tan	Cot	Cos		Deg.	Sin	Tan	Cot	Cos	
36.0	0.5878	0.7265	1.3764	0.8090	**54.0**	**40.5**	0.6494	0.8541	1.1708	0.7604	**49.5**
.1	.5892	.7292	1.3713	.8080	53.9	.6	.6508	.8571	1.1667	.7593	.4
.2	.5906	.7319	1.3663	.8070	.8	.7	.6521	.8601	1.1626	.7581	.3
.3	.5920	.7346	1.3613	.8059	.7	.8	.6534	.8632	1.1585	.7570	.2
.4	.5934	.7373	1.3564	.8049	.6	.9	.6547	.8662	1.1544	.7559	49.1
.5	.5948	.7400	1.3514	.8039	.5	**41.0**	0.6561	0.8693	1.1504	0.7547	**49.0**
.6	.5962	.7427	1.3465	.8028	.4	.1	.6574	.8724	1.1463	.7536	48.9
.7	.5976	.7454	1.3416	.8018	.3	.2	.6587	.8754	1.1423	.7524	.8
.8	.5990	.7481	1.3367	.8007	.2	.3	.6600	.8785	1.1383	.7513	.7
.9	.6004	.7508	1.3319	.7997	53.1	.4	.6613	.8816	1.1343	.7501	.6
37.0	0.6018	0.7536	1.3270	0.7986	**53.0**	.5	.6626	.8847	1.1303	.7490	.5
.1	.6032	.7563	1.3222	.7976	52.9	.6	.6639	.8878	1.1263	.7478	.4
.2	.6046	.7590	1.3175	.7965	.8	.7	.6652	.8910	1.1224	.7466	.3
.3	.6060	.7618	1.3127	.7955	.7	.8	.6665	.8941	1.1184	.7455	.2
.4	.6074	.7646	1.3079	.7944	.6	.9	.6678	.8972	1.1145	.7443	48.1
.5	.6088	.7673	1.3032	.7934	.5	**42.0**	0.6691	0.9004	1.1106	0.7431	**48.0**
.6	.6101	.7701	1.2985	.7923	.4	.1	.6704	.9036	1.1067	.7420	47.9
.7	.6115	.7729	1.2938	.7912	.3	.2	.6717	.9067	1.1028	.7408	.8
.8	.6129	.7757	1.2892	.7902	.2	.3	.6730	.9099	1.0990	.7396	.7
.9	.6143	.7785	1.2846	.7891	52.1	.4	.6743	.9131	1.0951	.7385	.6
38.0	0.6157	0.7813	1.2799	0.7880	**52.0**	.5	.6756	.9163	1.0913	.7373	.5
.1	.6170	.7841	1.2753	.7869	51.9	.6	.6769	.9195	1.0875	.7361	.4
.2	.6184	.7869	1.2708	7859	.8	.7	.6782	.9228	1.0837	.7349	.3
.3	.6198	.7898	1.2662	.7848	.7	.8	.6794	.9260	1.0799	.7337	.2
.4	.6211	.7926	1.2617	.7837	.6	.9	.6807	.9293	1.0761	.7325	47.1
.5	.6225	.7954	1.2572	.7826	.5	**43.0**	0.6820	0.9325	1.0724	0.7314	**47.0**
.6	.6239	.7983	1.2527	.7815	.4	.1	.6833	.9358	1.0686	.7302	46.9
.7	.6252	.8012	1.2482	.7804	.3	.2	.6845	.9391	1.0649	.7290	.8
.8	.6266	.8040	1.2437	.7793	.2	.3	.6858	.9424	1.0612	.7278	.7
.9	.6280	.8069	1.2393	.7782	51.1	.4	.6871	.9457	1.0575	.7266	.6
39.0	0.6293	0.8098	1.2349	0.7771	**51.0**	.5	.6884	.9490	1.0538	.7254	.5
.1	.6307	.8127	1.2305	.7760	50.9	.6	.6896	.9523	1.0501	.7242	.4
.2	.6320	.8156	1.2261	.7749	.8	.7	.6909	.9556	1.0464	.7230	.3
.3	.6334	.8185	1.2218	.7738	.7	.8	.6921	.9590	1.0428	.7218	.2
.4	.6347	.8214	1.2174	.7727	.6	.9	.6934	.9623	1.0392	.7206	46.1
.5	.6361	.8243	1.2131	.7716	.5	**44.0**	0.6947	0.9657	1.0355	0.7193	**46.0**
.6	.6374	.8273	1.2088	.7705	.4	.1	.6959	.9691	1.0319	.7181	45.9
.7	.6388	.8302	1.2045	.7694	.3	.2	.6972	.9725	1.0283	.7169	.8
.8	.6401	.8332	1.2002	.7683	.2	.3	.6984	.9759	1.0247	.7157	.7
.9	.6414	.8361	1.1960	.7672	50.1	.4	.6997	.9793	1.0212	.7145	.6
40.0	0.6428	0.8391	1.1918	0.7660	**50.0**	.5	.7009	.9827	1.0176	.7133	.5
.1	.6441	.8421	1.1875	.7649	49.9	.6	.7022	.9861	1.0141	.7120	.4
.2	.6455	.8451	1.1833	.7638	.8	.7	.7034	.9896	1.0105	.7108	.3
.3	.6468	.8481	1.1792	.7627	.7	.8	.7046	.9930	1.0070	.7096	.2
.4	.6481	.8511	1.1750	.7615	.6	.9	.7059	.9965	1.0035	.7083	45.1
40.5	0.6494	0.8541	1.1708	0.7604	**49.5**	**45.0**	0.7071	1.0000	1.0000	0.7071	**45.0**
	Cos	Cot	Tan	Sin	Deg.		Cos	Cot	Tan	Sin	Deg.

TABLE D. TRIGONOMETRIC FUNCTIONS (radians)

Rad.	Sin	Tan	Cot	Cos	Rad.	Sin	Tan	Cot	Cos
.00	.00000	.00000	∞	1.00000	**.50**	.47943	.54630	1.8305	.87758
.01	.01000	.01000	99.997	0.99995	.51	.48818	.55936	1.7878	.87274
.02	.02000	.02000	49.993	.99980	.52	.49688	.57256	1.7465	.86782
.03	.03000	.03001	33.323	.99955	.53	.50553	.58592	1.7067	.86281
.04	.03999	.04002	24.987	.99920	.54	.51414	.59943	1.6683	.85771
.05	.04998	.05004	19.983	.99875	.55	.52269	.61311	1.6310	.85252
.06	.05996	.06007	16.647	.99820	.56	.53119	.62695	1.5950	.84726
.07	.06994	.07011	14.262	.99755	.57	.53963	.64097	1.5601	.84190
.08	.07991	.08017	12.473	.99680	.58	.54802	.65517	1.5263	.83646
.09	.08988	.09024	11.081	.99595	.59	.55636	.66956	1.4935	.83094
.10	.09983	.10033	9.9666	.99500	**.60**	.56464	.68414	1.4617	.82534
.11	.10978	.11045	9.0542	.99396	.61	.57287	.69892	1.4308	.81965
.12	.11971	.12058	8.2933	.99281	.62	.58104	.71391	1.4007	.81388
.13	.12963	.13074	7.6489	.99156	.63	.58914	.72911	1.3715	.80803
.14	.13954	.14092	7.0961	.99022	.64	.59720	.74454	1.3431	.80210
.15	.14944	.15114	6.6166	.98877	.62	.60519	.76020	1.3154	.79608
.16	.15932	.16138	6.1966	.98723	.66	.61312	.77610	1.2885	.78999
.17	.16918	.17166	5.8256	.98558	.67	.62099	.79225	1.2622	.78382
.18	.17903	.18197	5.4954	.98384	.68	.62879	.80866	1.2366	.77757
.19	.18886	.19232	5.1997	.98200	.69	.63654	.82534	1.2116	.77125
.20	.19867	.20271	4.9332	.98007	**.70**	.64422	.84229	1.1872	.76484
.21	.20846	.21314	4.6917	.97803	.71	.65183	.85953	1.1634	.75836
.22	.21823	.22362	4.4719	.97590	.72	.65938	.87707	1.1402	.75181
.23	.22798	.23414	4.2709	.97367	.73	.66687	.89492	1.1174	.74517
.24	.23770	.24472	4.0864	.97134	.74	.67429	.91309	1.0952	.73847
.25	.24740	.25534	3.9163	.96891	.75	.68164	.93160	1.0734	.73169
.26	.25708	.26602	3.7591	.96639	.76	.68892	.95045	1.0521	.72484
.27	.26673	.27676	3.6133	.96377	.77	.69614	.96967	1.0313	.71791
.28	.27636	.28755	3.4776	.96106	.78	.70328	.98926	1.0109	.71091
.29	.28595	.29841	3.3511	.95824	.79	.71035	1.0092	.99084	.70385
.30	.29552	.30934	3.2327	.95534	**.80**	.71736	1.0296	.97121	.69671
.31	.30506	.32033	3.1218	.95233	.81	.72429	1.0505	.95197	.68950
.32	.31457	.33139	3.0176	.94924	.82	.73115	1.0717	.93309	.68222
.33	.32404	.34252	2.9195	.94604	.83	.73793	1.0934	.91455	.67488
.34	.33349	.35374	2.8270	.94275	.84	.74464	1.1156	.89635	.66746
.35	.34290	.36503	2.7395	.93937	.85	.75128	1.1383	.87848	.65998
.36	.35227	.37640	2.6567	.93590	.86	.75784	1.1616	.86091	.65244
.37	.36162	.38786	2.5782	.93233	.87	.76433	1.1853	.84365	.64483
.38	.37092	.39941	2.5037	.92866	.88	.77074	1.2097	.82668	.63715
.39	.38019	.41105	2.4328	.92491	.89	.77707	1.2346	.80998	.62941
.40	.38942	.42279	2.3652	.92106	**.90**	.78333	1.2602	.79355	.62161
.41	.39861	.43463	2.3008	.91712	.91	.78950	1.2864	.77738	.61375
.42	.40776	.44657	2.2393	.91309	.92	.79560	1.3133	.76146	.60582
.43	.41687	.45862	2.1804	.90897	.93	.80162	1.3409	.74578	.59783
.44	.42594	.47078	2.1241	.90475	.94	.80756	1.3692	.73034	.58979
.45	.43497	.48306	2.0702	.90045	.95	.81342	1.3984	.71511	.58168
.46	.44395	.49545	2.0184	.89605	.96	.81919	1.4284	.70010	.57352
.47	.45289	.50797	1.9686	.89157	.97	.82489	1.4592	.68531	.56530
.48	.46178	.52061	1.9208	.88699	.98	.83050	1.4910	.67071	.55702
.49	.47063	.53339	1.8748	.88233	.99	.83603	1.5237	.65631	.54869
.50	.47943	.54630	1.8305	.87758	**1.00**	.84147	1.5574	.64209	.54030
Rad.	Sin	Tan	Cot	Cos	Rad.	Sin	Tan	Cot	Cos

TABLE D. TRIGONOMETRIC FUNCTIONS (radians)

Rad.	Sin.	Tan	Cot	Cos	Rad.	Sin	Tan	Cot	Cos
1.00	.84147	1.5574	.64209	.54030	**1.50**	.99749	14.101	.07091	.07074
1.01	.84683	1.5922	.62806	.53186	1.51	.99815	16.428	.06087	.06076
1.02	.85211	1.6281	.61420	.52337	1.52	.99871	19.670	.05084	.05077
1.03	.85730	1.6652	.60051	.51482	1.53	.99917	24.498	.04082	.04079
1.04	.86240	1.7036	.58699	.50622	1.54	.99953	32.461	.03081	.03079
1.05	.86742	1.7433	.57362	.49757	1.55	.99978	48.078	.02080	.02079
1.06	.87236	1.7844	.56040	.48887	1.56	.99994	92.621	.01080	.01080
1.07	.87720	1.8270	.54734	.48012	1.57	1.00000	1255.8	.00080	.00080
1.08	.88196	1.8712	.53441	.47133	1.58	.99996	−108.65	−.00920	−.00920
1.09	.88663	1.9171	.52162	.46249	1.59	.99982	−52.067	−.01921	−.01920
1.10	.89121	1.9648	.50897	.45360	**1.60**	.99957	−34.233	−.02921	−.02920
1.11	.89570	2.0143	.49644	.44466	1.61	.99923	−25.495	−.03922	−.03919
1.12	.90010	2.0660	.48404	.43568	1.62	.99879	−20.307	−.04924	−.04918
1.13	.90441	2.1198	.47175	.42666	1.63	.99825	−16.871	−.05927	−.05917
1.14	.90863	2.1759	.45959	.41759	1.64	.99761	−14.427	−.06931	−.06915
1.15	.91276	2.2345	.44753	.40849	1.65	.99687	−12.599	−.07937	−.07912
1.16	.91680	2.2958	.43558	.39934	1.66	.99602	−11.181	−.08944	−.08909
1.17	.92075	2.3600	.42373	.39015	1.67	.99508	−10.047	−.09953	−.09904
1.18	.92461	2.4273	.41199	.38092	1.68	.99404	− 9.1208	−.10964	−.10899
1.19	.92837	2.4979	.40034	.37166	1.69	.99290	− 8.3492	−.11977	−.11892
1.20	.93204	2.5722	.38878	.36236	**1.70**	.99166	− 7.6966	−.12993	−.12884
1.21	.93562	2.6503	.37731	.35302	1.71	.99033	− 7.1373	−.14011	−.13875
1.22	.93910	2.7328	.36593	.34365	1.72	.98889	− 6.6524	−.15032	−.14865
1.23	.94249	2.8198	.35463	.33424	1.73	.98735	− 6.2281	−.16056	−.15853
1.24	.94578	2.9119	.34341	.32480	1.74	.98572	− 5.8535	−.17084	−.16840
1.25	.94898	3.0096	.33227	.31532	1.75	.98399	− 5.5204	−.18115	−.17825
1.26	.95209	3.1133	.32121	.30582	1.76	.98215	− 5.2221	−.19149	−.18808
1.27	.95510	3.2236	.31021	.29628	1.77	.98022	− 4.9534	−.20188	−.19789
1.28	.95802	3.3413	.29928	.28672	1.78	.97820	− 4.7101	−.21231	−.20768
1.29	.96084	3.4672	.28842	.27712	1.79	.97607	− 4.4887	−.22278	−.21745
1.30	.96356	3.6021	.27762	.26750	**1.80**	.97385	− 4.2863	−.23330	−.22720
1.31	.96618	3.7471	.26687	.25785	1.81	.97153	− 4.1005	−.24387	−.23693
1.32	.96872	3.9033	.25619	.24818	1.82	.96911	− 3.9294	−.25449	−.24663
1.33	.97115	4.0723	.24556	.23848	1.83	.96659	− 3.7712	−.26517	−.25631
1.34	.97348	4.2556	.23498	.22875	1.84	.96398	− 3.6245	−.27590	−.26596
1.35	.97572	4.4552	.22446	.21901	1.85	.96128	− 3.4881	−.28669	−.27559
1.36	.97786	4.6734	.21398	.20924	1.86	.95847	− 3.3608	−.29755	−.28519
1.37	.97991	4.9131	.20354	.19945	1.87	.95557	− 2.2419	−.30846	−.29476
1.38	.98185	5.1774	.19315	.18964	1.88	.95258	− 3.1304	−.31945	−.30430
1.39	.98370	5.4707	.18279	.17981	1.89	.94949	− 3.0257	−33.051	−.31381
1.40	.98545	5.7979	.17248	.16997	**1.90**	.94630	− 2.9271	−.34164	−.32329
1.41	.98710	6.1654	.16220	.16010	1.91	.94302	− 2.8341	−.35284	−.33274
1.42	.98865	6.5811	.15195	.15023	1.92	.93965	− 2.7463	−.36413	−.34215
1.43	.99010	7.0555	.14173	.14033	1.93	.93618	− 2.6632	−.37549	−.35153
1.44	.99146	7.6018	.13155	.13042	1.94	.93262	− 2.5843	−.38695	−.36087
1.45	.99271	8.2381	.12139	.12050	1.95	.92896	− 2.5095	−.39849	−.37018
1.46	.99387	8.9886	.11125	.11057	1.96	.92521	− 2.4383	−.41012	−.37945
1.47	.99492	9.8874	.10114	.10063	1.97	.92137	− 2.3705	−.42185	−.38868
1.48	.99588	10.983	.09105	.09067	1.98	.91744	− 2.3058	−.43368	−.39788
1.49	.99674	12.350	.08097	.08071	1.99	.91341	− 2.2441	−.44562	−.40703
1.50	.99749	14.101	.07091	.07074	**2.00**	.90930	− 2.1850	−.45766	−.41615
Rad.	Sin	Tan	Cot	Cos	Rad.	Sin	Tan	Cot	Cos

Answers to Odd-Numbered Problems

PROBLEM SET 1-1

1. Let x be one number and y the other; $x + \frac{1}{3}y$. **3.** Let x be one number and y the other; $2x/3y$.
5. Let x be the number; $0.10x + x$, or $1.10x$. **7.** Let x be one side and y the other; $x^2 + y^2$.
9. xy **11.** y/x **13.** $(30/x) + (30/y)$ **15.** x^2 **17.** $6x^2$ **19.** $4\pi(x/2)^2 = \pi x^2$ **21.** $10x^2$
23. $\frac{4}{3}\pi(x/2)^3 = \pi x^3/6$ **25.** $x^3 - 4\pi x$ **27.** $A = x^2 + \frac{1}{4}\pi x^2$; $P = 2x + \pi x$
29. Let x be the number; $x + \frac{1}{2}x = 45$; $x = 30$. **31.** Let x be the smaller odd number; $x + x + 2 = 168$; $x = 83$.
33. $\frac{9}{2}x = 252$; $x = 56$ **35.** 10.5 meters **37.** $V = 64$ cubic inches; $S = 96$ square inches
39. Volume is 8 times as large; surface area is 4 times as large. **41.** $(D/100) + (D/200) = 4$; $D = 266$ miles
43. $400/[(200/50) + (200/70)] = 58.3$ kilometers per hour

PROBLEM SET 1-2

1. 12 **3.** -31 **5.** -67 **7.** $-60 + 9x$ **9.** $14t$ **11.** $\frac{8}{9}$ **13.** $-\frac{3}{4}$ **15.** $(1 - 3x)/2$
17. $(-2x + 3)/2$ **19.** $\frac{21}{12} = \frac{7}{4}$ **21.** $\frac{19}{20}$ **23.** $\frac{23}{36}$ **25.** $\frac{106}{108} = \frac{53}{54}$ **27.** $\frac{1}{2}$ **29.** $\frac{3}{4}$ **31.** $\frac{5}{4}$
33. $\frac{3}{8}$ **35.** $\frac{54}{7}$ **37.** $\frac{17}{7}$ **39.** -17 **41.** $\frac{9}{17}$ **43.** $\frac{25}{18}$ **45.** $-\frac{7}{9}$ **47.** $\frac{1}{24}$ **49.** $\frac{22}{189}$
51. $\frac{4}{11}$ **53.** $-\frac{5}{16}$ **55.** $\frac{15}{4}$ **57.** $\frac{3}{11}$

PROBLEM SET 1-3

1. $2 \cdot 5 \cdot 5 \cdot 5$ **3.** $2 \cdot 2 \cdot 2 \cdot 5 \cdot 5$ **5.** $2 \cdot 2 \cdot 3 \cdot 5 \cdot 5 \cdot 7$ **7.** $2 \cdot 2 \cdot 2 \cdot 5 \cdot 5 \cdot 5 \cdot = 1000$ **9.** $2 \cdot 2 \cdot 3 \cdot 5 \cdot 5 \cdot 5 \cdot 7 = 10{,}500$
11. $2 \cdot 2 \cdot 2 \cdot 3 \cdot 5 \cdot 5 \cdot 5 \cdot 7 = 21{,}000$ **13.** 97/1000 **15.** 289/10,500 **17.** $-1423/21{,}000$ **19.** $.\bar{6}$
21. $.625\bar{0}$ **23.** $.\overline{461538}$ **25.** $\frac{7}{9}$ **27.** 235/999 **29.** 13/40 **31.** $318/990 = 53/165$ **33.** 21,000
35. (a) $.3125\bar{0}$ (b) $.\overline{27}$ **37.** $.\overline{637728}$ **39.** No; for example, $\sqrt{2} + (-\sqrt{2}) = 0$.
41. Follow hint; also, see solution to Problem 43. **43.** Suppose $\frac{3}{4}\sqrt{2} = r$, where r is rational. Then $\sqrt{2} = \frac{4}{3}r$; however, $\frac{4}{3}r$ is rational, which is a contradiction since $\sqrt{2}$ is irrational. **45.** a; b; d; e; g **47.** .000000001; no.
49. Suppose $\sqrt{3} = m/n$, where m and n are positive integers greater than 1. Then $3n^2 = m^2$. Both n^2 and m^2 must have an even number of 3's as factors. This contradicts $3n^2 = m^2$, since $3n^2$ must have an odd number of 3's.
51. It is nonrepeating. **53.** Negative.

PROBLEM SET 1-4

1. 1431 **3.** 1700 **5.** 61 **7.** 3 **9.** 2 **11.** Associative and commutative properties of multiplication.
13. Commutative property of addition.
15. Associative property of addition; additive inverse; zero is neutral element for addition. **17.** Distributive property.
21. True; $a - (b - c) = a + (-1)[b + (-c)] = a + (-1)b + (-1)(-c) = a - b + c$.
23. False; $1 \div (1 + 1) = \frac{1}{2}$, but $(1 \div 1) + (1 \div 1) = 2$. **25.** True; $ab(a^{-1} + b^{-1}) = aba^{-1} + abb^{-1} = b + a$.
27. False; $(1 + 2)(1^{-1} + 2^{-1}) = 3(1 + \frac{1}{2}) = 3(3/2) = \frac{9}{2}$. **29.** False; $(1 + 2)(1 + 2) = 3 \cdot 3 = 9$, but $1^2 + 2^2 = 5$.
31. False; $1 \div (2 \div 3) = 1/\frac{2}{3} = \frac{3}{2}$ but $(1 \div 2) \div 3 = \frac{1}{2}/3 = \frac{1}{6}$. **33.** (a) No; no. (b) Yes; yes. (c) Yes.
35. (a) Yes; yes. (b) Yes; yes. (c) Yes; 5. (d) Yes; 1. **37.** Additive inverse; zero is neutral element for addition; distributive property; associative property of addition; additive inverse; zero is neutral element for addition.
39. No; no.

PROBLEM SET 1-5

1. $>$ **3.** $>$ **5.** $>$ **7.** $>$ **9.** $<$ **11.** $=$ **13.** $-3\sqrt{2}/2$; $-\sqrt{2}$; $-\pi/2$; 3/4; $\sqrt{2}$; 43/24

15.
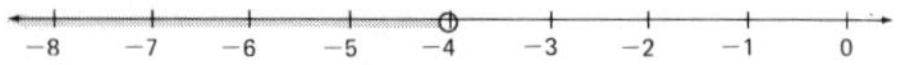

17.
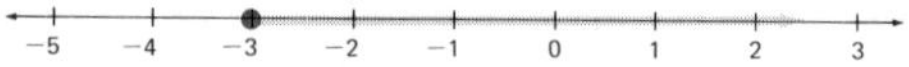

19.
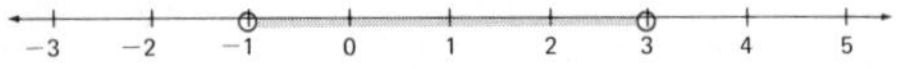

21.

23.
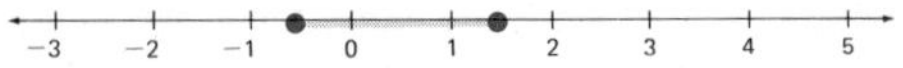

25. $-2 \le x \le 3$ **27.** $x \ge 2$ **29.** $-2 < x \le 3$

31. $-1 < x < 2$ **33.** $-4 \le x \le 4$
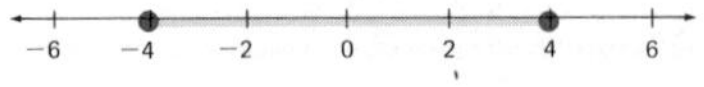

35. $1 < x < 5$ 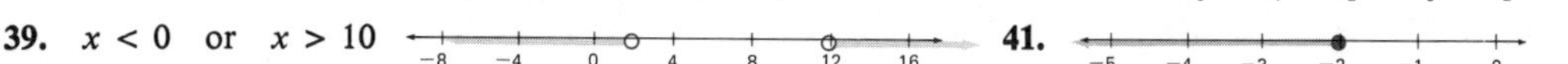 **37.**

39. $x < 0$ or $x > 10$ **41.**

43. **45.** No solutions. **47.**

49. No; $-2 < -1$ but $(-2)^2 > (-1)^2$. **51.** $1/a > 1/b$ **53.** (a) $|x - 6| < 4$ (b) $|x + 1| \le 5$ (c) $|x - 4.5| < 1.5$ (d) $|x - 1.5| < 4.5$

PROBLEM SET 1-6

1. $-2 + 8i$ **3.** $-4 - i$ **5.** $0 + 6i$ **7.** $-4 + 7i$ **9.** $11 + 4i$ **11.** $14 + 22i$ **13.** $16 + 30i$
15. $61 + 0i$ **17.** $\frac{3}{2} + \frac{7}{2}i$ **19.** $2 - 5i$ **21.** $\frac{11}{2} + \frac{3}{2}i$ **23.** $\frac{2}{5} + \frac{1}{5}i$ **25.** $\sqrt{3}/4 - i/4$ **27.** $\frac{1}{5} + \frac{8}{5}i$
29. $(\sqrt{3} - 1/4) + (-\sqrt{3}/4 - 1)i$ **31.** $i^4 = (i^2)(i^2) = (-1)(-1) = 1$. The four 4th roots of 1 are $1, -1, i$, and $-i$.
33. $(1 - i)^4 = (1 - i)^2(1 - i)^2 = (-2i)(-2i) = 4i^2 = -4$ **35.** (a) $4 - 12i$ (b) $14 + 10i$ (c) $\frac{7}{25} - \frac{51}{25}i$
37. $(1 + \sqrt{3}i)(1 + \sqrt{3}i)(1 + \sqrt{3}i) = (-2 + 2\sqrt{3}i)(1 + \sqrt{3}i) = -8$ **39.** $3 - \frac{1}{3}i$

CHAPTER 1. REVIEW PROBLEM SET

1. $V = x^2y$; $S = x^2 + 4xy$ **2.** $100/(x + y) + 100/(x - y)$ **3.** (a) $\frac{15}{24} = \frac{5}{8}$ (b) $\frac{1}{2}$ (c) $\frac{1}{9}$ (d) $\frac{1}{5}$ (e) $\frac{25}{7}$ (f) $\frac{27}{10}$
4. $2\cdot2\cdot2\cdot5\cdot5\cdot5$; $2\cdot2\cdot3\cdot3\cdot5$ **5.** $2\cdot2\cdot2\cdot3\cdot3\cdot5\cdot5\cdot5 = 9000$ **6.** $.\overline{384615}$; $1.5\overline{71428}$ **7.** 257/999, 122/99
8. Yes; no. **9.** $a(bc) = (ab)c$; $a + b = b + a$ **10.** 1.4; $\sqrt{2}$; $1.\bar{4}$; $\frac{29}{20}$, $\frac{13}{8}$

11. $-8 < x < 4$

$-1 \le x \le 2$

12. No. If $x < 0$, then $|-x| = -x$. For example, $|-(-2)| = -(-2)$. **13.** (a) $-6 + 7i$ (b) $17 - 18i$ (c) $\frac{29}{13} + \frac{28}{13}i$ (d) $2 + 11i$ (e) $\frac{5}{34} + \frac{3}{34}i$
14. Suppose that $5 + \sqrt{5} = r$, where r is rational. Then $\sqrt{5} = r - 5$, which is rational, a contradiction. Handle $5\sqrt{5}$ similarly. **15.** 320 square centimeters

PROBLEM SET 2-1

1. $1/5^2 = 1/25$ **3.** $-1/5^2 = -1/25$ **5.** $1/(-2)^5 = -1/32$ **7.** $-27/8$ **9.** $27/4$ **11.** $81/16$
13. $3/64$ **15.** $1/72$ **17.** $81x^4$ **19.** x^6y^{12} **21.** $16x^8y^4/w^{12}$ **23.** $27y^6/(x^2z^6)$ **25.** $25x^8$
27. $2z^3/(x^6y^2)$ **29.** $-x^3z^2/(2y^7)$ **31.** a^4b^3 **33.** $d^{40}/(32b^{15})$ **35.** $a^3/(a + 1)$ **37.** 0
39. $3x^4/y^4$ **41.** $3y^4/x^3$ **43.** $5184/(x^8z^4)$ **45.** $(x^4 + 1)/x^2$ **47.** $x/(x^2 + 1)$ **49.** $x + 1$
51. (a) 2^{-5} (b) 2^0 (c) 2^{-10} **53.** (a) $2^{19} = \frac{1}{2}(2^{20}) \approx \frac{1}{2}(1000)^2 = 500{,}000$¢ $= \$5000$ (b) $2^{49} = \frac{1}{2}(2^{50}) \approx \frac{1}{2}(1000)^5 = \frac{1}{2}(10)^{15}$¢ $= \$(5 \times 10^{12})$ **55.** About February 7

PROBLEM SET 2-2

1. 3.41×10^8 **3.** 5.13×10^{-8} **5.** 1.245×10^{-10} **7.** 9.2625×10^7 **9.** 8.4×10^{-17}
11. 4.1×10^{-1} **13.** 1.2×10^8 **15.** 4.132×10^4 **17.** 4×10^{11} **19.** 9.144×10^2 **21.** 1×10^4
23. 4×10^3 **25.** 4.63×10^{10} **27.** 7.2×10^{-4} **29.** 3.888×10^{11} **31.** 1.3717×10^{23}
33. About 1.20×10^{24} **35.** About 7.01×10^{23} **37.** About 26.1 liters

PROBLEM SET 2.3

(*Note*: Answers may vary slightly depending on the calculator used.)
1. 48.35 **3.** −2441.7393 **5.** 303.27778 **7.** 2.7721×10^{15} **9.** 1.286×10^{10} **11.** −13.138859
13. 1.7891883 **15.** 1.0498907 **17.** .90569641 **19.** .00000081 **21.** About 1.28 seconds
23. About 4.068×10^{16} **25.** About 9.3×10^{32} **27.** −657.42103 **29.** About 186 pounds
31. About 464.0 square millimeters **33.** About 1.61×10^5 **35.** About 1.01×10^7 **37.** About 107 meters

PROBLEM SET 2-4

1. Polynomial of degree 2. **3.** Polynomial of degree 3. **5.** Polynomial of degree 0. **7.** Not a polynomial. **9.** Not a polynomial. **11.** $-2x + 1$ **13.** $4x^2$ **15.** $-2x^2 + 5x + 1$ **17.** $6x - 15$ **19.** $-10x + 12$ **21.** $35x^2 - 55x + 19$ **23.** $t^2 + 16t + 55$ **25.** $x^2 - x - 90$ **27.** $2t^2 + 13t - 7$ **29.** $y^2 + 2y - 8$ **31.** $7.9794x^2 + 0.1155x - 4.4781$ **33.** $x^2 + 20x + 100$ **35.** $x^2 - 64$ **37.** $4t^2 - 20t + 25$ **39.** $4x^8 - 25x^2$ **41.** $(t + 2)^2 + 2(t + 2)t^3 + t^6 = t^6 + 2t^4 + 4t^3 + t^2 + 4t + 4$ **43.** $(t + 2)^2 - (t^3)^2 = -t^6 + t^2 + 4t + 4$ **45.** $5.29x^2 - 6.44x + 1.96$ **47.** $-3x^3 - 19x^2 + 2x + 5$ **49.** $2y^2 + y - 15$ **51.** $4y^2 + 20y + 25$ **53.** $4z^2 - 49$ **55.** $x^4 + x^2 - 12$ **57.** $6x^3 + 7x^2 - 13x - 10$ **59.** $x^2 - 2$ **61.** $x^2 - 4\sqrt{7}x + 28$ **63.** $8x$ **65.** $8x^3 + 12x^2 + 6x + 1$ **67.** $5.7121x^2 + 15.5636x + 16.9744$ **69.** $25x^2 - y^2$ **71.** $25x^2 - 10xy + 2y^2$ **73.** $8x^2 - 10xy - 3y^2$ **75.** $x^3 + y^3$ **77.** $x^3 + 6x^2y + 12xy^2 + 8y^3$ **79.** $x^4 - y^4$ **81.** $x^4 - x^2y^2 - 6y^4$ **83.** $6x^2y^2 - 7xyz - 3z^2$ **85.** $x^4 + 3x^3y + 2x^2y^2 + xy^3 - y^4$

PROBLEM SET 2-5

1. $x(x + 5)$ **3.** $(x + 6)(x - 1)$ **5.** $y^3(y - 6)$ **7.** $(y + 6)(y - 2)$ **9.** $(y + 4)^2$ **11.** $(2x - 3y)^2$ **13.** $(y - 8)(y + 8)$ **15.** $(1 - 5b)(1 + 5b)$ **17.** $(2z - 3)(2z + 1)$ **19.** $(5x + 2y)(4x - y)$ **21.** $(x + 3)(x^2 - 3x + 9)$ **23.** $(a - 2b)(a^2 + 2ab + 4b^2)$ **25.** $x^3(1 - y^3) = x^3(1 - y)(1 + y + y^2)$ **27.** Does not factor over integers. **29.** Does not factor over integers. **31.** $(y - \sqrt{5})(y + \sqrt{5})$ **33.** $(\sqrt{5}z - 2)(\sqrt{5}z + 2)$ **35.** $t^2(t - \sqrt{2})(t + \sqrt{2})$ **37.** $(y - \sqrt{3})^2$ **39.** Does not factor over real numbers. **41.** $(x + 3i)(x - 3i)$ **43.** $(x^3 + 7)(x^3 + 2)$ **45.** $(2x - 1)(2x + 1)(x - 3)(x + 3)$ **47.** $(x + 4y + 3)^2$ **49.** $(x^2 - 3y^2)(x^2 + 2y^2)$ **51.** $(x - 2)(x^2 + 2x + 4)(x + 2)(x^2 - 2x + 4)$ **53.** $x^4(x - y)(x + y)(x^2 + y^2)$ **55.** $(x^2 + y^2)(x^4 - x^2y^2 + y^4)$ **57.** $(x^2 + 1)(x - 4)$ **59.** $(2x - 1 - y)(2x - 1 + y)$ **61.** $(3 - x)(x + y)$ **63.** $(x + 3y)(x + 3y + 2)$ **65.** $(x + y + 2)(x + y + 1)$ **67.** $(x^2 - 4x + 8)(x^2 + 4x + 8)$ **69.** $(x^2 - x + 1)(x^2 + x + 1)$ **71.** $5x(x - 2)$ **73.** $(2x + 1)^2$ **75.** $(2y - 1)(2y + 1)$ **77.** $(2y - 5)(y - 1)$ **79.** $4(x^2 - 2)(x^2 + 2)$ **81.** $(a + b - 5)(a + b + 5)$ **83.** $(a + b)(a + b - 2)$ **85.** $(-2x)(y - 3)(y^2 + 3y + 9)$ **87.** $(x^2 - y + 2)(x^2 + y - 2)$ **89.** $(a + 2b - 3)(a + 2b + 2)$ **91.** Does not factor over integers. **93.** $(x^2 - x + 1)(x^2 + x + 1)$ **95.** $(x + 3)(x - 7)^2(x^2 - 4x - 20)$ **97.** $(x - 1.7674841)(x + 1.7674841)$ **99.** $(x + 1.69965391)(x^2 - 1.69965391x + 2.8888234)$

PROBLEM SET 2-6

1. $1/(x - 6)$ **3.** $y/5$ **5.** $(x + 2)^2/(x - 2)$ **7.** $z(x + 2y)/(x + y)$ **9.** $(9x + 2)/(x - 2)(x + 2)$ **11.** $2(4x - 3)/(x - 2)(x + 2)$ **13.** $(2xy + 3y - 1)/x^2y^2$ **15.** $(x^2 + 10x - 3)/(x - 2)^2(x + 5)$ **17.** $(4 - x)/(2x - 1)$ **19.** $(6y^2 + 9y + 2)/(3y - 1)(3y + 1)$ **21.** $(3m^2 + m - 1)/3(m - 1)^2$ **23.** $5x/(2x - 1)(x + 1)$ **25.** $1/(x - 3)(x - 2)$ **27.** $y^2(x^3 - y^3)$ **29.** $x/(x - 4)$ **31.** $5(x + 1)/x(2x - 1)$ **33.** $x/(x - 2)$ **35.** $(x - a)(x^2 + a^2)/(x + 2a)(x^2 + ax + a^2)$ **37.** $(y - 1)/2y$ **39.** $-2/(2x + 2h + 3)(2x + 3)$ **41.** $(-2x - h)/(x + h)^2x^2$ **43.** $(y - 1)(y + 2)/(7y + 19)$ **45.** $(a^2 + b^2)/ab$ **47.** 1 **49.** $(y - 1)/y$ **51.** $(2x^2 - y^2)/(x - y)$ **53.** $(-3x^2 - 26x + 2)/2(x + 10)(x - 6)$ **55.** $(2x - 3y)(x^2 + xy + y^2)$ **57.** $(13x + 15)/(2x + 3)(2x - 3)$ **59.** $(x - 2)/3x^2$

CHAPTER 2. REVIEW PROBLEM SET

1. $16/9$ **2.** $36/25$ **3.** $36/49$ **4.** $y^5/2x^3$ **5.** $x^6/27y^9$ **6.** $b^{11}/2a^8$ **7.** 1.382×10^6 **8.** 6.82×10^{-3} **9.** 5×10^7 **10.** Polynomial of degree 2. **11.** Not a polynomial. **12.** Not a polynomial. **13.** Not a polynomial. **14.** $2x^2 + x + 6$ **15.** $x^3 - x^2 - x + 7$ **16.** $2x^2 + 7x - 15$ **17.** $-4x + 1$ **18.** $z^6 - 16$ **19.** $2x^4 + 5x^2w - 12w^2$ **20.** $x^3 + 8a^3$ **21.** $6y^3 - 19y^2 + 27y - 18$ **22.** $9t^4 - 6t^3 + 7t^2 - 2t + 1$ **23.** $a^2 - b^2c^2d^2$ **24.** $x^2(2x^2 - x + 11)$ **25.** $(y - 4)(y - 3)$ **26.** $(3z - 1)(2z + 1)$ **27.** $(7a - 5)(7a + 5)$ **29.** $(a - 3)(a^2 + 3a + 9)$ **30.** $(2ab + 1)(4a^2b^2 - 2ab + 1)$ **31.** $x^4(x - y)(x + y)(x^2 + y^2)$ **32.** $(x + y - z^2)(x + y + z^2)$ **33.** $(2c + d)(2c - d - 3)$ **34.** $(3x - \sqrt{11})(3x + \sqrt{11})$ **35.** $(x - 4i)(x + 4i)$ **36.** $(x^2 + 2x + 4)/2$ **37.** $-2/(x - 1)$ **38.** $2/x$ **39.** $(x - 2)/(x + 1)$ **40.** $(x + 1)/10$

PROBLEM SET 3-1

1. Conditional equation. **3.** Identity. **5.** Conditional equation. **7.** Conditional equation. **9.** Identity. **11.** 2 **13.** $\frac{2}{5}$ **15.** $\frac{9}{2}$ **17.** $2/\sqrt{3}$ **19.** 7.57 **21.** -8.71×10^1 **23.** -24 **25.** $-\frac{1}{4}$ **27.** $\frac{22}{5}$ **29.** 3 **31.** 6 **33.** No solution (2 is extraneous). **35.** $-\frac{3}{4}$ **37.** $-\frac{13}{2}$ **39.** $P = A/(1 + rt)$ **41.** $r = (nE - IR)/nI$ **43.** $h = (A - 2\pi r^2)/2\pi r$ **45.** $R_1 = RR_2/(R_2 - R)$ **47.** 8 **49.** 60 **51.** -1.732 **53.** $\frac{19}{3}$ **55.** $-\frac{1}{3}$ **57.** No solution (-3 is extraneous). **59.** (a) $F = \frac{9}{5}C + 32$ (b) 95 (c) -40 (d) 72.5 (e) $\frac{80}{3} = 26.7$

PROBLEM SET 3-2

1. 9 **3.** 12 **5.** 23 **7.** 31 centimeters **9.** 89 **11.** 7 **13.** 12 **15.** 12:40 A.M. **17.** 4:20 P.M. **19.** 9 miles per hour **21.** 15 **23.** 6000 **25.** 142.86 liters **27.** \$145 **29.** 9 feet **31.** 7.5 days **33.** 4.6 feet from fulcrum **35.** 125 **37.** 44.44

PROBLEM SET 3-3

1. $x = -13; y = 13$ **3.** $u = \frac{3}{2}; v = -4$ **5.** $x = -1; y = 3$ **7.** $x = 4; y = 3$ **9.** $x = 6; y = -8$ **11.** $s = 1; t = 4$ **13.** $x = 3; y = -1$ **15.** $x = 4; y = 7$ **17.** $x = 9; y = -2$ **19.** $x = 16; y = -5$ **21.** $x = \frac{1}{2}; y = \frac{1}{3}$ **23.** $x = 4; y = 9$ **25.** $x = 1; y = 2; z = 0$ **27.** $x = \frac{4}{3}, y = -\frac{2}{3}; z = -\frac{5}{3}$ **29.** 7; 11 **31.** \$7200; \$5800 **33.** \$28,500 (certificates); \$19,000 (bonds) **35.** 32,500 (\$8 tickets); 12,500 (\$10 tickets) **37.** 20 nickels; 3 dimes **39.** 5 hours; 11 hours **41.** 62.79 pounds (\$1.69 coffee); 37.21 pounds (\$1.26 coffee) **43.** 28 coats; 50 dresses

PROBLEM SET 3-4

1. $5\sqrt{2}$ **3.** $\frac{1}{2}$ **5.** $3i/2$ **7.** 22 **9.** $(5 + 6\sqrt{2})/5$ **11.** $(2 + \sqrt{3})/4$ **13.** $(6 + i)/2$ **15.** ± 5 **17.** $-1; 7$ **19.** $-\frac{15}{2}; \frac{5}{2}$ **21.** $\pm 3i$ **23.** ± 3 **25.** $\pm .12$ **27.** $-2; 5$ **29.** $-2; \frac{1}{3}$ **31.** $-\frac{4}{3}; \frac{7}{2}$ **33.** $-9; 1$ **35.** $-\frac{1}{2}; \frac{3}{2}$ **37.** $-2 \pm \sqrt{5}i$ **39.** $-6; -2$ **41.** $(-5 \pm \sqrt{13})/2$ **43.** $(3 \pm \sqrt{42})/3$ **45.** $-2 \pm i$ **47.** $(3 \pm \sqrt{13}i)/2$ **49.** $-.2714; 1.8422$ **51.** $-1.6537; .8302$ **53.** $y = 2 \pm 2x$ **55.** $y = -6x$ or $y = 0$ **57.** $y = -2x + 3$ or $y = -2x + 5$ **59.** ± 12 **61.** $-\frac{3}{2}; 2$ **63.** $-3; \frac{1}{2}$ **65.** $-\frac{5}{4}, 0$ **67.** $-1 \pm \sqrt{6}$ **69.** $i; 4i$ **71.** 13 feet by 7 feet **73.** About 91 feet square **75.** $2 + 8\sqrt{2} \approx 13.31$ inches square **77.** (a) After 8 seconds. (b) After 2 seconds and after 6 seconds. (c) Never. **79.** $x = 3, y = 4; x = 4.8, y = -1.4$ **81.** $x = 2, y = -1; x = -2, y = -3$

PROBLEM SET 3-5

1. Conditional. **3.** Unconditional. **5.** Conditional. **7.** Conditional. **9.** Conditional.

11. Unconditional. **13.** $\{x : x < -6\}$

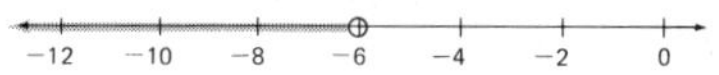

15. $\{x : x > -24\}$ 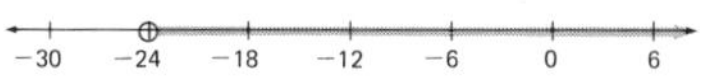 **17.** $\{x : x < \frac{30}{7}\}$

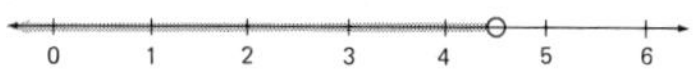

19. $\{x : -5 \le x \le 2\}$

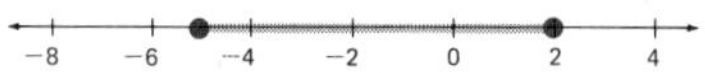

21. $\{x : x < -3 \text{ or } x > \frac{1}{2}\}$

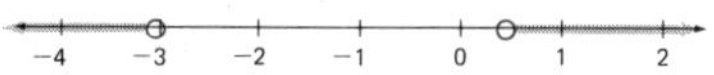

23. $\{x : x \le 1 \text{ or } x \ge 4\}$

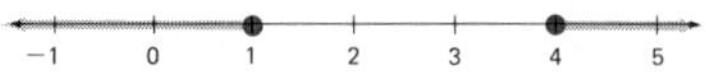

25. $\{x : \frac{1}{2} < x < 3\}$

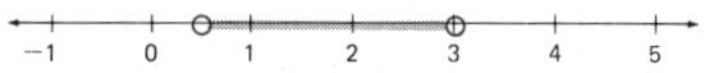

27. $\{x : -\frac{5}{2} < x < -\frac{1}{2}\}$

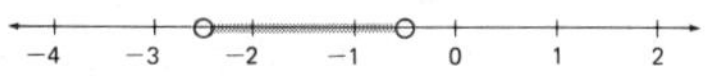

29. $\{x : -1 \le x \le 0\}$

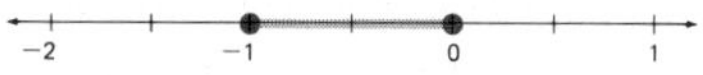

31. $\{x: -4 \le x \le 0 \text{ or } x \ge 3\}$

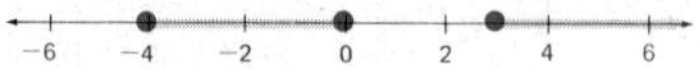

33. $\{x: x < 5 \text{ and } x \ne 2\}$

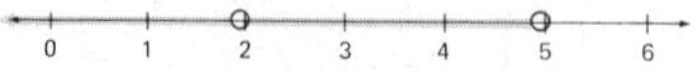

35. $\{x: -2 < x \le 5\}$

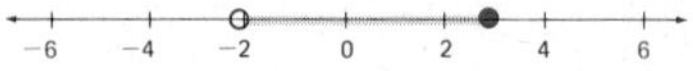

37. $\{x: -2 < x < 0 \text{ or } x > 5\}$

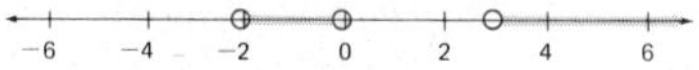

39. $\{x: -2 < x < 2 \text{ or } x > 3\}$

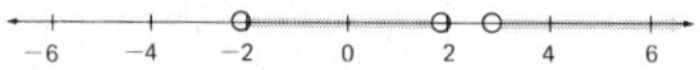

41. $|x - 3| < 3$ **43.** $|x - 3| \le 4$ **45.** $|x - 6.5| < 4.5$

47. $\{x: -\sqrt{7} < x < \sqrt{7}\}$

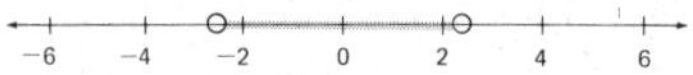

49. $\{x: x \le 2 - \sqrt{2} \text{ or } x \ge 2 + \sqrt{2}\}$

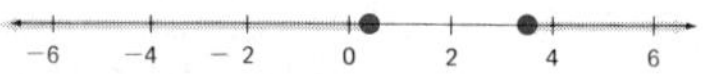

51. $\{x: x < -5.71 \text{ or } x > -.61\}$

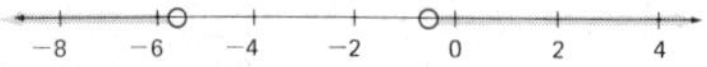

53. 4 **55.** 100

57. $\{x: x > -\frac{9}{2}\}$

59. $\{x: x < -4 \text{ or } x > 2\}$

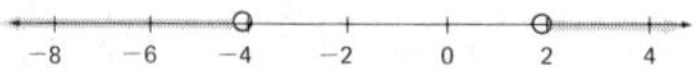

61. $\{x: -\frac{1}{3} \le x \le \frac{5}{3}\}$

63. $\{x: x < 0 \text{ or } 3 < x \le 5\}$

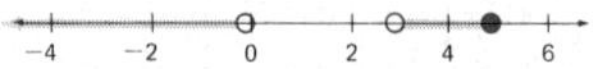

65. $k \le \frac{9}{4}$ **67.** All real numbers. **69.** (a) A score higher than 57. (b) A score between 132 and 192.
71. (a) 144 feet above the ground. (b) $0 < t < 4$ (c) After 5 seconds. (d) $4.55 < t \le 5$
73. At most 100 miles. **75.** $\{x: x \ge 3\}$ **77.** $\{x: x > -2\}$

PROBLEM SET 3-6

1. About 18,333 feet **3.** 525 miles **5.** About 21.82 minutes **7.** 4 hours and 48 minutes
9. About 4.15 hours after takeoff **11.** 12 standard; 20 deluxe **13.** 60 miles per hour **15.** 500 meters
17. 20 feet **19.** About 8.17 feet **21.** $\frac{5}{6}$ kilograms **23.** About 3.29 milligrams
25. .298 grams sodium chloride; .202 grams sodium bromide **27.** 36 shares
29. 14 pounds walnuts; 11 pounds cashews **31.** 23,500 **33.** \$785,714.29
35. 2340 undergraduate students; 864 graduate students **37.** 2 feet **39.** 5; 12; 13 **41.** $a = 8, b = 5, c = 10$
43. About 13.34 feet **45.** Two solutions: longer piece is 28 inches or $\frac{236}{7}$ inches.

CHAPTER 3. REVIEW PROBLEM SET

1. (a) Identity; (b) Conditional equation; (c) Conditional equation; (d) Identity.
2. (a) $-\frac{11}{12}$ (b) 7 (c) $\frac{5}{2}$ (d) No solution. **3.** $v_0 = (s - \frac{1}{2}at^2)/t$ **4.** $-320/13$
5. (a) $x = 2; y = -1$ (b) $x = 1; y = -2$
6. (a) $(2 + \sqrt{2})/4$ (b) $5 - 5\sqrt{3}$ (c) $(-1 + \sqrt{2}i)/3$ **7.** ± 7 **8.** $-5; 2$ **9.** -3 **10.** $-3; 7$
11. $-5; 4$ **12.** 4; 5 **13.** 0; 4 **14.** $-1 \pm \sqrt{5}$ **15.** $(-1 \pm \sqrt{13})/6$ **16.** $(-m \pm \sqrt{m^2 - 8n})/2$
17. $y = 2x \pm 2$ **18.** $y = (3 - 3x)/2$ or $y = (1 - 3x)/2$ **19.** $\{x: x \le \frac{12}{7}\}$ **20.** $\{x: x \le -6 \text{ or } x \ge 1\}$
21. $\{x: (-1 - \sqrt{13})/2 < x < (-1 + \sqrt{13})/2\}$ **22.** $\{x: x < -1 \text{ or } x > 4\}$
23. \$3500 in the bank; \$6500 in the credit union. **24.** 50 miles per hour **25.** 2 miles per hour

PROBLEM SET 4-1

1. 5 **3.** 4 **5.** 1.7 **7.** $2\pi - 5$ **9.** 3; 7 **11.** $-6; 2$ **13.** $\frac{7}{4}; \frac{13}{4}$ **15.** Rectangle.
17. Parallelogram. **19.** 5; $(\frac{7}{2}, 1)$ **21.** $2\sqrt{2}$; (3, 3) **23.** 3; $(\sqrt{3}/2, \sqrt{6}/2)$ **25.** 14.54; (3.974, 1.605)

27. (a) $\sqrt{10}$; $\sqrt{5}$; $\sqrt{10}$; $\sqrt{5}$ (b) Each midpoint has coordinates (5/2, 5). (c) Opposite sides have the same length; the two diagonals bisect each other. **29.** (a) $(-2, 3)$; $(4, 0)$ (b) $(2, 7)$; $(8, -1)$

31. If the points are labeled A, B, and C, respectively, then $d(A, B) = \sqrt{20}$, $d(B, C) = \sqrt{20}$, and $d(A, C) = \sqrt{40}$. Then note that $(\sqrt{40})^2 = (\sqrt{20})^2 + (\sqrt{20})^2$.

33. (a) The three distances are 5, 10, and 15, and $15 = 5 + 10$. (b) The distances are 5, 10, and 15. **35.** (9, 0)

37. (17, 35) **39.** $-\frac{3}{2}$; 1; $\frac{7}{2}$ **41.** (a) $(-1, 5)$ (b) $(-10, 11)$ **43.** 12 **45.** (1, 10), (6, 17), (11, 24), (16, 31)

47. \$1458.03 by road; \$1344.75 by air

PROBLEM SET 4-2

1.

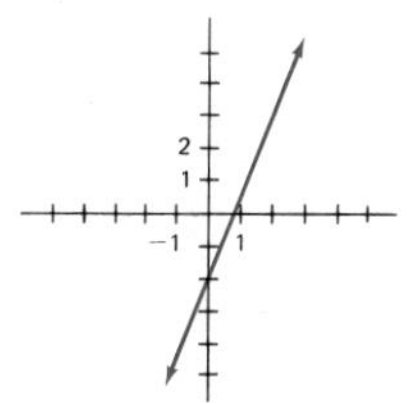

3.

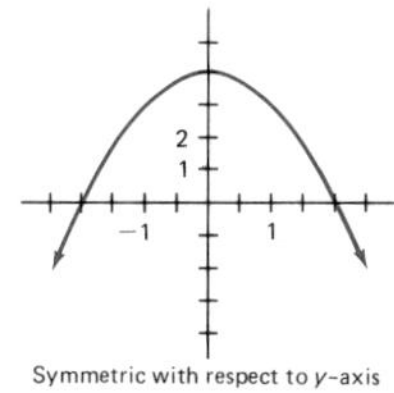

Symmetric with respect to y-axis

5.

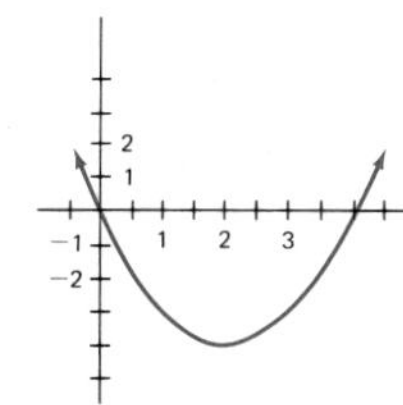

7.

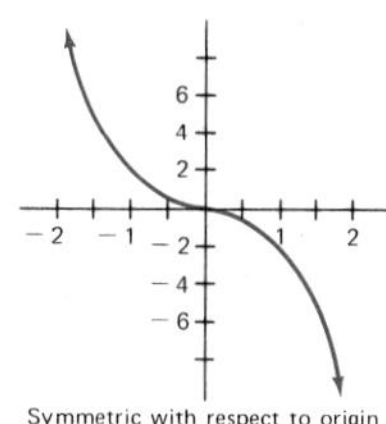

Symmetric with respect to origin

9.

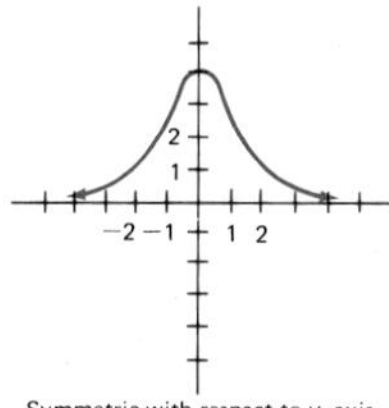

Symmetric with respect to y-axis

11.

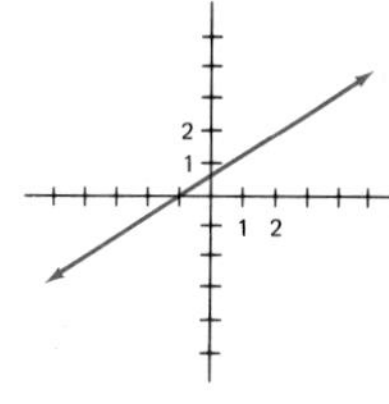

13.

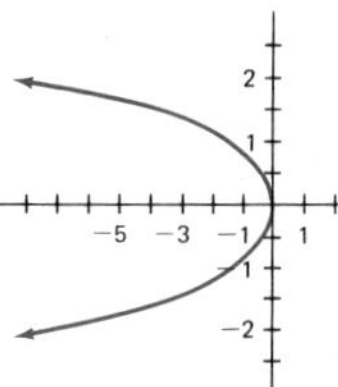

15.

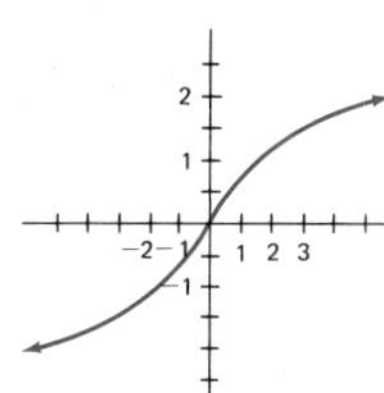

17. $x^2 + y^2 = 36$

19. $(x - 4)^2 + (y - 1)^2 = 25$

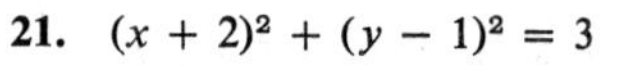

21. $(x + 2)^2 + (y - 1)^2 = 3$

23.

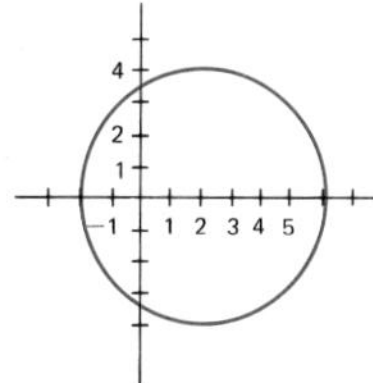

25.

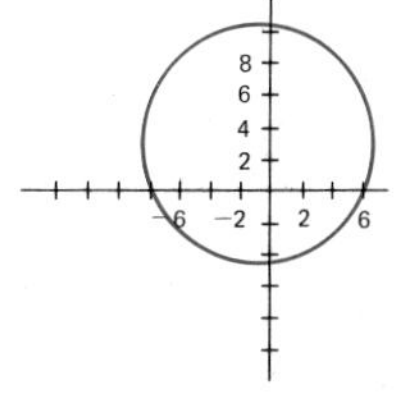

27. $(-1, 5)$; 1 **29.** (6, 0); 1 **31.** $(-\frac{1}{2}, \frac{3}{2})$; $\frac{3}{2}$

33.

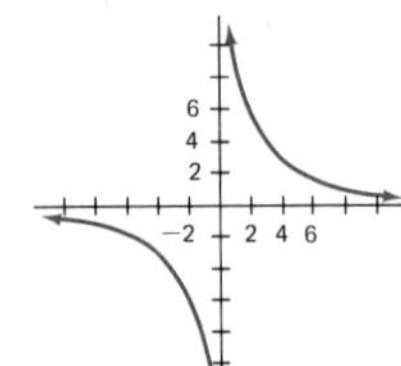

35.

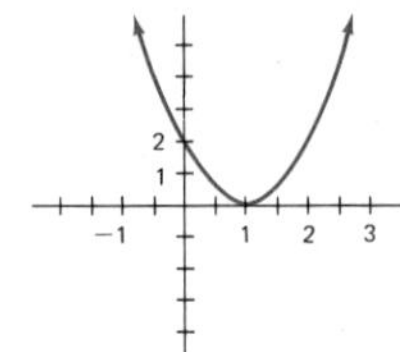

37.

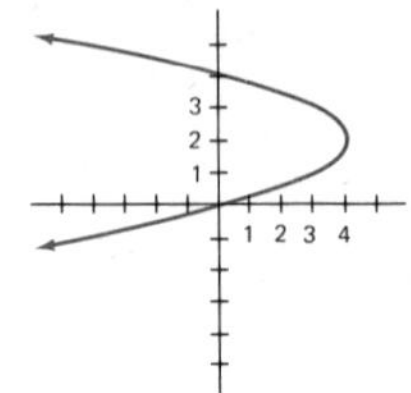

39. $x^2 + y^2 = 41$ 41. $x^2 + y^2 = 58$ 43. $(x - 6)^2 + (y - 6)^2 = 36$ 45. $(x - 4)^2 + (y - 1)^2 = 4$

PROBLEM SET 4-3

1. $\frac{5}{2}$ **3.** $-\frac{2}{7}$ **5.** $-\frac{5}{3}$ **7.** 0.1920 **9.** $4x - y - 5 = 0$ **11.** $2x + y - 2 = 0$ **13.** $2x + y - 4 = 0$ **15.** $y - 5 = 0$ **17.** $5x - 2y - 4 = 0$ **19.** $5x + 3y - 15 = 0$ **21.** $1.56x + y - 5.35 = 0$ **23.** $x - 2 = 0$ **25.** 3; 5 **27.** $\frac{2}{3}, -\frac{4}{3}$ **29.** $-\frac{2}{3}, 2$ **31.** $-4; 2$ **33.** (a) $y + 3 = 2(x - 3)$ (b) $y + 3 = -\frac{1}{2}(x - 3)$ (c) $y + 3 = -\frac{2}{3}(x - 3)$ (d) $y + 3 = \frac{3}{2}(x - 3)$ (e) $y + 3 = -\frac{3}{4}(x - 3)$ (f) $x = 3$ (g) $y = -3$ **35.** $y + 4 = 2x$ **37.** $(-1, 2); y - 2 = \frac{3}{2}(x + 1)$ **39.** $(3, 1); y - 1 = -\frac{4}{3}(x - 3)$ **41.** $\frac{7}{5}$ **43.** $\frac{18}{13}$ **45.** $\frac{6}{5}$ **47.** $-\frac{4}{5}$ **49.** $y - 2 = -(x - 4)$ **51.** (a) Parallel. (b) Neither. (c) Perpendicular. (d) Parallel. (e) Perpendicular. **53.** (a) $\frac{9}{5}$ (b) $\frac{9}{5}$; 32 (c) $F = -40$, $C = -40$. At 40° below zero, and only then, the two temperature scales give the same reading. **55.** Slope is 1600; A-intercept is 20,000. **57.** -9600. Each year the piece of equipment depreciates \$9600. **59.** $V = 80{,}000 - 3900n$ **61.** (a) .75 (b) Increasing production by one item increases the total cost by 75¢. **63.** The slope of the line through $(a, 0)$ and $(0, b)$ is $-b/a$. An equation of the line is $y = (-b/a)(x - a)$ which can be rewritten $x/a + y/b = 1$. **65.** $x + y = 5$

PROBLEM SET 4-4

1.

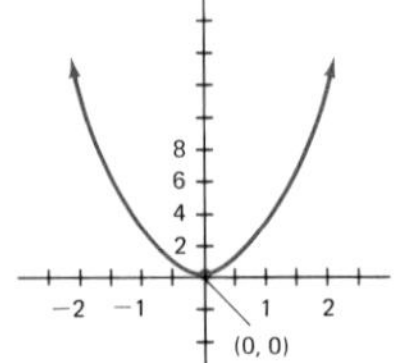

3.

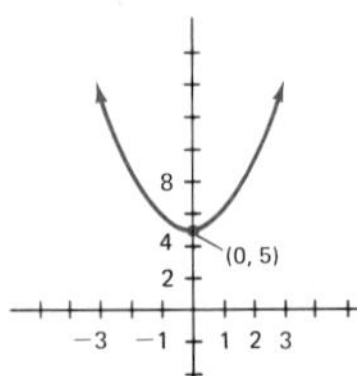

5.

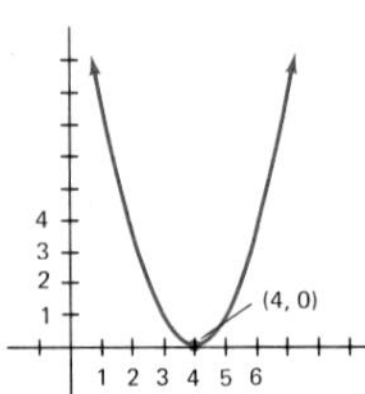

7.

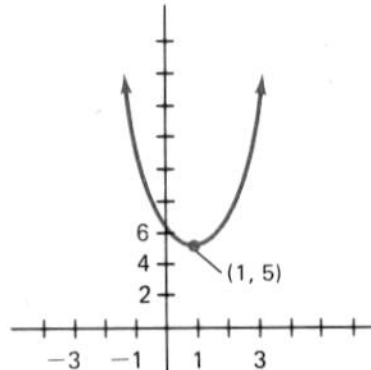

9.

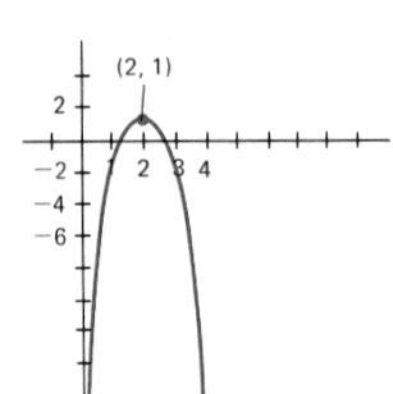

11. $y = 2x^2 - 4x + 9$

13. $y = -\frac{1}{2}x^2 + 5x - 10$

15.

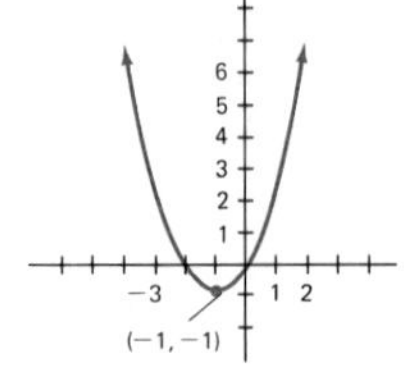

17.

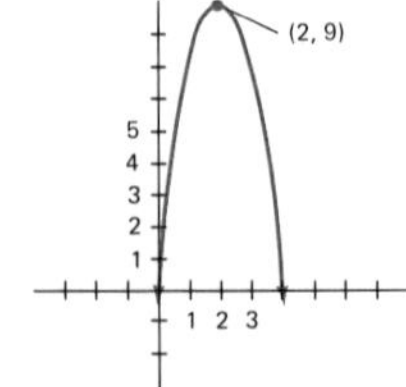

19.

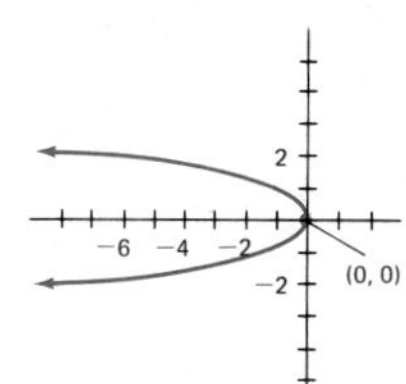

21.

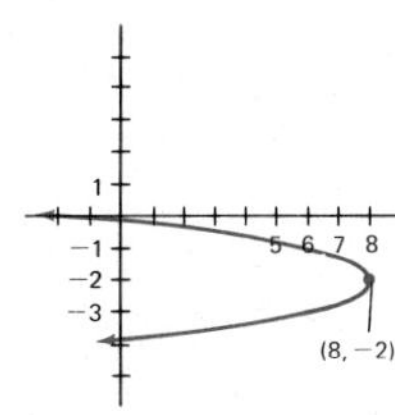

23.

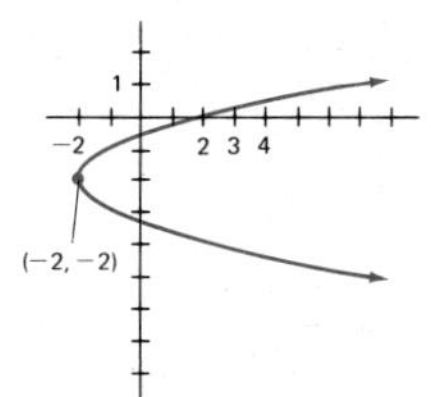

25. $(-3, 4); (0, 1)$ **27.** $(-1, 3); (2, -3)$ **29.** $(-.64, 2.25); (5.04, 10.75)$ **31.** $(x - 1)^2 = 8(y - 1)$
33. $(x + 3)^2 = -4(y - 5)$ **35.** $(x - 2)^2 = 16(y + 1)$ **37.** $x^2 = -24(y - 6)$
39. $(0, 0), (0, 2); y = -2, x = 0$ **41.** $(2, 4), (2, 0); y = 8, x = 2$ **43.** $(3, -5), (3, -\frac{19}{4}); y = -\frac{21}{4}, x = 3$

45.

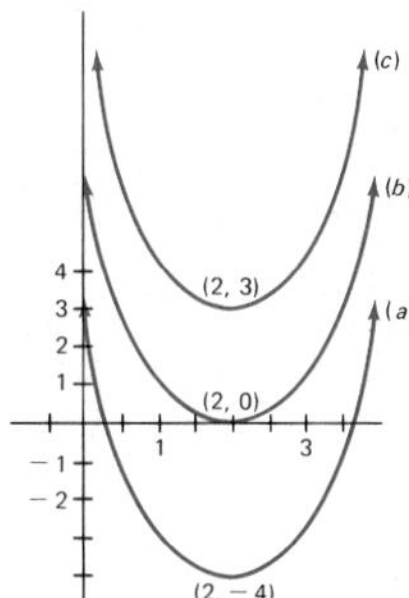

47. 32, 0, −24; two, one, none. **49.** $\frac{5}{4}$ **51.** $(-2, -8); (4, 4)$

53. Eliminating y gives $x^2 - 4x + 5 = 0$, which has no real solutions. **55.** $\frac{45}{4}$ **57.** $y = \frac{7}{6}x^2 - \frac{3}{2}x - \frac{2}{3}$
59. (a) $C = 12{,}000 + 80x$ (c) 334 (d) 1667; \$21,333

PROBLEM SET 4-5

1. Center: (0, 0);
Endpoints of major diameter: $(\pm 5, 0)$;
Endpoints of minor diameter: $(0, \pm 3)$.

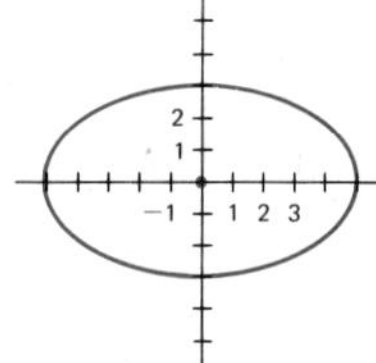

3. Center: (0, 0);
Endpoints of major diameter: $(0, \pm 5)$;
Endpoints of minor diameter: $(\pm 3, 0)$.

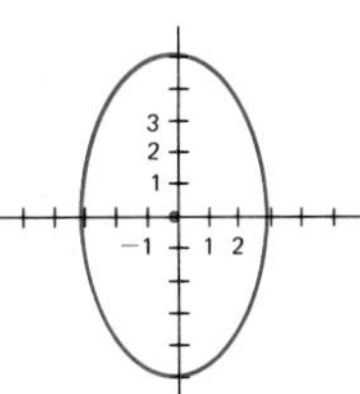

5. Center: $(2, -1)$;
Endpoints of major diameter: $(-3, -1), (7, -1)$;
Endpoints of minor diameter: $(2, -4), (2, 2)$.

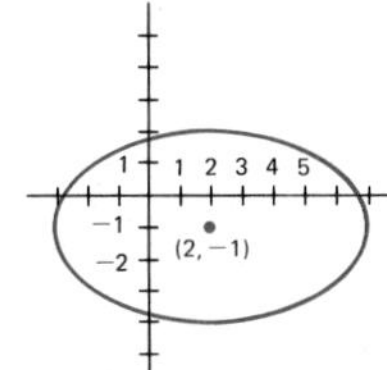

7. Center: $(-3, 0)$;
Endpoints of major diameter: $(-3, -4)$, $(-3, 4)$;
Endpoints of minor diameter: $(-6, 0)$, $(0, 0)$.

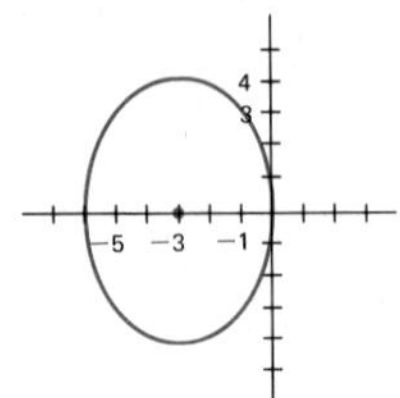

9. $(0, 0)$; $x^2/36 + y^2/9 = 1$ **11.** $(0, 0)$; $x^2/16 + y^2/36 = 1$ **13.** $(2, 3)$; $(x - 2)^2/36 + (y - 3)^2/9 = 1$
15. $(x + 2)^2/9 + (y - 3)^2/16 = 1$; vertical; $(-2, 3)$; 8 and 6. **17.** $(x - 4)^2/4 + y^2/9 = 1$; vertical; $(4, 0)$; 6 and 4.

19. $(x + 2)^2/16 + (y - 3)^2/9 = 1$; horizontal; $(-2, 3)$; 8 and 6. **21.**

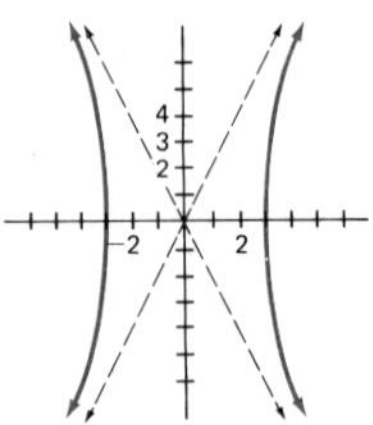

23.

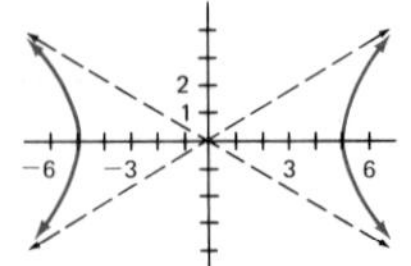

25.

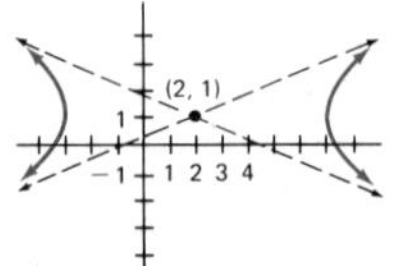

27.

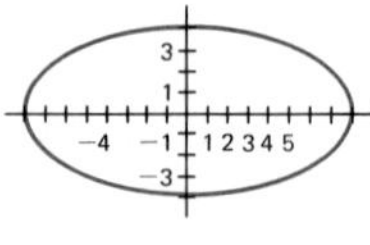

29.

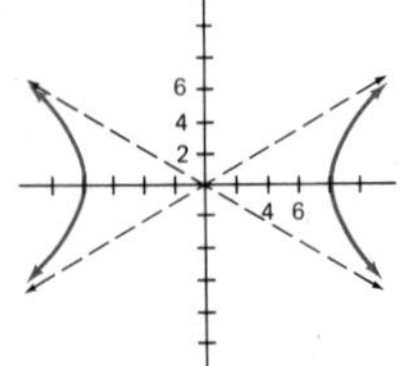

31.

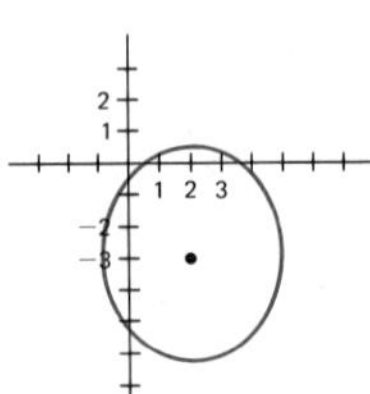

33.

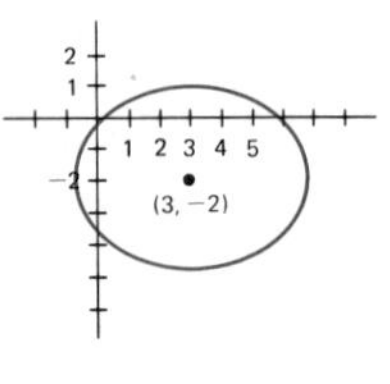

35. $(x - 3)^2/49 + (y - 2)^2/25 = 1$ **37.** $x^2/100 + y^2/64 = 1$ **39.** ± 3.7486 **41.** (a) 27.57 (b) 33.10
43. $x^2/16 - y^2/9 = 1$

CHAPTER 4. REVIEW PROBLEM SET

1. (a) Line. (b) Circle. (c) Line. (d) Parabola. (e) Parabola. (f) Circle. (g) Ellipse. (h) Hyperbola.

(i) Parabola. (j) Ellipse. (k) Hyperbola. (l) Hyperbola. **2.** (a)

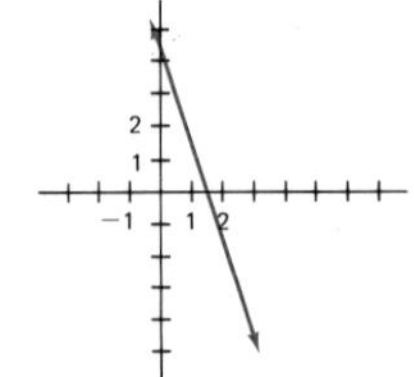

(b)

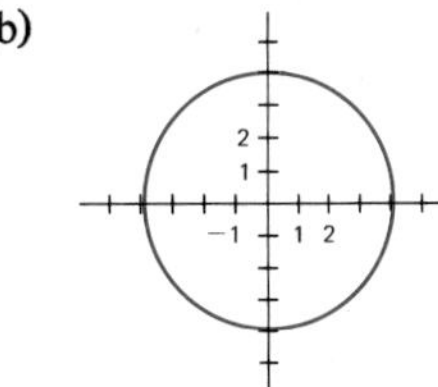

(c)

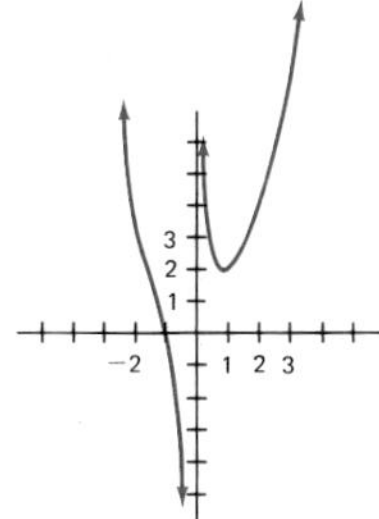

(d)

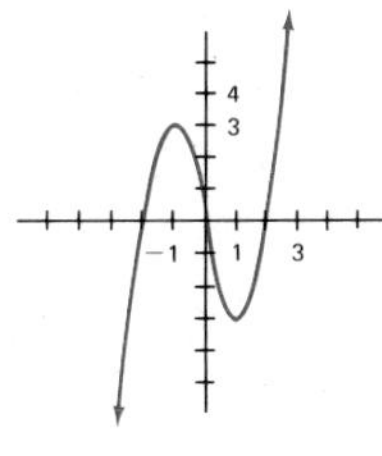

(e)

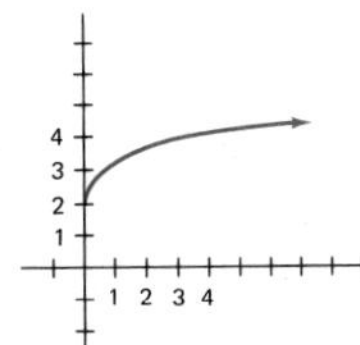

(f)

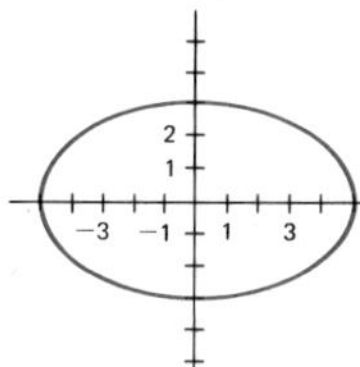

3. (a)

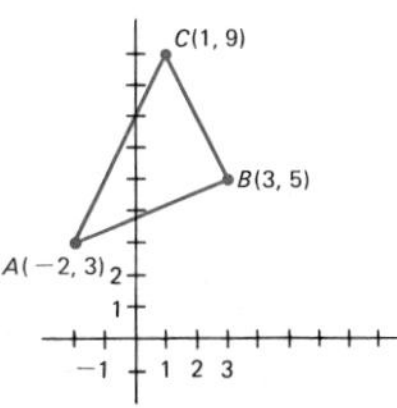

(b) $d(A, B) = \sqrt{29}$; $d(B, C) = 2\sqrt{5}$; $d(C, A) = 3\sqrt{5}$ (c) $\frac{2}{5}$; -2; 2 (d) $y - 3 = \frac{2}{5}(x + 2)$; $y - 5 = -2(x - 3)$; $y - 3 = 2(x + 2)$ (e) $(\frac{1}{2}, 4)$, $(2, 7)$, $(-\frac{1}{2}, 6)$ (f) $y - 3 = -2(x + 2)$ (g) $y - 3 = \frac{1}{2}(x + 2)$ (h) $12/\sqrt{5} \approx 5.37$ (i) 12

4. (a) $x - 4 = 0$ (b) $y + 1 = 0$ (c) $3x + 2y - 10 = 0$ (d) $5x - y - 15 = 0$ (e) $2x - 3y + 13 = 0$ (f) $4x - y - 1 = 0$ (g) $x + y - 10 = 0$ (h) $12x + 5y - 169 = 0$

5. (a) $(0, 0)$, $(0, -\frac{3}{2})$; $y = \frac{3}{2}$ (b) $(0, 0)$, $(4, 0)$; $x = -4$ (c) $(-2, 1)$, $(-2, \frac{7}{2})$; $y = -\frac{3}{2}$ (d) $(\frac{1}{4}, -\frac{1}{2})$, $(-\frac{1}{2}, -\frac{1}{2})$; $x = 1$ (e) $(2, -4)$, $(2, -\frac{15}{4})$; $y = -\frac{17}{4}$ (f) $(-17, -3)$, $(-16.875, -3)$; $y = -17.125$

6. (a) $y^2 = -20x$ (b) $x^2 = 32y$ (c) $y^2 = -72(x - 3)$ **7.** $(1, 2)$; $(4, 20)$

8. (a) $(x + 1)^2 + (y - 2)^2 = 7$; circle with center $(-1, 2)$ and radius $\sqrt{7}$. (b) $(x + 1)^2/10 + (y - 2)^2/20 = 1$; vertical ellipse with center $(-1, 2)$. (c) $(x + 1)^2/6 - (y + 2)^2/12 = 1$; horizontal hyperbola with center $(-1, -2)$. (d) $(x + 1)^2 = \frac{1}{2}(y + 16)$; vertical parabola opening up with vertex $(-1, -16)$.

9. (a) $x^2/9 + y^2/25 = 1$ (b) $x^2/25 + y^2/169 = 1$

PROBLEM SET 5-1

1. (a) 0 (b) -4 (c) $-15/4$ (d) -3.99 (e) -2 (f) $a^2 - 4$ (g) $(1 - 4x^2)/x^2$ (h) $x^2 + 2x - 3$
3. (a) $\frac{1}{4}$ (b) $-\frac{1}{2}$ (c) 2 (d) -8 (e) Undefined. (f) 100 (g) $x/(1 - 4x)$ (h) $1/(x^2 - 4)$ (i) $1/(h - 2)$ (j) $-1/(h + 2)$
5. All real numbers. **7.** $\{x : x \neq \pm 2\}$ **9.** $\{x : x \neq -2 \text{ and } x \neq 3\}$ **11.** All real numbers.
13. $\{x : x \geq 2\}$ **15.** $\{x : x \geq 0 \text{ and } x \neq 25\}$ **17.** (a) 9 (b) 5 (c) 3; the positive integers. **19.** 20
21. 5 **23.** 8 **25.** Undefined. **27.** $18xy - 10x$ **29.** $3 - 5x$ **31.** $y = 4x$ **33.** $y = 1/x$
35. $I = 324s/d^2$ **37.** 8.04 **39.** (a) $R = 2v^2/45$ (b) About 28,444 feet **41.** 8 gallons
43. (a) $\frac{5}{4}$ (b) 17 (c) 9 (d) 26 (e) 508.4 (f) $(5 + 8h^2 + 4h^3)/h^2$ **45.** $\{x : x \neq 4\}$ **47.** $\{x : x \geq 0 \text{ and } x \neq 4\}$
49. $y = 4000x^3$; 256,000 **51.** y varies inversely as x^3.
53. y varies jointly as x and y and inversely as the square root of w. **55.** 333.3 pounds
57. (a) .00123 (b) Decreases.
59. $T(x) = 2200 + 151x$, domain $\{x : x \geq 0\}$; $U(x) = (2200 + 151x)/x$, domain $\{x : x > 0\}$.

PROBLEM SET 5-2

1.

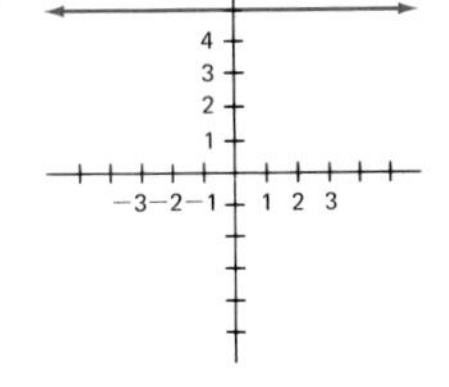

3.

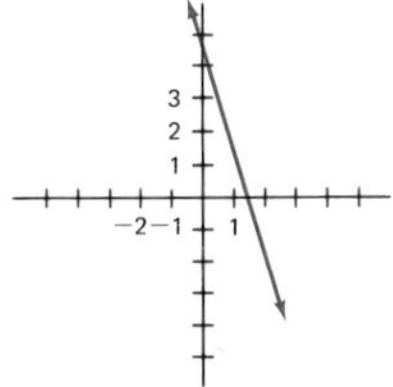

5.

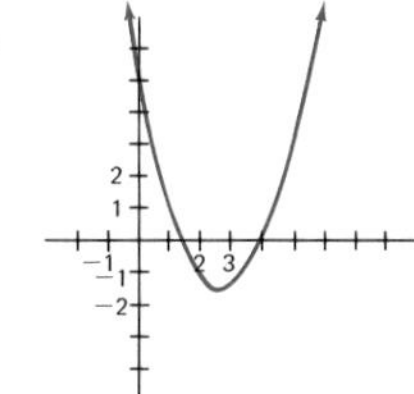

7. 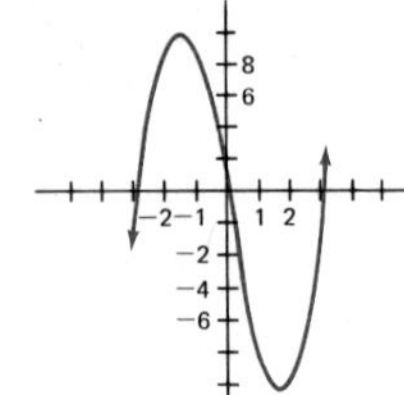

9. 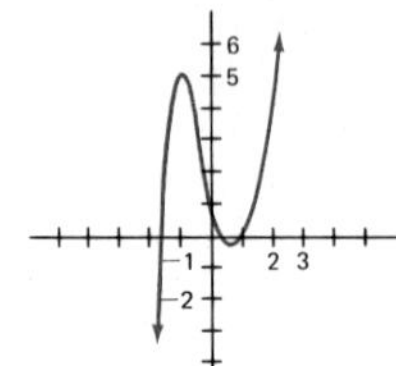

11. 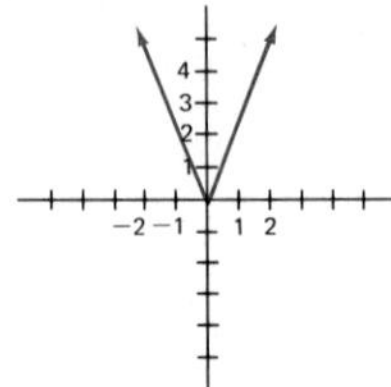

13. 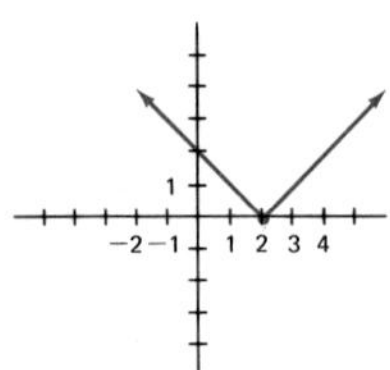

15. 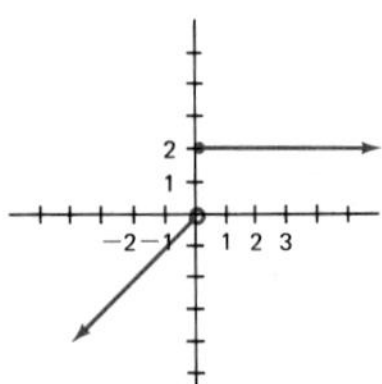

17. 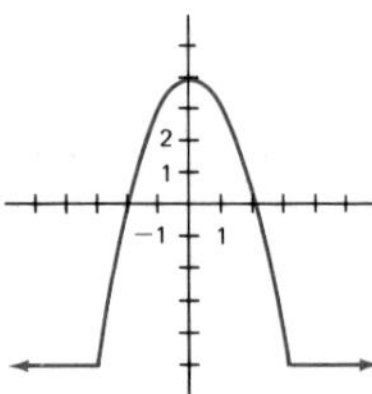

19. Even.

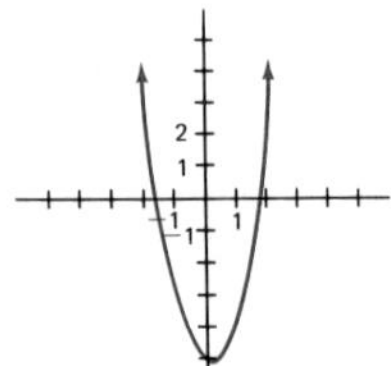

21. Neither even nor odd.

23. Odd.

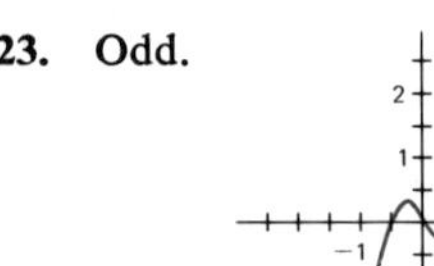

25. Even.

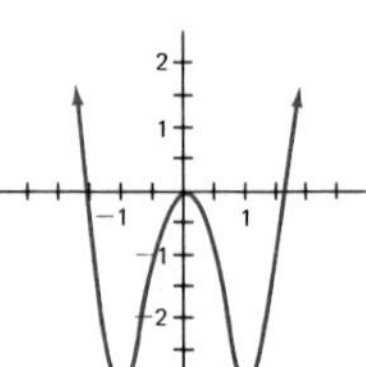

27. 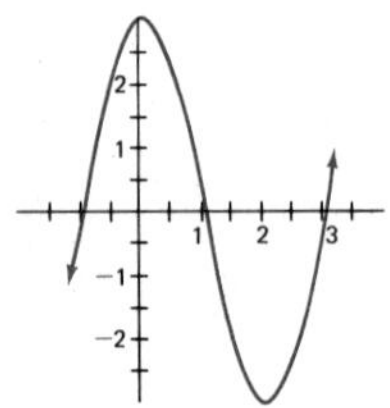

29. 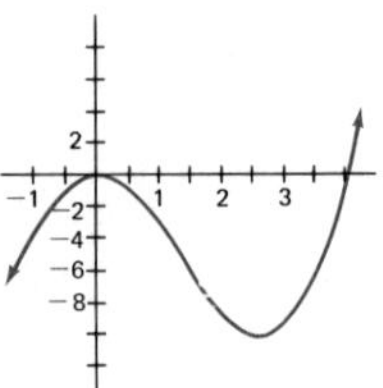

31. 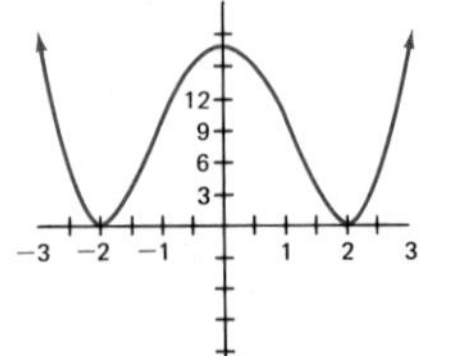

33. 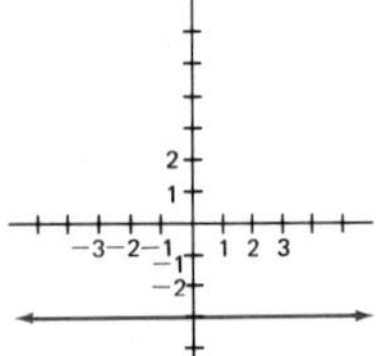

35. 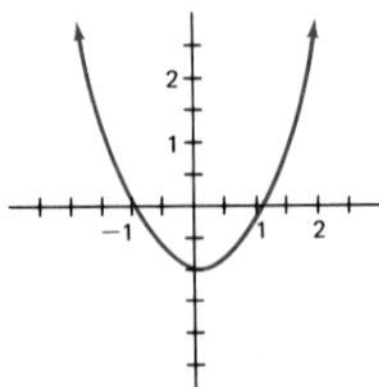

37. 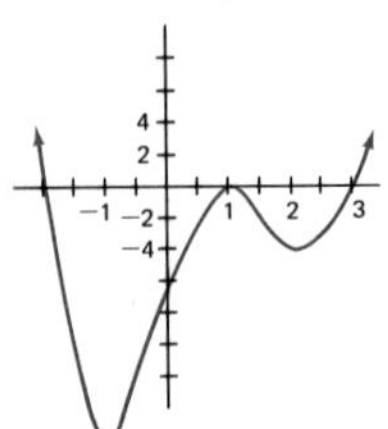

39. 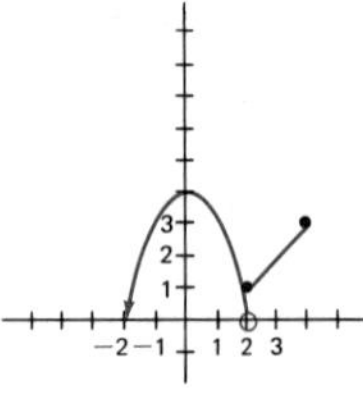

41. 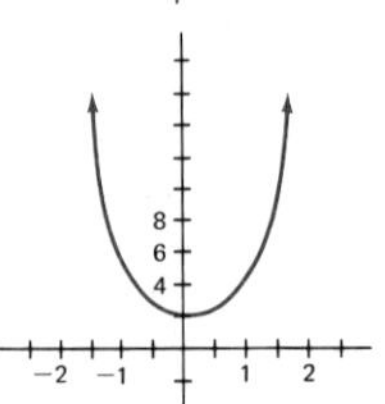

43. 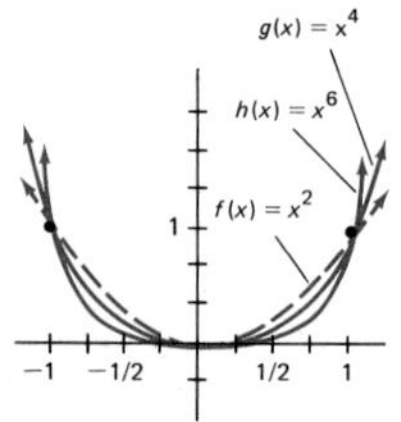

45. 55; 10.1923; 1033.9648

47. 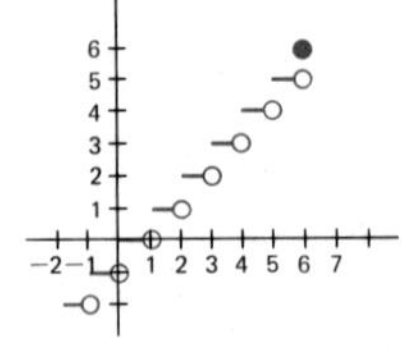

49. $V(x) = 8000 - 541.67x$

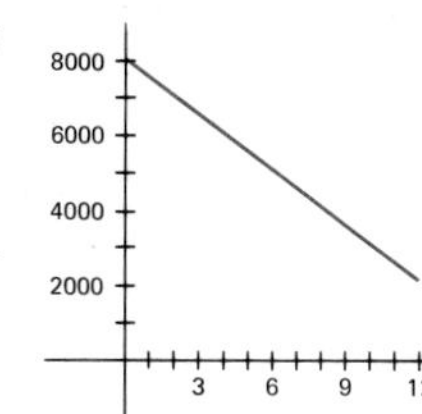

51. $C(x) = 15 + 10[x]$

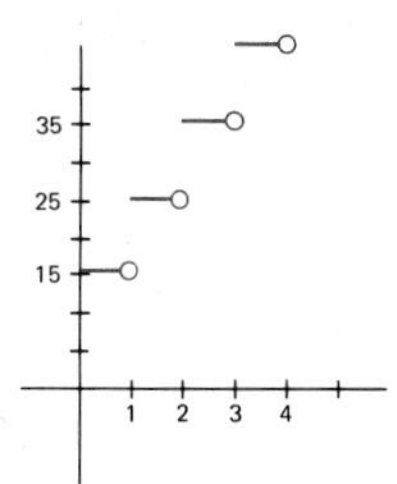

PROBLEM SET 5-3

1.

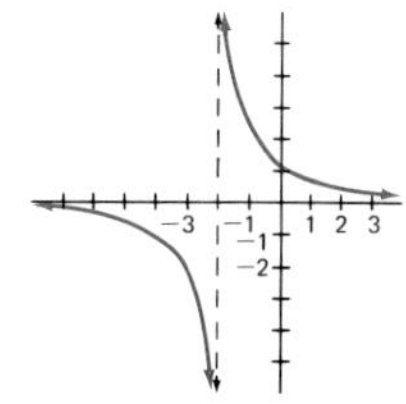

3.

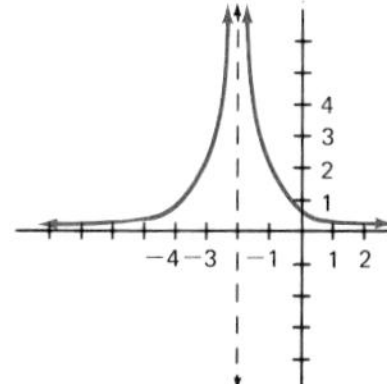

5.

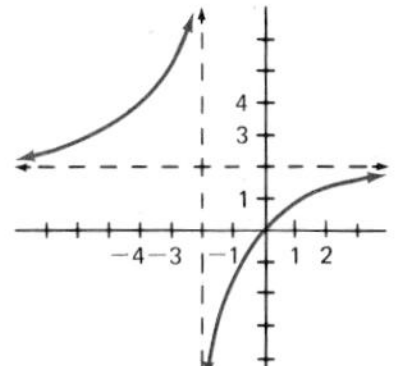

7.

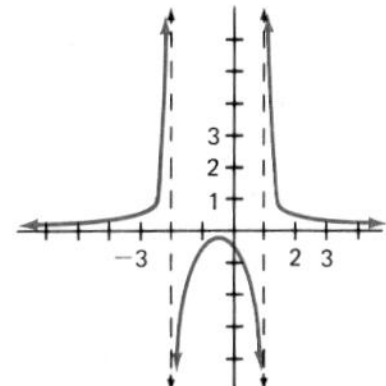

9.

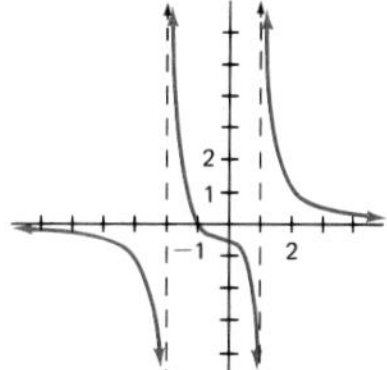

11.

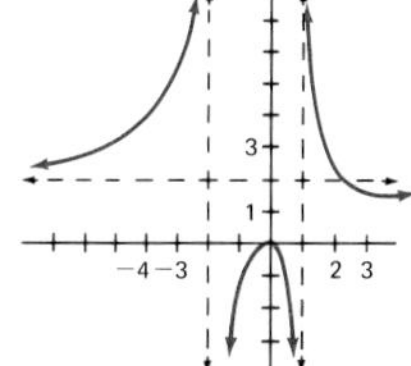

13.

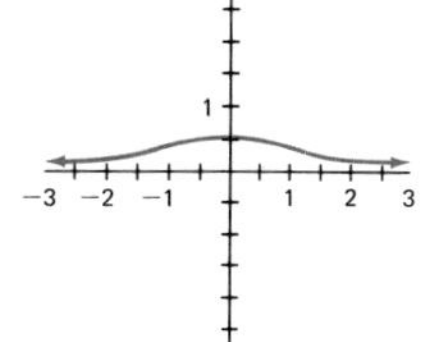

15.

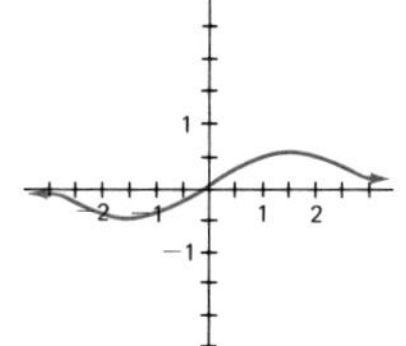

17.

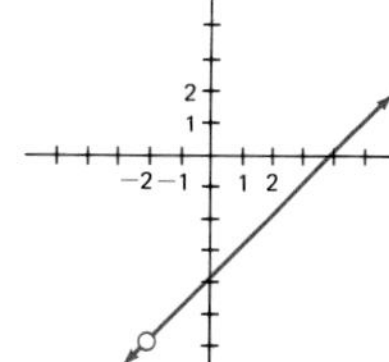

19.

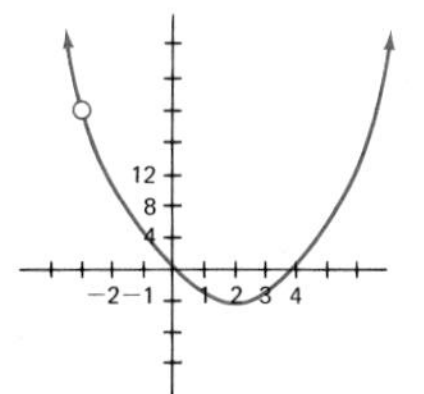

21.

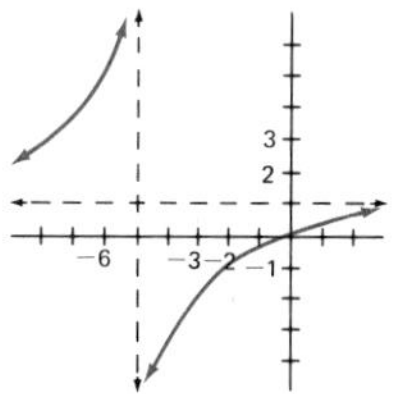

23.

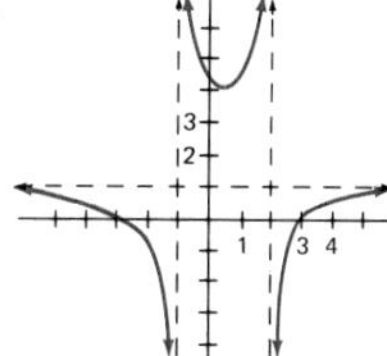

25.

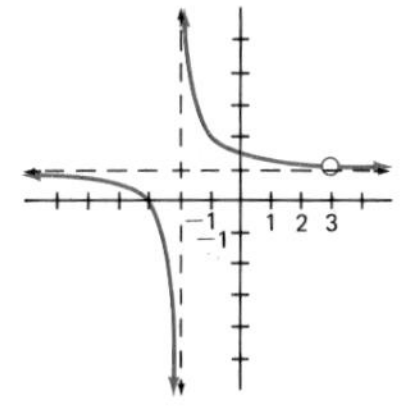

27.

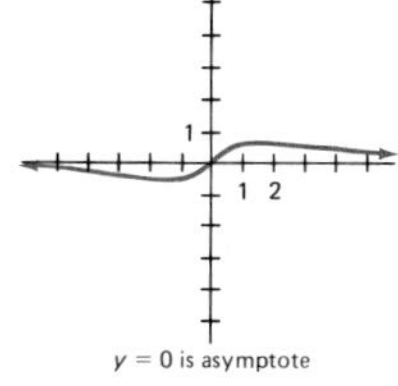

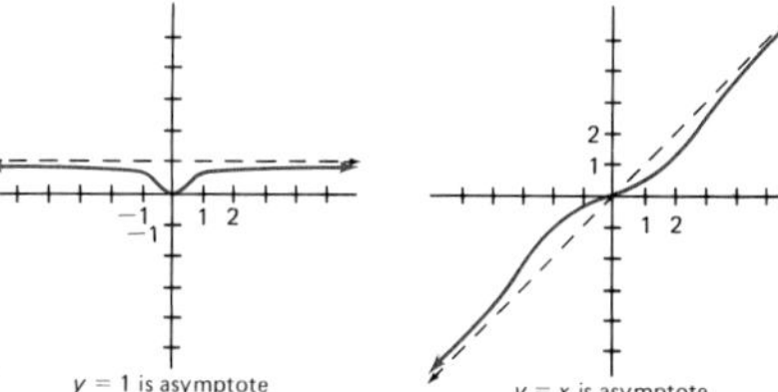

29. $U(x) = (20{,}000 + 50x)/x$

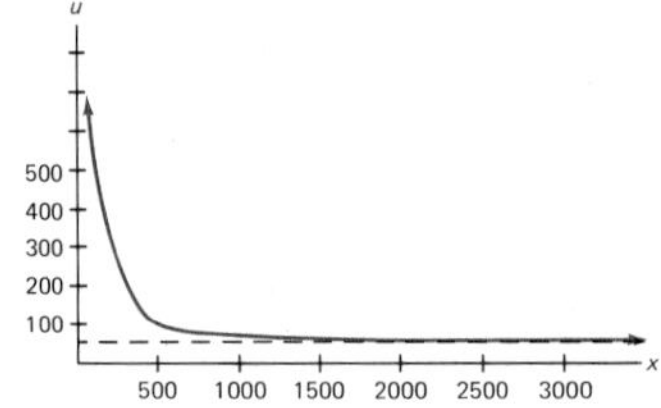

PROBLEM SET 5-4

1. (a) 3 (b) Undefined. (c) -1 (d) -10 (e) 2 (f) $\frac{1}{2}$ (g) 10 (h) $\frac{2}{5}$ (i) 3
3. $(f+g)(x) = x^2 + x - 2$, all real numbers; $(f-g)(x) = x^2 - x + 2$, all real numbers; $(f \cdot g)(x) = x^3 - 2x^2$, all real numbers; $(f/g)(x) = x^2/(x-2)$, $\{x : x \neq 2\}$
5. $(f+g)(x) = x^2 + \sqrt{x}$, $\{x : x \geq 0\}$; $(f-g)(x) = x^2 - \sqrt{x}$, $\{x : x \geq 0\}$; $(f \cdot g)(x) = x^2\sqrt{x}$, $\{x : x \geq 0\}$; $(f/g)(x) = x^2/\sqrt{x}$, $\{x : x > 0\}$
7. $(f+g)(x) = (x^2 - x - 3)/(x-2)(x-3)$, $\{x : x \neq 2 \text{ and } x \neq 3\}$; $(f-g)(x) = (-x^2 + 3x - 3)/(x-2)(x-3)$, $\{x : x \neq 2 \text{ and } x \neq 3\}$; $(f \cdot g)(x) = x/(x-2)(x-3)$, $\{x : x \neq 2 \text{ and } x \neq 3\}$; $(f/g)(x) = (x-3)/x(x-2)$, $\{x : x \neq 0,\ x \neq 2,\ x \neq 3\}$.
9. $(g \circ f)(x) = x^2 - 2$, all real numbers; $(f \circ g)(x) = (x-2)^2$, all real numbers.
11. $(g \circ f)(x) = (3x+1)/x$, $\{x : x \neq 0\}$; $(f \circ g)(x) = 1/(x+3)$, $\{x : x \neq -3\}$.
13. $(g \circ f)(x) = x - 4$, $\{x : x \geq 2\}$; $(f \circ g)(x) = \sqrt{x^2 - 4}$, $\{x : |x| \geq 2\}$.
15. $(g \circ f)(x) = x$, all real numbers; $(f \circ g)(x) = x$, all real numbers.
17. $g(x) = x^3$; $f(x) = x + 4$ **19.** $g(x) = \sqrt{x}$; $f(x) = x + 2$ **21.** $g(x) = 1/x^3$; $f(x) = 2x + 5$
23. $g(x) = |x|$; $f(x) = x^3 - 4$ **25.**

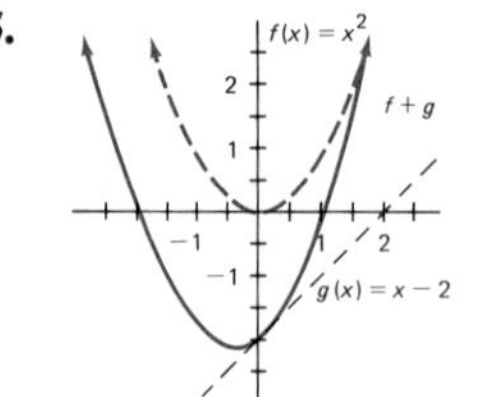

27.

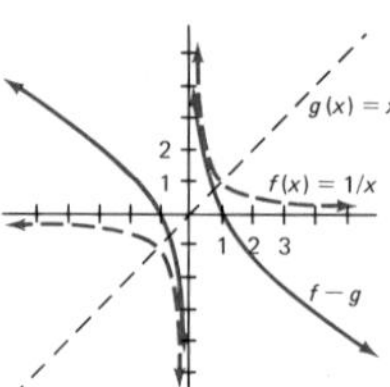

29.

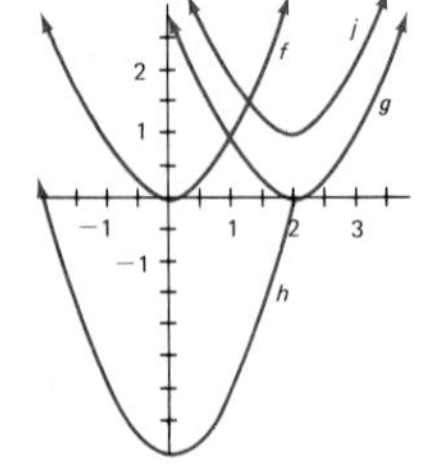

31.

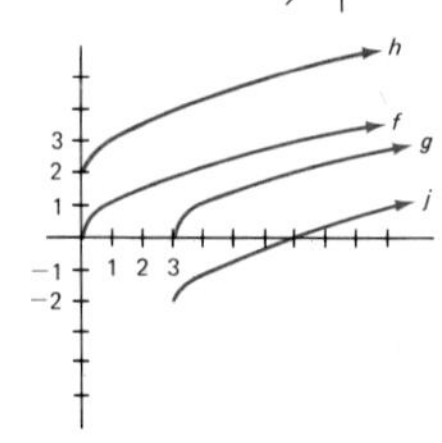

33. (a) $x^3 + 2x + 3$ (b) $x^3 - 2x - 3$ (c) $2x^4 + 3x^3$ (d) $(2x+3)/x^3$ (e) $2x^3 + 3$ (f) $(2x+3)^3$ (g) $4x + 9$ (h) x^{27} **35.** $1/|x^2 - 4|$; $\{x : x \neq -2 \text{ and } x \neq 2\}$ **37.** (a) $2x + h$ (b) 2 (c) $-1/x(x+h)$ (d) $-2/(x+2)(x+h-2)$ **39.** (a) .7053 (b) 1.6105 **41.** (a) $P = \sqrt{27 + \sqrt{t} + t}$ (b) 6.773

PROBLEM SET 5-5

1. (a) i; ii; iii; vii; viii (b) i; ii; viii; (c) i; ii; viii **3.** (a) 1 (b) $-\frac{1}{3}$ (c) $\frac{16}{3}$
5. (a) $f^{-1}(x) = -\frac{1}{3}(x-2)$ (b) $f^{-1}(x) = (2+2x)/x$ (c) $f^{-1}(x) = (x-2)^2$ **7.**

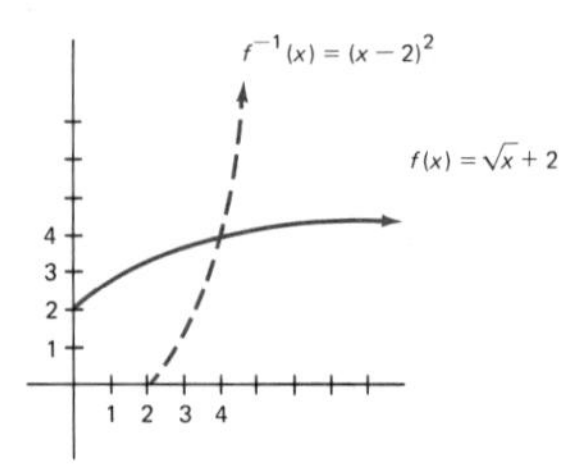

9. (a)

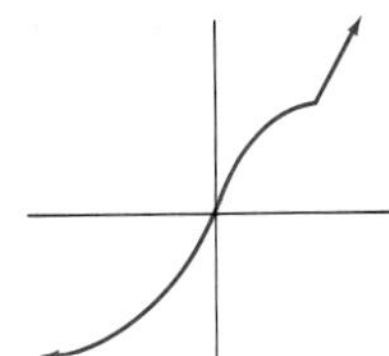

(b)

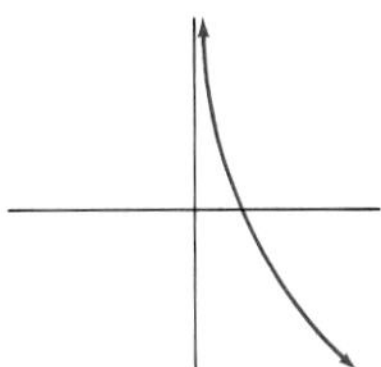

(c)

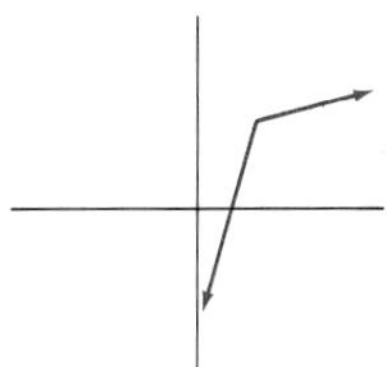

11. $(f \circ g)(x) = 3\left(\frac{2x}{3-x}\right)\Big/\left(\frac{2x}{3-x}+2\right) = \frac{6x}{3-x}\Big/\frac{6}{3-x} = x$; $(g \circ f)(x) = 2\left(\frac{3x}{x+2}\right)\Big/\left(3-\frac{3x}{x+2}\right) = \frac{6x}{x+2}\Big/\frac{6}{x+2} = x$

13. $\{x : x \geq 1\}$; $f^{-1}(x) = 1 + \sqrt{x}$ 15. $\{x : x \geq -1\}$; $f^{-1}(x) = -1 + \sqrt{x+4}$

17. $\{x : x \geq -3\}$; $f^{-1}(x) = -3 + \sqrt{x+2}$ 19. $\{x : x \geq -2\}$; $f^{-1}(x) = x - 2$

21. $\{x : x \geq -\frac{1}{2}\}$; $f^{-1}(x) = x + \sqrt{x^2 + x}$

23.

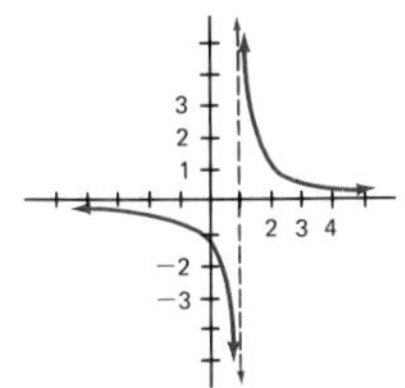

25. $f^{-1}(x) = (x+1)/x$

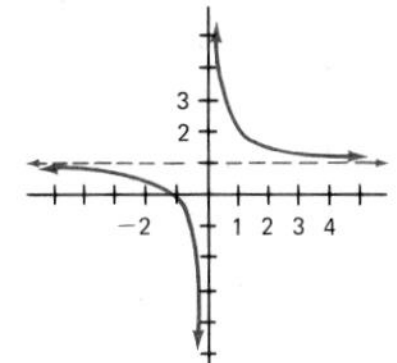

(a) $\frac{1}{2}$ (b) 3 (c) -1 (d) 0 (e) $\frac{4}{3}$ (f) $\frac{1}{2}$

27.

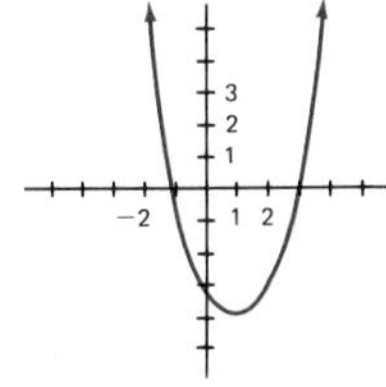

$\{x : x \geq 1\}$ $f^{-1}(x) = 1 + \sqrt{x+4}$

29. $f^{-1}(x) = \frac{1}{2}(x-5)$; $g^{-1}(x) = 1/x$; $(g \circ f)(x) = 1/(2x+5)$; $(g \circ f)^{-1}(x) = (1-5x)/2x$; $(f^{-1} \circ g^{-1})(x) = \frac{1}{2}(1/x - 5) = (1-5x)/2x$ 31. They are equal.

CHAPTER 5. REVIEW PROBLEM SET

1. (a) 15 (b) 4 (c) Undefined. (d) 0 (e) $-\frac{3}{4}$ (f) $\frac{2}{15}$ 2. $\{x : x \geq -1$ and $x \neq 1\}$ 3. $y = \frac{1}{8}x^3$

4. 12 5. (a)

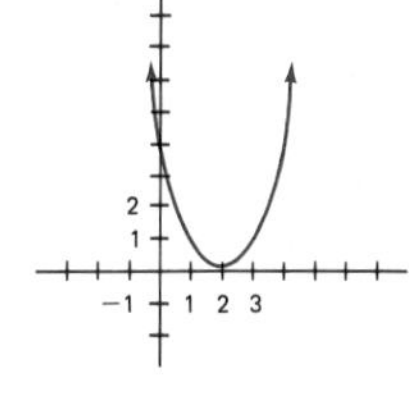

(b)

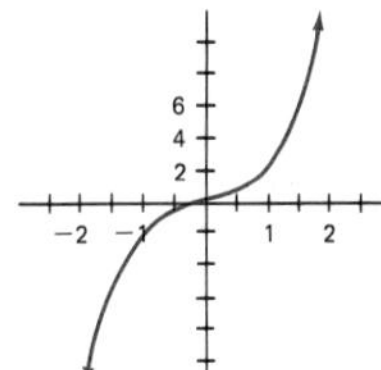

(c)

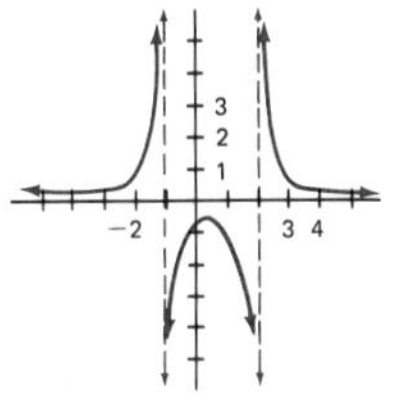

(d)

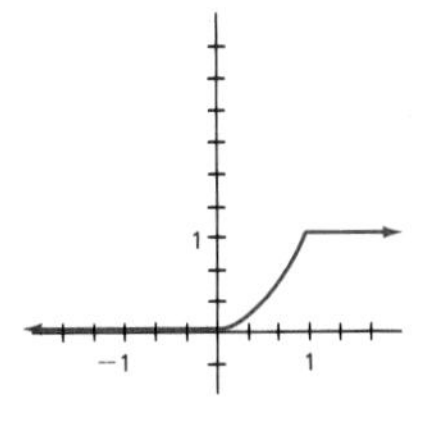

6.

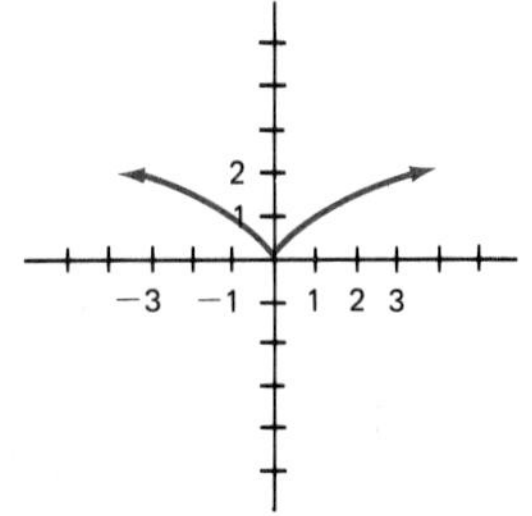

7. 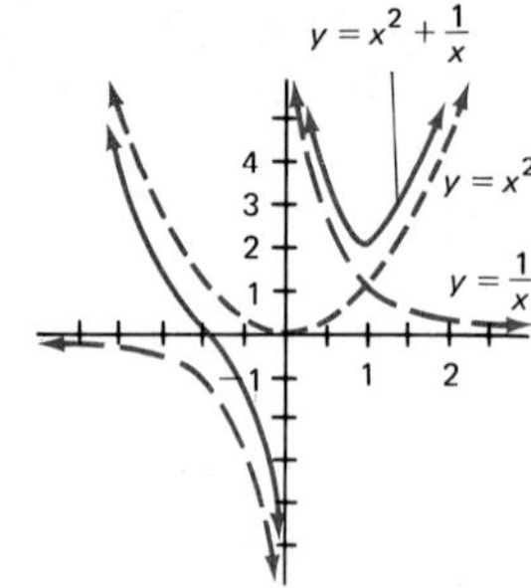

8. (a) $x^3 + 6x^2 + 12x + 9$ (b) $x^3 + 3$ (c) $x^9 + 3x^6 + 3x^3 + 2$ (d) $x + 4$ (e) $\sqrt[3]{x - 1}$ (f) $x - 2$ (g) $x + h + 2$ (h) 1 (i) $27x^3 + 1$ **9.** It is moved 2 units to the right and up 3 units.
10. (a) Even. (b) Odd, one-to-one. (c) Even. (d) One-to-one.

11. 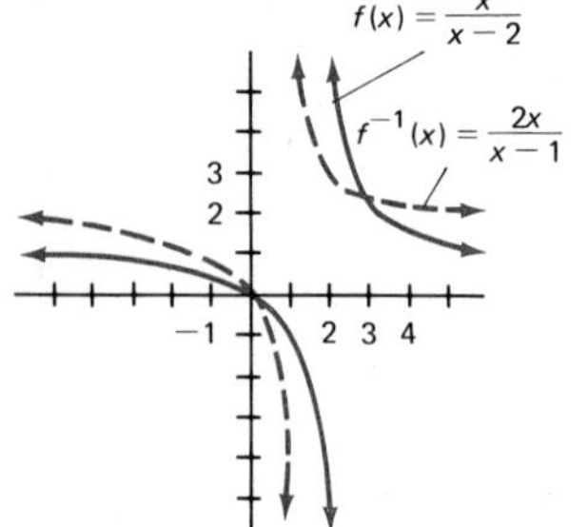

12. $\{x : x \geq -2\}$

PROBLEM SET 6-1

1. 3 **3.** 2 **5.** 7 **7.** $\frac{3}{2}$ **9.** 25 **11.** 9 **13.** 2 **15.** $\frac{1}{100}$ **17.** $\sqrt{2}/2$ **19.** $\sqrt{5}$
21. $3xy\sqrt[3]{2xy^2}$ **23.** $(x + 2)y\sqrt[4]{y^3}$ **25.** $x\sqrt{1 + y^2}$ **27.** $x\sqrt[3]{x^3 - 9y}$ **29.** $(xz^2/y^2)\sqrt[3]{x}$
31. $2(\sqrt{x} - 3)/(x - 9)$ **33.** $2\sqrt{x + 3}/(x + 3)$ **35.** $\sqrt[4]{2x}/2x$ **37.** $(2y/x)\sqrt[3]{x^2}$ **39.** $\sqrt{2}$ **41.** 26
43. $\frac{9}{2}$ **45.** $-\frac{32}{15}$ **47.** 0 **49.** 4 **51.** 2 **53.** $2y^2$ **55.** $-3x/y$ **57.** $3bc^2\sqrt{6b}$
59. $x\sqrt[4]{1 + x^4y^4}$ **61.** $2\sqrt{a - 1}/(a - 1)$ **63.** $(a^2 + 1)/a$ **65.** 6 **67.** 13 **69.** 0; 2
71. (a) 2.44289 (b) 5.97209 (c) 1.09648 (d) 5 **73.** Separate into two cases: first $x \geq 0$, then $x < 0$.
75. They are reflections of each other in the line $y = x$. **77.** $0 \leq x \leq 1$

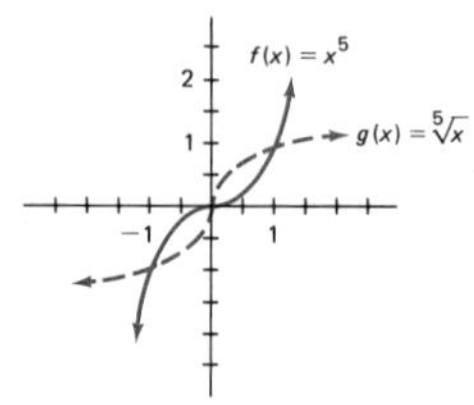

79. (a) 3.875; 3.8730 (b) 6.333; 6.3246 (c) 6.357; 6.3246

PROBLEM SET 6-2

1. $7^{1/3}$ **3.** $7^{2/3}$ **5.** $7^{-1/3}$ **9.** $7^{-2/3}$ **11.** $x^{2/3}$ **13.** $x^{5/2}$ **15.** $(x + y)^{3/2}$ **17.** $(x^2 + y^2)^{1/2}$
19. $\sqrt[3]{16}$ **21.** $1/\sqrt{8^3} = \sqrt{2}/32$ **23.** $\sqrt[4]{x^4 + y^4}$ **25.** $y\sqrt[5]{x^4y}$ **27.** $\sqrt{\sqrt{x} + \sqrt{y}}$ **29.** 5 **31.** 4
33. $\frac{1}{27}$ **35.** .04 **37.** .000125 **39.** $\frac{1}{5}$ **41.** $\frac{1}{16}$ **43.** $\frac{1}{16}$ **45.** $-6a^2$ **47.** $8/x^4$ **49.** x^4
51. $4y^2/x^4$ **53.** y^9/x^{30} **55.** $(2y^3 - 1)/y$ **57.** $x + y + 2\sqrt{xy}$ **59.** $\sqrt[6]{32}$ **61.** $\sqrt[12]{8x^2}$ **63.** $\sqrt{x}$
65. 2.53151 **67.** 4.6364 **69.** 1.70777 **71.** .0050463

73. 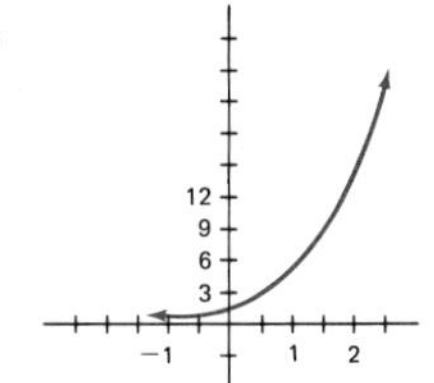

75. 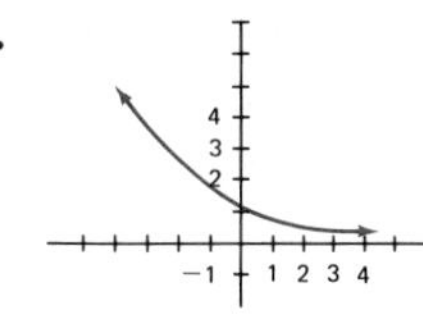

77. 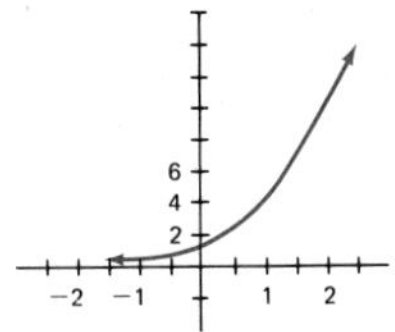

79. $a^{2/3}$ **18.** $(a + 2b)^{3/5}$ **87.** $\sqrt[4]{216}$ **85.** $1/\sqrt{4 + \sqrt{x}}$ **87.** 2 **89.** $\frac{1}{4}$ **91.** .0625 **93.** 5

95. $(1 + x^2)/x$ **97.** 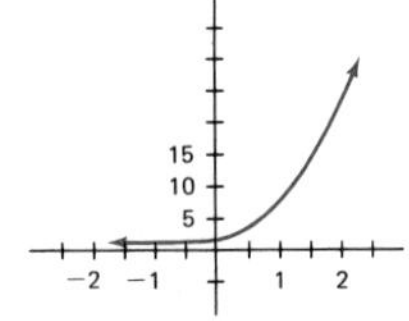

99. 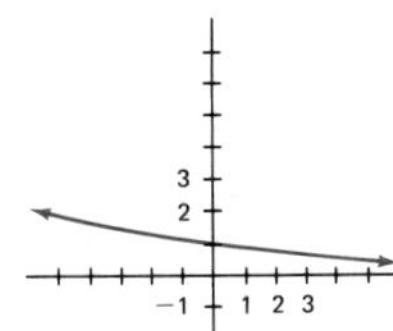

101. $5^{5/3}$; $5^{\sqrt{3}}$; $5^{7/4}$ **103.** 1; 8

PROBLEM SET 6-3

1. (a) Decays; (b) Grows; (c) Grows; (d) Decays. **3.** 7; 11; 17 **5.** 10^{11}
7. (a) 5.384 billion (b) 6.564 billion (c) 23.772 billion **9.** (a) 4.66095714 (b) 17.00006441 (c) 4.80102063 (d) 4.87543916 **11.** (a) \$76,035.82 (b) \$325,678.40 **13.** $p(1 + r/100)^n$ **15.** \$7401.22 **17.** \$4440.73
19. \$4466.59 **21.** $\$1000(1 + .08/12)^{120} = \$2219.64$ **23.** 1600; 1200; 900; 675 **25.** 24,000
27. (a) \$8635.70 (b) \$8832.16 (c) $\$4000(1 + .08/12)^{120} = \$8878.56$ (d) $\$4000(1 + .08/365)^{3650} \approx \$8901$
29. (a) $\$2500(1 + .0525/4)^{18} \approx \$3161.38$ (b) $\$4175(1 + .0776/12)^{141} \approx \$10,360$ **31.** 54,000; 18,000; 2000
33. (a) 5000 (b) 7071 (c) 40,000 (d) 14,142 **35.** 1.25 **37.** (a) 842.433 (b) 503.664 (c) 40.43
39. About \$122,390 **41.** (a) \$2621.44 (b) \$4000 (c) No.

PROBLEM SET 6-4

1. $\log_4 64 = 3$ **3.** $\log_{27} 3 = \frac{1}{3}$ **5.** $\log_4 1 = 0$ **7.** $\log_{125}(1/25) = -\frac{2}{3}$ **9.** $\log_{10} a = \sqrt{3}$
11. $\log_{10} \sqrt{3} = a$ **13.** $5^4 = 625$ **15.** $4^{3/2} = 8$ **17.** $10^{-2} = .01$ **19.** $5^{5/2} = 25\sqrt{5}$ **21.** $b^0 = 1$
23. $b^x = N$ **25.** 2 **27.** −1 **29.** $\frac{1}{2}$ **31.** −6 **33.** 0 **35.** $\frac{3}{2}$ **37.** 5 **39.** $\frac{1}{2}$ **41.** 5.6
43. .778 **45.** 1.204 **47.** −.602 **49.** 1.380 **51.** −.051 **53.** .699 **55.** $\log_{10}[(x + 1)^3(4x + 7)]$
57. $\log_2[8x(x + 2)^2/(x + 8)^2]$ **59.** $\log_6(\sqrt{x}\sqrt[3]{x^3 + 3})$ **61.** 47 **63.** $-\frac{11}{4}$ **65.** $\frac{16}{7}$
67. 5; 2 is extraneous. **69.** (a) $\log_{16}(\frac{1}{8}) = -\frac{3}{4}$ (b) $\log_{10} y = 3.24$ (c) $\log_{1.4} b = a$ **71.** (a) 2 (b) $\frac{1}{2}$ (c) 125 (d) 32 (e) 10 (f) 4 **73.** $\{x : x > 3\}$
77. Since $b = a^x$ and $a = b^y$, $b = (b^y)^x = b^{yx}$. Therefore $1 = yx$ and so $x = 1/y$.
79. Set $\log_b x = c$ so $x = b^c$. Taking $\log_a$ of both sides gives the result.

81. 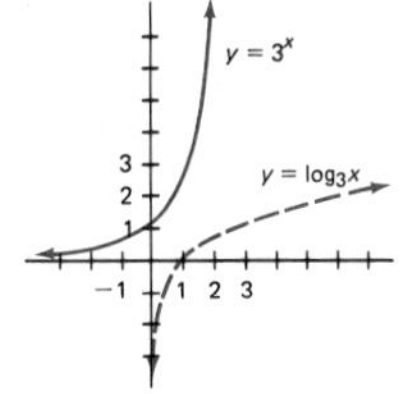

83. (a) 1 (b) $x > 1$ (c) $0 < x < 1$

PROBLEM SET 6-5

1. 1 **3.** 0 **5.** $\frac{1}{2}$ **7.** −3 **9.** 3.5 **11.** −.2 **13.** −7.5 **15.** 4.787 **17.** 6.537
19. .183 **21.** 9.1 **23.** .9 **25.** 90 **27.** About 2.73 **29.** About −.737 **31.** About 6.84

33. (a) 5^{10} (b) 9^{10} (c) 10^{20} (d) 10^{1000} **35.**

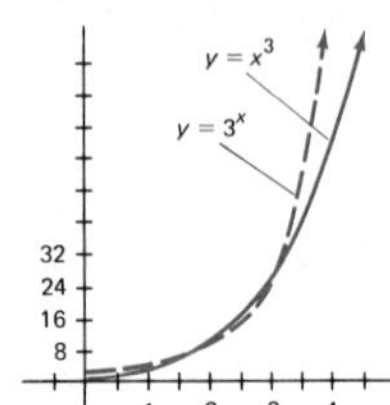

37. $y = ba^x$; $a \approx 1.5$, $b \approx 64$

39. $y = bx^a$; $a \approx 4$, $b \approx 12$ **41.** 3.5 **45.** 3.303 **47.** 7.844 **49.** (a) About -1.05 (b) About 26
51. (a) 0 (b) $2C \ln r$ (c) $C \ln 2$ **53.** About 0.231 years **55.** $x^x = e^{x \ln x}$
57. $(x - 5)^{1/x} = e^{(1/x)\ln(x-5)}$, $x > 5$ **59.** (a) 6.29209 (b) 10.1148

PROBLEM SET 6-6

1. 4 **3.** -2 **5.** $\frac{11}{2}$ **7.** -15 **9.** 10,000 **11.** .01 **13.** $10^{3/2}$ **15.** $10^{-3/4}$ **17.** .6355
19. 2.1987 **21.** .5172 − 2 **23.** 5.7505 **25.** 8.9652 **27.** 32.8 **29.** .0101 **31.** 3.98×10^8
33. 166 **35.** .838 **37.** .7191 **39.** 3.8593 **41.** .0913 − 3 **43.** 7.075 **45.** 8184
47. .03985 **49.** $\frac{5}{4}$ **51.** $\frac{5}{6}$ **53.** 1 **55.** $10^{2/3}$ **57.** .9926 **59.** .3404 − 3 **61.** .9833
63. 856.0 **65.** 5.370 **67.** .003570 **69.** 2.75 **71.** .78 **73.** -6

PROBLEM SET 6-7

1. 128 **3.** .0959 **5.** .0208 **7.** 7.12×10^7 **9.** 3.50 **11.** .983 **13.** 6.05 **15.** 4762
17. 6.143 **19.** 3.530×10^{-6} **21.** 8.90 **23.** 18.2 **25.** 5.19 **27.** $-.5984$ **29.** 2.24
31. .0120 **33.** .00632 **35.** 101 **37.** 1.81 **39.** $\frac{2}{9}$ **41.** -4; 1 **43.** (a) 33.2 grams (b) 391
45. (a) About 6.61 billion (b) In the year 2013. **47.** 21.7 feet

CHAPTER 6. REVIEW PROBLEM SET

1. (a) $-2y^2\sqrt[3]{z}/z^5$ (b) $2xy^2\sqrt[4]{2x}$ (c) $2\sqrt[6]{5}$ (d) $2(\sqrt{x} + \sqrt{y})/(x - y)$ (e) $5\sqrt{2 + x^2}$ (f) $6\sqrt{2}$ **2.** (a) 12 (b) 4

3. (a) $125a^3$ (b) $1/a^{1/2}$ (c) $1/5^{7/4}$ (d) $3y^{13/6}/x^6$ (e) $x - 2x^{1/2}y^{1/2} + y$ (f) $2^{7/6}$ **4.**

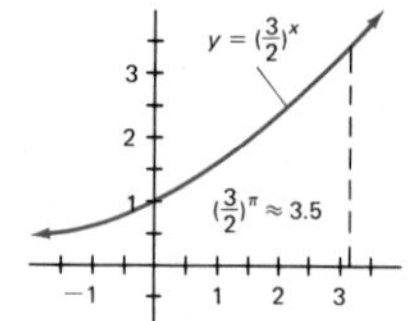

5. $(\frac{1}{2})^{81} \approx 4.14 \times 10^{-25}$ **6.** 16 million **7.** \$220.80 **8.** (a) 3 (b) $\frac{1}{8}$ (c) 7 (d) 1 (e) $\frac{3}{2}$ (f) 5 (g) 10
(h) 1.14 **9.** $\log_4[(3x + 1)^2(x - 1)/\sqrt{x}]$ **10.** (a) $\frac{3}{2}$ (b) $\frac{4}{3}$ (c) $\frac{1}{2}$ (d) -2.773 **11.** (a) 1.680 (b) 9.3

(c) 3.517 (d) .9 **12.** 1.807 **13.** 13.9 years **14.** .1204 **15.**

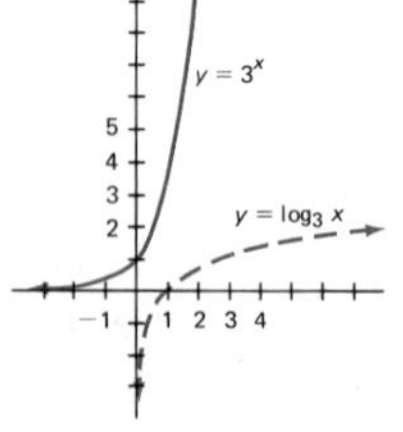

16. 3999

PROBLEM SET 7-1

1. .6600 **3.** .6534 **5.** 3.133 **7.** 12.5° **9.** 66.6° **11.** 69.3° **13.** $16.97 \approx 17$
15. $41.34 \approx 41$ **17.** $66.60 \approx 67$ **19.** $\beta = 48°$; $a = 23.42 \approx 23$; $b = 26.01 \approx 26$ **21.** $\alpha = 33.8°$; $a = 50.8$; $b = 75.9$ **23.** $\beta = 50.6°$; $b = 146$; $c = 189$ **25.** $c = 15$; $\alpha = 36.9°$; $\beta = 53.1°$
27. $b = 30$; $\alpha = 53.1°$; $\beta = 36.9°$ **29.** $\alpha = 26.7°$; $\beta = 63.3°$; $b = 29.0$ **31.** $\alpha = 32.9°$; $\beta = 57.1°$; $c = 17.5$
33. 14.6° **35.** 7.0° **37.** 31.2 feet **39.** (a) .7581 (b) .8545 (c) 1.6842
41. $\beta = 48.7°$; $b = 388$; $c = 517$ **43.** 12.5° **45.** About 1510 yards **47.** 12.25

PROBLEM SET 7-2

1. $2\pi/3$ **3.** $4\pi/3$ **5.** $7\pi/6$ **7.** $7\pi/4$ **9.** 3π **11.** $-7\pi/3$ **13.** $8\pi/9$ **15.** $\frac{1}{9}$ **17.** 240°
19. $-120°$ **21.** 540° **23.** 259.0° **25.** 18.2° **27.** (a) 2 (b) 3.14
29. (a) 3 centimeters (b) 5.5 inches **31.** II **33.** III **35.** II **37.** IV **39.** (a) $-3\pi/4$ (b) $25\pi/3$ (c) $\frac{1}{4}$
41. (a) 10 centimeters (b) 5.89 centimeters **43.** (a) 2.5 radians per second (b) .6 feet **45.** 672
47. About 1 minute **49.** (a) 1 (b) 5.5 **51.** 14.14 centimeters **53.** $A = \pi r^2\theta/360$ **55.** 2675 miles
57. 8.6×10^5 miles

PROBLEM SET 7-3

1. $\sqrt{2}/2$ **3.** $-\sqrt{2}/2$ **5.** $-\sqrt{2}/2$ **7.** $\sqrt{2}/2$ **9.** 1 **11.** 0 **13.** $-\sqrt{3}/2$ **15.** $\frac{1}{2}$
17. $-\sqrt{2}/2$ **19.** $\frac{1}{2}$ **21.** $-\frac{1}{2}$ **23.** $-\sqrt{3}/2$ **25.** $-.95557$; $-.29476$ **27.** (a) $(1/\sqrt{5}, 2/\sqrt{5})$ (b) $2/\sqrt{5}$, $1/\sqrt{5}$ **29.** (a) $\sin(\pi + t) = -y = -\sin t$ (b) $\cos(\pi + t) = -x = -\cos t$ **31.** (a) $\sqrt{3}/2$ (b) $-\sqrt{2}/2$ (c) $\sqrt{2}/2$ (d) $\frac{1}{2}$ (e) $-\sqrt{2}/2$ (f) $-\frac{1}{2}$ (g) $\sqrt{2}/2$ (h) $\frac{1}{2}$ **33.** (a) $\frac{3}{5}$ (b) $-\frac{4}{5}$ (c) $-\frac{4}{5}$ (d) $\frac{3}{5}$ (e) $\frac{3}{5}$ (f) $\frac{4}{5}$ **35.** (a) 0°; 180°; 360° (b) 90°; 270°; 450° **37.** $\pi/6 < t < 5\pi/6$
39. (a) $(2/\sqrt{5}, -1/\sqrt{5})$ (b) $(1/\sqrt{5}, 2/\sqrt{5})$ (c) $(1/\sqrt{5}, -2/\sqrt{5})$ (d) $(-1/\sqrt{5}, -2/\sqrt{5})$

PROBLEM SET 7-4

1. (a) $-\frac{4}{3}$ (b) $-\frac{3}{4}$ (c) $-\frac{5}{3}$ (d) $\frac{5}{4}$ **3.** $\sqrt{3}/3$ **5.** $2\sqrt{3}/3$ **7.** 1 **9.** $2\sqrt{3}/3$ **11.** $-\sqrt{3}/2$
13. $\sqrt{3}$ **15.** 0 **17.** $-\sqrt{3}/3$ **19.** -2 **21.** (a) $\pi/2$; $3\pi/2$; $5\pi/2$; $7\pi/2$ (b) $\pi/2$; $3\pi/2$; $5\pi/2$; $7\pi/2$ (c) 0; π; 2π, 3π; 4π (d) 0; π; 2π, 3π; 4π **23.** $-12/13$; $-12/5$; $13/5$ **25.** $-2\sqrt{5}/5$; 2; $-\sqrt{5}$ **27.** $\frac{3}{5}$; $\frac{5}{4}$
29. $-\frac{12}{13}$; $-\frac{12}{5}$ **31.** $(\frac{5}{13}, -\frac{12}{13})$ **33.** (a) $-\sqrt{2}/2$ (b) $-\sqrt{2}/2$ (c) 1 (d) $-\sqrt{2}$ (e) $-\sqrt{3}$ (f) -1 (g) 0 (h) 0 (i) $-\sqrt{2}$ **35.** $-2\sqrt{5}/5$; $-3\sqrt{5}/5$ **37.** (a) $\tan(-t) = \sin(-t)/\cos(-t) = -\sin t/\cos t = -\tan t$ (b) $\sec(-t) = 1/\cos(-t) = 1/\cos t = \sec t$ (c) $\csc(-t) = 1/\sin(-t) = 1/(-\sin t) = -\csc t$ **39.** .5592
41. II and III; I and III **43.** The point (1, 1000) is on the terminal side of such an angle.
45. $\tan(t + \pi) = \sin(t + \pi)/\cos(t + \pi) = -\sin t/(-\cos t) = \tan t$
47. (a) 3.039; 2.007; 2.000067; 2.0000006; 2.0000000 (b) 2

PROBLEM SET 7-5

(*Note*: Some of the answers may not be very accurate due to the use of 3.14 for π.)
1. 1.30 **3.** .40 **5.** 1.10 **7.** 1.06 **9.** 1.12 **11.** .50 **13.** $3\pi/8$ **15.** $\pi/3$ **17.** .24
19. .24 **21.** $\pi/2$ **23.** .15023 **25.** 5.4707 **27.** .84147 **29.** -1.2885 **31.** $-.82534$
33. 1.25; 1.89 **35.** 1.65; 4.63 **37.** 1.37; 4.51 **39.** 1.84; 4.98 **41.** .4051 **43.** $-.1962$
45. .4051 **47.** .15126 **49.** .9657 **51.** 21.3°; 158.7° **53.** 26.3°; 206.3° **55.** 155.3°; 204.7°
57. .15023 **59.** $-.54030$ **61.** $-.6009$ **63.** (a) 1.20 (b) .60 (c) 1.27 (d) .37 **65.** $1.57 < \pi/2 < 1.58$
67. $-.61806$; $-.06655$; .62052 **69.** (a) .09983 (Table .09983) (b) .38942 (Table .38942) (c) .84167 (Table .84147)

PROBLEM SET 7-6

1.

t	0	$\frac{\pi}{6}$	$\frac{\pi}{4}$	$\frac{\pi}{3}$	$\frac{\pi}{2}$	$\frac{3\pi}{4}$	π	$\frac{5\pi}{4}$	$\frac{3\pi}{2}$	$\frac{7\pi}{4}$	2π
$\cos t$	1	$\frac{\sqrt{3}}{2}$	$\frac{\sqrt{2}}{2}$	$\frac{1}{2}$	0	$-\frac{\sqrt{2}}{2}$	-1	$-\frac{\sqrt{2}}{2}$	0	$\frac{\sqrt{2}}{2}$	1

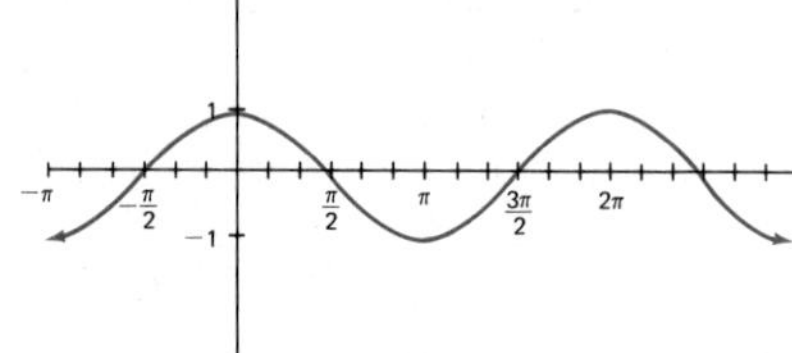

3.

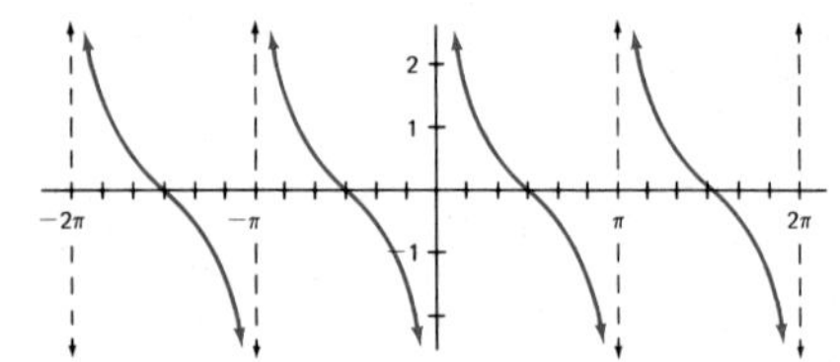

5.

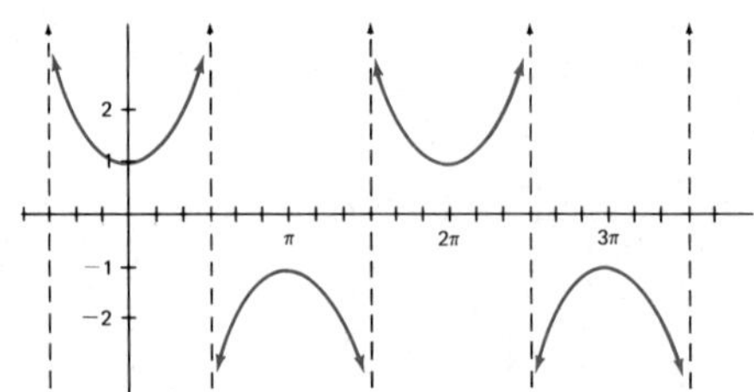

7. Domain: $\{t: t \neq \pi/2 + k\pi$, k any integer); range: $\{y: |y| \geq 1\}$. **9.** π; 2π **11.** $\cot(-t) = -\cot t$

13.

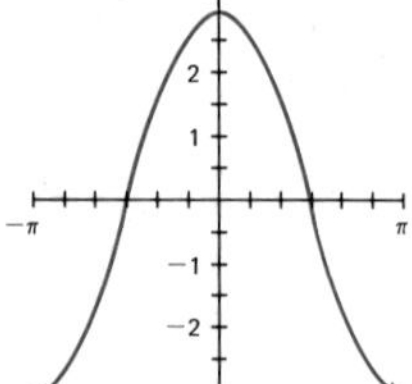

15.

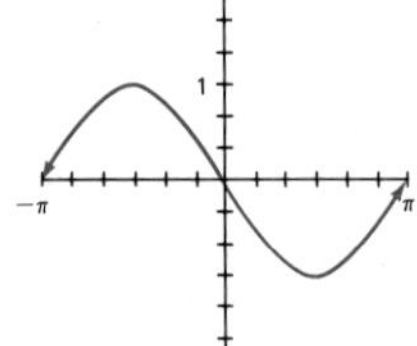

17.

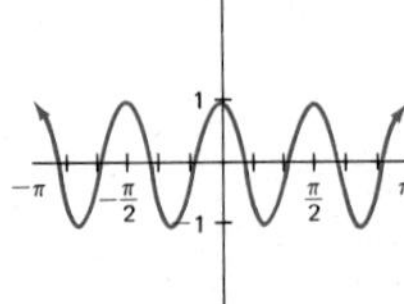

19.

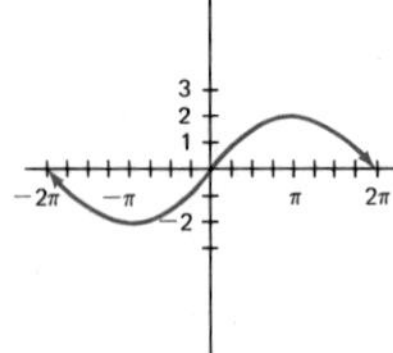

21.

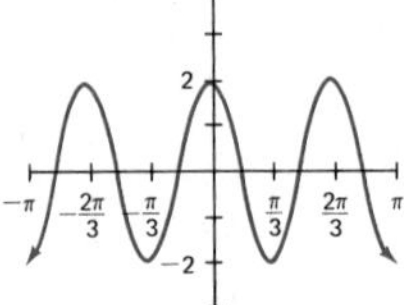

23.

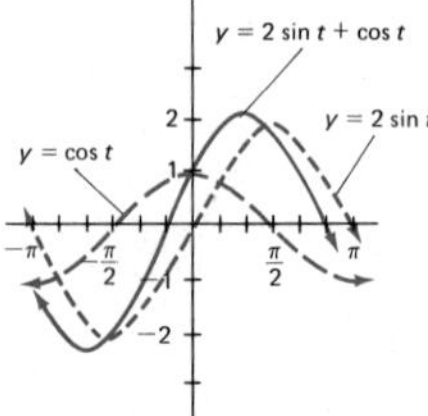

25.

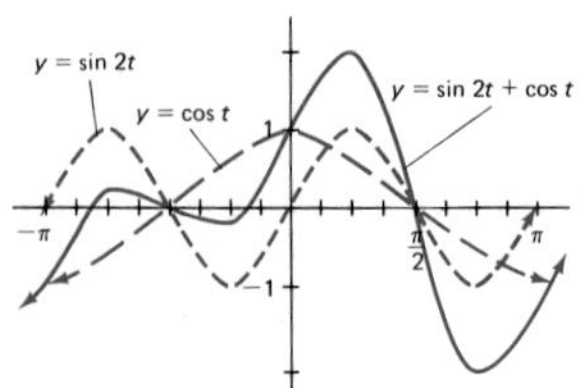

27.

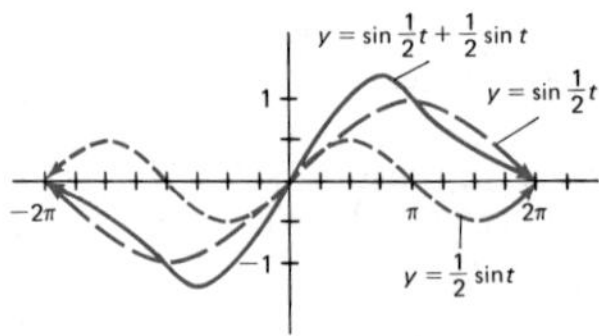

29.

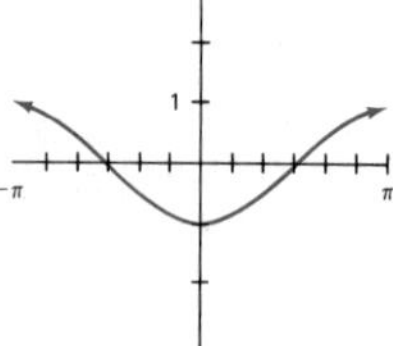

31.

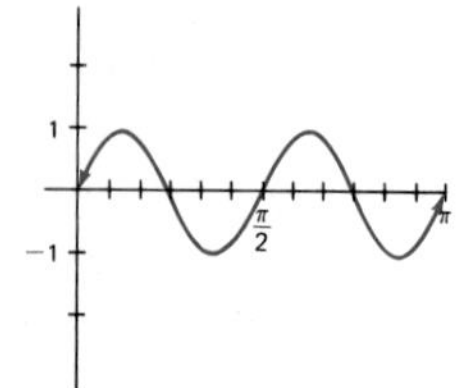

33. 2π; $\pi/2$

35.

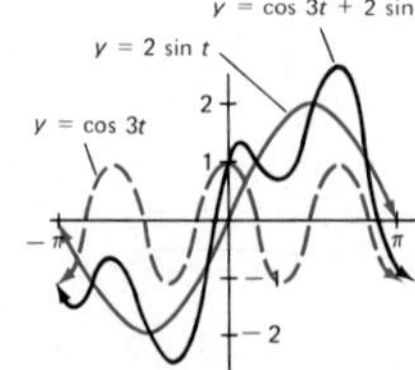

37. 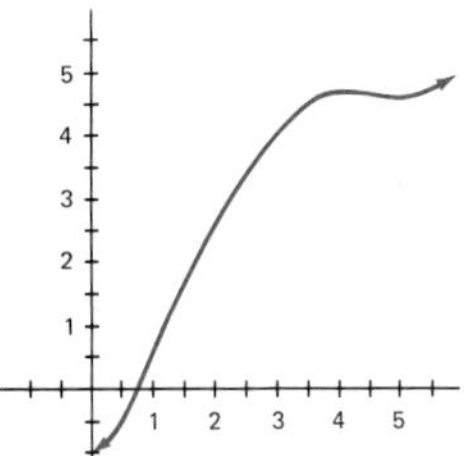

39. (a) 1/60 seconds (b) 60 (c) 30 amperes

41. (a) $1/\pi, 1/2\pi, 1/3\pi, 1/4\pi, \ldots$ (b) $-1, 1, 1, -1, \ldots$ (c) 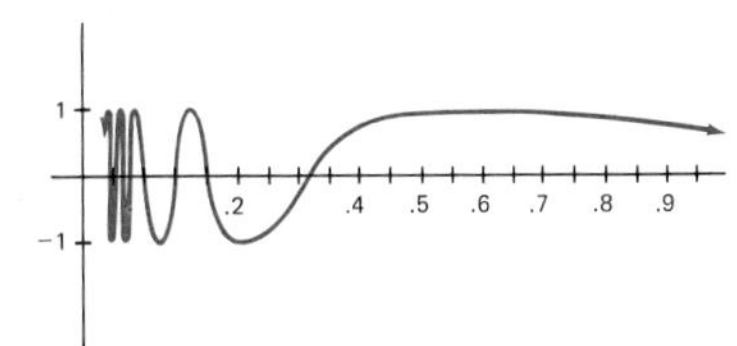

CHAPTER 7. REVIEW PROBLEM SET

1. (a) $\beta = 42.9°$; $a = 27.0$; $b = 25.1$ (b) $\alpha = 46.7°$; $\beta = 43.3°$; $b = 393$ 2. 5.01 feet 3. .576; 405° 4. 18,850 centimeters 5. (a) $-\frac{1}{2}$ (b) $\sqrt{3}/2$ (c) 1 (d) $\frac{1}{2}$ 6. (a) .7771 (b) $-.6157$ (c) $-.5635$ (d) .5258 7. (a) $-\sin t$ (b) $\sin t$ (c) $-\sin t$ (d) $\sin t$ 8. (a) $\{t: 0 \le t < \pi/2$ or $3\pi/2 < t \le 2\pi\}$ (b) $\{t: 0 \le t < \pi/4$ or $3\pi/4 < t < 5\pi/4$ or $7\pi/4 < t \le 2\pi\}$ 9. (a) $\frac{5}{12}$ (b) $-\frac{13}{5}$ 10. $-2/\sqrt{21}$

11. 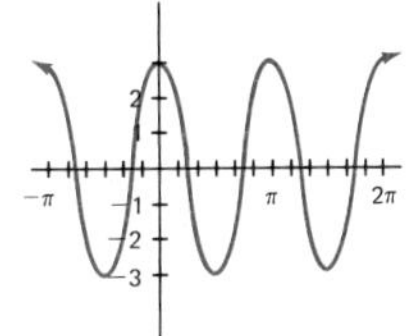

12. 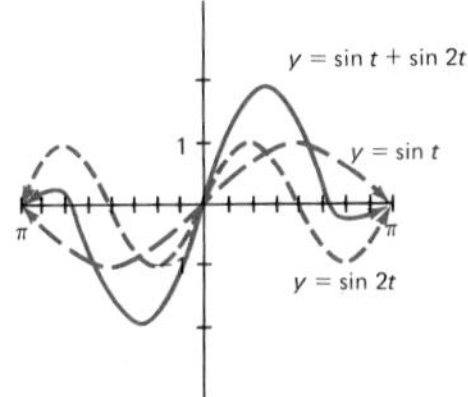

13. $\{y: -1 \le y \le 1\}$; $\{y: y \le -1$ or $y \ge 1\}$ 14. $\cot(-t) = \cos(-t)/\sin(-t) = \cos t/(-\sin t) = -\cot t$
15. See opening display of Section 7.3.

PROBLEM SET 8-1

1. (a) $1 - \sin^2 t$ (b) $\sin t$ (c) $\sin^2 t$ (d) $(1 - \sin^2 t)/\sin^2 t$ 3. (a) $1/\tan^2 t$ (b) $1 + \tan^2 t$ (c) $\tan t$ (d) 3
5. $\cos t \sec t = \cos t(1/\cos t) = 1$ 7. $\tan x \cot x = \tan x(1/\tan x) = 1$ 9. $\cos y \csc y = \cos y(1/\sin y) = \cot y$
11. $\cot \theta \sin \theta = (\cos \theta/\sin \theta) \sin \theta = \cos \theta$ 13. $\tan u/\sin u = (\sin u/\cos u)(1/\sin u) = 1/\cos u$
15. $(1 + \sin z)(1 - \sin z) = 1 - \sin^2 z = \cos^2 z = 1/\sec^2 z$
17. $(1 - \sin^2 x)(1 + \tan^2 x) = \cos^2 x \sec^2 x = \cos^2 x(1/\cos^2 x) = 1$

19. $\sec t - \sin t \tan t = \dfrac{1}{\cos t} - \dfrac{\sin^2 t}{\cos t} = \dfrac{(1 - \sin^2 t)}{\cos t} = \dfrac{\cos^2 t}{\cos t} = \cos t$

21. $\dfrac{(\sec^2 t - 1)}{\sec^2 t} = 1 - \dfrac{1}{\sec^2 t} = 1 - \cos^2 t = \sin^2 t$

23. $\cos t(\tan t + \cot t) = \sin t + \dfrac{\cos^2 t}{\sin t} = \dfrac{(\sin^2 t + \cos^2 t)}{\sin t} = \csc t$

25. $\sin t = (1 - \cos^2 t)^{1/2}$; $\tan t = (1 - \cos^2 t)^{1/2}/\cos t$; $\cot t = \cos t/(1 - \cos^2 t)^{1/2}$; $\sec t = 1/\cos t$; $\csc t = 1/(1 - \cos^2 t)^{1/2}$
27. $\cos t = -3/5$; $\tan t = -4/3$; $\cot t = -3/4$; $\sec t = -5/3$; $\csc t = 5/4$

29. $\dfrac{\sec t - 1}{\tan t} \cdot \dfrac{\sec t + 1}{\sec t + 1} = \dfrac{\sec^2 t - 1}{\tan t(\sec t + 1)} = \dfrac{\tan^2 t}{\tan t(\sec t + 1)} = \dfrac{\tan t}{\sec t + 1}$

31. $\dfrac{\tan^2 x}{\sec x + 1} = \dfrac{\sec^2 x - 1}{\sec x + 1} = \dfrac{(\sec x - 1)(\sec x + 1)}{\sec x + 1} = \sec x - 1 = \dfrac{1 - \cos x}{\cos x}$

33. $\dfrac{\sin t + \cos t}{\tan^2 t - 1} \cdot \dfrac{\cos^2 t}{\cos^2 t} = \dfrac{(\sin t + \cos t)\cos^2 t}{\sin^2 t - \cos^2 t} = \dfrac{\cos^2 t}{\sin t - \cos t}$

35. (a) $(3 - 2\cos^2 t)/\cos^2 t$ (b) $3\tan^2 t + 1$ 37. $\frac{1}{4}$; $\frac{17}{16}$

39. $(\tan x + \cot x)(\cos x + \sin x) = \left(\dfrac{\sin x}{\cos x} + \dfrac{\cos x}{\sin x}\right)(\cos x + \sin x)\ \dfrac{\sin^2 x + \cos x^2}{\cos x \sin x}(\cos x + \sin x)$

$$= \frac{1}{\cos x \sin x}(\cos x + \sin x) = \frac{1}{\sin x} + \frac{1}{\cos x} = \csc x + \sec x$$

41. $\dfrac{(1 + \cos^2 y)}{\sin^2 y} = \dfrac{(1 + 1 - \sin^2 y)}{\sin^2 y} = 2\csc^2 y - 1$

43. $\dfrac{\sin t - \cos t}{\sec t - \csc t} = \dfrac{\sin t - \cos t}{1/\cos t - 1/\sin t} = \dfrac{\sin t - \cos t}{(\sin t - \cos t)/\cos t \sin t} = \cos t \sin t = \dfrac{\cos t}{\csc t}$

45. $\dfrac{(\cos^3 t - \sin^3 t)}{(\cos t - \sin t)} = \dfrac{(\cos t - \sin t)(\cos^2 t + \cos t \sin t + \sin^2 t)}{(\cos t - \sin t)} = 1 + \sin t \cos t$

47. $(1 - \sin^4 t)(\sec^4 t) = \sec^4 t - \tan^4 t = (\sec^2 t + \tan^2 t)(\sec^2 t - \tan^2 t) = \sec^2 t + \tan^2 t$

49. $\dfrac{(\tan t)^{\sin^2 t}}{(\cot t)^{\cos^2 t}} - \tan t = \dfrac{(\tan t)^{\sin^2 t}}{(\tan t)^{-\cos^2 t}} - \tan t = (\tan t)^{\sin^2 t + \cos^2 t} - \tan t = \tan t - \tan t = 0$

PROBLEM SET 8-2

1. (a) $(\sqrt{2} + 1)/2 \approx 1.21$ (b) $(\sqrt{2}\sqrt{3} + \sqrt{4})/4 \approx .97$ 3. (a) $(\sqrt{2} + \sqrt{3})/2 \approx 1.57$ (b) $(\sqrt{2}\sqrt{3} - \sqrt{2})/4 \approx .26$
5. $\cos 2$ 7. $\sin \pi$ 9. $\cos 60°$ 11. $\sin(t + \pi) = \sin t \cos \pi + \cos t \sin \pi = -\sin t$
13. $\sin(t + 3\pi/2) = \sin t \cos(3\pi/2) + \cos t \sin(3\pi/2) = -\cos t$
15. $\sin(t - \pi/2) = \sin t \cos(\pi/2) - \cos t \sin(\pi/2) = -\cos t$
17. $\cos(t + \pi/3) = \cos t \cos(\pi/3) - \sin t \sin(\pi/3) = (1/2)\cos t - (\sqrt{3}/2)\sin t$
19. (a) $\frac{3}{5}$ (b) 12/13 (c) 56/65 (d) −33/65 (e) −16/65 (f) 63/65 (g) −56/33 (h) −16/63
21. (a) $\sin 64°$ (b) $\sin(\pi/6) = \frac{1}{2}$ (c) $\cos(\pi/6) = \sqrt{3}/2$ (d) $\cos 82°$ (e) $\cos^2 26°$ (f) $\sin^2 32°$
23. (a) $-\frac{4}{5}$ (b) $-\frac{24}{25}$ (c) $\frac{7}{25}$ (d) $-\frac{24}{7}$ 25. (a) $1/\sqrt{10}$ (b) $3/\sqrt{10}$ (c) $\frac{1}{3}$ 27. $\pm\sqrt{[1 - \cos(t/2)]/2}$
29. $\tan(s - t) = \tan(s + (-t)) = (\tan s + \tan(-t))/(1 - \tan s \tan(-t)) = (\tan s - \tan t)/(1 + \tan s \tan t)$
31. $\tan 2t = \tan(t + t) = (\tan t + \tan t)/(1 - \tan t \tan t) = (2\tan t)/(1 - \tan^2 t)$
33. $\cos 3t = \cos(2t + t) = \cos 2t \cos t - \sin 2t \sin t$
$= (2\cos^2 t - 1)\cos t - 2\sin^2 t \cos t = (2\cos^2 t - 1)\cos t - 2(1 - \cos^2 t)\cos t = 4\cos^3 t - 3\cos t$
35. $\csc 2t + \cot 2t = (1 + \cos 2t)/(\sin 2t) = (2\cos^2 t)/(2\sin t \cos t) = \cot t$
37. $\sin\theta/(1 - \cos\theta) = 2\sin(\theta/2)\cos(\theta/2)/2\sin^2(\theta/2) = \cot(\theta/2)$
39. $2\tan\alpha/(1 + \tan^2\alpha) = 2\tan\alpha/\sec^2\alpha = 2(\sin\alpha/\cos\alpha)\cos^2\alpha = 2\sin\alpha\cos\alpha = \sin 2\alpha$
41. (a) $(\sqrt{6} - \sqrt{2})/4 \approx .2588$ (b) $(\sqrt{6} + \sqrt{2})/4 \approx .9659$ 43. (a) $(\sqrt{6} - \sqrt{2})/4 \approx .2588$ (b) $\sqrt{2 - \sqrt{3}}/2 \approx .2588$
45. (a) $2\sqrt{2}/3$ (b) $-\sqrt{15}/4$ (c) $(1 - 2\sqrt{30})/12$ (d) $(2\sqrt{2} - \sqrt{15})/12$ (e) $4\sqrt{2}/9$ (f) $-\sqrt{10}/4$
47. (a) Add the two identities and divide by 2. (b) Subtract the two identities and divide by −2.
49. Follow the procedure of Problem 47.
51. (a) $\frac{1}{2}(\cos 3x + \cos x)$ (b) $-\frac{1}{2}(\cos 3x - \cos x)$ (c) $\frac{1}{2}(\sin 8\alpha + \sin 2\alpha)$ (d) $\frac{1}{2}(\sin 8\alpha - \sin 2\alpha)$
53. (a) $2\cos(x + h/2)\sin(h/2)$ (b) $-2\sin(x + h/2)\sin(h/2)$
55. $(\cos 6\theta + \cos 4\theta)/(\sin 6\theta - \sin 4\theta) = (2\cos 5\theta \cos\theta)/(2\cos 5\theta \sin\theta) = \cot\theta$
57. The given expression is equal to 1. To prove it, write $\cos 6t = 2\cos^2 3t - 1$ and use Problem 33.

PROBLEM SET 8-3

1. $\{0, \pi\}$ 3. $\{3\pi/2\}$ 5. No solution. 7. $\{5\pi/6, 7\pi/6\}$ 9. $\{\pi/4, 3\pi/4, 5\pi/4, 7\pi/4\}$
11. $\{\pi/4, 2\pi/3, 3\pi/4, 4\pi/3\}$ 13. $\{0, \pi, 3\pi/2\}$ 15. $\{0, \pi/3, \pi, 4\pi/3\}$ 17. $\{\pi/3, \pi, 5\pi/3\}$
19. $\{\pi/2, 7\pi/6, 3\pi/2, 11\pi/6\}$ 21. $\{0, \pi/2\}$ 23. $\{\pi/6, \pi/2\}$ 25. $\{0\}$
27. $\{\pi/6 + 2k\pi, 5\pi/6 + 2k\pi: k \text{ is an integer}\}$ 29. $\{k\pi: k \text{ is an integer}\}$
31. $\{\pi/6 + k\pi, 5\pi/6 + k\pi: k \text{ is an integer}\}$ 33. $\{0, \pi/2, \pi, 3\pi/2\}$ 35. $\{\pi/8, 5\pi/8, 9\pi/8, 13\pi/8\}$
37. $\{3\pi/8, 7\pi/8, 11\pi/8, 15\pi/8\}$ 39. $\{\pi/6, 11\pi/6\}$ 41. $\{\pi/2, 3\pi/2\}$ 43. $\{\pi/6, \pi/2, 5\pi/6, 3\pi/2\}$
45. $\{0, \pi/3, \pi, 5\pi/3\}$ 47. $\{.8213, 2.3203\}$ 49. $\{\pi/4 + k\pi, 3\pi/2 + 2k\pi: k \text{ is an integer}\}$ 51. $\{1.10, 2.04\}$
53. (a) 15 inches (b) $\tan\theta = \frac{2}{3}$ (c) 33.7° 55. (a) 26.6° (b) 10.3°
57. $\{k\pi/3, 2\pi/3 + 2k\pi, 4\pi/3 + 2k\pi: k \text{ is an integer}\}$ 59. 6, 4.4721, 6; $\{.4645, 1.1059\}$

PROBLEM SET 8-4

1. $\pi/3$ **3.** $\pi/4$ **5.** 0 **7.** $\pi/3$ **9.** $2\pi/3$ **11.** .2200 **13.** $-.2200$ **15.** .2037 **17.** .3486
19. 1.2803 **21.** 2/3 **23.** 10 **25.** $\pi/3$ **27.** $\pi/4$ **29.** $\frac{3}{5}$ **31.** $2/\sqrt{5}$ **33.** .9666 **35.** .4508
37. 24/25 **39.** $\frac{7}{25}$ **41.** 56/65 **43.** $\tan(\sin^{-1} x) = \sin(\sin^{-1} x)/\cos(\sin^{-1} x) = x/\sqrt{1 - x^2}$
45. $\tan(2 \tan^{-1} x) = 2 \tan(\tan^{-1} x)/[1 - \tan^2(\tan^{-1} x)] = 2x/(1 - x^2)$
47. $\{x: 0 \leq x \leq \pi/2\}$; $f^{-1}(x) = \frac{1}{2} \cos^{-1}(x/3)$ **49.** $\{x: x \leq -2/\pi$ or $x \geq 2/\pi\}$; $f^{-1}(x) = 1/\sin^{-1} x$
51. $\pi/6$ **53.** (a) .2037 (b) .8200 (c) -1.0800 **55.** $-.9$ **57.** $(\sqrt{21} - 4)/(5\sqrt{5}) \approx .0521$

59. Domain: all real numbers; range: $\{y: -\pi/2 < y < \pi/2\}$

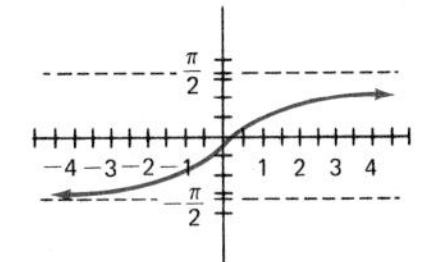

61. (a) 1.5706713 (b) 1.5707963. As x gets large, $\tan^{-1} x$ approaches $\pi/2$.
63. .1309 **65.** (a) 0 (b) $\pi/3$ (c) $2\pi/3$ (d) $2\pi/3$
67. (a) $\sin^{-1}(x/5)$ (b) $\tan^{-1}(x/3)$ (c) $\sin^{-1}(3/x)$ (d) $\tan^{-1}(3/x) - \tan^{-1}(1/x)$ **69.** $.2556 \approx .26$ radians

CHAPTER 8. REVIEW PROBLEM SET

1. (a) $\cot \theta \cos \theta = \cos^2 \theta/\sin \theta = (1 - \sin^2 \theta)/\sin \theta = 1/\sin \theta - \sin \theta = \csc \theta - \sin \theta$
(b) $(\cos x \tan^2 x)(\sec x - 1)/(\sec x + 1)(\sec x - 1) = \cos x \tan^2 x(\sec x - 1)/\tan^2 x = 1 - \cos x$
2. (a) $-\sin^3 x/(1 - \sin^2 x)$ (b) $1 - \sin x$ **3.** (a) $\cos 45° = \sqrt{2}/2$ (b) $\sin 90° = 1$ (c) $\cos 45° = \sqrt{2}/2$
4. (a) $24/25 = .96$ (b) $3/\sqrt{10} \approx .95$ **5.** (a) $\sin 2t \cos t - \cos 2t \sin t = \sin(2t - t) = \sin t$
(b) $\sec 2t + \tan 2t = (1 + \sin 2t)/\cos 2t = (\cos t + \sin t)^2/(\cos^2 t - \sin^2 t) = (\cos t + \sin t)/(\cos t - \sin t)$
(c) $\cos(\alpha + \beta)/\cos \alpha \cos \beta = \cos \alpha \cos \beta/\cos \alpha \cos \beta - \sin \alpha \sin \beta/\cos \alpha \cos \beta = 1 - \tan \alpha \tan \beta = \tan \alpha(\cot \alpha - \tan \beta)$
6. (a) $\{5\pi/6, 7\pi/6\}$ (b) $\{0, \pi, 7\pi/6, 11\pi/6\}$ (c) $\{0\}$ (d) $\{\pi/2, \pi\}$ (e) $\{\pi/6, 5\pi/6, 3\pi/2\}$
7. $-\pi/2 \leq t \leq \pi/2$; $0 \leq t \leq \pi$; $-\pi/2 < t < \pi/2$ **8.** (a) $-\pi/3$ (b) $5\pi/6$ (c) $-\pi/3$ (d) 6 (e) π (f) $\sqrt{5}/3$
(g) $-.02$ (h) 120/169 **9.** See the answer to Problem 59 of Section 8-4. **10.** -1.57

PROBLEM SET 9-1

1. $\gamma = 55.5°$; $b \approx 20.9$; $c \approx 17.4$ **3.** $\beta = 56°$; $a = c \approx 53$ **5.** $\beta \approx 42°$; $\gamma \approx 23°$; $c \approx 20$
7. $\beta \approx 18°$; $\gamma \approx 132°$; $c \approx 12$ **9.** Two triangles: $\beta_1 \approx 53°$, $\gamma_1 \approx 97°$, $c_1 \approx 9.9$; $\beta_2 \approx 127°$, $\gamma_2 \approx 23°$, $c_2 \approx 3.9$
11. 93.7 meters **13.** 44.7° **15.** $\gamma = 35°$; $b \approx 216$; $c \approx 221$ **17.** 27 feet **19.** .17 kilometers
21. 21.2°

PROBLEM SET 9-2

1. $a \approx 12.5$; $\beta \approx 76°$; $\gamma \approx 44°$ **3.** $c \approx 15.6$; $\alpha \approx 26°$; $\beta \approx 34°$ **5.** $\alpha \approx 44.4°$; $\beta \approx 57.1°$; $\gamma \approx 78.5°$
7. $\alpha \approx 30.6°$; $\beta \approx 52.9°$; $\gamma \approx 96.5°$ **9.** 98.8 meters **11.** 24 miles **13.** 106°
15. $a \approx 54$; $\beta \approx 23°$; $\gamma \approx 36°$ **17.** 120°; 26.5 **19.** 922 meters **21.** $A = \frac{1}{2}$(base)(height) $= \frac{1}{2}ch = \frac{1}{2}cb \sin \alpha$
23. 239 **25.** 5.78

PROBLEM SET 9-3

1. (a) (b) 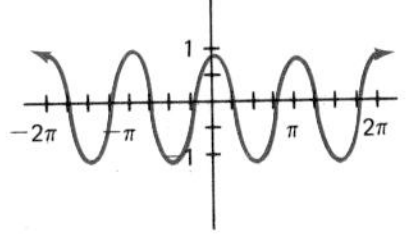 (c)

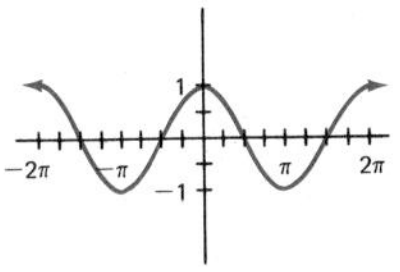

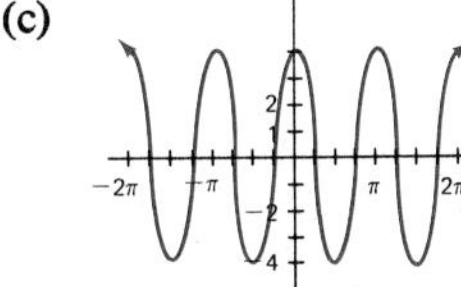

(d)

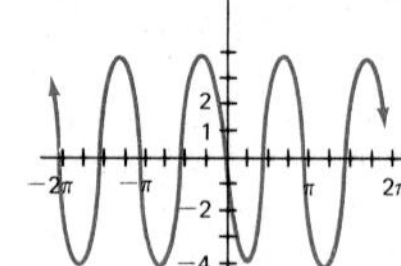

3. (a) $\pi; 4; 0$ (b) $2\pi; 3; -\pi/8$ (c) $\pi/2; 1; -\pi/32$ (d) $2\pi/3; 3; \pi/6$

5. (a)

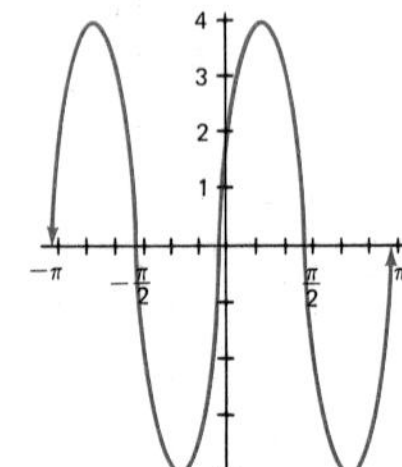

(b)

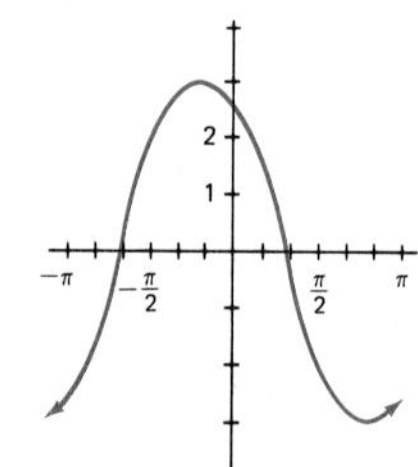

(c)

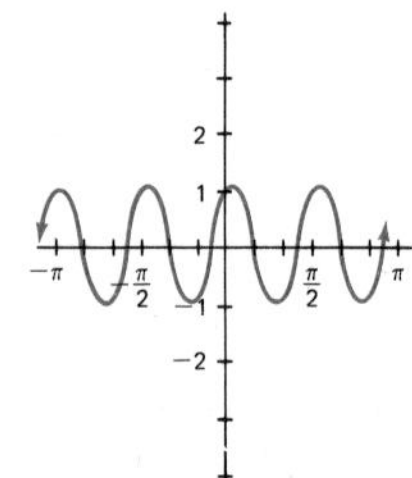

(d)

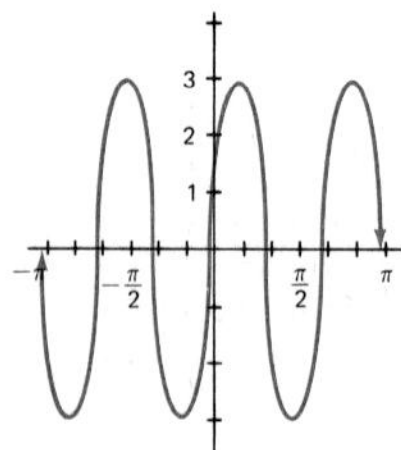

7.

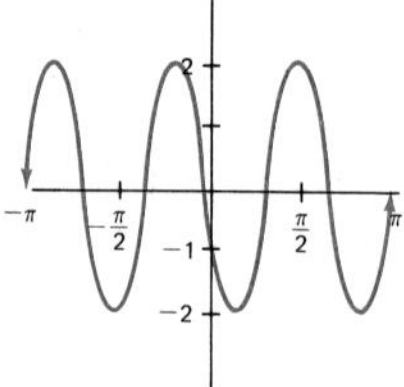

9.

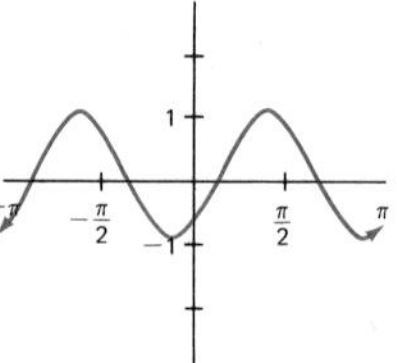

11.

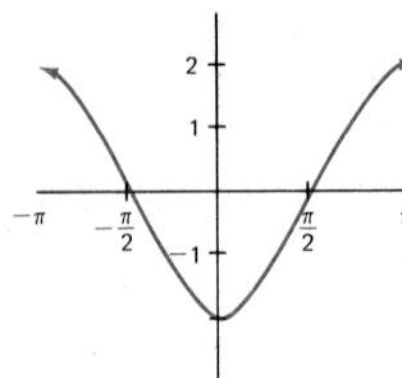

13. $(5\cos 4t, 5\sin 4t)$ 15. $(5\cos 4t, -8 + 5\cos 4t)$

17. (a) $2\pi/5; 1; 0$ (b) $4\pi; \frac{3}{2}; 0$ (c) $\pi/2; 2; \pi/4$ (d) $2\pi/3; 4; -\pi/4$ 19. 15 feet; 9 feet; 12 feet
21. $\sin 120\pi t + \sqrt{25 - \cos^2 120\pi t}$ 23. $2\cos(3t - \pi)$ 25. $A_1 = 0; A_2 = -2$
27. $A\sin(Bt + C) = A[\cos C \sin Bt + \sin C \cos Bt] = A[\sqrt{1 - (A_2/A)^2}\sin Bt + (A_2/A)\cos Bt]$
$= \sqrt{A^2 - A_2^2}\sin Bt + A_2\cos Bt = A_1 \sin Bt + A_2 \cos Bt$
29. 156; 55 31. (a) 0 (b) 0 (c) 53.53 33.

$R = 1000 + 150 \sin 2t$

$C = 200 + 50 \sin(2t - .7)$

PROBLEM SET 9-4

1–11.

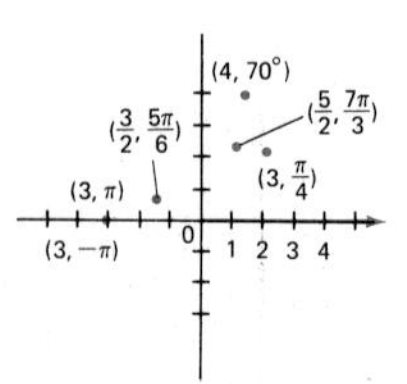

13. $(2\sqrt{2}, 2\sqrt{2})$ 15. $(-3, 0)$ 17. $(-5, -5\sqrt{3})$ 19. $(\sqrt{2}, -\sqrt{2})$

21. $(4, 0)$ **23.** $(2, \pi)$ **25.** $(2\sqrt{2}, \pi/4)$ **27.** $(2\sqrt{2}, 3\pi/4)$ **29.** $(2, -\pi/3)$ **31.** $(2\sqrt{3}, 11\pi/6)$

33.

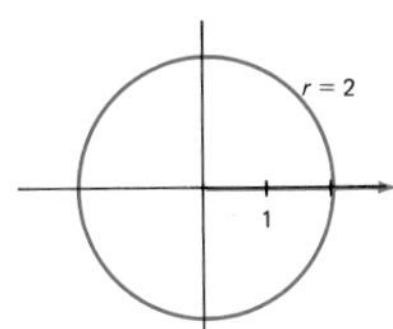

35.

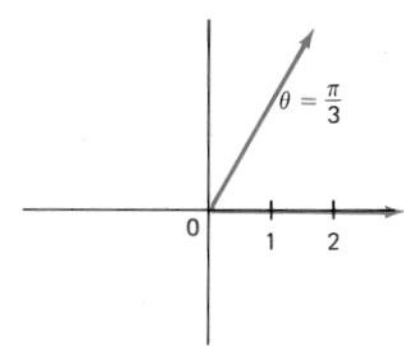

37.

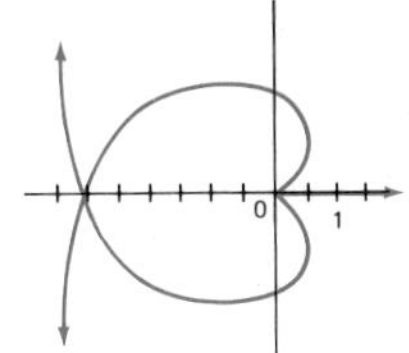

39.

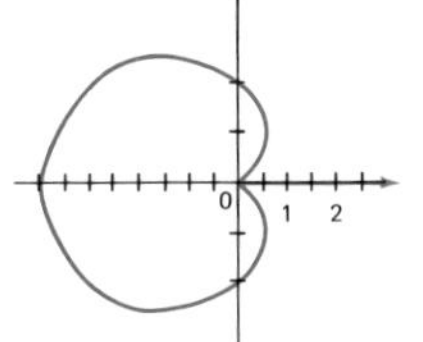

41.

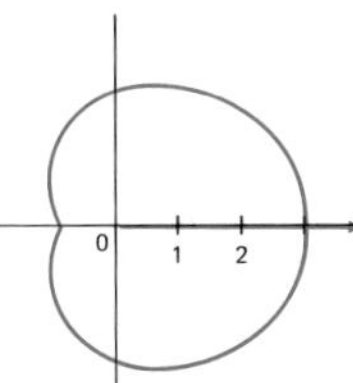

43. $r = 2$ **45.** $r = \tan\theta \sec\theta$

47. $y = 2x$ **49.** $(x^2 + y^2)^{3/2} = x^2 - y^2$ **51.**

53.

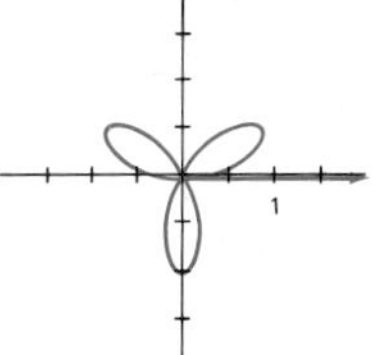

55.

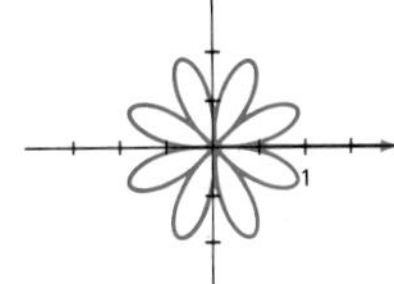

57. (a) $(-3/2, -3\sqrt{3}/2)$ (b) $(-\sqrt{2}, \sqrt{2})$ **59.** $r = 6\cos\theta - 8\sin\theta$

61.

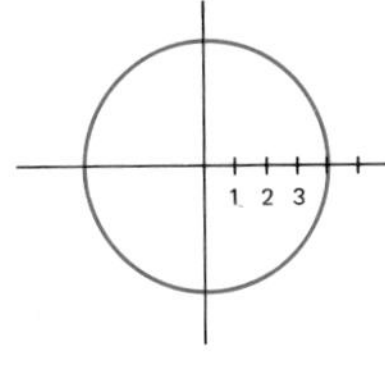

63.

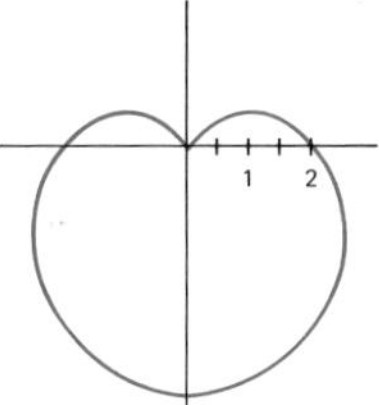

65.

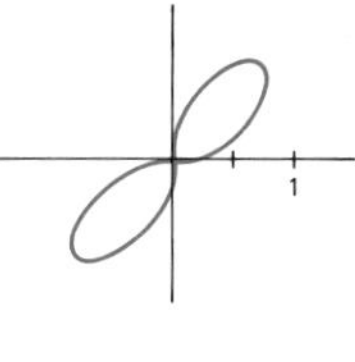

67. (a) $(2\sqrt{2}, \pi/4)$ (b) $(-3, \pi/2)$; $(-3, 3\pi/2)$; $(-3/2, \pi/3)$; $(-3/2; 5\pi/3)$

PROBLEM SET 9-5

1–11.

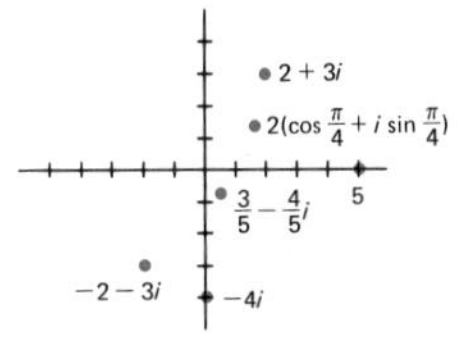

13. $\sqrt{13}$; $\sqrt{13}$; 5; 4; 1; 2 **15.** $0 - 4i$ **17.** $-\sqrt{2} - \sqrt{2}i$

19. $4(\cos \pi + i \sin \pi)$ **21.** $5(\cos 270° + i \sin 270°)$ **23.** $2\sqrt{2}(\cos 315° + i \sin 315°)$
25. $4(\cos \pi/6 + i \sin \pi/6)$ **27.** $6.403(\cos .6747 + i \sin .6747)$ **29.** $6(\cos 210° + i \sin 210°)$
31. $\frac{3}{2}(\cos 125° + i \sin 125°)$ **33.** $\frac{2}{3}(\cos 70° + i \sin 70°)$ **35.** $2(\cos 305° + i \sin 305°)$ **37.** $16 + 0i$
39. $-2 + 2\sqrt{3}i$ **41.** $16(\cos 60° + i \sin 60°)$ **43.** $1(\cos 120° + i \sin 120°)$ **45.** $16(\cos 270° + i \sin 270°)$

47. 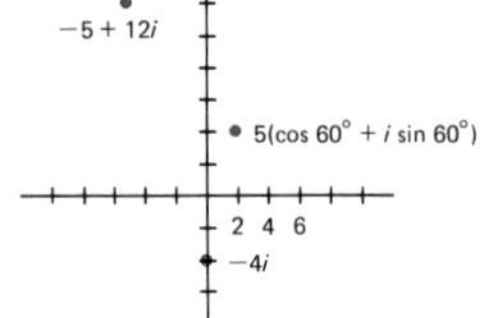

(a) $|-5 + 12i| = 13$ (b) $|-4i| = 4$ (c) $|5(\cos 60° + i \sin 60°)| = 5$

49. (a) $12(\cos 180° + i \sin 180°)$ (b) $3\sqrt{2}(\cos 225° + i \sin 225°)$ (c) $1(\cos 270° + i \sin 270°)$ (d) $2(\cos 315° + i \sin 315°)$ (e) $10(\cos 300° + i \sin 300°)$ (f) $5(\cos 323.13° + i \sin 323.13°)$
51. $20(\cos 255° + i \sin 225°)$ **55.** $\sqrt{5} + 2\sqrt{5}i;\ -\sqrt{5} - 2\sqrt{5}i$
57. (c) $(\cos(k\pi/4) + i \sin(k\pi/4))^8 = \cos 8(k\pi/4) + i \sin 8(k\pi/4) = \cos 2\pi k + i \sin 2\pi k = 1$

PROBLEM SET 9-6

1. $8[\cos(3\pi/4) + i \sin(3\pi/4)]$ **3.** $125(\cos 66° + i \sin 66°)$ **5.** $16(\cos 0° + i \sin 0°)$ **7.** $1 + 0i$
9. $-16\sqrt{3} + 16i$

11. $5(\cos 15° + i \sin 15°)$;
$5(\cos 135° + i \sin 135°)$;
$5(\cos 255° + i \sin 255°)$

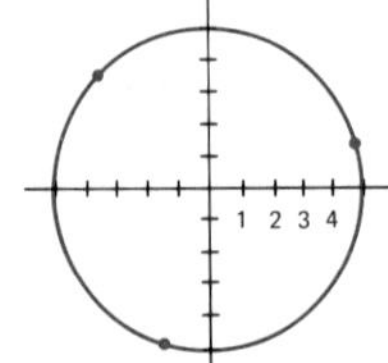

13. $2[\cos(\pi/12) + i \sin(\pi/12)]$; $2[\cos(5\pi/12) + i \sin(5\pi/12)]$; $2[\cos(9\pi/12) + i \sin(9\pi/12)]$; $2[\cos(13\pi/12) + i \sin(13\pi/12)]$; $2[\cos(17\pi/12) + i \sin(17\pi/12)]$; $2[\cos(21\pi/12) + i \sin(21\pi/12)]$

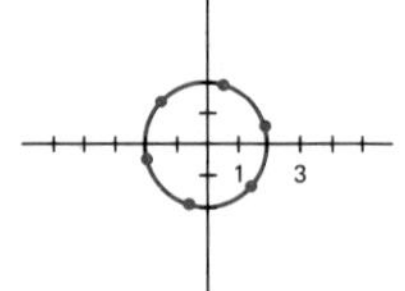

15. $\sqrt{2}(\cos 28° + i \sin 28°)$; $\sqrt{2}(\cos 118° + i \sin 118°)$; $\sqrt{2}(\cos 208° + i \sin 208°)$; $\sqrt{2}(\cos 298° + i \sin 298°)$

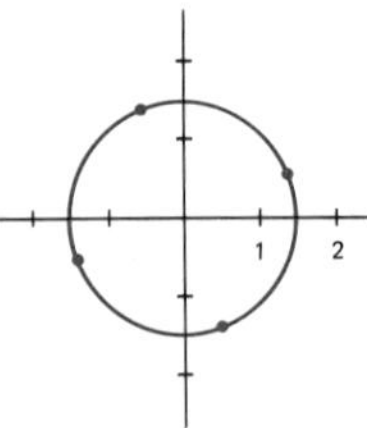

17. $\pm 2;\ \pm 2i$

19. $\pm(\sqrt{2} + \sqrt{2}i)$ **21.** $\pm(\sqrt{2} + \sqrt{6}i)$

23. $\pm 1;\ \pm i$

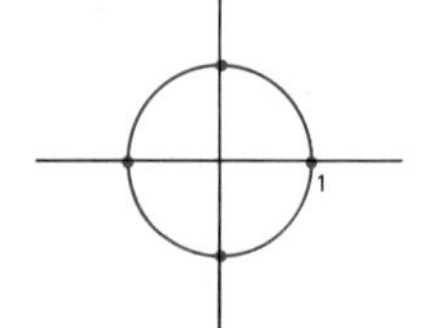

25. $\cos(k \cdot 36°) + i \sin(k \cdot 36°)$, $k = 0, 1, \ldots, 9$

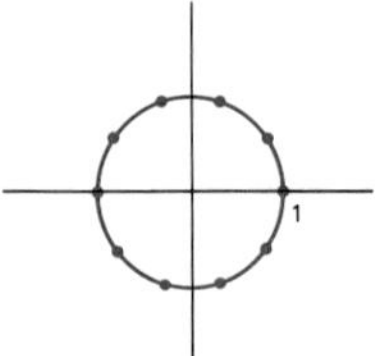

27. $81(\cos 104° + i\sin 104°)$ **29.** $1024(\cos 60° + i\sin 60°)$ **31.** $512 + 512\sqrt{3}i$
33. $2(\cos 55° + i\sin 55°)$; $2(\cos 145° + i\sin 145°)$; $2(\cos 235° + i\sin 235°)$; $2(\cos 325° + i\sin 325°)$
35. Roots: $\sqrt{2}/2 \pm (\sqrt{2}/2)i$, $-\sqrt{2}/2 \pm (\sqrt{2}/2)i$; sum: 0; product: 1.
37. $[\cos(\pi/4) + i\sin(\pi/4)]^4 + [\cos(\pi/4) + i\sin(\pi/4)]^2 + 1 = \cos\pi + i\sin\pi + \cos(\pi/2) + i\sin(\pi/2) + 1 = -1 + i + 1 = i$
39. $(\cos\theta + i\sin\theta)^{-n} = 1/(\cos\theta + i\sin\theta)^n = (\cos 0 + i\sin 0)/(\cos n\theta + i\sin n\theta) = \cos(-n\theta) + i\sin(-n\theta)$
41. $(\cos\theta + i\sin\theta)^3 = \cos 3\theta + i\sin 3\theta$; $(\cos\theta + i\sin\theta)^3 = \cos^3\theta + (3\cos^2\theta\sin\theta)i - 3\cos\theta\sin^2\theta - (\sin^3\theta)i = (\cos^3\theta - 3\cos\theta\sin^2\theta) + (3\cos^2\theta\sin\theta - \sin^3\theta)i$; Thus $\cos 3\theta = \cos^3\theta - 3\cos\theta\sin^2\theta$ and $\sin 3\theta = 3\cos^2\theta\sin\theta - \sin^3\theta$.

CHAPTER 9. REVIEW PROBLEM SET

1. (a) $\beta = 39.1°$; $a \approx 228$; $c \approx 139$ (b) $\alpha \approx 4.0°$; $\beta \approx 154.5°$; $\gamma \approx 21.5°$ (c) $c \approx 13.2$; $\alpha \approx 37.3°$; $\beta \approx 107.7°$ (d) $\gamma \approx 30.7°$; $\alpha \approx 100.7°$; $a \approx 75.5$ **2.** $x \approx 37.1$; $A \approx 281$ **3.** (a) π; 1; 0 (b) $\pi/2$; 3; 0 (c) $2\pi/3$; 2; $\pi/6$ (d) 4π; 2; -2π **4.** (a)

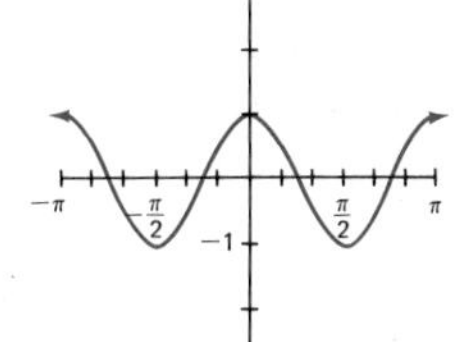

(b)

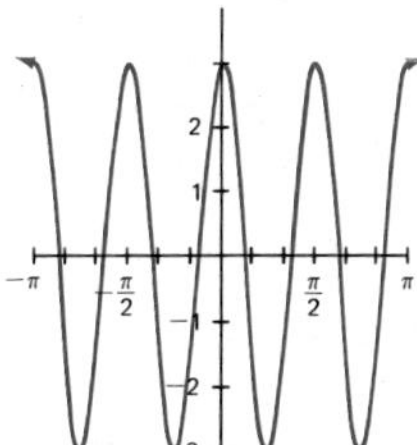

(c)

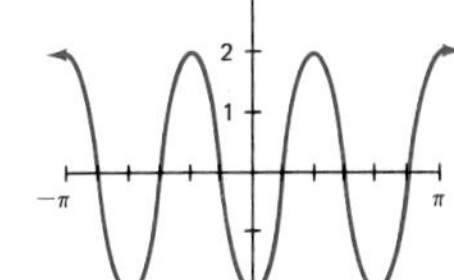

(d)

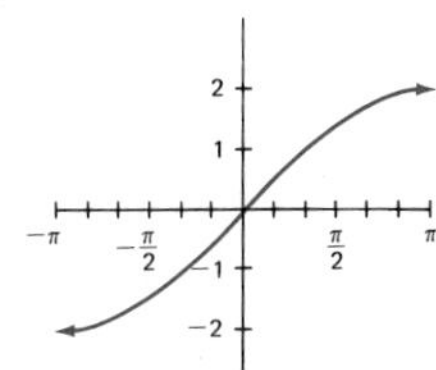

5. $(4\cos(3\pi t/4 + \pi), 4\sin(3\pi t/4 + \pi))$

6. (a) $2\pi/5$ seconds (b) At $x = -3$ feet. (c) When $t = \pi/5$ seconds. **7.**

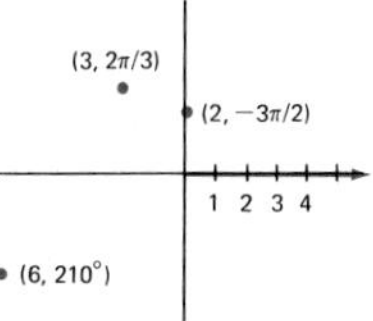

8. (a) $(-3/2, 3\sqrt{3}/2)$ (b) $(0, 2)$ (c) $(-3\sqrt{3}, -3)$ **9.** (a) $(5, 0)$ (b) $(4, 7\pi/4)$ (c) $(4, 5\pi/6)$

10. (a)

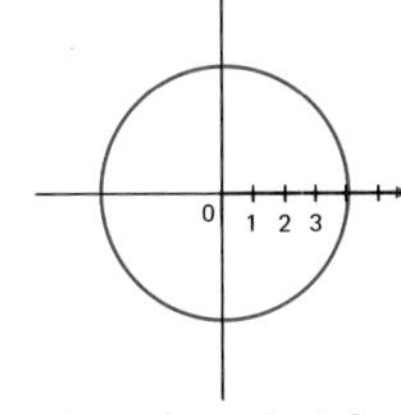

(b)

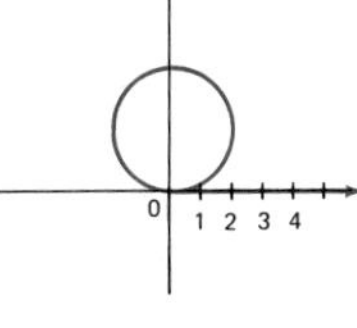

(c)

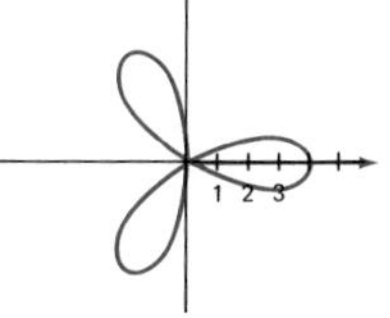

11. $r^2 = 4\sec\theta\csc\theta$; $(x^2 + y^2)^{3/2} = 2xy$ **12.**

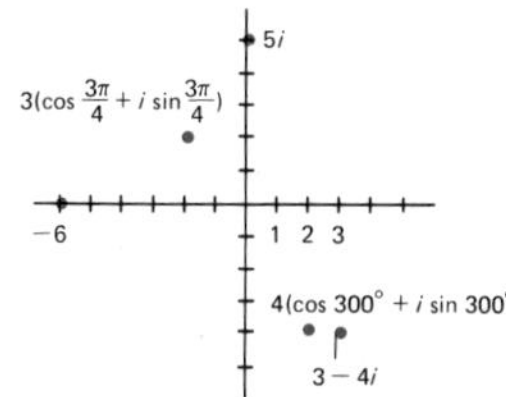

13. (a) 5 (b) 6 (c) 5 (d) 3 (e) 4

14. $-2\sqrt{3} + 2i$

15. (a) $3[\cos(\pi/2) + i\sin(\pi/2)]$ (b) $6(\cos\pi + i\sin\pi)$ (c) $\sqrt{2}[\cos(5\pi/4) + i\sin(5\pi/4)]$ (d) $4[\cos(11\pi/6) + i\sin(11\pi/6)]$
16. (a) $32(\cos 145° + i\sin 145°)$ (b) $2(\cos 65° + i\sin 65°)$ (c) $512(\cos 315° + i\sin 315°)$ (d) $4096(\cos 330° + i\sin 330°)$
17. $2(\cos 20° + i\sin 20°)$; $2(\cos 80° + i\sin 80°)$; $2(\cos 140° + i\sin 140°)$; $2(\cos 200° + i\sin 200°)$; $2(\cos 260° + i\sin 260°)$; $2(\cos 320° + i\sin 320°)$
18. $\cos 0° + i\sin 0°$; $\cos 72° + i\sin 72°$; $\cos 144° + i\sin 144°$; $\cos 216° + i\sin 216°$; $\cos 288° + i\sin 288°$

PROBLEM SET 10-1

1. $x + 1; 0$ **3.** $3x - 1; 10$ **5.** $2x^2 + x + 3; 0$ **7.** $x^2 + 4; 0$ **9.** $x + 2 + 5/x^2$
11. $x - 1 + (-x + 3)/(x^2 + x - 2)$ **13.** $2x^2 + x + 2; -2$ **15.** $3x^2 + 8x + 10; 0$
17. $x^3 + 3x^2 + 7x + 21; 62$ **19.** $x^2 + x - 4; 6$ **21.** $x^3 + x^2 + 3x + 4; 0$ **23.** $x^2 - 3ix - 4 - 6i; 4$
25. $x^3 + 2ix^2 - 4x - 8i; -1$ **27.** $x^2 + 8x + 11; 2x - 48$ **29.** $2x^2 - 4x + 13; -28$
31. $x^4 + 2x^3 + 4x^2 + 8x + 16; 0$ **33.** $Q = x^4 + 3x^3 + 6x^2 + 12x + 8; R = 0$
35. $Q = x^4 + (1 - 2i)x^3 + (-4 - 2i)x^2 + (-4 + 8i)x + 8i; R = 0$ **37.** 11 **39.** 25

PROBLEM SET 10-2

1. -2 **3.** $-\frac{9}{4}$ **5.** -6 **7.** 14 **9.** 1, -2, and 3, each of multiplicity 1.
11. $\frac{1}{2}$(multiplicity 1); 2(multiplicity 2); 0(multiplicity 3) **13.** $1 + 2i$ and $-\frac{2}{3}$, each of multiplicity 1.
15. $P(1) = 0$ **17.** $P(3) = 0$ **19.** $(5 \pm \sqrt{7}i)/4$ **21.** 3; 2 (multiplicity 2) **23.** $(x - 2)(x - 3)$
25. $(x - 1)(x + 1)(x - 2)(x + 2)$ **27.** $(x - 6)(x - 2)(x + 5)$ **29.** $(x + 1)(x + 1 - \sqrt{13})(x + 1 + \sqrt{13})$
31. $(x - 2)(x - 1)(x - 2i)$ **33.** $(x + 1)(x + 1)(x - 1 - i)$ **35.** $x^3 + x^2 - 10x + 8$ **37.** $12x^2 + 4x - 5$
39. $x^3 - 2x^2 - 5x + 10$ **41.** $4x^5 + 20x^4 + 25x^3 - 10x^2 - 20x + 8$
43. $x^3 + (1 - 3i)x^2 - (16 + 9i)x + 20 + 30i$ **45.** Remaining zeros: $-3, -2$. **47.** Remaining zeros: $\pm i$.
49. 22 **51.** 1 and 2, each of multiplicity 2. **53.** (a) $12x^3 - 25x^2 - 4x + 12$ (b) $x^3 + x^2 - 8x - 12$

55.

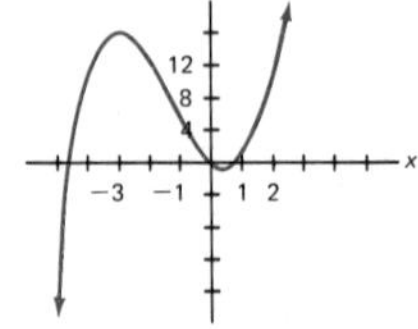

$y = 8$ at $x = -4, -\sqrt{2}, \sqrt{2}$

57. 3, 2.35 **59.** $16x^4 - 32x^3 + 24x^2 - 8x + 1$

61. $P(x) \geq 15$ for all real x.
63. If the coefficients are not all zero, then $P(x)$ can have at most n distinct zeros.

PROBLEM SET 10-3

1. $2 - 3i$ **3.** $-4i$ **5.** $4 + \sqrt{6}$ **7.** $(2 + 3i)^8$ **9.** $2(1 - 2i)^3 - 3(1 - 2i)^2 + 5$ **11.** $5 + i$
13. $3 + 2i; 5 - 4i$ **15.** $-i; \frac{1}{2}$ **17.** $1 - 3i; -2; -1$ **19.** $x^2 - 4x + 29$ **21.** $x^3 + 3x^2 + 4x + 12$
23. $x^5 - 2x^4 + 18x^3 - 36x^2 + 81x - 162$ **25.** $-1; 1; 3$ **27.** $\frac{1}{2}; -1 \pm \sqrt{2}$ **29.** $\frac{1}{2}; (1 \pm \sqrt{5})/2$
31. $3 + 2i; -\frac{1}{2}$ **33.** $-1; -2; 3 \pm \sqrt{13}$ **35.** (a) False. (b) False. (c) False. (d) True. (e) False.
37. $x - 3i = 0$ has no real solution and is of odd degree.
39. If c/d is a solution in reduced form, then d divides the leading coefficient 1. Thus $d = 1$ and c/d is an integer.
41. Remaining solution: -10.
43. (a) $(\sqrt{2} \pm \sqrt{2}i)/2, (-\sqrt{2} \pm \sqrt{2}i)/2$ (b) Rational solution: 2; Remaining solutions as in Part (a).

PROBLEM SET 10-4

1. $4x - 5; -1$ **3.** $4x + 1; 5$ **5.** $10x^4 + 4x^3 - 6x^2 + 8; 16$

7.

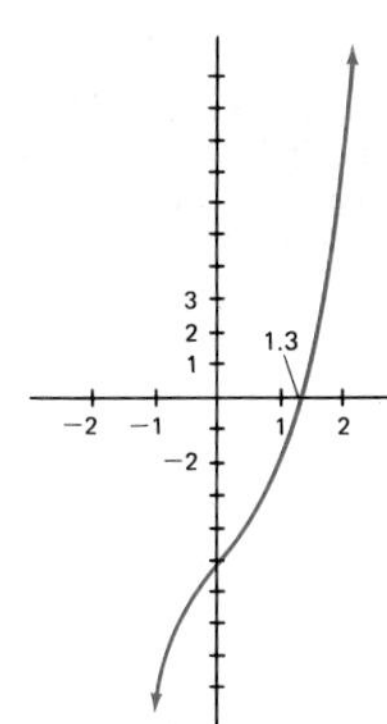

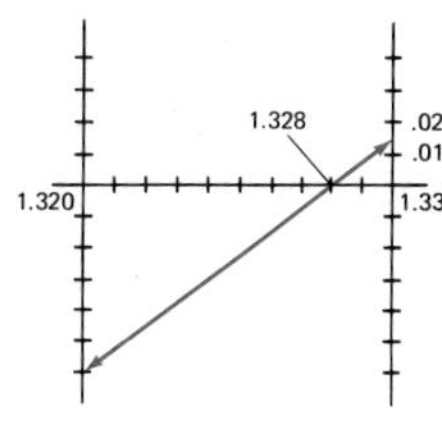

9.

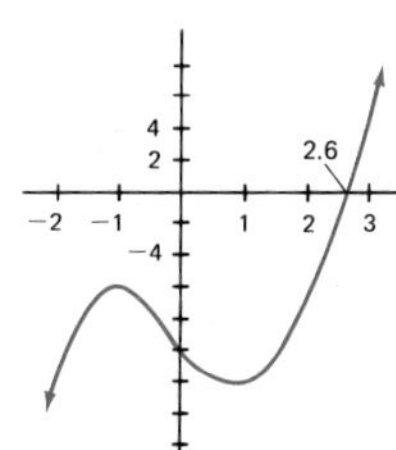

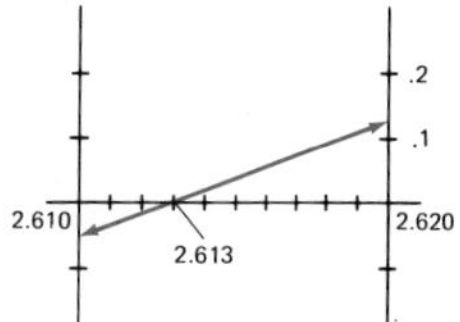

11. $x_1 = 1.3$; $x_2 = 1.33$; $x_3 = 1.328$; $x_4 = 1.3283$ 13. $x_1 = 2.6$; $x_2 = 2.61$; $x_3 = 2.613$; $x_4 = 2.6129$

15.

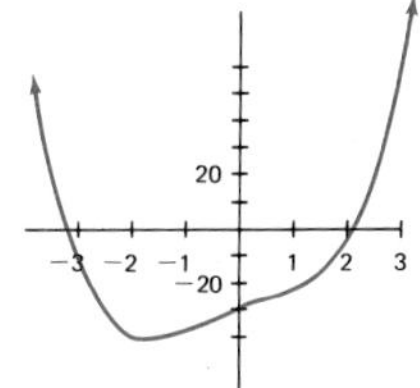

First estimate	Solution to thousandths
−3.2	−3.193
2.2	2.193

17.

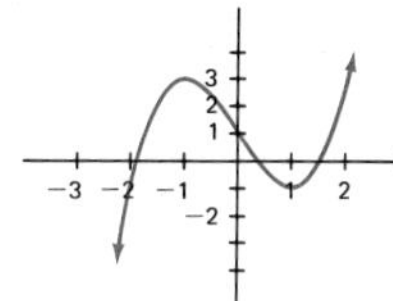

First estimate	Solution to thousandths
−1.9	−1.879
.4	.347
1.5	1.532

19. −2.11; .25; 1.86 21. $r^3 - 9r^2 + 9r - 3 = 0$; $r = 7.91$ 23. 9.701 percent 25. $y - 12 = 12(x - 2)$

27.

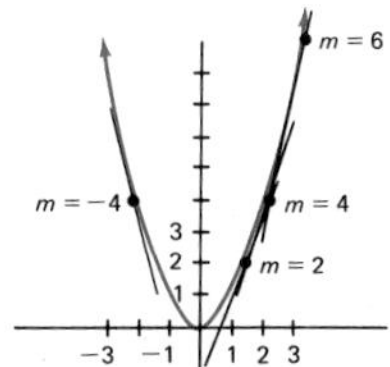

CHAPTER 10. REVIEW PROBLEM SET

1. (a) $2x - 5$; $20x - 20$ (b) $x^2 - 3x + 1$; $-3x + 5$ 2. (a) $x^2 - 4$; -1 (b) $2x^3 - 6x^2 + 3x - 5$; 12 (c) $x^2 + 5x + 1$; -1 3. 1; 97

4. (a) −1 (two); 1 (two); i (one); $-i$ (one) (b) 0 (one); $1 + \sqrt{3}i$ (one); $1 - \sqrt{3}i$ (one); $-\pi$ (three)

5. $2(x-3)(x-\frac{1}{2})(x+3)$ **6.** $(x+4)(x-\sqrt{7})(x+\sqrt{7})$ **7.** $x^4-9x^3+18x^2+32x-96$
8. $x^4-10x^3+31x^2-38x+10$ **9.** Remaining zeros: $-2, 4$. **10.** $6x^3-25x^2+3x+4$ **11.** -6
12. $x^3-6x^2+9x+50$ **13.** $2\pm 5i$; $\pm\sqrt{5}$ **14.** $\frac{3}{2}$, $3\pm 2\sqrt{2}$
15. A cubic equation has 3 solutions (counting multiplicities) and the nonreal solutions for an equation with real coefficients occur in conjugate pairs. The only possible rational solutions are ± 1 and ± 7, but none of them work.

16.

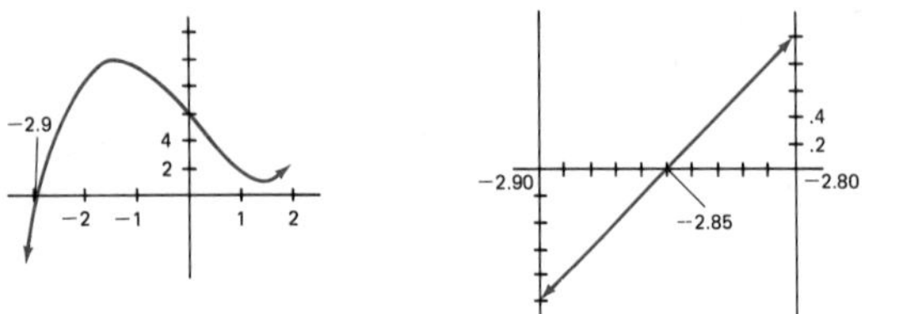

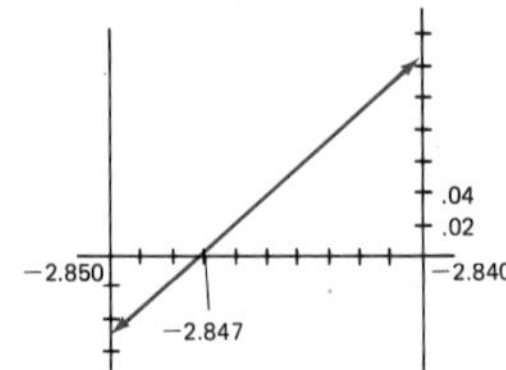

17. $x_1=-2.9$; $x_2=-2.85$; $x_3=-2.847$

PROBLEM SET 11-1

1. $(2,-1)$ **3.** $(-2,4)$ **5.** $(1,-2)$ **7.** $(0,0,-2)$ **9.** $(1,4,-1)$ **11.** $(2,1,4)$ **13.** $(0,0,0)$
15. $(5,6,0,-1)$ **17.** $(15z-110, 4z-32, z)$ **19.** $(2y-3z-2, y, z)$ **21.** $(-z, 2z, z)$ **23.** Inconsistent.
25. $(-z+\frac{2}{5}, z+\frac{16}{5}, z)$ **27.** $(2,4)$; $(10,0)$ **29.** $(5,-7)$; $(6,0)$ **31.** $(-1,2)$; $(1,2)$ **33.** $(\frac{23}{11}, -\frac{9}{11})$
35. $(0,-6)$ **37.** $(0,2,1)$ **39.** $a=-6$; $b=9$ **41.** 10 nickels; 6 dimes; 12 quarters **43.** 375
45. 13 inches by 10 inches **47.** 60; 30; 10

PROBLEM SET 11-2

1. $\begin{bmatrix}2 & -1 & 4\\ 1 & -3 & -2\end{bmatrix}$ **3.** $\begin{bmatrix}1 & -2 & 1 & 3\\ 2 & 1 & 0 & 5\\ 1 & 1 & 3 & -4\end{bmatrix}$ **5.** $\begin{bmatrix}2 & -3 & -4\\ 3 & 1 & -2\end{bmatrix}$ **7.** $\begin{bmatrix}1 & 0 & 0 & 5\\ 1 & 2 & -1 & 4\\ 3 & -1 & -5 & -13\end{bmatrix}$
9. Unique solution. **11.** No solution. **13.** Unique solution. **15.** Infinitely many solutions.
17. No solution. **19.** $(1,2)$ **21.** $(x, \frac{3}{2}x-\frac{1}{2})$ **23.** $(1,4,-1)$ **25.** $(\frac{16}{3}z+\frac{32}{3}, -\frac{7}{3}z-\frac{2}{3}, z)$
27. $(3,0,0)$ **29.** $(4.36, 1.26, -.97)$ **31.** $\begin{bmatrix}2 & -1 & -1 & -1\\ 4 & -1 & 0 & 3\\ 2 & -1 & 0 & 7\end{bmatrix}$ **33.** No solution.
35. Infinitely many solutions. **37.** No solution. **39.** $a=-4$; $b=8$; $c=0$
41. $D=-6$; $E=-4$; $F=-12$

PROBLEM SET 11-3

1. $\begin{bmatrix}8 & 4\\ 1 & 10\end{bmatrix}$; $\begin{bmatrix}-4 & -6\\ 5 & 4\end{bmatrix}$; $\begin{bmatrix}6 & -3\\ 9 & 21\end{bmatrix}$ **3.** $\begin{bmatrix}5 & 4 & 4\\ 8 & 3 & -6\end{bmatrix}$; $\begin{bmatrix}1 & -8 & 6\\ 0 & -3 & 0\end{bmatrix}$; $\begin{bmatrix}9 & -6 & 15\\ 12 & 0 & -9\end{bmatrix}$
5. $\begin{bmatrix}14 & 7\\ 4 & 36\end{bmatrix}$; $\begin{bmatrix}27 & 29\\ 5 & 23\end{bmatrix}$ **7.** $\begin{bmatrix}-3 & -4 & 2\\ 8 & 22 & -13\\ -2 & 0 & 9\end{bmatrix}$; $\begin{bmatrix}0 & 5 & -17\\ 13 & 10 & 3\\ 1 & -3 & 18\end{bmatrix}$ **9.** **AB** not possible; **BA** $=\begin{bmatrix}7 & 2 & -7 & 6\\ 15 & 2 & -11 & 16\end{bmatrix}$
11. **AB** $=\begin{bmatrix}2\\ 16\\ -2\end{bmatrix}$; **BA** not possible **13.** **AB** $=$ **BA** $=\begin{bmatrix}0 & 0\\ 0 & 0\end{bmatrix}$ **15.** $\begin{bmatrix}-4 & 7 & 9\\ -5 & -5 & 8\end{bmatrix}$
17. **A(B + C)** $=$ **AB** $+$ **AC** $=\begin{bmatrix}-7 & -3\\ 39 & 34\end{bmatrix}$; the distributive property. **19.** 93.5917
21. $\begin{bmatrix}2 & 5 & -1\\ -8 & 5 & -3\\ 16 & -2 & -1\end{bmatrix}$; $\begin{bmatrix}-16 & -6 & 8\\ 12 & 0 & 22\\ 11 & -16 & 19\end{bmatrix}$; $\begin{bmatrix}32 & -3 & 12\\ 36 & 29 & 24\\ 34 & 6 & 25\end{bmatrix}$
23. $(\mathbf{A}+\mathbf{B})^2=\mathbf{A}^2+\mathbf{AB}+\mathbf{BA}+\mathbf{B}^2$; not for matrices. **25.** It consists of zeros.

27. If **A** is $m \times n$ and **AB** and **BA** both make sense, then **B** is $n \times m$ and so **AB** is $m \times m$, **BA** is $n \times n$.

29. (a) $\begin{bmatrix} 16 \\ 13 \\ 14 \end{bmatrix}$ $\begin{matrix} \to \text{Art's wages on Monday.} \\ \to \text{Bob's wages on Monday.} \\ \to \text{Curt's wages on Monday.} \end{matrix}$ (b) $\begin{bmatrix} 18 \\ 14 \\ 16 \end{bmatrix}$ Each man's corresponding wages on Tuesday.

(c) $\begin{bmatrix} 7 & 9 & 3 \\ 9 & 3 & 4 \\ 8 & 5 & 4 \end{bmatrix}$ The combined output for Monday and Tuesday. (d) $\begin{bmatrix} 34 \\ 27 \\ 30 \end{bmatrix}$ Each man's combined wages for the two days.

31. It multiplies the first row of **B** by 3, the second row by 4, and the third by 5. It multiplies the first column of **B** by 3, the second column by 4, and the third by 5.

PROBLEM SET 11-4

1. $\begin{bmatrix} -1 & -3 \\ 1 & 2 \end{bmatrix}$ **3.** $\begin{bmatrix} \frac{1}{6} & \frac{7}{6} \\ 0 & \frac{1}{2} \end{bmatrix}$ **5.** $\begin{bmatrix} 1 & 0 \\ 0 & 1 \end{bmatrix}$ **7.** $\begin{bmatrix} 1/a & 0 \\ 0 & 1/b \end{bmatrix}$ **9.** $\begin{bmatrix} -2 & \frac{3}{2} \\ 1 & -\frac{1}{2} \end{bmatrix}$

11. $\begin{bmatrix} -\frac{4}{7} & \frac{2}{7} & \frac{3}{7} \\ \frac{6}{7} & -\frac{3}{7} & -\frac{1}{7} \\ \frac{5}{7} & \frac{1}{7} & -\frac{2}{7} \end{bmatrix}$ **13.** $\begin{bmatrix} -\frac{1}{9} & \frac{1}{9} & \frac{8}{9} \\ \frac{10}{9} & -\frac{1}{9} & -\frac{26}{9} \\ \frac{1}{9} & -\frac{1}{9} & \frac{1}{9} \end{bmatrix}$ **15.** $\begin{bmatrix} 1 & -1 & 2 & -\frac{5}{4} \\ 0 & \frac{1}{2} & -\frac{3}{2} & \frac{7}{8} \\ 0 & 0 & 1 & -\frac{3}{4} \\ 0 & 0 & 0 & \frac{1}{4} \end{bmatrix}$ **17.** $(\frac{5}{7}, \frac{10}{7}, -\frac{1}{7})$

19. $(\frac{47}{9}, -\frac{128}{9}, \frac{7}{9})$ **23.** Inverse does not exist. **25.** $\begin{bmatrix} -\frac{14}{57} & \frac{22}{57} & -\frac{27}{57} \\ -\frac{9}{57} & \frac{6}{57} & \frac{3}{57} \\ \frac{11}{57} & -\frac{1}{57} & \frac{9}{57} \end{bmatrix}$ **27.** $\begin{bmatrix} \frac{1}{2} & 0 & 0 \\ 0 & \frac{1}{3} & 0 \\ 0 & 0 & -\frac{1}{4} \end{bmatrix}$

29. $x = (-14a + 22b - 27c)/57$; $y = (-9a + 6b + 3c)/57$; $z = (11a - b + 9c)/57$

PROBLEM SET 11-5

1. -8 **3.** 22 **5.** -50 **7.** 0 **9.** (a) 12 (b) -12 (c) 36 (d) 12 **11.** -30 **13.** -16
15. 7 **17.** 42.3582 **19.** $(2, -3)$ **21.** $(2, -2, 1)$ **23.** 0 **25.** 0 **27.** $(1, 2, 3)$

29. $\begin{vmatrix} ka & kb \\ c & d \end{vmatrix} = kad - kbc = k(ad - bc) = k\begin{vmatrix} a & b \\ c & d \end{vmatrix}$ **31.** $D = 0$

PROBLEM SET 11-6

1. -20 **3.** -1 **5.** 39 **7.** 4 **9.** 57 **11.** $x = 2$ **13.** 6 **15.** -72 **17.** -960
19. $aehj$ **21.** 156.8659
23. Add $-r$ times the first row to the third row; then add $-s$ times the second row to the third row. You will get a row of zeros. **27.** They are rational numbers.

PROBLEM SET 11-7

1.

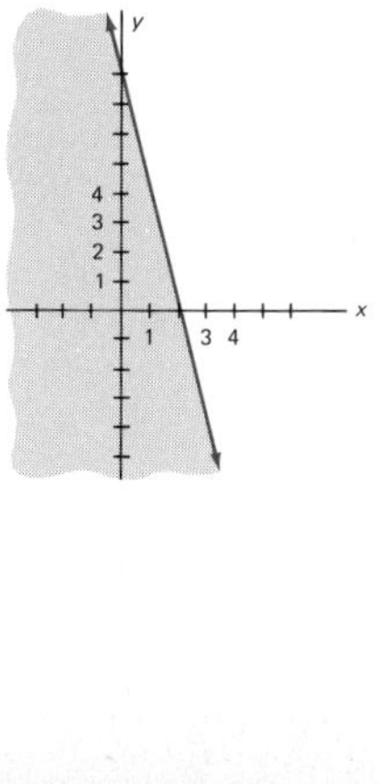

3.

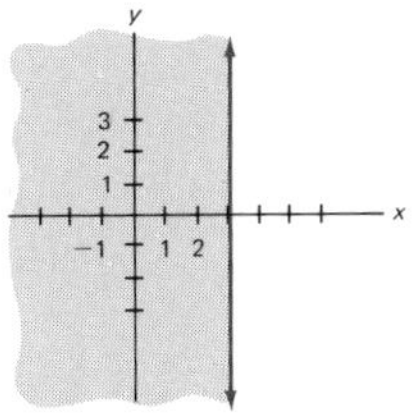

5.

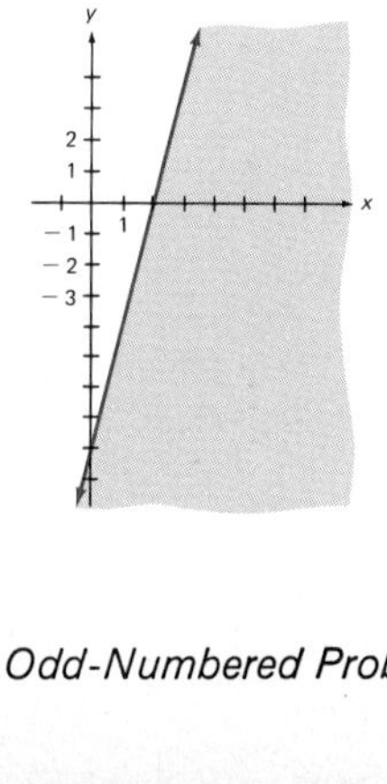

7.

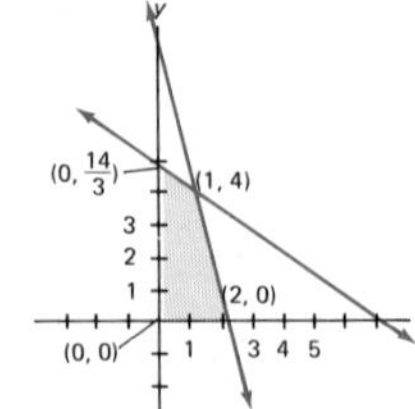

9.

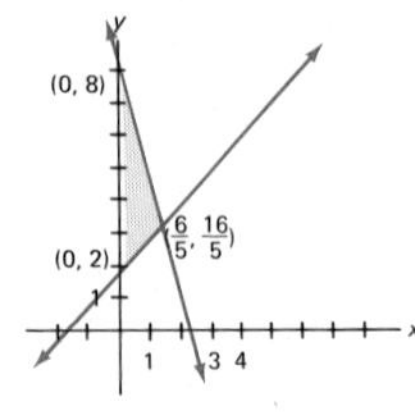

11. Maximum value: 6; minimum value: 0. 13. Maximum value: $-\frac{4}{5}$; minimum value: -8.
15. Minimum value of 14 at (2, 2). 17. Minimum value of 4 at $(\frac{3}{2}, 1)$.

19.

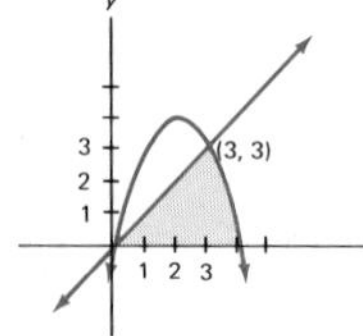

21.

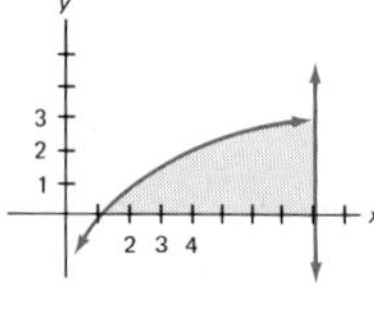

23.

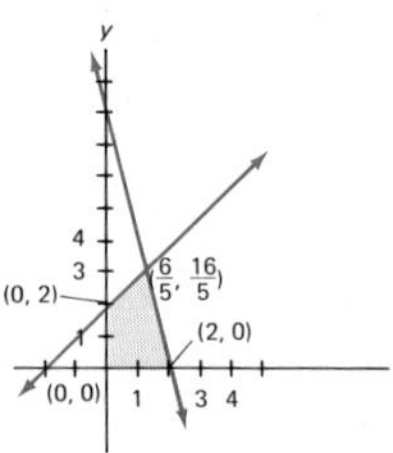

25. Maximum value of 4 at (2, 0); minimum value of -2 at (0, 2).
27. Maximum value of $\frac{11}{2}$ at $(\frac{9}{4}, \frac{13}{4})$; minimum value of 0 at (0, 0).
29. \$5700 (1700 barrels of type A, 300 barrels of type B). 31. \$9000 (150 pairs of shoes, 50 pairs of boots).
33. 60 acres of oats and 40 acres of wheat.

CHAPTER 11. REVIEW PROBLEM SET

1. (2, 6) 2. (5, 2, 3) 3. No solution. 4. No solution. 5. $(20w - 36, -9w + 22, -2w + 4, w)$
6. (a) $\begin{bmatrix} -1 & 8 & -2 \\ 8 & -1 & 3 \\ 5 & 13 & 0 \end{bmatrix}$ (b) $\begin{bmatrix} 7 & -11 & 9 \\ -1 & 7 & -11 \\ 0 & -16 & -5 \end{bmatrix}$ (c) $\begin{bmatrix} 0 & 12 & 3 \\ -8 & 6 & -9 \\ -3 & -3 & 0 \end{bmatrix}$ (d) $\begin{bmatrix} 7 & -5 & -5 \\ 3 & 9 & 0 \\ 32 & 14 & -10 \end{bmatrix}$ 7. $\begin{bmatrix} \frac{1}{6} & \frac{1}{2} & -\frac{1}{3} \\ \frac{1}{6} & -\frac{1}{2} & \frac{2}{3} \\ \frac{2}{3} & 0 & -\frac{1}{3} \end{bmatrix}$
8. $(-1, 2, 3)$ 9. 2 10. 0 11. -114 12. 0 13. -144 14. $(1, -2, 4)$

15.

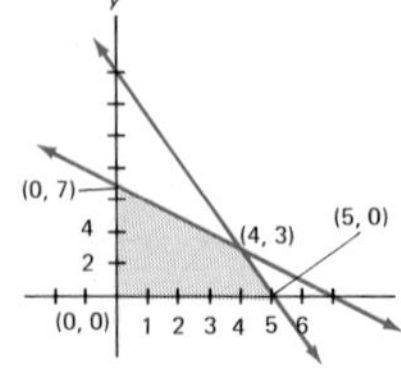

16.

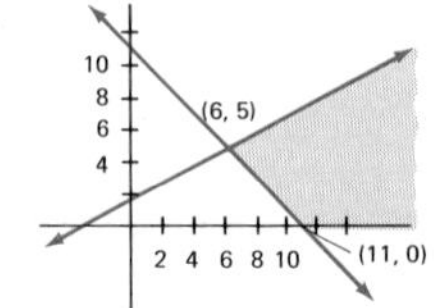

17. 14 18. 11 19. 24

20. 30 suits; 65 dresses

PROBLEM SET 12-1

1. (a) 9; 11 (b) 5; 2 (c) $\frac{1}{16}$; $\frac{1}{32}$ (d) 81; 121 3. (a) 11; 43 (b) $\frac{5}{6}$; $\frac{9}{10}$ (c) 49; 81 (d) -27; 81
5. (a) $a_n = 2n - 1$ (b) $b_n = 20 - 3n$ (c) $c_n = (\frac{1}{2})^{n-1}$ (d) $d_n = (2n - 1)^2$ 7. (a) 14 (b) 162 (c) $\frac{1}{2}$ (d) 81
9. (a) $a_n = a_{n-1} + 2$ (b) $b_n = b_{n-1} - 3$ (c) $c_n = c_{n-1}/2$ (d) $d_n = d_{n-1} + 8(n - 1)$ 11. 48 13. 42
15. 79 17. 69 19. (a) 11; 59 (b) 16; 1,048,576 21. (a) 27 (b) $\frac{1}{6}$ 23. (a) 40 (b) 62
25. (a) $a_n = 2 \cdot 3^{n-1}$ (b) $b_n = 2 + 4(n - 1) = -2 + 4n$ (c) $c_n = n(n - 1) + 2 = n^2 - n + 2$ (d) Hard.
(e) $e_n = 2(\frac{1}{2})^{n-1} = 2^{2-n}$ (f) $f_n = n^2 + 1$ 27. 1; 2; 3; 5; $(1/\sqrt{5})[((1 + \sqrt{5})/2)^{n+1} - ((1 - \sqrt{5})/2)^{n+1}]$
29. $F_n = f_{n+2} - 1$ 31. (a) 4; 8 (b) There are no primes between $m - 1$ and $m + 20$.
33. (a) $s_n = n^2$ (b) $s_n = s_{n-1} + 2n - 1$ (c) $t_n = t_{n-1} + n$ (d) $t_n = n(n + 1)/2$ 35. $a_8 = 4$; $a_{27} = 2$; $a_{53} = 5$
37. The number of letters in the English word for n is a_n.

PROBLEM SET 12-2

1. (a) 13; 16 (b) 3.2; 3.5 (c) 12; 8 **3.** (a) 3; 88 (b) .3; 10.7 (c) -4; -88 **5.** (a) 1335 (b) 190.5 (c) -900 **7.** .5 **9.** (a) 10,100 (b) 10,000 (c) 6,633 **11.** 382.5 **13.** 3.8; 4.6; 5.4; 6.2 **15.** 11; 2,772 **17.** (a) 90 (b) $\frac{25}{6}$ (c) 15,350 (d) 9,801 **19.** (a) $\sum_{i=3}^{20} b_i$ (b) $\sum_{i=1}^{19} i^2$ (c) $\sum_{i=1}^{n} 1/i$ **21.** (a) 2.5 (b) 130 (c) 3442.5 **23.** 923 **25.** 11,400 **27.** (a) $\sum_{i=1}^{112} b_i$ (b) $\sum_{i=0}^{100} (19 + 7i)$, or $\sum_{i=1}^{101} (12 + 7i)$ **29.** 2419 **31.** (a) 1; 7; 19; 37; 61 (b) 6; 12; 18; 24; Arithmetic.

PROBLEM SET 12-3

1. (a) 8; 16 (b) $\frac{1}{2}$; $\frac{1}{4}$ (c) .00003; .000003 **3.** (a) 2; 2^{28} (b) $\frac{1}{2}$; $(\frac{1}{2})^{26}$ (c) .1; $.3(.1)^{29} = 3 \times 10^{-30}$ **5.** (a) 31/2 (b) 31/2 (c) .33333 **7.** $100(2)^{10} = 102{,}400$ **9.** $\$(2^{32} - 1)$, which is over \$4 billion **11.** (a) $\frac{1}{2}$ (b) $\frac{4}{15}$ **13.** 30 feet **15.** $\frac{1}{9}$ **17.** 25/99 **19.** 611/495 **21.** 625; 125; 25; 5; 1 **23.** $a_n = 120(.2)^{n-1}$; $c_n = 6\pi - 2\pi n$; $e_n = 100(1.08)^n$; $f_n = 106 + 2n$ **25.** (a) 4429.124 (b) .467 (c) 46.522 (d) 2.414 **27.** $100(1.02)^{40}$; \$220.80 **29.** $25(1.02)(1.02^{40} - 1)/.02$; \$1540.25

29. If $r \geq 1$, the sum S_n of the first n terms grows large without bound as n increases. If $r \leq -1$, the sum S_n oscillates in value and therefore does not approach a fixed value.

PROBLEM SET 12-4

(*Note*: In the text, several proofs by mathematical induction are given in complete detail. To save space, we show only the key step here, namely, that P_{k+1} is true if P_k is true.)

1. $(1 + 2 + \cdots + k) + (k + 1) = k(k + 1)/2 + k + 1 = [k(k + 1) + 2(k + 1)]/2 = (k + 1)(k + 2)/2$

3. $(3 + 7 + \cdots + (4k - 1)) + (4k + 3) = k(2k + 1) + (4k + 3) = 2k^2 + 5k + 3 = (k + 1)(2k + 3)$

5. $(1\cdot2 + 2\cdot3 + \cdots + k(k + 1)) + (k + 1)(k + 2) = \frac{1}{3}k(k + 1)(k + 2) + (k + 1)(k + 2) = \frac{1}{3}(k + 1)(k + 2)(k + 3)$

7. $(2 + 2^2 + \cdots + 2^k) + 2^{k+1} = 2(2^k - 1) + 2^{k+1} = 2^{k+1} - 2 + 2^{k+1} = 2(2^{k+1} - 1)$ **9.** P_n is true for $n \geq 8$.

11. P_1 is true. **13.** P_n is true whenever n is odd. **15.** P_n is true for every positive integer n.

17. P_n is true whenever n is a positive integer power of 4.

19. $n = 4$. If $k + 5 < 2^k$, then $k + 6 < 2^k + 1 < 2^k + 2^k = 2^{k+1}$.

21. $n = 1$. Since $k + 1 < 10k$, $\log(k + 1) < 1 + \log k < 1 + k$.

23. $n = 1$. Multiply both sides of $(1 + x)^k \geq 1 + kx$ by $(1 + x)$: $(1 + x)^{k+1} \geq (1 + x)(1 + kx) = 1 + (k + 1)x + kx^2 > 1 + (k + 1)x$.

25. $x^{2k+2} - y^{2k+2} = x^{2k}(x^2 - y^2) + (x^{2k} - y^{2k})y^2$. Now $(x - y)$ is a factor of both $x^2 - y^2$ and $x^{2k} - y^{2k}$, the latter by assumption.

27. $(k + 1)^2 - (k + 1) = k^2 + 2k + 1 - k - 1 = (k^2 - k) + 2k$. Now 2 divides $k^2 - k$ by assumption and clearly divides $2k$.

29. $(5 + 15 + \cdots + (10k - 5)) + (10k + 5) = 5k^2 + 10k + 5 = 5(k + 1)^2$

31. $(1 + 2 + \cdots + k) + (k + 1) = (k^2 + k - 6)/2 + (k + 1) = (k^2 + 3k - 4)/2 = [(k + 1)^2 + (k + 1) - 6]/2$. No.

33. $(1^3 + 2^3 + \cdots + k^3) + (k + 1)^3 = [k(k + 1)/2]^2 + (k + 1)^3 = (k + 1)^2(k^2 + 4k + 4)/4 = [(k + 1)(k + 2)/2]^2$

35. (a) $[a + (a + d) + \cdots + (a + (k - 1)d)] + (a + kd) = (k/2)[2a + (k - 1)d] + (a + kd) = \frac{1}{2}[k(2a + kd - d) + 2a + 2kd] = \frac{1}{2}[(k + 1)2a + (k^2 + k)d] = [(k + 1)/2][2a + kd]$ (b) $(a + ar + \cdots + ar^{k-1}) + ar^k = a[(1 - r^k)/(1 - r)] + ar^k = (a - ar^k + ar^k - ar^{k+1})/(1 - r) = a[(1 - r^{k+1})/(1 - r)]$

37. The statement is true when $n = 3$ since it asserts that the angles of a triangle have a sum of 180°. Now any $(k + 1)$-sided convex polygon can be dissected into a k-sided polygon and a triangle. Its angles add up to $(k - 2)180° + 180° = (k - 1)180°$.

39. $[(1 - \frac{1}{4})(1 - \frac{1}{9})\cdots(1 - 1/k^2)](1 - 1/(k + 1)^2) = [(k + 1)/2k][((k + 1)^2 - 1)/(k + 1)^2] = (k + 2)/2(k + 1)$

41. $F_{k+1} = F_k + f_{k+1} = f_{k+2} - 1 + f_{k+1} = f_{k+3} - 1$

43. Assume the equality holds for a_k and a_{k+1}. Then $a_{k+2} = (a_k + a_{k+1})/2 = \frac{2}{3}[(1 - (-\frac{1}{2})^k + 1 - (-\frac{1}{2})^{k+1})/2] = \frac{2}{3}[1 - \frac{1}{2}(-\frac{1}{2})^k - \frac{1}{2}(-\frac{1}{2})^{k+1}] = \frac{2}{3}[1 - (-\frac{1}{2})^{k+2}]$.

PROBLEM SET 12-5

1. (a) 6 (b) 12 (c) 90 **3.** (a) 20 (b) 3024 (c) 720 **5.** 30 **7.** (a) 6 (b) 5 **9.** (a) 36 (b) 72 (c) 6 **11.** 1680 **13.** 25; 20 **15.** 240 **17.** 300 **19.** (a) 30 (b) 180 (c) 120 (d) 120 (e) 450 **21.** (a) $12\cdot11\cdot10\cdot9\cdot8\cdot7\cdot6\cdot5\cdot4$ (b) $2\cdot10\cdot9\cdot8\cdot7\cdot6\cdot5\cdot4\cdot3$ (c) $2\cdot11\cdot10\cdot9\cdot8\cdot7\cdot6\cdot5\cdot4$ **23.** 210 **25.** 34,650 **27.** 840 **29.** 126 **31.** (a) 5!/2! (b) 10!/2! (c) 8!/6! **33.** (a) 720 (b) 120 **35.** (a) 120 (b) 24 **37.** 1024 **39.** 9,000,000 **41.** 100 **43.** 144

PROBLEM SET 12-6

1. (a) 720 (b) 120 (c) 120 (d) 1 (e) 6 (f) 6 **3.** (a) 1140 (b) 161,700 **5.** 56 **7.** 495 **9.** 720
11. 63 **13.** (a) 36 (b) 100 (c) 24 **15.** (a) 84 (b) 39 (c) 130 **17.** ${}_{26}C_{13}$ **19.** 4^{13}
21. ${}_4C_2 \cdot {}_{48}C_{11} + {}_4C_3 \cdot {}_{48}C_{10} + {}_4C_4 \cdot {}_{48}C_9$ **23.** ${}_{52}C_5$ **25.** ${}_{13}C_2 \cdot {}_4C_2 \cdot {}_4C_2 \cdot 44$ **27.** 35 **29.** 126
31. (a) 10 (b) 10 (c) 16 **33.** 1680 **35.** (a) 4 (b) 8 (c) 16; ${}_nC_0 + {}_nC_1 + \cdots + {}_nC_n = 2^n$ **37.** 5040
39. (a) 45 (b) 90 (c) 100 (d) 55 **41.** (a) 36 (b) 6 (c) 15 **43.** $(6!)^2 = 518{,}400$

PROBLEM SET 12-7

1. $x^3 + 3x^2y + 3xy^2 + y^3$ **3.** $x^3 - 6x^2y + 12xy^2 - 8y^3$ **5.** $c^8 - 12c^6d^3 + 54c^4d^6 - 108c^2d^9 + 81d^{12}$
7. $x^{20} + 20x^{19}y + 190x^{18}y^2$ **9.** $x^{20} + 20x^{14} + 190x^8$ **11.** (a) 16 (b) 32 (c) 64 **13.** $405x^2y^{24}$
15. $10x^4$ **17.** 104.076 **19.** 294.4 **21.** 1219 (using 3 terms)
23. $32x^{15} - 80x^{11} + 80x^7 - 40x^3 + 10/x - 1/x^5$ **25.** $b^{36} + 36b^{33}c + 594b^{30}c^2$
27. There are n decisions to make and 2 ways of making each decision; hence there are 2^n ways of selecting a subset.
29. $1 + 10(.0003) + 45(.0003)^2 \approx 1.0030041$ **31.** $2^8 - 1 = 255$ **33.** 105

CHAPTER 12. REVIEW PROBLEM SET

1. (a) and (c) are arithmetic; (b) and (e) are geometric.
2. (a) $a_n = a_{n-1} + 3$ (b) $b_n = 3b_{n-1}$ (c) $c_n = c_{n-1} - .5$ (d) $d_n = d_{n-1} + d_{n-2}$ (e) $e_n = e_{n-1}/3$
3. (a) $a_n = 2 + (n-1)3 = 3n - 1$ (b) $b_n = 2 \cdot 3^{n-1}$ **4.** 6767 **5.** $3^{100} - 1$ **6.** 3 **7.** 89
8. 40 **9.** 250,500 **10.** $\frac{5}{9}$ **11.** $100(1.02)^{48}$ **13.** P_n is true for n a multiple of 3. **14.** 60 **15.** 64
16. (a) 5040 (b) 210 (c) 30 (d) 1225 (e) 28 (f) 2520 **17.** (a) 504 (b) 224 (c) 84 (d) 40 **18.** 15,120
19. 56 **20.** 256 **21.** $x^{10} + 20x^9y + 180x^8y^2 + 960x^7y^3$ **22.** $-20a^3b^6$ **23.** 1.0408

Index

Abscissa, 134
Absolute value, 29
Absolute value properties, 30
Addition
    complex numbers, 34, 356
    fractions, 11
    functions, 196
    matrices, 417
    polynomials, 62
    rational expressions, 76
Addition law for cosines, 305
Addition law for sines, 305
Addition of ordinates, 197, 293
Addition principle in counting, 488
Additive inverse, 22, 417
Algebra, 2
Algebraic logic in calculators, 56
Ambiguous case, 331
Amplitude, 342
Analytic geometry, 138
Angle, 266
Antilogarithm, 248
Arc length, 268
Area of a sector, 271
Area of a triangle, 339
Argand, Jean-Robert, 353
Argand diagram, 353
Arithmetic sequence, 465
Associative property, 21, 417, 419

Back substitution, 404
Bernoulli, Daniel, 460
Binomial coefficient, 502
Binomial Formula, 501
Bombelli, Raffael, 33
Bounded region, 447
Brahmagupta, 85
Briggs, Henry, 247

Cajori, F., 252
Calculators, 56
Calculus, 79, 132, 198, 247, 392, 395
Cardano, Hieronimo, 214
Carroll, Lewis, 7, 424
Cartesian coordinates, 133
Cayley, Arthur, 409, 416
Center of ellipse, 163
Change of base formula, 239
Characteristic, 248
Circle, 141
Closed half-plane, 446
Coefficients, 61
Cofunction identities, 279
Combination, 494
Common difference, 465
Common logarithms, 247
Common ratio, 472
Commutative property, 21, 417
Compatible, 419
Complete Factorization Theorem, 378
Completing the square, 108
Complex numbers, 34, 353
Complex plane, 353
Composition of functions, 198
Compound interest, 229
Compound interest table, 232
Condorcet, Marquis de, 329
Conic sections, 166
Conjugate, 35, 384
Conjugate Pair Theorem, 386
Constraints, 445
Convex region, 447
Coordinate, 132
Coordinate axes, 133
Cosecant, 278
Cosine, 272
Cotangent, 278
Cramer, Gabriel, 434
Cramer's rule, 434
Curve fitting, 243

DaVinci, Leonardo, 369
Davis, Philip J., 1
Decimals, 15
    nonrepeating, 17
    repeating, 16, 476

Decimals (*cont.*)
    terminating, 16
    unending, 16
Decomposing functions, 198, 200
Degree measurement, 266
Degree of a polynomial, 61
DeMoivre, Abraham, 360
DeMoivre's Theorem, 360
Descartes, René, 33, 93, 132, 457
Determinant, 432, 433, 440
Determinant properties, 434
Diameter, 162
    major, 162
    minor, 163
Directrix, 155
Dirichlet, P. G. L., 174
Discriminant, 110
Distance formula, 134
Distance from point to line, 150
Distributive property, 22, 419
Division
    by zero, 24
    complex numbers, 35, 355
    fractions, 11
    functions, 196
    polynomials, 370
    rational expressions, 77
    real numbers, 23
Division algorithm, 370
Division Law for Polynomials, 371
Domain, 176
Double-angle formulas, 306
Doubling time, 227

Einstein, Albert, 2
Ellipse, 162
Equality, 86
    complex numbers, 34
    matrices, 416
Equations, 86
    exponential, 241, 255
    linear, 88, 148
    logarithmic, 238
    polynomial, 384
    quadratic, 106
    system of, 100, 402
    trigonometric, 312
    with radicals, 217
Equivalent systems, 402
Equivalent matrices, 410
Euclid, 13, 86, 145
Euler, Leonhard, 240
Eves, Howard, 106
Expansion by minors, 440
Explicit formula for sequences, 459
Exponent, 42
    integral, 42
    negative, 44
    rational, 220
    real, 221
    zero, 44
Exponential decay, 228
Exponential function, 222
Exponential growth, 226
Extraneous solution, 89

Factor, 67
Factor completely, 67
Factorial, 488
Factor over, 71
Factor Theorem, 377
Fermat, Pierre de, 132
Fibonacci, Leonardo, 460
Focus, 155, 165
Fractions
    number, 10
    polynomial, 75
    signs of, 78
Frequency, 346
Functional notation, 175
Functions, 174, 196
    absolute value, 184
    composite, 198
    even, 186
    exponential, 222
    greatest integer, 188
    inverse, 203
    inverse trigonometric, 318
    logarithmic, 235
    odd, 186
    of two variables, 178
    one-to-one, 204
    periodic, 275
    polynomial, 181
    power, 241
    rational, 189
    translations of, 199
    trigonometric, 272
Fundamental Theorem of Algebra, 378
Fundamental Theorem of Arithmetic, 14

Galileo, 259, 297
Gauss, Carl, 376, 466, 509
Geometric sequences, 472
Graphs
    equations, 138
    functions, 181
    inequalities, 445
    inverse functions, 206
    polar equations, 349
Grouping symbols, 10, 58

Half-angle formulas, 306
Half-life, 228
Hardy, G. H., 14
Heisenberg, Werner, 416
Hyperbola, 164

Identities, 86, 298
Identity element, 22, 424
Imaginary axis, 353
Imaginary part, 34
Inconsistent, 407
Inequalities, 27
    conditional, 114
    linear, 114
    quadratic, 115
    systems of, 445
    unconditional, 114
Inequality properties, 28, 115
Integers, 8
Interpolation, 249, 510
Intersection of a line and a parabola, 158
Intersection of two lines, 150
Intervals, 29, 128
Inverse
    additive, 22, 417
    multiplicative, 22, 425
    of a function, 204, 236, 257
Irrational number, 14, 17

Jefferson, Thomas, 75

Kline, Morris, 2, 131, 181
Kōwa, Seki, 431

Lagrange, Joseph-Louis, 138
Laplace, P. S., 213
Law of cosines, 335
Law of sines, 330
Least common denominator, 18
Least common multiple, 17
Leibniz, Gottfried, 33, 431
Line, 145
Linear
    equation, 88
    inequality, 445
    interpolation, 249, 510
    programming problem, 447
Logarithm, 233
    common, 247
    general, 234
    natural, 240
Logarithmic equations, 238
Logarithmic function, 235
Logarithm properties, 234

Mach, Ernst, 278
Mantissa, 248
Marginal cost, 152
Marginal profit, 152
Mathematical induction, 478
Matrix, 409
Matrix of a system, 409
Metric system, 52
Midpoint formula, 134
Minor, 439
Monomial, 62
Multiplication
    complex numbers, 35, 355
    fractions, 11
    functions, 196
    matrices, 418
    polynomials, 63
    rational expressions, 76
    scalar, 419
Multiplication principle in counting, 487
Multiplicative identity, 22, 424
Multiplicative inverse, 22, 37, 425, 426
Multiplicity of zeros, 379

Napier, John, 233
Natural domain, 176
Natural logarithm, 240
Negative exponent, 44
Negative integers, 7
Neutral element, 22, 417, 425

Newton, Isaac, 33, 392
Newton's method, 394
Nonlinear systems, 407
Numbers
- complex, 34, 353
- integer, 8
- irrational, 14, 17
- prime, 14
- rational, 8
- real, 15
- whole, 7

Number sequence, 459

Oblique triangle, 330
Open half-plane, 446
Order, 27
Ordered pair, 134
Order properties, 28
Ordinate, 134
Origin, 133

Parabola, 153
Parallel lines, 148
Pascal, Blaise, 502
Pascal's Triangle, 502
Period, 275, 291, 342
Permutation, 487
Perpendicular lines, 148
Phase shift, 342
Poincaré, Henri, 173
Point-slope form, 147
Polar axis, 347
Polar coordinates, 347
Polar form, 354
Pole, 347
Polya, George, 99
Polygonal region, 447
Polynomials, 61
- complex, 61
- in several variables, 64
- linear, 62
- quadratic, 62
- real, 61

Power, 42, 360
Prime factorization, 17
Prime number, 14
Product formulas, 310
Proper rational expressions, 373
Pythagorean Theorem, 13

Quadrants, 133
Quadratic equation, 106
Quadratic Formula, 109

Radian measurement, 266
Radicals, 214
Range, 176
Rational expression, 75
Rationalizing denominators, 216
Rational numbers, 8
Rational Solution Theorem, 387
Real axis, 353
Real line, 15
Real number, 15
Real part, 34
Recursion formula, 460
Reduced form, 9, 75
Reducing fractions, 10
Reference angle, 284
Reference number, 284
Remainder Theorem, 376
Restricting domains, 209, 318
Reverse Polish logic, 56
Right triangle, 260
Rise, 145
Roots, 214, 361, 363
Roots of unity, 365
Rules for exponents, 45
Rules for modifying equations, 87, 100
Rules for radicals, 215
Run, 145

Scientific notation, 50
Secant, 278
Sector of a circle, 271
Sequence, 458
Sigma notation, 468
Significant digits, 51
Simple harmonic motion, 291, 340
Sine, 272
Slope, 145, 395
Slope-intercept form, 147
Slope Theorem, 395
Solving
- equations, 87
- inequalities, 114

Special angles, 261, 273
Split-point method, 116
Square roots, 107
Standard equation
- circle, 141
- ellipse, 163
- parabola, 156

Standard position
- angle, 272
- decimal point, 248

Statement, 478
Subtraction
- complex numbers, 35
- fractions, 11
- functions, 196
- matrices, 417
- polynomials, 63
- rational expressions, 76
- real numbers, 23

Successive approximations, 392
Successive enlargements, 392
Sum and difference formulas, 311
Sylvester, James Joseph, 401, 439
Symmetry, 139
Synge, J. L., 162
Synthetic division, 371

Table of values, 138
Tait, P. G., 416
Tangent function, 278
Tangent line, 394
Translations, 199
Triangular form, 404
Trigonometry, 260

Unbounded region, 449
Unit circle, 268

Variable
- dependent, 176
- independent, 176

Variation
- direct, 176
- inverse, 176
- joint, 177

Vertex
- parabola, 153, 155
- polygon, 447

Weber, E., 243
Weber-Fechner Law, 244
Whitehead, Alfred North, 41
Whole numbers, 7

Zero of a polynomial, 376

GRAPHS OF TRIGONOMETRIC FUNCTIONS

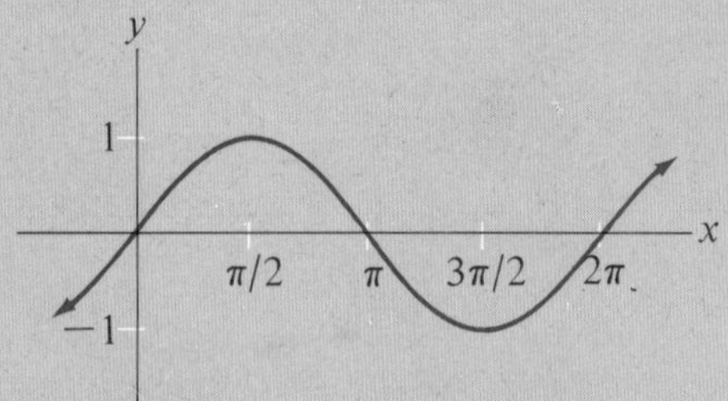

$y = \sin x$

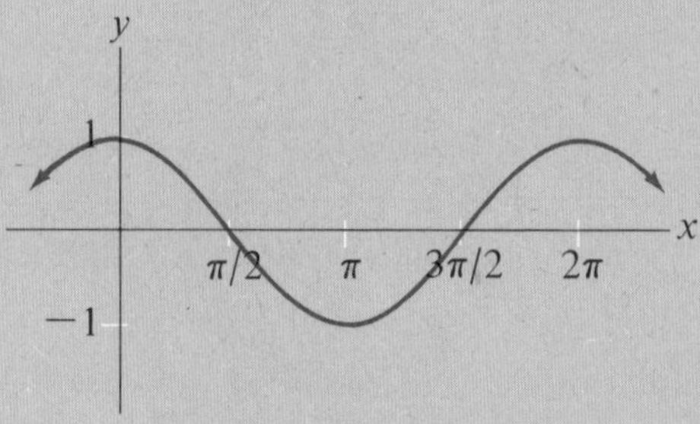

$y = \cos x$

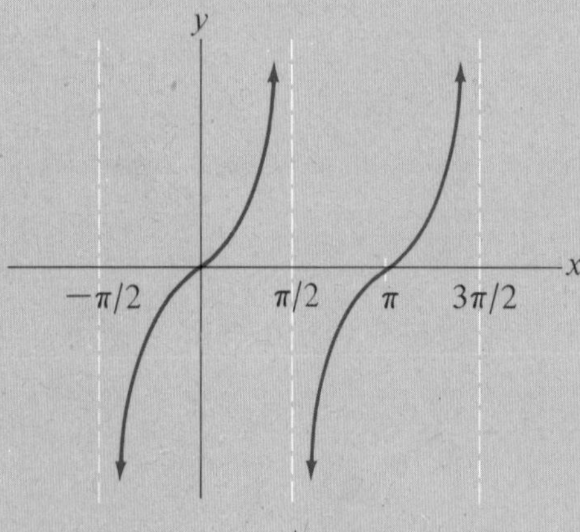

$y = \tan x$

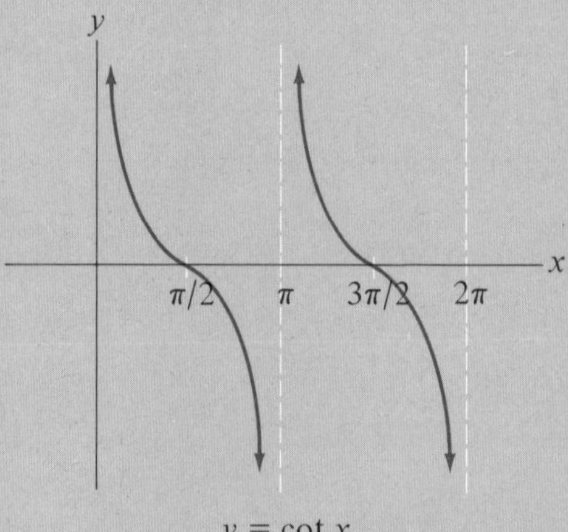

$y = \cot x$

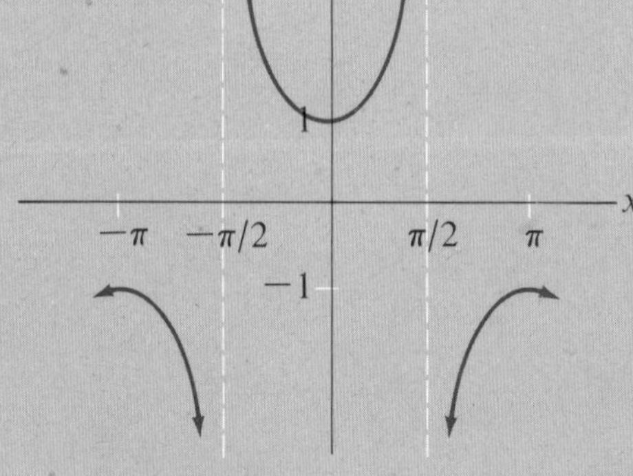

$y = \sec x$

y

1

x

−π −π/2 π/2 π

−1

$y = \csc x$